Multiphase Catalytic Reactors

Multiphase Catalytic Reactors

Theory, Design, Manufacturing, and Applications

Edited by

Zeynep Ilsen Önsan
Department of Chemical Engineering
Boğaziçi University
Istanbul, Turkey

Ahmet Kerim Avci
Department of Chemical Engineering
Boğaziçi University
Istanbul, Turkey

WILEY

Published by John Wiley & Sons, Inc., Hoboken, New Jersey

Published simultaneously in Canada

Library of Congress Cataloging-in-Publication Data:

Names: Önsan, Zeynep Ilsen, editor. | Avci, Ahmet Kerim, editor.
Title: Multiphase catalytic reactors : theory, design, manufacturing, and applications / edited by Zeynep Ilsen Önsan, Ahmet Kerim Avci.
Description: Hoboken, New Jersey : John Wiley & Sons Inc., [2016] | Includes bibliographical references and index.
Identifiers: LCCN 2016009674 | ISBN 9781118115763 (cloth) | ISBN 9781119248477 (epub) | ISBN 9781119248460 (epdf)
Subjects: LCSH: Phase-transfer catalysis. | Chemical reactors.
Classification: LCC TP159.C3 M85 2016 | DDC 660/.2832–dc23
LC record available at https://lccn.loc.gov/2016009674

Set in 9.5/12pt Minion by SPi Global, Pondicherry, India

10 9 8 7 6 5 4 3 2 1

Contents

List of contributors

Anton Alvarez-Majmutov
Instituto Mexicano del Petróleo, Management of Products for the Transformation of Crude Oil, Mexico City, Mexico

Jorge Ancheyta
Instituto Mexicano del Petróleo, Management of Products for the Transformation of Crude Oil, Mexico City, Mexico

Ahmet Kerim Avci
Department of Chemical Engineering, Boğaziçi University, Istanbul, Turkey

Vivek V. Buwa
Department of Chemical Engineering, Indian Institute of Technology-Delhi, New Delhi, India

Joaquim M.S. Cabral
Department of Bioengineering and IBB-Institute for Bioengineering and Biosciences, Instituto Superior Técnico, Universidade de Lisboa, Lisboa, Portugal

Sourav Chatterjee
School of Chemistry and Chemical Engineering, Queen's University Belfast, Belfast, UK

Burtron H. Davis
Center for Applied Energy Research, University of Kentucky, Lexington, KY, USA

Pedro Fernandes
Department of Bioengineering and IBB-Institute for Bioengineering and Biosciences, Instituto Superior Técnico, Universidade de Lisboa; Faculdade de Engenharia, Universidade Lusófona de Humanidades e Tecnologias, Lisboa, Portugal

Rebecca R. Fushimi
Materials Science & Engineering Department, Idaho National Laboratory, Idaho Falls, ID, USA

John T. Gleaves
Department of Energy, Environmental and Chemical Engineering, Washington University, St. Louis, MO, USA

John R. Grace
Department of Chemical and Biological Engineering, The University of British Columbia, Vancouver, British Columbia, Canada

Ion Iliuta
Chemical Engineering Department, Laval University, Québec City, Québec, Canada

Gary Jacobs
Center for Applied Energy Research, University of Kentucky, Lexington, KY, USA

Seda Keskin
Department of Chemical and Biological Engineering, Koc University, Istanbul, Turkey

Teuvo Kilpiö
Process Chemistry Centre, Åbo Akademi University, Turku/Åbo, Finland

Gunther Kolb
Fraunhofer ICT-IMM, Decentralized and Mobile Energy Technology Department, Mainz, Germany

Faïçal Larachi
Chemical Engineering Department, Laval University, Québec City, Québec, Canada

Carolina Leyva
Centro de Investigación en Ciencia Aplicada y Tecnología Avanzada, Unidad Legaria, Instituto Politécnico Nacional, Mexico City, Mexico

João P. Lopes
Department of Chemical Engineering and Biotechnology, University of Cambridge, Cambridge, UK

Päivi Mäki-Arvela
Process Chemistry Centre, Åbo Akademi University, Turku/Åbo, Finland

Dmitry Yu. Murzin
Process Chemistry Centre, Åbo Akademi University, Turku/Åbo, Finland

Zeynep Ilsen Önsan
Department of Chemical Engineering, Boğaziçi University, Istanbul, Turkey

Vivek V. Ranade
Chemical Engineering & Process Development Division, National Chemical Laboratory, Pune, India

Evgeny Rebrov
School of Engineering, University of Warwick, Coventry, UK

Alírio E. Rodrigues
Laboratory of Separation and Reaction Engineering,
Associate Laboratory LSRE/LCM, Department of Chemical Engineering,
Faculty of Engineering, University of Porto, Porto, Portugal

Shantanu Roy
Department of Chemical Engineering, Indian Institute of Technology-Delhi,
New Delhi, India

Tapio Salmi
Process Chemistry Centre, Åbo Akademi University,
Turku/Åbo, Finland

Gregory S. Yablonsky
Parks College of Engineering, Aviation and Technology,
Saint Louis University, St. Louis, MO, USA

Preface

The single irreplaceable component at the core of a chemical process is the chemical reactor where feed materials are converted into desirable products. Although the essential variables by which chemical processes can be controlled are reaction temperature, pressure, feed composition, and residence time in the reactor, two technological developments of major consequence starting with 1960s have made possible cost-effective operation under less severe conditions; these are the extensive use of efficient catalysts and the introduction of improved or innovative reactor configurations. The impact of heterogeneous catalysis is significant in this respect since petroleum refining, manufacturing of chemicals, and environmental clean-up, which are the three major areas of the world economy today, all require the effective use of solid catalysts. The challenges involved in the design of novel solid catalysts and modification of many existing ones for higher selectivity and stability have also prompted the development of "engineered" catalysts befitting novel reactor configurations, requiring the use of new supports such as monolithic or foam substrates as well as the establishment of new techniques for coating surfaces with diverse catalyst components in order to ensure longevity particularly in cyclic processes.

In industrial practice, the composition and properties of the complex feed mixtures that are processed for producing a range of valuable chemicals generally necessitate the use of heterogeneous catalytic reactors. Numerous chemical and physical rate processes take place in a heterogeneous reactor at different length and time scales and frequently in different phases. The prerequisite for the successful design and operation of catalytic reactors is a thorough microkinetic analysis starting from intrinsic kinetic models of the steady-state chemical activity and leading to global rate expressions obtained by overlaying the effects of physical rate phenomena occurring at the particle scale. Kinetic models of increasing complexity may be required depending on the variety of components and number of reactions involved. The second critical stage in reactor modeling and design is a macrokinetic analysis including the detailed description of physical transport phenomena at the reactor scale and utilizing the global rate expressions of the microkinetic analysis. The final catalytic reactor model which integrates these essential stages can successfully predict the performance and dynamics of plant-scale industrial reactors as well as simulating their start-up, shutdown, and cyclic operation. Taking into account engineered catalysts and new reactor configurations, the modeling and scaling up of reactions conducted at the bench-scale to pilot plant and industrial-scale reactor levels have to be modified in order to include simultaneous multiscale approaches along with the conventional sequential modes.

Multiphase Catalytic Reactors: Theory, Design, Manufacturing, and Applications is a comprehensive up-to-date compilation on multiphase catalytic reactors which will serve as an excellent reference book for graduate students, researchers, and specialists both in academia and in industry. The content of the book is planned to cover topics starting from the first principles involved in macrokinetic analysis of two- and three-phase catalytic reactors to their particular industrial applications. The main objective is to provide definitive accounts on academic aspects of multiphase catalytic reactor modeling and design along with detailed descriptions of some of the most recent industrial applications employing multiphase catalytic reactors, in such a way as to balance the academic and industrial components as much as possible. Accordingly, seven chapters are included in Parts II, III, and IV to review the relevant mathematical models and model equations utilized in the fundamental analysis and macroscopic design of specific reactor types together with some useful approximations for their design and scale-up from a practical standpoint, while the four chapters in Part VI describe specific industrial applications and contain pointers that tie in with the modeling and design approaches presented for the particular multiphase catalytic reactor types discussed in Parts II, III, and IV. Furthermore, the chapters included in Parts I and V of the book contain detailed reviews of the basic principles and essential tools of catalytic reaction engineering that are crucial for the successful design and operation of catalytic reactors. All chapters of the book are contributed by experts distinguished in their respective fields.

The total of 15 chapters included in *Multiphase Catalytic Reactors: Theory, Design, Manufacturing, and Applications* are organized in six parts. Part I is an overview of the principles of catalytic reaction engineering, embracing Chapter 1 which is a survey of multiphase catalytic reactor types and their industrial significance as well as Chapter 2 on the microkinetic analysis of heterogeneous catalytic systems which surveys the formulation of intrinsic rate equations describing chemical rate processes and the construction of global rate expressions that include the effects of physical mass and heat transport phenomena occurring at the particle scale. Chapters 3 through 9 in

Parts II, III, and IV discuss individual two- and three-phase catalytic reactor types and provide design equations and empirical relationships that characterize different multiphase reactors; mathematical modeling is an integral part of these chapters. In Part II, two-phase catalytic reactors are grouped as fixed-bed gas–solid catalytic reactors (Chapter 3) and fluidized-bed catalytic reactors (Chapter 4). Part III deals exclusively with three-phase catalytic reactors and includes Chapter 5 on three-phase fixed-bed reactors as well as Chapter 6 on three-phase slurry reactors, both of which find significant industrial applications; moreover, multiphase bioreactors are also included in Part III as Chapter 7. Part IV is devoted to the discussion of the more recent state-of-the-art structured reactors; the theoretical aspects and examples of structured reactors enabling process intensification in multiphase operation are treated in Chapter 8 on monolith reactors and in Chapter 9 on microreactors of different configurations including microstructured packed beds and microchannel reactors. Part V of the book is specifically designed for surveying the essential tools of catalytic reactor modeling and design and comprises two chapters. Chapter 10 discusses the recent developments and experimental techniques involved in lab-scale testing of catalytic reactions, including steady-state and transient flow experiments as well as the micro-kinetic and TAP approaches to kinetic analysis, while Chapter 11 surveys the numerical solution techniques that are frequently used in catalytic reactor analysis and demonstrates with some case studies. The capstone section of the book, Part VI, contains four chapters devoted to specific industrial applications of multiphase catalytic reactors and includes the recent developments and practices in Fischer–Tropsch technologies (Chapter 12); a thorough discussion of reactor modeling, simulation, and scale-up approaches involved in the hydrotreating of oil fractions (Chapter 13); a detailed assessment of the performances of various reactor configurations used for fuel processing (Chapter 14); and a comprehensive discussion of catalytic deoxygenation of fatty acids in a packed-bed reactor as case study in production of biofuels (Chapter 15).

It is indeed a pleasure to thank all of the contributors who have made this challenging task achievable. The editors are sincerely grateful for their willingness to devote their valuable time and effort to this project, for their readiness in sharing their vision, knowledge, years of experience, and know-how, and also for their patience in tolerating various expected or unexpected extensions arising from the busy schedules of different contributors. It has definitely been a privilege to work with the authors, coauthors, and reviewers involved in this book. The editors would also like to extend their thanks to Wiley-Blackwell for their commitment to this project and to Michael Leventhal for his organization and management of the publication process.

On a more personal note, the editors would like to take this opportunity to express their sincere gratitude to the late Professor David L. Trimm, who has inspired their research in catalysis and catalytic reaction engineering through many years as supervisor, mentor, colleague, and friend.

Zeynep Ilsen Önsan,
Ahmet Kerim Avci,
Istanbul, October 2015

PART 1

Principles of catalytic reaction engineering

CHAPTER 1

Catalytic reactor types and their industrial significance

Zeynep Ilsen Önsan and Ahmet Kerim Avci
Department of Chemical Engineering, Boğaziçi University, Istanbul, Turkey

Abstract

The present chapter is aimed to provide a simplified overview of the catalytic reactors used in chemical industry. Each reactor type is described in terms of its key geometric properties, operating characteristics, advantages, and drawbacks among its alternatives and typical areas of use. The significance of the reactors is explained in the context of selected industrial examples. Industrial reactors that do not involve the use of solid catalysts are also discussed.

1.1 Introduction

Today's chemical markets involve many different products with diverse physical and chemical properties. These products are produced in chemical plants with different architectures and characteristics. Despite these differences, general structure of a chemical plant can be described by three main groups of unit operations, namely, upstream operations, downstream operations, and the reaction section, as shown in Figure 1.1. Among these groups, the reactor is the most critical section that determines the plant profitability via metrics such as reactant conversion, product selectivity, and yield: high per-pass conversions will reduce the operating expenses involved in product separation and purification steps as well as the recycling costs (Figure 1.1). At this stage selection of the appropriate reactor type and ensuring their efficient operation become critical issues to be addressed.

In almost all reactors running in the chemical industry, the desired product throughput and quality are provided by catalysts, the functional materials that allow chemical synthesis to be carried out at economic scales by increasing the reaction rates. Owing to this critical feature, more than 98% of the today's industrial chemistry is involved with catalysis. Since catalysts have direct impact on reactor performance, they have to be operated at their highest possible effectiveness, which is determined by the degree of internal and external heat and mass transport resistances defined and explained in detail in Chapter 2. At this stage, the function of the reactor is to provide conditions such that the catalyst particles can deliver the best possible performance (e.g., activity, selectivity, yield) at sufficient stability. For example, for a highly exothermic reaction system such as Fischer–Tropsch (FT) synthesis, heat transport/removal rates within the reactor should be very high to prevent undesired temperature elevations that can negatively affect product distribution and, more importantly, cause thermally induced deactivation of the catalysts. Considering the fact that transport rates are favored by good mixing of the reactive fluid at turbulent conditions, the selected reactor type should allow a wide operating window in terms of pressure drop, which is a limit against the occurrence of well-mixed conditions. The possibility of integration and operation of effective external heat exchange systems should also be taken into account in the selected reactor type. The final selection is carried out in the context of fixed capital investment, operating expenses, and profitability of the technically feasible solutions.

Synthesis of commercial chemical products having different physical and chemical functional properties involves the existence of different combinations of catalytic chemistry, thermodynamic properties, and heat and mass transport conditions (e.g., nature of the catalyst and fluids) within the reactor volume. As a result, several reactor types are being proposed. Classification of the reactors can be carried out based on various criteria such as compatibility with the operating mode (batch vs. continuous reactors) and the number of phases (homogeneous vs. heterogeneous reactors). In this chapter, reactors are classified according to the position of the catalyst bed, that is, whether it is fixed or mobile. In packed-bed, trickle-bed, and structured (i.e., monolith and microchannel) reactors, catalyst bed is fixed, while it is mobile in fluidized-bed, moving-bed, and slurry reactors. The descriptions of these reactor types are summarized in the following sections.

1.2 Reactors with fixed bed of catalysts

1.2.1 Packed-bed reactors

In packed-bed reactors (PBRs), the solid particulate catalyst particles forming the bed are fixed in an enclosed volume. The

Multiphase Catalytic Reactors: Theory, Design, Manufacturing, and Applications, First Edition. Edited by Zeynep Ilsen Önsan and Ahmet Kerim Avci.

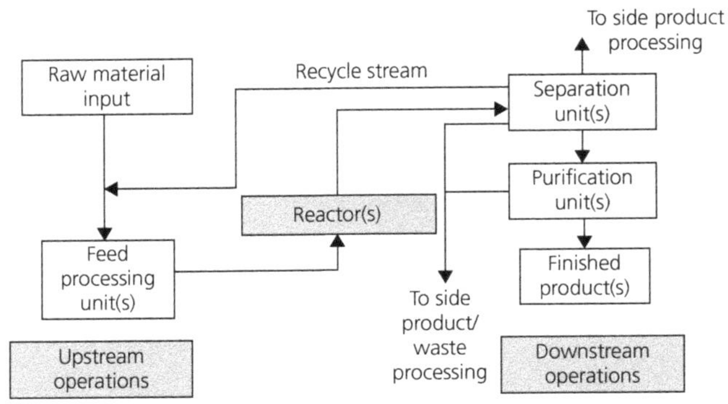

Figure 1.1 Outline of a chemical process.

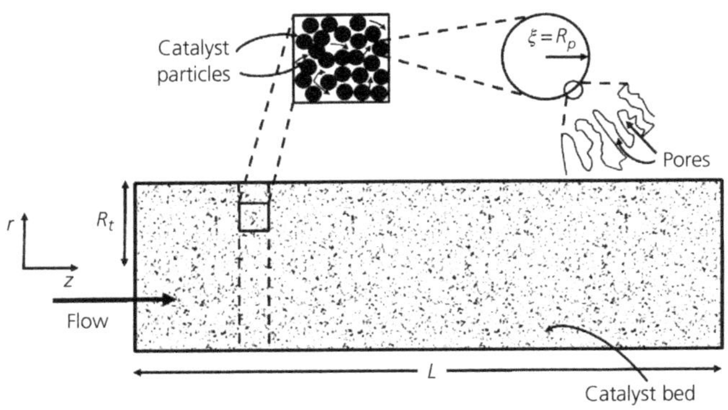

Figure 1.2 Schematic presentation of a packed-bed reactor. (Source: Onsan and Avci [1]. Reproduced with permission of Elsevier.)

particles are randomly packed, so there is not a regular structure, and, as a result, fluid flow takes place through irregular, random paths. Reactions take place over the active sites that are buried within the pores of the catalyst particles. A simple description of the PBR operation is shown in Figure 1.2 [1]. Owing to their relatively simple configuration and operation, PBRs are widely used in the chemical industry. They are used in high-throughput, continuous operations. Since the catalyst is considered as a separate solid phase and the fluid types are either gas only or gas–liquid mixtures, PBRs are classified as heterogeneous reactors. In the case of coexistence of three phases with concurrent downflow of liquid and gas over the solid packing, the reactor is called as a trickle-bed reactor (see Section 1.2.4). The geometry of the catalyst-containing volume, which can be either a tube or a vessel, dictates the type of the PBR. Descriptions of the so-called tubular and vessel-type PBRs are given later.

1.2.1.1 Tubular PBRs

PBRs are known to have inherently weak heat transfer properties due to the presence of voids within the catalyst bed (Figure 1.2 [1]) that act as resistances against the transport of heat along the reactor. The tubular PBR geometry, which involves the location of catalyst-containing tubes in a particular pattern within a shell, is preferred over a regular vessel when high rates of heat input or removal are essential for highly endothermic or exothermic reactions, respectively. This advantage of the tubular configuration, however, comes at the expense of higher pressure drop. It is also worth noting that the process of catalyst packing and unloading in tubular geometry is more difficult than that involved in vessels. Therefore, catalyst lifetime in tubular PBRs should be long enough to minimize the downtimes for and costs associated with catalyst changeover.

The shell/tube configuration of tubular PBRs depends on the nature of the catalytic reaction. For highly endothermic reactions such as catalytic steam reforming, the reactor geometry is similar to that of a fired furnace in which the catalyst-packed tubes are heated by the energy released by the combustion of a fuel on the shell side. Catalytic steam reforming involves the conversion of a hydrocarbon to a hydrogen-rich mixture in the presence of steam:

$$C_mH_nO_k + (m-k)H_2O = mCO + (m-k+n/2)\,H_2,\ (m>k) \qquad \Delta H^{\circ} > 0 \tag{1.1}$$

The process is known as the conventional method of producing hydrogen for meeting the hydrogen demands of the refining and petrochemical industry. The most widely used fuel in steam reforming is natural gas, which is mostly composed of methane:

$$CH_4 + H_2O = CO + 3H_2, \qquad \Delta H^{\circ} = 206\ \text{kJ/mol} \tag{1.2}$$

Methane steam reforming is conventionally carried out over Ni-based catalysts. Owing to the high endothermicity and slow kinetics, the process depends strongly on the input of external energy at high rates for ensuring commercially viable throughput of hydrogen. The critical energy demand of the reaction is met in a reactor (also called as the reformer) where multiple Ni-based catalyst-packed tubes are heated mainly via radiative heat generated by homogeneous combustion of a fuel, typically natural gas, in a process furnace. This configuration sets the basis for the development and use of various types of commercial steam reforming reactors described in Figure 1.3 [2], which differ in the positions of heat source and the degree of delivery of the combustion energy to the so-called reformer tubes. A further detailed representation of a tubular reformer is provided in Figure 1.4 [2]. Depending on the capacity of the reactor, the number of tubes can be increased up to 1000, each having outer diameter, wall thickness, and heated length ranges of 10–18 cm, 0.8–2.0 cm and 10–14 m, respectively. The degree of furnace-to-tube heat transfer affecting the rate of Reaction 1.2 and hydrogen production capacity of the reactor is limited by thermal stability of the tube material which is found to decrease significantly with temperature above ca. 850°C [3]. Therefore special alloys, particularly microalloys, composed of 25Cr 35Ni Nb Ti are used to improve the operating window of the reactor [3].

The multitubular PBR configuration is preferred when convection is not sufficient for delivering the necessary heat flux to sustain the operation. However, in most of the exothermic and endothermic reactions, the temperature of the catalyst bed can be regulated by convective external heat transfer. In

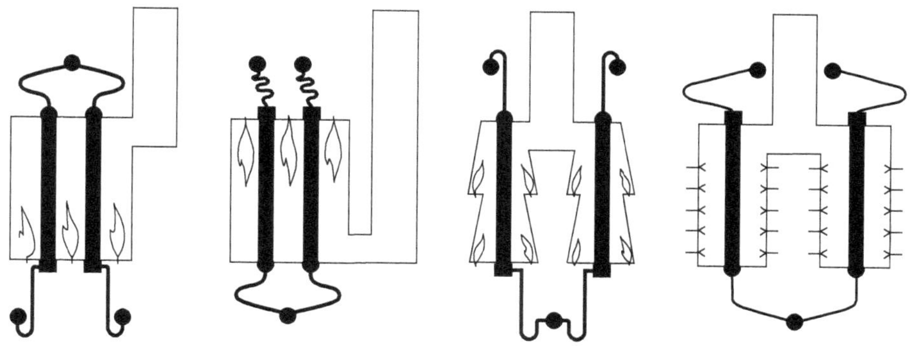

Figure 1.3 Furnace configurations for multitubular packed-bed reformers. (Source: Dybkjaer [2]. Reproduced with permission of Elsevier.)

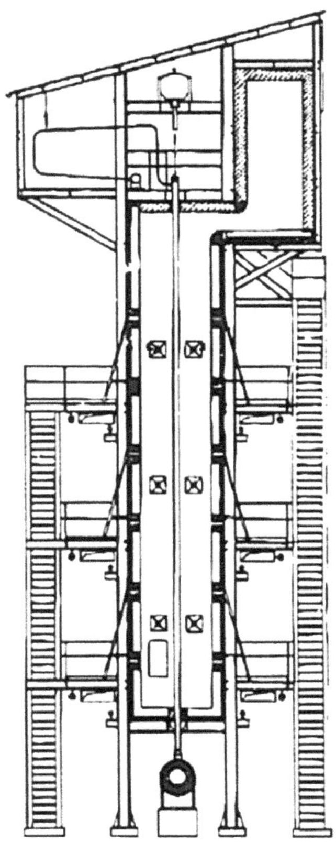

Figure 1.4 Side-fired tubular reformer design by Haldor-Topsøe. (Source: Dybkjaer [2]. Reproduced with permission of Elsevier.)

such cases, the catalyst-containing tubes are bundled in a shell-and-tube heat exchanger like configuration involving circulation of the heat transfer fluid on the shell side. This PBR concept is described in Figure 1.5 [4] in which alternative methods of circulation of the heat transfer fluid around the packed tubes are introduced. In mildly endothermic or exothermic reactions, heat transfer can be realized to provide nearly isothermal conditions in cross-flow and parallel flow configurations shown in Figure 1.5a and b [4], respectively. In such reactors, inside diameters and lengths of the tubes are reported to vary between 2–8 cm and 0.5–15 m, respectively [4]. For endothermic cases, the heating medium can be a gas or a liquid, with the latter offering better heat transfer rates due to higher convective heat transfer coefficients of liquids. Cooling in exothermic reactions is carried out either by circulation of a heat transfer fluid or by boiling heat transfer. In the former case, fluids such as molten salts are force-circulated around the tube bundle. The heated liquid leaving the reactor is then passed through an external steam generator and cooled for the next cycle. In the case of boiling heat transfer (Figure 1.5c [4]), however, the cooling fluid that is fed from the bottom of the reactor rises up due to natural circulation induced by the decreasing density profile that is caused by continuous heat absorption from the tubes. Partial evaporation of the cooling water is also observed. Vapor bubbles agitate the liquid and increase the convective heat transfer coefficient. The resulting vapor–liquid mixture is then let to settle in a steam drum where steam is separated, and the remaining liquid sent back to the cooling cycle together with some makeup water. Even though this configuration eliminates the need for cooling fluid transportation equipment, the tubes may be overheated if heat generation in the tubes becomes excessive to evaporate cooling water on the shell side. In such a case, the rate of convective heat removal will be less than the rate of catalytic heat generation, and the tubes are subjected to the risk of burning out.

In multitubular PBRs heat management can be improved by increasing the heat transfer area per catalyst volume, which is possible by using tubes with smaller diameters. In this case, definite amounts of catalyst will be packed into a higher number of tubes, which will offer increased external tube surface area for heat transfer. Due to the reduced tube cross-sectional area, smaller tube diameters will also increase the linear flow rate of the reactive mixture and favor well-mixed conditions that increase the heat transport rates. However, these advantages are naturally limited by pressure drop, as higher flow rates will cause increased frictional loss of mechanical energy of the reactive fluid and will require increased pumping/compression costs. Nevertheless, the trade-off between heat transfer rates and pressure drop can be relaxed by the possibility of using different combinations of size, shape, and material of the catalyst pellets [4, 5]. For example, pellet shapes offering higher void fractions and larger hydraulic diameters allow lower pressure drop operations. It is worth noting that the rate of catalytic reactions increases with the surface area of the catalyst bed that necessitates the use of smaller pellets. Therefore pellet size also requires careful optimization.

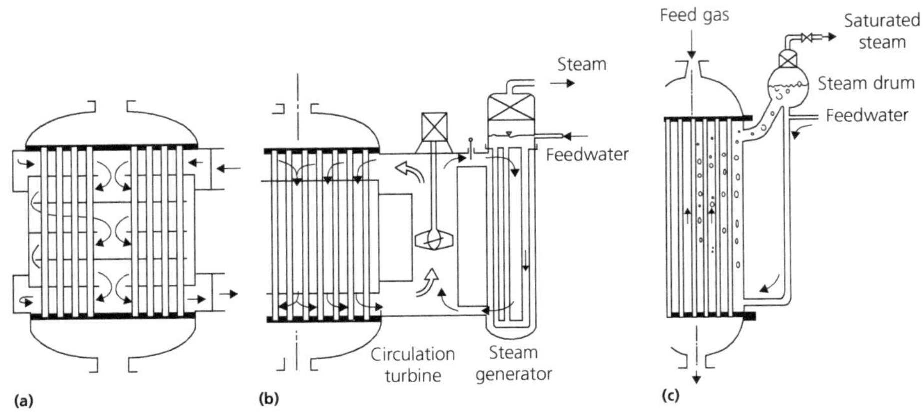

Figure 1.5 Heat transfer strategies in multitubular packed-bed reactors. (a) Cross-flow, (b) parallel flow, and (c) boiling-water cooling. (Source: Eigenberger [4]. Reproduced with permission of John Wiley & Sons, Inc.)

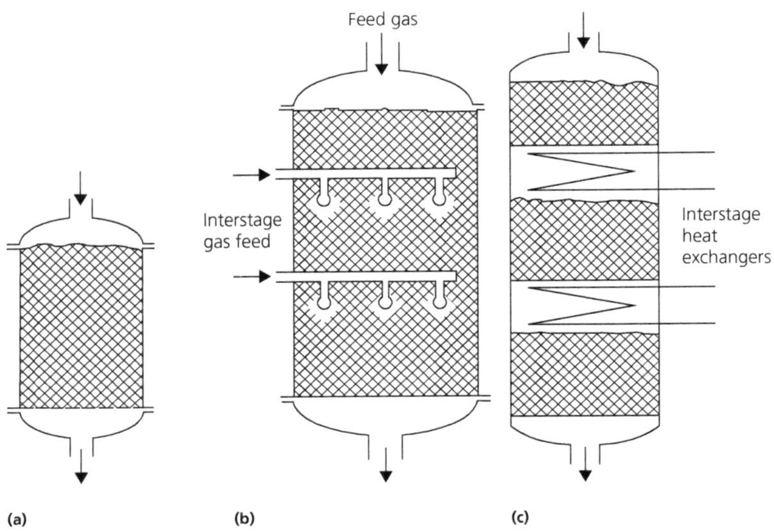

Figure 1.6 Various configurations of vessel-type packed-bed reactors. (a) Single-bed adiabatic packed-bed reactor, (b) adiabatic reactor with interstage gas injection, and (c) multiple adiabatic beds with interstage heat exchange. (Source: Eigenberger [4]. Reproduced with permission of John Wiley & Sons, Inc.)

The length and diameter of the tube and the particle size (hydraulic diameter) also affect flow distribution within the packed tube. If the ratio of the tube diameter to that of the particle diameter is above 30, radial variations in velocity can be neglected, and plug (piston) flow behavior can be assumed. The ratio of the tube length to particle diameter is also important; if this ratio exceeds 50, axial dispersion and axial heat conduction effects can be ignored. These effects bring notable simplifications into the modeling of PBRs, which are discussed in Chapter 3.

1.2.1.2 Vessel-type PBRs

The design and operational requirements explained for tubular PBRs are also valid for PBRs in which the catalyst bed is packed in one vessel as described schematically in Figure 1.6a [4]. This reactor configuration is preferred when the reaction is carried out at adiabatic conditions. However, as demonstrated in Figure 1.6b and c [4], bed temperature can be changed by heat addition to/removal from the bed for obtaining a temperature profile as close as possible to that of the optimum. Figure 1.6b [4] is a representation of addition or removal of heat to/from the catalyst bed by direct injection of hot or cold feed to the bed. This heat management strategy can be used where the heats of reactions are low. Successful implementation of this strategy depends on careful consideration of mixing and redistribution of the injected fluid with that of the reactive mixture and of the adiabatic temperature change upon injection, which should be within acceptable limits. A better regulation of the bed temperature is possible by the use of interstage heat exchangers between multiple adiabatic beds (Figure 1.6c [4]). This configuration is more suitable for improving conversions or product selectivities in reactions limited by chemical equilibrium. The possibility of using different heat exchange equipment between the stages helps in handling high reaction enthalpies. For endothermic reactions, interstage heating is usually carried out by means of fired heating, in which the heat transfer fluid is heated in a fired furnace and then circulated between the beds to provide heat to the reactive fluid. Adiabatic heat generated during exothermic reactions is removed by contacting the hot bed effluent with interstage heat exchange tubes in which a coolant, for example, water, is circulated for steam generation purposes.

Multiple adiabatic beds with interstage heat exchange configuration compete with tubular PBR geometry, as both configurations provide regulation of the bed temperature to improve reactant conversion and product selectivity. In this respect, the tubular PBR alternative is better, because it offers continuous control over the bed temperature. However, although temperature regulation is only possible through a stepwise pattern in the multiple adiabatic beds, they do offer several practical advantages such as the possibility of (i) changing the catalyst bed in individual stages at different times, (ii) distributed stagewise feeding of a reactant instead of its total feeding at the inlet, and (iii) drawing a limiting product from an intermediate stage in case of reactions limited by equilibrium [4, 5].

Vessel-type PBRs are widely used in chemical industry. A descriptive example is ammonia synthesis, which is an exothermic equilibrium reaction:

$$N_2 + 3H_2 = 2NH_3, \quad \Delta H^\circ = -92.4\ \text{kJ/mol} \tag{1.3}$$

The reaction is carried out in a multistage PBR with interstage cooling (Figure 1.7 [4]) in the 400–500°C range and involves the use of iron-based catalysts. In order to favor ammonia production by shifting the chemical equilibrium to the product side, pressures up to 300 bar are required. As adiabatic temperature rise hinders conversion due to the equilibrium limit, the reactive mixture is cooled down between the beds, and the recovered heat is used for steam generation. The resulting conditions deliver a product mixture including ca. 20% NH_3 which is separated by a series of condensers. Upon separation, unreacted mixture of N_2 and H_2 is combined with fresh makeup feed and recycled to the first stage of the reactor.

Another commercial example involving the use of a vessel-type PBR is autothermal reforming (ATR) of natural gas. It is a key step in gas-to-liquid (GTL) processes and is used to produce synthesis gas ($CO + H_2$) for FT synthesis in which a mixture of hydrocarbons in the C_1–C_{30+} range is synthesized [6]. In ATR, noncatalytic oxidation (Reaction 1.4) and Ni-catalyzed steam reforming of natural gas (Reaction 1.2) are combined, and product distribution is affected by water–gas shift (Reaction 1.5), an important side reaction of steam reforming [3, 7]:

$$CH_4 + 1.5O_2 \rightarrow CO + 2H_2O, \quad \Delta H^\circ = -519\ \text{kJ/mol} \tag{1.4}$$

$$CH_4 + H_2O = CO + 3H_2, \quad \Delta H^\circ = 206\ \text{kJ/mol} \tag{1.2}$$

$$CO + H_2O = CO_2 + H_2, \quad \Delta H^\circ = -41\ \text{kJ/mol} \tag{1.5}$$

ATR is carried out in an adiabatic PBR as described in Figure 1.8 [7]. Natural gas, steam, and oxygen (or enriched air) are cofed to a mixer–burner unit for ensuring combustion of the homogeneous mixture of reactants taking place in the combustion chamber. Heat produced in the combustion zone, where temperature can be well above ca. 1500°C, is then transferred to the Ni-based catalyst bed on which Reactions 1.2 and 1.5 take place to produce a mixture of H_2 and CO at molar ratios close to 2 at temperatures above ca. 1000°C and at pressures up to ca. 30 bar [3, 7]. Success of the reactor depends on keeping the exothermic heat within the vessel, that is, operating the reactor adiabatically. For this purpose, the inner wall of the steel pressure vessel is lined with multiple layers of refractory insulation. A special catalyst pellet shape including numerous holes is used to minimize pressure drop along the bed and to avoid bypass of gas through the refractory layer.

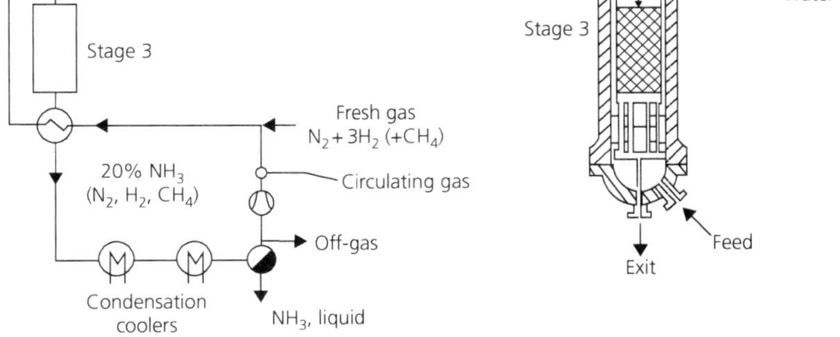

Figure 1.7 Packed-bed reactor with multiple adiabatic beds for ammonia synthesis.
(Source: Eigenberger [4]. Reproduced with permission of John Wiley & Sons, Inc.)

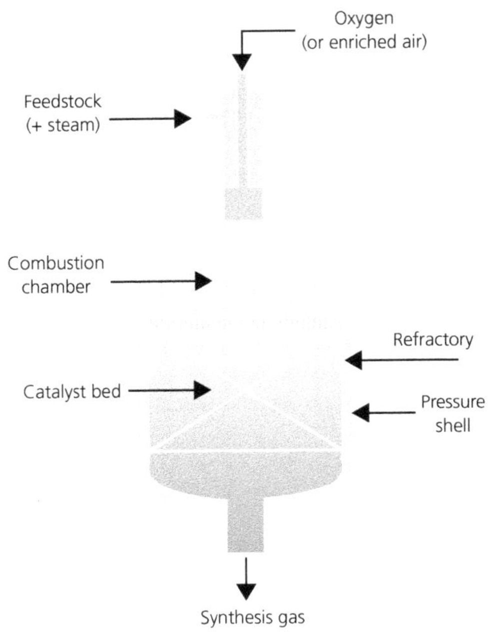

Figure 1.8 Packed-bed reactor configuration for autothermal reforming of methane to synthesis gas.
(Source: Aasberg-Petersen et al. [7]. Reproduced with permission of Elsevier.)

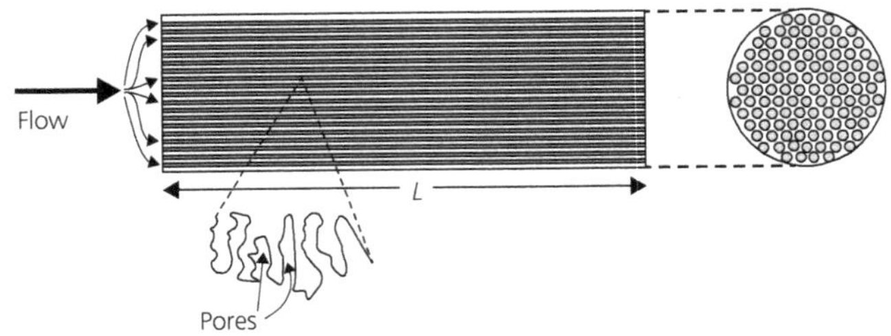

Figure 1.9 Schematic presentation of a monolith reactor.
(Source: Onsan and Avci [1]. Reproduced with permission of Elsevier.)

1.2.2 Monolith reactors

Monolith reactors are composed of a large number of parallel channels, all of which contain catalyst coated on their inner walls (Figure 1.9 [1]). Depending on the porosity of the monolith structure, active metals can be dispersed directly onto the inner channel walls, or the catalyst can be washcoated as a separate layer with a definite thickness. In this respect, monolith reactors can be classified among PBR types. However, their characteristic properties are notably different from those of the PBRs presented in Section 1.2.1. Monolith reactors offer structured, well-defined flow paths for the reactive flow, which occurs through random paths in PBRs. In other words, the residence time of the reactive flow is predictable, and the residence time distribution is narrow in monoliths, whereas in a PBR, different elements of the reactive mixture can pass through the bed at different rates, resulting in a wider distribution of residence times. This is a situation that is crucial for reactions where an intermediate species is the desired product and has to be removed from the reactor before it is converted into an undesired species.

Hydraulic diameters of monolithic channels range between ca. 3×10^{-4} m and 6×10^{-3} m [8]. Combination of such small diameter channels leads to surface areas per reactor volume in the order of ~10^4 m^2/m^3 (which is ~10^3 m^2/m^3 for PBRs) and void fractions up to ~75% (which is ~40% for PBRs). As shown in Figure 1.10 [9], these design properties allow monolith reactors to operate with pressure drops that are up to three orders of magnitude less than those observed in PBRs.

Monolith reactors differ from PBRs in terms of transport properties. Owing to the small channel diameters, the flow regime is laminar. In this case, channel shape and diameter dictate the values of heat and mass transfer coefficients according to the definitions of the Nusselt ($Nu = h_f d_h / \lambda_f$) and Sherwood ($Sh = k_g d_h / D_{AB}$) numbers, respectively. Assuming that the flow is fully developed, values of Nu and Sh are constant for a given channel shape [10]. However, in the case of PBRs, where turbulent flow conditions are valid, transport coefficients improve with the degree of turbulence and mixing within the reactor. It is worth noting that transport coefficients in monolith channels can be slightly affected by the flow rate if the surface of the channel is tortuous. The reader is directed to Chapter 8 for a detailed analysis and discussion of monolith reactors.

Heat management in monolith reactors via external heating or cooling is not as effective as in PBRs due to lack of convective heat transport in the radial direction. At this point, the material of construction of the monolithic structure affects the overall performance. Monolith reactors can be made of metals or ceramics. In case of nonadiabatic reactions, metallic monoliths are preferred due to their higher thermal conductivity which partially eliminates the lacking convective contribution. Ceramic monoliths, on the other hand, have very low thermal conductivities (e.g., 3 W/m.K for cordierite [11]) and are suitable for use in adiabatic operations.

Despite their notable advantages in terms of residence time distribution and pressure drop, the operating windows of monolith reactors are narrower than those of PBRs. As the catalyst is integrated to the monolithic structure, replacement of the catalyst bed in case of its irreversible deactivation becomes a serious issue. Moreover, small channels are subject to the risk of plugging either by the dirt and scale that can come together with the feed stream or by phenomena such as coking that may occur during reactions involving hydrocarbons conducted at high temperatures. In such as case, flow distribution and residence time in the channels will be disturbed, and product distribution will be adversely affected. Prevention of these risks is possible by careful selection and control of the operating conditions, which in turn put some limitations on the versatility of using monolith reactors.

The capability of offering high surface area-to-volume ratios together with low pressure drop makes monolith reactors the

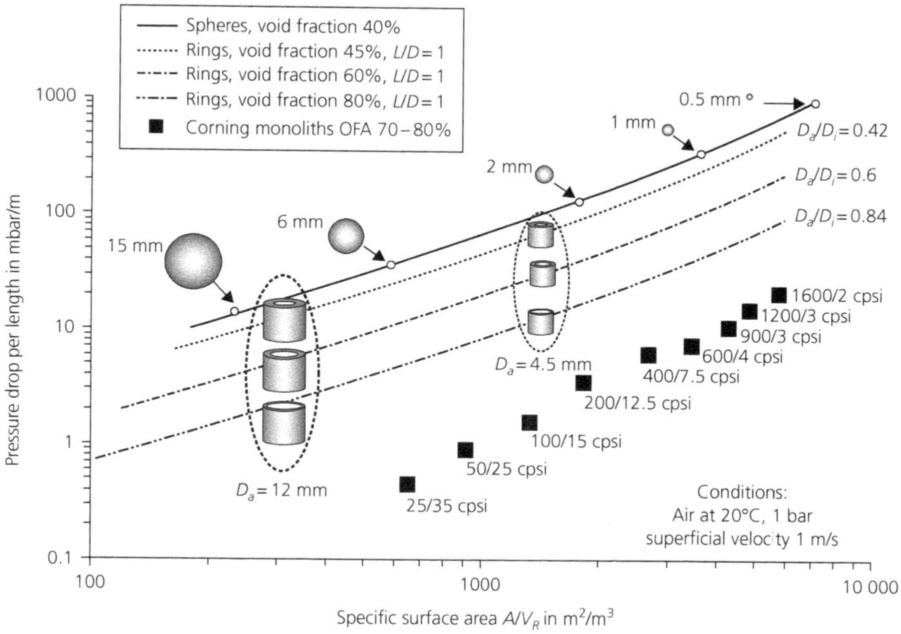

Figure 1.10 Comparison of pressure drop in various configurations of monoliths and packing structures.
(Source: Boger et al. [9]. Reproduced with permission of American Chemical Society.)

unique choice for use as three-way catalytic converters in vehicles to regulate the emission levels. The compact nature of the monolithic catalytic converters allows their integration into the exhaust gas aftertreatment zone of the vehicles. These converters involve washcoated layers of precious metal catalysts that are capable of reducing the NO_x, CO, and unburned hydrocarbon content of the exhaust gas below the legislative limits. Apart from vehicular use, monolith reactors are also used in NO_x removal from flue gases in power stations because of their capability of providing adiabatic conditions with low pressure drop. It is worth noting that monolith reactors are not limited for use only in gas-phase reactions and can also be used for handling gas–liquid-type reactive mixtures [10].

1.2.3 Radial flow reactors

In addition to monolith reactors, pressure drop in fixed-bed operation can be reduced by employing radial flow reactors. These units are essentially packed-bed type, with gaseous reactive flow being in the radial direction, that is, perpendicular to the catalyst bed, instead of being in the axial direction (Figure 1.11 [4]). The radial flow pattern is achieved by directing the flow to the catalyst pellets that are packed between two perforated cylinders or concentric screens. The flow orientation is flexible, that is, can be either from outside cylinder to inside cylinder or vice versa. In this design, radial flow distance along the catalyst bed is constant and is independent of the amount of catalyst packed. This unique feature makes radial flow reactors suitable for use in cases where large catalyst volumes are needed in high-pressure operations with strict pressure drop limitations. During operation, however, the catalyst bed settles down and causes a gap for bypassing of the fresh feed through the upper part of the perforated cylinder. This issue can be addressed by refining the design of the upper closure [4]. Radial flow reactors are used in such applications as the synthesis of ammonia (Figure 1.12 [12]) and methanol.

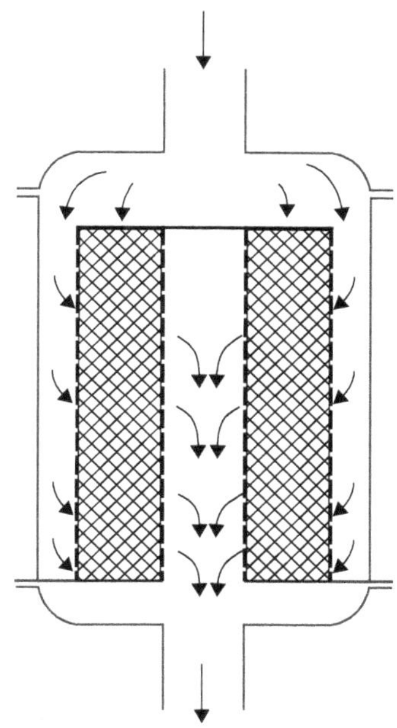

Figure 1.11 Radial flow reactor concept.
(Source: Eigenberger [4]. Reproduced with permission of John Wiley & Sons, Inc.)

1.2.4 Trickle-bed reactors

Trickle-bed reactors are similar to the PBR geometry described in Section 1.2.1.2, with the main difference being the coexistence

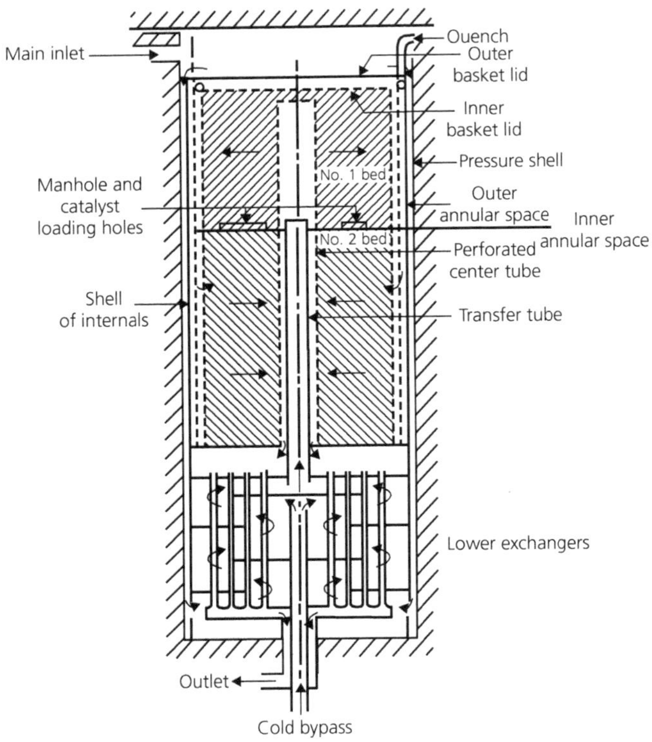

Figure 1.12 Radial flow ammonia synthesis converter by Haldor-Topsøe. (Source: Couper et al. [12]. Reproduced with permission of Elsevier.)

of gas and liquid phases in the reactive mixture and putting trickle-bed reactors among those classified as three-phase (gas–liquid–solid) reactors. In gas–solid PBRs described in Section 1.2.1.2, headspace above the catalyst bed is usually filled with inert ceramic balls to ensure uniform distribution of the gaseous feed over the entire bed. Cocurrent feeding of gas and liquid phases, however, calls for using a more sophisticated distributor design that is expected to mix the two phases and then distribute them uniformly across the catalyst bed to ensure sufficient wetting of the catalyst pellets and to prevent channeling of the gas and liquid components in the feed. The requirement of sophisticated distributors such as bubble cap trays is another factor that differentiates trickle-bed reactors from gas–solid PBRs. Status of feed mixture distribution to the catalyst bed dictates the diameter of the reactor, which is usually under 5 m. Height-to-diameter ratio is usually in the range of 5 and 25 [13]. Typical sizes of the catalyst pellets, which can be cylinder, sphere, extrudate, needle, or bead in shape, range between 1 and 5×10^{-3} m and give bed void fractions between ~0.35 and 0.40 [13]. Details on the design, analysis, and operation of trickle-bed reactors are provided in Chapters 5 and 13.

Trickle-bed reactors are mainly used in key petroleum refining applications such as hydrocracking, hydrodesulfurization, and hydroisomerization. The process involves the combination of hydrogenation/hydrotreating and cracking of vacuum gas oil and residues (liquid phase) to produce lighter hydrocarbons such as gasoline in the presence of hydrogen (gas phase) over a catalyst (solid phase) in the 300–600°C range and at pressures up to ~150 atm to ensure high solubility of the gaseous phase in the liquid. Conventional hydrocraking catalysts, such as Pt on aluminosilicates or zeolites, involve two components, namely, an acidic component for cracking and isomerization reactions and a noble metal component for the hydrogenation reactions [14]. The trickle-bed reactor involves the presence of up to six successive catalyst beds. Since hydrocracking reactions are exothermic, adiabatic temperature rise in each bed is regulated by interstage cooling enabled by the injection of cold hydrogen quenches; the gas–liquid mixture is remixed and redistributed prior to its entrance to the succeeding bed. In hydrodesulfurization, which is an important operation in crude oil refining, the organic sulfur components, that is, sulfides, disulfides, thiols, and thiophenes existing in crude oil (liquid phase), are converted to hydrogen sulfide in the presence of hydrogen (gas phase) over alumina-supported Co–Mo or Ni–Mo catalysts (solid phase) in the 350–400°C range. The resulting H_2S is then removed by processing over beds of ZnO. In hydroisomerization, on the other hand, the light alkanes in the C_4–C_6 range are converted to branched-chain isomers in the presence of hydrogen for producing high-octane component additives for being blended into gasoline. The process, carried out in trickle-bed reactors, involves the use of catalysts such as Pt supported on chlorinated alumina or on acidic zeolites. In contrast with hydrocrackers, interstage heat exchange is not used in hydroisomerization reactors which involve milder conditions, with temperatures and pressures ranging between ca. 110–180°C and 20–70 atm, respectively. As exothermic equilibrium reactions are involved in hydroisomerization, the catalyst should be able to operate at low temperatures to favor the desired conversions.

1.2.5 Short contact time reactors

Pressure drop in fixed beds can be reduced by minimizing the amount of catalyst used, which leads to the existence of short contact times. In addition to reduction of pressure drop, these reactors are ideal for carrying out reactions whose extent and product distribution depend strongly on the contact time (e.g., direct partial oxidation of hydrocarbons to synthesis gas). A typical concept of such a reactor, called the disk reactor, is shown in Figure 1.13 [4]. The reactor involves a thin layer of catalyst in the form of wire gauzes or pellets, whose height and diameter are in the orders of centimeters and meters, respectively. Quenching at the downstream of the catalyst bed helps in halting further conversion of the products into other unwanted species.

In addition to the disk reactor, short contact times can also be achieved in monolith reactors (Section 1.2.2) and in microchannel reactors (Section 1.2.5), the latter involving fluid mechanical

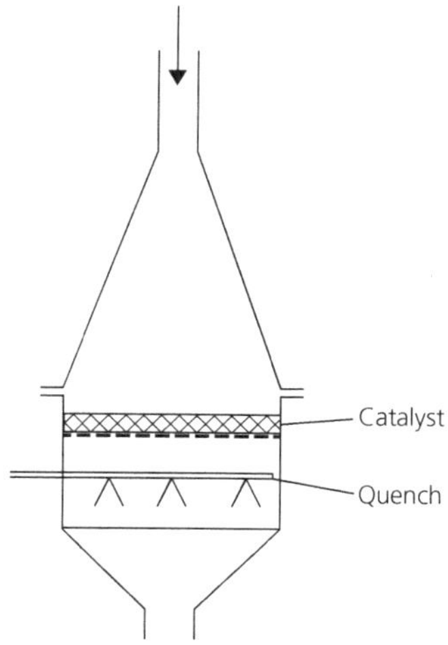

Figure 1.13 Disk reactor concept.
(Source: Eigenberger [4]. Reproduced with permission of John Wiley & Sons, Inc.)

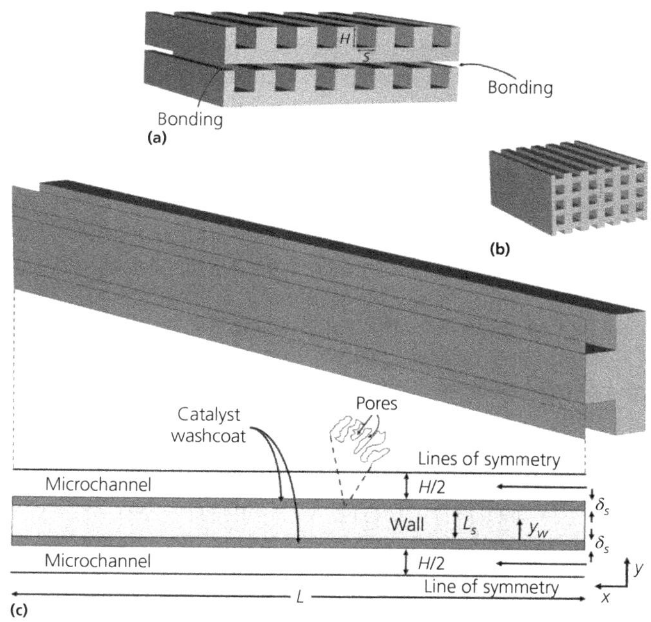

Figure 1.14 Schematic presentation of a microchannel reactor. (a) Machined plates with microchannels, (b) microchannel reactor block obtained after bonding the plates, and (c) characteristic section of the multichannel reactor. (Source: Onsan and Avci [1]. Reproduced with permission of Elsevier.)

properties and architectures similar to those of monoliths, where the existence of thin layers of washcoated porous catalysts together with high fraction of void space allows fast fluid flow almost without compromise from pressure drop (Figure 1.14 [1]). These factors lead to the occurrence of contact times in the order of milliseconds, whereas it is in the order of seconds in PBRs. Like in the case of monoliths, the existence of a structured flow pattern in microchannel units leads to precise control of residence times that promotes selective productions. Even though such similarities exist between monolith and microchannel reactors, they differ in certain aspects. Microchannel units have channel diameters in the submillimeter range, whereas larger diameter channels up to 6×10^{-3} m are used in monoliths. Owing to the constant *Nu* and *Sh* numbers per cross-sectional channel shape, higher heat and mass transport coefficients can be obtained in microchannels as a result of the smaller hydraulic diameters which also lead to higher surface area-to-volume ratios (i.e., up to ~5×10^4 m^2/m^3) than those of monoliths. These factors favor precise regulation of reaction temperature, an important benefit for strongly exothermic reactions. Due to their special manufacturing techniques involving micromachining and bonding of the plates (Figure 1.14 [1]), various nonlinear patterns (e.g., wavy shapes) along the channel length, which induce static mixing and improve heat transport, can be implemented in microchannels [15]. On the other hand, in monoliths, channels are limited to have straight axial patterns. Finally, the range of materials of construction is versatile (e.g., various metals and ceramics, polymers, silicon) in microchannels, whereas monoliths can be made of ceramics and metals only.

In addition to their advantages stated earlier, compact dimensions of the microchannel reactors allow inherently safe productions, as the risks associated with reactions (e.g., thermal runaway) are not significant due to the small quantities in the order of microliters processed in each channel. Even though small throughput is a disadvantage of short contact time reactors, the capacity of the microchannel reactors can be rapidly increased through the so-called numbering-up approach, which is much simpler than the traditional scaling-up approach. The resulting capacities are expected to be suitable for small-scale throughput industries such as pharmaceuticals and fine chemical productions. Applications of microchannel reactors in these industries are provided by Hessel et al. [16]. Nevertheless, production capacities of the microchannel units and other short contact time reactors are far from being able to compete with those of the continuously operating commercial reactors involved in the petroleum and petrochemical industries. The reader is directed to Chapters 9 and 14 for more detailed information about the microchannel reactors.

1.3 Reactors with moving bed of catalysts

1.3.1 Fluidized-bed reactors

Fluidized-bed reactors (FBRs) are continuously operating units of the gas–solid type, involving a catalyst bed which is fluidized when the volumetric flow rate of the gaseous feed stream exceeds a limiting value called the *minimum fluidization* flow rate. The resulting degree of mixing between the gas and solid phases in the FBR brings several operational advantages over a gas–solid PBR (Section 1.2.1). FBRs offer uniform temperature distribution due to intensive mixing, which minimizes the chance of hot spot formation in exothermic reactions. Heat management in FBRs is conventionally carried out by the heat transfer surfaces that are immersed into the reactor vessel. In this respect, fluidization favors heat transfer coefficients and subsequent fast heat exchange between the bed and immersed heat transfer surfaces. Mobility of the catalyst phase widens the operating window for allowable pressure drop. Therefore, pellet sizes smaller than those involved in PBRs can be used in FBRs, and higher reaction rates can be obtained due to increased catalytic surface area per unit bed volume. Even though higher heats of reactions evolve with increased rates, the possibility of fast heat exchange helps in effective regulation of the bed temperature. FBRs also allow constant catalytic activity either by online addition of fresh catalyst or by its continuous regeneration in a separate zone, like in the case of the fluidized catalytic cracking (FCC) operation described later. Modeling and design aspects of FBRs are explained in detail in Chapter 4.

The advantages listed previously for FBRs, however, have to be considered together with several operational limitations. Fluidization of the catalyst pellets at high velocities can cause unavoidable acceleration of the erosion of both reactor vessel and heat exchange surfaces, and their undesirable breakdown into smaller particle sizes eventually calls for the need of cost-intensive catalyst separation/gas purification equipment. In contrast with breakdown, the pellets can also merge into each other, and the resulting increase in particle weights can cause defluidization, which can seriously disturb the reactor operation. Moreover, residence time distribution is not narrow in FBRs due to the chaotic movement of reactive fluid inside the vessel. Another operational drawback of FBRs is linked with their high sensitivity against the presence of sulfur in the gaseous feed mixture. Once they enter the reactor, sulfur-containing molecules can immediately poison the entire bed due to intense mixing of the phases and the highly exposed surface area of small catalyst particles and can eventually cause a sudden drop in pressure. This serious drawback, however, is less serious in gas–solid PBRs as sulfur poisoning moves like a wave front. In other words, at the beginning of the operation, only the section of the packed bed near the inlet will be poisoned, while pellets at the downstream will remain active until the ones at the upstream are saturated with sulfur.

Apart from the operational drawbacks stated earlier, capital and operating expenses involved in an FBR exceed those of a PBR of equivalent capacity due to requirements of larger vessel volume for handling fluidization and of installing gas purification and solid circulation components. Chaotic nature of the operation also calls for a tedious preliminary study of the process of interest at the pilot scale that should be followed by a labor and cost-intensive scaling-up stage, all of which eventually increase the capital cost of the commercial FBR unit.

Although not as widely used as a gas–solid PBR, FBR remains as the only choice for processes such as FCC and high-temperature Fischer–Tropsch (HTFT) synthesis, both of which have key roles in the petroleum processing and petrochemical industries. FCC is a critical step in petroleum refining and involves catalytic breakdown of heavy gas oil molecules into

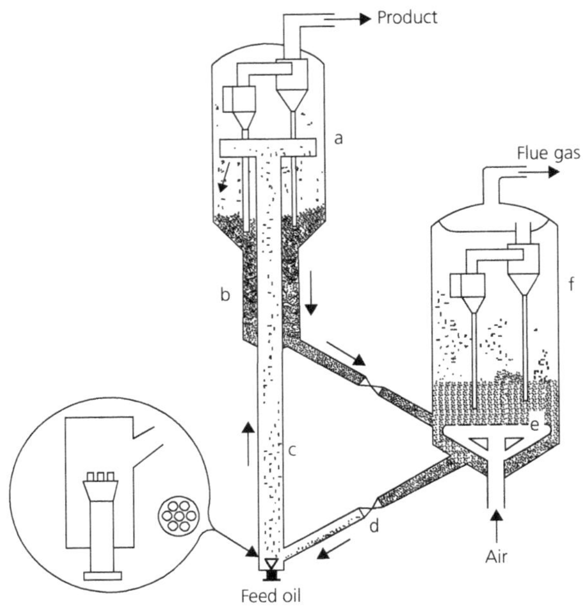

Figure 1.15 Riser cracking process by UOP. (a) Reactor, (b) stripper, (c) riser, (d) slide valve, (e) air grid, and (f) regenerator.
(Source: Werther [17]. Reproduced with permission of John Wiley & Sons, Inc.)

commercially valuable products such as gasoline, diesel, and olefins. The FBR reactor, shown in Figure 1.15 [17], is composed of a riser and a regenerator between which the catalyst is circulated continuously at rates that can exceed 100 tons/min. Endothermic cracking reactions that take place in the riser at temperatures of 500–550°C unavoidably deposit coke on the surface of the zeolite-based catalyst pellets [17]. Spent catalysts are continuously transported to the regenerator in which coke is burned off with hot air at ca. 730°C for the restoration of the catalytic activity. The cycle is completed when the regenerated catalysts are conveyed back to the riser unit. Heat needed to drive the endothermic cracking reactions is supplied by the hot catalysts that come from the regenerator. HTFT synthesis, on the other hand, involves catalytic conversion of synthesis gas into a hydrocarbon mixture rich in olefins and gasoline. The process is carried out at 340°C and 20 atm over iron-based catalysts. As FT synthesis is strongly exothermic and the product distribution is a strong function of temperature, the catalyst bed should be maintained at isothermal conditions. This requirement is met by the circulating fluidized-bed (CFB) reactor, known as the Sasol Synthol reactor, shown in Figure 1.16a [12], in which heat released during reactions is absorbed by the cooling coils immersed into the reactor vessel to produce steam [18, 19]. These reactors can operate with capacities up to 8×10^3 barrels/day (3.3×10^5 tons/year). CFB reactors are then replaced by turbulent FBRs, known as Sasol Advanced Synthol reactors (Figure 1.16b [19]), due to their smaller size, lower capital expense requirements and maintenance costs, and their ability to operate at higher conversions and capacities up to 2×10^4

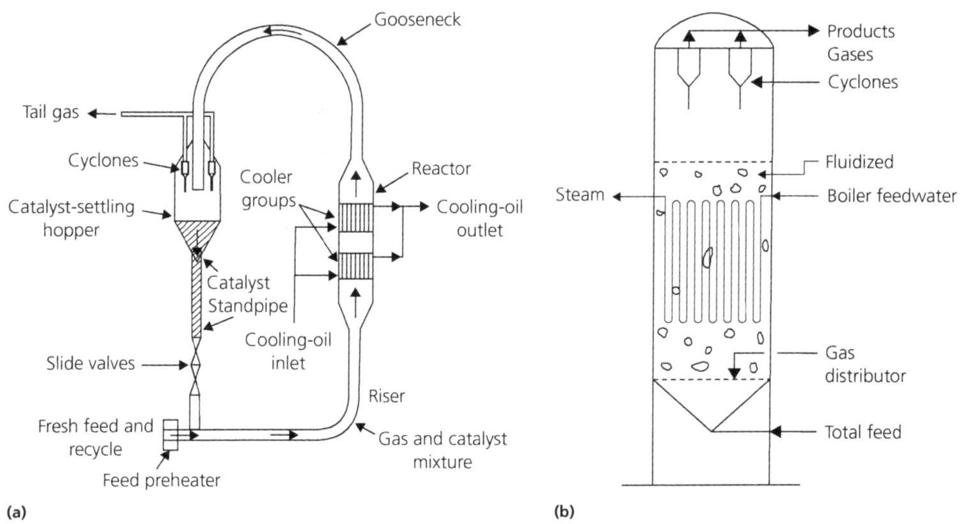

Figure 1.16 High-temperature Fischer–Tropsch synthesis reactors. (a) Sasol Synthol circulating fluidized-bed reactor. (Source: Couper et al. [12]. Reproduced with permission of Elsevier.) (b) Sasol Advanced Synthol turbulent fluidized-bed reactor. (Source: Steynberg et al. [19]. Reproduced with permission of Elsevier.)

barrels/day (8.5×10^5 tons/year) with lower pressure drop [18, 19]. The use of FBRs in HTFT is extensively discussed in Chapter 12.

1.3.2 Slurry reactors

Slurry reactors involve the coexistence and intense mixing of gas, liquid, and solid phases in the same volume. The possibility to run slurry reactors in the batch, semibatch, or continuous modes differentiates these reactors from others in terms of operational flexibility. In slurry reactors, the roles of the three phases can be different, that is, liquid can be a reactant, a product, or an inert that serves as a contacting medium for gas and solids. Similarly, dissolved gas can either be a reactant or an inert for inducing mixing of liquid and solids via bubbling. The solid phase usually corresponds to the finely dispersed catalyst particles with diameters lower than 5×10^{-3} m [20].

Slurry reactors are typically used for highly exothermic reactions. Heat removal from the reaction mixture is provided by cooling coils immersed into the reactor vessel. Intense mixing, which is enabled either by gas bubbling or by a mechanical agitator, increases the heat transfer coefficient between the reaction mixture and coils and improves the rate of heat removal. High heat capacity and heat transfer coefficients of the slurries are other factors that further promote heat transport and temperature control. Excellent heat management capabilities of slurry reactors make them promising candidates for several processes, with the most popular one being the low-temperature Fischer–Tropsch (LTFT) synthesis that involves conversion of syngas into a hydrocarbon mixture heavier than that synthesized in HTFT. LTFT is carried out in the ~190–250°C range and at pressures between 20 and 40 atm over Co-based catalysts [6, 18]. As Co is more active than the Fe catalyst of HTFT [21], exothermic heat generation is higher, and the demand for fast heat removal becomes more critical. The reaction starts in the gas–solid mode, where the synthesis gas with a molar H_2/CO ratio of ~2 contacts the Co-based catalyst pellets. In the course of reaction, the liquid phase, called wax, is produced first in the pores of the pellets and then in the entire reactor. These conditions can be handled in a slurry bubble column reactor (SBCR), a special version of the slurry reactor, described in Figure 1.17 [21]. The same process can also be carried out in a multitubular PBR involving trickle flow. However, the slurry bubble column offers several advantages such as lower pressure drop (ca. 1 atm in SBCR vs. 4 atm in PBR), higher intrinsic catalytic activity due to the possibility of using small particle sizes that minimize intraparticle diffusion limitations, higher mass transfer coefficients due to well mixing, longer runs due to possibility of online addition/removal of the catalyst, better temperature control improving reactant conversion and product selectivity, and lower capital expenditure requirements [21]. Nevertheless, the drawbacks brought by the mobility of the catalyst phase, that is, the need for catalyst–wax separation and the risk of immediate catalyst poisoning, should not be underestimated in SBCR operation. Apart from LTFT synthesis, slurry reactors are used in other applications such as oxidation and hydroformylation of olefins, methanation and polymerization reactions, and ethynylation of aldehydes [20]. Further information regarding the modeling and design of the slurry reactors is presented in Chapter 6. The reader is also directed to Chapter 12 for a detailed discussion about the use of slurry reactors in LTFT.

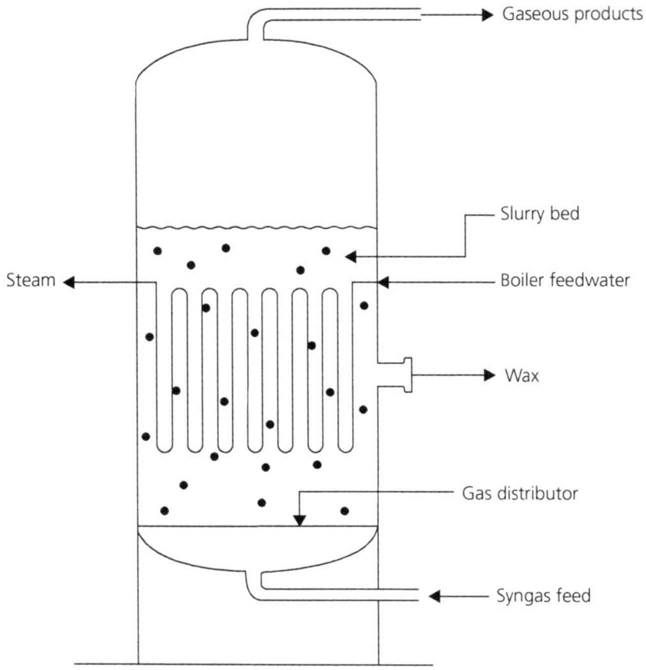

Figure 1.17 Slurry bubble column reactor for low-temperature Fischer–Tropsch synthesis.
(Source: Espinoza et al. [21]. Reproduced with permission of Elsevier.)

1.3.3 Moving-bed reactors

Moving-bed reactors are preferred when there is a need for continuous catalyst regeneration. In this operation, fresh catalyst is fed from the top of the reactor, and it moves in the downflow direction by gravitational forces. Spent catalyst leaving the reactor at the bottom is usually replaced in the continuous mode. While the catalyst movement is downward, reactive mixture flow can be cocurrent or countercurrent to that of the catalyst flow.

Moving-bed reactors do not involve intense mixing of the catalyst bed with the reaction mixture. In this respect, heat management within the bed is not as efficient as that involved in FBRs or in slurry reactors. High heat capacity of the circulating catalyst pellets dictates the heat transport in the moving-bed reactors. As described in Chapter 13, these reactors are used in catalytic hydrotreating of heavy oils, in which the moving bed ensures steady conditions for the catalyst and therefore minimizes the need for periodic shutdowns.

1.4 Reactors without a catalyst bed

The reactor types introduced in Sections 1.2 and 1.3 depend on the existence of a catalyst bed, either fixed or moving, for the operation. However, there are multiphase reactions, such as the gas–liquid type, which do not involve the use of a solid catalyst. Gas cleaning/purification applications, such as removal of CO_2 or H_2S from gas streams via mono-/diethanolamine or di-/triethylene glycol solutions and removal of nitrogen oxides by water; liquid-phase processes of oxidation, nitration, alkylation, hydrogenation, or manufacturing of products such as sulfuric acid, nitric acid, and adipic acid; and biochemical processes such as fermentation and oxidation of wastewater are examples of industrial applications of gas–liquid reactions [22]. Depending on factors such as residence time distribution of the phases, throughput demand of the process, and heat transfer requirements, gas and liquid phases can be contacted in various configurations; that is, gas can be distributed into the bulk liquid in the form of bubbles (bubble columns, plate columns), liquid can be sprayed to the bulk gas in the form of droplets (spray columns), or both phases can be contacted as thin films over an inert packing or on the reactor wall (packed columns, wetted wall columns). The common direction for liquid flow is from the top to the bottom of the reactor, and gas flow is usually in the opposite direction. Column-type reactors presented here involve a vessel and the particular components required to introduce or contact the phases (e.g., spargers for gas bubbling, spraying equipments for showering down the liquid, packing materials for contacting gas and liquid films, liquid distributors for ensuring uniform wetting of the packings, sieve plates for directing the liquid flow and for providing cross-contact with the rising gas). In general, reactor performance is affected by the gas solubility, which is expected to be high for improved rates. Operating temperature should be low, while pressure should be high for increasing gas solubility in the reactor. Depending on the heat of reaction, heat transfer equipment can be integrated to the reactor structure for regulating the temperature in the desired limits.

In some gas–liquid reactions, a mechanical agitator can be integrated into the reactor for improving mixing and mass transfer between the phases. In this case, the reactor is called as a stirred-tank reactor (Figure 1.18 [12]). The agitator is composed of an impeller that is mounted on a mechanically rotated shaft. Rotation and desired level of fluid mixing are provided by a variable speed electric motor that is placed on the reactor vessel. Gas–liquid stirred-tank reactors are also equipped by spargers for dispersing the gas bubbles into the liquid and by baffles to minimize swirl and vortex formations. In general, four baffles, each of which is one-tenth of the vessel diameter, are placed into the inner perimeter of the vessel. Aspect ratio, which is defined as the ratio of the liquid height in the tank to the tank diameter, is usually set up to be ~3 for increasing the residence time of the gas and improving the extent of reaction between phases. In such configurations, mixing is provided by multiple impellers mounted on the same shaft with distances up to one tank diameter [23].

In stirred-tank reactors, the possibility of regulating the agitation speed and the selection of various impeller types and diameters allow control over the degree of mixing of different fluids, which is quantified by the impeller Reynolds number ($Re = D^2 S\rho/\mu$; D, impeller diameter; S, speed of agitation; ρ, fluid density; μ, fluid viscosity). The impeller types not only affect the mixing characteristics but also the power consumption determined by the dimensionless power number ($Po = P/\rho S^3D^5$; P: power consumption). Plots of Po versus Re define the power characteristics of the impeller which is affected by factors such as its position in the tank and its diameter. In the laminar regime, characterized by $Re < 10$, Po decreases linearly with Re, whereas in the turbulent regime ($Re > 10^4$), Po remains constant and reaches an asymptotic value which is a function of the impeller type [23].

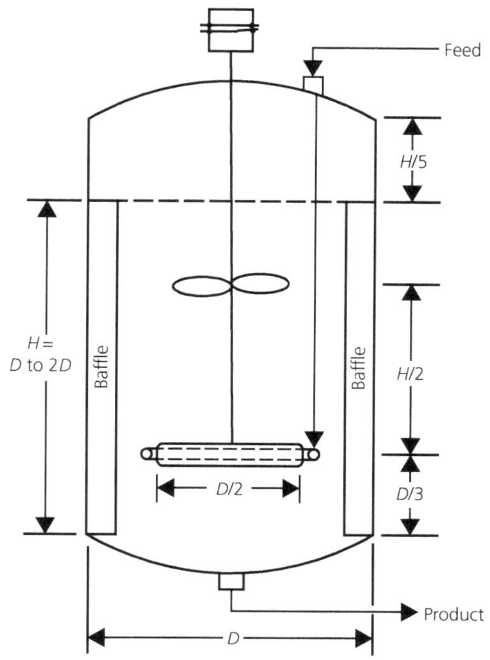

Figure 1.18 Stirred-tank reactor with typical dimensions.
(Source: Couper et al. [12]. Reproduced with permission of Elsevier.)

Heat transfer into/from the stirred-tank reactors is made possible by various configurations (Figure 1.19 [12]). Low heat duties can be realized by the heat transfer fluid flowing in a jacket surrounding the vessel (Figure 1.19a). For higher heat duties coils (Figure 1.19b) or internal tubes (Figure 1.19c) are immersed into the vessel for heat transfer fluid circulation. Heating/cooling of the reactive mixture in an external heat exchanger via a circulating loop (Figure 1.19d) is also possible. Other possible heat transfer configurations are shown in Figure 1.19e and f. In all cases, heat transfer coefficient on the reactor side is known to increase with the degree of mixing.

In addition to processes involving gas–liquid reactions, stirred-tank reactors can also be used for single (liquid)-phase reactions. Moreover, their operation is not limited to the continuous mode, and they can be easily adapted for use in semibatch and batch modes. The absence of a gas phase does not pose important structural and operational differences from those stated earlier for multiphase systems. However, in the case of single-phase operation, the aspect ratio is usually kept lower (~1) to ensure well mixing of the reactive liquid. Regardless of the number of phases involved, stirred-tank reactors can approach their ideal states if perfect mixing is established. Under such conditions, it is assumed that reaction takes place immediately just

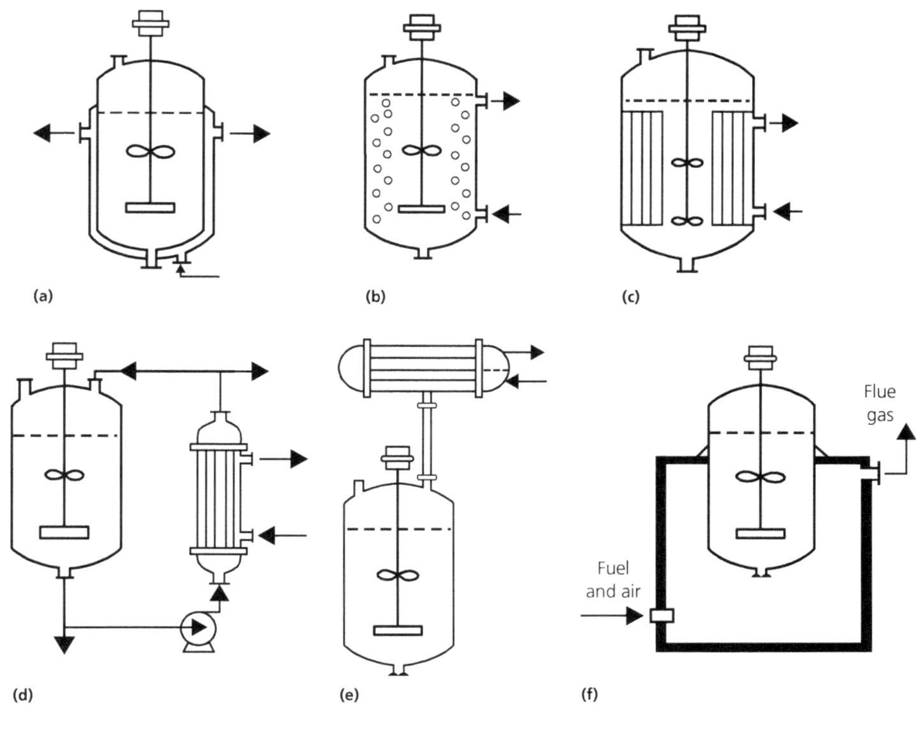

Figure 1.19 Heat transfer strategies in stirred-tank reactors. (a) Jacket, (b) internal coils, (c) internal tubes, (d) external heat exchanger, (e) external reflux condenser, and (f) fired heater.
(Source: Couper et al. [12]. Reproduced with permission of Elsevier.)

after the entrance of the reactants, and the properties of the exit stream are the same with those of the reactive mixture. Nevertheless, depending on the fluid properties and the specific internal geometry of the vessel, poorly mixed zones causing selectivity issues may develop in real operations.

1.5 Summary

Catalytic reactors are critical processing units of industrial chemistry. The complex combination of several factors such as conditions of the key reactions, requirements, and limitations of the catalytic chemistry and the demand for meeting the commercial targets for conversion and yield have led to the evolution of numerous catalytic reactor types. Besides technical requirements, fixed and operating capital expenses of the reactors determine the final decision for the selection of the appropriate reactor type. This chapter aims to provide an overview of all the factors involved that may help readers in understanding the key features of these complex reactors and their significance in chemical industry. The contents of this chapter are prepared to set the basis for the following chapters, each of which provides detailed information about the analysis, design, and modeling of the multiphase reactors covered in this book.

References

1 Onsan ZI, Avci AK. Reactor design for fuel processing. In: Shekhawat D, Spivey JJ, Berry DA, editors. Fuel cells: technologies for fuel processing. Amsterdam: Elsevier Science; 2011. p. 451–516.

2 Dybkjaer I. Tubular reforming and autothermal reforming of natural gas – an overview of available processes. Fuel Processing Technology 1995;42:85–107.

3 Aasberg-Petersen K, Christensen TS, Dybkjaer I, Sehested J, Ostberg M, Coertzen RM, Keyser MJ, Steynberg AP. Synthesis gas production for FT synthesis. In: Steynberg AP, Dry ME, editors. Fischer-Tropsch technology. Amsterdam: Elsevier; 2004. p. 258–405.

4 Eigenberger G. Catalytic fixed-bed reactors. In: Ertl G, Knözinger H, Schüth F, Weitkamp J, editors. Handbook of heterogeneous catalysis. Weinheim: Wiley-VCH; 2008. p. 2075–2106.

5 Eigenberger G. Fixed-bed reactors. In: Ullmann's processes and process engineering. Weinheim: Wiley VCH; 2004. p. 1983–2023.

6 Dry ME. The Fischer-Tropsch process: 1950–2000. Catalysis Today 2002;71:227–241.

7 Aasberg-Petersen K, Christensen TS, Nielsen CS, Dybkjaer I. Recent developments in autothermal reforming and pre-reforming for synthesis gas production in GTL applications. Fuel Processing Technology 2003;83:253–261.

8 Cybulski A, Moulijn JA. The present and the future of structured catalysts: an overview. In: Cybulski A, Moulijn JA, editors. Structured catalysts and reactors. Boca Raton, FL: CRC Press; 2006. p. 1–17.

9 Boger T, Heibel AK, Sorensen CM. Monolithic catalysts for the chemical industry. Industrial & Engineering Chemistry Research 2004;43:4602–4611.

10 Shah RK, London AL. Laminar flow forced convection in ducts. Advances in heat transfer, Supplement 1. New York: Academic Press; 1978.

11 Lide DR, editor. CRC handbook of chemistry and physics. Boca Raton, FL: CRC Press; 2003.

12 Couper JR, Penney WR, Fair JR, Walas SM. Chemical process equipment: selection and design. Boston: Butterworth-Heinemann; 2010.

13 Westerterp KR, Wammes WJA. Three-phase trickle-bed reactors. In: Ullmann's reaction engineering. Weinheim: Wiley-VCH; 2013. p. 651–663.

14 Dingerdissen U, Martin A, Herein D, Wernicke HJ. The development of industrial heterogeneous catalysis. In: Ertl G, Knözinger H, Schüth F, Weitkamp J, editors. Handbook of heterogeneous catalysis. Weinheim: Wiley-VCH; 2008. p. 37–56.

15 Hardt S, Ehrfeld W, Hessel V, Vanden Bussche KM. Strategies for size reduction of microreactors by heat transfer enhancement effects. Chemical Engineering Communications 2003;190:540–559.

16 Hessel V, Löb P, Löwe H. Industrial microreactor process development up to production. In: Hessel V, Renken A, Schouten JC, Yoshida J, editors. Micro process engineering, Vol. 3: System, process and plant engineering. Weinheim: Wiley-VCH; 2009. p. 183–247.

17 Werther J. Fluidized-bed reactors. In: Ertl G, Knözinger H, Schüth F, Weitkamp J, editors. Handbook of heterogeneous catalysis. Weinheim: Wiley-VCH; 2008. p. 2106–2132.

18 Steynberg AP, Dry ME, Davis BH, Breman BB. Fischer-Tropsch reactors. In: Steynberg AP, Dry ME, editors. Fischer-Tropsch technology. Amsterdam: Elsevier; 2004. p. 64–195.

19 Steynberg AP, Espinoza RL, Jager B, Vosloo AC. High temperature Fischer-Tropsch synthesis in commercial practice. Applied Catalysis A: General 1999;186:41–54.

20 Nedeltchev S, Schumpe A. Slurry reactors. In: Ertl G, Knözinger H, Schüth F, Weitkamp J, editors. Handbook of heterogeneous catalysis. Weinheim: Wiley-VCH; 2008. p. 2132–2156.

21 Espinoza RL, Steynberg AP, Jager B, Vosloo AC. Low temperature Fischer-Tropsch synthesis from a Sasol perspective. Applied Catalysis A: General 1999;186:13–26.

22 Walas SM. Chemical reactors. In: Perry RH, Green DW, Maloney JO, editors. Perry's chemical engineers' handbook. New York: McGraw-Hill; 1999. p. 23-1–23-61.

23 Nienow AW. Stirred tank reactors. In: Ullmann's reaction engineering. Weinheim: Wiley-VCH; 2013. p. 623–640.

CHAPTER 2

Microkinetic analysis of heterogeneous catalytic systems

Zeynep Ilsen Önsan
Department of Chemical Engineering, Boğaziçi University, Istanbul, Turkey

Abstract

This chapter deals with the microkinetics of gas–solid catalytic reaction systems. An applied approach is adopted in the discussion, which starts with the formulation of *intrinsic* rate equations that account for chemical processes of adsorption and surface reaction on solid catalysts and then proceeds with the construction of global rate expressions that include the individual and simultaneous effects of physical external and internal mass and heat transport phenomena occurring at the particle scale.

2.1 Heterogeneous catalytic systems

The task of the chemical reaction engineer is generally completed in two consecutive phases: (i) measurement and evaluation of the chemical kinetic behavior of a reaction system (microkinetic analysis and modeling) and (ii) use of this information in the design of equipment in which the reaction will be conducted (macrokinetic analysis and reactor design). Without underestimating the importance and complexity of the second phase, it can be said that the first phase of the task is by far the more critical, since it has to be completed correctly before the second phase is tackled. Chemical kinetic models, which are essential for efficient reactor design and scale-up, need to be based on experimental data that reflect steady-state chemical activity, that is, chemical events only. In solid-catalyzed heterogeneous systems, physical processes such as mass and energy transport at the particle scale may interfere with chemical (intrinsic) rates to modify the overall (global) reaction rates observed. These physical transport phenomena are analyzed depending on the characteristics of the particular catalyst/reactor system used and are then superimposed on the chemical kinetic model.

Accordingly, in order to arrive at the rate equation(s) appropriate for macrokinetic analysis at the reactor scale, microkinetic analysis has to take into account several chemical and physical rate processes at the particle scale:

1 Transfer of reactant(s) from the bulk gas stream to the exterior catalyst surface
2 Diffusion of reactant(s) from the exterior surface into the interior surface
3 Chemisorption of reactant(s) on the inner surface of the pores
4 Surface chemical reaction to form product(s)
5 Desorption of product(s) from the surface of the pores
6 Diffusion of product(s) from the pores to the exterior catalyst surface
7 Transfer of product(s) from the exterior catalyst surface to the bulk gas stream

In this sequence, steps 3–5 are the chemical rate processes; laboratory analysis of these steps in the absence of physical effects yields the intrinsic reaction rate. Steps 1 and 7 are external physical rate processes separated from and in series with the chemical rate processes, while steps 2 and 6 are internal physical rate processes occurring simultaneously with chemical rate processes. The external and internal physical transport effects existing in a particular system are superimposed on the intrinsic reaction rate to obtain the global reaction rate, which is used in the macroscopic mass and energy transport equations required for reactor design.

In the intrinsic heterogeneous catalytic cycle, the reactants are adsorbed on the catalyst surface at specific locations called active sites, and they are activated by chemical interaction with these sites to form the catalyst–reactant complex, thus rapidly transforming on the active site to adsorbed products which subsequently desorb from these sites allowing them to momentarily return to their original state until other reactant molecules adsorb. The simple hypothesis initiating from Langmuir's work on chemisorption [1, 2] forms the basis of the modern theory used in the interpretation of the kinetics of reactions at the catalyst surface:

$$\text{Adsorption of reactants} \rightarrow \text{Surface reaction} \rightarrow \text{Desorption of products}$$

This postulation has been useful in correlating a wide variety of kinetic results as well as in predicting the effects of new

Multiphase Catalytic Reactors: Theory, Design, Manufacturing, and Applications, First Edition. Edited by Zeynep Ilsen Önsan and Ahmet Kerim Avci.

conditions imposed on reacting systems. There are, however, some conceptual difficulties arising from experimental results which suggest that only a small fraction of the surface is active and that active sites for chemisorption are not the same for all species. The simple physical model of the catalyst surface proposed later by Taylor [3] has the following features: (i) the catalyst surface can offer a variety of sites where molecules can adsorb with various bond strengths, (ii) the structure of the adsorbed species depends on bond strength, (iii) for a particular surface reaction to happen, bond strengths must be within specific limits, and (iv) sites that meet these bond energy requirements are called the active sites for the reaction. In short, there are a "fixed number of active sites" that account for the catalytic activity of a solid catalyst.

While the basic variables by which chemical processes can be controlled are temperature, pressure, inlet reactant concentrations, and residence time in the reactor, two technological developments of major consequence starting with 1960s have made possible cost-effective operation under less severe conditions: the prevalent use of efficient catalysts and improved reactor configurations. The impact of heterogeneous catalysis is significant, since three major areas of the world economy, namely, petroleum refining, chemicals manufacturing, and environmental cleanup, all require the use of efficient solid catalysts.

The general definition of a catalyst is common to homogeneous, heterogeneous, and enzyme catalysis. A catalyst is a substance that increases the rate at which a chemical reaction approaches equilibrium without itself suffering permanent chemical change. This description indicates that a catalyst gets temporarily involved in the chemical reaction, changes chemical reaction rates, but does not disturb chemical reaction equilibrium. Catalysts can only accelerate reactions that are thermodynamically feasible, that is, only those with negative Gibbs free energy change, $\Delta G^\circ < 0$, at a specified temperature. For a given reaction, the chemical equilibrium reached in the absence and presence of a catalyst is the same equilibrium:

$$\Delta G^\circ = -R_g T \ln K \tag{2.1}$$

Since the overall reaction equilibrium constant K is also equal to the quotient of the velocity constants for the forward and reverse reactions ($K = k_f/k_r$), both reactions are accelerated by the same factor. This does not, however, suggest that all the reactions in a multiple reaction system are accelerated to the same extent; quite the reverse, the merit of a successful catalyst is to accelerate only the desirable reaction(s).

In solid-catalyzed reactions, the reactant binds to an active site on the catalyst surface where an intermediate catalyst–reactant complex is formed, and reaction occurs on the active site to form products which are then released into the gas. Transformation of the reactant into product is expedited, because the role of the catalyst is to convert reactant(s) into a form in which conversion to product(s) is easier, and by this means, the catalyst provides a new reaction path that is energetically more beneficial than the uncatalyzed path (Figure 2.1).

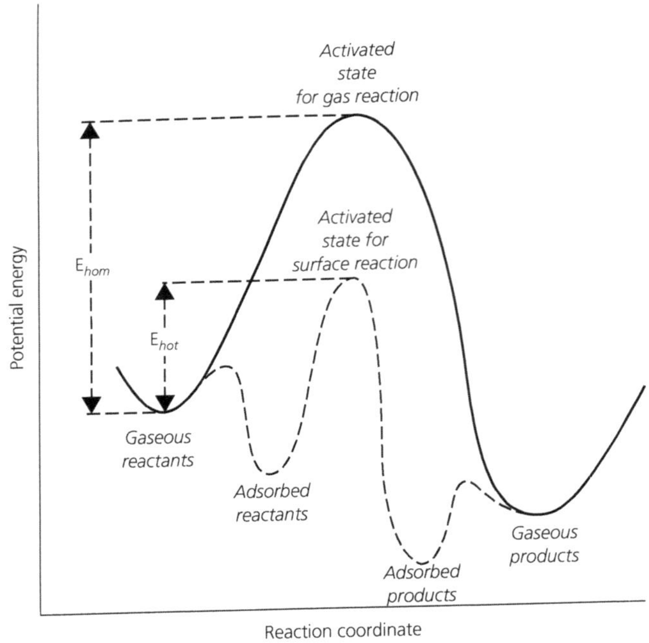

Figure 2.1 Potential energy curves representing the action of a solid catalyst. (Source: Davis [4]. Reproduced with permission of John Wiley & Sons.)

Chemical reactions, catalyzed or uncatalyzed, take place in accordance with the Arrhenius equation:

$$k = A \exp\left(-\frac{E_A}{R_g T}\right) \tag{2.2}$$

The preexponential or the frequency factor A is catalyst dependent, that is, it varies with the extent of surface and has the same units as the rate constant k. On the basis of the collision theory, it can be estimated that the frequency factor of a unimolecular heterogeneous reaction is smaller than that of its homogeneous counterpart by a factor of 10^{12}. It follows that, for efficient catalysis, the activation energy E_A of the catalyzed reaction should be at least 80 kJ/mol lower than that of the uncatalyzed one at 298 K. At higher reaction temperatures, the difference in E_A must also be higher in order to keep the advantage of the catalyzed reaction rate. E_A and A usually tend to compensate the change in one another; hence, the compensation (or theta) effect between A and E_A has to be taken into account [4].

2.1.1 Chemical and physical characteristics of solid catalysts

In heterogeneous catalysis, the reaction takes place at the interface between the catalyst and the less dense phase. Adsorption is defined as the preferential concentration of gas molecules at a fresh solid surface, caused by the existence of a field force that attracts molecules of the contacting fluid. Two major types of adsorption have been recognized, namely, physical adsorption and chemisorption [5, 6].

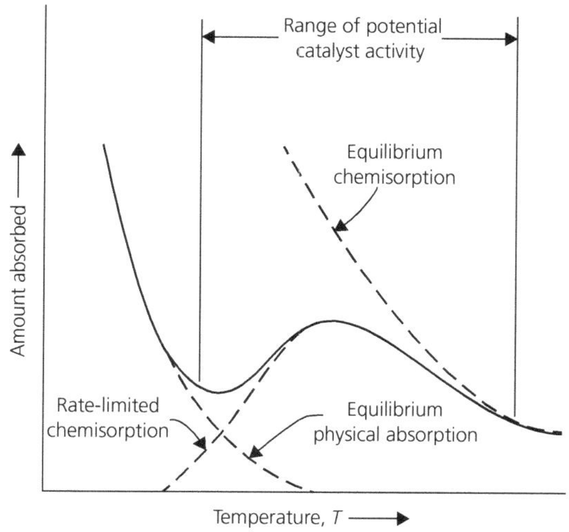

Figure 2.2 Effect of temperature on amount of gas adsorbed for simultaneous physical adsorption and activated chemisorption. (Source: Hill [7]. Reproduced with permission of John Wiley & Sons.)

Physical adsorption, which is similar to the condensation of vapor molecules onto a liquid surface of the same composition, (i) is due to weak attractive forces of the van der Waals type, (ii) is multilayer and nonspecific, (iii) occurs at temperatures close to the boiling point of the adsorbate, and (iv) has low heats of adsorption close to the heats of condensation of the adsorbate involved.

Chemical adsorption (chemisorption), on the other hand, is similar to a chemical reaction resulting in the formation of an intermediate compound restricted to the surface layer of the adsorbent and, unlike physical adsorption, it (i) involves chemical bonding and exchange of electrons between the adsorbate and the partially uncoordinated active sites of the adsorbent, (ii) is monolayer and highly specific, (iii) occurs at temperatures well above the boiling point of the adsorbate, and (iv) has much higher heats of adsorption close to the heats of chemical reactions. Conditions required for catalysis designate chemisorption as the essential precursor to surface reaction. Physical adsorption may, nonetheless, facilitate the transition of reactants from the gaseous to the chemisorbed state (Figure 2.2).

2.1.1.1 Quantitative treatment of chemisorption

The key concept in the quantitative treatment of chemical adsorption is due to Langmuir [1, 2] in his pioneering work aiming to find "a relation between the quantity of gas adsorbed by a solid and the pressure of the gas over the solid when equilibrium is reached." His original derivation was a kinetic one, with the implicit assumptions of (i) monolayer adsorption, taking place through the collision of gaseous adsorbate molecules with vacant active sites on the surface, (ii) one site–one entity interaction, with each surface site accommodating only one entity (i.e., one atom or one molecule), and (iii) energetic uniformity of the entire active surface.

Langmuir used fractional surface coverage by the adsorbate gas, θ_A, as a measure of the amount of gas adsorbed and envisaged a dynamic equilibrium between the adsorption and desorption rates of the adsorbate, $R_{ads} = R_{des}$. The original form of the Langmuir isotherm for molecular adsorption of the adsorbate gas, $A + S \leftrightarrow A - S$, was obtained as

$$\theta_A = \frac{K_A P_A}{1 + K_A P_A} \tag{2.3}$$

$$K_A = \frac{k_{ads}}{k_{des}} = \text{adsorption equilibrium coefficient}$$

$$R_{ads} = k_{ads} P_A (1 - \theta_A) \quad \text{and} \quad R_{des} = k_{des} \theta_A$$

For dissociative adsorption, $A_2 + 2S \leftrightarrow 2A - S$, the Langmuir isotherm becomes

$$\theta_A = \frac{(K_A P_A)^{\frac{1}{2}}}{1 + (K_A P_A)^{\frac{1}{2}}} \tag{2.4}$$

For multicomponent adsorption on similar sites, $A + S \leftrightarrow A - S$ and $B + S \leftrightarrow B - S$,

$$\theta_A = \frac{K_A P_A}{1 + K_A P_A + K_B P_B}$$

$$\theta_A = \frac{K_A P_A}{1 + \sum K_i P_i} \quad \text{for } i \text{ number of components} \tag{2.5}$$

The monolayer assumption of the Langmuir treatment is valid since exchange of electrons and chemical bonding are involved in chemisorption, and the usual range of chemical bond distances would indicate the formation of only a monolayer restricted to the surface. The second one site–one entity assumption is not always true, since chemisorption of more molecules on one site or one molecule on two or more sites is possible depending on the coordination between active sites and adsorbate molecules. The third assumption regarding energetic equivalence of the active surface contains an important weak point; experimental observations clearly indicate decreasing heats of adsorption ΔH_{ads} with increasing surface coverage θ. The major reasons for the decline in ΔH_{ads} are listed as surface heterogeneity and lateral interaction between adjacent species; that is, highly active sites are covered first and adsorption on neighboring sites increases surface repulsions.

The early work of Beeck in 1950 shows isosteric heats of adsorption for hydrogen as a function of surface coverage on several metal films, exhibiting their dependence on surface coverage [8, 9]. These data also indicate that there is a common region corresponding to intermediate surface coverages ($0.2 < \theta < 0.8$) that are essential for efficient catalysis, where the decline in the heats of adsorption is linear and an average ΔH_{ads} value may be used with some approximation if the fall is not appreciable. The distinct advantage of the Langmuir isotherm is that it readily describes multicomponent chemisorption in all partial pressure ranges and also predicts the two limiting conditions of $\theta_A \to 0$ when $P_A \to 0$ and $\theta_A \to 1$ when $P_A \to \infty$; as a result, it forms the basis of the modern treatment of heterogeneous reaction kinetics in the formulation of rate equations.

Two other well-known isotherms that do not involve an assumption regarding energetic equivalence of the active surface are the Temkin isotherm and the Freundlich isotherm [6]. The Temkin isotherm takes into account a linear fall in ΔH_{ads} with increasing θ and permits its interpretation in terms of surface heterogeneity as well as lateral repulsion between adsorbed species:

$$\theta_A = k_1 \ln(k_2 P_A) \tag{2.6}$$

This isotherm may be derived from kinetic considerations for intermediate surface coverages ($0.2 < \theta < 0.8$), but it does not lend itself to multicomponent adsorption and also fails to predict the limiting conditions of $\theta_A \to 0$ when $P_A \to 0$ and $\theta_A \to 1$ when $P_A \to \infty$. Even though it was used for correlating the kinetics of ammonia synthesis, the Temkin isotherm has not found much use in the kinetic analysis of solid-catalyzed gas-phase reactions.

Originally postulated as an empirical equation, the Freundlich isotherm with two constants, k and n, can be derived from thermodynamic or statistical considerations with the assumptions that ΔH_{ads} decreases exponentially with increasing surface coverage and that this decrease is due to surface heterogeneity:

$$\theta_A = k(P_A)^{\frac{1}{n}}; \quad n > 1 \tag{2.7}$$

The statistical derivation shows that the Freundlich isotherm is expected to be valid at low surface coverages; in fact, the isotherm successfully predicts that $\theta_A \to 0$ when $P_A \to 0$ but fails to predict $\theta_A \to 1$ when $P_A \to \infty$. The Freundlich isotherm can handle multicomponent adsorption to some extent, and in some cases, the Langmuir isotherm can be reduced to the power function form of the Freundlich isotherm.

2.1.1.2 BET treatment of physical adsorption

The Langmuir approach was extended to multilayer adsorption by Brunauer, Emmett, and Teller in the form of the BET equation with two constants. The linearized form of the BET equation is important in the measurement of total surface areas of porous solid catalysts [5]:

$$\frac{P}{V_{ads}(P_0 - P)} = \frac{1}{V_m C} + \frac{C-1}{V_m C}\frac{P}{P_0} \tag{2.8}$$

Here, P is the pressure of adsorbate (N_2), in mmHg; P_0 is the saturation or vapor pressure of adsorbate, in mmHg; V_{ads} is the volume of adsorbed gas, in cm^3; V_m is the volume of monolayer, also in cm^3; and C is a constant for the particular gas–solid system used and temperature. Utilizing the P versus V_{ads} data obtained on a constant-volume or constant-pressure BET equipment, the volume of the monolayer is easily calculated from the slope and intercept of the BET equation. The specific surface area S_g of the catalyst is then calculated in a simple sequence of steps using the ideal gas law, Avogadro's number N_0, and the cross-sectional area A_m of one molecule of the adsorbate:

$$S_g(\text{m}^2/\text{g}) = \text{total surface area per unit weight of catalyst sample}$$

$$V_m = \text{monolayer volume } (\text{cm}^3)$$

$$\frac{V_m}{22414} = \frac{\text{cm}^3}{\text{cm}^3/\text{mol at STP}} = \text{moles of gas in monolayer}$$

$$\left(\frac{V_m}{22414}\right)\left(6.02 \times 10^{23}\frac{\text{molecules}}{\text{mol}}\right) = \text{number of molecules in monolayer}$$

$$\left(\frac{V_m}{22414}\right)(6.02 \times 10^{23})(A_m) = \text{area covered by the molecules in monolayer}$$

$$A_m = \text{cross-sectional area of one molecule} = 16.2\,\text{Å}^2 \text{ for } N_2$$

$$S_g(\text{m}^2/\text{g}) = \left(\frac{V_m}{22414}\right)(6.02 \times 10^{23})(A_m)\left(\frac{1}{w_{cat}}\right)(10^{-20}) \tag{2.9}$$

including the conversion of (Å)^2 to m^2 as well as the weight of the solid catalyst sample.

2.1.1.3 Catalyst physical properties

The physical properties of solid catalysts have a pronounced effect on their catalytic performance and are also used in geometric models of catalyst particles as well as in expressing effectiveness factors. The more frequently used properties are listed in the following.

S_g (m^2/g), total surface area per gram of catalyst, or specific surface area, is a measure of the extent of surface available for adsorption and determines the amount of gas adsorbed.

V_g (cm^3/g), void volume or pore volume per gram of catalyst particle, is a measure of the effectiveness of the internal surface and is calculated from

$$V_g(\text{cm}^3/\text{g}) = \frac{V_{Hg} - V_{He}}{m_p} \tag{2.10}$$

Here, V_{Hg} and V_{He} (both in cm^3) are the volumes of Hg and He displaced by the particle as measured by pycnometry, respectively, and m_P is the mass of the catalyst sample.

$\bar{a}$ (Å), mean pore radius, is roughly estimated by assuming all pores are cylindrical, straight, and parallel with the same radius and length:

$$\bar{a}\,(\text{Å}) = \frac{2V_g}{S_g} \tag{2.11}$$

ρ_S (g/cm^3), density of the solid phase in the particles, is calculated using V_{He}:

$$\rho_S(\text{g/cm}^3) = \frac{m_P}{V_{He}} \tag{2.12}$$

ρ_P (g/cm^3), density of the porous particles, is calculated using V_{Hg}:

$$\rho_P(\text{g/cm}^3) = \frac{m_P}{V_{Hg}} \tag{2.13}$$

ϵ_P, void fraction or porosity of the particles, is calculated using the difference between V_{Hg} and V_{He} or from ($V_g\rho_P$); voidage of most industrial catalysts is in the range of 0.40–0.60:

$$\epsilon_P = \frac{V_{Hg} - V_{He}}{V_{Hg}} \quad \text{or } V_g\rho_P \tag{2.14}$$

Pore-volume distribution, distribution of void volume according to pore size or radius of pore mouth, is measured by N_2 adsorption–desorption experiments [5, 8].

2.1.2 Activity, selectivity, and stability

The three fundamental properties inherent in the actual definition of a catalyst are activity, selectivity, and stability. Moreover, for successful industrial applications, catalysts must be regenerable, reproducible, mechanically and thermally stable with suitable morphological characteristics, and also economical.

2.1.2.1 Catalyst activity

Activity is a measure of the rate at which the catalyst causes the chemical reaction to arrive at equilibrium. In terms of kinetics, the reaction rate defines catalyst activity as the quantity of reactant consumed per unit time per unit volume or mass of catalyst:

$$(-R_A)_V = \text{mol/L/h} \quad \text{or} \quad (-R_A)_P = \text{mol/kg/h}$$

In industrial practice, it is more practical to use readily measured parameters such as

$$STY = \text{space-time yield} = \text{mol/L/h} \quad \text{or} \quad (R_B)_V = \text{mol/L/h}$$

Space-time yield is the quantity of product formed per unit time per unit volume of reactor or catalyst, since reactor volume is taken as the catalyst-packed volume.

Space time is defined as the time required for processing one reactor volume of feed and is calculated by dividing the reactor volume by the volumetric flow rate of feed. The reciprocal of space time is defined as space velocity, with units of reciprocal time, and signifies the number of reactor volumes of feed processed per unit time. The phase and the conditions at which the volumetric flow rate of feed is measured have to be specified.

High activity is reflected in either high space-time yield from comparatively small catalyst volumes or mild operating conditions that enhance selectivity and stability. Catalyst activity defined as $(-R_A)_V$ or $(R_B)_V$ depends on pressure, temperature, and reactant concentrations.

In the screening of a range of solid catalysts in order to select the best candidate(s), a correct comparison of their catalytic activity is possible by determining one of the following, under otherwise similar reaction conditions [10, 11]: (i) their conversion levels, x, (ii) the space velocity required in each case for achieving a given constant conversion level, x, (iii) the space-time yield, or (iv) the temperature necessary for reaching a given conversion level, x.

2.1.2.2 Catalyst selectivity

In complex reaction systems, several stable products are produced by more than one reaction, and some of the products are not desirable. Selectivity is a measure of the extent to which the catalyst accelerates the formation of desired product(s) and is usually a function of the degree of conversion of reactant and reaction conditions, particularly temperature. A number of different definitions of selectivity are used according to purpose.

The basic concept is overall selectivity, defined as the ratio of the quantity of desired product to the quantity of reactant converted, (mol/mol) or (mol%). For parallel (competing) reactions, rate selectivity is defined as the ratio of the rate of formation of desirable product B to the rate of formation of another product C, $(R_B)_V/(R_C)_V$, as in the case of the simultaneous reactions, $A \rightarrow B$ and $A \rightarrow C$.

2.1.2.3 Turnover frequency and turnover number

Turnover frequency (TOF) quantifies the number of molecules converted or formed per catalytic site per second at specified conditions of temperature, pressure, and conversion:

$$TOF = \frac{(-R_A)_V}{\left(\frac{\text{number of centers}}{\text{volume}}\right)}; \quad \frac{\text{mol/L/s}}{\text{mol/L}} = \text{s}^{-1}$$

For most relevant industrial applications, TOF values in the range of 10^{-2}–$10^{3}\ \text{s}^{-1}$ have been observed. For enzyme-catalyzed reactions, TOF levels are much higher at 10^{3}–$10^{7}\ \text{s}^{-1}$. TOF is limited by the difficulty in determining the number of active centers for multimetallic, nonmetallic, and mixed oxide catalysts used more frequently in large-scale operations.

Turnover number (TON) specifies the number of catalytic cycles for which the catalyst is effective up to the decline in activity. For most industrial applications, TON values are in the range of 10^{6}–10^{8}.

2.1.2.4 Catalyst stability

The stability of a catalyst is determined by its ability to withstand changes in physical and chemical properties that take place during use, leading to catalyst deactivation. Chemical, thermal, and mechanical stability of a catalyst determines its lifetime in industrial reactors. Total catalyst lifetime is usually crucial for the economics of a catalytic process.

A catalyst with good stability will change only very slowly over the course of time under conditions of use and regeneration. Catalyst stability is influenced by numerous factors, including decomposition, coke formation, poisoning, and sintering. The priority of target properties in catalyst design and development for industrial applications is commonly given in the following order: selectivity > stability > activity.

2.1.2.5 Catalyst deactivation

It is misleading to say that a catalyst is totally unchanged by the reaction it catalyzes. Gradual physical and chemical alterations may take place during catalysis or with usage. Industrial catalysts are slowly deactivated by phenomena that accompany the main catalytic process. Catalyst aging, or deactivation, is indicated by the decrease in catalyst activity with time. It introduces additional complexity to the determination of rate parameters and has to be considered in macrokinetic analysis, that is, in catalytic reactor design.

The most common causes of catalyst deactivation are [12] (i) poisoning by strong chemisorption of impurity chemicals on active sites, (ii) fouling or coking by the deposition of carbon on active sites, (iii) sintering due to loss of active surface by the agglomeration of metals, narrowing or closing of pores of the solid support, (iv) chemical decomposition due to loss of active components by vapor transport, and (v) mechanical failure caused by the attrition and crushing of catalysts.

Poisoning is a chemical effect, and catalyst poisons are extraneous materials forming strong adsorptive bonds with the active sites on the catalyst surface. Adsorbed poisons physically block adsorption sites and may also induce changes in the electronic or geometric structure of active surfaces. For instance, sulfur adsorbs strongly on metals such as Ni and prevents or modifies the adsorption of reactant molecules; its presence causes substantial or complete loss of activity. Sulfur poisoning is a major problem in the industrial processes of steam reforming, hydrogenation, methanation, and Fischer–Tropsch synthesis. The order of decreasing toxicity for sulfur is given as $H_2S > SO_2 > SO_4^{2-}$, which results from the increased shielding by oxygen.

Frequently, reaction products may adsorb more strongly than reactants; reaction products that desorb slowly from the active sites and thereby reduce reaction rates are generally termed as inhibitors, not as poisons, and are taken into account in reaction rate equations.

Fouling, coking, and carbon formation are used interchangeably and refer to the physical deposition of species from the fluid phase onto the catalyst surface, resulting in activity loss due to blockage of sites and/or pores. Coke-forming processes may also be accompanied by the chemisorption of condensed hydrocarbons which act as poisons.

On nonmetallic catalysts, coke formation is a result of cracking reactions involving alkenes and aromatics. On metallic catalysts, depending on temperature, carbon deposits may contain little or no hydrogen. Carbon is formed either as graphite or as filaments growing out from metal surfaces, causing metal dispersion and deterioration. Coke formation processes can be attributed to the following reactions:

$$2CO \rightarrow C + CO_2; \quad CH_4 \rightarrow C + 2H_2; \quad 2H_2 + CO_2 \rightarrow C + H_2O$$

Thermally induced catalyst deactivation may result from (i) loss of catalytic surface due to metal crystallite growth, that is, metal(s) present in the form of separate dispersed atoms or small clusters rearrange to form larger crystallites, (ii) loss of support area due to support collapse or pore collapse on metal crystallites, resulting in pore closure and encapsulation of metals, or (iii) transformation of catalytic phases to noncatalytic phases, as in the solid-state reaction of NiO with Al_2O_3 to form a stable but inactive $NiAl_2O_4$ under steam-containing or oxidizing conditions at temperatures above 400–500 °C. The first two processes described in (i) and (ii) here are typically called sintering.

2.1.2.6 Measures against catalyst deactivation

A brief synopsis of commonly used measures is given here to emphasize their significance both in catalyst development and in processing strategies [12, 13]. Poisoning of metal catalysts can be avoided by the incorporation of suitable promoters in catalyst formulations as well as pretreatment of feed mixtures to remove impurities. A good example of increasing the sulfur resistance of Ni or Co catalysts is the addition of Mo in the hydrogenation of CO_x or in hydrotreatment processes. Coke formation can be reduced substantially by increasing the hydrogen partial pressure, by partial neutralization of acid sites with promoters, and by additives such as SiO_2, Al_2O_3, TiO_2, MoO_3, or WO_3 to prevent filamentous carbon in the case of Ni–Fe catalysts. In steam reforming processes, the steam to carbon ratio is increased to inhibit carbon formation and/or to gasify the carbon deposited on the surface. Coke already formed is removed by periodic regeneration of the catalyst by combustion (burning off) of the deposited carbon layer in a controlled manner to avoid local sintering of the active phase or carrier. In sintering, catalyst stabilization is increased by using particles with lower densities and narrow pore-size distributions. For a given reactant, the stability of active metals against sintering increases as follows, with *Re* being the most stable:

$$Ag < Cu < Au < Pd < Ni < Co < Pt < Rh < Ru < Ir < Os < Re$$

Addition of higher melting noble metals like Rh and Ru to base metals such as Ni also improves thermal stability.

Considering some of the commonly used support materials, the stability against sintering increases in the following order:

$$TiO_2 < SiO_2 < Al_2O_3 < MgO$$

TiO_2 is an exception, since it is the typical support for strong metal–support interactions (SMSI). Addition of Ba, Zn, La, Si, and Mn promoters improves the thermal stability of Al_2O_3 supports and hinders loss of total surface through extended use at relatively high temperatures.

2.2 Intrinsic kinetics of heterogeneous reactions

The intrinsic catalytic cycle contains only the chemical steps 3–4–5 of the 7-step sequence listed in the so-called continuous reaction model. It is necessary to make the assumption of zero gradients with respect to heat and mass transport both outside and within the catalyst particle. Therefore, experimental conditions in the laboratory have to be adjusted to ensure that (i) external transport processes (steps 1 and 7 of the sequence) are very rapid compared to chemical steps and (ii) internal transport processes (steps 2 and 6 of the sequence) are negligible, that is, particle sizes are small enough to ignore pore structure. In Figure 2.3, the reactant concentration profile labeled as IV represents the case for intrinsic kinetics.

In the interpretation of the intrinsic kinetics of catalytic reactions, the simple scheme based on Langmuir's work including chemisorption of reactants, surface chemical reaction, and desorption of products provides the framework together with Taylor's physical surface model postulating a fixed number of active surface sites. This analysis has been successful in correlating a wide range of kinetic results and also in predicting possible effects of new reaction conditions.

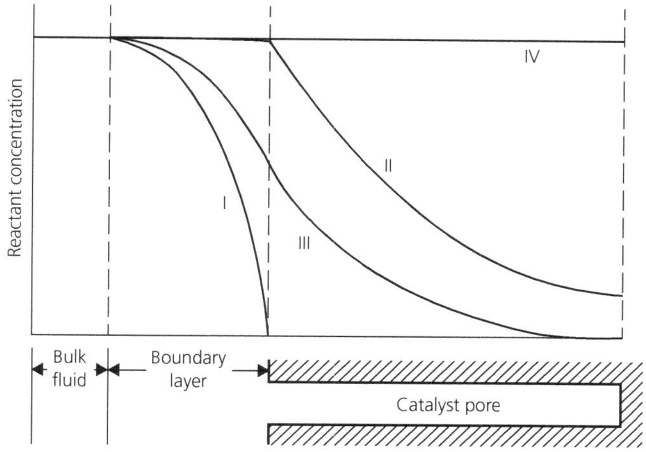

Figure 2.3 Reactant concentration profiles in different global rate regimes: I, external mass transfer limitation; II, pore diffusion limitation; III, both external and internal mass transfer limitations; IV, no mass transfer limitations on the intrinsic rate.
(Source: Hill [7]. Reproduced with permission of John Wiley & Sons.)

The three key principles used in the formulation of intrinsic rate equations originate from these surface model concepts [14]:

1 Constancy of the total number of active sites, which is *a priori* assumption based on the physical surface model
2 Quasi- or pseudo-steady-state approximation, which assumes that concentrations of intermediate complexes formed on the surface are small and time invariant
3 Presence of rate-controlling or slow step(s) in the reaction mechanism comprising adsorption, reaction, and desorption steps, which establishes the final functional form of the intrinsic rate equation.

2.2.1 Kinetic models and mechanisms

Consider a single reaction of the form

$A \leftrightarrow B$

Since the reaction is solid catalyzed, it is clear that it does not take place as written by the stoichiometric equation. Postulating a possible reaction mechanism in terms of elementary reaction steps and intermediate complexes of the type described by Langmuir,

$A + S \leftrightarrow A - S$ adsorption; k_1, k_{-1}

$A - S \leftrightarrow B - S$ surface reaction; k_2, k_{-2}

$B - S \leftrightarrow B + S$ desorption; k_{-3}, k_3

If we let $S = S_V$, $A–S = S_A$, and $B - S = S_B$, then S_V, S_A, and S_B refer to the chemical forms of unoccupied and occupied active surface sites, respectively. Reactant A adsorbs on vacant site S_V to form the catalyst–reactant complex S_A which is converted to adsorbed product complex S_B, and finally, adsorbed product desorbs to give gaseous product B and also regenerate the vacant active site S_V so that a cyclic reaction pattern repeats itself and a large number of product molecules can be formed by each active site. The important point is that the vacant site S_V consumed by the first step is regenerated in the third step of the reaction mechanism, leading to a closed sequence.

In order to proceed with the kinetic analysis, the *a priori* assumption that the number of active sites is a constant proportional to the mass of catalyst (S_O) is utilized in writing a "site balance":

$$S_O = S_V + S_A + S_B \tag{2.15a}$$

Considering that the number of active sites on the surface is small compared to the number of reactant molecules in the gas phase, a dynamic steady state is readily established between gaseous and adsorbed species if the intermediate steps are reactive enough. Under conditions where the quasi- or pseudo-steady-state approximation is applicable, the distribution of active sites between occupied and unoccupied forms does not change with respect to time, and thus, surface concentrations of intermediate species can be related to their gas-phase concentrations:

$$\frac{dS_A}{dt} = k_1 C_A S_V - k_{-1} S_A - k_2 S_A + k_{-2} S_B = 0 \tag{2.15b}$$

$$\frac{dS_B}{dt} = k_2 S_A - k_{-2} S_B - k_{-3} S_B + k_3 C_B S_V = 0 \tag{2.15c}$$

Since the net rates of all the consecutive steps in the mechanism are identical under the steady-state approximation, the net steady-state rate for the overall reaction may be evaluated from any one of the steps. However, considering that most reactions involve more than one reactant and/or product, the resulting sizeable rate equations are cumbersome and tend to correlate virtually any set of data with little distinction.

Simplification of the rate expression is possible if the rate constants corresponding to one of the elementary steps in the reaction mechanism can be identified as being small compared to others. This is called the "slow step" or the rate-controlling/rate-limiting/rate-determining step in the overall reaction mechanism. In the limiting case, all elementary reaction steps of the mechanism are essentially at equilibrium except the rate-determining slow step; therefore, the net steady-state rate can be expressed in terms of the slow step, and equilibrium statements can directly be written for all other steps in the mechanism.

The slow step may be any one of the three steps in the reaction mechanism; so, the limiting case may be that of (I) surface reaction controlling, (II) adsorption of reactant controlling, or (III) desorption of product controlling. Since a large majority of heterogeneous reactions are surface reaction controlled, the Langmuir–Hinshelwood approach to kinetics of fluid–solid catalytic reactions [15] based on fractional surface coverages of reactants was restricted only to this particular rate-limiting step. The Langmuir–Hinshelwood formulation is a special case of the comprehensive approach put forward later by Hougen and Watson [16] for deriving rate expressions when adsorption, surface reaction, or desorption is controlling the rate; the latter treatment provides a rational and structured approach to catalytic kinetics, despite the restrictions of the Langmuir isotherm on which it is based [13, 17]. Furthermore, parameters accounting for catalytic activity, catalyst effectiveness as a result of diffusion, and/or activity decay may also be included in the Hougen-Watson derivations. It must, however, be kept in mind that the equations obtained are kinetic models, not mechanistic descriptions, and they will only indicate that the proposed sequence of steps are plausible.

Case I Surface reaction controlling
This limiting case corresponds to the assumption that the adsorption and desorption steps of the reaction mechanism are fast and essentially at equilibrium, while the surface reaction step is slow and far from equilibrium:

$$k_2S_A, k_{-2}S_B \ll k_1C_AS_V,\ k_{-1}S_A,\ k_{-3}S_B,\ k_3C_BS_V$$

$$(-R_A)_{SI} = k_2S_A - k_{-2}S_B = \text{net steady-state rate for Case I} \tag{2.16a}$$

$$S_A = \frac{k_1}{k_{-1}}C_AS_V = K_AC_AS_V = \text{adsorption equilibrium statement for } A \tag{2.16b}$$

$$S_B = \frac{k_3}{k_{-3}}C_BS_V = K_BC_BS_V = \text{desorption equilibrium statement for } B \tag{2.16c}$$

$$S_O = S_V + S_A + S_B = S_V + K_AC_AS_V + K_BC_BS_V = \text{site balance} \tag{2.16d}$$

$$S_V = \frac{S_O}{1 + K_AC_A + K_BC_B}$$

$$(-R_A)_{SI} = \frac{(k_2K_AS_O)\left(C_A - \frac{C_B}{K}\right)}{1 + K_AC_A + K_BC_B} \tag{2.17}$$

$$K = \frac{K_{sr}K_A}{K_B} \tag{2.18}$$

Case II Adsorption of reactant controlling
This case relates to the situation where the adsorption of reactant A on a vacant active site to form the active complex S_A is slow, while both the surface reaction converting S_A to adsorbed product S_B and the desorption of B are fast:

$$k_1C_AS_V,\ k_{-1}S_A \ll k_{-3}S_B,\ k_3C_BS_V,\ k_2S_A,\ k_{-2}S_B$$

$$(-R_A)_{SII} = k_1C_AS_V - k_{-1}S_A = \text{net steady-state rate for Case II} \tag{2.19a}$$

$$S_B = \frac{k_3}{k_{-3}}C_BS_V = K_BC_BS_V = \text{desorption equilibrium statement for } B \tag{2.19b}$$

$$k_2S_A = k_{-2}S_B;\quad S_B = \frac{k_2}{k_{-2}}S_A = K_{sr}S_A = \text{reaction equilibrium statement}$$

$$S_A = \frac{K_B}{K_{sr}}C_BS_V \quad \text{since } S_B = K_BC_BS_V = K_{sr}S_A \tag{2.19c}$$

$$S_O = S_V + S_A + S_B = S_V + \frac{K_B}{K_{sr}}C_BS_V + K_BC_BS_V = \text{site balance} \tag{2.19d}$$

$$S_V = \frac{S_O}{1 + \frac{K_B}{K_{sr}}C_B + K_BC_B}$$

$$(-R_A)_{SII} = \frac{k_1S_O\left(C_A - \frac{C_B}{K}\right)}{1 + \frac{K_A}{K}C_B + K_BC_B} \tag{2.20}$$

The denominator of this equation does not explicitly contain the reactant concentration C_A.

Case III Desorption of product controlling
This is the step where S_B decomposes to give gaseous product B and regenerates S_V. Both the adsorption of A and the surface reaction converting S_A to S_B are fast:

$$k_{-3}S_B, k_3C_BS_V \ll k_2S_2, k_{-2}S_3, k_1C_AS_1, k_{-1}S_2$$

$$(-R_A)_{SIII} = k_{-3}S_B - k_3C_BS_V = \text{net steady-state rate for Case III} \tag{2.21a}$$

$$S_A = \frac{k_1}{k_{-1}}C_AS_V = K_AC_AS_V = \text{adsorption equilibrium statement for } A \tag{2.21b}$$

$$k_2S_A = k_{-2}S_B;\ S_B = \frac{k_2}{k_{-2}}S_A = K_{sr}S_A = \text{reaction equilibrium statement}$$

$$S_B = K_AK_{sr}C_AS_V \tag{2.21c}$$

$$S_O = S_V + S_A + S_B = S_V + K_AC_AS_V + K_AK_{sr}C_AS_V = \text{site balance} \tag{2.21d}$$

$$S_V = \frac{S_O}{1 + K_AC_A + K_AK_{sr}C_A}$$

$$(-R_A)_{SIII} = \frac{k_1S_OK\left(C_A - \frac{C_B}{K}\right)}{1 + K_AC_A + KK_BC_A} \tag{2.22}$$

The denominator of this expression does not explicitly contain the product concentration C_B.

2.2.1.1 Langmuir–Hinshelwood–Hougen–Watson rate equations

The intrinsic Langmuir–Hinshelwood–Hougen–Watson (LHHW) rate expressions (Eqs. 2.17, 2.20, and 2.22) derived for various reactions with different or similar postulated slow steps are of the following general form:

$$\text{Rate} = \frac{(\text{kinetics term})(\text{driving potential term})}{(\text{adsorption term})^n} \tag{2.23}$$

Here, the exponent n shows the number of sites involved per catalytic reaction cycle, and its value can be 1 or 2, very rarely 3, for surface reaction-controlling cases. Since one active site is involved per reaction cycle in the example discussed earlier, the exponent of the adsorption term for the surface reaction limiting case is unity.

The individual terms appearing in LHHW rate expressions describing different kinetic schemes were prepared in the form of tables first by Yang and Hougen [18] for four different reactions that cover nearly all possible types of catalytic reactions [5, 13, 19]:

Table 2.1 Individual terms of LHHW rate equations for the surface reaction-controlling cases of various catalytic reactions.

Reaction type	Postulated reaction steps	Kinetics term	Driving potential term	Adsorption term	n
$A \leftrightarrow B$	$A+S \leftrightarrow A-S$ $A-S \leftrightarrow B-S$ $B-S \leftrightarrow B+S$	$k_{sr}K_A$	$P_A - P_B/K$	$(1+K_AP_A+K_BP_B)$	1
$A \leftrightarrow B+C$	$A+S \leftrightarrow A-S$ $A-S+S \leftrightarrow B-S+C-S$ $B-S \leftrightarrow B+S$; $C-S \leftrightarrow C+S$	$k_{sr}K_A$	$P_A - P_BP_C/K$	$(1+K_AP_A+K_BP_B+K_CP_C)$	2
$A+B \leftrightarrow C$	$A+S \leftrightarrow A-S$ $B+S \leftrightarrow B-S$ $A-S+B-S \leftrightarrow C-S+S$ $C-S \leftrightarrow C+S$	$k_{sr}K_AK_B$	$P_AP_B - P_C/K$	$(1+K_AP_A+K_BP_B+K_CP_C)$	2
$A(g)+B \leftrightarrow C$ (one gaseous reactant)	$B+S \leftrightarrow B-S$ $A(g)+B-S \leftrightarrow C-S$ $C-S \leftrightarrow C+S$	$k_{sr}K_B$	$P_AP_B - P_C/K$	$(1+K_BP_B+K_CP_C)$	1
$A+B \leftrightarrow C+D$	$A+S \leftrightarrow A-S$ $B+S \leftrightarrow B-S$ $A-S+B-S \leftrightarrow C-S+D-S$ $C-S \leftrightarrow C+S$; $D-S \leftrightarrow D+S$	$k_{sr}K_AK_B$	$P_AP_B - P_CP_D/K$	$(1+K_AP_A+K_BP_B+K_CP_C+K_DP_D)$	2
$½A_2+B \leftrightarrow C+D$ (with dissociation)	$A_2+2S \leftrightarrow 2A-S$ $B+S \leftrightarrow B-S$ $A-S+B-S \leftrightarrow C-S+D-S$ $C-S \leftrightarrow C+S$; $D-S \leftrightarrow D+S$	$k_{sr}(K_A)^{½}K_B$	$(P_A)^{½}P_B - P_CP_D/K$	$(1+(K_AP_A)^{½}+K_BP_B+K_CP_C+K_DP_D)$	2
$A+B \leftrightarrow C+D$ (unlike active sites)	$A+S_1 \leftrightarrow A-S_1$ $B+S_2 \leftrightarrow B-S_2$ $A-S_1+B-S_2 \leftrightarrow C-S_1+D-S_2$ $C-S_1 \leftrightarrow C+S_1$; $D-S_2 \leftrightarrow D+S_2$	$k_{sr}K_{A1}K_{B2}$	$P_AP_B - P_CP_D/K'$	$(1+K_{A1}P_A+K_{C1}P_C)\,(1+K_{B2}P_B+K_{D2}P_D)$	1 each

$A \leftrightarrow B$; $A \leftrightarrow B+C$; $A+B \leftrightarrow C$; $A+B \leftrightarrow C+D$

The rate equation for any specific situation is easily assembled using these tables. Surface reactions of molecularity greater than two are not known. Since the surface reaction limiting case is the most important for industrial-scale reactions, the specific terms and exponent n values corresponding to this particular case are formulated in Table 2.1 for various reactions and mechanisms. The surface reaction rate constants (k_{sr}) appearing in the kinetic terms of the various cases are lumped parameters including the total number of active sites S_O or the number of adjacent sites in some form, since the latter is generally unknown or not independently measurable.

2.2.1.2 Mechanisms of bimolecular surface reactions

There are two possible mechanisms for solid-catalyzed reactions that involve two reactants. The Langmuir–Hinshelwood mechanism postulates that the surface reaction takes place only between two adjacently adsorbed reactants, while the Rideal-Eley mechanism hypothesizes that the surface reaction can occur between an adsorbed reactant and a gaseous reactant. In both cases the stoichiometric equation is the same:

$A + B \longrightarrow \text{Products}$

In the Langmuir–Hinshelwood mechanism, it is assumed that the chemisorptions of both A and B are fast and essentially at equilibrium, while the irreversible surface reaction between adsorbed reactants is the rate-determining step (rds):

$A + S \leftrightarrow A-S$ adsorption; k_1, k_{-1}

$B + S \leftrightarrow B-S$ adsorption; k_2, k_{-2}

$A-S+B-S \longrightarrow \text{Product} + 2S$ reaction; k_3

The surface reaction is considered as giving directly the gaseous product, as desorption rates are unknown in most cases. The Langmuir adsorption isotherm is used for obtaining surface concentrations of reactants, and the probability of gaseous product formation is taken to be proportional to the product of fractional surface coverages, $\theta_A\theta_B$:

$$\theta_A = \frac{K_AP_A}{1+K_AP_A+K_BP_B} \quad \text{and} \quad \theta_B = \frac{K_BP_B}{1+K_AP_A+K_BP_B}$$

The net steady-state rate $(-R_A)_S$ is then formulated as

$$(-R_A)_{SI} = k\theta_A\theta_B = \frac{kK_AP_AK_BP_B}{(1+K_AP_A+K_BP_B)^2} \tag{2.24}$$

In the Rideal–Eley mechanism, on the other hand, it is assumed that the chemisorption of A is fast and basically at equilibrium, while the irreversible surface reaction that occurs between adsorbed A and gaseous B is the rate-determining step (rds):

$A + S \leftrightarrow A - S$ adsorption; k_1, k_{-1}

$A - S + B(g) \rightarrow$ Products $+ S$ reaction; k_2

The probability of surface reaction is proportional to the product of the fractional surface coverage of A and the partial pressure of B in the gas phase, $\theta_A P_B$, giving directly the gaseous product; θ_A is then given by

$$\theta_A = \frac{K_A P_A}{1 + K_A P_A}$$

$$(-R_A)_{SI} = k\theta_A P_B = \frac{kP_B K_A P_A}{1 + K_A P_A} \tag{2.25}$$

Kinetic studies indicate whether a surface reaction-controlled bimolecular reaction proceeds by an L–H or R–E mechanism. Equation 2.24 indicates that in the L–H mechanism $(-R_A)$ passes through a maximum when either P_A or P_B is increased while the other is fixed. The decrease in the rate at high P_A or P_B is rationalized by supposing that the more strongly adsorbed reactant displaces other species from the surface as its partial pressure is increased. This type of behavior was observed in the transition metal-catalyzed reaction of cyclopropane with hydrogen, where the strongly adsorbed hydrogen displaced cyclopropane from the surface [20]. In the R–E mechanism, on the other hand, $(-R_A)$ tends to become independent of reactant partial pressure when P_A is steadily increased (Eq. 2.25).

Example 1 Consider a bimolecular reversible reaction of the general type

$A + B \leftrightarrow C + D$

Case 1: A simple Langmuir–Hinshelwood sequence of steps is postulated for this reaction:

$A + S \leftrightarrow A - S$ adsorption

$B + S \leftrightarrow B - S$ adsorption

$A - S + B - S \rightarrow C - S + D(g) + S$ reaction (rds)

$C - S \leftrightarrow C + S$ desorption

The rate equation for the surface reaction-controlling case is assembled by making use of the corresponding terms listed in Table 2.1, bearing in mind that D is a gaseous product:

$$(-R_A)_{SI} = \frac{k_{sr} K_A K_B (P_A P_B - P_C P_D / K)}{(1 + K_A P_A + K_B P_B + K_C P_C)^2} \tag{2.26}$$

Several special forms of this rate equation are likely, depending on the assumptions made on the basis of experimental observations:

(a) If the overall equilibrium constant K is large, or when there is a gaseous product, the reaction is considered irreversible:

$$(-R_A)_{SI} = \frac{k_{sr} K_A K_B P_A P_B}{(1 + K_A P_A + K_B P_B + K_C P_C)^2} \tag{2.26a}$$

(b) If initial rate data are being analyzed, then $P_C \approx 0$, and $(-R_A)_S = (-R_A)_0$:

$$(-R_A)_0 = \frac{k_{sr} K_A K_B P_A P_B}{(1 + K_A P_A + K_B P_B)^2} \tag{2.26b}$$

(c) If reactant A is weakly adsorbed,

$$(1 + K_A P_A + K_B P_B) \approx 1 + K_B P_B$$

$$(-R_A)_0 = \frac{k_{sr} K_A K_B P_A P_B}{(1 + K_B P_B)^2} \tag{2.26c}$$

(d) When both A and B are weakly adsorbed,

$$K_A P_A + K_B P_B \ll 1; \quad (1 + K_A P_A + K_B P_B) \approx 1$$

$$(-R_A)_0 = k_{sr} K_A K_B P_A P_B = k' P_A P_B \tag{2.26d}$$

(e) When B is strongly adsorbed and A is weakly adsorbed,

$$(1 + K_B P_B) \approx K_B P_B$$

$$(-R_A)_0 = k_{sr} \frac{K_A}{K_B} P_A P_B^{-1} = k'' P_A P_B^{-1} \tag{2.26e}$$

(f) With C in the feed, if A and B are weakly adsorbed and C is strongly adsorbed,

$$(1 + K_A P_A + K_B P_B + K_C P_C) \approx K_C P_C$$

$$(-R_A)_0 = k_{sr} \frac{K_A K_B}{K_C^2} P_A P_B P_C^{-2} = k''' P_A P_B P_C^{-2} \tag{2.26f}$$

Under these conditions, the products slow down the surface reaction considerably; this is the case of product inhibition.

Case 2: A Rideal–Eley mechanism is assumed where adsorbed A reacts with gaseous B:

$A + S \leftrightarrow A - S$ adsorption

$A - S + B(g) \rightarrow C - S + D(g)$ reaction (rds)

$C - S \leftrightarrow C + S$ desorption

The rate expression based on this mechanism is

$$(-R_A)_S = \frac{k_{sr} K_A P_A P_B}{1 + K_A P_A + K_C P_C} \tag{2.27}$$

Special forms of this equation are also possible. For instance,

(a′) If both reactant A and product C are weakly adsorbed,

$$(1 + K_A P_A + K_C P_C) \approx 1$$

$$(-R_A)_S = k_{sr} K_A P_A P_B = k P_A P_B \tag{2.27a′}$$

(b′) If A is strongly adsorbed and C is weakly adsorbed,

$$(1 + K_A P_A + K_C P_C) \approx K_A P_A$$

$$(-R_A)_S = \frac{k_{sr} K_A P_A P_B}{K_A P_A} = k_{sr} P_A^{0} P_B = k_{sr} P_B \tag{2.27b′}$$

It is safe to state that when weak adsorption or low surface coverage of reactant A is involved, the reaction rate tends toward first-order dependence on P_A, while strong adsorption and high surface coverage of reactant A leads to zeroth-order dependence on P_A because of its abundance on the surface.

2.2.1.3 Activation energies of surface reactions

The reaction rate constant of a single-stage process varies with temperature according to the Arrhenius functionality. In the analysis of solid catalyzed reactions, the differential form of Equation 2.2 is used to describe the exponential dependence of k_{sr} on temperature:

$$\frac{d(\ln k_{sr})}{dT} = \frac{E_A}{R_g T^2} = \frac{E_{true}}{R_g T^2}$$

The temperature dependence of the reaction equilibrium coefficient K_T is obtained from the second law of thermodynamics using $\Delta G^\circ = \Delta H^\circ - \Delta S^\circ$ and Equation 2.1.

The resulting van't Hoff equation gives the temperature dependence of K:

$$\frac{d(\ln K)}{dT} = \frac{-\Delta H_r}{R_g T^2} \tag{2.28}$$

Considering that chemisorption is a chemical reaction restricted to the surface layer of the catalyst, the dependence of adsorption equilibrium coefficients, K_i, is also expressed in terms of the van't Hoff equation, with ΔH_{ads} being used instead of ΔH_r:

$$\frac{d(\ln K_i)}{dT} = \frac{-\Delta H_{ads,i}}{R_g T^2} \tag{2.28a}$$

In heterogeneous kinetics, the experimentally observed rate constant is an apparent rate constant, k_{app}, and the activation energy obtained from it is also apparent activation energy, E_{app}, which is a relatively complex combination of E_{true} with ΔH_{ads} of reactants and products. It is difficult to identify the separate components of E_{app} unless a specific experimental design is formulated.

In part (d) of Example 1, the experimentally observed apparent rate constant, k', is equal to $(k_{sr}K_AK_B)$. Therefore,

$$\frac{d(\ln k')}{dT} = \frac{d(\ln k_{sr})}{dT} + \frac{d(\ln K_A)}{dT} + \frac{d(\ln K_B)}{dT}$$
$$= \frac{E_{apparent}}{R_g T^2} = \frac{E_{true}}{R_g T^2} + \frac{\Delta H_A}{R_g T^2} + \frac{\Delta H_B}{R_g T^2}$$

Since chemisorption is mostly exothermic, the heats of adsorption for reactants A and B are negative, and the measured E_{app} is lower than E_{true} by the sum of $\Delta H_A + \Delta H_B$:

$$E_{app} = E_{true} + \Delta H_A + \Delta H_B$$

Considering part (e) of the same example, k'' is equal to $(k_{sr}K_A/K_B)$, and since B is adsorbed strongly and A is adsorbed weakly, the E_{app} calculated is higher than E_{true} by an amount equal to the difference between ΔH_B and ΔH_A:

$$E_{app} = E_{true} + \Delta H_A - \Delta H_B$$

Similarly, k''' in part (f) of the example is equal to $(k_{sr}K_AK_B/K_C{}^2)$, and because product inhibition is involved, E_{app} is higher than E_{true} by the difference $(\Delta H_A + \Delta H_B - 2\Delta H_C)$:

$$E_{app} = E_{true} + \Delta H_A + \Delta H_B - 2\Delta H_C$$

In this example, the only case where E_{true} is measured is that of the Rideal–Eley mechanism in part (b′) involving strong adsorption of reactant A and weak adsorption of product C.

2.2.2 Analysis and correlation of rate data

The mathematical models generally used for correlation of rate data on solid-catalyzed reactions fall into two broad classifications:

1 LHHW models

$$\text{Rate} = \frac{(\text{kinetics term})(\text{driving potential term})}{(\text{adsorption term})^n} \tag{2.23}$$

2 Power function (or power law) models

$$\text{Rate} = k\,C_A{}^\alpha\, C_B{}^\beta\, C_C{}^\gamma \tag{2.29}$$

Providing the full equation is correct, LHHW equations can be extrapolated to calculate reaction rates at other conditions not included in the kinetic study. They give a general idea about the reaction mechanism postulated for deriving the model equation(s); nevertheless, good fit of data to the model is only a necessary but not sufficient condition for deciding on a particular reaction mechanism. LHHW equations usually complicate the mathematics of reactor design and reactor control, particularly if diffusion effects are present.

Power function models, on the other hand, directly utilize the concept of reaction order. Unlike homogeneous reactions, the reaction orders encountered in solid-catalyzed reactions can be negative or positive, integer, fractional, or zero; moreover, product concentrations may also appear in the rate equation. Due to the simplicity of their form, power function models are considerably easier to handle and integrate than the full LHHW expressions and are preferred especially if the reaction is affected by diffusional limitations. These models cannot, however, be used to discriminate between other than grossly different mechanisms, and they are reliable only within the limits of the reaction conditions used to obtain the kinetic data.

The definite advantages to the use of both formulations under appropriate conditions are evident; however, care must be taken not to apply either of the models arbitrarily. Estimation of reaction orders is desirable in a number of applications, and LHHW equations can be reduced to power law form by making the most abundant surface intermediate (MASI) approximation, if the fractional surface coverage of one adsorbed intermediate is much greater than all others under reaction conditions [21]. Alternatively, catalytic rate data analysis can be started by expressing the rate in terms of a power function model and then translating this model into a plausible LHHW model and evaluating its parameters.

2.2.2.1 Model discrimination and parameter estimation

In a kinetic investigation, the rate-determining step and, hence, the functional form of the rate model are not known *a priori*; also unknown are the rate constants and adsorption equilibrium coefficients. Hence, the aim of data procurement and correlation is both model discrimination and parameter estimation which are completed in tandem [17]. The critical problem at this point is to obtain reliable experimental data from which kinetic models that reflect steady-state chemical activity can be extracted and evaluated. In order to measure correctly the rates of chemical events only, (i) external and internal mass and heat transport resistances at the particle scale have to be eliminated, (ii) an ideal flow pattern and isothermal operation have to be established to reduce transport effects at the reactor scale, and (iii) possibility of catalyst deactivation during experiments should be minimized. The measures to be taken for overcoming these problems at the laboratory level depend on the careful selection of experimental reactors, operating conditions, and kinetic analysis methods [13]. A series of preliminary diagnostic experiments need to be conducted to ensure conditions typical of heterogeneous catalysis before the kinetic investigation is launched.

The full conventional kinetic study conducted at steady-state conditions for determining intrinsic rate equations consists of various consecutive steps:

1 Collection of concentration versus space-time data at a single temperature and under isothermal conditions
2 Numerical differentiation of these data to obtain reaction rate data
3 Utilization of LHHW and/or power function models together with linear and nonlinear regression techniques to fit rate data for purposes of model discrimination and parameter estimation
4 Validation of the model and its parameters by testing against new independent experimental data
5 Repeating these isothermal experiments at several other temperatures to find the temperature dependences of the rate and adsorption parameters

Although kinetic models, numerical methods, and present-day computational capabilities can very well handle nonisothermal conditions in reactor design and dynamic simulations, determination of the functional form of the rate equation is accurate only on the basis of isothermal data [10, 13, 17].

More recently, specific reactors are used for nonsteady-state studies of catalytic kinetics which also allow observation of the intermediate steps in a reaction mechanism [22]. Transient studies of catalytic reaction schemes and kinetics are discussed in Chapter 10.

Example 2 Linear and nonlinear regression techniques are utilized to fit experimental data to an LHHW or a power function rate expression in order to decide which equation best describes the data. Consider the bimolecular reaction $A + B \leftrightarrow C + D$ in Example 1 and part (b) in particular. The surface reaction-controlled expression for the initial reaction rate $(-R_A)_0$ given in Equation 2.26b may well be written in terms of reactant concentrations C_A and C_B instead of partial pressures. The units and hence numerical values of the adsorption equilibrium coefficients will change when C_A and C_B are used.

The initial reaction rates $(-R_A)_0$ calculated by numerical differentiation of concentration versus space-time data at a single temperature can be directly fitted to this equation by nonlinear regression to estimate the model parameters k_{sr}, K_A, and K_B:

$$(-R_A)_0 = \frac{k_{sr} K_A K_B \, C_A C_B}{(1 + K_A C_A + K_B C_B)^2} \tag{2.30}$$

Concurrently, linear regression techniques may be used to provide (i) the first estimates of the model parameters and also (ii) useful information on goodness of fit, both of which may be used in making the initial guesses required by nonlinear regression. The rearranged and linearized form of the rate expression is as follows:

$$\left\{\frac{(C_A C_B)}{(-R_A)_0}\right\}^{\frac{1}{2}} = \frac{(1 + K_A C_A + K_B C_B)}{(k_{sr} K_A K_B)^{\frac{1}{2}}}$$

$$\left\{\frac{(C_A C_B)}{(-R_A)_0}\right\}^{\frac{1}{2}} = \frac{1}{(k_{sr} K_A K_B)^{\frac{1}{2}}} + \frac{K_A^{\frac{1}{2}} C_A}{(k_{sr} K_B)^{\frac{1}{2}}} + \frac{K_B^{\frac{1}{2}} C_B}{(k_{sr} K_A)^{\frac{1}{2}}}$$

$$(y = m + nx + qz) \quad \text{or} \quad (y = m + nC_A + qC_B)$$

If the model tested is to be acceptable, all of the rate parameters evaluated by regression analysis should be positive and the statistical fit of data should be good. A disadvantage of linearization is that independent and dependent variables of the rate expression are grouped together in the y term. However, if kinetic experiments are designed carefully and the quantity of data generated is sufficient, both linear and nonlinear regression analyses should give compatible results.

Example 3 Experimental data on the effect of total pressure P_T on the initial reaction rate $(-R_A)_0$ provide information on the functional form of the rate equation, and hence on the rate-limiting step of the reaction, as well as on estimates of some of the rate parameters if not all. The requirement to the use of this technique is that all other conditions such as feed composition, temperature, and space time must be kept constant in all runs while only the total pressure P_T in the reactor is gradually increased.

Consider the single reaction $A \leftrightarrow B$ discussed in Section 2.2.1. The rate equations derived for the three different rate-limiting steps of this reaction may also be written in terms of P_T. For the surface reaction-controlling case, Case 1,

$$(-R_A)_{SI} = \frac{(k_2 K_A S_0)\left(C_A - \frac{C_B}{K}\right)}{(1 + K_A C_A + K_B C_B)} \tag{2.17}$$

$$(-R_A)_0 = \frac{(k_2K_AS_0)(C_A)}{(1+K_AC_A)} = \frac{(k_2K_AS_0)\left(\frac{P_Ty_A}{R_gT}\right)}{\left[1+K_A\left(\frac{P_Ty_A}{R_gT}\right)\right]}$$

$$(-R_A)_0 = \frac{aP_T}{(1+bP_T)} \text{ (saturation curve)} \tag{2.17a}$$

For the adsorption of reactant-controlling case, Case II,

$$(-R_A)_{SII} = \frac{(k_1S_0)\left(C_A - \frac{C_B}{K}\right)}{\left[1+\left(\frac{K_A}{K}\right)C_B + K_BC_B\right]} \tag{2.20}$$

$$(-R_A)_0 = \frac{(k_1S_0)(C_A)}{(1+0+0)} = k_1S_0C_A = \left(k_1S_0\frac{y_A}{R_gT}\right)(P_T) = a'P_T \text{ (linear)} \tag{2.20a}$$

For the desorption of product-controlling case, Case III,

$$(-R_A)_{SIII} = \frac{(k_3S_0K)\left(C_A - \frac{C_B}{K}\right)}{(1+K_AC_A+KK_BC_A)} \tag{2.22}$$

$$(-R_A)_0 = \frac{(k_3S_0K)(C_A)}{[(K_A+KK_B)C_A]} = \frac{(k_3S_0K)}{(K_A+KK_B)} = a'' \text{ (independent of } P_T) \tag{2.22a}$$

This example demonstrates that the dependence of the initial rate $(-R_A)_0$ on total pressure P_T gives a clear indication of the rate-controlling step and hence of the form of the LHHW equation. The linear P_T dependence observed for adsorption-controlling cases and the independence from P_T of desorption-controlling cases are similar in all reaction types. The P_T dependence of $(-R_A)_0$ for surface reaction-controlling cases of dual-site or bimolecular reactions is generally expressed by rate equations with a squared term in the denominator:

$$(-R_A)_0 = \frac{aP_T}{(1+bP_T)^2} \text{ (dual-site, } A \rightarrow B + C) \tag{2.31}$$

$$(-R_A)_0 = \frac{aP_T^2}{(1+bP_T)^2} \text{ (bimolecular, } A + B \rightarrow C) \tag{2.32}$$

The P_T dependence of $(-R_A)_0$ for $A \rightarrow B + C$ is given in Figure 2.4 for surface reaction-controlling (Eq. 2.31) as well as adsorption- and desorption-controlling cases.

The three principal criteria of model discrimination and parameter estimation are:

1 Reaction rate constants, k, and adsorption coefficients, K_i, obtained from linear and nonlinear regression analyses must be positive.
2 Good statistical fit of the model equation to the data should be demonstrated.
3 Total pressure dependence of the initial rate should be in agreement with (1) and (2).

In the best strategy of experimentation for model discrimination and parameter estimation, experimental studies are started by determining the P_T dependence of $(-R_A)_0$, which will indicate the functional form of the rate expression. Secondly, pure feed experiments are conducted both at space times within the initial rates region and at longer space times to determine the reaction rate constant and adsorption equilibrium coefficients of reactants. Thirdly, mixed feed experiments carried out in the initial rates region and at longer space times are used in the calculation of adsorption equilibrium coefficients of products [17].

2.2.2.2 Laboratory catalytic reactors

Kinetic studies on solid-catalyzed gas-phase reactions are generally conducted in flow reactors. Continuous, steady-state, isothermal, isobaric reactor operation with nearly ideal flow pattern and no concentration gradients is ideal for kinetic data procurement [10, 13]. Continuous stirred tank reactors (CSTRs) and plug-flow reactors (PFRs) are commonly used, since they represent the two limiting cases of complete mixing and no mixing, respectively, and thus the complication of data evaluation by axial and radial dispersion terms of the continuity equation is avoided (Figure 2.5). The most practical flow reactor is the PFR packed with catalyst particles and operated in a single pass with fresh catalyst being used in each run. PFRs may be operated in the differential or integral modes. When either a CSTR or a differential PFR is used, it is possible to measure the global rate directly. If an integral PFR is used, data are obtained at higher conversion levels and can be analyzed by either the differential or the integral methods of data analysis. However, care must be taken at all times to eliminate physical transport effects at the particle and reactor scales as well as minimizing the probability of deactivation.

CSTR: One type of catalytic CSTR is the spinning basket reactor in which each experiment is carried out at a given inlet

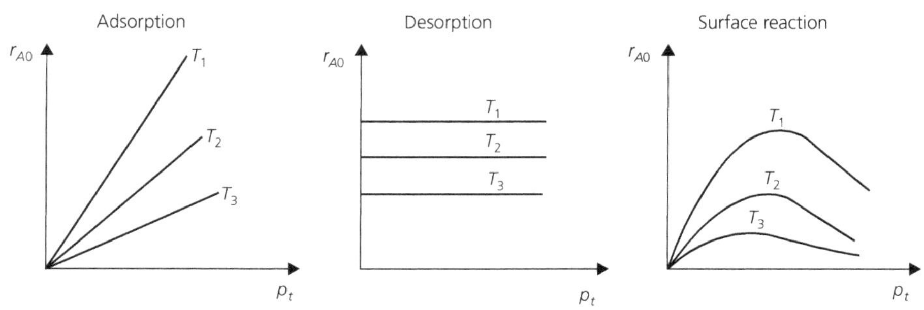

Figure 2.4 Total pressure dependence of initial rates for the reaction $A \rightarrow B + C$. (Source: Froment [17]. Reproduced with permission of John Wiley & Sons.)

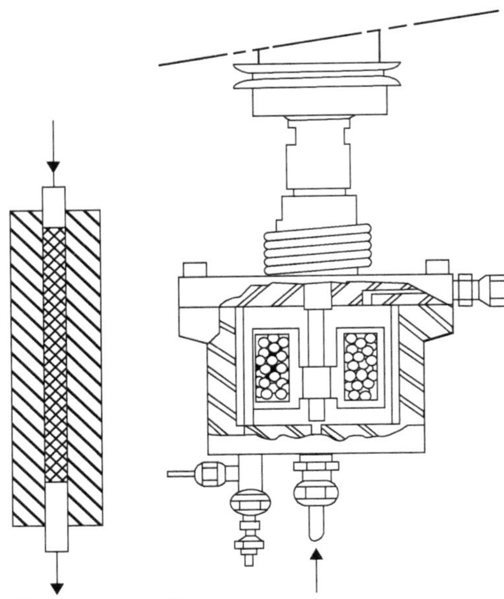

Figure 2.5 Types of laboratory catalytic reactors mostly used: (a) plug-flow packed-bed reactor and (b) spinning basket reactor mimicking CSTR performance.
(Source: Froment [17]. Reproduced with permission of John Wiley & Sons.)

composition and space time to provide only one value of the reaction rate (Figure 2.5). The ideal CSTR is characterized by complete mixing of fluid elements so that all reaction parameters in the reactor are uniform and equal to those of the exit stream, thus enabling gradientless operation as well as mathematically simpler data evaluation. The design equation for an ideal CSTR is algebraic, and the rate $(-R_{Ae})_P$ is evaluated at exit conditions:

$$(-R_{Ae})_P \rho_B V = F_{A0} x_{Ae} \tag{2.33}$$

In CSTR experiments, inlet concentrations and space times have to be varied over a wide range, using several different nonstoichiometric feed compositions. If reactant concentrations are varied in stoichiometric ratios for a reaction with two reactants, the overall reaction order $(\alpha + \beta)$ is determined from experimental data rather than the individual orders of α and β for reactants A and B, respectively. The disadvantage of a CSTR is that the exit reactant concentrations have to be controlled by manipulating the inlet concentration and the space time by a trial-and-error routine, which is difficult at times when more than one reactant is involved [23].

PFR: The one-dimensional design equation for an ideal catalyst-packed PFR (called a packed-bed reactor (PBR)) is obtained by making a component mass balance over a volume element extending over the entire cross section of the reactor:

$$\frac{dW}{F_{A0}} = \frac{d(\rho_B V)}{F_{A0}} = \frac{dx_A}{(-R_A)_P} \tag{2.34}$$

A differential PFR is a laboratory microreactor operated at very low fractional conversions of reactant(s), preferably not more than 5 mol% but certainly lower than 10 mol%. Differential operation occurs at low space times achieved by keeping the catalyst weight small, the gas flow rate high, and the reaction temperature moderate. At very low conversion levels, it can be assumed that the reaction rate is constant at an average value in the direction of flow within the small catalyst bed and

$$\frac{W}{F_{A0}} = \frac{x_A}{(-R_A)_{Pavg}} \tag{2.34a}$$

The average reaction rate may then be calculated directly from $(-R_A)_{P,avg} = x_A F_{A0}/W$. The average of the inlet and exit reactant concentrations is normally used in calculations. The differential PFR is convenient and versatile for generating substantial amounts of conversion versus space-time data by monitoring the inlet composition only. In order to find the individual concentration dependencies of $(-R_A)_P$ accurately in the case of power law models, it is necessary to obtain data in as wide a range as possible by using several different nonstoichiometric feed compositions at constant temperature.

In an integral PFR, on the other hand, the fractional conversion of reactant is generally much higher; the reaction rate decreases in the direction of flow, and therefore, the conversion versus space-time data should be differentiated in order to calculate the rate values at each conversion level. The slope of the tangent to the x_A versus W/F_{A0} curve is equal to $(-R_A)_P$ at any selected value of x_A:

$$(-R_A)_P = \frac{dx_A}{d(W/F_{A0})} \tag{2.34b}$$

Although a single set of experiments at one feed composition may, in principle, be sufficient to determine the rate equation, this is possible only if the reactants are not in stoichiometric ratio in the feed. The kinetic analysis is much more reliable if data are taken in several runs at different nonstoichiometric feed compositions [13, 17, 23].

Transport Criteria in PBRs: In laboratory catalytic reactors, basic problems are related to scaling down in order to eliminate all diffusional gradients so that the reactor performance reflects chemical phenomena only [24, 25]. Evaluation of catalyst performance, kinetic modeling, and hence reactor scale-up depend on data that show the steady-state chemical activity and selectivity correctly. The criteria to be satisfied for achieving this goal are defined both at the reactor scale (macroscale) and at the catalyst particle scale (microscale). External and internal transport effects existing around and within catalyst particles distort intrinsic chemical data, and catalyst evaluation based on such data can mislead the decision to be made on an industrial catalyst or generate irrelevant data and false rate equations in a kinetic study. The elimination of microscale transport effects from experiments on intrinsic kinetics is discussed in detail in Sections 2.3 and 2.4 of this chapter.

Since the fluid flow pattern defines the performance of continuous reactors, the two crucial prerequisites for the operation of laboratory PBRs are (i) to ensure an ideal plug-flow pattern for enabling clear-cut data analysis and (ii) to establish fully isothermal conditions for getting reliable quantitative data on catalyst performance.

Deviations from an ideal plug-flow pattern are caused by either wall flow or axial gradients that develop in the direction of flow. The bed void fraction at the reactor wall is likely to be somewhat higher than the void fraction in the catalyst bed, ϵ_B. In order to eliminate the wall effect on the flow pattern, the tube diameter to particle diameter ratio is chosen to be greater than 10 as a general rule, but since the microreactors used for kinetic studies have very small diameters (4–10 mm), and for reasons discussed in the following, a higher ratio of 15 is indicated to minimize the wall effect in laboratory PBRs:

$$\frac{d_{tube}}{d_{particle}} > 15 \tag{2.35}$$

Axial gradients may arise as a result of reactant conversion along the catalyst bed, which may be important in integral reactors. The rule of thumb for minimizing axial dispersion effects is concerned with the reactor tube length to particle diameter ratio, which is reported as being at least 50 for first-order reactions and particle Reynolds numbers greater than 10:

$$\frac{L_{tube}}{d_{particle}} > 50; \text{ for } (Re)_P > 10 \text{ or } (Pe)_P = 2 \tag{2.36}$$

It is difficult to achieve $(Re)_P > 10$ in laboratory PBRs with very small particle sizes of the order of microns or millimeters; hence, a higher L_{tube}/d_P ratio is suggested for minimizing axial concentration gradients in these reactors [13, 24]:

$$\frac{L_{tube}}{d_{particle}} \approx 100; \text{ for } (\mathrm{Re})_P \approx 0.1-1 \text{ or } 2 \tag{2.36a}$$

$$(\mathrm{Re})_P = \frac{u d_P \rho}{\mu}; \quad (Pe)_P = \frac{u d_P}{D_{axial}} \tag{2.37}$$

Deviations from the second important criterion regarding isothermality are caused by temperature gradients that develop in the axial flow direction as a result of reactant conversion. The rate constant of a chemical reaction is exponentially dependent on temperature as given by the Arrhenius equation; therefore, small changes in temperature may lead to large changes especially in the rates of exothermic reactions. Minimization of axial temperature gradients is subject to the same rule about the L_{tube}/d_P ratio as in axial concentration gradients. Moreover, radial temperature gradients may result from the low effective thermal conductivity of the catalyst bed. The radial aspect ratio of particle diameter to tube radius (d_P/r_{tube}) must be decreased in order to minimize these radial gradients, and radial aspect ratios of 1/8–1/5 have been recommended.

Large temperature gradients can develop in the catalyst bed even for moderately exothermic reactions under typical testing conditions. Although differential reactors are characterized by low reactant conversions, the same measures as in integral reactors must be taken to guarantee completely isothermal operation [11, 13, 24]. These measures may be listed as follows: (i) small reactor diameter, d_{tube}, (ii) small catalyst particle diameter, d_P, to decrease bed voidage, (iii) catalyst support having high thermal conductivity, (iv) bed dilution with inert particles having high thermal conductivity, (v) high gas flow rates, (vi) feed dilution with inert gas having high thermal conductivity, such as H_2 or He, (vii) differential reactor operation with low reactant conversion levels, and (viii) an appropriate reactor furnace with a flat temperature profile to avoid temperature gradients along the catalyst bed. The use of a CSTR-type reactor, if possible, is also recommended.

2.2.2.3 Methods of data analysis

The data obtained in laboratory PFRs are invariably concentration versus space-time data. There are four major methods of data analysis and one method of partial analysis which may be used, in principle, for analyzing data on homogeneous or heterogeneous reactions. The major methods may be listed as the differential method, the integral method, the method of half-lives, and the method of initial rates; the method of partial analysis is called the isolation method or the method of excess which is used together with one of the major methods when more than one reactant is involved. More recently, linear and nonlinear regression are also listed among methods for analyzing rate data [17, 23, 26]. Several powerful software packages, including MATLAB, are available for facilitating data analysis by regression techniques.

The two major methods used predominantly in the kinetic analysis of isothermal data on solid-catalyzed reactions conducted in plug-flow PBRs are the differential method and the method of initial rates. The integral method is less frequently used either when data are scattered or to avoid numerical or graphical differentiation. Linear and nonlinear regression techniques are widely used in conjunction with these major methods.

Differential Method: In order to use the differential method of data analysis, it is necessary to differentiate the reactant concentration versus space-time data obtained in a plug-flow PBR. There are three methods of differentiation that are commonly used: (i) graphical equal-area differentiation, (ii) numerical differentiation or finite difference formulas, and (iii) polynomial fit to the data followed by analytical differentiation. The aim of differentiation is to obtain point values of the reaction rate $(-R_A)_P$ at each reactant concentration C_A or conversion x_A or space time (W/F_{A0}), as required. All three differentiation methods can introduce some error to the evaluation of $(-R_A)_P$. Information on and illustration of the various differentiation techniques are available in the literature [23, 26].

The differential method of data analysis is convenient since it requires only one experiment for rate equations containing only one reactant and is readily applied to determine reaction parameters of power law or LHHW rate equations. The rate equations may be linearized to allow the use of linear regression (least squares) analysis, or nonlinear regression analysis can be applied to the rate equation as such. Consider the reaction $A \rightarrow B$, power law or LHHW models, and the linearized forms of these models are given in the following as examples:

$$(-R_A)_P = k_\alpha C_A^\alpha \quad \text{or} \quad \ln(-R_A)_P = \ln k_\alpha + \alpha \ln C_A$$

$$(-R_A)_0 = \frac{k_{sr} K_A C_{A0}}{1 + K_A C_{A0}} \quad \text{or} \quad \frac{C_{A0}}{(-R_A)_0} = \frac{C_{A0}}{k_{sr}} + \frac{1}{(k_{sr} K_A)}$$

When more than one reactant is present in the rate equation, the isolation method can be used together with the differential method. The isolation method requires the use of one reactant in large excess while the concentration of the isolated reactant is varied; in this case, kinetic analysis is conducted in stages using one isolated reactant at a time. However, this method of partial analysis has to be employed with caution in solid-catalyzed reactions since, in certain reactions, a large excess of any one reactant may have an adverse effect on the active sites of the catalyst. Linearized equations of the form presented in Example 2 can easily be handled by the computational methods available even if reactant concentrations are comparable but present in non-stoichiometric proportions:

$$(-R_A)_P = k_\alpha C_A{}^\alpha C_B{}^\beta$$

$$\ln(-R_A)_P = \ln k_\alpha + \alpha \ln C_A + \beta \ln C_B$$

$$\ln(-R_A)_P = \ln\left(k_\alpha C_{B0}{}^B\right) + \alpha \ln C_A \text{ (when } C_{B0} \text{ is large and constant)}$$

$$\ln(-R_A)_P = \ln(k_\alpha C_{A0}{}^\alpha) + \beta \ln C_B \text{ (when } C_{A0} \text{ is large and constant)}$$

Initial Rates Method: The method of initial rates is also a differential method in essence, the main difference being the number of experiments to be conducted. Several experiments are carried out using different initial concentrations of reactant, C_{A0}, in each run. The initial rate values $(-R_A)_0$ are calculated by differentiating each set of data and extrapolating to zero space time. The same rate equations are valid as in other methods, but they are expressed in terms of initial concentrations, C_{A0} and C_{B0}. In view of the fact that the product concentrations are negligible in the initial rates region, the initial rates method is used very effectively in the analysis of (i) reversible reactions, for studying only the forward or the reverse reaction, and (ii) solid-catalyzed reactions, for avoiding the effects of product inhibition or possible catalyst deactivation. In this method, the number of experiments determines the number of data points, $[C_{A0}, (-R_A)_0]$, to be used in the regression analyses. The accuracy of $(-R_A)_0$ values at each C_{A0} is increased if an adequate number of data points collected at low space times are included in the differentiation step.

Integral Method: Alternatively, the integral method of data analysis may be used with a trial procedure, in which a particular rate equation is assumed and integrated to yield a relationship between space time (W/F_{A0}) and concentration (or conversion, in the case of more than one reactant) of the following form:

$$\frac{W}{F_{A0}} = f(x_A, k_{sr}, K_A, K_B, \ldots)$$

Nonlinear regression can be used to minimize either

$$\sum\left[\frac{W}{F_{A0}} - \left(\frac{W}{F_{A0}}\right)_{calc}\right]^2 \quad \text{or} \quad \sum\left[X_A - (X_A)_{calc}\right]^2$$

Illustrations of this approach are presented in Froment et al. [17], and the results obtained from the integral method with nonlinear regression are compared with the results found by the differential method using linear regression.

In power function rate expressions, the relationship between space time (W/F_{A0}) and x_A can be linearized to estimate the values of the unknown rate parameters. The calculated reaction rates are then compared with experimentally measured ones using the sum of squares to find out if the model assumption is correct.

2.3 External (interphase) transport processes

For the heterogeneous catalytic process to be effective, the reactants present in the surrounding fluid phase must be transported to the surface of the solid catalyst, and after the reaction, the products formed must be carried back from the surface to the bulk fluid. The path of the physical rate processes at the particle scale is divided into two parts, as depicted in the 7-step sequence of the continuous reaction model used in microkinetic analysis:

1 Transfer of reactants from the bulk fluid to the exterior surface of the catalyst particle (external or interphase transport)
2 Diffusion of reactants from the exterior surface to the active interior surface of the porous particle (internal or intraparticle transport)

The last two steps, steps 6 and 7, of the sequence involve the analogous transport of products back to the bulk fluid. Steps 1 and 7 are external physical rate processes which are in series with chemical steps 3–4–5 of adsorption–reaction–desorption; in other words, they are separated from the chemical steps. Steps 2 and 6, on the other hand, are internal physical rate processes which are concurrent with chemical steps and require simultaneous treatment of chemical and physical rate processes. Heat transfer between bulk fluid and outer catalyst surface and within the porous particle is also treated by the same reasoning.

In this section, the catalyst particle or the catalytic surface is assumed to be nonporous, and the 7-step sequence is reduced to a 5-step sequence with steps 1, 3–4–5, and 7. In this case, the only mass transfer resistance involved is between the fluid and the outer surface of the particle. The rates of external mass transfer depend on (i) the temperature, pressure, and physical properties of the fluid phase under these conditions, (ii) the gas velocity relative to the solid surface, and (iii) the intrinsic rate of the surface reaction. In other words, the rate at which mass is transported from the fluid to the surface is determined by the relative magnitudes of:

1 The mass transfer coefficient, k_m, between bulk fluid and pellet surface
2 The intrinsic rate constant, k_{sr}, of the surface reaction

When conditions are not turbulent, the rate of external mass transfer may be relatively slow leading to surface concentrations that are lower than the reactant concentrations in the bulk fluid, which actually slows down the progress of the surface reaction.

Similarly, there is a temperature difference between the bulk fluid and the particle surface, depending on:

1 The heat transfer coefficient, h, between the bulk gas and the solid surface
2 The intrinsic rate constant of surface reaction, k_{sr}
3 The heat released or absorbed by the chemical reaction (heat of reaction, ΔH_r)

External mass transfer and heat transfer processes are significant in some reactor types and negligible in others. For example, mass and heat transfer features of two-phase fluidized beds are excellent partly because of the turbulent mixing but mostly because of the large solid surface exposed per unit volume by the finely divided catalyst particles. Fixed-bed (or packed-bed) reactors, on the other hand, are characterized by larger catalyst pellets packed in a stationary position with the fluid phase being forced through the packing. In two- or three-phase fixed-bed reactors, both external and internal mass and heat transport processes are existent, and they affect the reactor performance significantly under some operating conditions [7, 27]. The discussion in this chapter pertains to external transport resistances prevailing in two-phase fixed-bed reactors.

2.3.1 External mass transfer: Isothermal conditions

A fluid passing over a solid surface develops a boundary layer in which the velocity parallel to the surface varies rapidly over a very short distance normal to the direction of flow. The velocity is zero at the solid surface but approaches the bulk-stream velocity at a distance less than a millimeter from the surface. Mixing occurs in the main fluid stream where reactants and products are transported at rates that depend on the nature of flow. The fluid velocity near the surface is low with little mixing; therefore, mass transport perpendicular to the surface is by molecular diffusion. Although mass transport in the main stream is essentially independent of the molecular diffusion coefficient, $D_{A\text{-}mix}$, it is proportional to $D_{A\text{-}mix}$ very near the surface. Since the flow passages between the particles/pellets packed in fixed beds are quite complex, the approach generally used is to define transport rates in terms of "average" mass transfer and heat transfer coefficients and then to use semiempirical correlations for estimating these transport coefficients.

Consider the unimolecular, irreversible solid-catalyzed gas-phase reaction, $A \longrightarrow B$, carried out in a fixed-bed reactor packed with completely nonporous particles. Assume that the chemical steps of adsorption–reaction–desorption are represented by first-order kinetics and that the bulk temperature, T_b, and surface temperature, T_S, are the same around a particle located at any point along the length of the reactor.

Under steady-state conditions, the rate of reaction will be balanced by the rates of mass transfer of reactants and products. If $(-R_A)_P$ is defined as the experimentally measured global rate of disappearance of A per unit mass of catalyst (mol/g-s), the rates of steps 1, 3–4–5, and 7 in the 5-step sequence will all be equal to $(-R_A)_P$.

Step 1: Mass transfer of reactant A from the bulk fluid to the catalyst surface

$$(-R_A)_P = (k_m)_A a_m (C_{Ab} - C_{AS}) \tag{2.37a}$$

Step 3–4–5: Surface reaction $A \longrightarrow B$

$$(-R_A)_P = (k_{sr})\, a_m C_{AS} \tag{2.38}$$

Step 7: Mass transfer of B from the catalyst surface to the bulk fluid

$$(-R_A)_P = (R_B)_P = (k_m)_B a_m (C_{BS} - C_{Bb}) \tag{2.39}$$

In these equations, a_m is the external surface area per unit mass of catalyst (m^2/g), k_m is the gas-phase mass transfer coefficient based on unit external surface area (m/s), k_{sr} is the apparent rate constant of the surface reaction per unit external surface area (m/s), and C_{AS} and C_{BS} are the surface concentrations of A and B (mol/m^3), respectively.

Since an irreversible reaction is considered here, step 7 does not affect the reaction rate; hence, $(-R_A)_P$ can be formulated from steps 1 and 3-4-5. In principle, for reactant A to be transported from the bulk gas to the surface, $C_{Ab} > C_{AS}$; therefore, $(-R_A)_P$ is normally less than what it would be if all the outer surface were at bulk conditions, that is, if $C_{Ab} = C_{AS}$.

There are three possible cases that can be encountered, depending on the relative magnitudes of the mass transfer coefficient, k_m, and the intrinsic reaction rate constant, k_{sr}.

Case 1: The general case of comparable k_m and k_{sr},

$$(k_m)_A a_m (C_{Ab} - C_{AS}) = (k_{sr}) a_m C_{AS}; \quad C_{AS} = \frac{(k_m)_A C_{Ab}}{\left[k_{sr} + (k_m)_A\right]} \tag{2.40}$$

$$(-R_A)_P = \left\{ \frac{k_{sr}(k_m)_A}{\left[k_{sr} + (k_m)_A\right]} \right\} a_m C_{Ab} = k_o a_m C_{Ab}$$

$$\frac{1}{k_o} = \frac{1}{k_{sr}} + \frac{1}{(k_m)_A} \tag{2.41}$$

Here, k_o is the overall reaction rate constant that will be experimentally observed.

Case 2: Slow surface reaction and fast mass transfer,

$k_{sr} \ll (k_m)_A$ intrinsic kinetics is observed (surface reaction-controlling)

$$C_{AS} \longrightarrow C_{Ab}; \quad (-R_A)_P = (k_{sr}) a_m C_{Ab} \tag{2.38a}$$

Case 3: Fast surface reaction and slow mass transfer,

$k_{sr} \gg (k_m)_A$ mass transfer is observed (diffusion-controlling)

$$C_{AS} \longrightarrow 0; \quad (-R_A)_P = (k_m)_A a_m C_{Ab} \tag{2.37b}$$

When conducting intrinsic kinetic studies in laboratory PBRs, the situation described by the limiting Case 2 is desirable. The conditions for ensuring small k_{sr} and relatively large k_m are achieved by keeping the reaction temperature at moderate levels and by increasing the linear velocity of the fluid through the catalyst bed. The limiting Case 3 of fast surface reaction is observed at higher temperatures, since the reaction rate constant k_{sr} increases rapidly with temperature according to the Arrhenius equation, while the mass transfer coefficient k_m is dependent on temperature only through fluid physical properties which are not as responsive to temperature changes as k_{sr}. In Figure 2.3, the concentration profile labeled as I shows the limiting case of fast surface reaction and slow mass transfer.

When data are taken in a laboratory PBR system where external transport effects are present and if this possibility is arbitrarily neglected, experimental results will lead to falsified reaction kinetics:

1 When the surface reaction is fast, k_m determines the overall kinetics, and since mass transfer is a first-order process, this dependence is observed instead of the true order of the surface reaction. This is a serious mistake if intrinsic kinetic data are sought, but if the intrinsic kinetics is already known, this strategy can be used for an experimental evaluation of the mass transfer coefficient, k_m.
2 Even if the reaction orders coincide as in the case of a first-order surface reaction, the apparent activation energy E_A calculated in the limiting case of $k_{sr} \gg k_m$ will be extremely low (just a few kJ/mol) since the actual temperature dependence of k_m is reflected in the data; very low E_A values are a clear indication of mass transfer limitations and must be treated with caution.
3 Similarly, the apparent E_A calculated from the experimentally measured value of k_o will also be different from the apparent E_A of the chemical steps:

$$k_o = \left\{ \frac{A \exp\left(-E_A/R_g T\right)(k_m)_A}{\left[A \exp\left(-E_A/R_g T\right) + (k_m)_A\right]} \right\} \tag{2.41a}$$

At low temperatures, $k_o \approx k_{sr} \approx A \exp(-E_A/R_g T)$. As the temperature is gradually increased, k_{sr} increases exponentially with temperature whereas k_m is relatively insensitive to temperature; hence, k_o approaches a nearly constant value equal to k_m.

The general procedure applicable in laboratory experiments for first-order reactions when external diffusion effects are not known includes (i) measurement of $(-R_A)_P$, (ii) calculation of k_o from the global rate data, (iii) estimation of k_m using empirical correlations based on dimensionless numbers, and (iv) calculation of k_{sr} from Equation 2.41. For higher surface reaction orders, it was shown that consecutive rate processes of different orders are difficult to combine in an overall rate expression and numerical techniques have to be used for distinguishing between transport and reaction rate coefficients when their values are comparable [17].

2.3.1.1 Diagnostic experiments and criteria

The presence or significance of interphase transport resistances is confirmed either by conducting diagnostic experiments or by utilizing a well-established criterion. A preliminary set of experiments that must be conducted in laboratory PBRs in order detect the effect of external mass transport is to measure the dependence of the exit conversion x_A on the linear velocity of the fluid through the catalyst bed, u (m/s), which is in fact the velocity of the fluid relative to the catalyst particles [7, 23].

In this set of experiments, all operating conditions such as the reaction temperature and the feed composition are fixed, and the gas flow rate is increased at constant space time, which means that the catalyst amount, W, and the total feed flow rate, F_{total}, are increased by the same factor. Since W/F_{A0} is an indirect measure of space time, and the feed composition is kept constant, the diagnostic experiment can be conducted according to the following plan:

$$\frac{W_1}{F_{A01}} = \frac{W_2}{F_{A02}} = \frac{W_3}{F_{A03}}$$

Since the particle Reynolds numbers in laboratory PBRs are very low, as stated in Section 2.2.2.2, the range of flow rates to be covered for producing adequate increases in transport coefficients has to be rather wide. It is important to determine the minimum gas flow rate after which the exit conversions x_A or the global rates $(-R_A)_P$ remain constant (Figure 2.6). All subsequent kinetic experiments must be conducted at flow rates equal to or above this minimum. When the overall process is surface reaction controlled, that is, if intrinsic kinetics is being observed, neither x_A nor $(-R_A)_P$ will change with increasing linear velocity of the fluid.

Mears' Criterion for External Mass Transfer Effects: In a number of cases, it may be necessary to rapidly estimate whether transport limitations are present or not, and Mears' criterion is

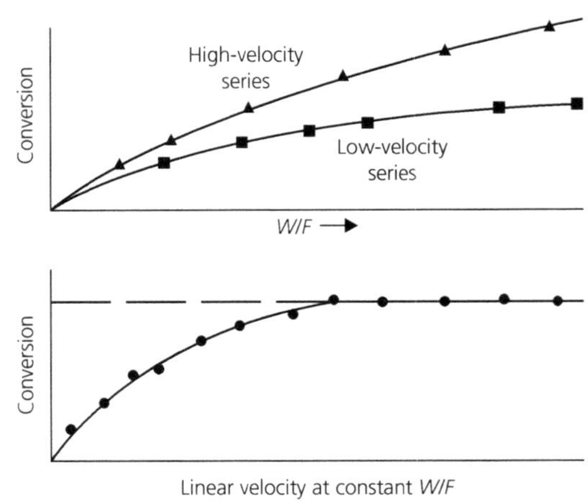

Figure 2.6 The effect of external mass transport limitations on conversion rates.
(Source: Hill [7]. Reproduced with permission of John Wiley & Sons.)

used both in laboratory experiments and for large-scale operations to predict if external diffusion effects are negligible:

$$\left\{\frac{\left[(-R_A)_P \rho_B \left(\frac{d_P}{2}\right) n\right]}{k_m C_{Ab}}\right\} < 0.15 \qquad (2.42)$$

The criterion makes use of $(-R_A)_P$, the measured global rate (kmol/kg-s); n, order of reaction; d_P, the particle diameter (m); ρ_B, the bed density (kg/m^3); k_m, mass transfer coefficient (m/s); and C_{Ab} = bulk concentration of A (kmol/m^3). The intrinsic kinetics of the reaction must be known in order to use Mears' criterion. The mass transfer coefficient, k_m, can be estimated from existing empirical correlations [26–28].

2.3.1.2 Correlations for mass transfer coefficients

Since the flow channels between the particles in packed beds are irregular and intricate, transport rates are generally expressed in terms of average mass and heat transfer coefficients that can be estimated from semiempirical correlations that are developed using mass and heat transfer data in packed beds together with dimensionless groups describing the flow conditions in the bed.

The mass transfer coefficient is defined by the Sherwood number, *Sh*, which represents the ratio of total mass transfer to diffusive mass transfer. *Sh* is a function of the dimensionless Reynolds and Schmidt numbers. *Re* is the ratio of inertial forces to viscous forces expressed in terms of the extensive factor characteristic of the system, d_P, the superficial velocity, and physical properties of the fluid. *Sc* correlates fluid physical properties and the diffusivity of the transferred species in the fluid mixture:

$$Sh = \frac{k_m d_P}{D} = f(Re, Sc)$$

$$(Re)_P = \frac{d_P G}{\mu} = \frac{d_P(\rho u)}{\mu}$$

$$Sc = \frac{\mu}{\rho D}$$

The following relationship is suggested as a basis for the correlation of mass transfer data [9, 17, 23]:

$$j_D = \frac{k_m \rho}{G}\left(\frac{\mu}{\rho D}\right)^{\frac{2}{3}} = f\left(\frac{d_P G}{\mu}\right) \qquad (2.43)$$

The experimental results obtained by many investigators for mass transfer between a fluid and a bed of particles are reported as plots of j_D or $\epsilon_B j_D$ as a function of *Re*, where ϵ_B is the void fraction of the bed of particles. All data can almost be summarized in almost a single curve of j_D versus *Re* (Figure 2.7). After calculating *Re*, the j_D factor is read from the plot in order to calculate k_m. Numerical expressions allowing the calculation of j_D as a function of *Re* are also reported for packed beds of spherical particles with $\epsilon_B = 0.37$, and they describe the results plotted in Figure 2.7 reasonably well [17]:

$$(Re)_P < 190; \; j_D = 1.66\left[(Re)_P\right]^{-0.51} \qquad (2.43a)$$

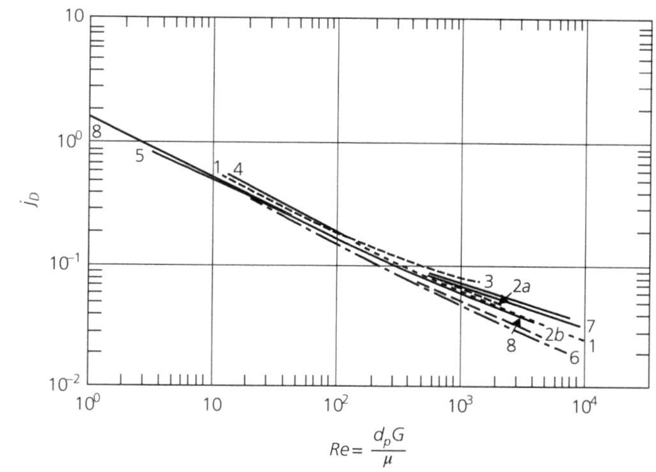

Figure 2.7 Correlation of experimental data on external mass transport between bulk fluid and particle surface in packed beds by various researchers. (Source: Froment [17]. Reproduced with permission of John Wiley & Sons.)

$$(Re)_P > 190; \; j_D = 0.983\left[(Re)_P\right]^{-0.41} \qquad (2.43b)$$

2.3.2 External temperature effects

The temperature differences between the bulk fluid and the catalyst surface originate from the heat(s) of reaction associated with surface reaction(s) and are significant particularly when highly exothermic chemical reactions are involved. External temperature gradients can be large even when mass transfer limitations are negligible, which disguises the actual reaction kinetics occurring at T_S and not at the measured T_b. At the particle scale, the major resistance to heat transfer is the laminar film next to the catalyst surface rather than intraparticle heat transfer; external mass transfer limitations are generally less important than external heat transfer resistance.

2.3.2.1 Diagnostic experiments and criteria

The measures taken in laboratory experiments for minimizing temperature gradients are similar to those for avoiding concentration gradients: the usage of high fluid linear velocities and dilution of the feed mixture with inerts having high thermal conductivity in order to reduce heat generation per unit packed-bed volume.

Mears' Criterion for External Heat Transfer Effects: There is an analogous Mears' criterion for predicting possible heat transfer effects [26–28]:

$$\left\{\frac{\left[(-\Delta H_r)(-R_A)_P \rho_B \left(\frac{d_P}{2}\right)(E_A)\right]}{h T^2 R_g}\right\} < 0.15 \qquad (2.44)$$

This criterion also makes use of $(-R_A)_P$, the measured global rate (kmol/kg-s); d_P, the particle diameter (m); and ρ_B, the bed density (kg/m^3). Other parameters included are h, heat transfer coefficient (kJ/m^2-s-K); $(-\Delta H_r)$, heat of reaction (kJ/mol); E_A, apparent activation energy (kJ/kmol); and R_g = gas constant (kJ/mol-K). The apparent activation energy of the surface reaction must be known to use this criterion. The heat transfer coefficient, h, can be estimated from existing empirical correlations.

2.3.2.2 Correlations for heat transfer coefficients

Heat and mass are transferred between a solid surface and a fluid by similar mechanisms; heat transfer by radiation is not considered unless reactions occur well above 400 °C in industrial reactors run with pellet sizes larger than 6 mm [7, 9]. For present purposes, the addition of radiation effects is not necessary.

The heat transfer coefficient is defined by the Nusselt number, Nu, which is a function of the dimensionless Reynolds and Prandtl numbers. Pr is the analog of Sc and correlates fluid physical properties and thermal conductivity of the fluid mixture:

$$Nu = \frac{hd_P}{\lambda} = f(Re, Pr)$$

$$Pr = \frac{c_P\mu}{\lambda}$$

Here, μ is the fluid viscosity, c_P is the constant-pressure heat capacity per unit mass of fluid, λ is the thermal conductivity of the fluid, and h is the film heat transfer coefficient.

Experimental data on heat transfer in packed beds are correlated in much the same way as data on mass transfer, that is, in terms of a j_H factor. The ratio (j_D/j_H) was initially reported as 0.70; upon further evaluation of new data available, this number was revised to 0.93 [9, 17, 29]. The reason for the disparity is unclear in the absence of heat transfer by radiation. Hence, for many flow geometries, the Chilton–Colburn analogy between heat and mass transfer leads to $j_H \approx j_D$ which agrees well with experimental results, and Equations 2.43a and 2.43b can be used to calculate j_H values [19, 30]:

$$j_H = \frac{h}{c_p G}\left(\frac{c_P\mu}{\lambda}\right)^{\frac{2}{3}} = f\left(\frac{d_P G}{\mu}\right) \tag{2.45}$$

2.3.2.3 Effect of ΔC_A on ΔT

The relationship between the degree of mass transfer control of a surface reaction and the temperature difference between particle surface and the main fluid can easily be derived for steady-state conditions. Under these conditions, the rate of mass transfer of reactant from the fluid to the solid surface must be equal to the rate of reactant conversion by surface reaction:

$$(k_m)_A a_m (C_{Ab} - C_{AS}) = (k_{sr}) a_m C_{AS}{}^n \tag{2.40a}$$

The implicit assumption made here is that the outer surface of the catalyst particle is uniformly accessible to the reactant(s): that is, the thickness of the concentration and thermal boundary layers over the particle surface has constant values. Since each section of the outer surface behaves kinetically the same as all other parts, steady-state analysis of such a system is essentially one-dimensional [14]. Hence, even when the functional form of the rate equation or the reaction order is not known, the heat generated by the surface reaction can be calculated by multiplying the mass transfer rate with the heat of reaction per mol of reactant. At steady state, the rate of energy generation by the surface reaction must be equal to the rate of energy removal by heat transfer from the solid back to the bulk fluid:

$$(k_m)_A a_m (C_{Ab} - C_{AS})(-\Delta H_r) = h a_m (T_S - T_b)$$

Solving for the temperature difference and considering the j-factor correlations,

$$(T_S - T_b) = \left(\frac{k_m}{h}\right)(C_{Ab} - C_{AS})(-\Delta H_r) \tag{2.46}$$

$$k_m = \frac{j_D G}{\rho (Sc)^{\frac{2}{3}}};\ h = \frac{j_H G c_p}{(Pr)^{\frac{2}{3}}};\ j_D \approx j_H;\ (Sc)^{\frac{2}{3}} \approx (Pr)^{\frac{2}{3}}$$

(for simple gas mixtures)

$$(T_S - T_b) = \left(\frac{-\Delta H_r}{\rho c_p}\right)(C_{Ab} - C_{AS}) \tag{2.46a}$$

If a fraction f is so defined as to express the degree of mass transfer control, $f = 1$ indicates total diffusion control and $f = 0$ indicates complete surface reaction control:

$$f = \{\text{extent of } MT \text{ control}\} = \frac{(C_{Ab} - C_{As})}{C_{Ab}}$$

Then the energy balance equation becomes

$$(T_S - T_b) = \left(\frac{-\Delta H_r C_{Ab}}{\rho c_p}\right) f \tag{2.46b}$$

The temperature difference is directly proportional to (i) the heat of reaction per mol of diffusing reactant and (ii) the fractional drop in concentration between the bulk fluid and the solid surface. The terms appearing on the right-hand side of the equation are

$(-\Delta H_r C_{Ab})$ = heat release by complete reaction of unit volume of reacting gas mixture, and

(ρc_p) = volumetric heat capacity of the reacting gas mixture

The quotient of these two terms gives the temperature rise equivalent to complete adiabatic conversion of the reacting mixture, and $(T_S - T_b) = (\Delta T)_{max}$ when $f = 1$, since $C_{AS} \to 0$. This equation also shows that heat transfer limitation and ΔT may be significant if $(-\Delta H_r)$ values are large, even when concentration gradients are small. For example, hydrogen oxidation over Pt-Al_2O_3 was reported to give a ΔT of 115 °C even though the degree of diffusion control was less than 5% [9]; such $(T_S - T_b)$ values cause a large increase in the observed global rate over the rate which would be measured if the surface were indeed at T_b. A significant error is made in calculating reaction rates even when a much smaller ΔT is neglected at random.

2.3.3 Nonisothermal conditions: Multiple steady states

An interesting feature of systems involving exothermic chemical reactions accompanied by external transport resistances is that of multiple steady states [9, 13, 19]. The simplest case exists when internal transport resistances are unimportant but external transport resistances are present. At steady state, the heat released by the chemical reaction on any element of the outer catalyst surface must be transported from the solid to the bulk fluid. Let Q_R = heat liberated by the surface reaction per unit time and per unit mass of catalyst and Q = heat transported away from the surface to the bulk gas per unit time per unit mass of catalyst. Assuming an nth order single-reactant surface reaction, steady-state analysis gives

$$Q_R = Q$$

$$Q_R = (-R_A)_P(-\Delta H_r) \tag{2.47}$$

$$Q = ha_m(T_S - T_b) \tag{2.48}$$

$$(-R_A)_P = ka_m C_{AS}{}^n = \left[A\exp\left(\frac{-E_A}{R_g T_s}\right)\right] a_m C_{AS}{}^n$$

Then,

$$Q_R = a_m A(-\Delta H_r)\exp\left(\frac{-E_A}{R_g T_S}\right) C_{AS}{}^n$$

Rearranging and defining new dimensionless parameters,

$$Q_R = a_m A(-\Delta H_r)\exp\left(\frac{-E_A}{R_g T_S}\right) C_{AS}{}^n \tag{2.47a}$$

$$\alpha = \frac{E_A}{R_g T_b} \quad \text{and} \quad \theta = \frac{(T_S - T_b)}{T_b} \tag{2.49}$$

$$\left(\frac{-E_A}{R_g T_s}\right) = \exp\left[\frac{-\alpha}{(\theta+1)}\right]$$

$$Q_R = a_m A(-\Delta H_r)\exp\left[\frac{-\alpha}{(\theta+1)}\right] C_{AS}{}^n \tag{2.47b}$$

$$Q = ha_m(T_S - T_b) = ha_m T_b \theta \tag{2.48a}$$

The equations for Q_R and Q define the temperature difference $(T_S - T_b)$ in terms of parameters including a_m, A, ΔH_r, h, and C_{AS}. The requirement that $Q_R = Q$ at steady state leads to multiple steady-state solutions, and it is typical that the particular steady state which is physically realized during operation depends on the conditions under which the reaction is started. There are two possible situations to be considered, depending on whether external concentration gradients are (i) negligible or (ii) substantial.

Negligible $(C_{Ab} - C_{AS})$: This case is simpler, because the unknown surface concentration C_{AS} is replaced by C_{Ab}, since $C_{AS} \rightarrow C_{Ab}$ for slow surface reaction and fast mass transfer. The Q_R equation is directly written in terms of C_{Ab}:

$$Q_R = a_m A(-\Delta H_r)\exp\left[\frac{-\alpha}{(\theta+1)}\right] C_{Ab}{}^n \tag{2.47c}$$

An analytical solution for $Q_R = Q$ is not feasible; however, when both Q_R and Q are plotted versus the dimensionless temperature rise θ, the intersection point(s) of the two curves will determine the steady-state value of T_S for a given C_{Ab} and T_b.

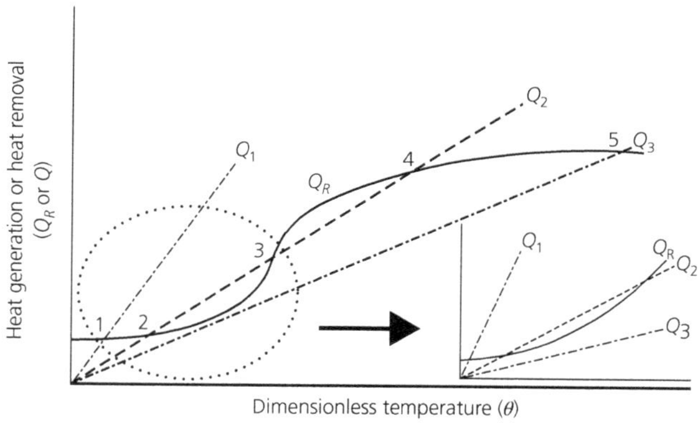

Figure 2.8 Steady-state multiplicity caused by external temperature gradients in simple nonporous surface-catalyzed exothermic reactions when external mass transfer resistance is either considerable or negligible.

The Q_R equation is an exponential curve in θ while the equation for Q is linear in θ, as also depicted in the insert in Figure 2.8. In this case, there can be one or two intersection points depending on the shape and location of the Q_R curve and the magnitude of the heat transfer coefficient, h. In Figure 2.8, Q_1, corresponding to rather fast heat removal rate, gives rise to one stable solution, θ_1. In case of Q_2, even though there are two intersection points indicating two solutions, there is in fact only one truly stable θ value, θ_2, and hence only one stable T_S value. The second intersection point, θ_3, is pseudostable because small perturbations will force the system either into the unstable region between θ_2 and θ_3 where $Q_2 > Q_R$ leading to cooling of the surface with stabilization at θ_2 or into the unstable region above θ_3 where $Q_R > Q_2$ leading to temperature escalation with no stabilization. The particular merit of the graphical method is to clearly demonstrate which values of the parameters will lead to multiple steady-state solutions and which solution is actually stable.

Substantial $(C_{Ab} - C_{AS})$: In this case, the equality of rates of surface reaction and mass transfer given by Equation 2.40a and the equality of Q_R and Q have to be solved simultaneously to determine the steady-state values of C_{AS} and T_S. For the sake of simplicity, $n = 1$ is assumed for the reaction order:

$$C_{AS} = C_{Ab}\left\{\frac{k_m}{(k_m + k_{sr})}\right\} = C_{Ab}\left\{\frac{k_m}{\left[k_m + A\exp\left(\frac{-E_A}{R_g T_S}\right)\right]}\right\}$$

$$Q_R = \frac{\{a_m(-\Delta H_r)C_{Ab}\}}{\left\{\left(\frac{1}{A}\right)\exp\left[\frac{\alpha}{(\theta+1)}\right] + \frac{1}{k_m}\right\}} \tag{2.47d}$$

$$Q = ha_m(T_S - T_b) = ha_m T_b \theta \tag{2.48a}$$

An analytical solution for $Q_R = Q$ is not achievable. When Q_R and Q are plotted versus θ, the intersection points of the two curves determine the steady-state values of C_{AS} and T_S. Lower values of θ are more likely to fall into the kinetically controlled region, and hence, the Q_R curve starts out in the exponential form. As $(T_S - T_b)$ increases, the Q_R curve tends to level out on account of the $(1/k_m)$ term in the denominator becoming more significant, thus giving rise to an S-shaped Q_R curve.

Three heat removal cases corresponding to different heat transfer coefficient (h) values are demonstrated in Figure 2.8. Q_1, corresponding to rapid heat removal rate, gives rise to one stable solution, θ_1, in the kinetics-controlled regime. Q_3, representing the lower heat removal rate, also leads to one stable solution, θ_5, which is in the diffusion-controlled regime. Q_2, typifying moderate heat removal rates, results in three mathematical steady-state solutions, θ_2, θ_3, and θ_4. Two solutions, θ_2 and θ_4, are truly stable in the kinetics-controlled and diffusion-controlled regime, respectively, while θ_3 is not actually sustainable since small perturbations will force T_S either into the unstable region between θ_2 and θ_3 where $Q_2 > Q_R$ leading to cooling of the surface with stabilization at θ_2 or into the unstable region between θ_3 and θ_4 where $Q_R > Q_2$ leading to heating of the surface with stabilization at θ_4.

2.3.4 External effectiveness factors

The external effectiveness factor is defined as the ratio of the global rate to the rate evaluated at bulk conditions:

$$\gamma = \frac{(-R_A)_P}{(-R_A)_b};\ (-R_A)_P = (-R_A)_S = f(C_{AS}, T_S) \text{ and}$$
$$(-R_A)_b = f(C_{Ab}, T_b) \tag{2.50}$$

In general, it can be stated that $\gamma \le 1$ for isothermal conditions and $\gamma \ge 1$ for nonisothermal conditions arising from exothermic surface reactions. Power function rate equations are used in deriving an expression for γ and recalling that

$$\alpha = \frac{E_A}{R_g T_b} \text{ and } \theta = \frac{(T_S - T_b)}{T_b}$$

External effectiveness factor expressed for the case of negligible $(C_{Ab} - C_{AS})$ carries only the ΔT effect; that is, it is based on $(-R_A)_P = f(C_{Ab}, T_S)$ and $(-R_A)_b = f(C_{Ab}, T_b)$:

$$(-R_A)_P = a_m A \exp\left(\frac{-E_A}{R_g T_S}\right) C_{Ab}{}^n$$
$$(-R_A)_b = a_m A \exp\left(\frac{-E_A}{R_g T_b}\right) C_{Ab}{}^n$$
$$\gamma = \exp\left[\frac{\alpha\theta}{(\theta+1)}\right] \tag{2.50a}$$

The expression obtained for γ in the case of substantial $(C_{Ab} - C_{AS})$ in addition to $(T_S - T_b)$ reflects the effects of both ΔT and ΔC_A. The numerator of the equation for γ represents the influence of ΔT while the denominator represents that of ΔC_A. A comparison of the two equations clearly indicates that the presence of diffusional limitations reduces the external effectiveness factor [9]:

$$(-R_A)_P = a_m A \exp\left(\frac{-E_A}{R_g T_S}\right) C_{AS}$$
$$(-R_A)_b = a_m A \exp\left(\frac{-E_A}{R_g T_b}\right) C_{Ab}$$
$$\gamma = \frac{\left\{\exp\left[\frac{\alpha\theta}{(\theta+1)}\right]\right\}}{\left\{1 + \left(\frac{A}{k_m}\right)\right\}\exp\left[\frac{-\alpha}{(\theta+1)}\right]} \tag{2.50b}$$

The isothermal external effectiveness factor stated for the case of negligible $(T_S - T_b)$ carries only the ΔC_A effect and is based on $(-R_A)_P = f(C_{AS}, T_b)$ and $(-R_A)_b = f(C_{Ab}, T_b)$. Keeping the first-order surface reaction assumption, the following simple expression is found for γ:

$$(-R_A)_P = a_m A \exp\left(\frac{-E_A}{R_g T_b}\right) C_{As}$$
$$(-R_A)_b = a_m A \exp\left(\frac{-E_A}{R_g T_b}\right) C_{Ab}$$
$$\gamma = \frac{(-R_A)_P}{(-R_A)_b} = \frac{C_{As}}{C_{Ab}} = \frac{k_m}{[k_{sr} + k_m]} = \frac{1}{\left[1 + \left(\frac{k_{sr}}{k_m}\right)\right]} \tag{2.50c}$$

The Damköhler number is defined as the ratio (k_{sr}/k_m) or $(k_V/k_m a_m \rho_B)$ for linear kinetics and as the ratio of the chemical reaction rate $(k_V C_{Ab}{}^{n-1})$ to the external mass transfer rate $(k_m a_m \rho_B)$ for nth order reactions. Isothermal external effectiveness factors have been reported for one-half and second-order surface reactions in terms of a Damköhler number [13]. In isothermal systems, mass transfer limitations become more influential at higher reaction orders. For fast mass transfer and slow surface reaction, $Da \to 0$ and the intrinsic rate is observed. It should be remembered that k_m determines the overall kinetics if the surface reaction is fast, and no matter what the order of the chemical reaction is, first-order dependence will be observed since mass transfer itself is a first-order process, that is, Da will be large and in the limit the global rate will approach $(k_m a_m \rho_B C_{Ab})$.

It is also worth mentioning at this point that, in isothermal operation with porous particles, external mass transfer limitations are likely to exist only when internal concentration gradients are present [14]. However, if the active phase is not impregnated on a porous solid, as in the case of metal wires and foils or as in wire gauzes and extremely thin surface coatings, only external mass transport can modify the intrinsic reaction rate [19, 28]. On the other hand, external temperature gradients can be significant even when external concentration gradients are small; therefore, under practical operating conditions, one of the major transport resistances is external heat transport.

2.4 Internal (intraparticle) transport processes

A major segment of catalytic processes conducted on an industrial scale employ fixed-bed (or packed-bed) reactors, and most PBRs do not operate in a regime where intrinsic kinetics is applicable. On the contrary, it is generally desirable to use solid catalysts with an intricate pore structure, since these pores can (i) provide hundreds of square meters of catalytic surface per unit weight of catalyst, (ii) disperse active metal components to avoid sintering, (iii) impart mechanical strength thus hindering attrition, and (iv) minimize pressure drop along the packed bed. A descriptive diagram of the pore structure of a catalyst particle having both macropores and micropores is presented in Figure 2.9.

The most important mass transfer limitation in packed beds is the diffusion of components into and out of the catalyst pore structure. In the 7-step sequence of the continuous reaction model used in microkinetic analysis, steps 2 and 6 are the physical rate processes that occur simultaneously with the chemical steps 3–4–5 within the pore structure.

2 Diffusion of reactants from the exterior to the interior surface of the porous particle

6 Diffusion of products from the interior to the exterior surface

In this section, the porous catalyst particle is studied on its own in order to evaluate the global rate by accounting for intraparticle concentration and temperature profiles over the entire particle with reference to the concentrations and temperature at its exterior surface.

In a porous catalyst, as reactants diffuse in the radial direction toward the center of the particle, reaction occurs on the pore walls, releasing or absorbing heat as required by the reaction. The interior surface of a porous catalyst is not as effective as its exterior because each point on the inner surface is exposed to a lower reactant concentration than that of the exterior (C_{AS}). The net effect of intrapellet mass transfer resistance is to reduce the global rate beneath the rate evaluated at surface conditions. The net effect of intrapellet heat transfer resistance, on the other hand, depends on the exothermicity–endothermicity of the surface reaction and on the relative significance of concentration and temperature gradients. If the gas-phase reaction involves a change in the total number of moles, intraparticle pressure gradients affecting diffusion rates may also develop at steady state; however, these changes are not large and are rarely taken into account.

The concentration and temperature gradients resulting from the diffusion–reaction process lead to the existence of different point rates at different locations along the pores and within the particle. Differential mass and energy balance equations are needed for obtaining the radial concentration and temperature profiles describing these variations. In order to arrive at the global rate, these profiles have to be integrated over the entire particle by making use of the intrinsic rate equation. The radial concentration and temperature profiles are based on characteristic physical parameters of the porous particle, such as the "effective diffusivity" and the "effective thermal conductivity." Hence, theories and correlations pertaining to the estimation of these parameters are prerequisite for establishing the global rate expressions for the entire particle.

2.4.1 Intraparticle mass and heat transfer

In order to predict the effective diffusivity in the porous catalyst particle, D_e, the following steps are taken:

1 The equations for diffusion in a single cylindrical pore are established.

2 These equations are combined with an appropriate geometric model of the pore structure of the particle.

The mechanisms of mass transfer in the pore volume have to be considered for the development of diffusion equations for the single cylindrical pore [9, 16, 29]. Pore diffusion may occur by one or more of three mechanisms:

1 Bulk diffusion, which takes place in the free space of the pore due to collisions of reactant molecules with each other

2 Knudsen diffusion, which is caused by the collisions of reactant molecules with the pore walls

3 Surface diffusion, which is the migration of adsorbed reactant molecules on the pore walls as a result of concentration gradients

Surface Diffusion: Molecules adsorbed on solid surfaces may show considerable mobility. Transport of adsorbed molecules by

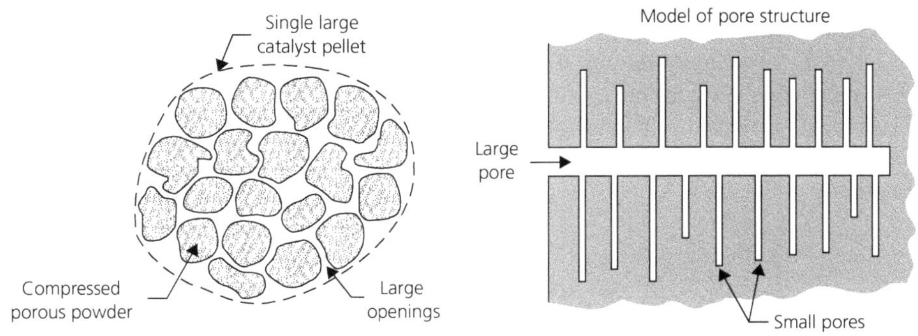

Figure 2.9 Representation of a catalyst particle having micro- and macropore structure. (Source: Levenspiel [31]. Reproduced with permission of John Wiley & Sons.)

movement on a surface is known as surface diffusion, with the coefficient $(D_S)_A$. Its direction is that of decreasing surface concentration. Since equilibrium adsorption is a function of the adsorbate partial pressure in the gas phase adjacent to the surface, both surface diffusion and bulk diffusion proceed in parallel. The effective diffusivity of a porous solid cannot be predicted if surface diffusion effects are significant. In general, surface diffusion is considered unimportant unless the amount of reactant adsorbed is appreciable; moreover, chemisorption is virtually immobile as molecules are detained on the surface by chemical bonds. Bulk diffusion and Knudsen diffusion are the only mechanisms that prevail in the catalysts and operating conditions used in practice.

Bulk Diffusion: Bulk or ordinary diffusion, with the diffusion coefficient D_{AB} or $D_{A\text{-}mix}$, occurs in the void space of the pores as a result of collisions between gas molecules and is likely to dominate when (i) the pores are macropores larger than 100 Å in radius, (ii) the gas is relatively dense, that is, at high pressures, and (iii) the pores are filled with a liquid.

Knudsen Diffusion: What is known as Knudsen flow or Knudsen diffusion in fine pores is described with the coefficient $(D_K)_A$ and occurs in gases when (i) the pores are micropores smaller than ~20 Å in radius and (ii) the gas density is low. Under these conditions, the molecules collide with the pore walls much more frequently than with each other. Molecules striking the pore wall are adsorbed for a moment and then reflected in arbitrary directions. Mass transfer is slowed down to some extent as a consequence of both the random reflection and the brief instant the molecules stay adsorbed.

2.4.1.1 Combined diffusivity D_{comb} in a single pore

Many reaction conditions, particularly the pressure, lead to the existence of diverse molecular concentrations in pore spaces where both bulk diffusion and Knudsen diffusion can prevail concurrently. In the so-called transition region, both mass diffusivities, (D_{AB}) and $(D_K)_A$, are included in the formulation of a combined mass diffusivity, D_{comb}. In the analysis of mass diffusivities, the key parameter to be considered is the size of the pore with respect to the mean free path of the molecules.

Considering a stagnant binary gas mixture, the molar flux (rate per unit area) is proportional to the concentration gradient in the direction of diffusion. Fick's law is used to describe the diffusion in pores, where D_{comb} is the combined mass diffusivity based on the cross-sectional area of a single pore, x is the direction of diffusion along the pore, and y_A is the mole fraction of A. The factor α is related to the diffusion fluxes of components A and B and is determined by the stoichiometry for diffusion with reaction:

$$N_A = -D_{comb}\left(\frac{dC_A}{dx}\right) \tag{2.51}$$

$$\frac{1}{D_{comb}} = \frac{(1-\alpha y_A)}{D_{AB}} + \frac{1}{(D_K)_A} \tag{2.52}$$

$$\alpha = 1 + \frac{N_B}{N_A}$$

For the reaction $A \longrightarrow B$ at steady state, diffusion in the pore is equimolar counterdiffusion, $N_A = -N_B$, and $\alpha = 0$. The composition dependence in Equation 2.52 is passed on to particle diffusivities (D_e) and presents an important drawback in the integration of intraparticle diffusion–reaction equations. However, diffusion fluxes in multicomponent systems may be diverse, and it has been indicated that, on the whole, the dependence of D_{comb} on composition is rather weak, and therefore, common practice has been to use the simpler composition-independent form of the equation [7, 9, 32]; this additive resistance relationship is the Bosanquet formula [17, 33]:

$$\frac{1}{D_{comb}} = \frac{1}{D_{AB}} + \frac{1}{(D_K)_A} \tag{2.52a}$$

$$D_{AB} \propto \left(\frac{1}{P_T}\right) \quad \text{and} \quad (D_K)_A \propto \bar{a}$$

The two special cases are the dominance of (i) bulk diffusion or (ii) Knudsen diffusion, and the parameters of importance in this respect are not only the pore size and the pressure but also the magnitude of the $(D_{AB})/(D_K)_A$ ratio.

The bulk diffusivity D_{AB} is inversely proportional to total pressure and hence becomes significant as pressure increases but does not depend on the pore size. The temperature dependence of D_{AB} is $T^{3/2}$. The Chapman–Enskog equation is suitable for accurate estimations of D_{AB} at moderate temperatures and pressures [30]. Knudsen diffusivity $(D_K)_A$ is directly proportional to the pore radius, and the average molecular velocity as predicted by the kinetic theory of gases, and hence its temperature dependence is $T^{1/2}$. Its contribution to D_{comb} increases as pore size decreases, but $(D_K)_A$ is independent of total pressure [30].

2.4.1.2 Effective diffusivity D_e of the particle

The combined diffusivity, D_{comb}, calculated for a single cylindrical pore is based on the cross-sectional area of the pore perpendicular to the direction of diffusion. A catalyst particle consists of an assembly of single pores. Therefore, the ultimate aim is to find the effective diffusivity of the porous catalyst particle, D_e, based on the total area exposed by the cross sections of all the pores in the particle, which constitutes the total mass transfer area normal to the direction of diffusion.

Experimental evaluation of the effective diffusivity D_e is possible, if intrinsic kinetics of the reaction is known, and requires a comparison of the measured and predicted global rates which may also involve an iterative procedure.

If no such data are available, D_e is estimated according to the following approach by using a geometric model of the physical pore structure in the particle:

1 The combined diffusivity, D_{comb}, is calculated for the single cylindrical pore.
2 A simple geometric model of the particle described in terms of measurable physical properties is devised.

3 The geometric model is used to convert D_{comb} of the single pore to D_e of the entire particle comprising an array of cylindrical pores.

The diffusion flux is still represented by Fick's law but is based on the total cross-sectional area exposed by all the pores, and the diffusion path r is the thickness of the particle:

$$(N_A)_e = -D_e\left(\frac{dC_A}{dr}\right) \tag{2.53}$$

The measurable physical properties of catalyst particles commonly used in geometric models are the total surface area S_g (m^2/g), pore volume V_g (cm^3/g), solid density ρ_S (g/cm^3), void fraction or porosity ϵ_P, and occasionally pore-volume distribution.

The two major pore models that have been used extensively over the years for practical purposes are the "parallel-pore" model proposed by Wheeler in 1955 [5, 9] and the "random-pore" model proposed by Wakao and Smith in 1962 [34]. Among the more recent advanced models are the "parallel cross-linked pore" model [35] and "pore-network" models [36, 37].

The original semiempirical parallel-pore model represents a monodisperse pore-size distribution and makes use of the measurable physical properties, S_g, V_g, ρ_S, and ϵ_P. The complex particle with porosity ϵ_P is replaced by an array of straight and parallel cylindrical pores of radius $\bar{a}$, much like a honeycomb structure. The mean pore radius $\bar{a}$ is simply calculated by assuming that the sum of the inner surface areas of all the n pores in an array ($n2\pi\bar{a}L$) is equal to the total surface area S_g and the sum of all the pore volumes ($n\pi\bar{a}^2L$) is equal to the experimental pore volume V_g [5]:

$$\bar{a}\,(\AA) = \frac{2V_g}{S_g} \tag{2.11}$$

If the pore walls are tortuous with changing pore cross sections, the expression for the average pore radius is modified by including a tortuosity factor, δ, and the porosity of the particle, ϵ_P:

$$\bar{a}(\AA) = \frac{2V_g}{S_g}\delta(1-\epsilon_P) \tag{2.11a}$$

The only other variable required for predicting D_e is the length of the diffusion path, which is the thickness of the particle multiplied also by the adjustable tortuosity factor, δ, that accounts for distorted diffusion pathways and also for varying pore cross sections in interconnections and constrictions; the value of δ varies between $\sqrt{2}$ and 10 but is typically 3 or 4 in most industrial catalysts. The simplest geometric model which is still commonly used in practical applications for estimating D_e is the parallel-pore model:

$$D_e = \frac{\epsilon_P D_{comb}}{\delta} \tag{2.54}$$

The random-pore model describes a bidisperse pore-volume distribution and uses separate void fractions in macropores (ϵ_M) and micropores (ϵ_μ), as well as separate mean pore radii for macropores ($\bar{a}_M$) and micropores ($\bar{a}_\mu$). In the microregion, $\bar{a}_\mu$ is calculated in the same manner as in the parallel-pore model using S_g and V_g, while in the macroregion related to the void spaces between primary particles, $\bar{a}_M$ is obtained from pore-volume distribution. When $\epsilon_M = 0$, a monodisperse model similar to the parallel-pore model is obtained [9, 17, 34].

2.4.1.3 Intraparticle heat transfer

In intraparticle heat transfer, the parameter analogous to the effective diffusivity D_e is the effective thermal conductivity, λ_e. The defining equation for the heat flux is

$$Q_e = -\lambda_e\left(\frac{dT}{dr}\right) \tag{2.55}$$

Q_e is the energy transferred per unit total area of the particle normal to the direction of heat transfer. The effective thermal conductivities of catalyst pellets are remarkably low because of the pore structure. The contribution of the thermal conductivity of the solid skeleton is little, since the extremely small heat transfer areas existing at solid–solid contact points offer substantial resistance to heat transfer. The gas phase filling the void spaces in the pores also participates in hindering heat conduction; experimental results indicate that λ_e decreases as ϵ_P increases. At low pressures, when the mean free path of molecules is greater than or equal to pore size, λ_e increases with total pressure since free-molecule conduction starts to dominate. There are no general correlations for predicting λ_e from the physical properties of the solid and fluid phases involved. An approximate correlation based on the thermal conductivities of the individual phases and the porosity of the particle has been proposed:

$$\lambda_e = \lambda_S\left(\frac{\lambda_g}{\lambda_S}\right)^{(1-\epsilon_P)} \tag{2.56}$$

It is, however, possible to select fairly accurate λ_e values because it varies within a rather narrow range; experimental results reported vary between 0.1 and 0.4 Btu/h-ft-°F, excluding vacuum conditions [9]. Other order of magnitude values that have been reported as 4×10^{-3} J/s-cm-K [16] or 3×10^{-4} cal/s-cm-K [19] are also based on the latter results. Bearing in mind the lesser impact of internal heat transport compared with internal mass transfer limitations, relatively little attention has been paid to correlations of λ_e inside pellets.

2.4.2 Mass transfer with chemical reaction: Isothermal effectiveness

Since solid-catalyzed reactions occur on the surface of a catalyst, the greater the extent of surface available for adsorption of reactants, the higher is the rate of reaction. Hence, particularly in practical applications, it is desirable to disperse the active components on a support having small volume but high surface area. This is achieved by using support materials with highly microporous particles; additional macropore volume is also introduced by pressing the primary particles into pellets or extrudates. The global rate of reaction for the pellet is, in this case, determined by both the intrinsic kinetics and the diffusional rate processes through the pore structure.

The quantitative treatment of the effects of concurrent mass transfer and chemical reaction within porous structures was started separately by Thiele in the United States, Damköhler in Germany, and Zeldovich in Russia and reported separately between 1937 and 1939 [4, 13]. These analyses were further developed by Wheeler, Weisz, Wicke, and Aris [13].

The most significant result of these studies is the "internal effectiveness factor" which links the intrinsic rate to the actually measured global rate and is also a measure of the efficacy with which the available surface area of the catalyst is utilized. The internal effectiveness factor is defined as

$$\eta = \frac{(-R_A)_P}{(-R_A)_S} = \frac{(\text{actual or global rate in pellet})}{(\text{rate evaluated at exterior surface conditions})} \tag{2.57}$$

$$(-R_A)_P = f(C_A, T) = \eta(-R_A)_S = \eta f(C_{AS}, T_S)$$

C_A and T denote the surface concentration of A and the surface temperature at any point along the pore. The central problem in the quantitative analysis of the simultaneous diffusion–reaction problem is to find an expression for η as a function of the parameters involved:

$$\eta = f(k_{sr}, k_e, D_e)$$

Once the internal effectiveness factor η is known, $(-R_A)_P$ can be obtained in terms of (C_{AS}, T_S). Subsequently, using the methods outlined in Section 2.3, $(-R_A)_P$ can be obtained as a function of bulk concentrations and temperatures (C_{Ab}, T_b).

The evaluation of the isothermal internal effectiveness factor η includes (i) determining the concentration profile in the particle, (ii) establishing the global reaction rate through the particle, $(-R_A)_P$, using the concentration profile obtained, and (iii) taking the ratio of the global rate $(-R_A)_P$ to the rate that would be observed if all the interior surface were concentrated at the exterior, $(-R_A)_S$. The internal effectiveness factor for an isothermal particle is defined as

$$\eta = \frac{(-R_A)_P}{(-R_A)_S} = \frac{f(C_A, T_S)}{f(C_{AS}, T_S)} \tag{2.57a}$$

Since the reactant concentrations along the pores and within the particles are lower than the external surface concentrations, the overall effect of internal mass transfer resistances is to reduce the actually observed global rate below that measured at exterior surface conditions. It can be stated for isothermal effectiveness factors that $\eta \le 1$. The concentration profile showing the pore diffusion-affected surface reaction is labeled as II in Figure 2.3.

2.4.2.1 Single cylindrical pore

A single cylindrical pore of length L and radius of r $(=\bar{a})$ located in a microscopic section of the catalyst particle is generally used for modeling the diffusion–reaction process (Figure 2.3). The steady-state component mass balance for a control volume extending over the cross section of the pore includes diffusion of reactant into and out of the control volume as well as reaction on the inner wall surface. The simple case taken as an example is that of an isothermal, irreversible first-order reaction:

$$-\pi r^2 D_{comb}\left(\frac{dC_A}{dx}\right)_{out} + \pi r^2 D_{comb}\left(\frac{dC_A}{dx}\right)_{in} + k_{sr} C_A(2\pi r \Delta x) = 0$$

$$k = k_V = \frac{2k_{sr}}{r}$$

$$k_v = k_{sr}\left(\frac{\text{surface}}{\text{volume}}\right) = k_{sr}\left(\frac{2\pi r L}{\pi r^2 L}\right) = \frac{2k_{sr}}{r} \tag{2.58}$$

In the limit as Δx approaches zero, a second-order linear homogeneous equation with constant coefficients is obtained:

$$\frac{d^2 C_A}{dx^2} - \frac{k_v}{D_{comb}} C_A = 0 \tag{2.59}$$

The ordinary differential equation is solved by conventional methods to give

$$C_A = M_1 e^{mx} + M_2 e^{-mx} \tag{2.59a}$$

$$m = \sqrt{\frac{k_v}{D_{comb}}} = \sqrt{\frac{2k_{sr}}{r D_{comb}}}$$

The boundary conditions specify (i) the reactant concentration at the pore mouth, C_{AS}, and (ii) the zero flux at the pore end since reactant does not diffuse into the bulk solid:

$$C_A = C_{AS}, \quad \text{at } x = 0$$

$$\frac{dC_A}{dx} = 0, \quad \text{at } x = L$$

The concentration profile of the reactant decreasing progressively as it diffuses along the pore length is given by the following equation:

$$\frac{C_A}{C_{As}} = \frac{e^{m(L-x)} + e^{-m(L-x)}}{e^{mL} + e^{-mL}} = \frac{\cosh m(L-x)}{\cosh mL} = \frac{\cosh \varphi_L\left(1 - \frac{x}{L}\right)}{\cosh \varphi_L} \tag{2.59b}$$

The drop in the concentration of reactant A through the pore is a function of the dimensionless parameter mL which is called the Thiele modulus φ_L for the single pore:

$$\varphi_L = mL = L\sqrt{\frac{k_v}{D_{comb}}} \quad (\text{for first-order reaction}) \tag{2.60}$$

The basic approach to evaluating the global rate in the pore would be to express the point rate prevailing at each point along the pore in terms of Equation 2.59b and then to integrate the rate equation over the entire volume of the pore. However, a mathematically much easier solution (especially for complex geometries) is offered by the fact that the overall reaction rate within the pore is actually equal to the rate of mass transfer into the pore. Therefore, differentiating Equation 2.59b with respect to distance along the pore, evaluating the derivative at $x = 0$ and substituting into the diffusion equation gives $(-R_A)_P$:

$$(-R_A)_P \approx \{\text{molar flow in at } x=0\} = -D_{comb}\pi r^2 \left(\frac{dC_A}{dx}\right)_{x=0}$$
$$= \pi r^2 D_{comb} C_{AS} \sqrt{\frac{k_v}{D_{comb}}} \tanh L \sqrt{\frac{k_v}{D_{comb}}}$$

If the entire active surface within the pore is concentrated at the exterior surface, no concentration gradients and hence no diffusional effects will exist. Then, the reaction rate evaluated at external surface conditions is simply obtained as

$$(-R_A)_S = 2\pi r L k_{sr} C_{AS} = \pi r^2 L k_v C_{AS}$$

The isothermal internal effectiveness factor, η, for the single cylindrical pore and first-order irreversible surface reaction is obtained from the aforementioned expressions for $(-R_A)_P$ and $(-R_A)_S$ using Equation 2.60:

$$\eta = \frac{(-R_A)_P}{(-R_A)_S} = \frac{(\tanh mL)}{mL} = \frac{\tanh \varphi_L}{\varphi_L} \tag{2.61}$$

The Thiele modulus φ_L includes an extensive factor, L, and the ratio of the intrinsic reaction rate constant to the effective mass diffusivity of the reactant through the pore structure. The two limiting cases are:

1 Severe diffusion limitation at large values of φ_L (for $\varphi_L > 3$), where tanh φ_L approaches unity, and therefore $\eta \to 1/\varphi_L$, which occurs if the pore is long (large L), the surface reaction is fast (large k_V), and/or the diffusion is slow (small D_{comb})

2 Diffusion-free surface reaction at small values of φ_L (for $\varphi_L <$ 0.3), where tanh φ_L approaches φ_L, and therefore $\eta \to 1$, which occurs when the pore is short (small L), the surface reaction is slow (small k_V), and/or the diffusion is rapid (large D_{comb})

The same expression as in Equation 2.61 is derived for the η of a catalyst with flat plate geometry, with the only difference that D_{comb} in Equation 2.60 is replaced by D_e of the flat plate having many cylindrical pores. For other particle geometries, such as cylindrical or spherical, the concentration profiles and the correlations for the isothermal internal effectiveness factor η are different, and the effective mass diffusivity, D_e, is used in all the analyses.

It can be shown that the η expression in Equation 2.61 holds approximately for irreversible nth order reactions, but the m in φ_L becomes concentration dependent [17, 23]:

$$\varphi_L = mL = L\sqrt{\frac{(n+1)k_v C_{AS}^{(n-1)}}{2D_e}} \quad (\text{for } n > -1) \tag{2.60a}$$

The relationship between the isothermal internal effectiveness factor η and the Thiele modulus φ_L for a flat plate is plotted in Figure 2.10 for φ_L values ranging from 0.1 to 20.

2.4.2.2 Porous spherical catalyst particle

Consider an isothermal porous spherical catalyst particle of radius R in which a single, irreversible first-order reaction takes place at steady state (Figure 2.11). Taking a spherical shell of thickness Δr at a radius r from the center, the steady-state component mass balance over a differential shell of volume $4\pi r^2 \Delta r$ includes diffusion of reactant into and out of the control volume in the radial direction as well as reaction on the inner surface of the particle:

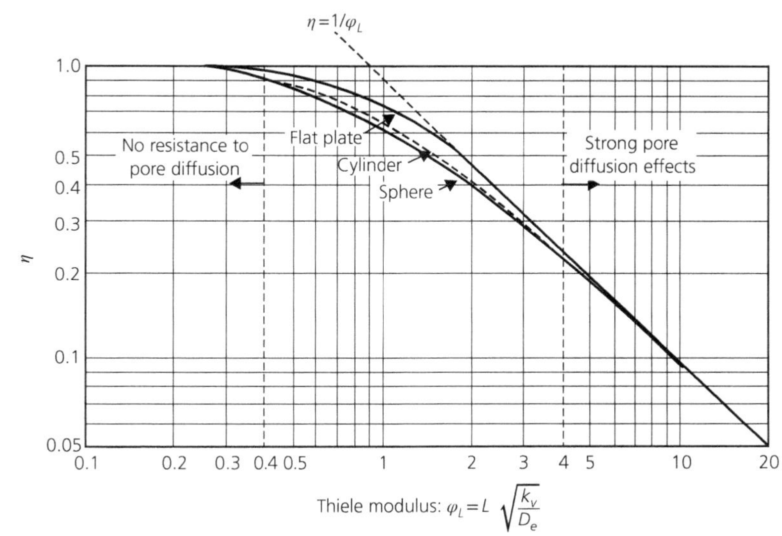

Figure 2.10 Internal effectiveness factor as a function of Thiele modulus for porous particles of various shapes. (Source: Levenspiel [31]. Reproduced with permission of John Wiley & Sons.)

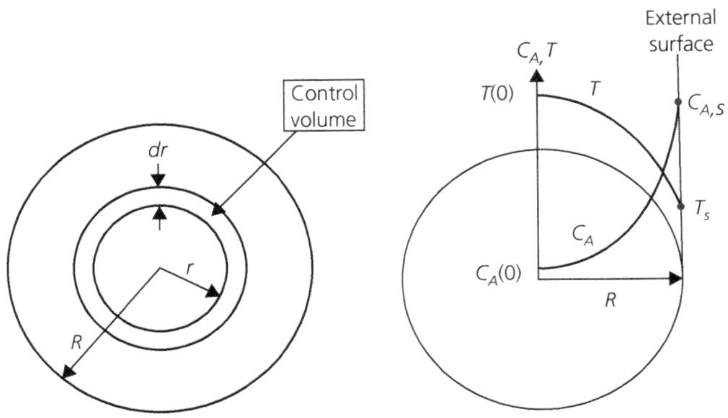

Figure 2.11 Reactant concentration and temperature profiles inside a porous spherical catalyst particle with exothermic reaction occurring at steady state. (Source: Roberts [23]. Reproduced with permission of John Wiley & Sons.)

$$\left\{4\pi(r+\Delta r)^2 D_e\left(\frac{dC_A}{dr}\right)\right\}_{in} - \left\{4\pi r^2 D_e\left(\frac{dC_A}{dr}\right)\right\}_{out}$$
$$= \left\{k_V\left(4\pi r^2\Delta r\right)C_A\right\}$$

Dividing the entire equation by $4\pi D_e \Delta r$ and taking the limit as Δr approaches zero, a second-order linear homogeneous ordinary differential equation is obtained:

$$\frac{d^2C_A}{dr^2} + \frac{2}{r}\frac{dC_A}{dr} = \frac{k_V C_A}{D_e} \tag{2.62}$$

The relevant boundary conditions are (i) the reactant concentration at the exterior surface, C_{AS}, and (ii) the zero flux at the center of particle, condition of particle symmetry:

$$r = R, \quad C_A = C_{AS}$$
$$r = 0, \quad \frac{dC_A}{dr} = 0$$

The concentration profile toward the interior of the particle in the radial direction (Figure 2.11) is given by the following equation written in terms of normalized distance (r/R):

$$\frac{C_A}{C_{AS}} = \frac{\left\{\sinh\left(\frac{r}{R}\right)\Phi_S\right\}}{\left\{\frac{r}{R}\sinh(\Phi_S)\right\}} \tag{2.62a}$$

It is convenient to define the Thiele modulus for the spherical catalyst particle, Φ_S, with radius of the particle R as the size parameter:

$$\Phi_S = R\sqrt{\frac{k_V}{D_e}} \text{ (for first order reaction)} \tag{2.63}$$

The global rate within the entire porous particle can be equated to the diffusion rate into the particle through the exterior surface. Differentiating the concentration profile (Eq. 2.62a) with respect to distance along the radius of the particle, evaluating the derivative at $r = R$ and substituting into the diffusion equation gives $(-R_A)_P$:

$$(-R_A)_P \approx \{\text{molar flow in at } r = R\} = -D_e 4\pi R^2\left(\frac{dC_A}{dr}\right)_{r=R}$$
$$= 4\Phi_S \pi R D_e C_{AS}\left\{\frac{1}{\tanh\Phi_S} - \frac{1}{\Phi_S}\right\}$$

If the entire active surface within the pore structure of the particle is exposed to exterior surface conditions, with no concentration gradients and hence no diffusional effects, the reaction rate at external surface conditions is simply calculated from

$$(-R_A)_S = \frac{4}{3}\pi R^3 k_V C_{AS}$$

The isothermal internal effectiveness factor, η, for a porous spherical particle and first-order irreversible surface reaction is obtained from the aforementioned expressions as follows:

$$\eta = \frac{(-R_A)_P}{(-R_A)_S} = \frac{D_e 4\pi R^2\left(\frac{dC_A}{dr}\right)_{r=R}}{\frac{4}{3}\pi R^3 k_V C_{AS}} = \frac{3D_e}{R k_V C_{AS}}\left(\frac{dC_A}{dr}\right)_{r=R} \tag{2.64}$$

$$\eta' = \frac{(-R_A)_P}{(-R_A)_S} = \frac{3}{\Phi_S}\left\{\frac{1}{\tanh\Phi_S} - \frac{1}{\Phi_S}\right\} \tag{2.64a}$$

The two limiting cases for porous particles are similar to those mentioned for the single pore:

1 Severe diffusion limitation at large values of Φ_S, where tanh Φ_S approaches unity, and $\eta' \to 3/\Phi_S$, which occurs if the particle is large (large R), the surface reaction is fast (large k_V), and/or the diffusion is slow (small D_e)
2 Diffusion-free surface reaction at small values of Φ_S, where tanh Φ_S approaches Φ_S, and $\eta' \to 1$, which occurs when the particle is small (small R), the surface reaction is slow (small k_V), and/or the diffusion is rapid (large D_e)

The Φ_S versus η' plot has the same shape as that of φ_L versus η for a flat plate but is shifted by a factor of 3 on the same log-log plot, if Equations 2.60 and 2.63 are used as basis. Aris [29, 38] has shown that, for a first-order reaction in various particle geometries, the plots between the Thiele modulus φ and the isothermal internal effectiveness factor η become identical at high and low values of φ (Figure 2.10), as long as the size parameter in the Thiele modulus is defined on a common basis. The characteristic size parameter L is, therefore, defined as the ratio of the particle volume to the external surface area available for reactant penetration, which enables its use for any arbitrary particle shape:

$$L = \frac{V_P}{S_{external}} \tag{2.65}$$

Then, the Thiele modulus for a sphere, φ_S, and the corresponding η expression become

$$\varphi_S = \frac{\Phi_S}{3} = \frac{R}{3}\sqrt{\frac{k_V}{D_e}} \text{ (for first order reaction)} \tag{2.63a}$$

$$\eta = \frac{(-R_A)_P}{(-R_A)_S} = \frac{1}{\varphi_S}\left\{\frac{1}{\tanh 3\varphi_S} - \frac{1}{3\varphi_S}\right\} \tag{2.64b}$$

In this case, as demonstrated in Figure 2.10, the φ versus η curves for all geometries coincide for very large and very small values of φ, and they are reasonably comparable for intermediate φ values. The numerical values of η and η' are the same for the spherical particle, as long as the appropriate size parameters defined by φ and Φ are used, respectively.

The characteristic size parameters for various particle geometries that are defined according to Equation 2.65 and used in Figure 2.10 may be listed as

$$L = \frac{\text{thickness}}{2}; \text{ for flat plates}$$

$$L = \text{thickness}; \text{ for flat plates sealed on one face}$$

$$L = \frac{R}{2}; \text{ for cylinders}$$

$$L = \frac{R}{3}; \text{ for spheres}$$

The analytical integration of the differential equation for diffusion with chemical reaction in a catalyst particle is achievable just for first-order reactions. A "generalized modulus" has been proposed for extending the use of the η expression in Equations 2.61 and 2.64a to any type of rate expression [17], at least approximately. For irreversible nth order reactions, the generalized modulus for a sphere becomes dependent on the exterior surface concentration:

$$\varphi_S = \frac{R}{3}\sqrt{\frac{(n+1)k_V {C_{AS}}^{n-1}}{2D_e}} \quad (\text{for } n > -1) \tag{2.63b}$$

In the case of more complicated kinetic expressions like LHHW equations, the effectiveness factor η can be determined by numerical integration of the differential equations for diffusion with chemical reaction. The complex kinetics of a number of reactions can practically be approximated by simpler power function expressions.

The Thiele moduli for first-order reactions in various geometries as well as the generalized moduli applicable for other reaction orders all assume that the chemical reaction is irreversible. Studies on the first-order reversible reaction $A \leftrightarrow B$ have shown that the same $\varphi_L - \eta$ function as in the irreversible case can be used, when the Thiele modulus is defined using the characteristic size factor L of Equation 2.65. The reaction equilibrium constant K is used, if the D_e value for A and B is more or less the same [13, 29]:

$$\varphi_L = L\sqrt{\frac{(K+1)k_{forward}}{KD_e}} \quad (\text{first order, reversible}) \tag{2.63c}$$

2.4.3 Heat and mass transfer with chemical reaction

The general approach in the evaluation of nonisothermal internal effectiveness factors is analogous to that outlined for determining isothermal η. However, since substantial temperature gradients can occur in the catalyst particles used in practice, the nonisothermal internal effectiveness factor is defined as

$$\eta = \frac{(-R_A)_P}{(-R_A)_S} = \frac{f(C_A, T)}{f(C_{AS}, T_S)} \tag{2.57}$$

For this reason, both a concentration and a temperature profile are required to describe the conditions at each point within the porous particle, before $(-R_A)_P$ can be determined. In the case of exothermic reactions, T at any point within the particle may be higher than T_S; hence, depending on the relative significance of ΔC_A and ΔT, the global rate $(-R_A)_P$ may be higher than the rate at surface conditions $(-R_A)_S$, and situations where $\eta > 1$ may be encountered.

A relationship may easily be derived between intraparticle temperature and concentration differences (ΔT and ΔC_A) just by considering a boundary surface enclosing some section or all of an arbitrary porous structure as the starting point. Under steady-state conditions, the diffusion of reactants across this boundary surface is equal to the global rate of reaction within the boundary surface. Then, the heat released by reaction within the surface can be expressed in terms of the diffusion rate across the boundary:

$$Q_R = -D_e\left(\frac{dC_A}{dx}\right)(-\Delta H_r)(Area) \tag{2.66}$$

Similarly, the heat released by reaction is transferred across the same boundary by conduction since this is the major mechanism of heat transfer inside catalyst particles:

$$Q = -\lambda_e\left(\frac{dT}{dx}\right)(Area) \tag{2.67}$$

At steady state, $Q_R = Q$, and therefore,

$$-D_e\left(\frac{dC_A}{dx}\right)(-\Delta H_r) = -\lambda_e\left(\frac{dT}{dx}\right)$$

In integrated form,

$$(T - T_S) = (-\Delta H_r)\frac{D_e}{\lambda_e}(C_{AS} - C_A) \tag{2.68}$$

Here, T and C_A are the temperature and concentration at any point within the particle, while T_S and C_{AS} are the boundary values at the outer surface, respectively. As evident in Equation 2.68, the heat of reaction ΔH_r and the transport properties D_e and λ_e are the essential parameters. This relationship, which was originally derived by Damköhler in 1943 [9], is valid for all kinetics and applies to all particle geometries. The only implicit assumption made is that of symmetry, that is, the assumption that T_S and C_{AS} are uniform over the entire boundary surface. Using this expression, it is possible to find the maximum temperature difference between the surface and the center of particle, which occurs when the reactant is used up before it reaches the center; ΔT_{max} is influenced by C_{AS} in addition to ΔH_r, D_e, and λ_e:

$$(T-T_S)_{max} = (-\Delta H_r)\frac{D_e}{\lambda_e}(C_{AS}) \tag{2.68a}$$

2.4.3.1 Nonisothermal spherical catalyst particle

Consider the nonisothermal porous spherical catalyst particle of radius R in which a single, irreversible, first-order reaction takes place at steady state (Figure 2.11). Taking the same spherical shell of thickness Δr at a radius r from the center, the steady-state energy balance over a differential shell of volume $4\pi r^2 \Delta r$ includes conduction into and out of the control volume in the radial direction as well as heat release by reaction within the control volume:

$$\left\{4\pi(r+\Delta r)^2\lambda_e\left(\frac{dT}{dr}\right)\right\}_{in} - \left\{4\pi r^2\lambda_e\left(\frac{dT}{dr}\right)\right\}_{out}$$

$$= \{k_V(4\pi r^2\Delta r)C_A\}(\Delta H_r)$$

Dividing the entire equation by $4\pi D_e \Delta r$ and taking the limit as Δr approaches zero, a second-order linear homogeneous ordinary differential equation is obtained in terms of temperature. Equating the mass balance and energy balance equations by elimination of the kinetic term and integrating twice using the appropriate boundary conditions also gives Equation 2.68:

$$\frac{d^2T}{dr^2}+\frac{2}{r}\frac{dT}{dr}=\frac{k_V C_A \Delta H_r}{\lambda_e} \tag{2.69}$$

$$\frac{d^2C_A}{dr^2}+\frac{2}{r}\frac{dC_A}{dr}=\frac{k_V C_A}{D_e} \tag{2.62}$$

The pertinent boundary conditions for solving the coupled equations are (i) the concentration and temperature at the exterior surface, C_{AS} and T_S, and (ii) the zero mass and heat fluxes at the center due to symmetry:

$$r=R,\ C_A=C_{AS},\ T=T_S$$

$$r=0,\ \frac{dC_A}{dr}=0,\ \frac{dT}{dr}=0$$

In order to obtain the concentration and temperature profiles in the particle, (i) the temperature dependence of the intrinsic rate constant k_V is expressed as an Arrhenius function, and (ii) the two differential equations (Eqs. 2.62 and 2.69) are solved by using Equation 2.68 to eliminate one variable. Numerical solution of the set of equations is necessary, because the differential equations are coupled through the exponential dependence of k_V on temperature.

Weisz and Hicks [39] solved these differential equations numerically to determine the concentration profile in the particle, using Equation 2.68 for eliminating the exponential temperature dependence. The internal effectiveness factor η was then calculated by using Equation 2.64, which is applicable to nonisothermal conditions if k_V is evaluated at the surface temperature, T_S:

$$\eta=\frac{(-R_A)_P}{(-R_A)_S}=\frac{3D_e}{R k_V C_{AS}}\left(\frac{dC_A}{dr}\right)_{r=R} \tag{2.64}$$

2.4.3.2 Nonisothermal effectiveness factor

In the results reported by Weisz and Hicks [39], the nonisothermal η was described in terms of three dimensionless parameters, that is, two new independent parameters were introduced in addition to the Thiele modulus:

1 The Thiele modulus evaluated at T_S:

$$\Phi_S = 3\varphi_S = R\sqrt{\frac{(k_V)_S}{D_e}} \tag{2.63c}$$

2 The Arrhenius number, exponent of the Arrhenius equation evaluated at T_S:

$$\gamma_s = \frac{E_A}{RT_S} \tag{2.70}$$

3 A heat generation function evaluated in terms of C_{AS} and T_S:

$$\beta = \left\{\frac{(-\Delta H_r)D_e C_{AS}}{\lambda_e T_S}\right\} = \frac{(\Delta T)_{max}}{T_S} \tag{2.71}$$

The temperature dependence of the transport parameters D_e and λ_e are assumed unimportant compared to the exponential temperature dependence of k_V. The parameter β represents the maximum ΔT that can exist in the particle relative to the exterior surface temperature, T_S. Under transient conditions, it is possible to exceed even $(\Delta T)_{max}$.

The internal effectiveness factor η is plotted as a function of Φ_S for a family of β curves at four different values of γ_s between 10 and 40 in the practical range [39]. Figure 2.12 shows η as a function of Φ_S and β at $\gamma_s = 20$ [40]. The $\beta = 0$ curve corresponds to the isothermal case and coincides with the curve in Figure 2.10.

For exothermic reactions, $\beta > 0$, and the β curves generated at each γ_s value show that η may indeed exceed unity, depending on whether or not the rate increase caused by temperature rise offsets the decrease resulting from the fall in concentration. When $\eta > 1$, $(-R_A)_P > (-R_A)_S$, but this situation may not always be beneficial because the increase in temperature toward the center of the particle may induce deactivation or may promote undesirable reactions with an overall decrease in the selectivity for the desirable product. At high values of the Thiele modulus, η becomes inversely proportional to Φ_S, as in the isothermal case; therefore, the reactant is rapidly consumed as it penetrates the particle so that the reaction takes place in a thin shell just underneath the particle surface while the interior is more or less isothermal at a higher temperature. On the other hand, at low values of Φ_S and for highly exothermic reactions, for example, $\beta > 0.3$, the value of η is not uniquely defined by the three dimensionless parameters of the analysis. Three possible values of η exist, each representing a different set of conditions at which the rate of heat release balances the rate of heat removal; the

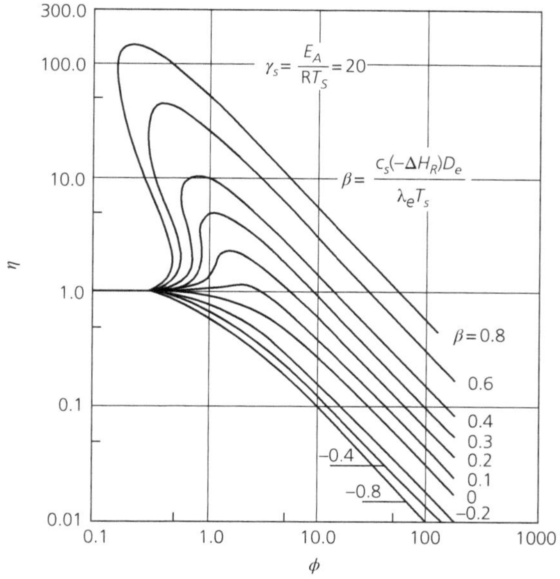

Figure 2.12 Nonisothermal internal effectiveness factors for first-order reactions in porous spherical particles.
(Source: Dittmeyer [40]. Reproduced with permission of John Wiley & Sons.)

intermediate η value is shown to be metastable, while the set of conditions physically attained is found to depend on the direction of approach to the steady state.

For endothermic reactions, $\beta < 0$, and the β curves at each γ_s value clearly indicate that the internal effectiveness factor η will always be less than unity, since both the temperature and the reactant concentration decline toward the center of the particle. In this case, the impact of heat transfer decreases, but the effect of mass transfer becomes almost negligible. An approximate solution can be obtained by ignoring the concentration profile and solving the differential energy balance (Eq. 2.69) by assuming that the reactant concentration is equal to C_{AS} within the particle:

$$\frac{d^2T}{dr^2} + \frac{2}{r}\frac{dT}{dr} = \frac{k_V \Delta H_r}{\lambda_e} C_{AS} \tag{2.69a}$$

The internal effectiveness factor η thus obtained is named a "thermal effectiveness factor" [9, 38].

2.4.4 Impact of internal transport limitations on kinetic studies

In a study of intrinsic kinetics, if data are taken in a laboratory PBR where intraparticle transport resistances are appreciable, and if this circumstance is arbitrarily neglected, experimental results will lead to falsified reaction kinetics:

1 For isothermal particles, under conditions when $\varphi > 3$, which is in the region with strong pore diffusion effects in Figure 2.10, the isothermal internal effectiveness factor η will be inversely proportional to φ. Considering an nth order reaction, and expressing φ with the characteristic size factor L as defined in Equation 2.65:

$$(-R_A)_P = \eta(-R_A)_S = \frac{1}{\varphi}(k_V C_{AS}{}^n) = \frac{\sqrt{k_V D_e}}{L\sqrt{(n+1)/2}} C_{AS}{}^{(n+1)/2} \tag{2.72}$$

Assuming that external transport resistances have been eliminated and $C_{AS} = C_{Ab}$:

$$(-R_A)_P = \frac{\sqrt{k_V D_e}}{L\sqrt{(n+1)/2}} C_{Ab}{}^{(n+1)/2} \tag{2.72a}$$

Although the reaction rate is nth order in the absence of internal mass transfer limitations, the order determined from global rate data affected by pore diffusion will be $[(n+1)/2]$ instead of the true order of the surface reaction, for example, 3/2 instead of 2. The intrinsic reaction order will be observed only when $n = 1$.

2 When data are taken under isothermal conditions at several temperatures in order to calculate the activation energy, the apparent rate constants evaluated will include the square root of $(k_V D_e)$ at each temperature. Since D_e is not very sensitive to temperature changes, the E_A calculated from the Arrhenius equation will be equal to one-half the E_A of the surface reaction, provided that external transport limitations do not exist:

$$(k_V)_{apparent} = \sqrt{k_V D_e} = A \exp\left[-\frac{(E_A)_{apparent}}{R_g T}\right]$$

$$(E_A)_{apparent} = \frac{(E_A)_{sr} + E_{Diff}}{2} \approx \frac{(E_A)_{sr}}{2}$$

Another point to be considered in the estimation of E_A is to select the temperature range carefully. At low temperatures, the reaction rate constant k_V is low so the value of φ is also low, leading to $\eta \approx 1$. However, k_V increases exponentially with temperature while D_e is not so responsive to temperature changes. Therefore, as the reaction temperature is gradually increased in successive runs, φ starts increasing and eventually η may drop below unity. Consider as an example [12] a flat plate catalyst with a 0.06 cm size factor and a D_e of 0.070 cm^2 s^{-1}; for a first-order reaction k_V of 0.84 s^{-1} at 499 K, η is calculated as 0.99. If the E_A of the surface reaction is 110 kJ mol^{-1}, k_V is calculated as 70.3 s^{-1} at 599 K; at this temperature, with the same size factor and D_e values, η is estimated as 0.50. This trend continues with further increases in temperature, until η becomes inversely proportional to φ.

3 Equation 2.72a shows that the global rate is also inversely proportional to L, the characteristic dimension, when $\varphi > 3$. In other words, if the experiments are being conducted in the region of strong diffusional effects, $(-R_A)_P$ increases with decreasing particle size.

The significance or even presence of intraparticle transport resistances must be demonstrated either by conducting diagnostic experiments or by utilizing well-established criteria.

2.4.4.1 Diagnostic experiments

The first set of preliminary experiments in a kinetic study carried out in laboratory PBRs should establish the flow conditions at which external transport effects are negligible. The second set of diagnostic experiments conducted should verify the particle size requirements under which intraparticle transport effects are eliminated. A study of intrinsic kinetics must make use of the flow rates and particle sizes that exclude both external and internal transport limitations, respectively.

The classical method of testing for the existence of pore diffusion is to measure the global rate $(-R_A)_P$ as a function of catalyst particle size. All operating conditions such as the reaction temperature, feed composition, gas flow rate, catalyst amount, and hence space time are fixed while only the particle size of the catalyst packed in the bed is systematically decreased in successive runs. This is equivalent to changing the characteristic size factor in the Thiele modulus, φ. If the particle sizes used are in the $\eta < 1$ range, $(-R_A)_P$ increases as catalyst particle size decreases; normally, particle size is decreased until $(-R_A)_P$ no longer changes. Under conditions where $\eta \approx 1$, $(-R_A)_P$ is not affected by φ. The use of this criterion does not necessitate knowledge of the rate equation or the rate constant of the reaction.

A variant of the aforementioned method is to compare the performance of two different particle sizes in terms of $(-R_A)_P$. It is assumed that k_V and D_e values are the same for both particle sizes:

1 In the diffusion-free region with $\varphi < 0.3$,

$$\frac{(-R_A)_{P1}}{(-R_A)_{P2}} = \frac{\eta_1 k_V C_{Ab}{}^n}{\eta_2 k_V C_{Ab}{}^n} = \frac{\eta_1}{\eta_2} \cong 1$$

2 In the intermediate diffusion region, for $d_{P2} > d_{P1}$,

$$\frac{(-R_A)_{P1}}{(-R_A)_{P2}} = \frac{\eta_1 k_V C_{Ab}{}^n}{\eta_2 k_V C_{Ab}{}^n} = \frac{\eta_1}{\eta_2} > 1$$

3 In the strong diffusion region, for $\eta = 1/\varphi$,

$$\frac{(-R_A)_{P1}}{(-R_A)_{P2}} = \frac{\eta_1}{\eta_2} = \frac{\varphi_2}{\varphi_1} = \frac{d_{P2}}{d_{P1}} = \frac{r_2}{r_1}$$

In general, $(-R_A)_P$ becomes inversely proportional to particle size after $\varphi > 3$, as indicated in Figure 2.10. If global rate data are procured using three catalyst particle sizes, as long as all three are not in the strong pore diffusion region, values of η and φ can be estimated for each particle size by an iterative method, starting with the assumption that $\eta = 1$ for the smallest particle and then correcting it at the end [23].

2.4.4.2 Weisz–Prater criterion for internal transport effects

In many cases, it is necessary to rapidly estimate whether internal transport limitations are present or not, without having to use the reaction rate constant. The Weisz–Prater criterion is used in laboratory experiments and for large-scale operations to confirm the absence of diffusion effects, and the criterion formulated for first-order reactions is [17, 26, 28]

$$\eta\varphi^2 = \left\{ \frac{\left[(-R_A)_P \rho_P \left(\frac{d_P}{2}\right)^2\right]}{D_e C_{AS}} \right\} \ll 1.0 \tag{2.73}$$

$$\varphi \ll 1, \quad \eta = 1 \text{ (no pore diffusion)}, \quad \eta\varphi^2 \ll 1.0$$

$$\varphi \gg 1, \quad \eta = \frac{1}{\varphi} \text{ (strong pore diffusion)}, \quad \eta\varphi^2 \gg 1.0$$

The Weisz–Prater criterion makes use of observable quantities like $(-R_A)_P$, the measured global rate (kmol/kg-s); d_P, the particle diameter (m); ρ_P, the particle density (kg/m^3); D_e, the effective mass diffusivity (m^2/s); and C_{AS}, the surface concentration of reactant (kmol/m^3). The intrinsic reaction rate constant k_V need not be known in order to use the Weisz–Prater criterion. If external mass transfer effects are eliminated, C_{Ab} can be used, and the effective diffusivity D_e can be estimated using catalyst and fluid physical properties. The criterion can be extended to other reaction orders and multiple reactions by using the generalized Thiele modulus, and various functional forms are quoted in the literature [17, 26, 28].

2.5 Combination of external and internal transport effects

The concept of effectiveness developed separately for external or internal transport resistances can be extended to an "overall effectiveness factor" for treating the general diffusion–reaction problem where both external and internal concentration and temperature gradients exist. The overall effectiveness factor, Ω, is defined for relating the actual global rate to the intrinsic rate, that is, $(-R_A)_P$ to $(-R_A)_b$. To sum up the definitions for γ, η, and Ω,

$$(-R_A)_S = \gamma(-R_A)_b; \quad (-R_A)_S = f(C_{AS}, T_S) \quad \text{and} \quad (-R_A)_b = f(C_{Ab}, T_b) \tag{2.50}$$

$$(-R_A)_P = \eta(-R_A)_S; \quad (-R_A)_P = f(C_{Ai}, T_i) \quad \text{and} \quad (-R_A)_S = f(C_{AS}, T_S) \tag{2.57}$$

$$(-R_A)_P = \Omega(-R_A)_b; \quad (-R_A)_P = f(C_{Ai}, T_i) \quad \text{and} \quad (-R_A)_b = f(C_{Ab}, T_b) \tag{2.74}$$

2.5.1 Isothermal overall effectiveness

Consider a catalyst particle located in a fluid medium where both external and internal mass transfer limitations affect the global rate. The reactant concentration profile for this situation is represented by curve III in Figure 2.3. The isothermal overall effectiveness factor, Ω, expressed in terms of bulk fluid conditions, is derived primarily for first-order reactions using

Equations 2.40 and 2.57a. At steady state, the rate of physical transport of reactants from the bulk fluid to the outer surface of the particles is equal to the total rate of chemical reaction on and within the catalyst particles:

$$(-R_A)_P = (k_m)_A a_m (C_{Ab} - C_{AS}) = \eta\, k_{sr} a_S C_{AS} \tag{2.40a}$$

Here, a_m and a_S represent the external and internal surface areas per unit weight of catalyst, respectively. The internal surface area a_S can safely be equated to the specific surface area, S_g, for all practical purposes. Solving for C_{AS} and substituting back,

$$(-R_A)_P = \eta k_{sr} S_g C_{AS} = \frac{\eta k_{sr} S_g k_m a_m C_{Ab}}{[\eta k_{sr} S_g + k_m a_m]} = \Omega k_{sr} S_g C_{Ab} = \Omega\,(-R_A)_b \tag{2.74a}$$

After rearrangement of terms, the overall effectiveness factor Ω for linear kinetics can be written as follows:

$$\Omega = \frac{\eta}{1+\eta \dfrac{k_{sr} S_g}{k_m a_m}} = \frac{\eta}{1+\eta \dfrac{k_{sr} S_g \rho_B d_P}{6(1-\epsilon_B)k_m}} \tag{2.75}$$

In order to express the parameters appearing in Ω in terms of measurable physical properties of the catalyst particles and the catalyst bed, the external surface area a_m is equated to

$$a_m = \frac{6(1-\epsilon_B)}{\rho_B d_P} \tag{2.76}$$

The porosity and density of the catalyst bed, ϵ_B and ρ_B, respectively, and the particle diameter, d_P, are the physical properties used in estimating a_m. The global rate can be expressed per unit volume, per unit weight, or per unit surface area, according to purpose:

$$(-R_A)_V = \rho_B\,(-R_A)_P = \rho_B S_g (-R_A)_A \tag{2.77}$$

The Damköhler number Da, which is a measure of the ratio of the surface chemical reaction rate to the external mass transfer rate, can also be defined as an observable quantity in terms of the global rate when multiplied by the internal effectiveness factor η [13]:

$$(\eta Da) = \frac{L\,(-R_A)_V}{k_m C_{Ab}} \tag{2.78}$$

External concentration gradients are dependent on the internal mass transfer process. The Biot number for mass $(Bi)_m$ is defined as a measure of the ratio of internal to external mass transport resistances and is expressed in terms of the size factor L of Equation 2.65 [13, 19]:

$$(Bi)_m = \frac{L k_m}{D_e} \tag{2.79}$$

Experimental results show that the mass Biot number $(Bi)_m$, which is sometimes called the "modified Sherwood number" [17], is much larger than unity, indicating that the major resistance to mass transfer resides in the internal pore diffusion process.

2.5.2 Nonisothermal conditions

The general problem of diffusion–reaction for the overall effectiveness factor Ω is rather complicated. However, the physical and chemical rate processes prevailing under practical conditions promote isothermal particles and negligible external mass transfer limitations. In other words, the key transport limitations are external heat transfer and internal mass transfer. External temperature gradients can be significant even when external mass transfer resistances are negligibly small.

The analogous Biot number for heat $(Bi)_h$ is defined as the ratio of internal to external heat transport resistances and is used together with $(Bi)_m$ to assess the relative importance of all heat and mass transport limitations:

$$(Bi)_h = \frac{hL}{\lambda_e} \tag{2.80}$$

$$(Bi)_r \equiv \frac{(Bi)_m}{(Bi)_h} = \left\{\frac{k_m}{h}\right\}\left\{\frac{\lambda_e}{D_e}\right\} = \frac{\beta_{ext}}{\beta} \tag{2.81}$$

Here, $(Bi)_r$ is the Biot number ratio. The external heat generation function β_{ext} is obtained from Equation 2.46 by dividing both sides by T_b and letting $C_{AS} \to 0$ for $(T_S - T_b) = (\Delta T)_{max}$. The internal heat generation function β (Eq. 2.71), the Arrhenius number γ_s (Eq. 2.70), and k_V of the Thiele modulus (Eq. 2.63c) are all evaluated at T_b [19].

The usual orders of magnitude for effective transport coefficients of gases, liquids, and porous solid particles are [13]:

For gases: $D_e = 1$–0.1 cm^2/s; $\lambda_e = 10^{-4}$–10^{-5} cal/cm-s-°C

For liquids: $D_e = 10^{-5}$–10^{-6} cm^2/s; $\lambda_e = 10^{-2}$–10^{-4} cal/cm-s-°C

For porous solids: $D_e = 10^{-1}$–10^{-3} cm^2/s; $\lambda_e = 10^{-3}$–10^{-4} cal/cm-s-°C

Then, the ranges of internal and external heat generation functions β and β_{ext} calculated on the basis of these coefficients show that the Biot number ratio is usually large (>10) in gas–solid catalytic reaction systems:

$$\beta = \frac{(-\Delta H_r) D_e C_{Ab}}{\lambda_e T_b} = 0.001 - 0.3$$

$$\beta_{ext} = \frac{(-\Delta H_r) k_m C_{Ab}}{h T_b} = 0.01 - 2.0$$

$$(Bi)_r = \frac{\beta_{ext}}{\beta} = \frac{(Bi)_m}{(Bi)_h} = 1.0 - 10^4$$

In reality, $(Bi)_r$ is in the range of hundreds under typical reaction conditions, and hence the particle can safely be considered isothermal [41]. If the catalyst particle is assumed isothermal, then a simple analysis can be made to find the relative significance of external to internal mass transfer limitations [19]:

$$\frac{(C_{Ab} - C_{AS})}{C_{Ab}} = \frac{[\eta \varphi^2]}{(Bi)_m} \tag{2.82}$$

The result in Equation 2.82 is significant because it shows that external concentration gradients are indeed affected by internal pore diffusion processes, and they are usually negligible in particulate systems unless the value of the Thiele modulus is close to the mass Biot number. The concentration and temperature differences between the exterior surface and the interior of a particle originate from the ΔH_r of highly exothermic reactions. Although steady-state multiplicity is of concern for both internal and external temperature gradients, it has been pointed out that multiple steady-state solutions resulting from internal temperature gradients occur at somewhat unrealistic values of the relevant parameters, and in the case of single reactions, it was shown that steady-state multiplicity is mostly caused by external and not internal gradients [19, 42]. This is supported by the typical results reported for $(Bi)_r$ values.

This discussion is not applicable to catalytic liquid–solid reactions since the range of β_{ext} values is much lower and the resulting Biot number ratios are also rather low [13, 19]:

$$\beta = \frac{(-\Delta H_r)D_e C_{Ab}}{\lambda_e T_b} = 0.001 - 0.1$$

$$\beta_{ext} = \frac{(-\Delta H_r)k_m C_{Ab}}{h T_b} = 0.001 - 0.05$$

$$(Bi)_r = \frac{\beta_{ext}}{\beta} = \frac{(Bi)_m}{(Bi)_h} = 10^{-4} - 10^{-1}$$

2.6 Summary

Successful reactor analysis and design necessitates chemical kinetic models based on steady-state chemical activity only. In solid-catalyzed heterogeneous systems, mass and energy transport processes at the particle scale can significantly disguise chemical reaction rates, thus altering experiential overall rates. Therefore, physical transport phenomena evaluated in accordance with the distinctive features of the actual catalyst/reactor system have to be superimposed on the chemical kinetic model to obtain global rate expressions suitable for macrokinetic analysis. In this chapter, firstly, a concise review of the basic concepts in heterogeneous catalytic kinetics is presented as well as the challenges involved in data acquisition and kinetic model development in the absence of physical transport effects. Secondly, the degree to which physical transport limitations can modify intrinsic reaction rates of solid-catalyzed gas-phase reactions is discussed for the simple case of external transport resistances in series with nonporous surface-catalyzed chemical reaction. It is underlined that external temperature gradients may be substantial even when external concentration gradients are small, and under practical operating conditions, one of the major transport resistances not to be overlooked is external heat transport. Thirdly, the case of mass and/or heat transport with simultaneous chemical reaction within porous catalyst particles is considered in order to evaluate the global rate by accounting for internal concentration and temperature profiles with reference to exterior surface conditions. It is concluded that the second major transport resistance not to be ignored is internal mass transport. Finally, the concept of overall effectiveness is considered for the general diffusion–reaction problem where both external and internal concentration and temperature gradients exist. Comparisons based on Biot number ratios confirm that physical and chemical rate processes that typically prevail under practical conditions lead to isothermal particles and negligible external mass transfer limitations.

Nomenclature

$\bar{a}$	mean pore radius, Å
A	preexponential or frequency factor (units of rate constant)
A_m	cross-sectional area of one adsorbate molecule, $Å^2$
a_m	external surface area per unit mass of catalyst, m^2/g
a_S	internal surface area per unit mass of catalyst, S_g, m^2/g
$(Bi)_h$	Biot number for heat, hL/λ_e
$(Bi)_m$	Biot number for mass, k_mL/D_e
$(Bi)_r$	Biot number ratio, $(Bi)_m/(Bi)_h$
C_{Ab}, C_{Bb}, ...	bulk gas concentrations of molecular species, mol/m^3
C_{AS}, C_{BS}, ...	surface concentrations of adsorbed molecular species, mol/m^3
c_p	heat capacity per unit mass of fluid, J/g-K
Da	Damköhler number, (k_{sr}/k_m) or $(k_V/k_m a_m \rho_B)$
D_{axial}	axial dispersion coefficient, cm^2/s
D_{comb}	combined diffusivity in pores, cm^2/s
D, $D_{A\text{-}mix}$ or D_{AB}	molecular diffusion coefficient, cm^2/s
D_e	effective diffusivity in particles, cm^2/s
D_K	Knudsen diffusivity in pores, cm^2/s
d_P	particle diameter, m or cm
D_S	surface diffusivity in pores, cm^2/s
d_{tube}	reactor tube diameter, m or cm
E_A	activation energy, kJ/mol
E_{app}	apparent activation energy, kJ/mol
E_{true}	true activation energy, kJ/mol
f	degree of mass transfer control, fraction
F_A	molar flow rate of species A, mol/s
G	fluid mass velocity, ρu, g/m^2-s
ΔG^0	standard Gibbs free energy change, kJ/mol
h	external heat transfer coefficient, kJ/m^2-s-K
ΔH_{ads}	heat of adsorption, kJ/mol
ΔH_r	heat of chemical reaction, kJ/mol
i	ith species
j_D, j_H	j factors for mass and heat transfer, respectively, dimensionless

K	overall reaction equilibrium constant, dimensionless
k_{ads}	rate constant of adsorption
K_A, K_B, ...	adsorption equilibrium coefficients of species, m^3/mol or 1/atm
k_{des}	rate constant of desorption
k_m	external mass transfer coefficient, m/s
K_{sr}	surface reaction equilibrium constant, dimensionless
k_{sr}	surface reaction rate constant per unit area, m/s
k_V	surface reaction rate constant per unit volume, 1/s
L	characteristic size factor, (V_P/S_{ext}), m or cm
L_{tube}	length of reactor tube, m
m_P	mass of catalyst particle, g
n	general reaction order
N_0	Avogadro's number, molecules/mol
N_A, N_B, ...	molar flux of species, mol/m^2-s
Nu	Nusselt number, hd_P/λ
P_A, P_B, ...	partial pressure of species, atm
$(Pe)_P$	particle Péclet number, ud_P/D_{axial}
P_o	saturation pressure of species, atm or mmHg
Pr	Prandtl number, $c_p\mu/\lambda$
P_T	total pressure, atm
Q	rate of heat removal per unit mass of catalyst, kJ/g-s
Q_R	rate of heat generation per unit mass of catalyst, kJ/g-s
r	radial coordinate
R	radius of particle, m or cm
R_{ads}	rate of adsorption
$(-R_A)_P$	global rate of reaction per unit mass of catalyst, mol/g-s
$(-R_A)_V$	global rate of reaction per unit volume of catalyst, mol/m^3-s
R_{des}	rate of desorption
$(Re)_P$	particle Reynolds number, d_PG/μ
R_g	gas constant, J/mol-K or cm^3-atm/mol-K
r_{tube}	radius of reactor tube, m or cm
Sc	Schmidt number, $\mu/\rho D$
S_g	total surface area per unit mass of catalyst, m^2/g
Sh	Sherwood number, k_md_P/D
S_V, S_i	chemical forms of vacant and occupied surface sites, respectively
T	temperature, °C or K
T_b	temperature of bulk fluid, °C or K
T_S	surface temperature of catalyst, °C or K
ΔT	temperature difference, °C or K
u	superficial fluid velocity, m/s
V	volume, m^3 or cm^3
V_{ads}	volume of adsorbed gas, cm^3
V_g	void volume per unit mass of catalyst, cm^3/g
V_{He}	volume of helium displaced, cm^3
V_{Hg}	volume of mercury displaced, cm^3
V_P	volume of particle, cm^3
w or W	weight of catalyst sample or bed, respectively, g
x, x_i	fractional conversion, conversion of ith species
y_i	mol fraction of ith species

Greek letters

α	Arrhenius number defined by Eq. 2.49, E_A/R_gT_b
α, β, γ, ...	reaction orders in power law models
β	heat generation function defined by Eq. 2.71
γ	external effectiveness factor defined by Eq. 2.50
γ_s	Arrhenius number defined by Eq. 2.70, E_A/R_gT_S
δ	tortuosity factor
ϵ_B	porosity of catalyst bed
ϵ_P	porosity of catalyst particle
η	internal effectiveness factor defined by Eq. 2.57
θ	dimensionless temperature defined by Eq. 2.49
θ_i	fractional surface coverage of ith species
λ, λ_g	fluid or gas thermal conductivity, J/m-s-K
λ_e	effective thermal conductivity of particle, J/m-s-K
λ_s	solid thermal conductivity, J/m-s-K
μ	fluid viscosity, g/m-s
ρ	fluid density, g/cm^3 or kg/m^3
ρ_B	bed density, g/cm^3 or kg/m^3
ρ_P	particle density, g/cm^3 or kg/m^3
ρ_S	solid density, g/cm^3 or kg/m^3
φ_L	Thiele modulus for flat plate defined by Eq. 2.60
φ_S	Thiele modulus for spherical particle defined by Eq. 2.63a
Φ_S	Thiele modulus for spherical particle defined by Eq. 2.63
Ω	overall effectiveness factor defined by Eq. 2.74

References

1 Langmuir I. The constitution and fundamental properties of solids and liquids. Part I. Solids. Journal of American Chemical Society 1916; 38: 2221–2295.

2 Langmuir I. The adsorption of gases on plane surfaces of glass, mica and platinum. Journal of American Chemical Society 1918; 40(9): 1361–1403.

3 Taylor HS. A theory of the catalytic surface. Proceedings of the Royal Society Series A: Mathematical Physical and Engineering Sciences 1925; 108A: 105–111.

4 Davis BH. Development of the science of catalysis. In: Handbook of heterogeneous catalysis, 2nd ed. Eds. Ertl G, Knözinger H, Schüth F, Weitkamp J. New York: Wiley; 2008; Ch 1, Section 1.2: 16–37.

5 Thomas JM, Thomas WJ. Principles and practice of heterogeneous catalysis. Weinheim: VCH; 1996.

6 Masel RI. Principles of adsorption and reaction on solid surfaces. New York: Wiley-Interscience; 1996.

7 Hill Jr. CG. An introduction to chemical engineering kinetics and reactor design. New York: Wiley; 1977.

8 Beeck O. Hydrogenation catalysts. Discussions of the Faraday Society 1950; 8: 118–128.

9 Smith JM. Chemical engineering kinetics, 3rd ed. New York: McGraw-Hill; 1981.
10 Dautzenberg FM. Ten guidelines for catalyst testing. In: Characterization and catalyst development: An interactive approach. Eds. Bradley SA, Gattuso MJ, Bertolacini RJ. ACS Symposium Series 411. Washington, DC: American Chemical Society; 1989; Ch 11: 99–119.
11 Moulijn JA, Tarfaoui A, Kapteijn F. General aspects of catalyst testing. Catalysis Today 1991; 11: 1–12.
12 Bartholomew CH, Farrauto RJ. Fundamentals of industrial catalytic processes, 2nd ed. Oxford: Blackwell Science; 2006.
13 Carberry JJ. Chemical and catalytic reaction engineering. New York: McGraw-Hill; 1976.
14 Petersen EE. Chemical reaction analysis. Englewood Cliffs, NJ: Prentice-Hall; 1965.
15 Hinshelwood CN. Kinetics of chemical change. London: Oxford University Press; 1940.
16 Hougen OA, Watson KM. Chemical process principles. Part 3. Kinetics and catalysis. New York: Wiley; 1947.
17 Froment GF, Bischoff KB, DeWilde J. Chemical reactor analysis and design, 3rd ed. New York: Wiley; 2011.
18 Yang KH, Hougen OA. Determination of mechanism of catalyzed gaseous reactions. Chemical Engineering Progress 1950; 46(3): 146–157.
19 Lee HH. Heterogeneous reactor design. Butterworths Series in Chemical Engineering. Boston: Butterworths; 1985.
20 Thomson SJ, Webb G. Heterogeneous catalysis. Edinburgh: Oliver & Boyd; 1969.
21 Boudart M, Djega-Mariadassou G. Kinetics of heterogeneous catalytic reactions. Princeton, NJ: Princeton University Press; 1984.
22 Gleaves JT, Yablonskii GS, Phanawadee P, Schuurman Y. TAP-2: An interrogative kinetics approach. Applied Catalysis A. General 1997; 160(1): 55–88.
23 Roberts GW. Chemical reactions and chemical reactors. New York: Wiley; 2009.
24 Carberry JJ. Designing laboratory catalytic reactors. Industrial and Engineering Chemistry 1964; 56(11): 39–46.
25 Weekman Jr VW. Laboratory reactors and their limitations. AIChE Journal 1974; 20(5): 833–840.
26 Fogler HS. Elements of chemical reaction engineering, 4th ed. Englewood Cliffs, NJ: Prentice-Hall; 2006.
27 Rase HF. Fixed-bed reactor design and diagnostics. London: Butterworths; 1990.
28 Onsan ZI, Avci AK. Reactor design for fuel processing. In: Fuel cells: Technologies for fuel processing. Eds. Shekhawat D, Spivey JJ, Berry DA. New York: Elsevier Science; 2011; Ch 14: 451–516.
29 Satterfield CN, Sherwood TK. The role of diffusion in catalysis. Reading, MA: Addison-Wesley; 1963.
30 Bird RB, Stewart WE, Lightfoot EN. Transport phenomena, 2nd ed. New York: Wiley; 2002.
31 Levenspiel O. Chemical reaction engineering, 3rd ed. New York: Wiley; 1999.
32 Nauman EB. An introduction to chemical reactor design, optimization and scaleup. New York: McGraw-Hill; 2002.
33 Zalc M, Reyes SC, Iglesia E. The effects of diffusion mechanism and void structure on transport rates and tortuosity factors in complex porous structures. Chemical Engineering Science 2004; 59: 2947–2960.
34 Wakao N, Smith JM. Diffusion in catalyst pellets. Chemical Engineering Science 1962; 17(11): 825–834.
35 Feng CF, Stewart WE. Practical models for isothermal diffusion and flow of gases in porous solids. Industrial & Engineering Chemistry Fundamentals 1973; 12: 143–146.
36 Beeckman JW, Froment GF. Deactivation of catalysts by coke formation in the presence of internal diffusion limitation. Industrial & Engineering Chemistry Fundamentals 1982; 21(3): 243–250.
37 Keil FJ. Modelling of phenomena within catalyst particles. Chemical Engineering Science 1996; 51(10): 1543–1567.
38 Aris R. On shape factors for irregular particles. 1. The steady state problem of diffusion and reaction. Chemical Engineering Science 1957; 6(6): 262–268.
39 Weisz PB, Hicks JS. The behavior of porous catalyst particles in view of internal mass and heat diffusion effects. Chemical Engineering Science 1962; 17(4): 265–275.
40 Dittmeyer R, Emig G. Simultaneous heat and mass transfer and chemical reaction. In: Handbook of heterogeneous catalysis, 2nd ed. Eds. Ertl G, Knözinger H, Schüth F, Weitkamp J. New York: Wiley; 2008; Ch 6, Section 6.3: 1741.
41 Butt JB, Downing DM, Lee JW. Inter-intraphase temperature gradients in fresh and deactivated catalyst particles. Industrial & Engineering Chemistry Fundamentals 1977; 16(2): 270–272.
42 Luss D. Diffusion-reaction interactions in catalyst pellets. In: Chemical reaction and reactor engineering. Eds. Carberry JJ, Varma A. New York: M. Dekker; 1987; Ch 4: 239–292.

PART 2

Two-phase catalytic reactors

CHAPTER 3

Fixed-bed gas–solid catalytic reactors

João P. Lopes[1] and Alírio E. Rodrigues[2]

[1]Department of Chemical Engineering and Biotechnology, University of Cambridge, Cambridge, UK

[2]Laboratory of Separation and Reaction Engineering, Associate Laboratory LSRE/LCM, Department of Chemical Engineering, Faculty of Engineering University of Porto, Porto, Portugal

Abstract

Fixed-bed reactors are a core technology in chemical engineering, which appear in a number of well-established processes or in new configurations with superior thermal and functional integration. Over the past decades, the theory associated with the modeling of these units has grown substantially. In this chapter, a review of the main transport and reaction phenomena occurring in a packed bed is provided. The commonly used models for design and simulation of packed beds are obtained from a sufficiently generic set of conservation equations, assuming dominance of dispersion and interphase transfer compared to the other processes present in the reactor. The analysis becomes complete with approximate solutions for the effectiveness factor at the particle level. This represents a methodology for consistent model hierarchization and selection, determination of equivalent formulations, and improvement of simplified descriptions.

3.1 Introduction and outline

Fixed-bed reactors have become an example of what can be considered a "conventional" technology in chemical engineering, since most processes involving solid catalysts require its implementation. Indeed they are used in many operations involving heterogeneous gas–solid chemical reactions, which are ubiquitous in the chemical industry for the large-scale production of commodities (e.g., petroleum refining and petrochemical industries) and fine chemicals, environmental protection, and bioprocessing. In general, the flow of reactants through a bed packed with pellets of supported catalyst is involved. In practice, however, several versions of this simplistic picture exist. They can be distinguished by their geometry, operating mode, temperature control and reactant dosing methods, and functional integration. Figure 3.1 presents a very schematic representation of some types of fixed-bed reactors, namely:

1 *Axial flow fixed-bed reactor (usually downflow): single* (Figure 3.1-a.1), *multistage* (Figure 3.1-a.2), *or multitubular* (Figure 3.1-a.3). Here, the different configurations are related with the requirements for appropriate heat management [1]. While a large diameter reactor may be used to approach adiabatic operation, higher conversions can be obtained by using several beds with intermediate cooling/heating or cold/hot feed injection. However, if reactions with selectivity constraints and with appreciable heat generation/consumption occur, it may be useful to adopt a shell and tube design (where the catalyst is placed in several tubes with diameters around a few centimeters while a cooling/heating medium circulates in the shell) [2–4].

2 *Radial flow fixed-bed reactor* (Figure 3.1-b). If a large quantity of catalyst is required, it is convenient to separate the length scales that determine catalyst volume (axial length) and pressure drop (flow-through transverse length). Since the bed length is typically several times the thickness of the catalyst packing, higher pressures and fluid velocities may be employed. Several configurations can appear regarding the radial and relative axial flow direction between the periphery annulus and the inner concentric channel. Inward flow occurs when the fluid travels from the annular channel to the inner tube (as depicted in Figure 3.1-b), while outward flow is verified in the opposite case. The relative direction of the axial flow that is established in both chambers also leads to a distinction between straight and reverse flow (the former corresponds to the situation where the points of reactant feeding and product collection are separated by the reactor length, whereas the latter implies that inlet and outlet are on the same side). Kareeri et al. [5] simulated the flow distribution in the four operating modes concerning the fluid flow direction in the two sides. Vek [6] presented a radial flow converter for ammonia synthesis operating adiabatically, as a way to overcome internal transport limitations in highly active catalysts. Other examples include styrene production [7, 8] and highly exothermic selective oxidations [9]. In the latter case, increased productivity, better thermal management and lower pressure drop are reported, in comparison with the axial flow counterpart. The UOP Isomar™ process for isomerization of xylenes also makes use of a radial-flow reactor [10].

Multiphase Catalytic Reactors: Theory, Design, Manufacturing, and Applications, First Edition. Edited by Zeynep Ilsen Önsan and Ahmet Kerim Avci.

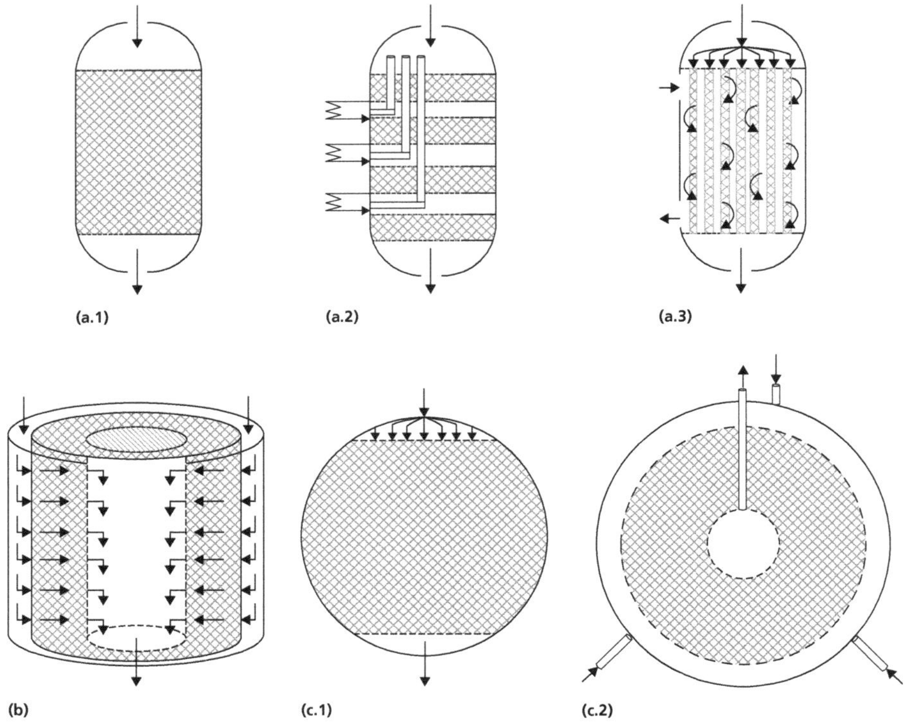

Figure 3.1 Fixed-bed reactors. Schematic representation of several operating configurations and arrangements: (a) Cylindrical with axial flow (a.1-single, a.2-multistage, a.3-multitubular); (b) cylindrical with radial flow; (c) spherical reactor (c.1-axial and c.2-radial flow).

3 *Spherical fixed-bed reactor: axial* (Figure 3.1-c.1) *or radial flow* (Figure 3.1-c.2). The increase of the cross-sectional area over the bed's length is expected to lead to lower pressure drops, lower investment and operation costs, and consequently smaller particles with higher effectiveness can be used. This geometry has been investigated in the production of methanol with radial [11, 12] and axial flow [13] configurations. The naphtha reforming process has been also studied [14]. In this case, axial flow was found to be advantageous over the radial configuration, due to improved feed distribution and the possibility of introducing membrane-based technologies [15].
4 *Multifunctional reactors.* Apart from the applications exclusively in the reaction engineering field, fixed beds can also be found integrated in apparatus with separation purposes. Two major classes can be identified: membrane [16, 17] and chromatographic (adsorptive) reactors [18]. In both cases, a familiar configuration involves a packed bed with a permeable, selective wall and/or packed with a catalyst/adsorbent. The latter is found in simulated moving bed reactors [19, 20], which can be seen as an arrangement of several of these beds. The overall effect from the coexistence of reaction and separation functionalities is the increase of conversion in equilibrium-limited systems, which are shifted due to the simultaneous separation of the products.

These designs have been implemented mostly at the macroscale (industrial or laboratory scale), but many developments at the microscale have also been reported under the trendy headline of "chemical microprocess engineering," not only for laboratory-scale measurements but also at the production scale [21, 22]. Guettel and Turek [23] compared microstructured and multitubular fixed-bed reactors for highly exothermic partial oxidations such as butane to maleic anhydride and *o*-xylene to phthalic anhydride. The microfixed beds showed a productivity increase between 2.5 and 7 times higher than the conventional technology. Recently, Bakhtiary-Davijany et al. [24] have proposed a multi-slit packed bed microstructured reactor with an integrated heat exchanger for the synthesis of methanol from synthesis gas over a $Cu/ZnO/Al_2O_3$ catalyst. Some of the benefits in the miniaturization of packed beds include higher catalyst loading (in comparison with wall coatings) and better temperature control (in comparison with the macroscale). This is especially attractive for the measurement of kinetic data and catalyst screening. Moreover, fixed or moving packed beds can be created using monoliths in a single pellet string configuration [25].

In this chapter, we revisit some of the main topics in modeling and analysis of two-phase (gas–solid) catalytic reactors. First, we consider the usual set of assumptions and simplifying concepts that constitute the theoretical framework that describes the transport–reaction processes occurring in the reactor (Section 3.2.1). Then, the generic continuous model that arises in this context is written and the timescales for the different processes are identified (Section 3.2.2). The remaining part of the chapter is concerned with the different reductions and simplified formulations that derive from the most complete model. Emphasis is given on the use of approximate analytical techniques in the treatment of phenomena at the particle scale (Section 3.3) or at the reactor level, in particular concerning the relative rate of mass and heat interphase transport (Sections 3.4 and 3.5, respectively), or dispersion (Section 3.6), compared with the remaining mechanisms.

3.2 Modeling of fixed-bed reactors

The formulation of conservation equations and its numerical solution have been the object of several treatments over the past decades. In the last years, application of computational fluid dynamics (CFD) to fixed-bed reactors [26, 27] has been explored as an improved alternative to the classical plug-flow models with effective transport quantities [28]. This has been studied mainly for low tube-to-particle diameter ratios, owing to the lower calculation effort involved when compared with packings with a large number of particles. However, increasing advances in computational resources are allowing numerical simulations to approach realistic ratios between bed and pellet typical dimensions. The results from CFD calculations in a geometrically representative mesh can be used to evaluate the bed hydrodynamics (e.g., velocity profiles can then be incorporated into reaction–dispersion–convection fixed-bed models) and transport properties in a more realistic way. Magnico [29] considered the wall-to-fluid heat transfer in tubular beds with small tube-to-particle-diameter ratio. The effect of turbulence models on the same phenomenon was also studied [30]. Dixon et al. [31, 32] used CFD to couple the fluid flow in a fixed bed with the transport and reaction in a solid catalyst particle with no-slip surface flow boundary condition. Other computational techniques include cell models [33, 34], network analysis [35, 36], or the Lattice-Boltzmann approach [37]. Freund et al. [38] described the use of numerical techniques to generate random packings (with Monte Carlo processes), calculate the flow field (with lattice Boltzmann methods), and simulate mass transport (using a particle tracking algorithm for the pore scale simulation of a tracer). Computational developments have also been made in the incorporation of detailed mechanistic descriptions of the surface kinetics into classical fixed-bed models [39]. At the same time, experimental techniques, such as magnetic resonance imaging [37, 40, 41] and spectroscopic profiling [42], have been applied to these studies. Nevertheless, continuum models with effective transport quantities still dominate the studies that are presented in the literature, and most computationally demanding approaches still make use of them at some point (e.g., with parameters that are more accurately estimated).

3.2.1 Description of transport–reaction phenomena

The behavior of a fixed-bed reactor comprises phenomena of different nature, occurring at several scales (Figure 3.2). In fact, flow (hydrodynamics) and mass/heat transfer, within and between the two phases, are coupled with chemical reactions taking place in the catalyst. These mechanisms are referred to a particular domain (interparticular or intraparticular) or to the interfaces between different phases. The simplest formulation of mass and heat fluxes in the conservation equations is usually based on constitutive equations of the same form of Fick's and Fourier's laws of diffusion and conduction with effective dispersion coefficients. Transfer terms are written by making use of concepts such as interphase transfer coefficients associated with appropriate driving forces. Experimental data or estimation methods for these quantities are required in order to simulate the behavior of a system with a given mathematical model. Extensive reviews on this subject have been provided in the literature, gathering a large amount of experimental data and proposing correlations to fit it as accurately as possible (e.g., in Ref. [43]). Nevertheless, new measurements for specific cases have been reported [44–46], while expressions to describe these quantities with as much system independence as possible are sought.

The form in which the different processes are described reflects the assumptions of the model, which we briefly review in the succeeding text.

1 *Plug-flow with superficial velocity, u.* The most convenient approach to describe the flow through the packed bed is to consider a uniform fluid velocity and to account for possible deviations in the dispersion term discussed later. The actual velocity profile $u(r, z)$ can be simulated with CFD packages, including models for turbulence. For most reaction engineering applications, the momentum equation is decoupled from the other model equations, as long as dependence of properties on the variables can be neglected. Attempts to account for velocity variations due to nonuniform radial porosity have also been presented [42, 47, 48]. Only some cases have treated the gas phase as compressible in a fixed-bed reactor [49, 50].
2 *Mass dispersion effective coefficients.* Effects that result in a more complex flow pattern, such as higher fluxes near the wall due to higher void fraction, turbulent transverse velocity

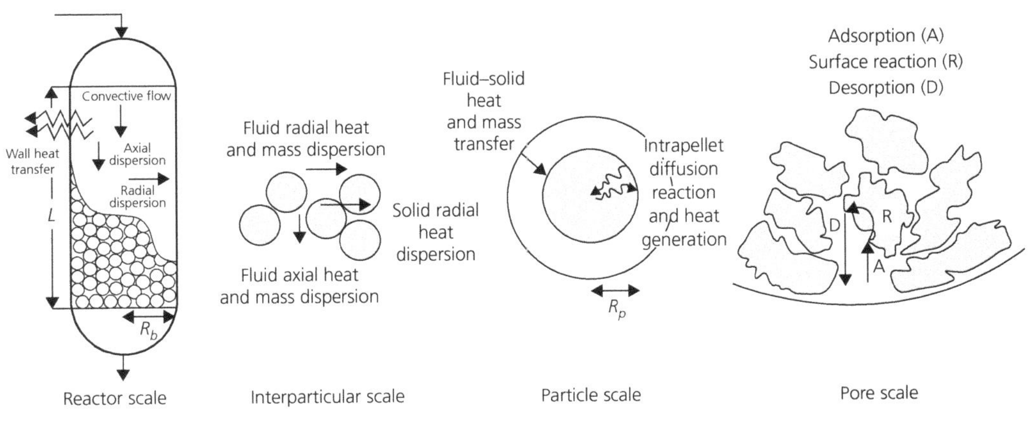

Figure 3.2 Transport–reaction mechanisms in a fixed-bed reactor at different scales.

variations, or mixing due to the presence of packing, are captured by axial and radial dispersion coefficients (D_{ae} and D_{re}, respectively). These coefficients are effective quantities (expressed in m^2_{bed}/s) and appear in the dispersion flux as a diffusivity in Fick's law. They are usually correlated in terms of the particle Peclet number:

$$Pe_{ma} = \frac{ud_p}{\varepsilon_b D_{ae}} = \frac{u_i d_p}{D_{ae}} \tag{3.1}$$

$$Pe_{mr} = \frac{ud_p}{\varepsilon_b D_{re}} = \frac{u_i d_p}{D_{re}} \tag{3.2}$$

Froment [51] compiled several previous experimental works in $Pe_{ma} - Re$ and $Pe_{mr} - Re$ diagrams, where the Reynolds number is defined as $Re = d_p G/\mu$. In general, one may say that for $0.1 < Re < 1000$, Pe_{ma} will take values between 0.5 and 2. Some correlations for the radial dispersion coefficient are of the form

$$Pe_{mr} = A\left[1 + B\left(\frac{d_p}{d_b}\right)^2\right]. \tag{3.3}$$

For $1 < Re < 1000$, A should lie between 6 and 40, although for practical purposes, it is reasonable to assume [51] that Pe_{mr} varies in a range between 6 and 8. Experiments for a given system may be used to obtain the values for the coefficients. Froment [51] collected data from other authors that suggests $A \approx 7-15$ and $B = 19.4$. Rase [52] also proposed that $B = 19.4$ for $d_p/d_b < 0.1$, while $B = 0$ for larger values of this ratio. Moreover, a dependence on the Reynolds number is introduced:

$$\frac{1}{A} = \frac{1}{m'} + \frac{0.38}{Re}, \tag{3.4}$$

where in general m' is a function of Re, even though for $Re > 400$, we can consider $m' = 11$. Gunn [53] presented a comparison between experimental data [54] on axial dispersion in fixed beds and correlations. In gas phase, the dependence on the Peclet number and the $ReSc$ product increases linearly (with a slope around 1) in the small $ReSc$ region (corresponding to direct proportionality between dispersion and molecular diffusion coefficients) up to a maximum and then falls to a plateau around 2 (the theoretical limit of Aris and Amundson [55]) as Re increases. A typical form of these correlations is given by Ref. [56]:

$$\frac{1}{Pe_{ma}} = \frac{A}{ReSc} + \frac{B}{1 + C/(ReSc)}. \tag{3.5}$$

Other expressions are available in Ref. [57]. An alternative to the diffusion-like treatment of dispersion, in terms of a nonequilibrium approach, was given by Kronberg and Westerterp [58].

3 *Fluid and solid heat dispersion coefficients.* The description of heat transfer follows the same lines, with two main distinctions concerning mostly the processes occurring in the transverse direction: (i) the bed central core and the region near the wall are significantly different, and (ii) fluid and solid phases should be treated separately. The first aspect is discussed later when transfer coefficients to/from the wall are introduced. The heterogeneous character of the phenomenon was approached by de Wasch and Froment [59] through the consideration of two coefficients (one for each phase). These quantities act as effective conductivities, expressed in $W/(m_{bed}K)$, in the fluid and solid dispersive heat fluxes. The formulation is similar to Fourier's law, with spatial gradients of the interparticular fluid temperature and the solid phase temperature. Park [60] took the latter as an averaged value over a catalyst pellet, while de Wasch and Froment [59] considered unitary effectiveness factor. These assumptions are reflected in the energy balance given by Equation 3.34. Most models following this treatment, however, ignore temperature gradients inside the catalyst particles [61]. De Wasch and Froment [59] showed that ignoring solid heat transfer leads to incorrect observations in fixed-bed reactor modeling, namely, in the prediction of runaway behavior. It was estimated that typically 25% of the radial heat transfer proceeds through the solid [62].

Several correlations for the heat Peclet numbers as a function of Reynolds and Prandtl numbers are available. For the fluid effective radial heat conductivity λ^f_{re}, they usually take the following form [63]:

$$\frac{1}{Pe^f_{hr}} = \frac{1}{Pe^f_{hr}(\infty)} + f\left(\frac{\varepsilon_b}{\tau_b}\frac{1}{RePr}\right), \tag{3.6}$$

where $Pe^f_{hr} = \rho_g c_{pg} d_p u/\lambda^f_{re}$ and the limiting Peclet number $Pe^f_{hr}(\infty)$ has been evaluated in the literature in a number of ways, including in some cases a dependence on the ratio between particle and bed diameters (A and B are constants, but only A depends on the packing geometry): $Pe^f_{hr}(\infty) = A\left[1 + B(d_p/d_b)^2\right]$. For spherical packings, the following values have been proposed [64]: $A = 10$ and $B = 19.4$, or $A = 7$ and $B = 2$, or $A = 12$ and $B = 0$. Values for other geometries are also available.

An alternative to (3.6) is to consider a sum of static and dynamic contributions (the latter from an analogy with mass transfer) using the following equation:

$$\lambda^f_{re} = \frac{\varepsilon_b}{\tau_b}\lambda^f + \rho_g c_{pg}\varepsilon_b D_{re}. \tag{3.7}$$

Heat transfer in the solid proceeds across the bed in the radial direction by conduction through the particles and contact points and by radiation. The two latter mechanisms are usually negligible. Under these conditions, the effective conductivity may be estimated [59, 65] according to

$$\frac{\lambda^s_{re}}{\lambda^f} = \frac{A(1-\varepsilon_b)}{B + C\lambda^f/\lambda_{cat}}. \tag{3.8}$$

Equation 3.8 combines the effects of conduction through particles (with characteristic length given by Cd_p) and in the stagnant fluid film (with thickness Bd_p) in the vicinity of the contact surface between two neighboring particles (with centers distanced by Ad_p) as two resistances in series. The reader is referred to the references cited earlier for further details and other expressions [66]. Experimental measurements with air flowing over vanadium pentoxide catalyst are reported in Ref. [67].

The axial heat Peclet number Pe^f_{ha} (with a definition similar to Pe^f_{hr}) varies from about 0.1 at $Re \sim 1$ to about 0.5–1 at $Re > 100$ [68]. A review of correlations for estimating the axial and radial effective thermal conductivities is given in Ref. [69]. They present the common form of $\lambda_e/\lambda_f = A + BRe_P Pr^C$.

4 *Fluid and solid heat transfer to the wall.* If the heterogeneous model is two-dimensional (2D), a boundary condition of third kind for heat transfer appears at the wall of the packed bed, and again heat transfer coefficients for the fluid and solid phases are required. Effective parameters may lump these processes together as discussed in Section 3.3.

Concerning the fluid phase, most of the published correlations are of the form [69, 70]

$$\frac{\alpha_{wf} d_p}{\lambda^f} = Re^a Pr^b f\left(\frac{d_b}{d_p}\right). \tag{3.9}$$

The values of a, b, and $f(d_b/d_p)$ are fitted in several ranges of Re. Pereira Duarte et al. [61] estimated α_{wf} from Yagi and Wakao correlation [71] for mass transfer. According to this reference, $b = 1/3$, while $a = 0.5$ and $f = 0.6$ for $Re < 40$, but $a = 0.8$ and $f = 0.2$ for $Re > 40$. There is recognizably a lack of experimental data on the wall–solid transfer coefficient α_{ws}. The following correlation is sometimes used [61]: $\alpha_{ws} = 2.12\left(\lambda^s_{re}/d_p\right)$.

5 *Interphase mass and heat transfer coefficients k_g and h.* Describing the fluid–solid heat and mass transfer rigorously is a very complex task, due to the nonuniformity of conditions around each pellet in a flow field, which is uncertain, over a packed bed characterized by a certain degree of randomness. The process is usually envisioned as occurring through a symmetric boundary layer, whose thickness depends on the hydrodynamic conditions (it decreases as the fluid velocity past the catalytic surface increases). The commonly adopted mass (k_g) and heat (h) transfer coefficients are actually averaged values of the position-dependent quantities, over the pellet surface. Correlations for these quantities are available from experimental results under reactive and nonreactive conditions or from theoretical analyses. They are usually included in dimensionless parameters, such as Sherwood and Nusselt numbers or j-factor expressions, defined in the succeeding text:

$$j = \frac{k_g}{u} Sc^{2/3} = \frac{h}{\rho_g c_{pg} u} Pr^{2/3} = \frac{A}{Re_i^{\,B}}. \tag{3.10}$$

Carberry [72] compiled several experimental data on fixed-bed mass transfer and derived a boundary layer model with $A = 1.15$, $B = 1$, and $Re_i = \rho_g u_i d_p/\mu_g$. Wakao et al. [73] proposed the following correlation:

$$Nu = 2 + 1.1Re^{0.6}Pr^{1/3} \tag{3.11}$$

for $15 < Re < 8500$. A similar correlation for mass transfer was obtained. Other correlations are given in Satterfield [74], Gunn [57], and Welty et al. [75]. The latter reference reports the following correlations:

$$\frac{hd_p}{\lambda_g} = 2 + 0.6Re_p^{0.5}Pr^{1/3} \tag{3.12}$$

$$\frac{k_g d_p}{D_g} = 1.17Re_p^{0.585}Sc^{1/3}. \tag{3.13}$$

3.2.2 Mathematical model

The previous description of the processes considered in a fixed-bed reactor leads to a heterogeneous non-isothermal 2D model (symmetry with respect to the angular coordinate is assumed). The unsteady state mass and energy balances for a single reactant species with dimensional concentration $\hat{c}_i$ (subscript i is omitted for convenience) and appropriate boundary conditions are given in Table 3.1. The independent variables are normalized as follows (see Figure 3.2):

$$z = \frac{\hat{z}}{L},\; r = \frac{\hat{r} - \hat{r}_0}{R_b} = \frac{\hat{r}}{R_b} - r_0, \text{ and } t = \frac{\hat{t}}{t_P}. \tag{3.14}$$

The radius of the cross section of the downflow cylindrical bed is R_b, which is also the radial direction dimension of cylindrical or spherical beds. The reference radial distance may be $\hat{r}_0 = 0$ for an axial flow reactor or equal the inner radius of radial flow reactors (Figure 3.1-b). For axial flow spherical reactors (with radius R_{sph} and distance H_S between reactor's and bed's inlets), R_b is a function of the axial position [76] (Figure 3.1-c.1):

$$R_b = \sqrt{R_{sph}^2 - \left(R_{sph} - H_S - z\right)^2}. \tag{3.15}$$

The dependent variables in the fluid phase are made dimensionless by

$$c = \frac{\hat{c} - \hat{c}_{ref}}{\Delta\hat{c}} \text{ and } T = \frac{\hat{T} - \hat{T}_{ref}}{\Delta\hat{T}}. \tag{3.16}$$

The reference values of concentration and temperature ($\hat{c}_{ref}$ and $\hat{T}_{ref}$) may be chosen as the values at feed conditions. A scale for the reaction rate term evaluated at pellet surface conditions (a nonlinear function of both concentration and temperature, in general) is also introduced as follows:

$$R_j\left(c^s_{surf}, T^s_{surf}\right) = \frac{\hat{R}_j\left(\hat{c}^s_{surf}, \hat{T}^s_{surf}\right)}{R_{scale,j}}, \tag{3.17}$$

Table 3.1 Fixed-bed reactor mathematical model (definition of timescales τ in Table 3.2).

Two-dimensional heterogeneous nonisothermal packed bed reactor

1. Interparticular fluid phase

1.1. *Mass conservation*

$$\frac{\partial}{\partial z}\left(\frac{c}{\tau_{CA}}\right)-\frac{\partial}{\partial z}\left(\frac{1}{\tau_{DA}}\frac{\partial c}{\partial z}\right)+\frac{1}{(r_0+r)^S}\frac{\partial}{\partial r}\left[(r_0+r)^S\frac{c}{\tau_{CR}}\right]-\frac{1}{(r_0+r)^S}\frac{\partial}{\partial r}\left[\frac{(r_0+r)^S}{\tau_{DR}}\frac{\partial c}{\partial r}\right]+\frac{1}{\tau_{FA}}\frac{\partial c}{\partial t}=\frac{c_{surf}^s-c}{\tau_{MT}} \quad (3.22)$$

Feed at $\hat{z}=0$:

$$\left.\frac{\partial c}{\partial z}\right|_{z=0}=-\frac{\tau_{DA}}{\tau_{CA}}\left(c_{in}-c|_{z=0}\right) \quad (3.23a) \qquad \left.\frac{\partial c}{\partial z}\right|_{z=1}=0 \quad (3.23b)$$

$$\left.\frac{\partial c}{\partial r}\right|_{r=0}=0 \quad (3.23c) \qquad \left.\frac{\partial c}{\partial r}\right|_{r=1}=0 \quad (3.23d)$$

Feed at $\hat{r}=\hat{r}_0$:

$$\left.\frac{\partial c}{\partial r}\right|_{r=0}=-\frac{\tau_{DR}}{\tau_{CR}}\left(c_{in}-c|_{r=0}\right) \quad (3.24a) \qquad \left.\frac{\partial c}{\partial r}\right|_{r=1}=0 \quad (3.24b)$$

Feed at $\hat{r}=R_b$:

$$\left.\frac{\partial c}{\partial r}\right|_{r=0}=0 \quad (3.25a) \qquad \left.\frac{\partial c}{\partial r}\right|_{r=1}=\frac{\tau_{DR}}{\tau_{CR}}\left(c_{in}-c|_{r=1}\right) \quad (3.25b)$$

Initial condition: $c|_{t=0}=c^0$ (3.26)

1.2. *Energy conservation*

$$\frac{1}{\tau_{CA}}\frac{\partial T}{\partial z}-\frac{\partial}{\partial z}\left(\frac{1}{\tau_{HDA}^f}\frac{\partial T}{\partial z}\right)+\frac{1}{\tau_{CR}(r_0+r)^S}\frac{\partial}{\partial r}\left[(r_0+r)^S T\right]-\frac{1}{\tau_{HDR}^f(r_0+r)^S}\frac{\partial}{\partial r}\left[(r_0+r)^S\frac{\partial T}{\partial r}\right]+\frac{1}{\tau_{FA}}\frac{\partial T}{\partial t}=\frac{T_{surf}^s-T}{\tau_{HT}} \quad (3.27)$$

Feed at $\hat{z}=0$:

$$\left.\frac{\partial T}{\partial z}\right|_{z=0}=-\frac{\tau_{HDA}^f}{\tau_{CA}}\left(T_{in}-T|_{z=0}\right) \quad (3.28a) \qquad \left.\frac{\partial T}{\partial z}\right|_{z=1}=0 \quad (3.28b)$$

$$\left.\frac{\partial T}{\partial r}\right|_{r=0}=0 \quad (3.28c) \qquad \left.\frac{\partial T}{\partial r}\right|_{r=1}=-\frac{\tau_{HDR}^f}{\tau_{WF}}\left(T|_{r=1}-T_{wall}\right) \quad (3.28d)$$

Feed at $\hat{r}=0$:

$$\left.\frac{\partial T}{\partial r}\right|_{r=0}=\frac{\tau_{HDR}^f}{\tau_{CR}}\left(T_{in}-T|_{r=0}\right) \quad (3.29a) \qquad \left.\frac{\partial T}{\partial r}\right|_{r=1}=0 \quad (3.29b)$$

Feed at $\hat{r}=R_b$:

$$\left.\frac{\partial T}{\partial r}\right|_{r=0}=0 \quad (3.30a) \qquad \left.\frac{\partial T}{\partial r}\right|_{r=1}=\frac{\tau_{HDR}^f}{\tau_{CR}}\left(T_{in}-T|_{r=1}\right) \quad (3.30b)$$

Initial condition: $T|_{t=0}=T^0$ (3.31)

2. Solid (catalyst) phase

2.1. *Mass conservation*

$$\frac{c-c_{surf}^s}{\tau_{MT}}+\sum_{j=1}^{N_{RXN}}\frac{1}{\tau_{R,j}}\nu_{ij}R_j\left(c_{i,surf}^s,T_{surf}^s\right)\eta_j=\frac{1}{\tau_{MA}}\frac{d\langle c^s\rangle}{dt} \quad (3.32)$$

Average pellet concentration: $\langle c^s\rangle=(\sigma+1)\int_0^1\xi^\sigma c^s d\xi$ (3.33)

2.2. *Energy conservation*

$$\frac{T_{surf}^s-T}{\tau_{HT}}+\frac{1}{\tau_{HA}}\frac{\partial\langle T^s\rangle}{\partial t}=\sum_{j=1}^{N_{RXN}}\frac{R_j\left(c_{i,surf}^s,T_{surf}^s\right)\eta_j}{\tau_{HG,j}}+\frac{1}{(r_0+r)^S}\frac{\partial}{\partial r}\left[\frac{(r_0+r)^S}{\tau_{HDR}^s}\frac{\partial\langle T^s\rangle}{\partial r}\right]+\frac{\partial}{\partial z}\left[\frac{1}{\tau_{HDA}^s}\frac{\partial\langle T^s\rangle}{\partial z}\right] \quad (3.34)$$

Average pellet temperature: $\langle T^s\rangle=(\sigma+1)\int_0^1\xi^\sigma T^s d\xi$ (3.35)

Wall–solid heat transfer: $\left.\frac{\partial\langle T^s\rangle}{\partial r}\right|_{r=1}=-\frac{\tau_{HDR}^s}{\tau_{WS}}\kappa\left(\langle T^s\rangle|_{r=1}-T_{wall}\right)$ (3.36)

where $\kappa=\frac{\rho_b c_{ps}}{\varepsilon_b\rho_g c_{pg}}$ is the fluid–solid heat capacity ratio

$$\left.\frac{\partial\langle T^s\rangle}{\partial r}\right|_{r=0}=0 \quad (3.37)$$

3. Ergun's equation

$$-\frac{d\hat{P}}{d\hat{z}}=150\frac{(1-\varepsilon_b)^2\mu_g}{d_p^2\varepsilon_b^2}u_i+1.75\frac{(1-\varepsilon_b)\rho_g}{d_p\varepsilon_b}u_i^2 \quad (3.38)$$

where R_j is the dimensionless rate of reaction j, defined so that $\nu_{ij}R_j$ represents the relative rate of formation of species i by the same reaction.

The term for flux in the transverse direction is written with a shape factor S, given by

$$S = \frac{R_b S_b}{V_b} - 1. \tag{3.18}$$

For cylindrical coordinates, $S = 1$, while for spherical coordinates, $S = 2$ and the axial component z should be ignored. For the case of the microslit fixed bed, $S = 0$. Note that dimensional variables are capped and the solid phase concentration (c^s) and temperature (T^s) are normalized by the same scales as in Eq. (3.16). In the governing equations at the particle level (discussed in detail in Section 3.3), the dimensionless spatial coordinate is

$$\xi = \frac{\hat{\xi} - \hat{\xi}_0}{R_p} \tag{3.19}$$

where the reference position $\hat{\xi}_0$ is 0 for solid pellets and equals the radius of the hole for a hollow cylinder, for example. The shape factor for the catalyst geometry is defined by

$$\sigma = \frac{R_p S_p}{V_p} - 1. \tag{3.20}$$

The total external surface area of catalyst particles per volume of reactor is given by

$$a_V = \frac{S_{surf}}{V_b} = \frac{(1 - \varepsilon_b)(\sigma + 1)}{R_p}. \tag{3.21}$$

For a description of the remaining symbols, the reader is referred to section "Nomenclature" at the end of this chapter. Admitting that physical properties, transfer coefficients, porosity, velocity, etc. can be assumed to be constant and uniform, or without loss of generality, represented by the same notation for scaling purposes, the order of magnitude of each term in the model equations is represented by the associated timescale. We will take advantage of this formulation in our analysis (Section 3.3). Some common dimensionless parameters appear when ratios between these timescales are taken, as shown in the following sections.

3.2.3 Model reduction and selection

Even though the model in Table 3.1 results from several assumptions (detailed in Section 3.2.1), it can be considered as quite comprehensive. In fact, what is commonly found in the literature is a simplified version of this model. The well-known classification of fixed-bed reactor models by Froment [51] and Froment and Bischoff [62] clearly exemplifies how a more general model unfolds into a hierarchy of several others with decreasing complexity. The dimensionality of the model (usually one- or two-dimensional) and the presence of interphase and intraparticular resistances to mass/heat transfer are the main basis for distinguishing between different categories.

For particular cases, it may be required to add more complex phenomena with additional effects or more evolved descriptions of the same mechanisms. In general, however, reduced models are appropriate and desirable. Historically, this stemmed from the shorter computational effort and time required for the numerical solution of such models. Today this is also an advantage for optimization, control, and real-time simulation applications, and reliable simplified models are still used for almost all purposes due to the lower number of dimensionless parameters requiring estimation and to the success found in the description of experimental results. On the other hand, complex detailed models fulfill the most generic purpose of reactor simulation, which is related to the prediction of the actual behavior from fundamental, independently measured parameters. Therefore, it is important to understand the equivalence and agreement between both detailed and reduced models, so as to take advantage of their predictive power without unnecessary effort.

In the following sections, we revisit and systematize models presented in the literature with different degrees of complexity, using scaling and perturbation analysis. Assuming the dominance of certain mechanisms, it is possible to estimate the error in using simplified models and to find equivalent formulations.

3.3 Averaging over the catalyst particle

The first reduction of the two-scale model that we consider is already included in the equations from Table 3.1 and refers to the reaction–diffusion problem inside catalytic particles. The treatment of this question based on the *effectiveness factor* concept (η) is widely generalized in the literature, after the seminal works of Damköhler [77], Thiele [78], and Zeldovitch [79]. It may be defined for a reaction j with respect to the conditions prevailing at the pellet surface by Ref. [80]:

$$\eta_j = \frac{\iiint R_j\left(c_i^s, T^s\right) d\hat{V}}{R_j\left(c_{i,surf}^s, T_{surf}^s\right) V_p} = \frac{(\sigma + 1)}{R_j\left(c_{i,surf}^s, T_{surf}^s\right)} \int_0^1 \xi^\sigma R_j\left(c_i^s, T^s\right) d\xi. \tag{3.39}$$

Equation 3.39 expresses an averaged reaction rate evaluated with the actual concentration and temperature profiles that are observed inside the pellet, in the presence of reaction superposed with the eventual presence of significant limitations from diffusion through the catalyst pores. If transport to the catalyst surface is also limited, conditions on the interface may differ significantly from those in the bulk fluid. In this case, it may be more useful to refer the observed average rate of reactant consumption to the conditions in the interparticular bulk fluid, defining a global effectiveness factor $\bar{\eta}$, which is related to (3.39) by

$$\eta_j R_j\left(c_{i,surf}^s, T_{surf}^s\right) = \bar{\eta}_j R_j(c_i, T). \tag{3.40}$$

The ratio $R_j\left(c^s_{i,surf}, T^s_{surf}\right)/R_j(c_i, T)$ is known as the external effectiveness factor. Additionally, as can be seen in Equations 3.32 and 3.34, the averages of the solid concentration and temperature distributions ($\langle c^s\rangle$ and $\langle T^s\rangle$) may be also required, especially for nonlinear kinetics in transient conditions or with appreciable heat effects. The reason behind the use of these averaged quantities is to naturally circumvent the solution of the possibly highly nonlinear partial differential conservation equations at the pellet scale, which are coupled between them by the source term and with the fluid phase by continuity conditions at each point of the reactor. Introducing the effectiveness factor, it is possible to avoid the simultaneous calculation at both domains, with values of η obtained from interpolation between a series of tabulated or plotted numerical results. In some cases, exact or approximate analytical solutions for the effectiveness factor are available [62, 78, 80–86]. Since the main obstacle for an analytical solution resides in the nonlinear reaction term, additional simplification is usually found in two asymptotes, corresponding to the timescale for reaction being much larger or smaller than the one for diffusion through the pores (e.g., this has given origin to the appearance of the so-called Aris' numbers from Wijngaarden et al. [87]).

Regarding the geometry of the catalyst pellet, the most well-known shape normalization is based on the behavior at high reaction rates [80]. However, over the past decades, a 1D generalized cylinder model [88, 89] has been explored to describe a wide range of shapes of solid and hollow particles [90, 91] and also of multihole and multilobe pellets [92]. This approximate treatment considers mass and heat transfer along a single coordinate, and the deviation from the actual geometry (2D or 3D) is reduced by fitting of a shape parameter σ. This quantity can be related to the ratio of the pellet volume per surface area and per particle characteristic length, according to Equation 3.20. The choice of values for σ has been proposed to ensure agreement with the behavior of the actual geometry at low reaction rates [93], high reaction rates [94], or both [95].

In the case of nonlinear kinetics, the two asymptotes of (3.39) (for small and large reaction rates) can be combined empirically to correctly describe the effectiveness factor over the whole range of timescales for reaction and transport. Examples can be found in Gottifredi and Gonzo [82], Wijngaarden et al. [87], and Lopes et al. [96]. When the reaction is exothermic, the dependence of the effectiveness factor on the scale for the reaction rate is not monotonous. In fact, an intermediate region may exist where the degree of utilization of the catalyst volume increases with the rate of consumption of reactant [97, 98]. Moreover, other effects such as steady-state multiplicity can be observed for certain combinations of parameters [99]. These behaviors result from thermal feedback but could also appear in isothermal kinetics with significant inhibition effects (concentration feedback). Analytical attempts to describe these phenomena are rather limited, but the asymptotes of the effectiveness factor are still interesting [100]. For a first-order non-isothermal reaction occurring in a catalyst slab (with fixed surface values), Cardoso and Rodrigues [101] have derived perturbation solutions for the temperature and concentration profiles and for η.

The relevance of interphase gradients distinguishes between two different classes of problems, and this is reflected on the type of boundary condition at the pellet's surface. It is known that specifying the value of the concentration (or temperature) at the surface (Dirichlet boundary condition) may not be realistic, and thus finite external transfer effects have to be considered (in a Robin-type boundary condition) [72]. Apart from these, a large number of additional effects have also been considered. Some examples include the nonuniformity of the porous pellet structure (distribution of pore sizes [102], bidisperse particles [103], etc.), nonuniformity of catalytic activity [104], deactivation by poisoning [105], presence of multiple reactions [106], and incorporation of additional transport mechanisms such as Soret diffusion [107] or intraparticular convection [108].

As an illustration of the application of approximate analytical techniques, we derive the asymptotic forms of the effectiveness factor and average solid concentration/temperature for a non-isothermal reaction in a catalyst particle with external heat/mass gradients. Adopting the normalization in Equations 3.16–3.20 for a solid porous particle ($\hat{\xi}_0 = 0$) with characteristic dimension in the direction of diffusion R_p, the mass and heat balances become

$$\frac{D_{cat}}{R_p^2\xi^\sigma}\frac{\partial}{\partial\xi}\left(\xi^\sigma\frac{\partial c^s}{\partial\xi}\right) + \sum_{j=1}^{N_{RXN}}\frac{\rho_{cat}R_{scale,j}}{\Delta\hat{c}}\nu_{ij}R_j = \frac{\varepsilon_P}{t_P}\frac{\partial c^s}{\partial t} \quad (3.41a)$$

$$\frac{\lambda_{cat}}{R_p^2\xi^\sigma}\frac{\partial}{\partial\xi}\left(\xi^\sigma\frac{\partial T^s}{\partial\xi}\right) + \sum_{j=1}^{N_{RXN}}\frac{\rho_{cat}(-\Delta H_j)R_{scale,j}}{\Delta\hat{T}}R_j = \frac{\rho_{cat}c_{pcat}}{t_P}\frac{\partial T^s}{\partial t}, \quad (3.41b)$$

where in our case the reaction scale is referred to $\hat{T}_{ref}$ and $\Delta\hat{c}$ and for an mth-order reaction of the type $A \longrightarrow$ Products:

$$R = -R_A = \exp\left[\frac{\gamma_{ref}\left(\Delta\hat{T}/\hat{T}_{ref}\right)T^s}{1+\left(\Delta\hat{T}/\hat{T}_{ref}\right)T^s}\right]\left(c^s_A\right)^m \quad (3.41c)$$

(as in Table 3.1, we will omit the subscript A for the sake of simplicity).

These reference conditions may be the ones at the reactor's inlet conditions, and therefore, $\hat{c}_{ref} = 0$, $\Delta\hat{c} = \hat{c}_{in}$, and $\hat{T}_{ref} = \Delta\hat{T} = \hat{T}_{in}$ in Equation 3.16. There is symmetry at $\hat{\xi} = 0$ ($\partial c^s/\partial\xi = 0$ and $\partial T^s/\partial\xi = 0$) and flux continuity at the surface expressed by

$$\left.\frac{D_{cat}}{R_p}\frac{\partial c^s}{\partial\xi}\right|_{\xi=1} = k_g\left(c - c^s_{surf}\right) \quad (3.42a)$$

$$a_V\left.\frac{\lambda_{cat}}{R_p}\frac{\partial T^s}{\partial\xi}\right|_{\xi=1} = a_V h\left(T - T^s_{surf}\right) + \frac{\lambda^s_{re}}{R_b^2 r^S}\frac{\partial}{\partial r}\left(r^S\frac{\partial\langle T^s\rangle}{\partial r}\right) + \frac{\lambda^s_{ae}}{L^2}\frac{\partial^2\langle T^s\rangle}{\partial z^2}. \quad (3.42b)$$

We note the presence of the reactor scale in the heat transfer boundary condition (3.42b). The formulation in terms of the average solid temperature $\langle T^s \rangle$ is an approximation, which should be reasonable given that these terms are likely to be subdominant in Equation 3.42b. As in Table 3.2, it is possible to identify the timescales for the different processes in Equations 3.41 and 3.42. The axial conduction term in (3.42b) can be ignored for beds with $R_b << L$. Combination of these scales yields some well-known parameters for the internal reaction–diffusion problem, such as

$$\text{Thiele modulus:}\quad \phi_j^2 = \frac{R_p^2 \rho_{cat} R_{scale,j}}{\Delta \hat{c}\, D_{cat}} \tag{3.43a}$$

$$\text{Prater's parameter:}\quad \beta_j = \frac{(-\Delta H_j) D_{cat} \Delta \hat{c}}{\lambda_{cat} \Delta \hat{T}} \tag{3.43b}$$

$$\text{Arrhenius parameter:}\quad \gamma_{j,ref} = \frac{E_j}{R_G \hat{T}_{ref}} \tag{3.43c}$$

Table 3.2 Timescales of the different processes in the model for a fixed bed.

Process *X*	**Timescale, τ_X**	
Convection (axial)	τ_{CA}	$\frac{\varepsilon_b L}{u_Z}$
Convection (radial)	τ_{CR}	$\frac{\varepsilon_b R_b}{u_R}$
Mass dispersion (axial)	τ_{DA}	$\frac{L^2}{D_{ae}}$
Mass dispersion (radial)	τ_{DR}	$\frac{R_b^2}{D_{re}}$
Fluid heat dispersion (axial)	τ_{HDA}^f	$\frac{\varepsilon_b L^2 \rho_g c_{pg}}{\lambda_{ae}^f}$
Fluid heat dispersion (radial)	τ_{HDR}^f	$\frac{\varepsilon_b R_b^2 \rho_g c_{pg}}{\lambda_{re}^f}$
Solid heat dispersion (axial)	τ_{HDA}^s	$\frac{\varepsilon_b \rho_g c_{pg} L^2}{\lambda_{ae}^s}$
Solid heat dispersion (radial)	τ_{HDR}^s	$\frac{\varepsilon_b \rho_g c_{pg} R_b^2}{\lambda_{re}^s}$
Fluid–solid mass transfer	τ_{MT}	$\frac{\varepsilon_b}{k_g a_V}$
Fluid–solid heat transfer	τ_{HT}	$\frac{\varepsilon_b \rho_g c_{pg}}{h a_V}$
Fluid accumulation	τ_{FA}	t_P
Solid mass accumulation	τ_{MA}	$\frac{\varepsilon_b t_P}{(1-\varepsilon_b)\varepsilon_p}$
Solid heat accumulation	τ_{HA}	$\frac{\varepsilon_b \rho_g c_{pg}}{\rho_b c_{ps}} t_P$
Reactant consumption	τ_R	$\frac{\varepsilon_b \Delta \hat{c}}{\rho_b R_{scale}}$
Heat generation by reaction	τ_{HG}	$\frac{\varepsilon_b \rho_g c_{pg} \Delta \hat{T}}{\rho_b (-\Delta H) R_{scale}}$
Wall–fluid heat transfer	τ_{WF}	$\frac{\varepsilon_b \rho_g c_{pg} R_b}{\alpha_{wf}}$
Wall–solid heat transfer	τ_{WS}	$\frac{\rho_b c_{ps} R_b}{\alpha_{ws}}$

$$\text{Lewis number:}\quad Le_{cat} = \frac{\lambda_{cat} \varepsilon_P}{\rho_{cat} c_{pcat} D_{cat}} \tag{3.43d}$$

$$\text{Mass Biot number:}\quad Bi_m = \frac{R_p k_g}{D_{cat}} \tag{3.43e}$$

$$\text{Heat Biot number:}\quad Bi_h = \frac{R_p h}{\lambda_{cat}}. \tag{3.43f}$$

Another parameter that also appears from the scaling of Equation 3.42b is

$$\Lambda = \frac{R_p}{a_V R_b^2} \frac{\lambda_{re}^s}{\lambda_{cat}}. \tag{3.43g}$$

Equation 3.43g compares the timescale for radial heat dispersion in the solid phase with the one for internal heat conduction. For catalysts with good heat conduction properties and low particle-to-bed diameter ratios, $\Lambda \ll 1$. In this case, the surface boundary condition is homogeneous and of Robin type, as given by the first terms on each side of (3.42b). A similar dimensionless number related with dispersion in the axial direction also appears, but its magnitude is considered much smaller than that of the other parameters in Equation 3.43, due to the geometrical reasons explained earlier. Note that Equations 3.32 and 3.34 are obtained by integrating Equation 3.41 with respect to ξ over the pellet domain and using Equation 3.42 as boundary conditions.

In non-isothermal reaction–diffusion systems described by Equation 3.41, it is possible to find a single conserved scalar for any geometry, simply by combining the steady-state mass and energy balances so that the source term (which may have an arbitrary form) is eliminated. This results in Damköhler's equation [109] in terms of Prater's parameter [110], from which the maximum temperature observable in a given system can be estimated (corresponding to concentration annulment inside the pellet). Even though the surface concentration and temperature are generally unknown, this is also valid here:

$$T^s = T_{surf}^s + \beta\left(c_{surf}^s - c^s\right) \tag{3.44a}$$

$$T_{max}^s = T_{surf}^s + \beta c_{surf}^s. \tag{3.44b}$$

It is also possible to obtain a relationship between bulk and surface conditions:

$$T = T_{surf}^s - \beta \frac{Bi_m}{Bi_h}\left(c - c_{surf}^s\right) - \frac{\Lambda}{Bi_h} \nabla^2 \langle T^s \rangle. \tag{3.45}$$

In terms of observable quantities and incorporating the effectiveness factor concept (note that $R_{bulk} = R(c, T)$), Equation 3.44b becomes

$$T_{max}^s = T + \beta c + \beta \frac{\bar{\eta}\phi^2}{Bi_m} \frac{R_{bulk}}{1+\sigma}\left(\frac{Bi_m}{Bi_h} - 1\right) + \frac{\Lambda}{Bi_h} \nabla^2 \langle T^s \rangle. \tag{3.46a}$$

Comparing the external ($\Delta T_{ext} = T^s_{surf} - T$) and the internal ($\Delta T^s_{max} = T^s_{max} - T^s_{surf}$) temperature rise [111],

$$\frac{\Delta T^s_{max}}{\Delta T_{ext}} = \frac{c(1-Ca)}{\frac{Bi_m}{Bi_h} cCa + \frac{\Lambda}{\beta Bi_h}\nabla^2\langle T^s\rangle}, \tag{3.46b}$$

where the Carberry number [111] is an observable quantity, which can be defined as

$$Ca = \frac{\bar{\eta}\phi^2}{Bi_m} \frac{R_{bulk}}{(1+\sigma)c} = \frac{\rho_{cat}\hat{R}_{bulk,obs}}{\hat{c}} \frac{V_P}{S_P k_g}. \tag{3.47}$$

It is also common to relate η with $\bar{\eta}$. For a non-isothermal mth-order reaction ($\hat{R} = k\,\hat{c}^m$), from (3.40),

$$\left(\frac{1}{\bar{\eta}\rho_{cat}k_{bulk}}\right)^{1/m} = \left(\frac{1}{\eta\rho_{cat}k_{surf}}\right)^{1/m} + \frac{V_P/S_P}{k_g}(\bar{\eta}\rho_{cat}k_{bulk}\hat{c}^m)^{\frac{m-1}{m}} \tag{3.48}$$

The global resistance given by Equation 3.48 can be interpreted as a sum of the series of resistances for external mass transfer and internal diffusion coupled with reaction (each term has units of $mol^{(m-1)/m}(m^3_{cat}s)^{1/m}/m^3_f$).

We will now find expressions for the effectiveness factor, with surface and average temperatures given in terms of the conditions in the bulk fluid. Considering internal thermal effects, approximate results are possible in the chemical and diffusional regimes.

3.3.1 Chemical regime

For slow reactions, mass and heat balances can be simplified in the limit of *small Thiele modulus* ($\phi^2 \ll 1$). At steady-state conditions, these profiles are flat inside the particles, due to the comparably fast transport. Solutions at the isothermal conditions are available elsewhere [80]. This means that as a first approximation,

$$c^s = c^s_{surf} + O(\phi^2) = c + O(\phi^2) \quad \text{and} \quad T^s = T^s_{surf} + O(\phi^2). \tag{3.49}$$

Equation 3.42b then becomes

$$\frac{1}{r^S}\frac{\partial}{\partial r}\left(r^S\frac{\partial T^s}{\partial r}\right) + \frac{Bi_h}{\Lambda}(T - T^s) = 0. \tag{3.50}$$

Equation 3.50 reproduces the solid-phase energy balance in Equation 3.34 in the absence of a chemical reaction (with negligible axial dispersion at steady-state). For large Bi_h/Λ, $T^s \sim T$ and this should be valid at least in the bed's core and everywhere if Equation 3.52 is verified. In general, close to the wall (in a layer with thickness in the order of $\sqrt{\Lambda/Bi_h}$), significant temperature gradients in the solid phase are observed when

$$\frac{\alpha_{ws}R_p}{\lambda^s_{re}}\sqrt{\frac{\Lambda}{Bi_h}} \sim 1 \text{ or } >> 1. \tag{3.51}$$

In these conditions, the profiles in each phase are related by Equation 3.50. Temperature in both phases is identical only if a nearly linear profile for T^s can be assumed near the wall ($\partial^2 T^s/\partial r^2 = 0$). This was found to be a reasonable approximation by Froment [51], which compared several boundary conditions for the wall–solid heat transfer in a 2D heterogeneous model for a fixed-bed reactor. This leads to the following equivalence between parameters:

$$\frac{\alpha_{ws}}{\lambda^s_{re}} = \frac{\alpha_{wf}}{\lambda^f_{re}} = \frac{\alpha_w}{\lambda_{re}}, \tag{3.52}$$

where α_w and λ_{re} refer to a model with lumped parameters [59].

The effects of a slow heat-generating chemical reaction with $\phi^2 \sim \Lambda << 1$ are visible as higher order corrections in $c^s = c + c^s_1\phi^2 + c^s_2\phi^4 + O(\phi^6)$ and $T^s = T + T^s_1\phi^2 + T^s_2\phi^4 + O(\phi^6)$. The particular forms of c^s_n and T^s_n can be obtained by collection of the $O(\phi^{2n})$ terms in Equations 3.41 and 3.42. Retaining the most important terms, the effectiveness factor and the averaged internal profiles become

$$\bar{\eta} = 1 - \left[\left(1 + \frac{3+\sigma}{Bi_m}\right)D_c - \left(1 + \frac{3+\sigma}{Bi_h}\right)\beta D_T\right] \frac{\phi^2}{(1+\sigma)(3+\sigma)} + \frac{\Lambda}{Bi_h}\frac{D_T}{R_{bulk}}\nabla^2 T + O(\phi^4) \tag{3.53a}$$

$$\langle c^s\rangle = c - \frac{3+\sigma+Bi_m}{(3+\sigma)(1+\sigma)}\frac{\phi^2}{Bi_m}\left(R_{bulk} + \frac{\Lambda}{Bi_h}\nabla^2 T D_T\right) + O(\phi^4) \tag{3.53b}$$

$$\langle T^s\rangle = T + \frac{3+\sigma+Bi_h}{(3+\sigma)(1+\sigma)}\frac{\beta\phi^2}{Bi_h}\left(R_{bulk} + \frac{\Lambda}{Bi_h}\nabla^2 T D_T\right) + \frac{\Lambda}{Bi_h}\nabla^2 T + O(\phi^4), \tag{3.53c}$$

where $R_{bulk} = R(c,T)$, $D_c = \partial R_{bulk}/\partial c$, and $D_T = \partial R_{bulk}/\partial T$. Moreover,

$$c^s_{surf} = c - \frac{\phi^2}{(1+\sigma)Bi_m}\left(R_{bulk} + \frac{\Lambda}{Bi_h}\nabla^2 T D_T\right) + O(\phi^4) \tag{3.53d}$$

$$T^s_{surf} = T + \frac{\beta\phi^2}{(1+\sigma)Bi_h}\left(R_{bulk} + \frac{\Lambda}{Bi_h}\nabla^2 T D_T\right) + \frac{\Lambda\nabla^2 T}{Bi_h} + O(\phi^4). \tag{3.53e}$$

Equation 3.53 reduces to several asymptotic results presented in the literature [80, 101], as they represent a shape and kinetic normalization at low reaction rates with nonzero interphase resistances. The application of these estimates for the effectiveness factor is shown in Figure 3.3 for negligible external mass transfer resistance (Dirichlet problem) and in Figure 3.4 for finite interphase resistance.

3.3.2 Diffusional regime

When the *reaction is fast* enough to cause abrupt variations of concentration and temperature in a boundary layer near the catalyst surface, several methods can be used to derive the form

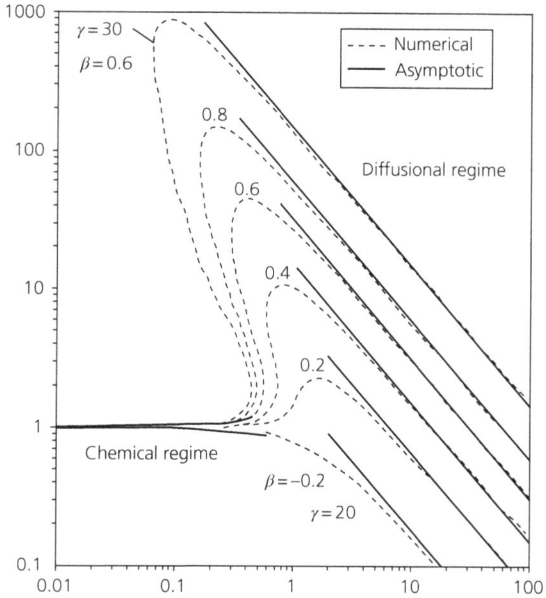

Figure 3.3 Effectiveness factor curve (η vs. ϕ) for a spherical catalyst particle with Dirichlet boundary condition and first-order exothermic reaction. Numerical results from Ref. [99]. A curve for $\gamma = 30$ is depicted (with $\beta = 0.6$), while the others were calculated for $\gamma = 20$ and several values of β. (Source: Weisz [99]. Reproduced with permission of Elsevier.)

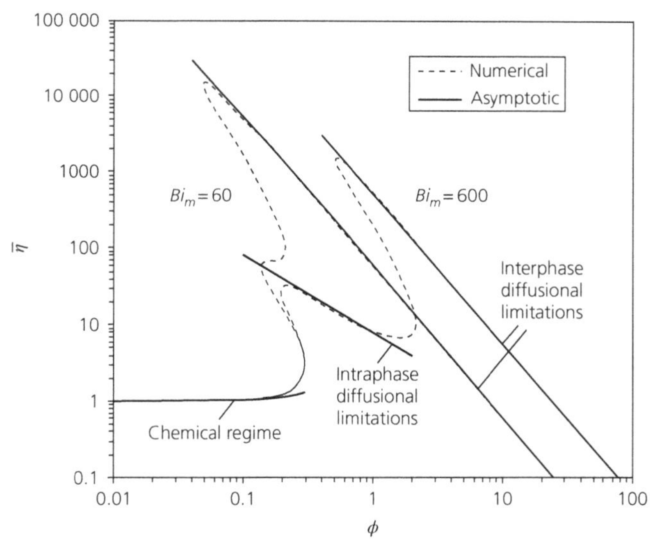

Figure 3.4 Asymptotic forms of the effectiveness factor ($\bar{\eta}$) as a function of the Thiele modulus (ϕ) for a slab catalyst with a first-order exothermic reaction ($\beta = 1/3$ and $\gamma = 27$) for $Bi_m/Bi_h = 6$. Numerical values in the intermediate region are given in Ref. [80].
(Source: Aris [80]. Reproduced with permission of Oxford University Press.)

of the effectiveness factor. Tavera [100] considered this limiting case (high reaction rates) and proposed an exact analytical solution for the non-isothermal effectiveness factor in a catalyst slab with known surface conditions. An approximation for low values of β, which are commonly observed in practice, was also presented. Previous work on this subject was also provided by Petersen [112]. More recently, an integral method was applied to a non-isothermal slab-shaped particle with Dirichlet boundary condition and zero [108] or first-order [101] kinetics. From these studies, it is possible to observe that the effectiveness factor can be written as

$$\eta = \frac{1+\sigma}{\phi^2 R\left(c^s_{surf}, T^s_{surf}\right)} \left.\frac{\partial c^s}{\partial \xi}\right|_{\xi=1} \simeq \frac{(1+\sigma)b}{\phi R\left(c^s_{surf}, T^s_{surf}\right)} + O(\phi^{-2}), \tag{3.54a}$$

$$\text{where } b = \sqrt{2\int_0^{c^s_{surf}} R(c^s)dc^s}. \tag{3.54b}$$

These expressions reflect a shape and a kinetic normalization in the diffusional limit [80, 81, 113]. The former is translated by the factor written as $1/(1+\sigma)$, which according to Equation 3.20, replaces the length scale in the Thiele modulus (R_p) by V_p/S_p. The latter is given by the parameter b in (3.54b), accounting for kinetic nonlinearities in concentration (for reactions which are not first-order) and temperature (a dependence of Arrhenius type is usually adopted for the nonisothermal cases). For an isothermal first-order reaction, $b = 1$. Note that using Equation 3.44a, the dimensionless reaction rate being integrated in the definition of b can be written in terms of c^s only. In some cases, it is possible to evaluate this integral analytically.

Using the conditions at the pellet surface as reference ($c^s_{surf} = 1$), the following result is obtained for isothermal kinetics of the form $R(c^s) = \frac{(1+k')^p (c^s)^m}{(1+k' c^s)^p}$:

$$b = \sqrt{2(1+k')^p \frac{\Gamma(m+1)}{\Gamma(m+2)}\, {}_2F_1(1+m, p, 2+m, -k')}. \tag{3.55a}$$

For negligible inhibition, the result for power-law kinetics is given by

$$b = \sqrt{\frac{2}{m+1}}. \tag{3.55b}$$

For non-isothermal kinetics, we adapt the result derived by Tavera [100] originally for Dirichlet conditions, and as expected b will be a function of kinetic descriptors (e.g., order of reaction m), nonisothermal parameters (γ and β), and surface conditions (c^s_{surf} and T^s_{surf}). The dependence on the latter quantities (generally unknown) leads to an implicit calculation of the effectiveness factor. Substituting Equation 3.42 into 3.54a,

$$c^s_{surf} = c - \frac{\bar{\eta} R_{bulk}}{(1+\sigma)} \frac{\phi^2}{Bi_m} = c - \frac{b\phi}{Bi_m} \tag{3.56a}$$

$$T^s_{surf} = T + \frac{\bar{\eta} R_{bulk}}{(1+\sigma)} \frac{\beta\phi^2}{Bi_h} = T + \beta \frac{Bi_m}{Bi_h} \frac{b\phi c}{Bi_m}, \tag{3.56b}$$

where $R_{bulk} = R(c,T)$ and $\bar{\eta}$ is the global effectiveness factor referred to bulk conditions. For the purpose of this discussion, the reactor-scale heat dispersion terms in Equation 3.42b were ignored. When external heat and mass transfer resistances can be ignored, the solution of the Dirichlet problem according to Equation 3.54 is $\eta \simeq (1+\sigma)\, b/\phi$, and this is plotted in Figure 3.3 for several values of γ and β. A simpler estimate was proposed by Carberry [72], which showed that the non-isothermal intraphase effectiveness factor for $\phi > 3$ could be well described by

$$\eta = \frac{\exp(\gamma\beta/5)}{\phi}. \tag{3.57}$$

The same asymptote is shown in Figure 3.4 for the region mainly limited by intraphase diffusion. The regime of strong interphase mass transfer resistance is also depicted in the same representation for two values of the mass Biot number. The asymptotic behavior of the effectiveness factor in this limit ($\phi/Bi_m \gg 1$ and low c^s_{surf}) is obtained from (power-law nonisothermal kinetics with conditions in the bulk taken as reference)

$$\begin{aligned} b &\simeq \sqrt{\frac{2}{m+1}}\exp\left[\frac{\gamma T^s_{max}}{2\left(1+T^s_{max}\right)}\right]\left(c^s_{surf}\right)^{\frac{m+1}{2}} \\ &\simeq \sqrt{\frac{2}{m+1}}\exp\left[\frac{\gamma T^s_{max}}{2\left(1+T^s_{max}\right)}\right]\left(1-\frac{b\phi}{Bi_m}\right)^{\frac{m+1}{2}}. \end{aligned} \tag{3.58a}$$

In order to observe low concentration values at the pellet surface, a first estimate for b is

$$b \simeq \frac{Bi_m}{\phi}. \tag{3.58b}$$

Replacing this result in the nondominant terms of Equation 3.58, the following improved expression is obtained:

$$b \simeq \frac{Bi_m}{\phi} - \left(\frac{Bi_m}{\phi}\right)^{\frac{m+3}{m+1}}\left(\frac{m+1}{2}\right)^{\frac{1}{m+1}}\exp\left[\frac{-\gamma T^s_{max}}{(m+1)\left(1+T^s_{max}\right)}\right], \tag{3.58c}$$

where

$$T^s_{max} = \beta + \frac{\beta b\phi}{Bi_m}\left(\frac{Bi_m}{Bi_h} - 1\right) \tag{3.58d}$$

is the maximum temperature inside the catalyst, changing from Prater's limit (for $\phi/Bi_m \to 0$) to $T^s_{max} = \beta Bi_m/Bi_h$, as ϕ/Bi_m increases. We will not extend our analysis to the more complex matter of steady-state multiplicity (numerical results are presented in Figure 3.3, as well as in Figure 3.4). Suitable descriptions of this topic can be found elsewhere [80, 114, 115]. Briefly, the maximum number of steady states and the extent of the multiplicity region depend on the kinetics, catalyst geometry, and boundary conditions. For a first-order reaction, the following criteria have been derived [116]:

$$\gamma\beta < 4\left(\frac{Bi_h}{Bi_m} + \beta\right) \quad \text{(uniqueness of the steady state)} \tag{3.59a}$$

$$\gamma\beta > 8\left(\frac{Bi_h}{Bi_m} + \beta\right) \quad \text{(existence of multiple steady states).} \tag{3.59b}$$

3.4 Dominant fluid–solid mass transfer

The ratio of timescales for internal diffusion within the catalyst pellet and for external mass transfer is of the order of the mass Biot number given in (3.43e). Since for many practical conditions Bi_m is reasonably high ($10^2 - 10^5$ as estimated in Ref. [72]), it is expected that the external mass transfer dominates over the internal one. If we extend this consideration to the reactor scale, then a very useful *pseudo-homogeneous model* [51] is obtained. The scaling condition that expresses fast fluid–solid mass transfer compared with other processes (X) in Equation 3.22 is

$$\frac{\tau_{MT}}{\tau_X} \ll 1. \tag{3.60}$$

We may take as a small parameter

$$\varepsilon = \frac{\tau_{MT}}{\tau_{CA}}, \tag{3.61}$$

where the axial convection timescale was chosen for comparison. This is appropriate for, but not restricted to, convective dominated processes since in this case all terms will be correctly scaled. The relationship of other timescales (except τ_{MT}) with τ_{CA} remains $\tau_{CA}/\tau_X \sim O(1)$.

The fluid mass balance with uniform physical properties becomes

$$\begin{aligned} c = c^s_{surf} &- \varepsilon\frac{\partial c}{\partial z} + \varepsilon\frac{\tau_{CA}}{\tau_{DA}}\frac{\partial^2 c}{\partial z^2} - \varepsilon\frac{\tau_{CA}}{\tau_{CR}}\frac{1}{(r_0+r)^S}\frac{\partial}{\partial r}\left[(r_0+r)^S c\right] \\ &+ \varepsilon\frac{\tau_{CA}}{\tau_{DR}}\frac{1}{(r_0+r)^S}\frac{\partial}{\partial r}\left[(r_0+r)^S\frac{\partial c}{\partial r}\right] - \varepsilon\frac{\tau_{CA}}{\tau_{FA}}\frac{\partial c}{\partial t}. \end{aligned} \tag{3.62}$$

A perturbation solution for concentration (in the fluid and at the particle surface) can be written as

$$c = \sum_{n=0}^{\infty} c_n \varepsilon^n \text{ and } c^s_{surf} = \sum_{n=0}^{\infty} c^s_{surf,n}\,\varepsilon^n. \tag{3.63}$$

Series solutions of the same type can be written for the average concentration. The leading-order problem that can be collected from (3.62) appears at $O(1)$ and simply expresses the absence of concentration gradients between the bulk fluid and the particle surface:

$$c_0 = c^s_{surf,0}. \tag{3.64}$$

Note that the scaling analysis hasn't involved the reaction timescale so far. This will be required to obtain a consistent balance at $O(1)$ from the steady state solid mass balance. Equation 3.32 can be written as

$$c = c^s_{surf} + \varepsilon\frac{\tau_{CA}}{\tau_{MA}}\frac{\partial\langle c^s\rangle}{\partial t} - \varepsilon\sum_{j=1}^{N_{RXN}}\frac{\tau_{CA}}{\tau_{Rj}}\nu_{ij}R_j\left(c^s_{i,surf}, T^s_{surf}\right)\eta_j. \tag{3.65}$$

At steady state and in the presence of reaction, the leading-order result is only consistent if

$$\frac{\tau_{MT}}{\tau_{Rj}} \ll 1. \tag{3.66}$$

For multiple reactions, a conservative approach would consist in the choice of the smallest τ_{Rj} (fastest kinetics). If this reaction scale fulfills Equation 3.66, then the whole system should satisfy the same requirement. In this case, the $O(1)$ result that could be extracted from Equation 3.65 reproduces Equation 3.64. As shown later, the so-called *pseudo-homogeneous model* [51] is only obtained if Equation 3.66 is also verified. A feature of this development is that the solution for c_0 is obtained at $O(\varepsilon)$, since no derivatives are retained in Equation 3.62 when $\varepsilon = 0$. At this higher order, the results from Equations 3.62 to 3.65 can be combined to yield

$$\frac{\tau_{CA}}{\tau_{DR}}\frac{1}{(r_0+r)^S}\frac{\partial}{\partial r}\left[(r_0+r)^S\frac{\partial c_0}{\partial r}\right] + \frac{\tau_{CA}}{\tau_{DA}}\frac{\partial^2 c_0}{\partial z^2} + \sum_{j=1}^{N_{RXN}}\frac{\tau_{CA}}{\tau_{Rj}}\nu_{ij}R_j\left(c_0, T^s_{surf,0}\right)\eta_j$$
$$= \frac{\partial c_0}{\partial z} + \frac{\tau_{CA}}{\tau_{CR}}\frac{1}{(r_0+r)^S}\frac{\partial}{\partial r}\left[(r_0+r)^S c_0\right] + \frac{\partial}{\partial t}\left(\frac{\tau_{CA}}{\tau_{FA}}c_0 + \frac{\tau_{CA}}{\tau_{MA}}\langle c^s_0\rangle\right) \tag{3.67}$$

For now we are not making any considerations regarding the fluid–solid heat transfer rate, and therefore we leave the reaction rate as a function of $T^s_{surf,0}$ (calculated from Eqs. 3.27 and 3.34 with c_0). In its current form, Equation 3.67 must be solved numerically with the energy balances in the two phases and the appropriate boundary conditions from Table 3.1. For our purposes of estimating the deviation from the pseudo-homogeneous model, the magnitude of the $O(\varepsilon)$ contribution is of interest. The relative error associated with this model is estimated by

$$\delta = \frac{c - c_0}{c} = \frac{\varepsilon c_1}{c_0 + \varepsilon c_1} + O\left(\varepsilon^2\right) \simeq \frac{\varepsilon c_1}{c_0} + O\left(\varepsilon^2\right) \tag{3.68}$$

The right-hand side of (3.68) assumed that the fluid concentration could be accurately described by (3.63). This should be reasonable for low values of δ, which is anyway the limit of interest for a regime boundary (small ε). As before, information on c_n is obtained at $O(\varepsilon^{n+1})$. Thus, c_1 is calculated from the problem at $O(\varepsilon^2)$, where the transfer term $\left(c_2 - c^s_{surf,2}\right)$ is eliminated from Equations 3.62 and 3.65. The following equation for c_1 is obtained:

$$\frac{\partial c_1}{\partial z} + \frac{\tau_{CA}}{\tau_{CR}}\frac{1}{(r_0+r)^S}\frac{\partial}{\partial r}\left[(r_0+r)^S c_1\right] + \frac{\partial}{\partial t}\left(\frac{\tau_{CA}}{\tau_{FA}}c_1 - \frac{\tau_{CA}}{\tau_{MA}}\langle c^s_1\rangle\right)$$
$$= \sum_{j=1}^{N_{RXN}}\frac{\tau_{CA}}{\tau_{Rj}}\nu_{ij}\partial\left.\frac{(R_j\eta_j)}{\partial c^s_{surf}}\right|_{\varepsilon=0} c^s_{surf,1} + \frac{\tau_{CA}}{\tau_{DR}}\frac{1}{(r_0+r)^S}\frac{\partial}{\partial r}\left[(r_0+r)^S\frac{\partial c_1}{\partial r}\right] + \frac{\tau_{CA}}{\tau_{DA}}\frac{\partial^2 c_1}{\partial z^2}. \tag{3.69}$$

Note that the reaction term $\eta_j R_j\left(c^s_{surf}, T^s_{surf}\right)$, where $c^s_{surf}(\varepsilon)$ is given by (3.63), was expanded in a Taylor series for small ε. Elimination of $c^s_{surf,1}$ occurs by substitution according to the $O(\varepsilon)$ subproblem from (3.65):

$$c^s_{surf,1} = c_1 - \frac{\tau_{CA}}{\tau_{MA}}\frac{\partial\langle c^s_0\rangle}{\partial t} + \sum_{j=1}^{N_{RXN}}\frac{\tau_{CA}}{\tau_{Rj}}\nu_{ij}\left.\left(\eta_j R_j\right)\right|_{c^s_{surf,0},\,T^s_{surf,0}}. \tag{3.70}$$

In general, Equation 3.69 can be solved numerically for c_1. However, for the purpose of criteria definition from this subdominant (relatively negligible) correction, an approximate solution of Equation 3.69 is enough. Note that the non-isothermal model is completed with Equations 3.27, 3.34 and respective boundary conditions. In the reaction term in the solid phase energy balance, c^s_{surf} should be replaced by c_0 to yield $T^s_{surf,0}$ and T_0. This arises from the fact that Equation 3.60 is also considered valid for the energy balance processes, inclusive $\tau_{MT}/\tau_{HT} \sim O(\varepsilon)$ (see Section 3.5). The result is a somewhat hybrid model in the sense that a pseudo-homogeneous description for mass transfer coexists with a heterogeneous one for temperature (see Section 3.4.2). Particular cases of this model are discussed in the succeeding text. The adequacy of pseudo-homogeneous descriptions in fixed-bed reactor modeling has been discussed in the literature [61, 117–119].

3.4.1 Isothermal axial flow bed

If the fixed bed can be considered to operate under isothermal conditions (as in the case of kinetic measurements [120], some laboratory-scale fixed beds, and packed microreactors [121]), the previously described mass balances are the only equations required. In this case, $T^s_{surf} = T$ and the particle is also considered internally isothermal. Specifically, we consider an axial flow fixed-bed reactor dominated by convection (i.e., with negligible axial dispersion for $L \gg R_b$), operating at the steady state. In the case of a single reaction, Equation 3.69 for the reactant concentration simplifies to

$$\frac{\partial c_1}{\partial z} + \omega_1 c_1 + \omega_0 = \frac{\tau_{CA}}{\tau_{DR}}\frac{1}{r^S}\frac{\partial}{\partial r}\left(r^S\frac{\partial c_1}{\partial r}\right) \tag{3.71a}$$

where

$$\omega_0 = -\left(\frac{\tau_{CA}}{\tau_R}\right)^2\left.\frac{\partial(\eta R)_{surf}}{\partial c^s_{surf}}\right|_{\varepsilon=0}(\eta R)|_{c_0,T} \tag{3.71b}$$

$$\omega_1 = \frac{\tau_{CA}}{\tau_R}\left.\frac{\partial(\eta R)_{surf}}{\partial c^s_{surf}}\right|_{\varepsilon=0}. \tag{3.71c}$$

Equation 3.71 must be solved with the following boundary conditions:

$$c_1\big|_{z=0} = 0 \tag{3.72a}$$

$$\left.\frac{\partial c_1}{\partial r}\right|_{r=0} = \left.\frac{\partial c_1}{\partial r}\right|_{r=1} = 0. \tag{3.72b}$$

For *linear kinetics* ($R(c) = c$), ϕ and consequently η are independent of c^s_{surf}; hence

$$\omega_0 = -\left(\eta\frac{\tau_{CA}}{\tau_R}\right)^2 c_0 \quad \text{and} \quad \omega_1 = \eta\frac{\tau_{CA}}{\tau_R}.$$

An estimate for c_0 is required (only for purposes of calculating a small correction), and it can be obtained from averaging Equation 3.67:

$$c_0 \simeq \langle c_0\rangle = \exp\left(-\frac{\tau_{CA}}{\tau_R}\eta z\right). \tag{3.73}$$

The perturbation series for the average fluid concentration over the reactor cross section $\langle c\rangle$ is obtained by averaging Equation 3.63:

$$\langle c\rangle = \sum_{n=0}^{\infty} \langle c_n\rangle\, \varepsilon^n. \tag{3.74}$$

Thus, the solution of Equations 3.71 and 3.72 for the cross-sectional averaged contribution to the concentration profile is given by

$$\langle c_1\rangle = \left(\eta\frac{\tau_{CA}}{\tau_R}\right)^2 \exp\left(-\frac{\tau_{CA}}{\tau_R}\eta z\right) z. \tag{3.75}$$

These results can be substituted into (3.68) to yield the following criterion for the validity of a pseudo-homogeneous mass transfer model:

$$\delta \simeq \frac{\varepsilon\langle c_1\rangle}{\langle c_0\rangle + \varepsilon\langle c_1\rangle} = \frac{\varepsilon(\eta Da)^2 z}{1 + \varepsilon(\eta Da)^2 z} \tag{3.76}$$

where the Damköhler number is here defined as $Da = \tau_{CA}/\tau_R$. Low values of δ are obtained for small Da, and in these conditions, $\delta \simeq \varepsilon(\eta Da)^2 z$. At an axial global scale ($z \sim 1$), δ equals the following timescale ratio: $\tau_{MT}\tau_{CA}/\tau^2_{R,obs}$ ($\tau_{R,obs}$ includes internal limitations and τ_{CA} is associated with the mechanism that dominates transport in the calculation of the correction). When $\tau_{MT}\tau_{CA}/\tau^2_{R,obs} \le \delta$, the simplified model has an associated relative error of δ or less. Note that in the special case of a first-order reaction, a 1D solution is given by

$$\langle c\rangle = \exp\left(-\frac{Da\eta z}{1 + \varepsilon Da\eta}\right) \simeq \langle c_0\rangle + \varepsilon\langle c_1\rangle + O(\varepsilon^2). \tag{3.77}$$

Expanding this result for small ε, consistency is obtained with the perturbation series derived earlier. In general, one might say that

$$\delta = \varepsilon(\eta Da)^2 \frac{z e^{-Da\eta z}}{1 - X_{obs}}, \tag{3.78}$$

where X_{obs} is the reactant conversion that is observed experimentally or calculated from a more complex model for an axial flow fixed bed. This represents a limit to the reactor's conversion represented by this model: $X_{obs} < 1 - \varepsilon(\eta Da)^2 e^{-Da\eta}/\delta$. The correction to the pseudo-homogeneous model is positive ($\langle c_1\rangle > 0$), which shows how the presence of isothermal interphase resistance decreases the reactor's conversion ($X_{PH} = 1 - \langle c_0\rangle$ is an upper bound).

Even though axial diffusion effects are not considered (making the analysis of the inlet region as $z \to 0$ incomplete), it is possible to obtain an estimate for the axial distance below which the pseudo-homogeneous approach is valid for a given δ, from (3.76):

$$z < \frac{\delta}{\varepsilon(\eta Da)^2}\frac{1}{1-\delta}. \tag{3.79}$$

According to Equation 3.79, the validity of Equation 3.67 increases as the inlet is approached (the relative error decreases linearly). The full model in Table 3.1 and its pseudo-homogeneous version described in Section 3.4 were solved numerically using gPROMS® [122]. The numerically determined cross-sectional concentration averages from both cases allowed the calculation of δ, and this is compared with Equation 3.76 in Figure 3.5.

When the reaction rate can be fitted by an expression of *power-law* type and the catalyst particle is free from internal diffusional limitations: $R_{surf} = c^m_{surf}$ and $\eta = 1$. Then, according to Equation 3.71,

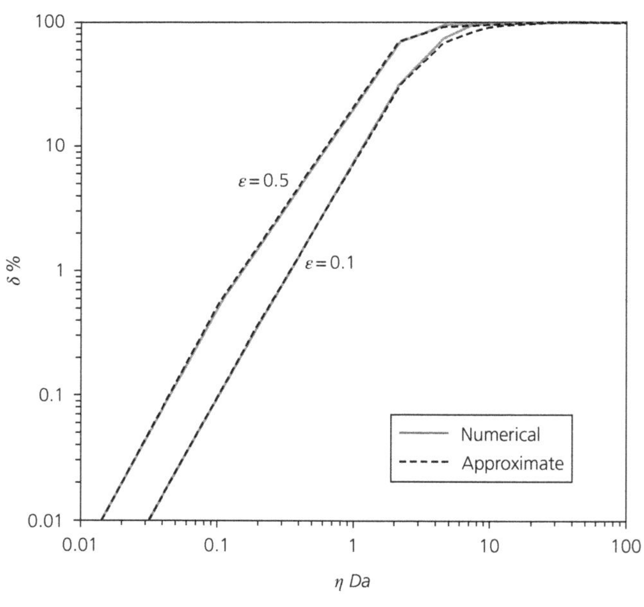

Figure 3.5 Relative error (δ %) associated with the calculation of conversion from a pseudo-homogeneous model for a first-order isothermal reaction as a function of $\eta\ Da = \tau_{CA}/\tau_{R,obs}$, with $\varepsilon = \tau_{MT}/\tau_{CA} = 0.1$ or 0.5.

$$\omega_0 = -Da^2 m c_0^{2m-1}$$
$$\omega_1 = mDac_0^{m-1},$$

where the 1D estimate for c_0 is taken from ($m \neq 1$):

$$c_0 = [1+(m-1)Daz]^{1/(1-m)}. \tag{3.80}$$

From (3.69) and (3.70), the convection-dominated solution for $\langle c_1 \rangle$ with arbitrary kinetics is (the condition for negligible radial dispersion is $\tau_{CA}/\tau_{DR} << 1$)

$$\langle c_1 \rangle = \exp\left[-\int_1^z \omega_1(z')dz'\right]\int_z^0 \exp\left[\int_1^{z*} \omega_1(z')dz'\right]\omega_0(z^*)dz^*. \tag{3.81}$$

For second-order reactions, it is possible to solve the integrals in (3.81) explicitly yielding

$$\langle c_1 \rangle = 2Da\frac{\ln(1+Daz)}{(1+Daz)^2} \quad \text{for } m=2. \tag{3.82}$$

In general for mth-order kinetics, the limit for small Da is important for the desirably low values of δ:

$$\langle c_1 \rangle \simeq mDa^2 z\left(1-\frac{m-1}{2}Daz\right)[1+(m-1)Daz]^{\frac{m}{1-m}}. \tag{3.83}$$

According to Equations 3.83, the criterion for a pseudo-homogeneous description of mass transfer is given by

$$\delta \simeq \frac{\varepsilon\langle c_1 \rangle}{c_0} = \frac{\varepsilon m Da^2 z}{1+(m-1)Daz}\left(1-\frac{m-1}{2}Daz\right) \quad \text{for } Da \to 0. \tag{3.84}$$

In particular for $m=2$, from Equation 3.82,

$$\delta \simeq \frac{\varepsilon\langle c_1 \rangle}{c_0} = 2Da\varepsilon\frac{\ln(1+Daz)}{1+Daz}. \tag{3.85}$$

These results are compared with the ones obtained numerically with gPROMS® in Figure 3.6.

When internal diffusional limitations are present in catalysts where a reaction with power-law kinetics occurs, information on the effectiveness factor is required. For isothermal systems, the chemical and diffusional limits are included into the results of Section 3.3. According to Equation 3.68,

$$\delta \simeq \varepsilon m Da^2 z\left[1-\frac{3m-1}{(\sigma+1)(\sigma+3)}\left(1+\frac{\sigma+3}{Bi_m}\right)\phi^2\right]+O(\varepsilon Da^3 z^2) \quad (\phi \ll 0) \tag{3.86a}$$

$$\delta = \varepsilon Da^2 z(1+\sigma)\left[\sqrt{\frac{2}{1+m\phi}}\frac{1}{\phi}-\frac{1}{Bi_m}\right]+\cdots \quad (\phi \gg 1). \tag{3.86b}$$

Note that the Damköhler, Thiele, and Biot numbers are related to each other and that higher ϕ for the same ε and Da implies higher Bi_m and hence lower interphase resistance.

An interesting representation of the results in this section can be made by a regime diagram. For an isothermal axial flow bed with $R_b \ll L$, three dimensionless parameters are of relevance: the Damköhler number ($Da = \tau_{CA}/\tau_R$), the radial Peclet number ($\alpha_b P_{mR} = \tau_{DR}/\tau_{CA}$), and the perturbation parameter $\varepsilon = \tau_{MT}/\tau_{CA}$. The numerical simulations performed for $\alpha_b P_{mR} = 0.1$, 1, and 10 show insensitiveness of δ to these values for nonlinear kinetics. The axial Peclet number is kept as $P_{mA} = \tau_{DA}/\tau_{CA} = 1000$. Therefore, a $Da-\varepsilon$ regime diagram is appropriate to plot boundaries associated with small values of δ, delimiting the areas where the pseudo-homogeneous model for mass transfer is applicable. For a second-order reaction, this is depicted in Figure 3.7. Numerical simulations

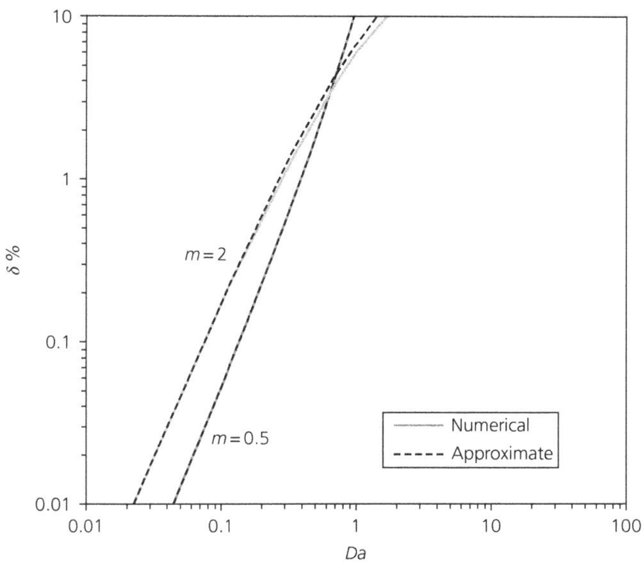

Figure 3.6 Relative error (δ %) associated with the calculation of conversion from a pseudo-homogeneous model for an mth-order isothermal reaction as a function of $Da = \tau_{CA}/\tau_R$, with $m = 2$ or 0.5.

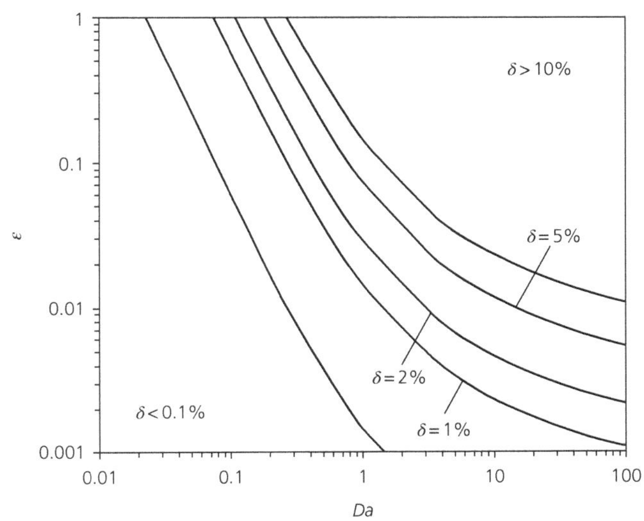

Figure 3.7 $Da-\varepsilon$ regime diagram for a second-order reaction occurring in an isothermal fixed bed. Iso-δ lines given by Equation 3.85.

in several points of the $Da-\varepsilon$ space are in excellent agreement with these predictions.

3.4.2 Non-isothermal non-adiabatic axial flow bed

The most general case involves the energy balances in Table 3.1 and the boundary conditions for heat transfer at the wall with their complete structures (e.g., as given by Eqs. 3.28d and 3.36). Numerical solution of this type of the model has been presented in several case studies in the literature (e.g., Refs. [60, 61, 123–126]). Pereira Duarte et al. [61] compared the performance of three 2D models with different degrees of complexity (I: pseudo-homogeneous; II: heterogeneous with heat transfer through the fluid and lumped thermal parameters; and III: heterogeneous with heat transfer through solid and fluid, comparable to the one presented in Table 3.1). The errors from the simplified models (I and II) relative to the more complex one (III) were calculated numerically for a first-order exothermic reaction. The deviation between the cross-sectional average concentrations from the different models (I or II relative to III) was evaluated at the axial position corresponding to the maximum temperature of the solid phase from model III. These were found to be proportional to the volume averaged rates of reaction, between the inlet and this position, which is subsequently replaced by the rate at feed conditions as an approximation. The influence of the Reynolds number, catalyst particle size, and reaction activation energy on the relative error was studied numerically. In particular, the positive error associated with the pseudo-homogeneous model decreases with the Reynolds number up to a certain value. Above this point (where the error is zero), the now negative deviation grows in absolute value again, due to the higher temperature in the solid phase compared to the one in the fluid. This effect is not compensated by the reduction of interphase resistance, and requiring a pseudo-homogeneous model for heat transfer may be too restrictive (see Section 3.5).

Adams II and Barton [127] used a dynamic heterogeneous model for water–gas shift reactors. The choice for this description was justified by previous reports of large (60°C) temperature differences between the catalyst bed and the gas. They also considered the internal diffusion–reaction problem in large catalyst pellets for industrial-scale applications. The difference between the catalyst core temperature and the one in the gas phase was simulated over the reactor axial position and time.

The results from Section 3.3 allow the formulation of the model equations only in terms of the concentration and temperature of the fluid phase, in the limits of small and large Thiele modulus. Considering weak internal diffusional limitations, Equation 3.53 can be written as follows:

$$\frac{\partial T}{\partial z}-\frac{1}{P_{ha}}\frac{\partial^2 T}{\partial z^2}-\frac{1}{\alpha_b P_{hr}}\frac{1}{r^S}\frac{\partial}{\partial r}\left(r^S\frac{\partial T}{\partial r}\right)$$
$$=Da_h R_{bulk}\bar{\eta}+\left(1+\frac{Bi_h}{3+\sigma}\right)\frac{H}{r^S}\frac{\partial}{\partial r}\left(r^S\frac{\partial R_{bulk}}{\partial r}\right). \tag{3.87}$$

For a first-order reaction, $R_{bulk}=c\exp\left(\frac{\gamma T}{1+T}\right)$ and $\bar{\eta}$ is given by Equation 3.53a. Several additional parameters enter into consideration now:

$$\frac{1}{P_{ha}}=\frac{\tau_{CA}}{\tau^f_{HDA}}=\frac{\lambda^f_{ae}}{u_Z\rho_g c_{pg}L} \tag{3.88a}$$

$$\frac{1}{\alpha_b P_{hr}}=\frac{1}{\alpha_b P^f_{hr}}+\frac{1}{\alpha_b P^s_{hr}}=\frac{L}{R_b\rho_g c_{pg}u_Z R_b}\frac{\lambda^f_{re}+\lambda^s_{re}}{} \tag{3.88b}$$

$$Da_h=\frac{\tau_{CA}}{\tau_{HG}}=\frac{L\rho_b R_{scale}(-\Delta H)}{\rho_g c_{pg}u_Z\Delta\hat{T}}=BDa \tag{3.88c}$$

$$H=\frac{\tau_{HT}}{\tau_{HG}}\frac{\tau_{CA}}{\tau^s_{HDR}}=\varepsilon\frac{Da\beta}{\alpha_b P^s_{hr}}\frac{Bi_m}{Bi_h} \tag{3.88d}$$

$$Nu_{wf}=\frac{\alpha_{wf}R_b}{\lambda^f_{re}}. \tag{3.88e}$$

Equation 3.88e appears in the boundary condition for the fluid phase, but as we have seen in Equation 3.52, the heat transfer with the wall in this limit can follow a lumped treatment between the two phases ($Nu_{wf}=Nu_{ws}=Nu_w$). For our purposes, axial diffusion may be considered negligible ($P_{ha}\gg 1$), and we seek to understand the influence of the non-isothermal parameters (H and Da_h) on δ. Even though H is expected to be small (for small ϕ^2), this dimensionless number also includes the effects from interphase thermal transfer and from the additional radial heat transfer through the solid, which could not be absorbed in the lumped parameter (3.88b). This term disappears whenever the following condition is satisfied:

$$\left(1+\frac{Bi_h}{3+\sigma}\right)H\to 0. \tag{3.89}$$

If smallness in comparison with the reaction term is demanded, the following condition is obtained:

$$\left(1+\frac{Bi_h}{3+\sigma}\right)\frac{\lambda^s_{re}}{a_V R_b^2 h}\ll 1. \tag{3.90}$$

The heterogeneous model for heat transfer (Table 3.1) can be solved with the heterogeneous and pseudo-homogeneous formulations for mass transfer detailed earlier. Numerical simulations were performed with gPROMS® for $Da=0.1$, $B=1$, $\gamma=10$, and no internal transfer limitations. The relative error of the pseudo-homogeneous mass transfer model at the exit increases from 0.11 to 1.1% when ε changes from 0.1 to 1. Therefore, the proposed scaling ($\delta\sim O(\varepsilon Da^2)$) remains valid.

3.5 Dominant fluid–solid mass and heat transfer

In the previous section, the case of fast interphase mass transfer was considered without any further simplification resulting from Equation 3.60 with respect to the fluid phase energy balance. Since

$$\frac{Bi_h}{Bi_m} = \frac{hD_{cat}}{k_g \lambda_{cat}} \sim \frac{Le_g^{2/3}}{Le_s}$$

is typically low for gas–solid systems [72], mass transfer is not limiting in the definition of a model where both mass and energy balances are described in a pseudo-homogeneous manner. Relatively fast interphase heat transfer compared with the other processes (X) in the energy balance is expressed by

$$\frac{\tau_{HT}}{\tau_X} \ll 1. \tag{3.91}$$

Assuming reasonably fast radial heat dispersion, the perturbation parameter can be written in this case as

$$\varepsilon' = \frac{\tau_{HT}}{\tau_{HDR}^f} = \varepsilon \frac{\tau_{HT}}{\tau_{MT}} \frac{\tau_{CA}}{\tau_{HDR}^f}. \tag{3.92}$$

Thus, it is expectable that $\varepsilon' \gg \varepsilon$ or at least $\varepsilon' \approx \varepsilon$. The first estimate in the fluid phase energy balance (for $\varepsilon' = 0$) expresses the absence of interphase temperature gradients:

$$T_0 = T_{surf,0}^s. \tag{3.93}$$

This term remains undetermined until we consider the second term in a perturbation series for temperature of the form

$$T = \sum_{n=0}^{\infty} T_n (\varepsilon')^n, \; T_{surf}^s = \sum_{n=0}^{\infty} T_{surf,n}^s (\varepsilon')^n \text{ and } \langle T^s \rangle = \sum_{n=0}^{\infty} \langle T_n^s \rangle (\varepsilon')^n. \tag{3.94}$$

At $O(\varepsilon')$, it is possible to collect from Equation 3.27:

$$T_1 = T_{surf,1}^s - \frac{\tau_{HDR}^f}{\tau_{CA}} \frac{\partial T_0}{\partial z} - \frac{\tau_{HDR}^f}{\tau_{CR}} \frac{1}{(r_0 + r)^S} \frac{\partial}{\partial r} \left[(r_0 + r)^S T_0 \right] + \frac{\tau_{HDR}^f}{\tau_{HDA}^f} \frac{\partial^2 T_0}{\partial z^2} + \frac{1}{(r_0 + r)^S} \frac{\partial}{\partial r} \left[(r_0 + r)^S \frac{\partial T_0}{\partial r} \right] - \frac{\tau_{HDR}^f}{\tau_{FA}} \frac{\partial T_0}{\partial t}. \tag{3.95}$$

The solid phase energy balance is also required to continue the calculation (a pseudo-homogeneous model for mass transfer is also considered):

$$T_1 = T_{surf,1}^s - \sum_{j=1}^{N_{RXN}} \frac{\tau_{HDR}^f}{\tau_{HGj}} R_j(c_0, T_0)\eta_j + \frac{\tau_{HDR}^f}{\tau_{HA}} \frac{\partial \langle T_0^s \rangle}{\partial t} - \frac{\tau_{HDR}^f}{\tau_{HDR}^s} \frac{1}{(r_0 + r)^S} \frac{\partial}{\partial r} \left[(r_0 + r)^S \frac{\partial \langle T_0^s \rangle}{\partial r} \right] - \frac{\tau_{HDR}^f}{\tau_{HDA}^s} \frac{\partial^2 \langle T_0^s \rangle}{\partial z^2}. \tag{3.96}$$

An equation for T_0 is obtained by elimination of the transfer term in Equations 3.95 and 3.96 and making $\langle T_0^s \rangle \approx T_0$ or using the results in Section 3.3. The interphase temperature gradient is given by $\Delta T_{ext} = \varepsilon' \left(T_{surf,1}^s - T_1 \right) + O(\varepsilon'^2)$. The deviation in the observed reaction rate can act as a criterion for negligible interphase thermal resistance. If the reaction term dominates in (3.96), then (note that $\hat{T}_{bulk} = \hat{T}_0$)

$$\left| \frac{R_{bulk} - R_{surf}}{R_{bulk}} \right| = \gamma \varepsilon' \frac{T_{surf,1}^s - T_1}{(1 + T_0)^2} = \gamma_{bulk} \frac{\rho_b (-\Delta H) R_{bulk} \eta}{h a_V \hat{T}_{bulk}} < 0.05. \tag{3.97}$$

Equation 3.97 reproduces Mears' criterion [128, 129]. A 2D non-isothermal pseudo-homogeneous model was used by Brandstädter and Kraushaar-Czarnetzki [49] to simulate a production-scale reactor for maleic anhydride from a C_4 mixture. The model included radial variation of the bed's porosity and a complex reaction network. They justify the choice for a pseudo-homogeneous model based on previous observations of small differences between the temperatures of catalyst grains and surrounding gas. Tye et al. [130] also used a pseudohomogeneous model to simulate a catalytic reactor for the oxidative coupling of methane, which was found to give reasonable results. Nevertheless, they point out the presence of mass and heat transfer resistances as a reason for the deviation with experimental results. Sharma et al. [3] described a fixed-bed reactor by a 2D pseudo-homogeneous model for the Fischer–Tropsch synthesis. Besides mass and heat dispersion, intraparticle diffusion and radial heat loss at the wall were also considered. They intended to obtain kinetic parameters from a laboratory-scale fixed bed and with these results and simulate a realistic industrial unit. Cornelio [131] compared heterogeneous and pseudohomogeneous models for the dynamic modeling of an ethylene oxide reactor. He claims that temperature differences between the catalyst surface and the gas phase are sufficiently small (<10 K) for the simplest model to be accurate enough.

Joshi and Doraiswamy [132] and Doraiswamy [133] report the use of a pseudo-homogeneous model for a non-isothermal non-adiabatic fixed-bed reactor (reduction of nitrobenzene to aniline). 1D and 2D models were compared in terms of the cross-sectional average concentration and temperature (Figure 3.8). From the presented values for the parameters [133] and for a reactor tube with $R_b = 2$ cm and $L = 2$ m, we estimate $Da = 3.3$, $B = 6.4$, $\alpha_b P_{hr} = 0.003$, $\alpha_b P_{mr} = 5410$, $\gamma = 3.1$, and $Nu_w = 0.14$. Though it is not specified, the global heat transfer coefficient U for the 1D model can be assumed to be calculated according to [134] as follows:

$$\frac{1}{U} = \frac{1}{\alpha_w} + \frac{R_b}{3\lambda_{re}} \frac{Nu_w + 3}{Nu_w + 4}. \tag{3.98}$$

The wall Nusselt number based on U calculates as $Nu_{w,1D} = (S+1)UR_b/\lambda_{re} = 0.26$. Due to the high values of the timescales for mass radial dispersion compared to the one for convection, a 1D model is expected to be appropriate for the concentration profile (Section 3.6).

Iordanidis [135] and Iordanidis et al. [136] simulated numerically a fixed bed for the partial oxidation of methanol to formaldehyde. In terms of the dimensionless parameters defined earlier, we estimate $\varepsilon = 0.02$, $\varepsilon' = 0.03$, $Da = 2.1$, $\alpha_b P_{hR} = 0.47$, $\alpha_b P_{mR} = 0.36$, $B = 0.55$, $\gamma = 18.5$, $Nu_{w,1D} = 4.57$ (based on U),

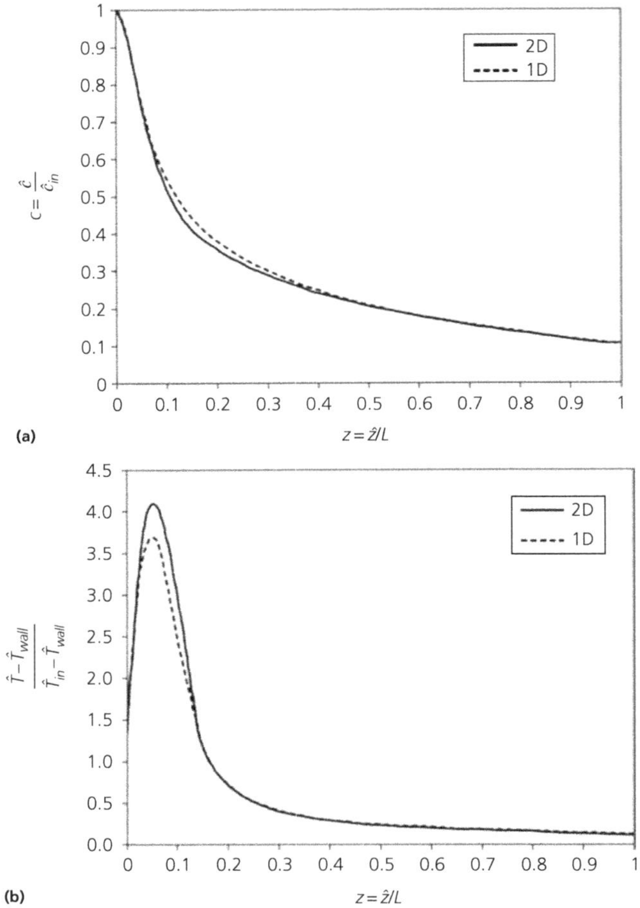

Figure 3.8 Cross-sectional average concentration (a) and temperature (b) profiles according to a pseudohomogeneous model (full lines: two-dimensional; dashed lines: one-dimensional). After Ref. [133] with $\hat{T}_{in} = 160°C$ and $\hat{T}_{wall} = 100°C$.
(Source: Doraiswamy [133]. Reproduced with permission of Oxford University Press.)

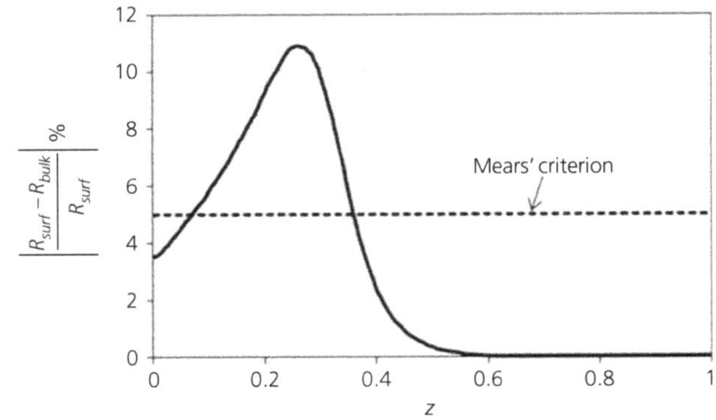

Figure 3.9 Deviation in the reaction rate between pseudo-homogeneous and heterogeneous models in the partial oxidation of methanol to formaldehyde, as calculated in Ref. [135].
(Source: Reproduced with kind permission of Iordanidis.)

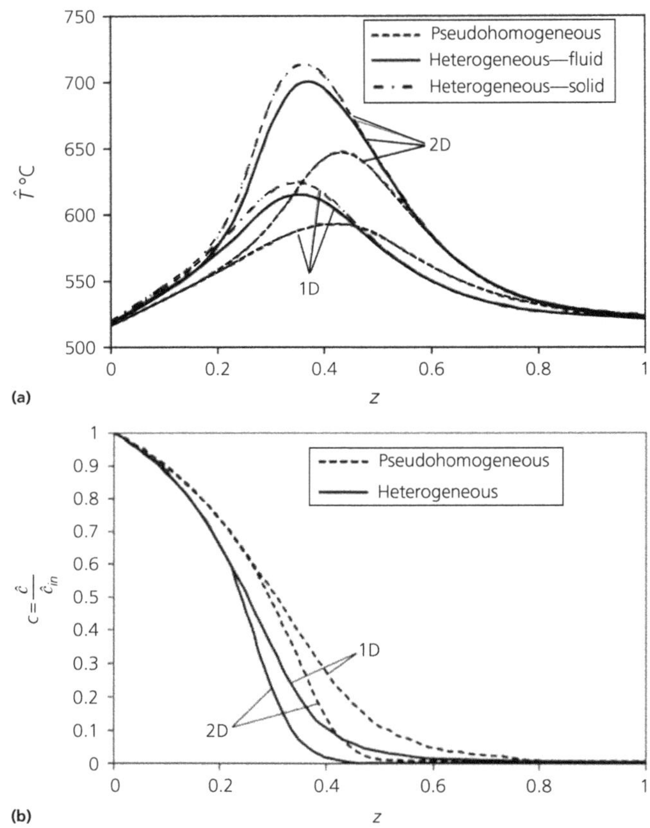

Figure 3.10 Comparison of the axial temperature (a) and methanol concentration (b) according to several models (cross-sectional averages for 2D models) for a fixed-bed reactor to produce formaldehyde from methanol [135].
(Source: Reproduced with kind permission of Iordanidis.)

and $Nu_w = 5.5$ (based on α_w). Pseudo-homogeneous and heterogeneous models were compared according to the deviation in the reaction rate, similar to Equation 3.97. Using 1D models, the result is reproduced in Figure 3.9. The models differ from over 10%, corresponding to about 30°C variation in the predicted temperatures by 1D models at the hot spot. The reactant concentration and temperature profiles calculated from several models are plotted in Figure 3.10.

In Figures 3.8 and 3.10, it is possible to observe a hot spot. This phenomenon is frequently observed for strong exothermic processes and can possibly lead to runaway. Several criteria have been put forward to predict this behavior and related thermal effects, such as steady-state multiplicity. Several references have been devoted to this complex subject [137], which is out of the scope of this chapter.

3.6 Negligible mass and thermal dispersion

The simplest manifestation of the nonisothermal nature of a packed bed is the presence of temperature gradients at the reactor scales (axially over the bed length L and radially across its radius R_b). The presence of these spatial variations may be detrimental to the reactor's performance and operation but is especially prohibitive if the purpose is to measure intrinsic reaction rates without interference of transport processes. Feed and/or catalyst dilution with highly conductive materials and miniaturization (increase of the L/R_b ratio) are some of the strategies that have been pursued to approach the isothermal behavior. On the other hand, 1D models are preferable for simplicity. Avci et al. [138] used a 1D heterogeneous model to simulate the catalytic oxidation and steam reforming of methane in a fixed-bed reactor.

Mears [68] derived criteria for negligible *axial* mass and heat transfer effects in nonisothermal reactors with uniform wall temperature (so that the rate deviates <5% from the one observed from a plug-flow model), extending the results from Young and Finlayson [139]. In terms of the bed length to particle diameter ratio, it writes as

$$\frac{L}{d_p} > 20\frac{\rho_b \hat{R}_{in} L}{\hat{c}_{in} u_Z}\left|\frac{m}{Pe_{mA}} - \frac{\gamma B}{Pe_{hA}}\right|. \tag{3.99}$$

Under isothermal conditions, the criterion includes previous results for negligible axial mass dispersion [128, 140]. In terms of the observed conversion, the criterion can be written as [128, 140]

$$\frac{L}{d_p} > 20\frac{\varepsilon_b D_{Ae}}{d_p u_Z}\ln\frac{1}{1-X}. \tag{3.100}$$

A comparison between thermal and mass dispersion effects in Equation 3.99 suggests that for typical parameter values ($\gamma\ |B|\ Pe_{mA}/Pe_{hA} > m$), thermal axial dispersion is more important than mass dispersion for gases [68]. Therefore, the more important terms in Equation 3.99 reflect a balance between reaction and axial heat dispersion. For example, according to Equation 3.87,

$$\frac{1}{P_{hA}} \ll Da_h\bar{\eta},\ 1,\ \frac{1}{\alpha_b P_{hR}}. \tag{3.101}$$

Nevertheless, the inclusion of axial dispersion may be interesting from the point of view of the numerical methods used to solve the conservation equations or in studies regarding the appearance of multiple steady-state solutions [141, 142]. Petersen [81] presented an analysis for a 1D reactor in terms of the dispersion factor F, which is the ratio between the length of a plug-flow reactor (no dispersion) and the one for a reactor with dispersion yielding the same conversion (L_M). Due to the coordinate transformations employed, F is a function of $\alpha^2 = kD_{Ae}/u^2$ (isothermal first-order reaction). Asymptotic solutions for the dispersion ratio were obtained and are given by

$$F = 1 \text{ as } \alpha^2 \to 0$$

$$F = \frac{1}{\alpha} + \frac{\ln(1+\alpha)}{\alpha^2 Pe} \text{ as } \alpha^2 \to \infty \quad (k \to \infty \text{ or } u \to 0)$$

where in this case, $Pe = uL_M/(\varepsilon_b D_{Ae})$. It was concluded that the axial diffusive transport rarely is important, a situation that corresponds to small values of α.

Concerning the *radial* heat dispersion, the criterion for negligible interparticle thermal resistance (applicable at the hot spot) is given by [128]:

$$\gamma_{wall}\frac{R_b^2\rho_b(-\Delta H)R_{bulk}\eta}{\lambda_{Re}^s\hat{T}_{wall}} < \frac{0.4}{1+8(R_p/R_b)Bi_{wall}}. \tag{3.102}$$

If this assumption is reasonable, it is common to use an overall heat transfer coefficient U, combining the individual resistances. Compared with axial interparticular effects, radial heat transfer at the hot spot is more significant. Moreover, axial and radial mass transfer effects are also less important in experimental reactors. A more generic statement of Equation 3.102 is based on the scaling from (3.87):

$$\frac{1}{\alpha_b P_{hR}} \ll 1,\ Da_h\bar{\eta}.$$

In general, a minimum value for d_b/d_p is usually reported, whereas for axial effects, a minimum L/d_p is required. However, some discrepancy in the values of these ratios exists.

The analysis for mass transfer follows the same lines. It is likely that if the previous criteria are observed for heat transfer, then concentration gradients can also be neglected. The ratios between timescales for axial and radial dispersion compared to the one for convection are the parameters of interest, as seen in previous examples. In Figure 3.10, the relatively low values of $\alpha_b P_{hR}$ and $\alpha_b P_{mR}$ justify the differences observed between 1D and 2D models.

3.7 Conclusions

The theory associated with modeling of fixed-bed reactors is still greatly based on simplified velocity profiles and on effective quantities to describe transport properties. A typical generic model constructed on this framework was presented. Approximate solutions to solid concentration/temperature profiles and effectiveness factor allow some developments in the analysis of effects depending on the coupling between both domains (e.g., radial solid heat dispersion with nonnegligible internal thermal effects). A perturbation procedure conceived for fast mass and heat interphase transfer yields the common pseudo-homogeneous model, while the negligibility of dispersive phenomena determines whether 1D models are appropriate or not. Classical criteria for these limits were presented, and the equations that need to be solved for improving the predictions at first approximation were derived (accounting for finite fluid–solid transfer). In particular, we consider a regime where mass transfer between fluid and catalyst is fast, but heat transfer may be finite. The relative error associated with the pseudo-homogeneous model for mass transfer can be expressed by the difference of concentration or average concentrations, instead of the typical deviation between observed reaction rates. A general procedure to obtain these estimates was given, and we derived its form for particular cases. These simple results are in excellent agreement with numerical calculations.

Nomenclature

a_V	external surface area of catalyst particles per volume of reactor, m^{-1}
B	dimensionless adiabatic temperature rise
b	nonisothermal coefficient in the diffusional regime's effectiveness factor
Bi_h	heat Biot number
Bi_m	mass Biot number
c	dimensionless fluid concentration
$\langle c \rangle$	dimensionless average fluid concentration over the reactor cross section
$\Delta\hat{c}$	scale for concentration change, mol/m^3_{fluid}
c^0	initial dimensionless concentration distribution
Ca	Carberry number
c_{in}	dimensionless inlet concentration
c_n	coefficients in perturbation series for concentration
c_{out}	dimensionless external radial concentration
c_{pg}	specific heat of the fluid (gas) phase, J/kg/K
c_{ps}	specific heat of the solid phase, J/kg/K
$\hat{c}_{ref}$	reference concentration, mol/m^3_{fluid}
c^s	dimensionless solid phase concentration
$\langle c^s \rangle$	dimensionless volume-averaged solid concentration
c^s_{surf}	dimensionless fluid–solid surface concentration
Da	mass Damköhler number
D_{ae}	axial effective mass dispersion coefficient, m^2_{bed}/s
Da_h	heat Damköhler number
d_b	catalyst bed diameter, m
D_{cat}	catalyst effective diffusivity, m^2_{cat}/s
D_g	gas molecular diffusivity, m^2_{fluid}/s
d_p	catalyst pellet diameter, m
D_{re}	radial effective mass dispersion coefficient, m^2_{bed}/s
${}_2F_1(a,b,c,z)$	hypergeometric function ${}_2F_1$
F	dispersion ratio
G	superficial mass flow velocity, kg/s
h	fluid–solid heat transfer coefficient, $W/m^2/K$
ΔH_j	heat of reaction j, J/mol
j	mass-heat analogy factor
k	kinetic constant at reference conditions, $mol^{1-m}\ m_f^{3m}/s/kg_{cat}$
k_{bulk}	kinetic constant at bulk conditions, $mol^{1-m}\ m_f^{3m}/s/kg_{cat}$
k_g	fluid–solid mass transfer coefficient, m/s
k_{surf}	kinetic constant at surface conditions, $mol^{1-m}\ m_f^{3m}/s/kg_{cat}$
L	length of the packed bed, m
Le	Lewis number
Le_{cat}	catalyst Lewis number
Le_g	fluid Lewis number
Le_s	solid Lewis number
m	order of reaction
n	order of term in perturbation series
N_{RXN}	number of reactions
Nu	Nusselt number
Nu_w	wall Nusselt number
$Nu_{w,1D}$	wall Nusselt number for 1D models (based on U)
Nu_{wf}	wall–fluid Nusselt number
Nu_{ws}	wall–solid Nusselt number
$\hat{p}$	pressure, Pa
P_{ha}	axial heat Peclet number (based on the bed length)
P_{hr}	radial heat Peclet number (based on the bed length)
Pe^f_{ha}	fluid axial heat Peclet number (based on particle diameter)
Pe^f_{hr}	fluid radial heat Peclet number (based on particle diameter)
Pe_{ma}	axial mass Peclet number (based on particle diameter)
Pe_{mr}	radial mass Peclet number (based on particle diameter)
Pr	Prandtl number
r_0	dimensionless reference radial distance
R_b	bed radial characteristic distance, m
R_{bulk}	dimensionless reaction rate at bulk conditions
R_j	dimensionless rate of reaction j
$R_j\left(c^s_{surf}, T^s_{surf}\right)$	dimensionless rate of reaction j evaluated at catalyst surface conditions
R_p	pellet characteristic length, m
$R_{scale,j}$	scale for the rate of reaction j, $mol/s/kg_{cat}$
R_{sph}	radius of the spherical reactor, m
Re	Reynolds number
S	bed shape factor
S_b	geometric external area of the bed, m^2_{bed}
S_p	geometric external area of the pellet, m^2_{part}
S_{surf}	fluid–particle interfacial surface area, m^2_{surf}
Sc	Schmidt number
Sh	Sherwood number
t	dimensionless time
T	dimensionless fluid temperature
T^0	initial dimensionless temperature distribution
T^s	dimensionless solid phase temperature
$\langle T^s \rangle$	dimensionless volume-averaged solid temperature
$\Delta\hat{T}$	scale for temperature change, K
ΔT_{ext}	dimensionless external temperature rise
T_{in}	dimensionless inlet temperature
T^s_{max}	maximum dimensionless temperature in the solid
ΔT^s_{max}	maximum dimensionless internal temperature rise
$\hat{T}_{ref}$	reference temperature, K

T^s_{surf} dimensionless fluid–solid surface temperature
T_{wall} dimensionless wall temperature
t_P characteristic time of transient effects in the process, s
u, u_Z superficial velocity of the fluid phase in the axial direction, m/s
U global heat transfer coefficient, W/m^2/K
u_i interstitial axial velocity, m/s
u_R superficial velocity of the fluid phase in the radial direction, m/s
$\hat{V}$ volume, m^3
V_b volume of bed, m^3_{bed}
V_p volume of solid pellet, m^3_{cat}
X reactant conversion
z dimensionless axial coordinate

Greek letters

α_b bed aspect ratio
α_{wf} heat coefficient for wall–fluid heat transfer, W/m^2/K
α_{ws} heat coefficient for wall–solid heat transfer, W/m^2/K
β_j Prater's parameter
γ Arrhenius number
δ relative error between different models for concentration
ε perturbation parameter for fast interphase mass transfer, $\tau_{MT}\tau_{CA}^{-1}$
ε' perturbation parameter for fast interphase heat transfer, $\tau_{HT}\left(\tau^f_{HDR}\right)^{-1}$
ε_b bed porosity
ε_P pellet porosity
ϕ_j^2 Thiele modulus
$\bar{\eta}$ global effectiveness factor
η_j effectiveness factor of reaction j
λ^f fluid thermal conductivity, W/m/K
λ^f_{ae} axial effective thermal dispersion coefficient in the fluid phase, W/m/K
λ^s_{ae} axial effective thermal dispersion coefficient in the solid phase, W/m/K
λ_{cat} catalyst thermal conductivity, W/m/K
λ^f_{re} radial effective thermal dispersion coefficient in the fluid phase, W/m/K
λ^s_{re} radial effective thermal dispersion coefficient in the solid phase, W/m/K
μ_g gas (dynamic) viscosity, Pa s
ν_{ij} stoichiometric coefficient of species i in reaction j
ρ_b solid phase density (on a bed volume basis), kg/m^3
ρ_g fluid (gas) phase density, kg/m^3
ρ_s, ρ_{cat} catalyst (solid phase) density, kg/m^3
σ pellet shape factor
τ_b bed tortuosity
τ_X timescale of process X, s
ξ dimensionless 1D spatial coordinate
$\hat{\xi}_0$ reference spatial position in catalyst pellet, m
∇^2 $\frac{1}{r^S}\frac{\partial}{\partial r}\left(r^S\frac{\partial}{\partial r}\right)$

References

1 Eigenberger G. Fixed-bed reactors. In: Elvers B, editor. Ullmann's Encyclopedia of Industrial Chemistry. Weinheim: Wiley-VCH Verlag GmbH & Co. KGaA; 2000.
2 Halloin VL, Wajc SJ. A multifunctional catalytic reactor suitable for the exothermal synthesis of condensable products. Chemical Engineering Science 1994;49: 4691–4698.
3 Sharma A, Philippe R, Luck F, Schweich D. A simple and realistic fixed bed model for investigating Fischer–Tropsch catalyst activity at lab-scale and extrapolating to industrial conditions. Chemical Engineering Science 2011;66: 6358–6366.
4 Song D, Cho W, Lee G, Park DK, Yoon ES. Numerical analysis of a pilot-scale fixed-bed reactor for dimethyl ether (DME) synthesis. Industrial and Engineering Chemistry Research 2008;47: 4553–4559.
5 Kareeri AA, Zughbi HD, Al-Ali HH. Simulation of flow distribution in radial flow reactors. Industrial and Engineering Chemistry Research 2006;45: 2862–2874.
6 Vek V. Optimization of large reactors with extremely active catalysts. Industrial & Engineering Chemistry Process Design and Development 1977;16: 412–424.
7 Abdollahi F, Mostoufi N, Sotudeh-Gharebagh R. Optimization of radial flow reactors of styrene production. International Journal of Chemical Reactor Engineering 2007;5: A75.
8 Lee WJ, Froment GF. Ethylbenzene dehydrogenation into styrene: Kinetic modeling and reactor simulation. Industrial and Engineering Chemistry Research 2008;47: 9183–9194.
9 Suter D, Bartroli A, Schneider F, Rippin DWT, Newson EJ. Radial flow reactor optimization for highly exothermic selective oxidation reactions. Chemical Engineering Science 1990;45: 2169–2176.
10 Silady PJ. UOP isomar process. In: Meyers RA, editor. Handbook of Petroleum Refining Processes, 3rd edition. New York: McGraw-Hill; 2004.
11 Hartig F, Keil FJ. Large-scale spherical fixed bed reactors: Modeling and optimization. Industrial & Engineering Chemistry Research 1993;32: 424–437.
12 Rahimpour MR, Parvasi P, Setoodeh P. Dynamic optimization of a novel radial-flow, spherical-bed methanol synthesis reactor in the presence of catalyst deactivation using differential evolution (DE) algorithm. International Journal of Hydrogen Energy 2009;34: 6221–6230.
13 Rahimpour MR, Pourazadi E, Iranshahi D, Bahmanpour AM. Methanol synthesis in a novel axial-flow, spherical packed bed reactor in the presence of catalyst deactivation. Chemical Engineering Research and Design 2011;89: 2457–2469.
14 Iranshahi D, Rahimpour MR, Asgari A. A novel dynamic radial-flow, spherical-bed reactor concept for naphtha reforming in the presence of catalyst deactivation. International Journal of Hydrogen Energy 2011;35: 6261–6275.
15 Iranshahi D, Pourazadi E, Paymooni K, Bahmanpour AM, Rahimpour MR, Shariati A. Modeling of an axial flow, spherical packed-bed reactor for naphtha reforming process in the presence of the catalyst deactivation. International Journal of Hydrogen Energy 2010;35: 12784–12799.

16 Sanchez Marcano JG, Tsotsis TT. Catalytic Membranes and Membrane Reactors. Weinheim: Wiley-VCH; 2002.
17 Bakhtiary-Davijany H, Dadgar F, Hayer F, Phan XK, Myrstad R, Venvik HJ, Pfeifer P, Holmen A. Analysis of external and internal mass transfer at low Reynolds numbers in a multiple-slit packed bed microstructured reactor for synthesis of methanol from syngas. Industrial and Engineering Chemistry Research 2012;51: 13574–13579.
18 Cho BK, Carr Jr RW, Aris R. A continuous chromatographic reactor. Chemical Engineering Science 1980;35: 74–81.
19 Minceva M, Gomes PS, Meshko V, Rodrigues AE. Simulated moving bed reactor for isomerization and separation of p-xylene. Chemical Engineering Journal 2008;140: 305–323.
20 Silva VMTM, Rodrigues AE. Novel process for diethylacetal synthesis. AIChE Journal 2005;51: 2752–2768.
21 Deshmukh SR, Tonkovich ALY, Jarosch KT, Schrader L, Fitzgerald SP, Kilanowski DR, Lerou JJ, Mazanec TJ. Scale-up of microchannel reactors for Fischer-Tropsch synthesis. Industrial and Engineering Chemistry Research 2010;49: 10883–10888.
22 Lerou JJ, Tonkovich AL, Silva L, Perry S, McDaniel J. Microchannel reactor architecture enables greener processes. Chemical Engineering Science 2010;65: 380–385.
23 Guettel R, Turek T. Assessment of micro-structured fixed-bed reactors for highly exothermic gas-phase reactions. Chemical Engineering Science 2010;65: 1644–1654.
24 Bakhtiary-Davijany H, Hayer F, Phan XK, Myrstad R, Venvik HJ, Pfeifer P, Holmen A. Characteristics of an integrated micro packed bed reactor-heat exchanger for methanol synthesis from syngas. Chemical Engineering Journal 2011;167: 496–503.
25 Kapteijn F, Nijhuis TA, Heiszwolf JJ, Moulijn JA. New non-traditional multiphase catalytic reactors based on monolithic structures. Catalysis Today 2001;66: 133–144.
26 Gunjal PR, Ranade VV, Chaudhari RV. Computational study of a single-phase flow in packed beds of spheres. AIChE Journal 2005;51: 365–378.
27 Gao X, Zhu YP, Luo ZH. CFD modeling of gas flow in porous medium and catalytic coupling reaction from carbon monoxide to diethyl oxalate in fixed-bed reactors. Chemical Engineering Science 2011;66: 6028–6038.
28 Dixon AG, Nijemeisland M. CFD as a design tool for fixed-bed reactors. Industrial & Engineering Chemistry Research 2001;40: 5246–5254.
29 Magnico P. Pore-scale simulations of unsteady flow and heat transfer in tubular fixed beds. AIChE Journal 2009;55: 849–866.
30 Guardo A, Coussirat M, Larrayoz MA, Recasens F, Egusquiza E. Influence of the turbulence model in CFD modeling of wall-to-fluid heat transfer in packed beds. Chemical Engineering Science 2005;60: 1733–1742.
31 Dixon AG, Taskin ME, Nijemeisland M, Stitt EH. CFD method to couple three-dimensional transport and reaction inside catalyst particles to the fixed bed flow field. Industrial and Engineering Chemistry Research 2010;49: 9012–9025.
32 Dixon AG, Ertan Taskin M, Hugh Stitt E, Nijemeisland M. 3D CFD simulations of steam reforming with resolved intraparticle reaction and gradients. Chemical Engineering Science 2007;62: 4963–4966.
33 Calo JM. Cell model studies of radial flow, fixed bed reactors. In: Weekman Jr VW, Luss D, editors. Chemical Reaction Engineering—Houston. ACS Symposium Series, Vol. 65. Washington, DC: American Chemical Society; 1978: 550–561.
34 Jiang Y, Khadilkar MR, Al-Dahhan MH, Dudukovic MP. Single phase flow modeling in packed beds: Discrete cell approach revisited. Chemical Engineering Science 2000;55: 1829–1844.
35 Chigada PI. Characterisation of flow and heat transfer patterns in low aspect ratio packed beds by a 3D network-of-voids model. Chemical Engineering Research and Design 2011;89: 230–238.
36 Thompson KE, Fogler HS. Modeling flow in disordered packed beds from pore-scale fluid mechanics. AIChE Journal 1997;43: 1377–1389.
37 Mantle MD, Sederman AJ, Gladden LF. Single-and two-phase flow in fixed-bed reactors: MRI flow visualisation and lattice-Boltzmann simulations. Chemical Engineering Science 2001;56: 523–529.
38 Freund H, Bauer J, Zeiser T, Emig G. Detailed simulation of transport processes in fixed-beds. Industrial and Engineering Chemistry Research 2005;44: 6423–6434.
39 Bizzi M, Saracco G, Schwiedernoch R, Deutschmann O. Modeling the partial oxidation of methane in a fixed bed with detailed chemistry. AIChE Journal 2004;50: 1289–1299.
40 Sharma S, Mantle MD, Gladden LF, Winterbottom JM. Determination of bed voidage using water substitution and 3D magnetic resonance imaging, bed density and pressure drop in packed-bed reactors. Chemical Engineering Science 2001;56: 587–595.
41 Lysova AA, Kulikov AV, Parmon VN, Sagdeev RZ, Koptyug IV. Quantitative temperature mapping within an operating catalyst by spatially resolved 27Al NMR. Chemical Communications 2012;48: 5763–5765.
42 Geske M, Korup O, Horn R. Resolving kinetics and dynamics of a catalytic reaction inside a fixed bed reactor by combined kinetic and spectroscopic profiling. Catalysis Science & Technology 2013;3: 169–175.
43 Schlünder EU. Transport phenomena in packed bed reactors. In: Luss D, Weekman Jr VW, editors. Chemical Reaction Engineering Reviews—Houston. Washington, DC: American Chemical Society; 1978.
44 Elsari M, Hughes R. Axial effective thermal conductivities of packed beds. Applied Thermal Engineering 2002;22: 1969–1980.
45 Gupta R, Bansal A. Effect of bed configuration on dispersion in a packed-bed reactor. Industrial and Engineering Chemistry Research 2010;49: 9525–9528.
46 Smirnov EI, Kuzmin VA, Zolotarskii IA. Radial thermal conductivity in cylindrical beds packed by shaped particles. Chemical Engineering Research and Design 2004;82: 293–296.
47 Vortmeyer D, Schuster J. Evaluation of steady flow profiles in rectangular and circular packed beds by a variational method. Chemical Engineering Science 1983;38: 1691–1699.
48 Hayer F, Bakhtiary-Davijany H, Myrstad R, Holmen A, Pfeifer P, Venvik HJ. Characteristics of integrated micro packed bed reactor-heat exchanger configurations in the direct synthesis of dimethyl ether. Chemical Engineering and Processing 2013;70: 77–85.
49 Brandstädter WM, Kraushaar-Czarnetzki B. Maleic anhydride from mixtures of n-butenes and n-butane: Simulation of a production-scale nonisothermal fixed-bed reactor. Industrial and Engineering Chemistry Research 2007;46: 1475–1484.
50 Jakobsen HA, Lindborg H, Handeland V. A numerical study of the interactions between viscous flow, transport and kinetics in fixed bed reactors. Computers and Chemical Engineering 2002;26: 333–357.
51 Froment GF. Analysis and design of fixed bed catalytic reactors. In: Bischoff KB, editor. Chemical Reaction Engineering. Washington, DC: American Chemical Society; 1974.

52 Rase HF. Fixed-Bed Reactor Design and Diagnostics: Gas Phase Reactions. Boston, MA: Butterworths; 1990.
53 Gunn DJ. An analysis of convective dispersion and reaction in the fixed-bed reactor. International Journal of Heat and Mass Transfer 2004;47: 2861–2875.
54 Gunn DJ, Pryce C. Dispersion in fixed beds. Transactions of the Institution of Chemical Engineers 1969;47: t341–t350.
55 Aris R, Amundson NR. Some remarks on longitudinal mixing or diffusion in fixed beds. AIChE Journal 1957;3: 280–282.
56 Edwards MF, Richardson JF. Gas dispersion in packed beds. Chemical Engineering Science 1968;23: 109–123.
57 Gunn DJ. Axial and radial dispersion in fixed beds. Chemical Engineering Science 1987;42: 363–373.
58 Kronberg AE, Westerterp KR. Nonequilibrium effects in fixed-bed interstitial fluid dispersion. Chemical Engineering Science 1999;54: 3977–3993.
59 de Wasch AP, Froment GF. A two dimensional heterogeneous model for fixed bed catalytic reactors. Chemical Engineering Science 1971;26: 629–634.
60 Park JH. Modeling of transient heterogeneous two-dimensional catalytic packed bed reactor. Korean Journal of Chemical Engineering 1995;12: 80–87.
61 Pereira Duarte SI, Barreto GF, Lemcoff NO. Comparison of two-dimensional models for fixed bed catalytic reactors. Chemical Engineering Science 1984;39: 1017–1024.
62 Froment GF, Bischoff KB. Chemical Reactor Analysis and Design. New York: John Wiley & Sons; 1990.
63 Ponzi PR, Kaye LA. Effect of flow maldistribution on conversion and selectivity in radial flow fixed-bed reactors. AIChE Journal 1979;25: 100–108.
64 Savoretti AA, Borio DO, Bucalá V, Porras JA. Non-adiabatic radial-flow reactor for styrene production. Chemical Engineering Science 1999;54: 205–213.
65 Kunii D, Smith JM. Heat transfer characteristics of porous rocks. AIChE Journal 1960;6: 71–78.
66 Gunn DJ, Ahmad MM, Sabri MN. Radial heat transfer to fixed beds of particles. Chemical Engineering Science 1987;42: 2163–2171.
67 de Wasch AP, Froment GF. Heat transfer in packed beds. Chemical Engineering Science 1972;27: 567–576.
68 Mears DE. On criteria for axial dispersion in nonisothermal packed-bed catalytic reactors. Industrial and Engineering Chemistry Fundamentals 1976;15: 20–23.
69 Wen D, Ding Y. Heat transfer of gas flow through a packed bed. Chemical Engineering Science 2006;61: 3532–3542.
70 Smirnov EI, Muzykantov AV, Kuzmin VA, Kronberg AE, Zolotarskii IA. Radial heat transfer in packed beds of spheres, cylinders and Rashig rings: Verification of model with a linear variation of λ_{er} in the vicinity of the wall. Chemical Engineering Journal 2003;91: 243–248.
71 Yagi S, Wakao N. Heat and mass transfer from wall to fluid in packed beds. AIChE Journal 1959;5: 79–85.
72 Carberry JJ. Chemical and Catalytic Reaction Engineering. New York: McGraw-Hill; 1976.
73 Wakao N, Kaguei S, Funazkri T. Effect of fluid dispersion coefficients on particle-to-fluid heat transfer coefficients in packed beds. Correlation of nusselt numbers. Chemical Engineering Science 1979;34(3): 325–336.
74 Satterfield CN. Mass Transfer in Heterogeneous Catalysis. Cambridge, MA: MIT Press; 1981.
75 Welty JR, Wicks CE, Wilson RE. Fundamentals of Momentum, Heat and Mass Transfer. New York: Wiley; 1976.
76 Fogler HS. Elements of Chemical Reaction Engineering. Englewood Cliffs, NJ: Prentice-Hall; 1999.
77 Damköhler G. The adsorption velocity of gases on porous adsorbents. Zeitschrift für Physikalische 1935;A174: 228–238.
78 Thiele EW. Relation between catalytic activity and size of the particle. Industrial and Engineering Chemistry 1939;31: 916–920.
79 Zeldovitch YB. The theory of reactions on powders and porous substances. Acta Physicochimica URSS 1939;10: 583–592.
80 Aris R. The Mathematical Theory of Diffusion and Reaction in Permeable Catalysts. London: Oxford University Press; 1975.
81 Petersen E. Chemical Reaction Analysis. Englewood Cliffs, NJ: Prentice-Hall; 1965.
82 Gottifredi JC, Gonzo EE. Application of perturbation and matching techniques to solve transport phenomena problems. In: Mujumdar A, Mashelkar RA, editors. Advances in Transport Processes. New Delhi: Wiley Eastern; 1986.
83 Gottifredi JC, Gonzo EE. Approximate expression for the effectiveness factor estimation and a simple numerical method for concentration profile calculation in porous catalyst. Chemical Engineering Journal 2005;109: 83–87.
84 Gottifredi JC, Gonzo EE. On the effectiveness factor calculation for a reaction—diffusion process in an immobilized biocatalyst pellet. Biochemical Engineering Journal 2005;24: 235–242.
85 Gottifredi JC, Gonzo EE, Quiroga OD. Isothermal effectiveness factor—I: Analytical expression for single reaction with arbitrary kinetics. Slab geometry. Chemical Engineering Science 1981;36: 713–719.
86 Magyari E. Exact analytical solution of a nonlinear reaction—diffusion model in porous catalysts. Chemical Engineering Journal 2008;143: 167–171.
87 Wijngaarden RJ, Kronberg A, Westerterp KR. Industrial Catalysis: Optimizing Catalysts and Processes. Weinheim: Wiley-VCH; 1998.
88 Datta R, Leung SWK. Shape generalized isothermal effectiveness factor for first-order kinetics. Chemical Engineering Communications 1985;39: 155–173.
89 Burghardt A, Kubaczka A. Generalization of the effectiveness factor for any shape of a catalyst pellet. Chemical Engineering and Processing: Process Intensification 1996;35: 65–74.
90 Keegan SD, Mariani NJ, Bressa SP, Mazza GD, Barreto GF. Approximation of the effectiveness factor in catalytic pellets. Chemical Engineering Journal 2003;94: 107–112.
91 Mariani NJ, Keegan SD, Martínez OM, Barreto GF. On the evaluation of effective reaction rates on commercial catalyst by means of a one-dimensional model. Catalysis Today 2008;133–135: 770–774.
92 Mariani NJ, Mocciaro C, Keegan SD, Martínez OM, Barreto GF. Evaluating the effectiveness factor from a 1D approximation fitted at high Thiele modulus: Spanning commercial pellet shapes with linear kinetics. Chemical Engineering Science 2009;64: 2762–2766.
93 Mariani NJ, Keegan SD, Martínez OM, Barreto GF. A one-dimensional equivalent model to evaluate overall reaction rates in catalytic pellets. Chemical Engineering Research and Design 2003;81: 1033–1042.
94 Keegan SD, Mariani NJ, Martínez OM, Barreto GF. Behaviour of smooth catalysts at high reaction rates. Chemical Engineering Journal 2005;110: 41–56.

95 Mocciaro C, Mariani NJ, Martínez OM, Barreto GF. A three-parameter one-dimensional model to predict the effectiveness factor for an arbitrary pellet shape. Industrial and Engineering Chemistry Research 2011;50: 2746–2754.
96 Lopes JP, Cardoso SSS, Rodrigues AE. Effectiveness factor for thin catalytic coatings: Improved analytical approximation using perturbation techniques. Chemical Engineering Science 2012;71: 46–55.
97 Tinkler JD, Pigford RL. The influence of heat generation on the catalyst effectiveness factor. Chemical Engineering Science 1961;15: 326–328.
98 Schilson RE, Amundson NR. Intraparticle diffusion and conduction in porous catalysts—I: Single reactions. Chemical Engineering Science 1961;13: 226–236.
99 Weisz PB, Hicks JS. The behaviour of porous catalyst particles in view of internal mass and heat diffusion effects. Chemical Engineering Science 1962;17: 265–275.
100 Tavera EM. Analytical expression for the non-isothermal effectiveness factor: The nth-order reaction in a slab geometry. Chemical Engineering Science 2005;60: 907–916.
101 Cardoso SSS, Rodrigues AE. Diffusion and reaction in a porous catalyst slab: Perturbation solutions. AIChE Journal 2006;52: 3924–3932.
102 Kočí P, Štěpánek F, Kubíček M, Marek M. Modelling of micro/nano-scale concentration and temperature gradients in porous supported catalysts. Chemical Engineering Science 2007;62: 5380–5385.
103 Wakao N, Smith JM. Diffusion and reaction in porous catalysts. Industrial and Engineering Chemistry Fundamentals 1964;3: 123–127.
104 Corbett Jr WE, Luss D. The influence of non-uniform catalytic activity on the performance of a single spherical pellet. Chemical Engineering Science 1974;29: 1473–1483.
105 Hwang S, Smith R. Heterogeneous catalytic reactor design with non-uniform catalyst considering shell-progressive poisoning behavior. Chemical Engineering and Technology 2008;31: 384–397.
106 Cukierman AL, Laborde MA, Lemcoff NO. Optimum activity distribution in a catalyst pellet for a complex reaction. Chemical Engineering Science 1983;38: 1977–1982.
107 Belfiore LA. Effects of the collision integral, thermal diffusion, and the Prater number on maximum temperature in macroporous catalysts with exothermic chemical reaction in the diffusion-controlled regime. Chemical Engineering Science 2007;62: 655–665.
108 Lopes JP, Cardoso SSS, Rodrigues AE. Convection, diffusion, and exothermic zero-order reaction in a porous catalyst slab: Scaling and perturbation analysis. AIChE Journal 2009;55: 2686–2699.
109 Damkohler G. The excess temperature in contact grains. Zeitschrift fur Physikalische Chemie-Leipzig 1943;193: 16–28.
110 Prater CD. The temperature produced by heat of reaction in the interior of porous particles. Chemical Engineering Science 1958;8: 284–286.
111 Carberry JJ. On the relative importance of external-internal temperature gradients in heterogeneous catalysis. Industrial and Engineering Chemistry Fundamentals 1975;14: 129–131.
112 Petersen EE. Non-isothermal chemical reaction in porous catalysts. Chemical Engineering Science 1962;17: 987–995.
113 Bischoff KB. An extension of the general criterion for importance of pore diffusion with chemical reactions. Chemical Engineering Science 1967;22: 525–530.
114 Marek M, Hlavacek V. Modelling of chemical reactors. VI. Heat and mass transfer in a porous catalyst particle; on the multiplicity of solutions for the case of an exothermic zeroth-order reaction. Collection of Czechoslovak Chemical Communications 1968;33: 506–517.
115 Drott DW, Aris R. Communications on the theory of diffusion and reaction—I: A complete parametric study of the first-order, irreversible exothermic reaction in a flat slab of catalyst. Chemical Engineering Science 1969;24: 541–551.
116 Van Den Bosch B, Luss D. Uniqueness and multiplicity criteria for an nth order chemical reaction. Chemical Engineering Science 1977;32: 203–212.
117 Balakotaiah V, Dommeti SMS. Effective models for packed-bed catalytic reactors. Chemical Engineering Science 1999;54: 1621–1638.
118 Vortmeyer D, Schaefer RJ. Equivalence of one- and two-phase models for heat transfer processes in packed beds: one dimensional theory. Chemical Engineering Science 1974;29: 485–491.
119 Dommeti SMS, Balakotaiah V, West DH. Analytical criteria for validity of pseudohomogeneous models of packed-bed catalytic reactors. Industrial & Engineering Chemistry Research 1999;38: 767–777.
120 Castaño P, Arandes JM, Pawelec B, Olazar M, Bilbao J. Kinetic modeling for assessing the product distribution in toluene hydrocracking on a Pt/HZSM-5 catalyst. Industrial and Engineering Chemistry Research 2008;47: 1043–1045.
121 Kulkarni AA, Zeyer KP, Jacobs T, Kienle A. Miniaturized systems for homogeneously and heterogeneously catalyzed liquid-phase esterification reaction. Industrial and Engineering Chemistry Research 2007;46: 5271–5277.
122 gPROMS® ModelBuilder 3.3.1. 2010. PSE – Process Systems Enterprise; London, UK.
123 Hoebink JHBJ, Couwenberg PM, Marin GB. Fixed bed reactor design for gas phase chain reactions catalysed by solids: The oxidative coupling of methane. Chemical Engineering Science 1994;49: 5453–5463.
124 Vasco De Toledo EC, Mariano AP, De Morais ER, Stremel DP, Meyer JFDCA, MacIel Filho R. Development of rigorous and reduced heterogeneous dynamic models for fixed bed catalytic reactor and three-phase catalytic slurry reactor. Chemical Product and Process Modeling 2008;3: A48.
125 Liu QS, Zhang ZX, Zhou JL. Steady-state and dynamic behavior of fixed-bed catalytic reactor for Fischer-Tropsch synthesis I: Mathematical model and numerical method. Journal of Natural Gas Chemistry 1999;8: 137–150.
126 Petera J, Nowicki L, Ledakowicz S. New numerical algorithm for solving multidimensional heterogeneous model of the fixed bed reactor. Chemical Engineering Journal 2013;214: 237–246.
127 Adams II TA, Barton PI. A dynamic two-dimensional heterogeneous model for water gas shift reactors. International Journal of Hydrogen Energy 2009;34: 8877–8891.
128 Mears DE. Tests for transport limitations in experimental catalytic reactors. Industrial and Engineering Chemistry: Process Design and Development 1971;10: 541–547.
129 Mears DE. Diagnostic criteria for heat transport limitations in fixed bed reactors. Journal of Catalysis 1971;20: 127–131.

130 Tye CT, Mohamed AR, Bhatia S. Modeling of catalytic reactor for oxidative coupling of methane using La2O3/CaO catalyst. Chemical Engineering Journal 2002;87: 49–59.

131 Cornelio AA. Dynamic modelling of an industrial ethylene oxide reactor. Indian Chemical Engineer. Section A 2006;48: 164–177.

132 Joshi JB, Doraiswamy LK. Chemical reaction engineering. In: Albright LF, editor. Albright's Chemical Engineering Handbook. Boca Raton: CRC Press; 2009.

133 Doraiswamy LK. Organic Synthesis Engineering. London: Oxford University Press; 2001.

134 Dixon AG. An improved equation for the overall heat transfer coefficient in packed beds. Chemical Engineering and Processing: Process Intensification 1996;35: 323–331.

135 Iordanidis AA. Mathematical modeling of catalytic fixed bed reactors [dissertation]. Twente: University of Twente; 2002. Available from: Twente.

136 Iordanidis AA, van Sint Annaland M, Kronberg AE, Kuipers JAM. A critical comparison between the wave model and the standard dispersion model. Chemical Engineering Science 2003;58: 2785–2795.

137 Hayer F, Bakhtiary-Davijany H, Myrstad R, Holmen A, Pfeifer P, Venvik HJ. Synthesis of dimethyl ether from syngas in a microchannel reactor—simulation and experimental study. Chemical Engineering Journal 2011;167: 610–615.

138 Avci AK, Trimm DL, Önsan ZI. Heterogeneous reactor modeling for simulation of catalytic oxidation and steam reforming of methane. Chemical Engineering Science 2001;56: 641–649.

139 Young LC, Finlayson BA. Axial dispersion in nonisothermal packed bed chemical reactors. Industrial & Engineering Chemistry Fundamentals 1973;12: 412–422.

140 Mears DE. The role of axial dispersion in trickle-flow laboratory reactors. Chemical Engineering Science 1971;26: 1361–1366.

141 Hlaváček V, Hofmann H. Modeling of chemical reactors—XVII: Steady state axial heat and mass transfer in tubular reactors. Numerical investigation of multiplicity. Chemical Engineering Science 1970;25: 187–199.

142 Luss D, Amundson NR. Uniqueness of the steady state solutions for chemical reactor occurring in a catalyst particle or in a tubular reaction with axial diffusion. Chemical Engineering Science 1967;22: 253–266.

CHAPTER 4

Fluidized-bed catalytic reactors

John R. Grace
Department of Chemical and Biological Engineering, The University of British Columbia, Vancouver, British Columbia, Canada

Abstract

Fluidized-bed reactors are used for solid-catalyzed reactions when their desirable features (favorable heat transfer, temperature uniformity, high effectiveness factors, low pressure drop, and ability to add/remove catalyst) outweigh their disadvantages (entrainment, attrition, wear, nonuniform residence time distributions, unpredictability). Because of their complexity and the large cast of variables which affect their performance, design and scale-up of fluidized-bed reactors rely on experience, as well as mechanistic understanding and models. This chapter provides an overview of key factors which affect the behavior and performance of catalytic fluidized-bed reactors operating in the flow regimes of practical interest.

4.1 Introduction

Fluidization of a bed of particles occurs when the pressure drop caused by upward flow of a gas or liquid through the bed is sufficient to support the bed weight minus buoyancy. The particles then begin to move relative to each other, acting somewhat like liquid elements of a boiling liquid, and hence the particles are said to be "fluidized." Given their many applications, challenges, and fascinating appearance, fluidized beds have attracted an enormous amount of attention since being introduced in the 1920s. Fluidization is now a mature technology, used widely, not only for catalytic reactions (the subject of this chapter) but also for gas–solid reactions like combustion and ore roasting and for physical operations like drying, coating, and granulation.

For comprehensive treatments of gas–solid fluidization, see Hetsroni [1], Yates [2], Davidson et al. [3], Geldart [4], Pell [5], Kunii and Levenspiel [6], Grace et al. [7], and Yang [8]. While fluidization can be applied to liquid–solid and three-phase (gas–liquid–solid) systems, the great majority of applications, especially in terms of catalytic processes, are for gas–solid systems, and this chapter is limited to this case. Readers interested in three-phase catalytic fluidized-bed reactors should consult Fan [9]. For a detailed review of spouted bed catalytic reactors, again not dealt with in this chapter but with linkages to it, see Rovero and Piccinini [10].

4.1.1 Advantages and disadvantages of fluidized-bed reactors

Since the 1940s, fluidized beds have provided a means of carrying out solid-catalyzed gas-phase reactions based on some significant features and advantages relative to competing packed-bed continuous flow reactors. In particular,

- Heat transfer is approximately an order of magnitude more favorable between fixed heat transfer surfaces (immersed tubes or wall of vessel) and gas-fluidized beds than between the same surfaces and packed beds. This is important for many reactions, exothermic and endothermic, where it is necessary to provide cooling or heating, respectively.
- Because of the rapid internal particle mixing in gas-fluidized beds, temperatures tend to be relatively uniform (typically not varying by more than 5–10°C within the bed itself), helping to avoid hot spots, explosions in the case of exothermic reactions, and undesirable side reactions.
- The catalyst particles used in most catalytic fluidized-bed reactor processes are small enough, typically less than 100 μm in diameter, that catalyst effectiveness factors are close to 1. As a result, intraparticle diffusional resistances tend to play a minor role in fluidized-bed catalytic processes, unlike in packed-bed reactors where catalyst particles are more than an order of magnitude larger.
- Pressure drops in fluidized beds are limited to those needed to support the weight of the bed minus buoyancy (wall friction playing a negligible role). Hence pressure drops tend to be much less than for packed-bed processes, resulting in substantial savings in compressor/blower operating costs.
- Due to the vigorous internal particle motion, it is possible to feed significant quantities of liquid into many fluidized beds without destroying the fluidization. This is important, for example, when it is desirable for hydrocarbons to be fed in liquid form into a reactor containing a catalyst.
- In cases where catalyst deactivation by coking occurs, catalyst particles can be readily withdrawn and replaced, or even

Multiphase Catalytic Reactors: Theory, Design, Manufacturing, and Applications, First Edition. Edited by Zeynep Ilsen Önsan and Ahmet Kerim Avci.

Table 4.1 Some typical key characteristics of catalytic fluidized-bed reactors compared with those of alternative types of reactor.

Characteristic (usual)	Moving packed bed	Turbulent fluidized bed	Circulating fluidized bed	Transported bed
Av. particle dia. (mm)	>1	0.04–0.1	0.05–0.3	<0.1
Particle size distribution	Usually narrow	Can be very broad	Intermediate	Limited
Sup. gas velocity (m/s)	1–2	0.3–0.6	3–12	10–16
Gas axial mixing	~ Plug flow	Axially dispersed plug flow	Axially dispersed plug flow	~ Plug flow
Part. axial mixing	~ Plug flow	~ Perfect mixing	~ Perfect mixing if particles do many cycles	~ Plug flow if once-through
Mean residence time of particles	Days to years depending on catalyst activity	Hours to months depending on catalyst activity	Seconds to days depending on catalyst activity	Seconds if once-through
Particle losses	None	Significant	Substantial	Substantial
Attrition/wear	Virtually none	Can be significant	Can be substantial	Can be significant

circulated to another vessel for online regeneration by burning off the coke, also taking advantage of recirculation of the heat generated by the coke oxidation.

- There is no upper limit on the scale of fluidized-bed reactors that can be constructed and operated. Hence reactors can be many meters in diameter, generally much larger than for competing types of reactor, where factors like radial temperature variations tend to limit the reactor diameter.

To set against these advantages, there are a number of inherent disadvantages of fluidized beds as catalytic reactors:

- Because of substantial axial mixing, both conversions and selectivities to intermediate products tend to be significantly less favorable than for packed-bed reactors of equal catalyst particle volume operating at similar temperatures and pressures.
- Further decreases in conversion and selectivity occur when gas bypasses contact with particles by traveling as bubbles or slugs in some flow regimes (see Sections 4.2.4 and 4.4.2).
- Catalyst particles tend to be entrained from the bed, requiring gas–solid separation equipment (e.g., cyclones, filters) and leading to catalyst losses. In particular, "fines" tend to be lost, since these are most readily entrained or "elutriated." The loss of fines can cause direct economic loss, pollution, and changes in reactor performance.
- Moving catalyst particles also tend to undergo attrition as they interact and collide with each other and with fixed surfaces.
- The moving catalyst particles also tend to cause wear of solid surfaces (heat exchanger tubes, baffles, piping, cyclones, vessel walls, etc.).
- The hydrodynamics, mixing, and contacting patterns in the fluidized bed are complex, with the result that it is difficult to predict accurately the performance of fluidized-bed reactors. The resulting uncertainty causes unwelcome risk relative to packed-bed reactors and other reactors which can be characterized more accurately and with greater security.
- If a makeup stream is needed to replace deactivated catalyst on a continuous or periodic basis, the nearly perfect mixing of the particles means that some of the replaced catalyst particles leave the reactor after only being present there for brief periods, leading to a suboptimal residence time distribution of the catalyst.

Typical properties of both turbulent and circulating fluidized-bed reactors are compared with those of moving (packed)-bed reactors and transported-bed reactors in Table 4.1.

4.1.2 Preconditions for successful fluidized-bed processes

Some of the conditions that are helpful, and in some cases required, to operate successful fluidized-bed catalytic processes are as follows:

- Particles must be robust enough to withstand collisions with each other and fixed surfaces without excessive breakage.
- Particle shapes must not be extreme (e.g., flaky or needlelike). Smooth rounded particle shapes are preferred.
- Sauter mean particle diameters should not be less than about 50 μm. It is helpful for there to be a broad range of sizes around the mean, for example, from approximately 5 to 150 μm.
- To avoid "bridging" of particles, that is, maintain smooth flow conditions, the minimum gap width inside the containing vessel (e.g., between heat transfer surfaces) should never be less than 20 times the mean particle diameter, with 50 times being preferred to give a margin of safety. Even larger ratios may be needed for unusually broad particle size distributions or extreme particle shapes.
- To assure uniform introduction of gas at the bottom of the bed, the fluidized bed should have a pressure drop across it of at least 30% of the pressure drop needed to support its weight. In addition, the number of gas-introduction points must be large enough to evenly distribute the incoming gas over the column cross section.
- Care is needed to ensure that the column is mounted truly vertically.
- Reentry of particles captured in cyclones and feeding of fresh particles should be well below the bed surface, normally just above the distributor. For vessels larger than about 1 m in diameter, multiple evenly spaced reentry points are desirable.
- If the walls of the vessel are tapered, the angle of taper to the horizontal should exceed the angle of repose of the particles.
- Any fixed surfaces immersed in the bed (such as heat transfer tubes) should be vertical or horizontal, not oblique.
- If there are multiple immersed open surfaces, particles should contact them on the outside, not the inside, to avoid stability issues leading to flow maldistribution and blockage of some passages [11].
- Sufficient height above the bed ("freeboard") is needed for particles to disengage from the gas stream. (See discussion of transport disengagement height (TDH) in Section 4.3.5.)

Table 4.2 Solid catalyzed gas-phase reactions which have been carried out in commercial fluidized-bed reactors.

Process	Comments
Fluid catalytic cracking	Widespread successful process
Acrylonitrile	Very successful SOHIO ammoxidation process
Ethylene oxide	Oxychlorination of ethylene successful process
Phthalic anhydride	Naphthalene partial oxidation; supplanted by fixed-bed process
Fischer–Tropsch synthesis	Several competing fluidized-bed processes
Polyolefin production	UNIPOL and BP processes compete
Maleic anhydride	Partial oxidation of butane; competes with fixed-bed process
Catalytic reforming	Supplanted by fixed-bed process
Aniline	Highly exothermic, well suited to fluid bed
Methanol to olefins	Reactor/regenerator system
Benzene reduction	Raises octane content while reducing benzene
Olefins to gasoline	Converts olefins to C5+ hydrocarbons
Isophthalonitrile	Ammoxidation of *m*-xylene
Catalytic oxidation	Means of low-temperature oxidation with favorable heat transfer
Steam methane reforming	So far unable to displace fixed-bed in-furnace process

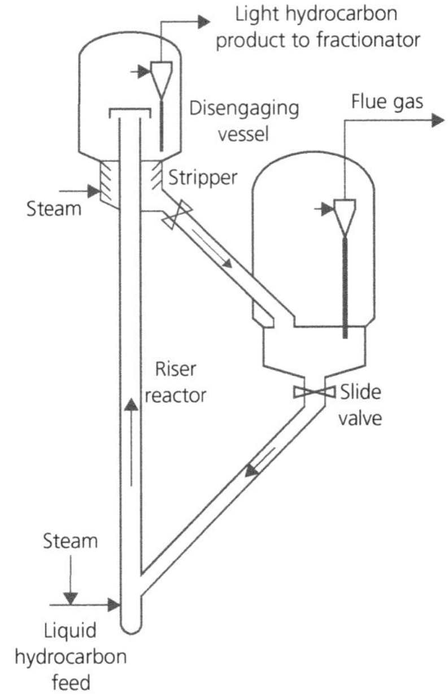

Figure 4.1 Schematic of principal features of a modern fluid catalytic cracking (FCC) reactor system.

4.1.3 Industrial catalytic processes employing fluidized-bed reactors

Reviews of catalytic applications of fluidized-bed reactors have been published by Geldart [12], Yerushalmi [13], Kunii and Levenspiel [6], and Jazayeri [14]. Major catalytic processes conducted in fluidized beds are identified, accompanied by brief commentary, in Table 4.2.

The most widespread fluidized-bed catalytic process is fluid catalytic cracking (FCC) of hydrocarbons [15, 16]. There are more than 400 FCC reactors operating worldwide, responsible for converting feedstocks of oil into liquid petroleum and heavier products. A key feature is that the FCC catalyst particles deactivate very quickly (in the order of seconds) due to coking on the surface; the coke is burned off in a fluidized-bed regenerator, both reactivating the catalyst and providing the heat needed for the endothermic cracking reactions when the hot catalyst particles are recirculated to the FCC reactor. Figure 4.1 shows a schematic which portrays the main features of modern FCC units, where catalyst is circulated through a riser reactor in which endothermic cracking occurs and then, via a steam stripper, to a regenerator in which coke is burned off the particle surface, before the hot catalyst particles repeat their circulation through the loop. In practice, there are many variants in configuration of FCC reactors [16].

In a manner similar to FCC, fluidized beds have also been utilized for catalytic reforming of straight-chain naphtha hydrocarbon compounds into branched-chain compounds of higher octane number. While most steam reforming of natural gas to produce hydrogen has been carried out in fixed-bed reactors, there have also been periodic efforts to conduct steam methane reforming (SMR) in fluidized beds.

One of the earliest successful catalytic fluidized-bed operations was to produce phthalic anhydride by partial oxidation of naphthalene. However, as xylene replaced naphthalene as the favored feedstock, these reactors were replaced by packed-bed reactors. Partial oxidation of butane to maleic anhydride has also been carried out in fluidized-bed reactors. On the other hand, nearly all acrylonitrile has been produced by fluidized-bed catalytic reactors, taking advantage of the excellent heat transfer capabilities of fluidized beds given the highly exothermic nature of the primary reaction. A schematic showing the main features of acrylonitrile synthesis reactors is provided in Figure 4.2. Aniline is sometimes made by hydrogenation of nitrobenzene in a fluidized-bed reactor. Ethylene dichloride is widely produced in fluidized-bed catalytic reactors by oxychlorination of ethylene. The catalytic methanol to olefins fluidized bed process has recently been commercialized in China [17]. In addition to these applications, synthesis of hydrocarbons from syngas can be carried out by Fischer–Tropsch synthesis in fluidized-bed reactors, utilizing different flow regimes [18]. South Africa has particular experience in this domain. One of the primary methods of producing polyethylene and polypropylene (polyolefin) particles involves dry polymerization around catalyst kernels in fluidized-bed reactors [19]. In this case the particles grow to be roughly 0.5–1 mm in diameter, and the reactor has more in common with gas–solid fluidized-bed reactors than with other catalytic fluidized-bed reactors.

Many other catalytic fluidized-bed processes have been tested at various scales. These include catalytic low-temperature oxidation, catalytic gasification and pyrolysis of biomass and waste plastic, production of carbon nanotubes, dry reforming of methane, hydrogenation and dehydrogenation of hydrocarbons, methanol-to-gasoline (MTG) process, synthesis of dimethyl ether (DME), and selective catalytic reduction of nitrogen

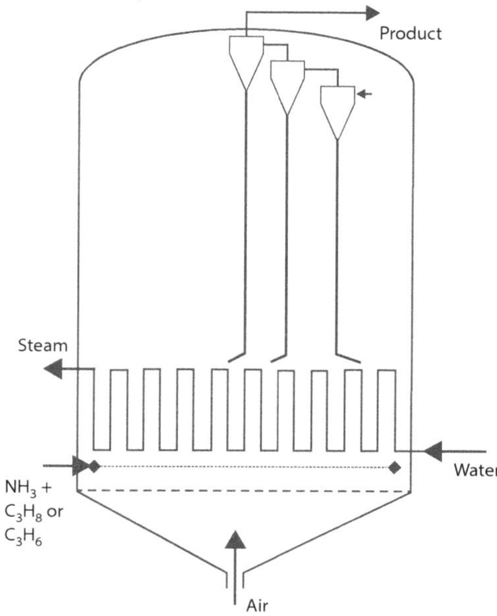

Figure 4.2 Schematic of main features of fluidized-bed reactor for production of acrylonitrile showing immersed serpentine heat transfer tubing, separate distributors for air and ammonia + propane or propene, and one representative cluster of internal cyclones in series.

Table 4.3 Typical operating ranges and features of catalytic fluidized-bed reactors.

Property	Normal range
Mean particle diameter	50–100 μm
Particle size distribution	Broad (e.g., 0–200 μm)
Reactor inner diameter	Up to ~7 m
Operating pressure	Up to ~80 bars
Temperature	Up to ~600°C
Superficial gas velocity	~0.3–0.6 m/s for bubbling or turbulent flow regime; 3–12 m/s for fast fluidization or dense suspension upflow
Static bed depth	~1–10 m
In-bed surfaces	Horizontal or vertical tubes may be inserted as heat transfer surfaces
Gas–solid separation	Heavily reliant on cyclones

oxides. Typical ranges of properties and operating conditions of the aforementioned catalytic gas-fluidized bed processes are identified in Table 4.3.

Table 4.4 compares the typical properties of major catalytic fluidized-bed processes, like those identified in Table 4.2, with corresponding practice for major gas–solid processes like combustion, gasification, roasting, incineration, and calcination.

4.2 Key hydrodynamic features of gas-fluidized beds

It is very important to understand key elements of the hydrodynamic behavior of gas-fluidized beds before planning new applications or undertaking design, since the flow characteristics affect gas–solid contacting, heat and mass transfer, dispersion of the gas as it passes through the bed, entrainment, attrition, wear, and so on, all of which are related to overall process and reactor performance.

4.2.1 Minimum fluidization velocity

When gas is passed upward through a bed of solid particles, a minimum flow of air is needed to fluidize the particles. The corresponding "minimum fluidization superficial velocity" (volumetric gas flow divided by total column cross-sectional area) is commonly predicted, with approximately ±20% accuracy, by an equation of the form

$$U_{mf} = \frac{\mu}{\rho d_p}\left\{\sqrt{C_1^2 + C_2 Ar} - C_1\right\} \tag{4.1}$$

$C_1 = 33.7$ and $C_2 = 0.0408$, fitted by Wen and Yu [20], are the most common values of these constants among many pairs proposed in the literature. Ar is the Archimedes number given by

$$Ar = \frac{\rho\left(\rho_p - \rho\right) g d_p^3}{\mu^2} \tag{4.2}$$

and d_p is the Sauter mean particle diameter, usually determined by screening the particles:

$$d_p = \left(\sum \frac{x_i}{d_{pi}}\right)^{-1} \tag{4.3}$$

Table 4.4 Comparison of typical properties of catalytic and gas–solid fluidized-bed reactors.

	Bubbling or turbulent bed		Circulating fluidized bed	
Feature	**Catalytic reaction**	**Gas–solid reaction**	**Catalytic reaction**	**Gas–solid reaction**
Cross-sectional shape	Circular	Rectangular or circular	Circular	Rectangular
Mean particle diameter (μm)	<100	>500	<100	>250
Geldart group	A	B or D	A	A or B
Temperature (°C)	Up to 600	Up to 1000	Up to 600	Up to 900
Pressure	Up to 80 bars	Up to 20 bars	Up to 5 bars	Atmospheric
Superficial gas velocity (m/s)	Up to 0.8	Up to 2	Up to 15	Up to 7
Overall voidage	0.7–0.8	0.65–0.75	0.8–0.95	0.9–0.98
Solid feeding	Minor makeup	Major factor	Minor makeup	Major factor

where d_{pi} is the mean screen opening size of the sieve which just retains the particles and the one with the smallest openings which just allows them to pass. When the catalyst particles are fine enough that $Ar < 1000$, then Equation 4.1 with the C_1 and C_2 values given earlier reduces to

$$U_{mf} = \frac{0.00061\left(\rho_p - \rho\right) g d_p^2}{\mu} \quad (4.4)$$

so that the minimum fluidization velocity is approximately 1/90th of the corresponding Stokes terminal velocity of single spherical particles of the same mean diameter. The corresponding void fraction at minimum fluidization (usually called minimum fluidization voidage and given the symbol ε_{mf} in the fluidization literature) is best determined experimentally. It is typically in the range of approximately 0.50–0.55 for catalyst particles of mean diameter <70 μm, but smaller for larger particles, for example, 0.40–0.43 for relatively coarse particles ($dp > 300$ μm), with broader particle size distributions tending to result in somewhat lower ε_{mf}.

4.2.2 Powder group and minimum bubbling velocity

Geldart [21] introduced a powder classification which is widely used to characterize particle behavior. For the particle mean diameters (typically 50–100 μm) and densities (typically 1200–2800 kg/m^3) of interest for catalytic fluidized-bed reactors, the particles fall into group A of the four Geldart classes of powder. Group A powders are generally well behaved. A notable feature is that U_{mb}, the minimum bubbling velocity (lowest superficial gas velocity at which bubbles form spontaneously), exceeds U_{mf} for these particles. In recent years there has been increasing interest in fluidization of ultrafine particles, including nanoparticles, with some of the intended applications being for catalytic processes. A useful review of this subject has been provided by Shabanian et al. [22].

4.2.3 Flow regimes and transitions

Figure 4.3 gives a diagrammatic representation of the sequence of flow regimes for gas-fluidized beds with increasing superficial gas velocity. Schematics of the most important flow regimes are provided in Figure 4.4. While the flow regimes are primarily determined by the Archimedes number and the superficial gas velocity, U, the column diameter plays a role in determining whether or not the slug flow regime is encountered, and the net upward flux of particles, G_s, is of central importance in determining the transition to dense suspension upflow (DSU). Five of these flow regimes are of particular importance for gas–solid fluidized-bed reactors—bubbling, slug flow, turbulent fluidization, fast fluidization, and DSU—and we deal with each of those in the next subsections.

4.2.4 Bubbling fluidized beds

The bubbling flow regime extends from the minimum bubbling velocity, U_{mb}, to the point where there is a transition either to slug flow (for bubbles which grow to be of similar size as

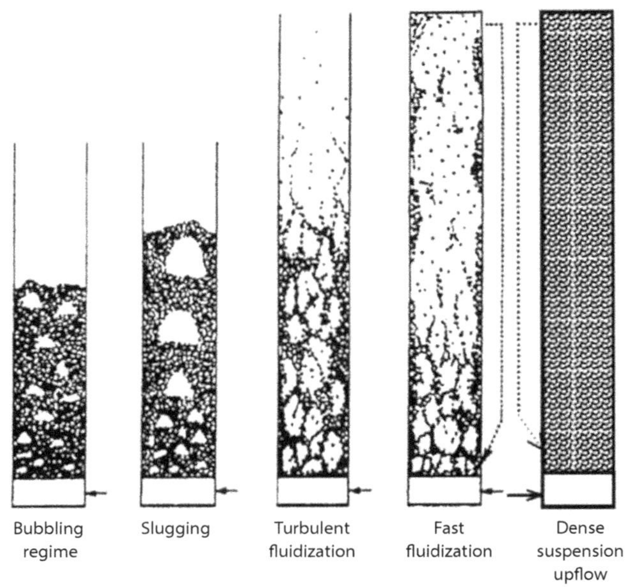

Figure 4.4 Pictorial representation of particles and void regions in flow regimes of principal interest for catalytic fluidized-bed reactors.

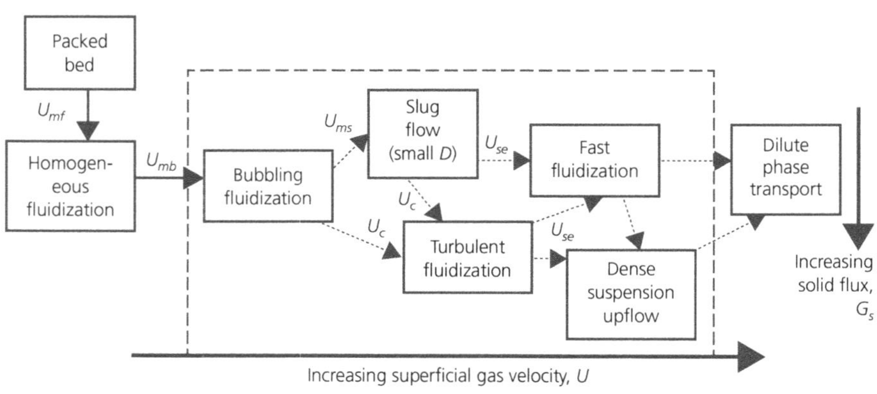

Figure 4.3 Diagrammatic representation of flow regimes of gas fluidization with increasing superficial gas velocity, with transition velocities indicated on the arrows between adjacent flow regimes. Dashed rectangular box encloses the five flow regimes of primary interest with respect to catalytic fluidized-bed reactors.

the column diameter, D) or to turbulent fluidization. In the bubbling regime, voids, commonly referred to as bubbles, occur due to instability of homogeneous fluidization. The bed is divided into two "phases": a *bubble phase*, denoted by subscript b, and the particles and interstitial gas outside bubbles making up the *dense phase*, denoted by subscript d. The flow associated with bubbles, Q_b, can, as a crude first approximation, be equated to the gas flow in excess of that needed for minimum fluidization, that is,

$$Q_b \approx \left(U - U_{mf}\right) A \tag{4.5}$$

However, one must be careful to realize that some of this flow passes upward through the bubbles, rather than being carried by translation of the void boundaries. To a first approximation, the dense phase is assumed to maintain a voidage equal to ε_{mf}, and the interstitial velocity in the dense phase is approximately U_{mf}/ε_{mf}.

Bubbles in fluidized beds tend to grow with height due to coalescence, countered to some extent by splitting. There is some evidence that bubbles approach a maximum stable size as they grow, a constraint that is particularly important for Geldart group A particles. A number of correlations have been proposed in the literature for the mean bubble diameter as a function of height. For one of the few such correlations that include the effects of both coalescence and splitting, as well as jet formation at the bottom of the bed, see Horio and Nonaka [23]. Alternatively, computational fluid dynamic (CFD) codes may be used to predict mechanistically the evolution of bubble size with height, with fair success, at least for beds of modest diameter, although there is a definite tendency for CFD codes to overpredict bed expansion for group A particles. For a review of the application of CFD to fluidized beds, see Lettieri and Luca [24].

In the bubbling regime, the bubbles are responsible for vigorous mixing of particles by transporting them in bubble wakes, as well as due to drift. Typically a volume of dense phase, with voidage ε_{mf}, approximately of the same order as the bubble volume, is displaced upward by each bubble and needs to be replaced by downward flux of particles elsewhere. This substantial induced particle motion is responsible for the excellent heat transfer to/from immersed or containing surfaces, as well as for the temperature uniformity and almost perfect mixing which characterize bubbling fluidized beds. On the other hand, the bubbles also lead to substantial interphase mass transfer resistances between the gas on the inside and outside of the bubbles as discussed in Section 4.3.4. The bubbles are sometimes assumed to contain no particles whatsoever, or they may be assumed to contain a small fraction (typically 0.1–1% by volume) of particles. Particles dispersed within the bubble phase play a very significant role for fast reactions.

While it is often helpful to consider the analogy between bubbles in liquids and in gas-fluidized beds, for example, affecting bubble rise velocities, shapes, and coalescence dynamics, one needs to recognize some very significant differences between gas–solid and gas–liquid systems, with profound impact:

- Gas–liquid systems have surface tension, affecting bubble shape and coalescence, whereas gas-fluidized beds are not subject to surface tension.
- Surface active agents play a major role in gas–liquid systems, but not in gas-fluidized beds.
- The boundaries of bubbles in fluidized beds are permeable, with the result that gas short-circuits through bubbles (entering the lower part and leaving through the roof). Gas can also be exchanged between bubbles without them coalescing. In addition, the total volume of bubbles involved in coalescence tends to grow (i.e., not be conserved) when coalescence is completed, presumably due to gaining gas flow from the dense phase.

4.2.5 Turbulent fluidization flow regime

When the overall bed voidage approaches a value of approximately 0.7–0.8, a transition takes place to a condition, analogous to the churn-turbulent flow regime in gas–liquid flow, where voids are in a continuous state of deforming, splitting, and coalescing. The onset superficial velocity to turbulent fluidization, designated U_c, is usually found experimentally by locating the superficial velocity corresponding to the maximum variance of differential pressure fluctuations. However, this transition, like other flow regime transitions in fluidized beds (see Figure 4.3), is gradual, rather than sharp, with U_c approximately corresponding to the onset of the transition range. Typically, $U_c \approx 0.3$–0.6 m/s for catalytic fluidized-bed reactors, with the level in the bed, as well as the particle size distribution, having some influence on the actual value.

The turbulent fluidization regime is transitional between bubbling, where gas bubbles containing very few particles are dispersed in a "continuous" dense phase composed of particles and interstitial gas, and the fast fluidization flow regime (see following text), where streamers of particles and interstitial gas are dispersed in a continuous phase, formed almost exclusively of gas. The turbulent regime is highly advantageous in terms of reactor characteristics, as the interphase mass transfer constraint that limits gas–solid contacting in the bubbling regime is largely obviated, while bed-to-surface heat transfer and temperature uniformity are excellent, even optimal. Hence, the turbulent flow regime is the flow regime of choice for many fluidized-bed catalytic reactors.

In the turbulent flow regime, voids are short-lived and distorted, darting to and fro, continuously coalescing and splitting. Hence the hydrodynamics are difficult to characterize, and, notwithstanding its importance, this flow regime has received much less attention from the research community than the bubbling flow regime. For an extensive review of the turbulent fluidization flow regime, see Bi et al. [25].

4.2.6 Fast fluidization and dense suspension upflow

If the superficial gas velocity is increased further than required for turbulent fluidization, a point is reached where particle entrainment is so large that there is no longer an identifiable bed surface. Instead there is a continuous suspension whose volumetric particle concentration varies continuously with height, decreasing over most of the column length, but possibly increasing toward the top, depending on the geometry of the exit there. The column is then often referred to as a "riser," and the entire system (riser, gas–solid separator(s), standpipe and nonmechanical device for returning entrained particles to the bottom of the riser, as well as the particles) is referred to as a "circulating fluidized bed" (CFB). For reviews of CFBs and fast fluidization, see Berruti et al. [26] and Grace et al. [27].

The minimum superficial velocity for fast fluidization can be predicted as

$$U_{FFmin} = 1.53\left(\frac{gd_p\left(\rho_p-\rho\right)}{\rho}\right)^{0.5} \tag{4.6}$$

In the "fast fluidization" flow regime, there is a core/annulus structure, with ascending flow of a dilute suspension of particles throughout the inner core of the column cross section and an outer annular region where particles on average are descending in concentrated streamers of voidage approximately 0.55–0.65. As a good approximation, frictional losses and acceleration effects on the pressure drop are of secondary importance, with the result that the pressure drop across the riser is directly related to the holdup of solids, that is,

$$(-\Delta P) = \rho_p g(1-\bar{\varepsilon})H \tag{4.7}$$

where $\bar{\varepsilon}$ is the overall voidage, that is, the fraction of the riser volume occupied by the gas. By determining the pressure profile along the riser, one can then estimate the profile of cross-sectionally averaged voidage, that is,

$$\varepsilon_{Av}(z) = 1 + \frac{1}{g\rho_p}\frac{dP}{dz} \tag{4.8}$$

The radial variation of voidage at any height z can then be estimated [28] by

$$\varepsilon(r,z) = \varepsilon_{mf} + \left[\varepsilon_{Av}(z)-\varepsilon_{mf}\right] \times \left[\varepsilon_{Av}(z)\right]^{\left(-1.5+2.1(r/R)^{3.1}+5.0(r/R)^{8.8}\right)} \tag{4.9}$$

The downward velocity at the wall is typically of order 1–2 m/s and has been correlated [29] by the simple empirical expression

$$v_{p(r=R)} = 36\sqrt{g\bar{d}_p} \tag{4.10}$$

The net upward flux of particles averaged over the column cross section is typically 20–200 kg/m^2 s in this fast fluidization flow regime.

Fast fluidization is commonly applied commercially for gas–solid reactions like combustion and gasification where the mean particle diameter is at least 200 μm, significantly larger than usual catalyst particles, and the superficial gas velocities are 4–8 m/s. This flow regime is less common for catalytic processes. A flow regime applicable to several catalytic processes including FCC is called "DSU" [30]. This regime is distinguished from fast fluidization by having particles ascending on average, not only in the core but also right to inner wall of the vessel, that is, upward flow through the entire cross section of the riser. This requires formidable solid fluxes, typically 300–1200 kg/m^2 s. To reach this flow regime, the superficial gas velocity is typically 8–15 m/s, and particle mean diameters are less than about 100 μm. The overall fraction of the reactor volume occupied by the particles is then typically 10–30%, with the system then often referred to as a "high-density CFB." In this flow regime, radial gradients are less severe than in the fast fluidization flow regime, though the time-mean solid volumetric concentration and particle velocity are again greater at the wall than in the interior of the riser. The major advantages of DSU over the fast fluidization regime are the greater utilization of the reactor volume, resulting in higher throughputs of gas and solids, and reduced axial dispersion of both gas and solids.

4.3 Key properties affecting reactor performance

Other important properties of catalytic fluidized-bed reactors—mixing characteristics, heat and mass transfer, entrainment, attrition, agglomeration, and wear—are outlined in this section.

4.3.1 Particle mixing

Particle mixing is important in processes where the ability of fluidized beds to add and remove particles on line is utilized to send spent (deactivated) catalyst to a separate regenerator column and to replace them by fresh and reactivated catalyst particles. The characteristic mixing time (often called "turnover time" for fluidized beds) is typically of order 20–200 s. If the mean residence time of the catalyst particles, τ, is much greater than this, the particles can be considered to be perfectly mixed. The exit age (or residence time) distribution of the particles is then given by

$$E(t) = \frac{e^{-t/\tau}}{\tau} \tag{4.11}$$

Perfect mixing will almost certainly be a very good approximation for solids in reactors operated in the bubbling and turbulent flow regimes, in which case the particle composition can be considered to be uniform over the entire reactor volume. However, for the DSU regime, where there is no downflow at the wall, there is substantially less dispersion, with mixing intermediate between plug flow and perfect mixing.

When there are broad particle size distributions, entrainment (see Section 4.3.5) of fines from the reactor (called "elutriation") may be contribute significantly to broadening of the catalyst particle residence time distribution. In these cases, it may be necessary to use a population balance (see Ref. [6]) to keep track of the various factors, including entrainment, which affect overall solid mixing and particle size distribution.

4.3.2 Gas mixing

Gas mixing is strongly related to particle mixing [31, 32]. For example, when particles descend at the wall of the column, they tend to drag gas down with them, contributing to "backmixing" and overall gas axial dispersion [33]. However, whereas most particles circulate multiple times through the reactor and its separation system (cyclones and standpipes), gas elements make a single pass. Radial mixing is also more likely to be important for the gas than for the particles in catalytic reactors, especially in reactors where gas or vaporizing liquid reactants are introduced as horizontal jets from the periphery of the vessel. Vertical gas mixing is often characterized by application of an axial dispersion coefficient, $D_{g,ax}$, and a corresponding dimensionless Peclet number, $Pe_{g,ax} = UH/D_{g,ax}$, where U is the superficial gas velocity and H is the bed depth for a bubbling, slugging, or turbulent bed, or column height in the case of a circulating fluidized bed riser. For fluidized beds, $Pe_{g,ax}$ is then intermediate between perfect mixing (for which $Pe_{g,ax} \to 0$) and plug flow (where $Pe_{g,ax} \to \infty$).

For bubbling beds, the dispersion model, which assumes small-scale random displacements, is not appropriate, given the large-scale and deterministic bubble displacements. Bubble rising velocities also differ from the velocity of interstitial gas percolating through the dense phase. In addition, there are velocity gradients associated with particle motion induced by bubbles. Notwithstanding the inappropriateness of the dispersion model, it is commonly applied in practice, resulting in a very wide range of values of dispersion coefficients [32]. In the turbulent fluidization flow regime, gas mixing results from more random and smaller-scale steps, with $Pe_{g,ax}$ typically in the range 5–20. In the DSU flow regime, there is substantially less axial dispersion, and it is likely that $Pe_{g,ax} \approx 10\text{–}50$.

4.3.3 Heat transfer and temperature uniformity

Catalyst particles in fluidized-bed processes tend to be small enough (mostly <100 μm in diameter) that their thermal Biot number <0.1, with the result that each particle can be treated as having a uniform temperature, that is, negligible internal temperature gradients. In addition, particle mixing within the bed for both the bubbling and turbulent flow regimes is normally rapid enough that the entire bed, beyond any jets or a region no deeper than about 10 mm at the bottom if there are no jets, can be treated as being of uniform temperature, with variations of temperature within the dense bed not exceeding about 10°C. Much larger temperature gradients and differences are, however, possible in the freeboard region.

Because fluidized-bed catalytic processes tend to have high heats of reaction, it is important to be able to take advantage of their favorable bed-to-surface heat transfer capability. Heat transfer surfaces should be either horizontal or vertical, not at any other orientation. Overall average heat transfer coefficients are nearly the same for vertical and horizontal tubes, despite gradients in local heat transfer coefficients in the latter case. The choice of whether to insert vertical or horizontal tubes therefore depends on other factors, such as the height/diameter ratio of the reactor and the degree of concern over attrition and tube wastage, both of which are bound to be greater for horizontal tubes than for vertical ones.

Excellent reviews of heat transfer in fluidized beds have been provided in books by Botterill [34] and Molerus and Wirth [35]. Many empirical and semiempirical correlations are available for predicting bed-to-surface heat transfer coefficients. The recommended one is that of Molerus and Wirth [35]:

$$\frac{h\lambda}{k_g} = \frac{0.125\left(1-\varepsilon_{mf}\right)\left[1+33.3\left(\sqrt[3]{\frac{U-U_{mf}}{U_{mf}}}\sqrt[3]{\frac{\rho_p c_p}{k_g g}}\left(U-U_{mf}\right)\right)^{-1}\right]^{-1}}{1+\frac{k_g}{2c_p\mu}\left[1+0.28\left(1-\varepsilon_{mf}\right)^2\left(\frac{\rho_g}{\rho_p-\rho_g}\right)^{0.5}\left(\sqrt[3]{\frac{\rho_p c_p}{k_g g}}\left(U-U_{mf}\right)^2\frac{U_{mf}}{\left(U-U_{mf}\right)}\right)\right]} + 0.165\ \mathrm{Pr}^{1/3}\left(\frac{\rho_g}{\rho_p-\rho_g}\right)^{1/3}\left[1+0.05\left(\frac{U_{mf}}{U-U_{mf}}\right)\right]^{-1} \tag{4.12}$$

$$\text{where}\ \lambda = \left[\frac{\mu}{\sqrt{g}\left(\rho_p-\rho_g\right)}\right]^{2/3} \tag{4.13}$$

These two equations mechanistically account for contributions from both particle convection to the fixed surface (large first term on right side of Eq. 4.12) and gas convection (second term commencing with 0.165), the former term being the dominant term for the relatively fine particles of interest in catalytic fluidized-bed reactors. For typical bubbling bed operating conditions at atmospheric pressure and for particles of mean size 50–100 μm, the overall bed-to-immersed-surface heat transfer coefficient (excluding radiation) can be expected to be approximately 250–400 W/m^2 K. Heat transfer is somewhat more favorable in the turbulent flow regime, with bed-to-surface coefficients below the bed surface of approximately 360–500 W/m^2 K. Somewhat lower values are expected in the DSU regime, on the order of 300–450 W/m^2 K.

In the freeboard region for bubbling and turbulent beds, the heat transfer coefficient falls off rapidly with increasing height as the particles, responsible for enhancing heat transfer, are disengaged and fall back onto the bed surface. If reactor temperatures exceed approximately 600°C, radiation also contributes appreciably to the overall bed-to-surface heat transfer. For start-up, shutdown, and changes of operating conditions, it is important to be able to estimate the bed temperature as a function of time. This requires an energy balance which must account for:

Heat transfer to reactor from heat transfer surfaces +
heat released by all reactions – heat loss to surroundings =
heat gain by all reactants and inerts in feed stream to bring
them to reactor temperature + heat gain by reactor shell (4.14)

When applying this macroscopic balance, one can generally include the simplifications that the dense bed remains essentially isothermal (spatially uniform temperature from the distributor up to the bed surface) and that the freeboard plays a secondary role in determining the evolution of the bed temperature. Note that the heat gain or loss by the vessel itself, especially if refractory-lined, can play a significant role in determining the transient behavior, especially in small vessels. Also, because the reaction rate is strongly temperature dependent, there is direct coupling between the temperature and the reaction rate.

4.3.4 Mass transfer

Because of their small diameters (<100 μm), fluidized-bed catalyst particles tend to have effectiveness factors close to unity, that is, it is possible to neglect intraparticle mass transfer resistances. These small particle sizes also lead to high surface-to-volume ratios, with one result being that external resistance to mass transfer can also commonly be neglected. For bubbling beds, the rate-determining mass transfer resistance is then generally that associated with interphase mass transfer, with bubbles having compositions which differ from that of the surrounding dense phase. Transfer across the bubble boundary takes place by two mechanisms which can be approximated as being additive. One is convective (associated with short-circuiting of gas from bottom to top through bubbles (often called "throughflow")), while the other is diffusional. To a good approximation, the overall interphase (bubble-to-dense-phase) mass transfer coefficient can then be estimated from the correlation of Sit and Grace [36]:

$$k_{bd} = \frac{U_{mf}}{3} + \sqrt{\frac{4D_{AB}\varepsilon_{mf}\bar{U}_b}{\pi \bar{d}_b}} \qquad (4.15)$$

D_{AB} is the molecular diffusivity of gaseous component A in gas B, whereas $\bar{d}_b$ and $\bar{U}_b$ are the average bubble diameter and bubble rising velocity, respectively. The first term on the right side of Equation 4.15 accounts for gas throughflow and is the dominant term for large particles, whereas the second term accounts for diffusion and is dominant for small particles. For catalytic fluidized-bed reactors, the interphase mass transfer resistance is almost always much greater than the external mass transfer resistance from interstitial gas to particles in the dense phase. From Equation 4.15, it is seen that keeping bubbles as small as possible, for example, by clever distributor design or by adding baffles, is helpful in promoting favorable interphase mass transfer. At atmospheric pressure and for typical fluidized-bed catalyst particles of size 50–100 μm, the resulting interphase coefficient is typically of order 0.01–0.03 m/s.

Given their lack of robust bubbles, interphase transfer resistances are much less likely to be rate limiting for either the turbulent fluidization or the DSU regimes. Hence chemical kinetics is usually rate controlling, at least for exothermic reactions, in these two flow regimes. For endothermic catalytic reactions, the supply of sufficient heat is likely to also play a significant role.

4.3.5 Entrainment

One disadvantage of fluidized-bed reactors is the loss of particles by carryover from the bed. In catalytic processes, this results in the loss of catalyst, as well as air pollution. Hence air separation devices, most commonly primary cyclones, usually backed up by one or more of secondary cyclones, filter bags, scrubbers, or electrostatic precipitators, are provided to capture entrained particles and return them to the bottom of the reactor vessel. The freeboard region of the reactor may also be enlarged in cross-sectional area to assist in separating of particles from the gas stream in the freeboard region. Since finer particles are more readily entrained than coarse ones, they are preferentially removed, a process called "elutriation." As a result, the particle size distribution tends to become coarser over time, with the loss of fines often negatively affecting the bed hydrodynamics. On the other hand, if the particles undergo significant attrition (see section 4.3.6), their size distribution tends to become finer over time, despite losses of fine particles due to entrainment.

In bubbling and slugging fluidized beds, clumps of particles are ejected into the freeboard by bubbles or slugs erupting at the bed surface. The bursting bubble and slugs also generate gas turbulence in the freeboard region. Many of the ejected particles return to the bed surface in what is commonly called the "splash zone." Disengagement tends to occur by particles reaching the reactor wall [37], where the upward gas velocity approaches 0, as required by the "no-slip" boundary condition. The distance from the expanded bed surface to the top of the splash zone is designated as the TDH. This height is the subject of a number of equations and is commonly adopted as the height of the freeboard region, as making the column taller would have negligible impact in terms of disengaging more particles. The graphical correlation of Zenz and Weil [38] is commonly used to provide conservative estimates of the TDH in reactor design.

Entrainment rates are notoriously difficult to predict accurately. There are many correlations (see Ref. [39] for a review), but they commonly give predictions that differ from each other by orders of magnitude, and no methods are widely accepted as being accurate or even "best available." To minimize losses, the design of effective particle separation equipment (mostly cyclones) for capture of entrained particles and the provision of return standpipes and pneumatic or mechanical valves for control of the returning particles to the bottom of the reactor are of critical importance.

4.3.6 Attrition

Particles in fluidized beds undergo collisions and frictional contacts with each other and with fixed surfaces, sometimes causing the particles to break [40, 41]. The most important mechanisms of attrition are impact attrition (also called fragmentation) and abrasion. Impacts can be especially energetic, and therefore likely to cause attrition, when particles are accelerated in distributor jets, feed jets, or cyclone entrances and then collide with fixed surfaces or stationary particles.

Impact attrition generates particles smaller than, but of the same order of size as the particles undergoing attrition. If impact attrition is dominant, the particle size distribution shifts gradually toward smaller sizes. On the other hand, abrasion occurs when the surfaces of rough particle rub together, releasing large numbers of much finer particles from the surface and resulting in a bimodal particle size distribution. A critical factor in fluidized-bed catalytic reactors is to assure that the catalyst particles are robust enough to withstand the normal and shearing forces that give rise to the attrition. Minimizing attrition is also favored by operating at modest superficial gas velocities and avoiding high-velocity jets interacting with fixed surfaces. Vertical surfaces also give much less attrition than horizontal ones.

4.3.7 Wear

Wear (also called wastage and erosion) of surfaces is a serious operational issue in some fluidized-bed reactors. Wear occurs when hard particles (e.g., silica-supported catalyst particles) continually strike fixed surfaces such as heat transfer tubes, reactor walls, or cyclone inner surfaces. The most damaging collisions tend to be those which are oblique (e.g., at 60°) to the surface, for example, at about the 5 and 7 o'clock positions, when bubble wakes slam into the underside of horizontal heat transfer tubes. If corrosion is also a factor, then the combined damage from erosion and corrosion can be considerably more extensive than estimated from the summation of the individual effects.

The material properties of the surface, as well as the hardness of the particles, influence the degree of wear. Careful design is needed to minimize impacts of particles traveling at high velocity on fixed surfaces. This can be accomplished by ensuring that grid jets and feed jets do not impinge directly on fixed (immersed or wall) surfaces. Another means of reducing wear is to cover endangered surfaces with refractory.

4.3.8 Agglomeration and fouling

Agglomeration and fouling of the reactor inner walls and fixed surfaces can also be challenging issues in some fluidized-bed processes, though this is more likely to be the case when the particles are themselves reacting rather than acting as catalysts. Particles can become sticky due to the presence of liquids or because of sintering as the temperature increases. Note that sintering can occur at temperatures well below the melting point [5]. To prevent agglomeration and minimize fouling, it is important to keep particles in motion throughout the entire reactor, for example, by avoiding corners, ledges, and upward-facing surfaces where particles tend to sit in a stagnant manner.

4.3.9 Electrostatics and other interparticle forces

Electrostatics can be a serious issue in fluidized beds, contributing to fouling and agglomeration, especially in polymerization processes [19, 42]. Electrostatic charges are generated when dielectric particles contact and then separate from each other and fixed surfaces. Bipolar charging can cause large particles to have different charge polarity than small ones, even when the particles are made of the same material. This can then affect entrainment, capture efficiency, and in-bed net charges as the fine particles are preferentially entrained. In some processes, antistatic agents or humidity are introduced to reduce the charging levels. Charges are recorded continuously in some reactors as a means of monitoring the process stability.

As noted earlier, most fluidized-bed catalysts belong to group A of the Geldart [21] powder classification scheme. These particles encounter significant interparticle forces (relative to gravity). Van der Waals forces affect fluidization properties, especially for fine particles and low excess superficial gas velocities ($U - U_{mf}$). For example, these forces can cause ultrafine particles to form clusters, increasing the effective particle size. If interparticle forces become excessive during operation, for example, due to liquid bridging on the surfaces of particle, they can lead to "bogging" (local or widespread defluidization).

4.4 Reactor modeling

4.4.1 Basis for reactor modeling

The complex fluid and particle dynamics of gas fluidization complicate the task of modeling fluidized-bed reactors. Reactor models serve as educational tools to show what is important, simulation tools for interpolation, interpretation of problems and process control, tools for carrying out "what-if" exercises, and predictive tools for design and scale-up. These models range from simple empirical models to sophisticated mechanistic, for example, based on CFD with reaction terms included. Mechanistic models that capture the essence of the rate-limiting processes, without being encumbered by factors of lesser significance, have the greatest chance of being applicable over a broad range of reactions and reactor configurations. For an extensive review of fluidized-bed reactor models, see Mahecha-Botero et al. [43].

In considering what to incorporate in models, one needs to keep in mind the long list of factors that can influence overall performance of catalytic fluidized-bed reactors:

- Hydrodynamics in the bed and freeboard
- Distributor effects; baffles and geometric effects
- Mass transfer (gas/particle, bed/surface)
- Heat transfer (bed/surface, gas/particle)

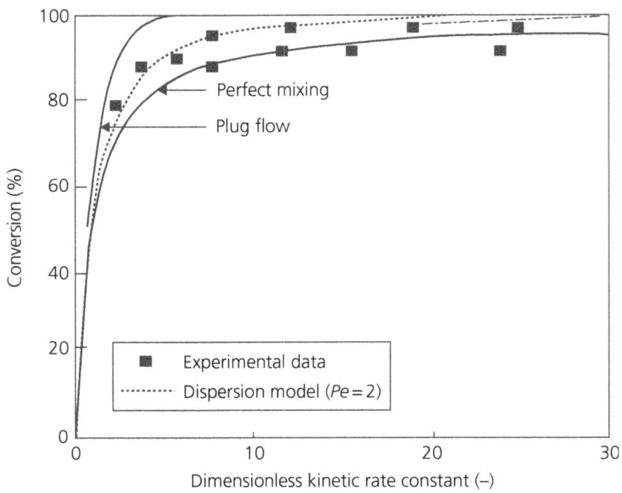

Figure 4.5 Simple single-phase model predictions for first-order irreversible catalytic ozone decomposition reaction in comparison with experimental fluidized-bed reactor data of Sun and Grace [44].

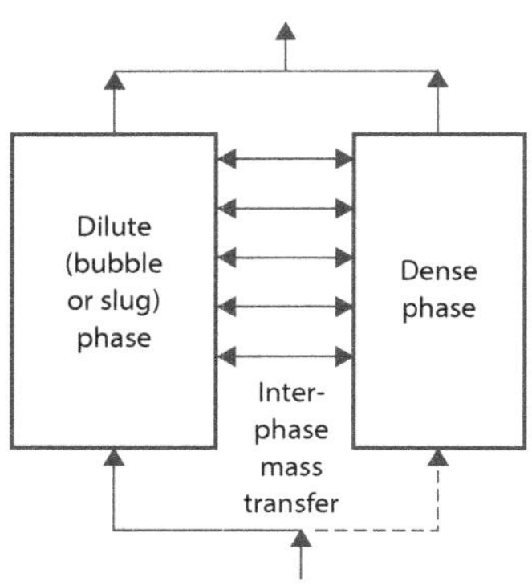

Figure 4.6 Schematic of two-phase model representation of bubbling or slugging fluidized-bed reactor.

- Mixing (axial and lateral, gas and solids)
- Chemical kinetics, including deactivation
- Approach to chemical equilibrium
- Thermochemical properties, such as equilibrium constants and heats of reaction
- Particle size distribution, attrition, and entrainment

Models typically focus on only a subset of the items in this list in order to include the factors of greatest influence.

4.4.2 Modeling of bubbling and slugging flow regimes

Figure 4.5 illustrates that simple single-phase models (plug flow and perfect mixing) do not give good predictions relative to experimental results obtained [44] in a laboratory reactor of 100 mm diameter. This is the case also in larger fluidized-bed reactors. To predict the performance of catalytic fluidized beds operating in the bubbling or slug flow regime, it is essential to consider their two-phase nature and, in particular, to make allowance for the interphase mass transfer resistance (see Eq. 4.15).

To formulate a two-phase model, one proceeds by writing separate species mole balances for the dilute phase (e.g., bubble or slug phase) and the dense phase, which are assumed to be in parallel, as shown in Figure 4.6, with exchange of gas by interphase mass transfer. Different models are available, depending on the assumptions adopted [45]. For a reactor at steady state, one must include terms for flow in and out of the control volume, consumption or production of the species by catalytic reaction, and transfer between the phases. For example, consider a simple case where there is an nth-order constant-volume irreversible solid-catalyzed reaction taking place at steady state, with plug flow in the dilute phase, and the dense phase treated as a stagnant region, exchanging mass with the dilute phase. Both temperature and pressure are assumed to be constant over the height interval of interest. Mole balances on reactant A in a differential slice of height dz then give

$$\text{Bubble phase:}\quad U\frac{dC_{Ab}}{dz}+k_{bd}a_b\varepsilon_b(C_{Ab}-C_{Ad})+k_r\varphi_b C_{Ab}^n=0 \tag{4.16}$$

$$\text{Dense phase:}\quad k_{bd}a_b\varepsilon_b(C_{Ad}-C_{Ab})+k_r\varphi_d C_{Ad}^n \tag{4.17}$$

where a_b is the bubble surface area/bubble volume, ε_b is the fraction of the bed occupied by bubbles, k_r is the reaction rate constant, and C_{Ab} and C_{Ad} are the local concentrations of A in the bubble and dense phases, respectively. Integration of these equations with the boundary condition C_{Ab} = entry concentration at $z = 0$ for a first-order reaction ($n = 1$) then results in prediction of the conversion within the dense bed:

$$\text{Conversion of component}\,A=1-\exp\left\{\frac{k_r^*\left[k_{bd}^*(\varphi_b+\varphi_d)+k_r^*\varphi_b\varphi_d\right]}{k_{bd}^*+k_r^*\varphi_d}\right\} \tag{4.18}$$

where $k_r^* = k_r H/U$ is a dimensionless first-order rate constant, $k_{bd}^* = k_{bd}a_b\varepsilon_b H/U$ is the dimensionless interphase mass transfer coefficient, whereas φ_b and φ_d are the fractions of bed volume occupied by particles associated with the bubble and dense phases, respectively.

For slow (rate-limiting) kinetics (i.e., low k_r^*), it can be shown from Equation 4.18 that the process is dominated by the chemical kinetics (k_r^*) group, with the bed hydrodynamics (e.g., bubble diameter) making very little difference. However, for relatively fast reactions, the interphase mass transfer group (k_{bd}^*) plays a preeminent role, with the fraction of solids within bubbles, φ_b, also becoming important.

Analytical solution of the mole balance equations is only likely to be possible when a number of simplifying assumptions can be made such as those adopted previously where we assumed a single irreversible first-order reaction, no change in molar flow due to reaction, isothermal reactor, negligible variation in pressure, plug flow of gas in the bubble phase, and either perfect mixing or plug flow in the dense phase (see Ref. [46]). Assumptions must also be made with respect to the respective flows through each phase and the extent to which particles are present within the dilute phase. When the various simplifying assumptions underlying Equation 4.18 do not apply, for example, when the reaction is not of first order, when there are changes in molar flow rate associated with the reaction and/or multiple reactions occur, it is then necessary to utilize numerical techniques (e.g., based on Runge–Kutta algorithms) to solve the ordinary differential equations arising from the phase mole balances. For unsteady-state processes, an accumulation (transient) term is also required in each mole balance equation, resulting in partial differential equations.

Similar two-phase modeling is required for the slug flow regime, with interphase mass transfer again playing a pivotal role (e.g., see Ref. [47]) and appropriate relationships for slug flow hydrodynamics, for example, for slug rise velocities, frequency, and spacing.

Both the distributor region (commonly called the "grid zone") and the freeboard region above the dense bed can also play major roles in determining overall reactor performance. These regions have been subjected to separate reactor models (as reviewed by Ho [45]).

4.4.3 Modeling of reactors operating in high-velocity flow regimes

For turbulent fluidization and DSU, segregation into separate phases is not such a serious issue as in the bubbling and slug flow regimes. Modeling may then proceed based on only a single phase, that is, neglecting interphase transfer resistance. On the other hand, axial dispersion needs to be included within the single phase, leading to a second-order ordinary differential equation for the steady state and the need for two (Danckwerts) boundary conditions, one at the entry and the other at the top of the dense phase region. In essence, the model then incorporates one or more reaction (consumption/generation) term into the widely used axially dispersed plug flow model.

In reality, the transition boundaries separating the flow regimes are not sharp. Instead there are gradual transitions with intermittent appearance (bursts) of behavior corresponding to the regime on either side of the transition. This has led to a probabilistic modeling approach (e.g., see Refs. [43, 48, 49]) where the regime-specific models are combined to provide a better representation of the full spectrum of superficial gas velocities and flow regimes, smoothly interpolating between the adjacent flow regimes.

4.5 Scale-up, pilot testing, and practical issues

Previous sections of this chapter have provided an overview of key issues affecting the performance of catalytic fluidized-bed reactors. In this section, we address more directly key challenges which affect the design, scale-up, and implementation of fluidized-bed processes.

4.5.1 Scale-up issues

Scale-up is a challenging issue for fluidized-bed reactors, as illustrated by several spectacular failures of commercial fluidized-bed processes which did not live up to the promise shown by laboratory- and pilot-scale reactors. With better understanding of the factors causing the worsening of performance, one can do a better job of avoiding the risks associated with scale-up. Useful tips were provided by Pell [5].

A major factor in the deterioration of reactor performance with increasing scale is that interphase mass transfer becomes less effective as a fluidized bed is scaled up (i.e., the interphase mass transfer coefficient given by Eq. 4.15 decreases as bubbles become larger and travel more quickly). Other contributing factors are likely to be greater axial dispersion, slugging in the small column giving way to bubbling in larger units, increased entrainment in larger columns for a given vessel height, more difficulty in distributing gas uniformly at the base of the larger reactor, and lack of uniformity when gas, vaporizing liquid, or solids are fed from a limited number of ports on the periphery of the reactor vessel.

When catalytic fluidized-bed reactors are scaled up or scaled down, there is no universally accepted scaling procedure. One approach is to maintain the H/D ratio constant; another is to hold H constant and only vary the diameter or cross-sectional area. In addition, there is no consensus on whether the distributor of a small facility should be a fully scaled geometrically identical version of the distributor intended for the full-scale reactor, or a portion of the full-scale version. Often the latter is impractical.

Hydrodynamic scaling by dimensional similitude, that is, by geometric scaling coupled with matching of all important dimensionless groups in small and large units [50], can be a useful technique for gaining information on industrial-scale reactors based on cold model experiments. However, this approach tends to be of limited applicability for the fine particle catalytic fluidized-bed reactors of interest in this chapter because interparticle forces, which play a significant role for fluidized catalyst (group A) particles, cannot be scaled simply in terms of hydrodynamic properties.

4.5.2 Laboratory and pilot testing

Intrinsic chemical kinetic data should not be obtained from fluidized beds, since the complex gas and particle dynamics in the fluidized bed make it difficult to separate the chemical factors from the interphase mass transfer and mixing constraints. Kinetics should instead be derived from fixed beds, Berty reactors, or other reactors where the mixing and mass transfer are well characterized and/or not rate limiting.

The minimum reactor diameter for pilot plant experiments on fluidized-bed reactors is generally recommended to be approximately 100 mm. Such small reactors can have many functions: measuring U_{mf}, ensuring that the catalyst particles can be fluidized smoothly, testing for catalyst attrition and deactivation, investigating fouling and similar issues, exploring operation over a wide range of gas velocities, and determining

Table 4.5 Common instrumentation required in fluidized-bed reactors.

Variable measured	Provision/instrument	Comments
Temperature	Thermocouples	Provide redundancy, response time of seconds
Pressure	Ports	Provide for continuous or periodic purging
Pressure	Transducers	Capable of measuring frequencies to 50 Hz
Gas composition (not always practicable)	Sampling ports	Maintain sufficient pressure drop to avoid sample bias; filter out solids, rapid quench
Bed level	Axial pressure profiles	Allow sufficient ports, in freeboard as well as in dense bed
Appearance	View ports	Useful if safety allows, and if they can be maintained clear
Electrical charge	Electrostatic probes	Signal difficult to interpret

baseline performance (through conversion and selectivity data). When reactors are scaled up stagewise, it is best not to increase linear dimensions in successive steps by more than a factor of 10. Avoid porous plate distributors, choosing instead a geometry of distributor similar to what is envisaged for the full-scale system. Operate if possible with the particles, temperature, pressure, and concentrations as similar as possible to those to be used in the commercial-scale unit. Also cover as broad a range of superficial gas velocity as possible. Operate for long enough to have as good an idea as possible of the long-term viability of the catalyst and fluidization stability.

4.5.3 Instrumentation

Instrumentation helpful in monitoring and interpreting fluidized-bed reactor performance is identified in Table 4.5.

Techniques for measuring hydrodynamic properties and gas concentrations in fluidized-bed reactors were reviewed by Werther et al. [51]. Essential instrumentation needed for industrial fluidized-bed catalytic reactors are likely to include:

- Gas flow measurement
- Catalyst addition (e.g., makeup) measurement
- Differential pressure measurements across the entire dense bed
- Thermocouples to monitor temperature at multiple positions
- Gas composition measurement at outlet

In addition, it is helpful to be able to make dynamic measurements of local gauge or differential pressure and temperature to check for upsets and defluidization. Continuous purging of pressure taps at a gas velocity close to 0.5 m/s is recommended. Since the frequencies of primary interest in fluidized beds are of order 1–10 Hz, sampling frequencies should be at least 50–100 Hz. Care must be exercised in setting up the pressure measurement system and in interpreting the signals [52]. Care is also needed in interpreting gas sampling concentrations [53].

Table 4.6 Complementary components of fluidized-bed reactor system.

Component	Comments
Catalyst preparation	May require drying, screening, or activation
Catalyst loading and makeup	Typically requires lock hopper system for feeding and an exit strategy to periodically discharge catalyst, recycle, or disposal facilities
Instrumentation	See Table 4.5
Steam	Boiler, feedwater treatment, condensate, traps
Gas feed	May require premixing or contact only in reactor
Liquid feed (if applicable)	Specify feed nozzles; choose pump and flow splitting system
Gas outlet stream	May require solids removal beyond cyclones; letdown temperature and pressure as needed
Heat transfer	Provision for thermal expansion, water treatment
Electrostatics	Ground the reactor vessel thoroughly; monitor; may require antistatic agent
Other safety	Conduct full HAZOP analysis; provide for emergency shutdowns, power failures, rupture disks, inert gas, quenching, and so on as required on a case-by-case basis
Shutdown	Manways and other provisions for access and cleaning; rapid removal and cleaning or replacement of inserts and distributor as required

4.5.4 Other practical issues

As listed in Table 4.6, many other components are required when installing fluidized-bed catalytic reactors. Provide a gas distributor with a pressure drop of at least 30% of the bed weight minus buoyancy and gas-entry holes no further apart than 100 mm. Avoid sintered plate distributors. Also avoid any fixed surfaces immersed in the bed which are other than horizontal or vertical in orientation. Gaps between adjacent fixed internal surfaces should be wider than the principal dimension of the largest particles and at least 20 times the Sauter mean particle diameter. For internal or external cyclones, maintain an inlet velocity of 15–25 m/s. To minimize attrition and wear, gas jets (and the particles they inevitably entrain) must not be allowed to impinge on fixed surfaces. Follow good practice [54, 55] with respect to the design and operation of cyclones, standpipes, and solid return systems. Be careful to maintain a substantial fraction of fines (particles smaller than ~40 μm in diameter) in the bed, as loss of fines can lead to marked deterioration in performance [5].

4.6 Concluding remarks

Fluidized-bed reactors can be very valuable because of certain advantages, like favorable heat transfer, uniform temperature, and the ability to add and remove catalyst particles online. On the other hand, fluidized beds are more risky and require attention to years of accumulated experience. This chapter provides an overview of key issues and pitfalls. Readers interested in fluidization as a possible tool for practical reactor design should consult the voluminous literature in the field and work with those with direct experience.

Nomenclature

A	Cross-sectional area of bed, m^2
a_b	Surface area of bubbles per unit bubble volume, m^{-1}
Ar	Archimedes number defined by Equation 4.2, —
C_1, C_2	Fitted constants, —
C_{Ab}	Local concentration of reactant A in bubble phase, $kmol/m^3$
C_{Ad}	Local concentration of reactant A in dense phase, $kmol/m^3$
c_p	Specific heat of particles, J/kg K
D_{AB}	Binary molecular diffusivity of gas A in gas B, m^2/s
d_b	Bubble diameter, m
$D_{g,ax}$	Effective gas-phase axial dispersion coefficient, m^2/s
d_p	Particle diameter, m
d_{pi}	Mid-point diameter of particles in i^{th} size fraction
$E(t)$	Exit age distribution function, m^{-1}
G	Acceleration of gravity, m/s^2
h	Bed-to-surface heat transfer coefficient, W/m^2 K
H	Expanded bed height, m
k_{bd}	Bubble-to-dense-phase interphase mass transfer coefficient, m/s
k_g	Thermal conductivity of gas, W/m K
k_r	Reaction rate constant, $mol^{1-n}/m^{3(1-n)}$ s
n	Order of reaction, —
$Pe_{g,ax}$	Peclet number for gas-phase axial mixing, —
Pr	Prandtl number, —
Q_b	Gas flow associated with bubble phase at any height, m^3/s
R	Column or riser radius, m
r	Radial coordinate, m
t	Time, s
U_b	Bubble velocity, m/s
U_c	Superficial gas velocity at onset of turbulent fluidization, m/s
U_{mb}	Superficial gas velocity at onset of bubbling, m/s
U_{mf}	Superficial gas velocity at minimum fluidization, m/s
x_i	Mass fraction of particle corresponding to ith size fraction, —
z	Height coordinate, m

Greek letters

$\bar{\varepsilon}$	Void fraction averaged over entire volume of riser, —
$\varepsilon_{Av}(z)$	Cross-sectional average void fraction at height z, —
ε_b	Volume fraction of bubbles, —
ε_{mf}	Bed void fraction at minimum fluidization, —
φ_b, φ_d	Fraction of bed volume occupied by bubble, dense-phase solids, —
μ	Gas absolute viscosity, Pa s
ρ_g, ρ_p	Density of gas, particles, kg/m^3
τ	Mean residence time of gas in bed, s

References

1 Hetsroni G, editor. Handbook of multiphase systems. Washington, DC: Hemisphere; 1982.
2 Yates JG. Fundamentals of fluidized-bed chemical processes. London: Butterworths; 1983.
3 Davidson JF, Clift R, Harrison D, editors. Fluidization, 2nd edition. London: Academic Press; 1985.
4 Geldart D, editor. Gas fluidization technology. New York: John Wiley & Sons, Inc.; 1986.
5 Pell M. Gas fluidization. Amsterdam: Elsevier; 1990.
6 Kunii D, Levenspiel O. Fluidization engineering, 2nd edition. Boston: Butterworth-Heinemann; 1991.
7 Grace JR, Avidan AA, Knowlton TM, editors. Circulating fluidized beds. London: Blackie; 1997.
8 Yang WC, editor. Handbook of fluidization and fluid-particle systems. New York: Marcel Dekker; 2003.
9 Fan LS. Gas–liquid-solid fluidization engineering. Boston: Butterworths; 1989.
10 Rovero G, Piccinini N. Catalytic reactors and their modeling. In: Epstein N, Grace JR, editors. Spouted and spout-fluid beds. Cambridge: Cambridge University Press; 2011. p 305–320.
11 Bolthunis CO, Silverman RW, Ferrari DC. Rocky road to commercialization: breakthroughs and challenges in the commercialization of fluidized bed reactors. In: Arena U, Chirone R, Miccio M, Salatino P, editors. Fluidization XI. Brooklyn: Engineering Conferences International; 2004. p. 547–554.
12 Geldart D. The fluidized bed as a chemical reactor: a critical review of the first 25 years. *Chemistry and Industry* 1967:1474–1481.
13 Yerushalmi J. Applications of fluidized beds. In: Hetsroni G, editor. Handbook of multiphase systems. Washington, DC: Hemisphere; 1982;8: p. 152–216.
14 Jazayeri B. Applications for chemical production and processing. In: Yang WC, editor. Handbook of fluidization and fluid-particle systems. New York: Marcel Dekker; 2003. p. 421–444.
15 Avidan AA. Fluid catalytic cracking. In: Grace JR, Avidan AA, Knowlton TM, editors. Circulating fluidized beds. London: Blackie; 1997. p. 466–488.
16 Sadeghbeigi R, editor. Fluid catalytic cracking handbook, 3rd edition. Amsterdam: Elsevier; 2012.
17 Tian P, Wei Y, Ye M, Liu Z. Methanol to olefins (MTO): from fundamentals to commercialization. ACS Catalysis 2015;5: 1922–1938.
18 Avidan AA. Future prospects. In: Grace JR, Avidan AA, Knowlton TM, editors. Circulating fluidized beds. London: Blackie; 1997. p. 568–577.
19 Burdett ID, Elsinger RS, Cai P, Lee KH. Gas-phase fluidization technology for production of polyolefins. In: Kwauk M, Li J, Yang WC, editors. Fluidization X. New York: United Engineering Foundation; 2001. p. 39–52.

20 Wen CY, Yu YH. A generalized method for predicting the minimum fluidization velocity. AIChE Journal 1966;12:610–612.
21 Geldart D. Types of gas fluidization. Powder Technology 1973;7:285–292.
22 Shabanian J, Jafari R, Chaouki J. Fluidization of ultrafine particles. International Review Chemical Engineering 2012;4:16–50.
23 Horio M, Nonaka A. A generalized bubble diameter correlation for gas-solid fluidized beds. AIChE Journal 1987;33:1865–1872.
24 Lettieri P, Luca M. Challenges and issues on the CFD modelling of fluidized beds: a review. Journal of Computational Multiphase Flows 2009;2:83–131.
25 Bi HT, Ellis N, Abba IA, Grace JR. A state-of-the-art review of gas-solid turbulent fluidization. Chemical Engineering Science 2000;55:4789–4825.
26 Berruti F, Chaouki J, Godfroy L, Pugsley TS, Patience GS. Hydrodynamics of circulating fluidized bed risers. Canadian Journal of Chemical Engineering 1995;73:579–602.
27 Grace JR, Bi HT, Golriz M. Circulating fluidized beds. In: Yang WC, editor. Handbook of fluidization and fluid-particle systems. New York: Marcel Dekker; 2003. p. 485–544.
28 Issangya AS, Grace JR, Bai D, Zhu JX. Radial voidage variation in CFB risers. Canadian Journal of Chemical Engineering 2001;79:279–286.
29 Griffith AF, Louge, MY. The scaling of cluster velocity at the wall of circulating fluidized bed risers. Chemical Engineering Science 1998;53:2475–2477.
30 Grace JR, Issangya AS, Bai D, Bi HT, Zhu J. Situating the high-density circulating fluidized bed. AIChE Journal 1999;45:2108–2116.
31 May WG. Fluidized bed reactor studies. Chemical Engineering Progress 1959;55(12):49–56.
32 Van Deemter JJ. Mixing patterns in large-scale fluidized beds. In: Grace JR, Matsen JM, editors. Fluidization. New York: Plenum; 1980. p. 69–89.
33 Li TW, Grace JR, Bi HT. Numerical investigation of gas mixing in gas-fluidized beds. AIChE Journal 2010;56:2280–2296.
34 Botterill JSM. Fluid-bed heat transfer. London: Academic Press; 1975.
35 Molerus O, Wirth KE. Heat transfer in fluidized beds. London: Chapman & Hall; 1997.
36 Sit SP, Grace JR. Effect of bubble interaction on interphase mass transfer in gas-fluidized beds. Chemical Engineering Science 1981;36:327–335.
37 Wen CY, Chen LH. Fluidized bed freeboard phenomena: entrainment and elutriation. AIChE Journal 1982;28:117–128.
38 Zenz FA, Weil NA. A theoretical-empirical approach to mechanism of particle entrainment from fluidized beds. AIChE Journal 1958;4:472–479.
39 Werther J, Hartge EU. Elutriation and entrainment. In: Yang WC, editor. Handbook of fluidization and fluid-particle systems. New York: Marcel Dekker; 2003. p. 129–154.
40 Bemrose CR, Bridgwater J. A review of attrition and attrition test methods. Powder Technology 1987;49:97–126.
41 Werther J, Reppenhagen J. Attrition. In: Wang WC, editor. Handbook of fluidization and fluid-particle systems. New York: Marcel Dekker; 2003. p. 201–237.
42 Hendrickson G. Electrostatics and gas phase fluidized bed polymerization reactor wall sheeting. Chemical Engineering Science 2006;61:1041–1064.
43 Mahecha-Botero A, Grace JR, Elnashaie SSEH, Lim CJ. Advances in modeling of fluidized-bed catalytic reactors: a comprehensive review. Chemical Engineering Communications 2009;196:1375–1405.
44 Sun G, Grace JR. The effect of particle size distribution on the performance of a catalytic fluidized bed reactor. Chemical Engineering Science 1990;45:2187–2194.
45 Ho TC. Modeling. In: Yang WC, editor. Handbook of fluidization and fluid-particle systems. New York: Marcel Dekker; 2003. p. 239–255.
46 Grace JR. Fluid beds as chemical reactors. In: Geldart D, editor. Gas fluidization technology. New York: John Wiley & Sons, Inc.; 1986. p. 287–341.
47 Hovmand S, Davidson JF. Chemical conversion in a slugging fluidized bed. Transactions. Institution of Chemical Engineers 1968;46:190–203.
48 Abba IA, Grace JR, Bi HT. Variable-gas-density fluidized bed reactor model for catalytic processes. Chemical Engineering Science 2002;57:4797–4807.
49 Abba IA, Grace JR, Bi HT, Thompson ML. Spanning the flow regimes: generic fluidized-bed reactor model. AIChE Journal 2003;49:1838–1848.
50 Glicksman LR, Hyre MR, Farrell PA. Dynamic similarity in fluidization. International Journal of Multiphase Flow 1994;20S:331–386.
51 Werther J, Hartge EU, Rensner D. Measurement techniques for gas-soli fluidized bed reactors. International Chemical Engineering 1993;33:18–27.
52 Brown RC, Brue E. Resolving dynamical features of fluidized beds from pressure fluctuations. Powder Technology 2001;119:68–80.
53 Grace JR, Bi HT, Zhang Y. Pitfalls in gas sampling from fluidized beds. Chemical Engineering Science 2009;64:2522–2524.
54 Knowlton TM. Cyclone separators. In: Yang WC, editor. Handbook of fluidization and fluid-particle systems. New York: Marcel Dekker; 2003. p. 599–617.
55 Knowlton TM. Standpipes and nonmechanical valves. In: Yang WC, editor. Handbook of fluidization and fluid-particle systems. New York: Marcel Dekker; 2003. p. 571–597.

PART 3

Three-phase catalytic reactors

CHAPTER 5

Three-phase fixed-bed reactors

Ion Iliuta and Faïçal Larachi
Chemical Engineering Department, Laval University, Québec City, Québec, Canada

Abstract

Hydrodynamics, mass and heat transfer in three-phase fixed bed reactors, scale-up rules and alternative ways to scale-down are briefly reviewed. Also, the recent advances in trickle-bed reactor modeling are examined, with no attempt made to systematically summarize this vast field. Two case studies with reference to modeling the simultaneous catalytic hydrodesulfurization/bed clogging in hydrotreating trickle-bed reactors and phenol biodegradation/biomass accumulation in trickle-bed bioreactors are considered. Additionally, the modeling of the integrated process of aqueous-phase glycerol reforming and dimethyl ether synthesis into an allothermal dual-bed reactor (membrane fixed-bed reactor/tickle-bed reactor) is analyzed.

5.1 Introduction

Three-phase fixed-bed reactors are frequently encountered in petroleum, petrochemical, chemical, and biochemical applications [1]. Fixed-bed reactors processing gas and liquid reactants can operate in downward cocurrent two-phase flow, in upward cocurrent flow, and in countercurrent flow (Figure 5.1). The criteria for choosing the proper flow direction have been established, and the evaluation of the effect of flow direction on the reactor performance has also been performed [2, 3]. Two-phase downflow fixed-bed reactors (trickle-bed reactors) are pervasive in industrial multiphase catalytic processes and span a broad range of applications from the manufacture of value-added products to the conversion of undesired chemicals into harmless and biocompatible species. Trickle-bed reactors are traditionally used in petroleum (hydrocracking, hydrodesulfurization, hydrodenitrogenation, hydrodeoxygenation, alkylation, etc.), petrochemical, and chemical industries (hydrogenation of higher aldehydes, reactive amination, liquid-phase oxidation, etc.); in wastewater and gas scrubbing biological treatments, such as filtration of sewage sludge through sand pack filters and removal of carbon disulfide, volatile organic compounds, or chlorinated aliphatic hydrocarbons; and in biochemical and electrochemical processing [1].

Two-phase upflow fixed-bed reactors (flooded-bed or packed-bed bubble column reactors) are sparingly practiced commercially even if they may outperform trickle-bed reactors with respect to catalyst life time and wetting efficiency, selectivity, liquid distribution, heat withdrawal, and liquid–solid mass transfer rates [1]. However, the interest in two-phase upflow fixed-bed reactors may be resurrected as portrayed by the last two decades' reported applications such as hydrogenation of nitrotoluenes in evaporating solvents [4, 5], partial hydrogenation of cyclododecatriene [6–8], selective hydrogenation of butadiene from C4 cuts [9], selective hydrogenation of pyrolysis gasoline, oligomerization [10], photocatalytic or catalytic wet air oxidation [11], etc. Three-phase fixed-bed reactors operating in countercurrent gas–liquid flow provide the opportunity for selective removal of by-products that may act as inhibitor, for example, in hydrodesulfurization where hydrogen sulfide may have an inhibitory effect [1]. Countercurrent flow achieves an intensive mixing of the phases, a more favorable flat axial temperature profile, a large gas–liquid interfacial area, and a large convective transport. Countercurrent operation is preferred over cocurrent when large heat of reaction is involved though at the expense of tighter control of the inception of bed flooding.

Two-phase flow in three-phase fixed-bed reactors makes the reactor design problem complex [12]. Interphase mass transfer can be important between gas and liquid as also between liquid and catalyst particle. Also, in the case of trickle-bed reactors, the rivulet-type flow of the liquid falling through the fixed bed may result (particularly at low liquid flow rates) in only part of the catalyst particle surface being covered with the liquid phase. This introduces a third mass transfer process from gas to the so-called "gas-covered" surface. Also, the reaction rates in three-phase fixed-bed catalytic reactors are highly affected by the heat transfer resistances: resistance to radial heat transfer and resistance to fluid-to-particle heat transfer. As a result of these and other factors, predicting the local (global) rate of reaction for a catalyst particle in three-phase fixed-bed reactors requires not only

Multiphase Catalytic Reactors: Theory, Design, Manufacturing, and Applications, First Edition. Edited by Zeynep Ilsen Önsan and Ahmet Kerim Avci.

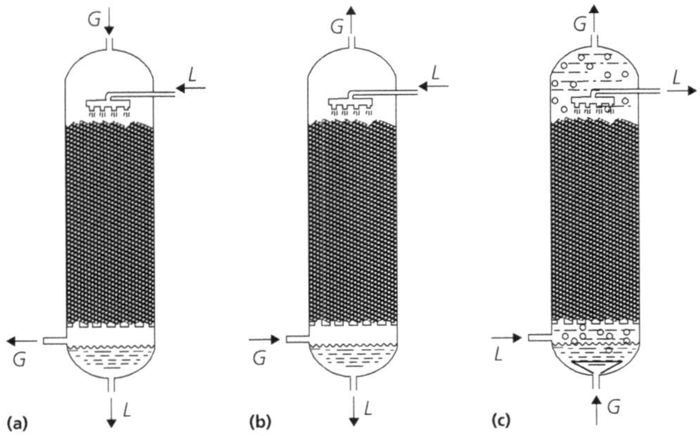

Figure 5.1 Fixed-bed reactors for gas–liquid–solid catalyzed systems. (a) Two-phase downflow fixed-bed reactor (trickle-bed reactor). (b) Fixed-bed reactor with countercurrent flow. (c) Two-phase upflow fixed-bed reactor (packed-bed bubble flow reactor).

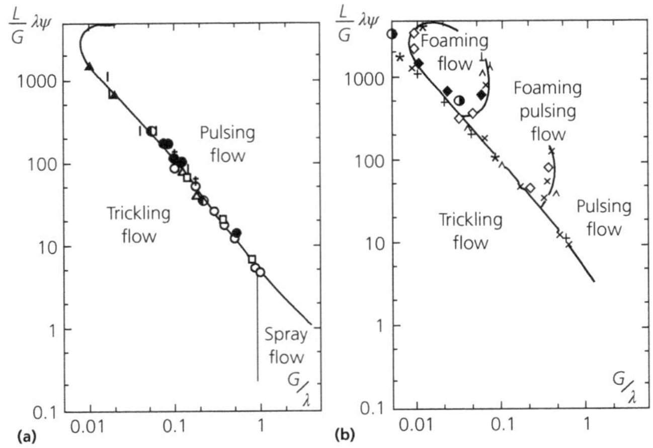

Figure 5.2 Hydrodynamic regimes for two-phase downflow in fixed beds—nonfoaming (a) and foaming (b) liquids.
(Source: Charpentier and Favier [16]. Reproduced with permission of John Wiley & Sons.)

intrinsic kinetics but also knowledge of the wetting efficiency and considerable mass and heat transfer rate information.

The aim of this chapter is to outline briefly the hydrodynamics, mass, and heat transfer in the commonly used three-phase fixed-bed reactors. In particular, scale-up rules and alternative ways to scale down trickle-bed reactors are discussed. Then recent developments in trickle-bed reactor design are analyzed, with no attempt made to systematically summarize this vast field. Two case studies with reference to modeling of simultaneous catalytic hydrodesulfurization/bed clogging in hydrotreating trickle-bed reactors and phenol biodegradation/biomass accumulation in trickle-bed bioreactors are considered. In addition, the modeling of the integrated process of aqueous-phase glycerol reforming and dimethyl ether (DME) synthesis into an allothermal dual-bed reactor (membrane fixed-bed reactor/tickle-bed reactor) is also summarized.

5.2 Hydrodynamic aspects of three-phase fixed-bed reactors

Extensive reviews of hydrodynamic of two-phase flow systems in fixed beds appeared in the last five decades [1, 12–15], and there is no point in repeating here the numerous tables and references that were provided in these reviews. We attempt here to summarize the key findings that ought to be of importance to the research and potentially plant engineers.

5.2.1 General aspects: Flow regimes, liquid holdup, two-phase pressure drop, and wetting efficiency

5.2.1.1 Flow regimes

Two-phase downflow fixed-bed reactors operate in a variety of flow regimes ranging from gas-continuous to liquid-continuous patterns. They usually fall into two broad categories referred to as low and high interaction regimes. The low interaction regime (trickling flow) manifests at low gas and liquid flow rates and is characterized by a weak gas–liquid interfacial activity and a gravity-driven liquid flow (Figure 5.2). Gas–liquid interactions in trickling flow regime increase at high gas and liquid flow rates and at elevated pressure (i.e., close to the transition to high interaction regimes). The high interaction regimes take place at moderate to high fluid flow rates due to the significant gas–liquid interfacial shears. As a result, various patterns arise depending on the gas-to-liquid holdup ratio and liquid tendency to foam. The pulsing and spray flow regimes were observed and depicted by Charpentier and Favier [16] (Figure 5.2), while Talmor [17] additionally noted bubble and dispersed bubble flow regimes. Low gas flow rates and sufficiently high liquid flow rates lead to the bubble flow regime with a continuous liquid phase which contains small bubbles. At medium gas flow rates and high liquid flow rates, the liquid phase remains continuous, but the bubbles coalesce and the gas flows in the form of elongated bubbles (dispersed bubble flow regime). Pulsing flow regime is observed at moderate liquid flow rates and moderate to high gas flow rates and may be approached either from the gas-continuous trickling flow or from the liquid-continuous coalesced bubble flow regime. Pulsing flow regime can be depicted as a macroscopic combination of dispersed bubble flow occurring in the liquid-rich slugs and trickling flow in the gas-rich slugs [14]. At still higher gas flow rates, a mist or spray flow is observed for which the gas becomes the continuous phase in which the liquid is entrained as droplets. These high interaction flow patterns apply for coalescing liquids, while for coalescence-inhibiting liquids two more flow regimes are encountered, namely, the foaming flow and foaming–pulsing flow regimes (Figure 5.2—[16]).

Because the trickling and pulsing flow regimes are of particular industrial interest, the majority of the literature work focused on the trickle-to-pulse flow transition. The following current trends emerge from the analysis of trickle-to-pulse flow regime transition literature [18]: (i) trickle-to-pulse flow transition is barely affected by intermediate-range bed porosity

(0.34 < ε < 0.40) and particle sizes up to 5 mm; for high-porosity beds or large-size particles, a shift of this transition toward higher liquid throughputs is observed; (ii) a significant drop of the liquid transitional velocity occurs with increasing liquid viscosity; (iii) low surface tension liquids exhibit a peculiar trend as higher liquid transitional velocities occur at higher rather than at lower superficial gas velocities; (iv) an increase in liquid density narrows down the trickling flow region; and (v) the trickling flow domain widens at elevated pressures.

Many flow regime charts and attempts at modeling trickle-to-pulse flow regime changeover have been proposed a summary of which is available in Saroha and Nigam [19] and Al-Dahhan et al. [14]. Because of the large number of variables affecting the trickle-to-pulse flow transition, compression into two or three flow chart coordinates seems illusive. Furthermore, since the choice of the compressing variables is not unanimous, the least to be said is that a universal flow chart/correlation is nonexisting and that each flow chart/correlation can only be employed interpolatively for conditions falling within those that served to its establishment. To circumvent these limitations, Iliuta et al. [18] developed a neural network model for trickle-to-pulse flow regime transition using an extensive historic fluid dynamic database and combined approaches relying on feed-forward neural networks, dimensional analysis, and phenomenological models. Also, some conceptual flow regime changeover models, based on microscopic and macroscopic approaches, were proposed in the literature [20–26]. Unfortunately, all recent experimental studies reach the same conclusion that none of them is yet entirely successful and no single unified approach can be recommended.

Two-phase upflow fixed-bed reactors operate in four flow regimes according to Shah [27]. A relatively low gas flow rate introduced in a single-phase liquid flow established in a fixed-bed reactor leads to bubble flow regime with a continuous liquid phase which contains small bubbles at slightly higher velocity than the liquid phase. Surging flow regime appears at moderate gas flow rates and low liquid flow rates and is characterized by the coalescence of the gas bubbles and formation of the large bubbles. As the gas flow rate is increased at a given liquid flow rate, pulsing flow occurs. This regime is characterized by alternate portions of liquid-rich and gas-rich mixtures of the two phases. With further increase of the gas flow rate, the density difference between the alternate slugs entirely disappeared, producing spray flow regime. These regimes were observed in fixed beds with large catalyst particles [27]. For fixed beds with smaller solid particles, only bubble flow, pulsing (slug) flow, and spray flow regimes were observed [28–30] (Figure 5.3).

Many flow regime charts and attempts at modeling flow regime changeover have been proposed. A list of the literature correlations for the interfaces between different flow regimes in two-phase upflow fixed-bed reactors is available in Rao et al. [31]. As in the case of two-phase downflow, because of the large number of variables affecting flow regime transition, a universal flow chart/correlation is nonexisting, and each flow chart/correlation can only be employed interpolatively for conditions falling within those that served to its establishment.

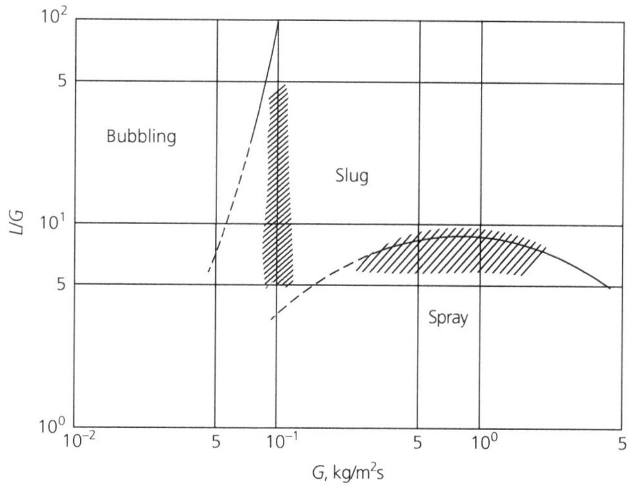

Figure 5.3 Hydrodynamic regimes for two-phase upflow in fixed beds: transition zones (/////) from Ref. [30] and transition lines from Ref. [28]. (Source: Mazzarino et al. [30]. Reproduced with permission of Elsevier.)

5.2.1.2 Liquid holdup and two-phase pressure drop

Liquid holdup is defined as the liquid volume fraction per unit reactor volume. It is indeed a basic parameter in reactor design owing to its relationship with other key hydrodynamic parameters such as two-phase pressure drop, interfacial mass transfer parameters, liquid mean residence time, catalyst loading-to-liquid volume ratio, liquid axial dispersion, radial effective thermal conductivity, convective wall heat transfer coefficient, etc. For two-phase downflow, under highly exothermic reaction conditions, the knowledge of the liquid holdup is essential for avoiding hot spots and for preventing reactor runaway. Liquid holdup also affects the catalyst external wetting efficiency that, in turn, has an impact on the reaction selectivity, depending on whether the reaction takes place exclusively on the wetted catalyst area or on the dry and wetted catalyst area alike.

Two-phase pressure drop represents the energy dissipated to offset the resistance to fluid motion through the reactor bed. It is important in determining energy losses, in the sizing of the compression and pumping devices, and, very often, in assessing the liquid holdup, the external wetting efficiency, the interfacial mass transfer coefficients levels, etc. [32]. Gas–liquid flow resistance in multiphase fixed-bed reactors is caused by friction forces due to the fluid viscosity at the gas–liquid, gas–solid (partial wetted conditions), and liquid–solid interfaces, inertial forces caused by successive acceleration and/or deceleration of the fluids, turbulent stress because of the local velocity field fluctuations in both gas and liquid phases, and capillary forces and gravity acting positively for two-phase downflow but exerting a resistance for two-phase upflow operation. The relative importance of these forces depends upon the flow regime in the reactor. For two-phase downflow, in the high interaction regime, two-phase pressure drop depends mainly on inertial forces and liquid-side shear stress, and in low interaction regime other forces (shear forces, gravity, and/or capillary forces) are important. For two-phase upflow, gravitational effects are important.

The following current trends emanate from the analysis of liquid holdup and two-phase pressure drop literature for two-phase downflow and upflow [18, 33]: (i) liquid holdup is an increasing function of liquid velocity, viscosity, and particle diameter; it is a decreasing function of gas superficial velocity and liquid surface tension; (ii) liquid holdup decreases as gas density increases, except for very low gas velocities where it is insensitive to gas density; (iii) noncoalescing liquids exhibit much smaller holdups (and much higher pressure drops) than coalescing liquids; (iv) gas viscosity appears to have marginal effect on liquid holdup and pressure drop; (v) mixing with fines improves liquid holdup at the expense of increased pressure drops; and (vi) increasing gas and liquid superficial velocities or mass fluxes, liquid viscosity, or gas density or decreasing particle diameters increases pressure drops.

Correlations for liquid holdup and two-phase pressure drop were summarized by Herskowitz and Smith [12], Al-Dahhan et al. [14], Saroha and Nigam [19], and Al-Dahhan and Dudukovic [34] for two-phase downflow and by Bensetiti et al. [33] and Larachi et al. [35] for two-phase upflow—the majority of them being mostly empirical and restricted to their specific narrow ranges of process conditions. Comparison of the two-phase downflow predictions with measurements [18] shows that no method emerges as clearly superior even though those relying on phenomenological considerations seem more reliable. To circumvent these setbacks, hydrodynamic models resulting from the equations of motion associated with the gas–liquid and liquid–solid drag forces were built considering two-phase trickle flow or bubble flow through a porous medium. These models are presented in Sections 5.2.2 and 5.2.3. More recently, the efforts have been routed for advanced CFD models based on the macroscopic mass and momentum conservation laws [36, 37].

5.2.1.3 Wetting efficiency

The rate of reaction in a single catalyst particle depends on intrinsic kinetics, intraparticle and interphase mass and heat transfer resistances, and effective wetting of the catalyst particle. While mass and heat transfer resistances are encountered in other three-phase fixed-bed reactors, the partial wetting of the catalyst particles, which leads to more direct contact between the gas and the solid catalyst, is a unique feature of trickle-bed reactors. Catalyst particle-scale incomplete external wetting is the cause of the deficient catalyst utilization. This in turn can lead to poor catalyst effectiveness factors and poor heat withdrawal from partially wetted catalyst pellets [38]. Particle-scale incomplete wetting arises when the tiny liquid irrigation rate feeding the bed is insufficient to ensure full coverage with a continuous liquid film of all pellets in the bed. Hence, the knowledge of the catalyst external wetting efficiency as a function of operating conditions is needed to provide a relationship between laboratory- and pilot-scale reactor data and the large-scale reactor operation for reliable scale-up/scale-down and design of commercial trickle-bed reactors. The reaction rate over externally incompletely wetted pellets can be greater or smaller than the rate observed over completely wetted ones. This depends upon whether the limiting reactant is present only in the liquid phase or in both fluids. If the reaction is liquid limited and the liquid reactant is nonvolatile, then a decrease in the external wetting efficiency reduces the surface for liquid–solid mass transfer, thereby causing a decrease of the reaction rate. If the liquid reactant is volatile and heat effects are significant, then an additional reaction may also occur at the dry catalyst surface. Higher reaction rates can also be achieved through gas-limited reactions where the gaseous reactant can access the catalyst pores from the externally dry area. Therefore, in order to predict trickle-bed reactor performance and behavior, it is crucial to precisely quantify the catalyst external wetting efficiency and the effectiveness factor of the resulting partially wetted catalyst [39, 40]. Numerous studies, summarized by Iliuta et al. [41], relative to the external wetting efficiency have been reported in the literature. Some deal with correlative and experimental methods of wetting efficiency, while others deal with demonstrating wetting effects on trickle-bed reaction performance. Furthermore, the majority of wetting efficiency correlations were developed based on observations taken at ambient conditions (room temperature and barometric pressure). Very often, these conditions did not reveal any dependence of wetting efficiency upon gas throughput and density. Only Al-Dahhan and Dudukovic [42, 43] developed an empirical correlation between the wetting efficiency and reactor pressure and gas flow rates. Also, some mechanistic models, based on macroscopic approach, were proposed in the literature. Pironti et al. [44] proposed a macroscopic semitheoretical approach to predict total wetting efficiency. Although the approach is phenomenological, it requires a priori measurements of liquid holdup, gas and liquid single-phase and two-phase pressure gradients to compute the wetting efficiency. Also, Iliuta et al. [41] and Iliuta and Larachi [45] proposed an implicit semitheoretical mechanistic two-fluid double-slit model for the prediction of the two-phase pressure drop, liquid holdup, external wetting efficiency, and gas–liquid interfacial area in trickle flow reactors operated under partially wetted conditions using an extensive trickle flow regime database.

5.2.2 Standard two-fluid models for two-phase downflow and upflow in three-phase fixed-bed reactors

Standard two-fluid models for two-phase flow are built considering a cocurrent two-phase trickle flow or bubble flow through a porous medium of uniform initial porosity and single-sized solid particles. Two-phase flow is isothermal, with no chemical reaction, and both phases are Newtonian. The liquid phase is incompressible and the gas phase is ideal. In the case of trickle flow regime, two-phase flow is assumed annular and separated. The packing surface is totally or partially covered by a liquid film and the gas flows in the remaining interstitial void. Under bubble flow the packing surface is totally covered by a continuous liquid film, and the gas is dispersed and consists of bubbles spherical in shape and uniform in size (the processes of coalescence and breakage are assumed to be absent). For both cases, each fluid phase is viewed as a continuum in any size of the domain under consideration for which the differential macroscopic balance equations can be applied. So the general structure of the continuity and momentum balance equations is the same for both two-phase flow problems, although the closure relations are system dependent. The physics conveyed by the standard two-phase flow models is minimalistic as it relies on a set of mass and momentum balance equations for the gas and liquid bulk phases as well as ad hoc drag force relationships. In standard two-phase flow models, mass and momentum conservation for the gas–liquid interface and mass exchange rates between interface and fluid bulks accounting for production and destruction of gas–liquid interfacial area are not taken into account. Two-fluid models are based on the general intrinsic α-phase average transport equation for multiphase systems [46] written for the conditions of no mass transfer between phases and no change of phases:

$$\frac{\partial}{\partial t}\varepsilon_\alpha\langle\psi_\alpha\rangle^\alpha + \nabla\cdot\varepsilon_\alpha\langle\underline{v}_\alpha\rangle^\alpha\langle\psi_\alpha\rangle^\alpha + \nabla\cdot\langle\xi_\alpha\rangle = \nabla\cdot\langle\Omega_\alpha\rangle + \varepsilon_\alpha\langle\sigma_\alpha\rangle^\alpha + \frac{1}{V}\oiint_{A_{\alpha\beta}}\Omega_\alpha\underline{n}_\alpha dA + \frac{1}{V}\oiint_{A_{\alpha\gamma}}\Omega_\alpha\underline{n}_\alpha dA \quad (5.1)$$

Here ε_α stands for α-phase volume fraction ($\varepsilon_\alpha = V_\alpha/V$), ψ_α represents an arbitrary tensor-valued function, $\langle\psi_\alpha\rangle^\alpha$ is the intrinsic α-phase average quantity normalized with respect to its own volume V_α, Ω_α is an arbitrary tensor-valued function representing a surface flux, $\langle\Omega_\alpha\rangle$ represents the value Ω_α averaged over the volume V, $\langle\sigma_\alpha\rangle^\alpha$ represents the intrinsic α-phase average of the tensor indicating the source term, $\langle\xi_\alpha\rangle$ is the dispersion vector, and $A_{\alpha\beta}$ is the interfacial area in the averaging volume for the α and β phases.

The mass and momentum balance equations can be obtained from Equation 5.1 upon substitution of the following quantities:

- $\langle\underline{v}_\alpha\rangle^\alpha = \underline{u}_\alpha$
- The momentum vector $\langle\psi_\alpha\rangle^\alpha = \rho_\alpha\underline{u}_\alpha$
- The body force vector $\langle\sigma_\alpha\rangle^\alpha = \rho_\alpha\underline{g}$
- The volume average stress tensor decomposed as the scalar tensor corresponding to pressure and the viscous stress tensor, that is, $\langle\Omega_\alpha\rangle^\alpha = -P_\alpha\underline{\underline{I}} + \underline{\underline{\tau}}_\alpha$
- The volume average force, $F_\alpha = F_{\alpha\beta}(+F_{\alpha\gamma})$, exerted on phase α by the other phases across their mutual interfaces $A_{\alpha\beta}$ and $A_{\alpha\gamma}$: $F_{\alpha\beta} = \frac{1}{V}\oiint_{A_{\alpha\beta}}\Omega_\alpha\underline{n}_\alpha dA$ and $F_{\alpha\gamma} = \frac{1}{V}\oiint_{A_{\alpha\gamma}}\Omega_\alpha\underline{n}_\alpha dA$
- The pseudoturbulence stress tensor $\langle\xi_\alpha\rangle^\alpha$ which is combined with the viscous stress tensor $\underline{\underline{\tau}}_\alpha$ and approximated by the phenomenological expression in the second term in Equation 5.3 RHS [47].

Using $\langle\psi_\alpha\rangle^\alpha = \rho_\alpha$, $\sigma_\alpha = 0$ (null sink in the absence of chemical reaction), $\xi_\alpha = 0$, and $\Omega_\alpha = 0$ results in the equation of conservation of mass which takes in 1D (z vertical direction) the following form:

$$\frac{\partial}{\partial t}(\varepsilon_\alpha\rho_\alpha) + \frac{\partial}{\partial z}(\varepsilon_\alpha\rho_\alpha u_\alpha) = 0 \quad (5.2)$$

Similarly, the 1D equation of conservation of momentum is

$$\frac{\partial}{\partial t}(\varepsilon_\alpha\rho_\alpha u_\alpha) + \frac{\partial}{\partial z}(\varepsilon_\alpha u_\alpha\rho_\alpha u_\alpha) = -\varepsilon_\alpha\frac{\partial P_\alpha}{\partial z} + \varepsilon_\alpha\mu_\alpha^e\frac{\partial^2 u_\alpha}{\partial z^2} + \varepsilon_\alpha\rho_\alpha g_z + F_\alpha \quad (5.3)$$

$g_z = -g$ (for two-phase upflow) and $g_z = +g$ (for two-phase downflow)

Equations 5.2 and 5.3 represent a set of four volume average momentum and mass conservation equations for the gas and the liquid phases. The assumption of bed full wetting entrains that the gas-phase drag will only have contributions due to effects located at the gas–liquid interface. The resultant of these forces denoted $F_{g\ell}$ is the drag force exerted on the gas phase as a result of the relative motion between the flowing phases to oppose slip. Similarly, the resultant of the forces exerted on the liquid phase involves two components: (i) the drag force $F_{\ell s}$ experienced by the liquid due to the shear stress nearby the liquid–solid boundary and (ii) the gas–liquid interfacial drag $F_{g\ell}$. In the case of trickle flow, it has been shown that under certain conditions the slit flow approximation yields a very satisfactory set of constitutive equations for the gas–liquid and liquid–solid drag forces [41, 48]. The slit flow becomes well representative of the trickle flow regime when the liquid texture is mainly contributed by catalyst-supported liquid films and rivulets, and the gas–liquid separated flow assumption holds. This generally occurs at small liquid irrigation rates that allow the transport of liquid in the form of a smooth and stable film [48]. However, there is no restriction that other constitutive formulations can be chosen instead [49, 50]. Under bubble flow regime, the liquid phase behaves as a continuous medium for which the Kozeny–Carman theory can be applied for predicting the liquid–solid drag forces [51]. The gas–liquid interaction force is obtained from the drag force exerted on a single bubble and the number of bubbles per unit fixed bed volume [51]. The 1D two-fluid model under fully wetted conditions is

$$\frac{\partial}{\partial t}\left(\varepsilon_g\rho_g\right) + \frac{\partial}{\partial z}\left(\varepsilon_g\rho_g u_g\right) = 0 \quad (5.4)$$

$$\frac{\partial}{\partial t}(\varepsilon_\ell\rho_\ell) + \frac{\partial}{\partial z}(\varepsilon_\ell\rho_\ell u_\ell) = 0 \quad (5.5)$$

$$\frac{\partial}{\partial t}\left(\rho_g\varepsilon_g u_g\right) + u_g\frac{\partial}{\partial z}\left(\rho_g\varepsilon_g u_g\right) = \varepsilon_g\mu_g^e\frac{\partial^2 u_g}{\partial z^2} - \varepsilon_g\frac{\partial P}{\partial z} + \varepsilon_g\rho_g g_z - F_{g\ell} \quad (5.6)$$

$$\frac{\partial}{\partial t}(\rho_\ell\varepsilon_\ell u_\ell) + u_\ell\frac{\partial}{\partial z}(\rho_\ell\varepsilon_\ell u_\ell) = \varepsilon_\ell\mu_\ell^e\frac{\partial^2 u_\ell}{\partial z^2} - \varepsilon_\ell\frac{\partial P}{\partial z} + \varepsilon_\ell\rho_\ell g_z + F_{g\ell} - F_{\ell s} \quad (5.7)$$

Note that in the formulation of the momentum balance equations, the capillary pressure between phase ℓ and phase g is neglected because the liquid holdup does not change significantly with z [50]. Equations 5.4–5.7 are coupled with the conservation of volume equation based on the assumption that any small volume of the bed void, at any particular time, can be regarded as containing a volume fraction ε_g of the gas phase and ε_ℓ of the liquid phase:

$$\varepsilon_\ell + \varepsilon_g = \varepsilon \quad (5.8)$$

The assumption of bed partial wetting under trickle flow conditions entrains that the gas-phase drag will have contributions due to effects located at the gas–liquid ($F_{g\ell}$) and gas–solid (F_{gs}) interfaces. Similarly, the resultant of the forces exerted on the liquid phase involves two components: (i) the drag force $F_{\ell s}$ experienced by the liquid due to the shear stress nearby the liquid–solid boundary and (ii) the gas–liquid interfacial drag due to the slip between fluids $F_{g\ell}$. Under these circumstances, the double-slit flow approach yields a satisfactory approximation of the constitutive equations needed for the gas–liquid and liquid–solid drag forces [41]. In this case the momentum balance equations for the gas and liquid phases are the following:

$$\frac{\partial}{\partial t}\left(\rho_g \varepsilon_g u_g\right) + u_g \frac{\partial}{\partial z}\left(\rho_g \varepsilon_g u_g\right) = \varepsilon_g \mu_g^e \frac{\partial^2 u_g}{\partial z^2} - \varepsilon_g \frac{\partial P}{\partial z} + \varepsilon_g \rho_g g - F_{g\ell} - F_{gs} \tag{5.9}$$

$$\frac{\partial}{\partial t}(\rho_\ell \varepsilon_\ell u_\ell) + u_\ell \frac{\partial}{\partial z}(\rho_\ell \varepsilon_\ell u_\ell) = \varepsilon_\ell \mu_\ell^e \frac{\partial^2 u_\ell}{\partial z^2} - \varepsilon_\ell \frac{\partial P}{\partial z} + \varepsilon_\ell \rho_\ell g + \frac{\varepsilon\eta_e - \varepsilon_\ell}{\varepsilon_g}\left[F_{g\ell} + F_{gs}\right] - F_{\ell s} \tag{5.10}$$

Equations 5.7 and 5.10 are valid only under the assumption of continuity of the velocity and shear stress profiles at the gas–liquid interface. If discontinuity in velocity and shear stress at gas–liquid interface is assumed, Equations 5.7 and 5.10 become

$$\frac{\partial}{\partial t}(\rho_\ell \varepsilon_\ell u_\ell) + u_\ell \frac{\partial}{\partial z}(\rho_\ell \varepsilon_\ell u_\ell) = \varepsilon_\ell \mu_\ell^e \frac{\partial^2 u_\ell}{\partial z^2} - \varepsilon_\ell \frac{\partial P}{\partial z} + \varepsilon_\ell \rho_\ell g + f_s F_{g\ell} - F_{\ell s} \tag{5.11}$$

$$\frac{\partial}{\partial t}(\rho_\ell \varepsilon_\ell u_\ell) + u_\ell \frac{\partial}{\partial z}(\rho_\ell \varepsilon_\ell u_\ell) = \varepsilon_\ell \mu_\ell^e \frac{\partial^2 u_\ell}{\partial z^2} - \varepsilon_\ell \frac{\partial P}{\partial z} + \varepsilon_\ell \rho_\ell g + \frac{\varepsilon\eta_e - \varepsilon_\ell}{\varepsilon_g} f_s \left[F_{g\ell} + F_{gs}\right] - F_{\ell s} \tag{5.12}$$

Two-fluid models presented earlier are applicable for two-phase flow of gas/Newtonian liquid systems. It has been successfully determined that the momentum balance equations for Newtonian fluids also apply to non-Newtonian fluids [52]. The only difference for the flow in porous media between a Newtonian fluid and a non-Newtonian fluid is that the effective viscosity is not the same. Hence, it becomes necessary that a suitable effective viscosity be defined for the flow of a non-Newtonian fluid [52]. Constitutive equations for the gas–liquid and liquid–solid drag forces are given by Iliuta and Larachi [45, 53] by upgrading the class of the slit models for rheologically complex fluids (Herschel–Bulkley fluid, Bingham plastic, and Ostwald–de Waele fluids).

5.2.3 Nonequilibrium thermomechanical models for two-phase flow in three-phase fixed-bed reactors

The physics conveyed by the standard two-phase flow models is minimalistic as it relies on a set of mass and momentum balance equations for the gas and liquid bulk phases as well as ad hoc drag force relationships. We state in the previous paragraph that such physics is deficient because it insufficiently describes the role and presence of interfaces and their thermodynamic properties.

Iliuta et al. [54] developed a rigorous mathematical for the description of two-phase flow in trickle beds by adapting the macroscopic nonequilibrium thermomechanical theory developed by Hassanizadeh and Gray [55] for multiphase flows in porous media. In this nonequilibrium thermomechanical theory, the interfaces and their thermodynamic properties are explicitly taken into account and treated as individual continua interacting and exchanging properties with the adjacent phase continua. The explicit inclusion of interfaces and interfacial properties is essential because they are known to have a significant role in determining the thermodynamic state of the whole system. This leads to the establishment of two categories of conservation equations whose macroscopic properties are defined at the scale of a representative elementary volume: one for the macroscopic bulk phase properties, that is, as in the standard model, and another for the macroscopic interfacial properties. The two sets of equations are coupled via terms accounting for exchange of thermodynamic properties between phase and interface continua.

The nonequilibrium two-phase flow model was built for permanently fully wetted beds when the system reduces to five superposed continua: three phases V_ℓ, V_g, and V_s and two interfaces $A_{g\ell}$ and $A_{\ell s}$ as sketched in Figure 5.4a. From this assumption, only $A_{g\ell}$ is subject to variations in time and position, whereas $A_{\ell s}$ exhibits a constant liquid–solid contacting area, $a_{\ell s}$. This view differs in many aspects from the standard Eulerian multifluid treatments of trickle-bed hydrodynamics. Here, the main feature of the two-phase flow model rests on the inclusion of (i) conservation equations for gas–liquid interfacial area and interfacial momentum, (ii) nonequilibrium capillary pressure, (iii) Helmholtz free energy gradients in the body supply of momentum for fluid bulk phases and gas–liquid interface, and (iv) mass exchange between interface and adjacent fluid bulk phases accounting for the production/destruction of gas–liquid interfacial area (Figure 5.4b). The mass and momentum conservation equations can be cast in the following general form:

Gas bulk conservation of mass:

$$\frac{\partial}{\partial t}\left(\rho_g \varepsilon_g\right) + \nabla \cdot \left(\rho_g \varepsilon_g \underline{u}_g\right) = \hat{e}_{g\ell}^g \tag{5.13}$$

Liquid bulk conservation of mass:

$$\frac{\partial}{\partial t}(\rho_\ell \varepsilon_\ell) + \nabla \cdot \left(\rho_\ell \varepsilon_\ell \underline{u}_\ell\right) = \hat{e}_{g\ell}^\ell \tag{5.14}$$

Gas–liquid interface conservation of mass:

$$\frac{\partial}{\partial t}(\Gamma_{g\ell} a_{g\ell}) + \nabla \cdot \left(\Gamma_{g\ell} a_{g\ell} \underline{w}_{g\ell}\right) = -\left(\hat{e}_{g\ell}^g + \hat{e}_{g\ell}^\ell\right) \tag{5.15}$$

Gas bulk conservation of momentum:

$$\rho_g \varepsilon_g \left(\frac{\partial \underline{u}_g}{\partial t} + \underline{u}_g \cdot \nabla \underline{u}_g\right) - \nabla \cdot \left\{\varepsilon_g \left[-P_g \underline{\underline{I}} + \mu_g \left(\nabla \underline{u}_g + {}^t\nabla \underline{u}_g\right) - \rho_g \overline{\widetilde{\underline{u}}_g \otimes \widetilde{\underline{u}}_g}^g\right]\right\} - \rho_g \varepsilon_g \underline{g} = \hat{e}_{g\ell}^g \underline{u}_g + \underline{\hat{T}}_{g\ell}^g \tag{5.16}$$

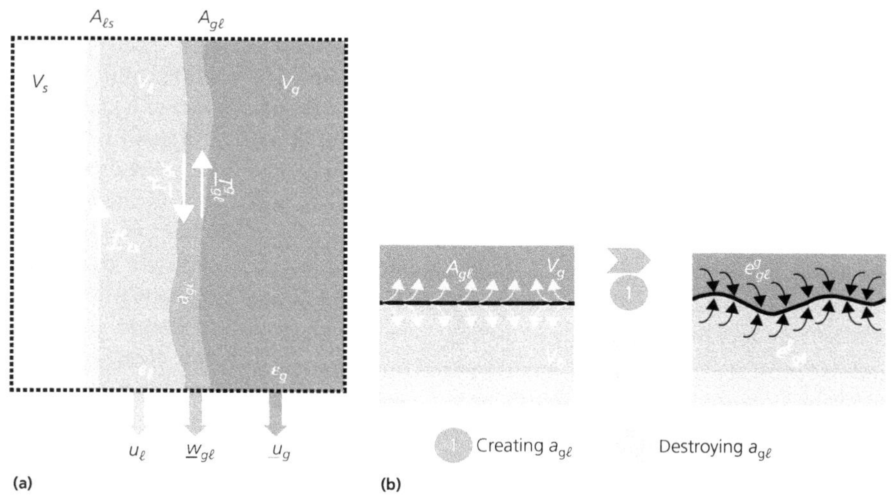

Figure 5.4 Representation of connectivity between subregions in the representative elementary volume: V_g, V_ℓ, and V_s for the bulk phases and $A_{g\ell}$ and $A_{\ell s}$ for the interfaces (a), interaction of interface $A_{g\ell}$ with V_g and V_ℓ bulk phases in creating and destroying gas–liquid interfacial area (b).
(Source: Iliuta et al. [54]. Reproduced with permission of John Wiley & Sons.)

Liquid bulk conservation of momentum:

$$\rho_\ell\varepsilon_\ell\left(\frac{\partial \underline{u}_\ell}{\partial t}+\underline{u}_\ell\cdot\nabla\underline{u}_\ell\right)-\nabla\cdot\left\{\varepsilon_\ell\left[-P_\ell\underline{\underline{I}}+\mu_\ell(\nabla\underline{u}_\ell+{}^t\nabla\underline{u}_\ell)\right.\right.$$
$$\left.\left.-\rho_\ell\overline{\underline{\tilde{u}}_\ell\otimes\underline{\tilde{u}}_\ell}^\ell\right]\right\}-\rho_\ell\varepsilon_\ell\underline{g}=\hat{e}_{g\ell}^\ell\underline{u}_\ell+\underline{\hat{T}}_{g\ell}^\ell+\underline{\hat{T}}_{\ell s}^\ell \tag{5.17}$$

Gas–liquid interface conservation of momentum:

$$\Gamma_{g\ell}a_{g\ell}\left(\frac{\partial \underline{w}_{g\ell}}{\partial t}+\underline{w}_{g\ell}\cdot\nabla\underline{w}_{g\ell}\right)-\nabla\cdot\left\{a_{g\ell}\left(\gamma_{g\ell}\underline{\underline{I}}-\Gamma_{g\ell}\overline{\underline{\tilde{w}}_{g\ell}\otimes\underline{\tilde{w}}_{g\ell}}^{g\ell}\right)\right\}$$
$$-a_{g\ell}\Gamma_{g\ell}\underline{g}=-\hat{e}_{g\ell}^g\left(\underline{u}_g-\underline{w}_{g\ell}\right)-\hat{e}_{g\ell}^\ell\left(\underline{u}_\ell-\underline{w}_{g\ell}\right)-\underline{\hat{T}}_{g\ell}^g-\underline{\hat{T}}_{g\ell}^\ell \tag{5.18}$$

Solid bulk conservation of momentum:

$$-\nabla\cdot\left\{(1-\varepsilon)P_s\underline{\underline{I}}\right\}-(1-\varepsilon)\rho_s\underline{g}+R_s=\underline{\hat{T}}_{\ell s}^s \tag{5.19}$$

Solid–liquid interface conservation of momentum:

$$-a_{\ell s}\Gamma_{\ell s}\underline{g}=-\left(\underline{\hat{T}}_{\ell s}^s+\underline{\hat{T}}_{\ell s}^\ell\right) \tag{5.20}$$

Volume conservation:

$$\varepsilon_\ell+\varepsilon_g=\varepsilon \tag{5.21}$$

where $\underline{u}_\alpha$ is the velocity vector for the bulk α-phase, $\underline{w}_{\alpha\beta}$ is the velocity of the $\alpha\beta$-interface, $a_{\alpha\beta}$ is the area of the $\alpha\beta$-interface per unit volume of porous medium, $\Gamma_{\alpha\beta}$ is the mass of $\alpha\beta$-interface per unit area of $\alpha\beta$-interface, $\gamma_{g\ell}$ is the macroscopic interfacial tension of the gas–liquid interface, $\hat{e}_{\alpha\beta}^\alpha$ is the rate of mass transfer from $\alpha\beta$-interface to the α-phase, $\underline{\hat{T}}_{\alpha\beta}^\alpha$ and is the body supply of momentum to the α-phase from $\alpha\beta$-interface.

Adapted to trickle-bed geometry assuming Newtonian unidirectional 1D (streamwise) and isothermal (local thermal equilibrium) immiscible (no mass transfer exchange between bulk phases) gas–liquid flows in a nondeformable and homogeneous (constant bed porosity) porous medium, the nonequilibrium thermomechanical model becomes as follows:

Gas bulk conservation of mass:

$$\frac{\partial}{\partial t}\left(\rho_g\varepsilon_g\right)+\frac{\partial}{\partial z}\left(\rho_g\varepsilon_g u_g\right)=\hat{e}_{g\ell}^g \tag{5.22}$$

Liquid bulk conservation of mass:

$$\frac{\partial}{\partial t}(\rho_\ell\varepsilon_\ell)+\frac{\partial}{\partial z}(\rho_\ell\varepsilon_\ell u_\ell)=\hat{e}_{g\ell}^\ell \tag{5.23}$$

Gas–liquid interface conservation of mass:

$$\frac{\partial}{\partial t}(\Gamma_{g\ell}a_{g\ell})+\frac{\partial}{\partial z}(\Gamma_{g\ell}a_{g\ell}w_{g\ell})=-\left(\hat{e}_{g\ell}^g+\hat{e}_{g\ell}^\ell\right) \tag{5.24}$$

Gas bulk conservation of momentum:

$$\rho_g\varepsilon_g\left(\frac{\partial u_g}{\partial t}+u_g\frac{\partial u_g}{\partial z}\right)+\frac{\partial}{\partial z}(\varepsilon_g P_g)-\frac{\partial}{\partial z}\left(\varepsilon_g\mu_g^e\frac{\partial u_g}{\partial z}\right)-\rho_g\varepsilon_g g=\hat{e}_{g\ell}^g u_g+T_{g\ell}^g \tag{5.25}$$

Liquid bulk conservation of momentum:

$$\rho_\ell\varepsilon_\ell\left(\frac{\partial u_\ell}{\partial t}+u_\ell\frac{\partial u_\ell}{\partial z}\right)+\frac{\partial}{\partial z}(\varepsilon_\ell P_\ell)-\frac{\partial}{\partial z}\left(\varepsilon_\ell\mu_\ell^e\frac{\partial u_\ell}{\partial z}\right)-\rho_\ell\varepsilon_\ell g=\hat{e}_{g\ell}^\ell u_\ell+T_{g\ell}^\ell+T_{\ell s}^\ell \tag{5.26}$$

Gas–liquid interface conservation of momentum:

$$\Gamma_{g\ell}a_{g\ell}\left(\frac{\partial w_{g\ell}}{\partial t}+w_{g\ell}\frac{\partial w_{g\ell}}{\partial z}\right)-\frac{\partial}{\partial z}\left(a_{g\ell}\gamma_{g\ell}\right)-a_{g\ell}\Gamma_{g\ell}g$$
$$=-\hat{e}_{g\ell}^g\left(u_g-w_{g\ell}\right)-\hat{e}_{g\ell}^\ell\left(u_\ell-w_{g\ell}\right)-T_{g\ell}^g-T_{g\ell}^\ell \tag{5.27}$$

Solid bulk + solid–liquid interface conservation of momentum:

$$-\frac{\partial}{\partial z}[(1-\varepsilon)P_s]-(1-\varepsilon)\rho_p g+R_s-a_{\ell s}\Gamma_{\ell s}g=-T_{\ell s}^\ell \tag{5.28}$$

Constitutive relations were established for the macroscopic solid pressure, the macroscopic Helmholtz free energies, and the mass exchange rates for production/destruction of interfacial area, leading to more complete formulations of the dynamic capillary pressure and the body supplies of momentum [54]. The structure of the momentum exchange terms, $T^{\alpha}_{\alpha\beta}$, was suggested from an analysis of the entropy inequality to fulfill the restrictions that at equilibrium no relative movement of phases and interfaces is possible, that is, u_α, $w_{g\ell}$, $\hat{e}^{\alpha}_{g\ell}$, $\partial\varepsilon_\alpha/\partial t = 0$, and whereby the system's net entropy generation is minimal [55]. Assuming nonequilibrium states are near equilibrium, Taylor expansion of nonequilibrium equations about the equilibrium relations is applicable [56]. It follows that $T^{\alpha}_{\alpha\beta}$ in nonequilibrium state is formulated as the sum of surviving equilibrium and nonsurviving nonequilibrium (i.e., vanishing as u_α, $w_{g\ell} \rightarrow 0$) terms for the body momentum supply. The equilibrium terms are expressed as a function of the macroscopic Helmholtz free energies per unit mass, A_g, A_ℓ, and $A_{g\ell}$ [55], whereas the nonequilibrium terms, $\tau^{\alpha}_{\alpha\beta}$, are closed using drag expressions formulated for trickle-bed reactors [50]:

$$T^{g}_{g\ell} = P_g \frac{\partial \varepsilon_g}{\partial z} - \rho_g \varepsilon_g \left(\frac{\partial A_g}{\partial a_{g\ell}} \frac{\partial a_{g\ell}}{\partial z} + \frac{\partial A_g}{\partial \varepsilon_\ell} \frac{\partial \varepsilon_\ell}{\partial z} \right) + \underbrace{\tau^{g}_{g\ell}}_{nonequilibrium} \tag{5.29}$$

$$T^{\ell}_{g\ell} + T^{\ell}_{\ell s} = P_\ell \frac{\partial \varepsilon_\ell}{\partial z} - \rho_\ell \varepsilon_\ell \left(\frac{\partial A_\ell}{\partial a_{g\ell}} \frac{\partial a_{g\ell}}{\partial z} + \frac{\partial A_\ell}{\partial \varepsilon_\ell} \frac{\partial \varepsilon_\ell}{\partial z} \right) + \underbrace{\tau^{\ell}_{g\ell} + \tau^{\ell}_{\ell s}}_{nonequilibrium} \tag{5.30}$$

$$T^{g}_{g\ell} + T^{\ell}_{g\ell} = \Gamma_{g\ell} a_{g\ell} \frac{\partial A_{g\ell}}{\partial \varepsilon_\ell} \frac{\partial \varepsilon_\ell}{\partial z} + \underbrace{\tau^{g}_{g\ell} + \tau^{\ell}_{g\ell}}_{nonequilibrium} \tag{5.31}$$

To solve the dynamic model, these constitutive relationships were calibrated using measurements of time-varying liquid holdup (from electrical capacitance tomography) and bed pressure drop. The developed constitutive equations verified the admissibility principle in accordance with the second law of thermodynamics.

Remarks. Close inspection of the nonequilibrium model outputs reveals that assumption of nonequilibrium capillary pressure in the studied range of experimental conditions was not necessary and static equilibrium described by $P_c \approx P_g - P_\ell$ was sufficient to account for the interfacial forces [54]. However, recourse to empirical capillary relationships, such as the Leverett *J*-function, is unnecessary as the nonequilibrium two-phase flow model enables access to capillary pressure via entropy-consistent constitutive expressions for the macroscopic Helmholtz free energies. Also, the role of mass exchange between bulk fluid phase holdups and gas–liquid interfacial area was shown to play a nonnegligible role in the dynamics of trickle-bed reactor [54]. By accounting for the production/destruction of interfacial area, they prompted much briefer response times for the system to attain steady state compared to the case without inclusion of these mass exchange rates.

5.3 Mass and heat transfer in three-phase fixed-bed reactors

For the design of the multiphase chemical reactors, one has to deal with hydrodynamics, mass and heat transfer, and catalyst activity. For heterogeneously catalyzed reactions, the rate of the reaction is often limited by interphase mass transfer instead of the intrinsic kinetics (Figure 5.5). The rate-limiting step can be the gas–liquid mass transfer, liquid–solid mass transfer, and

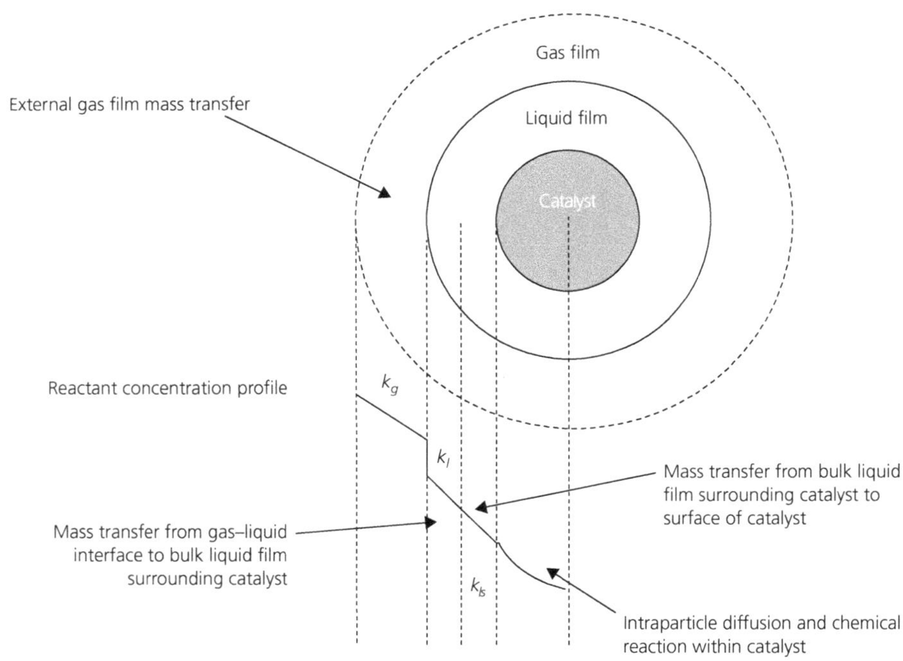

Figure 5.5 Mass transfer resistances in three-phase fixed-bed reactors.

intraparticle diffusion. These limitations reduce the overall conversion and selectivity of the process. Also, the reaction rates in three-phase fixed-bed catalytic reactors are highly affected by the heat transfer resistances: resistance to radial heat transfer and resistance to fluid-to-particle heat transfer. So a good knowledge of the packed-bed mass and heat transfer parameters is necessary for the design of the reactor and heat removal system.

5.3.1 Gas–liquid mass transfer

In two-phase downflow and upflow fixed-bed reactors, gas–liquid mass transfer resistance can be detrimental to the overall reactor performance [14, 32]. Therefore, accurate estimation of gas–liquid mass transfer parameters is important for achieving successful reactor design or scale-up. The overall gas–liquid mass transfer coefficient may be expressed, according to the two-film concept, in terms of the liquid-side and the gas-side mass transfer coefficients:

$$\frac{1}{K_\ell a_{g\ell}} = \frac{1}{He k_g a_{g\ell}} + \frac{1}{k_\ell a_{g\ell}} \tag{5.32}$$

In most cases, mass transfer resistance in the gas film is considerably smaller than the liquid-side resistance; therefore, the study of gas–liquid mass transfer has concentrated on the investigation of the mass transfer in the liquid film of the gas–liquid interface. Usually, gas–liquid mass transfer takes place across gas–liquid interfaces where the liquid can be stagnant or dynamic. While for slow reactions stagnant liquid affects very little the global reaction rate, fast reactions are characterized by a net contrast in reactant concentrations in the two regions resulting in reactionally ineffective stagnant zones.

The following current trends emanate from the analysis of the gas–liquid mass transfer literature: (i) liquid-side volumetric mass transfer coefficient, $k_\ell a_{g\ell}$, and gas–liquid interfacial area, $a_{g\ell}$, follow a trend qualitatively similar to pressure drops with regard to liquid and gas flow rates for a given pressure or gas density; (ii) for two-phase downflow, $k_\ell a_{g\ell}$ and $a_{g\ell}$ are higher for viscous liquids [57, 58], whereas, for two-phase upflow, $k_\ell a_{g\ell}$ decreases when liquid viscosity is increased [59], owing to a more important decrease of k_ℓ with regard to the increase of $a_{g\ell}$; (iii) $a_{g\ell}$ for noncoalescing liquids exceeds that for coalescing ones [57]; (iv) $a_{g\ell}$ for organic liquids does not differ significantly from that for ionic liquids [60]; (v) $a_{g\ell}$ is higher for nonspherical solid particles [35]; (vi) both $k_\ell a_{g\ell}$ and $a_{g\ell}$ improve in two-phase downflow fixed-bed reactors as gas reactor pressure increases, whereas $a_{g\ell}$ is not noticeably affected by pressure under two-phase upflow conditions [61, 62] because the coalescence/breakup process of the bubbles is dominated by the packing (any influence of pressure on the bubble formation at the gas distributor is masked by the coalescence/breakup process); and (vii) since all the packing is wetted, in upflow fixed-bed reactors, $a_{g\ell}$ is generated exclusively from interactions at the gas–liquid interface, whereas, in downflow trickle flow and partially wetted conditions, $a_{g\ell}$ is promoted by two simultaneous processes: the spreading of the liquid on the catalyst and the interaction between the gas and liquid at their boundary [59]. The scarcity of gas-side volumetric mass transfer coefficient is noteworthy; and to the best of our knowledge no experimental data on $k_g a_{g\ell}$ are available for high pressure conditions. Since k_g is expected to diminish with the increase in gas density, disregarding this fact can lead to overestimated overall gas–liquid mass transfer coefficients at high pressure, especially for a mixture of gases involving a highly soluble reactant. Gas–liquid mass transfer involving non-Newtonian liquids is also insufficiently addressed in the literature [58]. Considering the large number of biochemical processes that utilize fixed-bed reactors, this gap in knowledge needs to be filled.

In spite of the considerable research in the last 40 years, gas–liquid mass transfer in two-phase downflow fixed-bed reactors is still not well understood, and general correlations of mass transfer parameters, summarized by Saroha and Nigam [19], Al-Dahhan et al. [14], Larachi et al. [35], and Iliuta et al. [63], are still out of grasp. It is well recognized that the Achilles heel of the currently available fixed-bed reactor mass transfer design tools lies in their lack of generality inherent partly to the narrow experimental windows upon which validation of most correlations/models is carried out and partly to biased analysis and inappropriate mathematical expressions. Based on the largest gas–liquid interfacial mass transfer database (beyond 3200 data), Iliuta et al. [63] developed whole-flow-regime dimensionless correlations for gas–liquid mass transfer parameters in two-phase downflow fixed-bed reactors by combining dimensional analysis and artificial neural networks. The overall result was a significant improvement in predicting overall liquid-side and gas-side volumetric mass transfer coefficients and gas–liquid interfacial area. However, similarly to any other available correlation in literature, the neural network correlations are basically interpolation correlations. Hence, to prevent unacceptable deviations, it is suggested to verify a priori that the conditions to be predicted fall within the validity range.

In spite of the extensive studies on the gas–liquid mass transfer in two-phase upflow fixed-bed reactors, summarized by Shah [27], Ramachandran and Chaudhari [64], Gianetto and Silveston [65], and Molga and Westerterp [62], only few correlations are reported in the literature to evaluate gas–liquid mass transfer parameters under low and high pressure conditions [59, 66].

5.3.2 Liquid–solid mass transfer

Under certain conditions, for example, in the case of highly soluble gases, fast reactions of highly active catalysts, these reactions become partially or fully controlled by the rate of the liquid–solid mass transfer. Liquid–solid mass transfer resistance is significant when the following inequality is satisfied [13]:

$$\frac{10 d_p}{c_{sat}} r_{react}(1-\varepsilon) > K_{\ell s} \tag{5.33}$$

A significant amount of research has been performed on the measurement of liquid–solid mass transfer [67]. Generally, liquid–solid mass transfer in fixed-bed reactors has been studied by five methods: dissolution of slightly soluble solids into the liquid [68–73], chemical reaction with significant liquid–solid mass transfer resistance [74], ion exchange followed by an instantaneous irreversible reaction [75], dynamic absorption [76], and electrochemical technique [77–80]. The electrochemical method has certain advantages over the other: it facilitates direct and instantaneous measurements of solid–liquid mass transfer and is thus very useful to measure mass transfer fluctuations, especially under pulse flow conditions.

The following current trends emanate from the analysis of the liquid–solid mass transfer two-phase downflow fixed-bed literature: (i) volumetric liquid–solid mass transfer coefficient, $k_{\ell s}a_{\ell s}$, increases significantly with liquid flow rate in both low and high interaction flow regimes [67, 71, 81, 82]; (ii) liquid–solid mass transfer rate is slightly affected by the gas flow rate in trickle flow regime [71, 75, 83, 84] and increases in pulse flow regime [78]; (iii) transition from trickle flow to pulse flow regime causes an abrupt increase in the liquid–solid mass transfer [71]; (iv) in pulse flow regime higher mass transfer rates could be ascribed, in addition to the obvious influence of the liquid and gas velocities, to the improved wetting efficiency [77] and to the effect of the gas phase on the liquid–solid mass transfer mechanism through the local turbulence [78]; (v) volumetric liquid–solid mass transfer coefficient increases with a decrease in particle diameter [67, 70, 71, 73]; (vi) the effect of the reactor pressure on the liquid–solid mass transfer coefficient [67] follows the trend of reactor pressure on the pressure drop, liquid holdup, catalyst wetting efficiency, and gas–liquid interfacial area reported by Al-Dahhan and Dudukovic [42]; and (vii) trickle flow regime wetting efficiency acts on liquid–solid interfacial area ($a_{\ell s}$), while the liquid–solid mass transfer coefficient ($k_{\ell s}$) is mainly connected to flow turbulence through the interstitial liquid velocity [84].

Numerous correlations have been proposed for liquid–solid mass transfer coefficient in two-phase downflow fixed-bed reactors at atmospheric pressure [69–72, 74–78, 81–86]. A comparison of these correlations at zero gas flow rate in trickle flow and pulse flow regimes indicates a considerable scatter which can be attributed to various factors: particle diameter, interfacial area for liquid–solid mass transfer, and wetting efficiency of the packing [87]. Also, the current correlations for liquid–solid mass transfer coefficients developed based on atmospheric pressure data failed to predict the $k_{\ell s}$ values measured at high pressure and the trend of the effect of the pressure [67]. An attempt has been made to fill in this gap by Larachi et al. [87] who developed a neural network correlation based on a large database which includes the high-pressure data of Highfill and Al-Dahhan [67]. However, this correlation exhibits some deficiency in some cases which could be ascribed to the lack of a significant high-pressure $k_{\ell s}$ database [87]. Hence, further investigations are needed that cover a wide range of reactor pressures, different bed characteristics, physical properties, and operating conditions to develop reliable correlations for the prediction of the liquid–solid mass transfer coefficient in two-phase downflow fixed-bed reactors at high-pressure operation.

In two-phase upflow fixed-bed reactors operated in bubble flow regime, the enhanced liquid–solid mass transfer may be attributed to the following: (i) the presence of the gas bubbles in the liquid phase decreases the cross-sectional area available for liquid flow with a consequent increase in liquid velocity, (ii) the rising bubbles impart radial momentum to the surrounding liquid, (iii) the rising bubbles generate turbulence in their wakes by virtue of hydrodynamic boundary layer separation, (iv) bubble coalescence or breakdown in the vicinity of the solid particles generates turbulence which penetrates the diffusion layer, and (v) the collision of the rising bubbles with the solid particles disturbs the diffusion layer at the solid surface. For two-phase upflow fixed-bed reactors, the few correlations reported in the literature [88–91] have been developed based only on the liquid–solid mass transfer experiments performed at atmospheric pressure.

5.3.3 Heat transfer

Reactions carried out in three-phase fixed-bed reactors such as hydrogenation, oxidation, and hydrodesulfurization can be highly exothermic. Such situations require incorporation of an efficient heat removal system in order to avoid hot spots or catalyst deactivation as much as possible [13, 92]. A good knowledge of the packed-bed heat transfer parameters is necessary for the design of the reactor and heat removal system.

Several kinds of mathematical models have been used to analyze steady-state heat transfer experimental data. These models can be classified into basically two groups: heterogeneous models and pseudohomogeneous models. In heterogeneous models, separate energy balance equations are set up for the fluid phase and the solid phase. Pseudohomogeneous models make no distinction between the fluids and solid temperatures, and only one energy balance equation is required. Heat transfer in three-phase fixed-bed reactors has been mainly investigated using the two-dimensional homogeneous model [93]

$$\left(\rho_\ell c_{p,\ell} v_{s\ell} + \rho_g c_{p,g} v_{sg}\right)\frac{\partial T}{\partial z} = \Lambda_r\left(\frac{1}{r}\frac{\partial T}{\partial r} + \frac{\partial^2 T}{\partial r^2}\right) \tag{5.34}$$

$$r = R \quad T = T_w \tag{5.35}$$

$$r = 0 \quad \frac{\partial T}{\partial r} = 0 \tag{5.36}$$

In this heat transport model, the radial heat transfer is represented by a Fourier-like law, using an equivalent radial thermal conductivity of the medium with flowing fluids: $\dot{q}_r = \Lambda_r \partial T/\partial r$. The bed radial effective thermal conductivity was expressed as sum of two terms—conductive contribution and convective contribution: $\Lambda_r = \lambda_{s0} + \lambda_{r,g\ell}^{eff}$. In Hashimoto et al. [94] and Matsuura et al.'s [95] approach, the theory of single-phase gas flow heat transfer in fixed beds was extended to two-phase flow by adding the liquid convective contribution to heat transfer: $\Lambda_r = \lambda_{s0} + \lambda_{r,g}^{eff} + \lambda_{r,\ell}^{eff}$.

Also, heat transfer in three-phase fixed-bed reactors has been investigated using a two-dimensional homogeneous model with two parameters [94, 96]—the bed radial effective thermal conductivity and the heat transfer coefficient at the wall:

$$\left(\rho_\ell c_{p,\ell} v_{s\ell} + \rho_g c_{p,g} v_{sg}\right)\frac{\partial T}{\partial z} = \Lambda_r\left(\frac{1}{r}\frac{\partial T}{\partial r} + \frac{\partial^2 T}{\partial r^2}\right) \tag{5.37}$$

$$r = R \quad -\Lambda_r\frac{\partial T}{\partial r} = h_w(T - T_w) \tag{5.38}$$

$$r = 0 \quad \frac{\partial T}{\partial r} = 0 \tag{5.39}$$

The heterogeneous model for steady-state radial heat transfer assumes that the temperature variations within particles may be smoothed so that only large-scale changes in solid temperature in the axial and radial directions need to be considered. A similar approach is adopted for the fluids:

$$\left(\rho_\ell c_{p,\ell} v_{s\ell} + \rho_g c_{p,g} v_{sg}\right)\frac{\partial T}{\partial z} = \lambda_{g\ell,r}^{eff}\left(\frac{1}{r}\frac{\partial T}{\partial r} + \frac{\partial^2 T}{\partial r^2}\right) + \lambda_{g\ell,z}^{eff}\frac{\partial^2 T}{\partial z^2} - h_{\ell s} a_{\ell s}(T - T_s) \tag{5.40}$$

$$\lambda_{s,r}^{eff}\left(\frac{1}{r}\frac{\partial T_s}{\partial r} + \frac{\partial^2 T_s}{\partial r^2}\right) + \lambda_{s,r}^{eff}\frac{\partial^2 T_s}{\partial z^2} - h_{\ell s} a_{\ell s}(T - T_s) = 0 \tag{5.41}$$

$$z = 0 \quad \left(\rho_\ell c_{p,\ell} v_{s\ell} + \rho_g c_{p,g} v_{sg}\right)\left(T^{in} - T\right) = -\lambda_{g\ell,z}^{eff}\frac{\partial T}{\partial z} \quad T_s = T^{in} \tag{5.42}$$

$$z = H \quad \frac{\partial T}{\partial z} = \frac{\partial T_s}{\partial z} = 0 \tag{5.43}$$

$$r = 0 \quad \frac{\partial T}{\partial r} = \frac{\partial T_s}{\partial r} = 0 \tag{5.44}$$

$$r = R \quad -\lambda_{g\ell,r}^{eff}\frac{\partial T}{\partial r} = h_w(T - T_w) \quad -\lambda_{s,r}^{eff}\frac{\partial T_s}{\partial r} = h_{ws}(T_s - T_w) \tag{5.45}$$

Mears [97] found that the reaction rates in fixed-bed catalytic reactors are highly affected by two heat transfer resistances: resistance to radial heat transfer and resistance to particle-to-fluid heat transfer. Considerable effort has therefore been directed toward finding the effective radial conductivity and the fluid-to-wall heat transfer coefficient (which represents the radial heat transport) and particle–liquid heat transfer coefficient (which represents particle-to-fluid heat transport).

The following current trends emanate from the analysis of the radial heat transfer two-phase downflow and upflow fixed-bed literature [98]; (i) radial heat transfer is strongly influenced by the flow regime [96, 99, 100]; (ii) the bed radial effective thermal conductivity always increases with liquid flow rate for both two-phase downflow and upflow [96, 100]; (iii) Λ_r is very little dependent on gas flow rate in trickle flow, and it decreases with gas flow rate in pulsing flow regime and increases in dispersed bubble flow regime [99, 100]; (iv) Λ_r decreases with the increase of the liquid viscosity [101]; (v) the inhibition of coalescence induces higher Λ_r values [101]; (vi) Λ_r always increases with the liquid heat capacity [96]; (vii) Λ_r increases with the increase of the catalyst particle diameter [96]; (viii) for small particles the bed radial effective thermal conductivity is higher in two-phase upflow than in downflow (trickle flow or bubble flow regimes), and the result is a better thermal stability in two-phase upflow [96]; and (ix) for large particles, in the high interaction regime, the bed radial effective thermal conductivity in two-phase downflow is equal (or even slightly higher in pulsing flow) to bed conductivity in two-phase upflow [96]. Lamine et al. [96, 100] presents different correlations from the literature for the bed radial effective thermal conductivity in two-phase flow fixed-bed reactors (downflow and upflow), as well as their field of application.

For small catalyst particles the radial heat transfer resistance is well described by a single mechanism: the radial conduction through the solid bed [96]. When the particle diameter is bigger, the supplementary wall heat transfer resistance at the wall is no negligible. This mechanism is described by the wall heat transfer coefficient, h_w. In two-phase downflow fixed-bed reactors, the heat transfer coefficient at the wall, h_w, was found to be closely dependent on the hydrodynamic regimes. In the low interaction regime, h_w generally increases with the liquid and the gas flow rates, the effect of gas velocity being more marked at lower liquid velocity values [102]. In the high interaction regime, h_w are almost independent of both the fluid flow rates because the positive effect due to the increase of liquid and gas velocity is counterbalanced by the negative effect of the decrease of the radial spread [102]. The h_w values in two-phase flow are much higher (by at least one order of magnitude) than those for gas flow only, mainly due to the presence of a liquid film on the wall. The few correlations reported in the literature for heat transfer coefficient at the wall were summarized by Lamine et al. [96]. Also, in two-phase upflow fixed-bed reactors, heat transfer coefficient at the wall is strongly influenced by the flow regime. The relative importance of wall heat resistance decreases with decreasing packing diameter [100]. Only Lamine et al. [100] and Sokolov and Yablokova [103] have developed correlations for the heat transfer coefficient at the wall at atmospheric pressure.

As the catalytic reaction taking place inside the pellets is usually accompanied by heat effects, the particle–liquid heat transfer coefficient becomes a fundamental ingredient to be estimated for the assessment of the efficacy of the heat withdrawal from the particle level away to the reactor wall level. In particular, when highly exothermic reactions are in play, impediment of liquid replenishment over the dried spots on the catalyst surface may favor inception of hot spots that are responsible for reactor runaway. As a result, evacuation of heat across the liquid-covered pellet spots becomes a critical issue. Not many studies in literature deal with particle–liquid heat transfer rates in three-phase fixed-bed reactors. The main reason is probably the difficulty to find an accurate experimental method. The following current trends emanate from the analysis of the particle–liquid heat transfer two-phase downflow fixed-bed literature: (i) the transition from trickle to pulsing flow is accompanied by a substantial increase in particle–liquid heat transfer rates due to a large decrease in liquid holdup [104–106], (ii) no principal difference in particle–liquid heat transfer rates seems to exist between trickle and pulsing flow although the hydrodynamic behavior is totally different [106], (iii) particle–liquid heat transfer rates in both flow regimes are determined to be a function of only the linear liquid velocity [106], (iv) particle–liquid heat transfer rates in the pulses are 3–4 times higher than in between the pulses which probably means that the linear liquid velocity is very high inside the pulses [106], (v) the local structure of the packed bed has a considerable influence on local heat transfer rates, and (vi) the neighboring particles have an increasing effect on the particle–liquid heat transfer rates compared to a single particle in a liquid flow [107]. Larachi et al. [87] developed a neural network correlation for particle–liquid heat transfer coefficient in two-phase downflow fixed-bed reactors based on the measurements of Marcandelli et al. [105], Boelhouwer et al. [106], and Boelhouwer [108], obtained exclusively for the air–aqueous systems at room conditions.

Catalyst particles in three-phase fixed-bed reactors are usually completely filled with liquid. Then intraparticle temperature gradients are negligible due to the low effective diffusivities in the liquid phase, as pointed out by Satterfield [13] and Baldi [92]. However, if the limiting reactant and the solvent are volatile, vapor-phase reaction may occur in the gas-filled pores, causing significant intraparticle temperature gradients [109, 110]. In these conditions, intraparticle heat transfer resistance is necessary to describe the heat transfer.

5.4 Scale-up and scale-down of trickle-bed reactors

Scale rules of laboratory trickle-bed reactors are discussed. The scale rules are important in the scaling up of a novel process investigated in the laboratory to industrial process and in scaling down of laboratory test reactor with the aim of increasing cost effectiveness while maintaining the meaningfulness of the data generated in industrial practice [111].

5.4.1 Scaling up of trickle-bed reactors

A perennial problem in multiphase reactors is scale-up, that is, how to achieve the desired results in a large-scale reactor based on the observations made in the laboratory unit, which remains elusive due to complexities associated with transport-kinetic coupling [14]. The success of scale-up of trickle-bed reactors is based on the ability to understand and quantify the transport-kinetic interactions at the particle scale level (or single eddy scale), the interphase transport at the particle and reactor scales, and the flow pattern of each phase and phase contacting pattern and their changes with the changes in reactor scale and operating conditions [1].

The risks of using space velocity for scale-up of fixed-bed (gas–solid) catalytic reactors are well known [12]. Trickle-bed reactors introduce additional complications as liquid distribution and wetting efficiency, and it seems improbable that simple parameters, such as liquid hourly space velocity (defined as the ratio of the liquid flow rate to reactor volume), can be used for reliable scale-up, except in some special cases. Nowadays, it is understood that for liquid-limited reactions scale-up at constant space velocity is conservative since it results in improved wetting efficiency and better catalyst utilization. For gas-limited reactions scale-up at constant space velocity can lead to low performance [1] as the catalyst effectiveness factor drops with increased contacting efficiency due to a reduction in the gas reactant supply. Hence, for gas-limited reactions constant space velocity and constant reactor height are required in order to maintain the same performance upon scale-up. This leads sometimes to undesirable pan-cake reactor geometry which can be a problem in achieving uniform liquid distribution and hence model-based scale-up ought to be used [1].

The recommended procedure for scale-up of trickle-bed reactors is, first, to establish in the laboratory the rate of reaction for single catalyst pellets which will include the effect of wetting efficiency [12]. If there is a soluble gaseous reactant, the rate should account for mass transfer from both the gas- and liquid-covered surfaces of the pellet. This basic rate data then can be used with intrareactor mass and, if necessary, energy conservation expressions to design the large-scale reactor. This second step should include the liquid distribution. The required mass and energy transport rates will limit application of this approach because the majority of the literature is concerned exclusively with nearly atmospheric conditions.

However, with the available chemical engineering know-how embodied in correlations and models, a large-scale difference can generally be bridged: the design of the industrial unit is often possible on the basis of data from process development units without the need for larger intermediate-scale units such as demonstration units or semi-industrial plants. Sie and Krishna [111] showed that a large scale-up factor is possible in the case of gas oil hydroprocessing reactors due to the fact that mass transfer resistances do not play a crucial role in this application. Moreover, flow-related dispersions do not play a great role either as in properly designed laboratory- and industrial-scale trickle-bed reactors the gas and liquid flows are close to plug flow. Thus, on the basis of reliable data obtained in the laboratory process development units, the design of the industrial trickle-bed reactor is relatively straightforward. The desired capacity and conversion determine the total catalyst volume. Taking into account the heat generated in the process and the specific heats of the streams as well as the maximum allowed temperature rise per bed, the number of beds, the distribution of catalyst over the beds, and the quench gas flows are determined. The total reactor bed height and reactor diameter can be chosen given the total reactor volume. The linear flow rates are established, and the pressure drop is evaluated according to available models/correlations for the catalyst of the size and shape considered. In the design of the industrial reactors

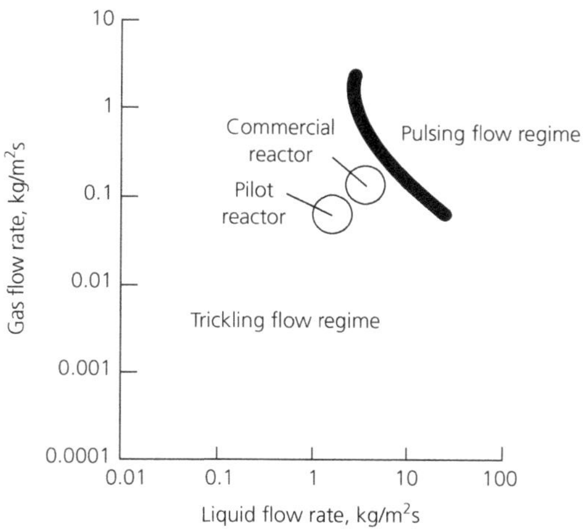

Figure 5.6 Liquid and gas loads in typical trickle-bed reactors for gas oil hydroprocessing.
(Source: Sie and Krishna [111]. Reproduced with permission of De Gruyter.)

for processes operating at elevated pressure, such gas oil hydroprocessing relatively high reactors are preferred within the constraint established by the permissible pressure drop. Tall trickle-bed reactors give rise to high velocities which may be close to the limits of transition to pulse flow and consequently ensure good wetting of the catalyst particles. Typical liquid and gas flow rates for commercial trickle-bed reactors in gas oil hydroprocessing are shown in Figure 5.6, which also shows the loads in typical process development units [111].

5.4.2 Scaling down of trickle-bed reactors

The process development can be seen, also as a scaling-down exercise. In the scaling-down process the problem is how to reduce the size of the reactor without degrading accuracy and meaningfulness of the process data. The main limiting factors for downsizing toward laboratory trickle-bed reactors are distortion of the packing in the wall region, deviation from plug flow, irrigation of catalyst particles, catalyst batch sample inhomogeneity, and interphase mass transfer and thermal radial gradients [112, 113].

It is possible to obtain kinetically representative data in very small laboratory reactors if the temperature and concentration gradients on the scale of the catalyst particle are absent. The bed needs to be isothermal and the reactor should behave as a plug-flow reactor. These requirements imply using small particles and a sufficient bed length. For flow through such small particles, surface tension forces become much more important, and established trickle-bed correlations cannot be extrapolated. For a reliable design of the reactor bed and for a choice of window of operation, the hydrodynamics of such fixed-bed microreactors have to be investigated.

In principle there are the following alternative ways to scale down the industrial trickle-bed reactors: down-scale according to geometric similarity, industrial reactor representation by different reactors, maintenance of the fluid dynamic similarity—industrial reactor representation by an integral laboratory reactor, and phenomenological approach.

5.4.2.1 Down-scale according to geometric similarity

In this approach the height and reactor diameter are reduced by the same factors leading to relatively wide laboratory reactors. The small length of the laboratory reactors leads to much lower liquid velocity than in the industrial reactor when operating at the same liquid space velocity [111]. These differences in velocities lead to differences in hydrodynamics (e.g., the catalyst particle wetting can be incomplete). Also, the relatively large diameters of the pilot- and bench-scale reactors may give undesired radial temperature profiles if the reactor is not operated under adiabatic conditions. So there are concrete outcomes nor incentives in scaling down while preserving geometric similarity, and this method is not suitable [111].

5.4.2.2 Industrial reactor represented by different reactors

If the reactor diameter and height are not reduced by the same factor but changed independently, one may chose between two options [111]: (i) maintaining the same fluid velocities but tolerating a reduction in the average residence time by increasing the space velocity with decreasing reactor bed height and (ii) maintaining the same average residence time without necessarily keeping the fluid velocities identical. The first possibility has the advantage that the hydrodynamics in the laboratory and industrial reactors are the same, but the reactant conversion is lower. In the extreme case, the laboratory reactor is a differential reactor rather than an integral one. The disadvantage of this approach is that it is necessary to study partial conversions at different conversion levels. To predict the integral conversion in the industrial reactor from a series of differential conversion experiments with feedstocks representing various degrees of conversion of the primary feed, a substantial experimental effort and a kinetic process modeling are needed. Although this approach is in essence consistent, it may not be practical all the times [111].

5.4.2.3 Maintenance of the fluid dynamic similarity: Industrial reactor represented by an integral laboratory reactor

As long as the fluid dynamics and reaction kinetics are closely interconnected so that their effects on the chemical reactions are inseparable, the only mode to scale down an industrial reactor to an integral laboratory reactor is to reduce the reactor diameter and keeping the same length. Thus, the fluid dynamic similarity is maintained (if the diameter is not reduced beyond the limit set by the wall effects), and the experiments are carried out not only at the same space velocity as in the industrial situation but also at the same liquid and gas linear velocities. This approach is based on the idea that in the industrial reactor the behavior in a horizontal plane within the bed is the same everywhere. Thus, the pilot plant reactor is a hypothetical narrow vertical column cut out of the industrial catalyst bed. The diameter of the representative pilot plant reactor is therefore determined by the catalyst particle diameter, and the height should be the same or not much smaller than that of the industrial reactor. Kallinikos and Papayannakos [114, 115] have performed diesel hydrodesulfurization studies using miniscale string bed reactors in a spiral form. They concluded that a governing parameter is the ratio between the gas and liquid superficial velocities and the reactor performance is independent of the liquid superficial velocity and of the reactor length in the range 2–6 m. Also, Hipolito et al. [116] proposed a single pellet string reactor to determine the kinetic of the fast exothermic reactions. In this geometry, the reactor diameter is close to the particle diameter, and it is possible to be operated with velocities similar to those of the industrial scale for a similar amount of catalyst as in a standard pilot plant. Single pellet string reactors present quite complex hydrodynamic patterns that cannot yet be explained with simple physical models. Obviously this approach is costly, and it is followed only if almost nothing is known of the process under investigation [113].

5.4.2.4 Phenomenological approach

In the phenomenological approach an attempt is made to isolate the various phenomena that affect the apparent reaction and to establish the appropriate correlations for each of them (assuming separable mechanisms). An example is the reactor models in which it is assumed that the reaction is pseudohomogeneous and the phases are in plug flow. For a first-order reaction one may write [113]

$$\ln(1-X)^{-1} = \frac{k_{app}}{s} \quad \text{where } k_{app} = \beta k \tag{5.46}$$

All scale-dependent factors are included in the parameter β. A vast amount of literature has been published on what the most appropriate correlation is for β. There is a disagreement over the validity of these relationships and especially about the prediction of the point from which $\beta = 1$. This piece of information is, however, essential as it is the reference point for the industrial-scale operation. The applicability of these types of reactor models is further limited by the fact that, in many situations, the intrinsic kinetics of the various reactions are not well enough known to be included in the model.

What the "smallest" scale of operation is will depend on the type of process to be investigated and on the kind of information required. Taking into account the unknown factors with regard to the reaction mechanism and intrinsic kinetics, the preferred strategy for process research is often to choose such experimental conditions which assure an independent composition of the reactor product as regards the scale of the operation. For trickle flow processes it is essential that complete wetting of each catalyst particle is ensured (as is the case of properly designed industrial reactor; nonwetted areas cause reduced, scattered conversion and may give rise to local hot spots). Deviation from plug flow should be minimal and in some instances mass and heat transfer resistances at a small scale have to be known.

5.4.3 Salient conclusions

Hydrodynamics, mass, and heat transfer in the commonly used three-phase fixed-bed reactors were briefly outlined. Also, scale-up rules and alternative ways to scale down trickle-bed reactors are discussed. In spite of the extensive studies on the hydrodynamics, mass, and heat transfer in three-phase fixed-bed reactors, clearly, a lot of work remains to be done in providing a fundamentally based description of the effect of pressure on the parameters of importance in three-phase fixed-bed reactors' operation, design, and scale-up or scale-down. It is evident that atmospheric data and models/correlations cannot, in general, be extrapolated to operation at elevated pressures. The physics conveyed by the standard two-phase flow models is minimalistic because it insufficiently describes the role and presence of interfaces and their thermodynamic properties. The explicit inclusion of interfaces and interfacial properties is essential because they are known to have a significant role in determining the thermodynamic state of the whole system.

5.5 Trickle-bed reactor/bioreactor modeling

The recent developments in trickle-bed reactor/bioreactor design are analyzed. Two case studies with reference to modeling of simultaneous catalytic hydrodesulfurization/bed clogging in hydrotreating trickle-bed reactors and phenol biodegradation/biomass accumulation in trickle-bed bioreactors are considered. In addition, the modeling of the integrated process of aqueous-phase glycerol reforming and DME synthesis into an allothermal dual-bed reactor (membrane fixed-bed reactor/tickle-bed reactor) is surveyed.

5.5.1 Catalytic hydrodesulfurization and bed clogging in hydrotreating trickle-bed reactors

One of the major unsolved complex problems confronting the chemical and petroleum industries at present is the deposition of fine particles dissolved or suspended in fluid flow systems [117–119]. When heavy gas oil suspensions containing a low concentration of nonfilterable fine solid impurities are treated in trickle-bed hydrotreaters, clogging develops and leads to the progressive obstruction of the bed that is accompanied with a rise in pressure drop. Although the fine particle concentration is in the 100 range weight ppm, the cumulative effect of several months of refining operation diverts the catalyst bed from its catalytic function to that of a huge filter where fine accumulation causes the pressure to rise by restricting the flow. In these conditions the unit must often be shut down to unload the physically deactivated catalyst for replacement with a pristine one. Additionally, fine deposition can provoke a possible unfavorable impact on the trickle-bed hydrotreating performance as a result of interparticle diffusional limitations by the fine deposits.

The knowledge on the fundamentals of the concomitant filtration and catalytic hydrotreating reactions could help improve understanding of these problems. An attempt has been made by Iliuta et al. [120] who studied the deposition of fine particles under chemical reaction conditions in a high-pressure/temperature trickle-bed reactor using a dynamic multiphase flow deep-bed filtration model coupled with heat and species balance equations in the liquid, gas, and solid (catalyst + deposit) phases. The high-pressure hydrodesulfurization process of dibenzothiophene (DBT) catalyzed by sulfided $CoO\text{-}MoO_3/\gamma\text{-}Al_2O_3$ catalyst was considered as a case study. The deep-bed filtration model incorporates the physical effects of porosity and effective specific surface area changes due to fine deposition/detachment, gas and suspension inertial effects, and coupling effects between the filtration parameters and the interfacial momentum exchange force terms. The detachment of the fine particles from the collector surface is assumed to be induced by the colloidal forces in the case of Brownian particles or by the hydrodynamic forces in the case of non-Brownian particles.

5.5.1.1 Modeling framework

In the model developed by Iliuta et al. [120], a cocurrent two-phase trickle flow through a porous medium of uniform initial porosity and single-sized catalytic particles is considered (Figure 5.7a and b). Two-phase flow is assumed unidirectional, and both the flowing phases are assumed as viscous Newtonian. The gas/liquid + fine/porous medium + fine multiphase system is viewed as a system of three interpenetrated continua: (i) a flowing gas, (ii) a dilute pseudohomogeneous suspension consisting of liquid and fine particles, and (iii) a stationary pseudocontinuous solid phase made up of the catalyst (collector) particles and the fine particles that get captured onto their surface. Hydrodesulfurization process by which the sulfur is removed from the hydrocarbon by reacting with the hydrogen in the presence of sulfided $CoO\text{-}MoO_3/\gamma\text{-}Al_2O_3$ catalyst at high temperature and pressure is considered. Even if the number of sulfur compounds present in gas oil can be as high as 70, DBT is used to represent the sulfur compounds since it is one of the less reactive sulfur organic compounds present in the light gas oil. DBT reacts with hydrogen via two parallel pathways, hydrogenolysis and hydrogenation [121]:

$$DBT + 2H_2 \Rightarrow BiPH + H_2S \tag{5.47}$$

$$DBT + 5H_2 \Rightarrow CHB + H_2S \tag{5.48}$$

The kinetics of DBT hydrogenolysis and hydrogenation reactions is summarized with the following rate equations recommended by Broderick and Gates [121], based on a Langmuir–Hinshelwood formulation:

$$r_{hs} = \frac{k_{hs}K_{DBT}K_{H_2}C_{DBT}C_{H_2}}{(1 + K_{DBT}C_{DBT} + K_{H_2S}C_{H_2S})^2(1 + K_{H_2}C_{H_2})} \tag{5.49}$$

$$r_{hn} = \frac{k_{hn}K'_{DBT}K'_{H_2}C_{DBT}C_{H_2}}{1 + K'_{DBT}C_{DBT}} \tag{5.50}$$

Each flowing phase is viewed as a continuum and the catalyst particles surface is completely covered by a liquid film and the gas flows in the remaining interstitial void. The properties of the fluid suspension (density, viscosity, holdup) are equal to those of the embracing liquid (influent fine volume fraction <0.1%). Only the influent liquid was considered as a source for (single-sized)

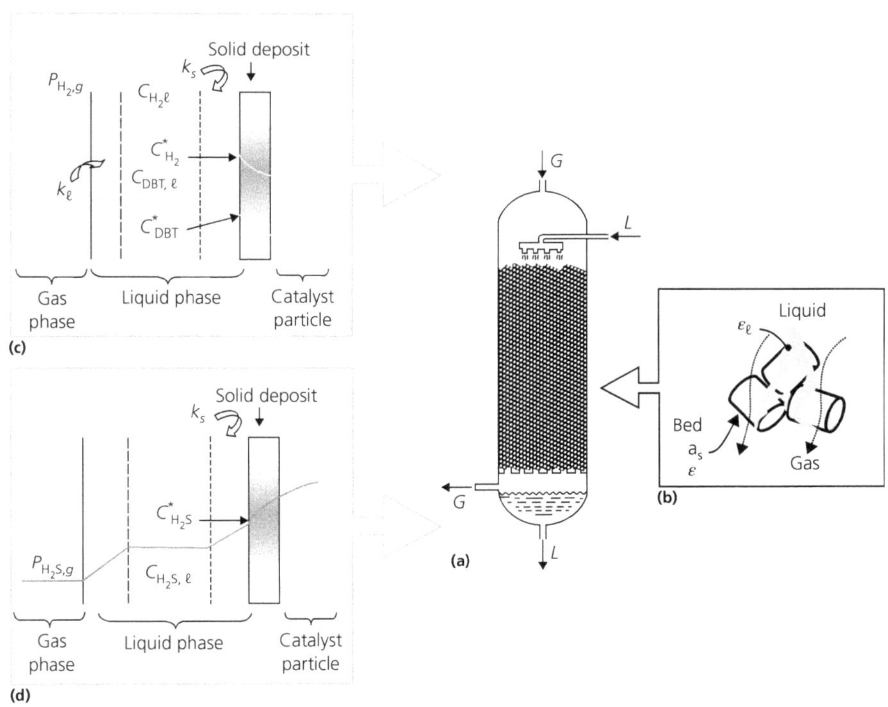

Figure 5.7 Sketch of trickle-bed reactor (a) and local-scale representation of the two-phase flow (b). Sketch showing the resistances involved in the G/L reaction on the catalyst particle (c, d). (Source: Iliuta et al. [120]. Reproduced with permission of Elsevier.)

fine particles. The filtration mechanism to occur is of the deep-bed filtration type and not of the cake filtration type [117]; in other words fine–fine attractive interactions within the fluid phase are assumed absent or at least small enough to preclude flocculation. Only the deposition under favorable surface interaction conditions is considered (the double layer force is attractive). The gas–liquid interface is impervious to fines, and the net sink in the fluid momentum balance due to fine mass transfer from fluid to the collector is neglected. In the case of colloidal (Brownian) particles ($d_f < 2$ μm), the release or detachment of fine particles from the collector surface is controlled by the colloidal forces (Brownian diffusion through the hydrodynamic boundary layer in the absence of energy barrier) [122, 123]. In the case of non-Brownian particles the detachment of fine particles is induced by hydrodynamic forces [123]. In formulating the mass conservation equations, the following additional assumption are made: the reactor is operated adiabatically and the possible temperature gradients inside the catalyst particles and between the catalyst surface and the liquid bulk are not considered, the heat consumed by vaporization is neglected, the catalyst activity is not declining with time, and chemical reactions only take place in the catalyst (no homogeneous gas- or liquid-phase reactions).

The model is based on the volume average form of the transport equations for multiphase systems [46]. The model equations consist of the conservation of volume, conservation of mass or continuity, conservation of momentum for the gas and liquid (with fines) phases, continuity for the solid stationary phase (i.e., the fixed bed), species balance in the gas and liquid phases, and heat balance [120]:

Conservation of volume:

$$\varepsilon_g + \varepsilon_\ell = \varepsilon \tag{5.51}$$

Continuity equations for the gas, the liquid (with fines), and the solid phases:

$$\frac{\partial}{\partial t}\left(\varepsilon_g\rho_g\right) + \frac{\partial}{\partial z}\left(\varepsilon_g\rho_g u_g\right) = 0 \tag{5.52}$$

$$\frac{\partial}{\partial t}(\varepsilon_\ell\rho_\ell) + \frac{\partial}{\partial z}(\varepsilon_\ell\rho_\ell u_\ell) = -\gamma_f\rho_f r_{dep,f} + \left(1-\gamma_f\right)\rho_f r^h_{rel,f} + \rho_f r^c_{rel,f} \tag{5.53}$$

$$\frac{\partial}{\partial t}\left[(1-\varepsilon^o)\rho_p + (1-\varepsilon_d)(\varepsilon^o-\varepsilon)\rho_f\right] = \gamma_f\rho_f r_{dep,f} - \left(1-\gamma_f\right)\rho_f r^h_{rel,f} - \rho_f r^c_{rel,f} \tag{5.54}$$

Species balance equation for the fines:

$$\frac{\partial}{\partial t}(\varepsilon_\ell c_f) + u_\ell\frac{\partial}{\partial z}(\varepsilon_\ell c_f) = -\gamma_f r_{dep,f} + \left(1-\gamma_f\right) r^h_{rel,f} + r^c_{rel,f} + D_\ell\frac{\partial^2}{\partial z^2}(\varepsilon_\ell c_f) \tag{5.55}$$

Momentum balance equations for the gas and liquid (with fines) phases:

$$\frac{\partial}{\partial t}\left(\rho_g\varepsilon_g u_g\right) + u_g\frac{\partial}{\partial z}\left(\rho_g\varepsilon_g u_g\right) = \varepsilon_g\mu^e_g\frac{\partial^2 u_g}{\partial z^2} - \varepsilon_g\frac{\partial P}{\partial z} + \varepsilon_g\rho_g g - F_{g\ell} \tag{5.56}$$

$$\frac{\partial}{\partial t}(\rho_\ell\varepsilon_\ell u_\ell) + u_\ell\frac{\partial}{\partial z}(\rho_\ell\varepsilon_\ell u_\ell) = \varepsilon_\ell\mu^e_\ell\frac{\partial^2 u_\ell}{\partial z^2} - \varepsilon_\ell\frac{\partial P}{\partial z} + \varepsilon_\ell\rho_\ell g + F_{g\ell} - F_{\ell s} \tag{5.57}$$

Deposition/detachment rate equation:

$$\frac{d\sigma_f}{dt} = r_{dep,f}\gamma_f - r^h_{rel,f}\left(1-\gamma_f\right) - r^c_{rel,f} \tag{5.58}$$

Species balance in the liquid phase (axial dispersion in the liquid phase):

$$\frac{\partial}{\partial t}(C_{\mathrm{DBT},\ell}\varepsilon_\ell) + u_\ell\frac{\partial}{\partial z}(C_{\mathrm{DBT},\ell}\varepsilon_\ell) = D_\ell\frac{\partial^2}{\partial z^2}(C_{\mathrm{DBT},\ell}\varepsilon_\ell) - r_{hs}\eta_{hs}\rho_p(1-\varepsilon) - r_{hn}\eta_{hn}\rho_p(1-\varepsilon) \tag{5.59}$$

$$\frac{\partial}{\partial t}(C_{\mathrm{H_2},\ell}\varepsilon_\ell) + u_\ell\frac{\partial}{\partial z}(C_{\mathrm{H_2},\ell}\varepsilon_\ell) = D_\ell\frac{\partial^2}{\partial z^2}(C_{\mathrm{H_2},\ell}\varepsilon_\ell) + (k_\ell a_{g\ell})_{\mathrm{H_2}}\left(\frac{P_{\mathrm{H_2},g}}{He_{\mathrm{H_2}}} - C_{\mathrm{H_2},\ell}\right) - 2r_{hs}\eta_{hs}\rho_p(1-\varepsilon) - 5r_{hn}\eta_{hn}\rho_p(1-\varepsilon) \tag{5.60}$$

$$\frac{\partial}{\partial t}(C_{\mathrm{H_2S},\ell}\varepsilon_\ell) + u_\ell\frac{\partial}{\partial z}(C_{\mathrm{H_2S},\ell}\varepsilon_\ell) = D_\ell\frac{\partial^2}{\partial z^2}(C_{\mathrm{H_2S},\ell}\varepsilon_\ell) + (k_\ell a_{g\ell})_{\mathrm{H_2S}}\left(\frac{P_{\mathrm{H_2S},g}}{He_{\mathrm{H_2S}}} - C_{\mathrm{H_2S},\ell}\right) + r_{hs}\eta_{hs}\rho_p(1-\varepsilon) + r_{hn}\eta_{hn}\rho_p(1-\varepsilon) \tag{5.61}$$

Species balance in the gas phase (no axial dispersion in the gas phase):

$$\frac{1}{RT}\frac{\partial}{\partial t}(P_{j,g}\varepsilon_g) + \frac{u_g}{RT}\frac{\partial}{\partial z}(P_{j,g}\varepsilon_g) = -(k_\ell a_{g\ell})_j\left(\frac{P_{j,g}}{He_j} - C_{j,\ell}\right) \quad j = \mathrm{H_2},\ \mathrm{H_2S} \tag{5.62}$$

Heat balance equation:

$$(\varepsilon_\ell\rho_\ell c_{p,\ell} + \varepsilon_s\rho_s c_{p,s})\frac{\partial T}{\partial t} + \varepsilon_\ell\rho_\ell u_\ell c_{p,\ell}\frac{\partial T}{\partial z} = \lambda^{eff}_{\ell,z}\frac{\partial^2 T}{\partial z^2} + \sum_i(-\Delta H_{Ri})r_i\eta_i\rho_p(1-\varepsilon) \tag{5.63}$$

The velocities at the inlet are specified in Dirichlet-type boundary conditions. At the outlet, an open boundary condition referred to as "outflow boundary condition" [124] is used. Outflow boundary conditions are usually appropriate to model exit flows where velocity and pressure distributions are not known a priori. They are appropriate when exit flows are fully developed or close to, which implies, except for pressure, zero gradients for all flow variables normal to the outflow boundary. The liquid holdup at the reactor inlet was evaluated assuming $\partial u_g/\partial z = \partial u_\ell/\partial z = 0$ and combining Equations 5.56 and 5.57:

$$\frac{F_{g\ell}}{\varepsilon_g} + \frac{F_{g\ell}}{\varepsilon_\ell} - \frac{F_{\ell s}}{\varepsilon_\ell} + \left(\rho_\ell - \rho_g\right)g = 0 \tag{5.64}$$

Initial and boundary conditions for Equations 5.59–5.63 are (j = DBT, H_2, H_2S)

$$t=0 \quad C_{j,\ell}=C_{j,\ell}^{in},\ P_{j,g}=P_{j,g}^{in},\ T=T^{in} \tag{5.65}$$

$$z=0 \quad P_{j,g}=P_{j,g}^{in} \quad u_\ell C_{j,\ell}^{in}=u_\ell C_{j,\ell}\big|_{z=0^+}-D_\ell\frac{\partial C_{j,\ell}}{\partial z} \tag{5.66}$$

$$u_\ell\rho_\ell c_{p,\ell}\varepsilon_\ell T^{in}=u_\ell\rho_\ell c_{p,\ell}\varepsilon_\ell T\big|_{z=0^+}-\lambda_{\ell,z}^{eff}\frac{\partial T}{\partial z} \tag{5.67}$$

$$z=H \quad \frac{\partial C_{j,\ell}}{\partial z}=0,\ \frac{\partial T}{\partial z}=0 \tag{5.68}$$

η_{hs} and η_{hn} are the effectiveness factors for the hydrogenolysis and hydrogenation reactions and are calculated as the ratio between the reaction rate with pore diffusion resistance and the reaction rate in absence of diffusion, namely, at liquid bulk concentrations:

$$\eta_i=\frac{\int_v r_i dV}{r_i|_\ell} \tag{5.69}$$

The concentration gradients inside the catalyst particles and fine solid deposit (Figure 5.7c and d) are given by the equations describing the simultaneous mass transport and reaction within the catalyst particles and the diffusion within the solid deposit considering that the reactants' diffusion occurs in the liquid phase by virtue of completely internally wetted catalysts:

$$\varepsilon_p\frac{\partial C_{j,p}}{\partial t}=\frac{1}{r^2}\frac{\partial}{\partial r}\left(r^2D_{j,p}^{eff}\frac{\partial C_{j,p}}{\partial r}\right)+\sum_i \upsilon_{ij}r_i\rho_p \quad j=\text{DBT, H}_2\text{, H}_2\text{S} \tag{5.70}$$

$$\varepsilon_d\frac{\partial C_{j,d}}{\partial t}=\frac{1}{r^2}\frac{\partial}{\partial r}\left(r^2D_{j,d}^{eff}\frac{\partial C_{j,d}}{\partial r}\right) \tag{5.71}$$

The corresponding boundary and initial conditions are given as (j = DBT, H_2, H_2S)

$$r=0 \quad \frac{\partial C_{j,p}}{\partial r}=0 \tag{5.72}$$

$$r=r_p^o \quad D_{j,d}^{eff}\frac{\partial C_{j,d}}{\partial r}\bigg|_{r=r_p^o}=D_{j,p}^{eff}\frac{\partial C_{j,p}}{\partial r}\bigg|_{r=r_p^o} \quad C_{j,d}\big|_{r_p^o}=C_{j,p}\big|_{r_p^o} \tag{5.73}$$

$r=r_p(t)$ external surface area of the catalyst particle is completely wetted

$$D_{j,d}^{eff}\frac{\partial C_{j,d}}{\partial r}\bigg|_{r=r_p(t)}=k_{\ell s,j}\left(C_{j,\ell}-C_j^*\right) \tag{5.74}$$

$$t=0 \quad C_{j,d}(r,0)=C_{j,\ell}^{in} \quad C_{j,p}(r,0)=C_{j,\ell}^{in} \tag{5.75}$$

The solution of the dynamic multiphase flow deep-bed filtration model requires the knowledge of the fine deposition and detachment processes kinetics. The deposition rate determines the degree of collection of fines. A simple logarithmic law is used to express the dependence of deposition rate versus fine concentration and superficial liquid velocity [125]:

$$r_{dep,f}=\lambda_f c_f v_{s\ell} \tag{5.76}$$

In Equation 5.76, λ_f is filter coefficient which can be thought of as the probability for a fine to be captured as it travels a unit distance through the bed [125]. The form of the filter coefficient is dictated by the nature of the capture phenomena in play and by the amount of capture as bed clogging proceeds. During the fine–collector capture step, the fines adhere individually as a monolayer on the collector surface via fine–collector interaction forces: fluid drag, gravity, van der Waals surface-interactive forces, and Brownian diffusion force [125]. For monolayer filtration, Logan et al. [126] and Nelson and Ginn [127] describe the attachment of suspended particles as a two-step process: the transport to the grain surface quantified by the collection efficiency (η^o) and the frequency at which particles in the aqueous phase come into contact with the collector quantified by the fractional collision efficiency (α):

$$\lambda_f=\frac{3}{2}(1-\varepsilon^o)^{1/3}\frac{\eta^o\alpha}{d_p} \tag{5.77}$$

Single-collector efficiency for monolayer filtration was estimated with the expression developed by Rajagopalan and Tien [128], obtained by the combination of the trajectory analysis of a spherical particle in the vicinity of a spherical collector with the contribution of the Brownian diffusion. For fine–fine capture step, filtration becomes driven by the fine–fine interaction forces yielding a multilayer deposit for which the filter coefficient no longer remains constant in time. The change of the filter coefficient as a function of the specific deposit was estimated using the correlation developed by Tien et al. [129]. Extra information about trickle-bed deep-bed filtration model is given in Iliuta and Larachi [130] and Iliuta et al. [119].

The release or detachment of the fine particles from the collector surface is assumed to be induced by the hydrodynamic forces in the case of non-Brownian particles or by the colloidal forces in the case of Brownian particles [131]. For non-Brownian fine particles, the rate of hydrodynamic particle entrainment is considered to be proportional to the difference between the wall shear stress and the critical shear stress [123]:

$$r_{rel,f}^h=\alpha_{rel,f}^h\bar{a}(1-\varepsilon)(\tau_w-\tau_{cr})\frac{1}{\rho_f} \quad \text{for } \tau_w>\tau_{cr} \tag{5.78}$$

$$r_{rel,f}^h=0 \quad \text{for } \tau_w\le\tau_{cr} \tag{5.79}$$

where the critical shear stress is estimated for the condition of equilibrium between the forces acting on an attached particle along the tangential direction [131]:

$$\tau_{cr}=\frac{k_f\frac{6(1-\varepsilon^o)}{d_p^o}\frac{Ha}{12\delta^2}}{2.551\pi d_f} \tag{5.80}$$

In the case of Brownian particles detachment by the hydrodynamic mechanism is not possible due to the much higher values of critical shear stress required for release [120]. For colloidally induced release, in the absence of an energy barrier, the rate of particle detachment is based on the assumption that the rate-limiting step is the diffusion of detached colloids across the boundary layer between collector surfaces and the liquid bulk [122]:

$$r^c_{rel,f} = \alpha^c_{rel,f}\bar{a}(1-\varepsilon)d_f\sigma_f \tag{5.81}$$

Note that in the formulation of the continuity, the species balance equation for the fine particles, and the deposition/detachment rate equation, the fraction of the collector surface area available/not available for fine particle adhesion/detachment (γ_f) corresponds to the regions where the shear stress acting on the collector is lower than the critical shear stress [131]. The expression for the fraction of the collector surface area available/not available for the fine particle adhesion/detachment is [131]

$$\gamma_f = \frac{\tau_{cr}}{\tau_w} \tag{5.82}$$

where the maximum shear stress acting on the collector surface was estimated using Happel's model:

$$\tau_w = 3\mu_\ell \frac{A_s}{d_p} u_\ell \quad A_s = \frac{2(1-p^5)}{w}, \quad p = (1-\varepsilon)^{1/3}, \quad w = 2 - 3p + 3p^5 - 2p^6 \tag{5.83}$$

In the case of Brownian particles, the critical shear stress is very high and all the particle surface area is available for particle adhesion ($\gamma_f = 1$).

The liquid–solid and gas–liquid drag forces are estimated using the slit model [24, 48, 132]. For the adaptation to the trickle flow clogging context, these drag equations are recast as a function of the local instantaneous values of porosity and effective specific surface area [119]:

$$F_{\ell s} = \left\{ \frac{E_1}{36} C_w^2 \frac{\bar{a}^2(1-\varepsilon)^2 \mu_\ell}{\varepsilon_\ell^3} + \frac{E_2}{6} C_{wi}\left(1+\psi_{g\ell}\right) \frac{\bar{a}(1-\varepsilon)}{\varepsilon_\ell^3} \rho_\ell |v_{s\ell}| \right\} v_{s\ell}\varepsilon_\ell \tag{5.84}$$

$$F_{g\ell} = \left\{ \frac{E_1}{36} C_w^2 \frac{\bar{a}^2(1-\varepsilon)^2 \mu_g}{\varepsilon_g^3} + \frac{E_2}{6} C_{wi}\left(1+\psi_{g\ell}\right) \frac{\bar{a}(1-\varepsilon)\rho_g}{\varepsilon_g^3} \left|v_{sg} - \varepsilon_g u_{g,i}\right| \right\} \left[v_{sg} - \varepsilon_g u_{g,i}\right]\varepsilon_g \tag{5.85}$$

The effective surface area is expressed as the summation of the surface area of the porous medium and the surface area of the fines [133]:

$$\bar{a} = \frac{N_c \xi \pi d_p^2(t) + N_c \partial N_f \left[\xi \pi d_f^2 - A_\Delta(t)\right]}{N_c \frac{\pi}{6}\left(d_p^o\right)^3 + N_f \frac{\pi}{6} d_f^3} \tag{5.86}$$

With the evolution of deposition, the solid–liquid surface area gets altered by two opposing phenomena [134]: an increase in surface area through addition of the area of the captured fines and a loss of area A_Δ due to the shadow effect [135] (the shadow effect refers to an exclusion collector area hidden nearby the fine once it gets captured forbidding thereafter access by future approaching fines).

5.5.1.2 Salient conclusions

The relevant criterion for clogging is the pressure drop in the reactor. Figure 5.8a shows the effect of the inlet fine particle concentration on the $\Delta P/\Delta P^o$ ratio for the operating conditions listed in Table 5.1 (ΔP^o is the two-phase pressure drop for the clean bed under otherwise identical conditions).

With the increase of the inlet fine particle concentration, the filtration rate increases yielding higher overall specific deposits at the same filtration time (Figure 5.8b). As a result, the bed porosity decreases and the two-phase pressure drop increases. So the major change resulting from fine particle deposition in the bed involves the effective porosity of the bed, and therefore the increase of two-phase pressure drop is the result of the decrease in the available free space. It is also to notice that inclusion in the model of the fine detachment mechanisms does not

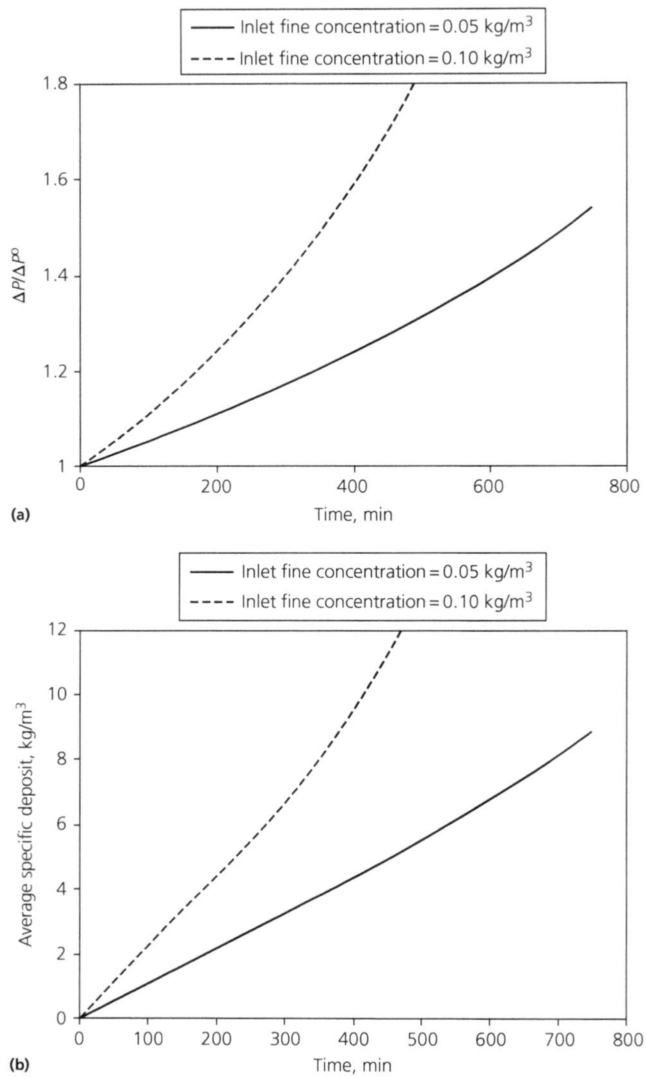

Figure 5.8 Two-phase pressure drop ratio (a) and bed volume average specific deposit (b) versus time at different values of fine particle concentration (d_f = 5 μm, T^{in} = 598 K, P^{in} = 10 MPa).

Table 5.1 Model parameters.

Liquid: carrier oil (*n*-hexadecane)	
Viscosity: 0.165×10^{-3} Pa s (598 K)	
Density: 610 kg/m^3	
Surface tension: 7.8×10^{-3} N/m	
Gas: hydrogen; hydrogen sulfide	
Fine particles: kaolinite	
Average diameter: 1–5 μm	
Density: 2000 kg/m^3	
Porosity of deposit layer: 0.8	
Influent concentration: 0–1 g/l	
Tortuosity of solid deposit: 3.7	
Catalyst packing: spherical catalyst particles	
Diameter: 0.003 m	
Porosity of catalyst: 0.65	
Bed porosity: 0.37	
Bulk density of the bed: 718 kg_{cat}/m_r^3	
Tortuosity: 3.7	
Geometry of fixed-bed reactor	
Diameter: 0.051 m	
Height: 1.0 m	
Operating conditions	
Inlet temperature: 598 K	
Inlet pressure: 10 MPa	
Superficial liquid velocity: 0.0038 m/s	
Superficial gas velocity: 0.035 m/s	
Inlet dibenzothiophene concentration: 0.03 kmol/m^3	
Gas hydrogen sulfide concentration: 0.025 kmol/m^3	
Fractional collision efficiency: $\alpha = 1$	
Coefficient of sliding friction: $k_f = 3.79 \times 10^{-6}$ m	
Hamaker constant: Ha = 1.4×10^{-20} J	
Hydrodynamic detachment rate coefficient: $\alpha^h_{rel,f} = 1.6 \times 10^{-8}$ kg/Ns	

eliminate deposition, but rather tends to stretch it in time. Said otherwise, when accounting for detachment, the model predicts that longer durations are needed to attain the same specific deposit values as if the detachment terms were to be excluded in the model equations.

Figure 5.9 shows the predicted outlet DBT conversions plotted as a function of inlet fine particle concentration for two different values of the fine particle diameter and constant filtration duration of 485 min. The larger the feed fine concentration, the lower the outlet DBT conversion. This can be rationalized as corresponding to larger overall specific deposits and, as a result, to higher external mass transfer resistances through the solid deposit of the reactants to reach the catalyst active core and of the back-diffusion of H_2S which further inhibits DBT hydrogenolysis. Nonetheless, Figure 5.9 shows that fine particle deposition does not afflict appreciably the trickle-bed reactor performance with a drop in conversion barely exceeding 1%.

The main conclusion of the Iliuta et al. [120] simulation results resides in the fact that the fine particle deposition process does not change appreciably the trickle-bed reactor performance, so the unique undesirable consequences of the fine particle deposition process reflects in an almost exclusive hydraulic effect of bed clogging.

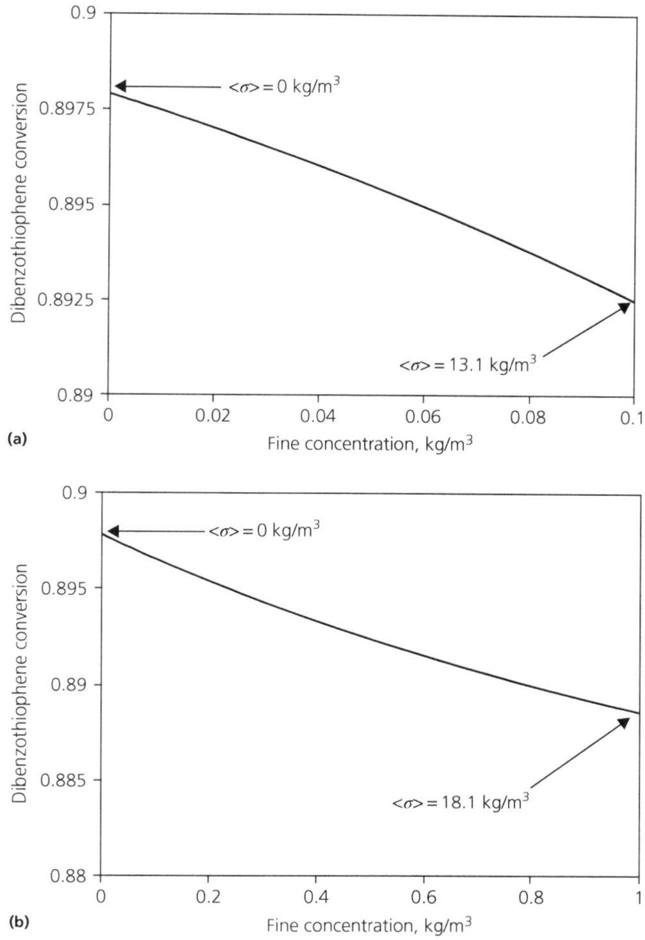

Figure 5.9 Effect of fine particle deposition process on dibenzothiophene conversion at $t = 485$ min ($T^{in} = 598$ K, $P^{in} = 10$ MPa): (a) $d_f = 5$ μm, (b) $d_f = 1$ μm.

5.5.2 Biomass accumulation and clogging in trickle-bed bioreactors for phenol biodegradation

Excessive biomass formation in two-phase flow trickle-bed bioreactors induces biological clogging and leads to the progressive obstruction of the bed that is accompanied with a buildup in pressure drop and flow channeling [132, 136]. One of the important aspects during biological clogging in trickle-bed bioreactors is the aggregation of cells and the detachment of cells and aggregates from pore bodies within the porous bed. In general, the theoretical models describing the transient behavior of biomass accumulation and the biological clogging in trickle-bed bioreactors neglect the cell aggregation process and the aggregate detachment. An attempt has been made by Iliuta and Larachi [137] who studied the dynamics of bacterial cell attachment, aggregation, growth, and detachment in trickle-bed bioreactors for wastewater treatment. Two-phase flow and space–time evolution of biological clogging were described using an Euler–Euler two-fluid dynamic model based on the volume average mass, momentum, and species balance equations, biomass dynamics equation, and filtration equations for the bacterial cells and the aggregates coupled with the discrete population balance equations for the cell agglomeration. The cell aggregation is described by the rate at which a given size aggregate is being formed by smaller aggregates minus the rate at which the aggregate combines to form a larger aggregate. The detachment of the cells or aggregates from the collector surface is assumed to be induced by the colloidal forces in the case of Brownian cells/aggregates or by the hydrodynamic forces in the case of non-Brownian aggregates. Coupling between the liquid suspension and solids is monitored via the cell and aggregate filtration rate equation, the cell and aggregate detachment rate equations, and the interaction drag or momentum exchange force terms. Phenol biodegradation by *Pseudomonas putida* as the predominant species immobilized on activated carbon is chosen as a case study to illustrate the consequences of the formation of excessive amounts of biomass.

5.5.2.1 Mathematical model

In the model developed by Iliuta and Larachi [137], a cocurrent downward gas–liquid trickle flow through a porous medium of uniform initial porosity and single-sized particles is considered. Two-phase flow is assumed unidirectional, and both the flowing phases are assumed as viscous Newtonian. The liquid is incompressible and the gas phase is ideal. The gas/liquid/biofilm/porous medium multiphase system is viewed as a system of three interpenetrated continua: (i) a flowing gas phase, (ii) a flowing liquid phase, (iii) and a stationary pseudocontinuous solid phase made up of the packing particles constituting the clean porous medium as well as of the biomass (cells and aggregates) captured onto their surface. The bioparticle surface is assumed partially wetted by a liquid film, and two-phase flow is annular and completely separated.

Phenol biodegradation by *P. putida* as the predominant species immobilized on activated carbon is considered. The cells seeded in the liquid are considered single sized and the cell aggregates are formed by lumps of bacterial cells at the surface of bioparticle. Cell attachment occurs via deep-bed filtration mechanism. The generation of cell aggregates was considered to take place by collision and subsequent bonding of the cells transported from the bulk liquid to the surface of the bioparticle in the liquid approaching bioparticle and by collision with the associated single-sized growth cells. The cell–cell attractive interactions within the bulk liquid phase are assumed absent or at least small enough to preclude the aggregation. Cells within the biofilm (including the cells associated with aggregate growth process) are individualized as in the case of individual-based modeling theory to describe the cell agglomeration process [138]. The detachment is induced by hydrodynamic forces (non-Brownian cells/aggregates) or colloidal forces (Brownian cells/aggregates).

The Euler–Euler model is based on the volume average form of the transport equation developed for multiphase systems [46] and is coupled with the population balance equations for the particle agglomeration [139]. The general model equations consist of the conservation of volume, the continuity and the Navier–Stokes equations for the gas and fluid phases, the continuity equation for the solid stationary phase and the species balance equation for the bacterial cells and detached aggregates undergoing migration from the fluid phase to the solid phase, and filtration equations for the bacterial cells [137, 140]:

Conservation of volume:

$$\varepsilon_\ell + \varepsilon_g = \varepsilon \tag{5.87}$$

Continuity equations for the gas, the liquid, and the solid phases:

$$\frac{\partial}{\partial t}\left(\varepsilon_g \rho_g\right) + \frac{\partial}{\partial z}\left(\varepsilon_g \rho_g u_g\right) = 0 \tag{5.88}$$

$$\frac{\partial}{\partial t}(\varepsilon_\ell \rho_\ell) + \frac{\partial}{\partial z}(\varepsilon_\ell \rho_\ell u_\ell) = -r_{dep,c} + r^c_{rel,c} + \sum_{k=2}^{max} r^c_{rel,a,k} + \sum_{k=2}^{max} r^h_{rel,a,k}\left(1-\gamma_{a,k}\right) \tag{5.89}$$

$$\frac{\partial}{\partial t}\left[(1-\varepsilon^o)\rho_p + (\varepsilon^o - \varepsilon)\rho_b\right] = \rho_c\left(\frac{d\sigma_c}{dt} + \sum_{k=2}^{max}\frac{d\sigma_{a,k}}{dt}\right) \tag{5.90}$$

Momentum balance equations for the gas and liquid phases:

$$\frac{\partial}{\partial t}\left(\rho_g \varepsilon_g u_g\right) + u_g \frac{\partial}{\partial z}\left(\rho_g \varepsilon_g u_g\right) = \varepsilon_g \mu^e_g \frac{\partial^2 u_g}{\partial z^2} - \varepsilon_g \frac{\partial P}{\partial z} + \varepsilon_g \rho_g g - F_{g\ell} - F_{gs} \tag{5.91}$$

$$\frac{\partial}{\partial t}(\rho_\ell \varepsilon_\ell u_\ell) + u_\ell \frac{\partial}{\partial z}(\rho_\ell \varepsilon_\ell u_\ell) = \varepsilon_\ell \mu^e_\ell \frac{\partial^2 u_\ell}{\partial z^2} - \varepsilon_\ell \frac{\partial P}{\partial z} + \varepsilon_\ell \rho_\ell g + \frac{\varepsilon \eta_e - \varepsilon_\ell}{\varepsilon_g}\left[F_{g\ell} + F_{gs}\right] - F_{\ell s} \tag{5.92}$$

Unsteady-state mass balance equation for planktonic cells (cells freely suspended in aqueous phase):

$$\frac{\partial}{\partial t}(\varepsilon_\ell c_c) + u_\ell \frac{\partial}{\partial z}(\varepsilon_\ell c_c) = D_\ell \frac{\partial^2}{\partial z^2}(\varepsilon_\ell c_c) + r_{gf} - r_{df} + r^c_{rel,c} - r_{dep,c} \tag{5.93}$$

Unsteady-state mass balance equation for κ-size aggregate in the liquid phase (without growth and death of detached cell *aggregates*):

$$\frac{\partial}{\partial t}(\varepsilon_\ell c_{a,k}) + u_\ell \frac{\partial}{\partial z}(\varepsilon_\ell c_{a,k}) = D_\ell \frac{\partial^2}{\partial z^2}(\varepsilon_\ell c_{a,k}) + r^c_{rel,a,k} + r^h_{rel,a,k}\left(1-\gamma_{a,k}\right) \quad k = 2,\ldots,max \tag{5.94}$$

Growth, attachment, and detachment equation for the bacterial cells:

$$\rho_c \frac{d\sigma_c}{dt} = r_{gs}\eta_G - r_{ds} + r_{dep,c} - \sum_{k=2}^{max}\frac{dn_k}{dt} v_c k \rho_c - r^c_{det,c}(\sigma_c) \tag{5.95}$$

Generation and detachment equation for the κ-size aggregate:

$$\rho_c \frac{d\sigma_{a,k}}{dt} = \frac{dn_k}{dt} v_c k \rho_c - r^c_{rel,a,k}(\sigma_{a,k}) - r^h_{rel,a,k}\left(1-\gamma_{a,k}\right) \quad k = 2,\ldots,max \tag{5.96}$$

Biomass dynamics equation:

$$\frac{d\sigma}{dt} = \frac{d\sigma_c}{dt} + \sum_{k=2}^{max} \frac{d\sigma_{a,k}}{dt} \tag{5.97}$$

Species balance in the liquid phase (axial dispersion in the liquid phase):

$$\frac{\partial}{\partial t}(C_{S,\ell}\varepsilon_\ell) + u_\ell \frac{\partial}{\partial z}(C_{S,\ell}\varepsilon_\ell) = D_\ell \frac{\partial^2}{\partial z^2}(C_{S,\ell}\varepsilon_\ell) - \frac{r_{gf}}{Y_{X/S}} - \frac{r_{gs}\eta_G}{Y_{X/S}} \tag{5.98}$$

$$\frac{\partial}{\partial t}(C_{O_2,\ell}\varepsilon_\ell) + u_\ell \frac{\partial}{\partial z}(C_{O_2,\ell}\varepsilon_\ell) = D_\ell \frac{\partial^2}{\partial z^2}(C_{O_2,\ell}\varepsilon_\ell) + (k_\ell a_{g\ell})_{O_2}\left(\frac{C_{O_2,g}}{He_{O_2}} - C_{O_2,\ell}\right) - \frac{r_{gf}}{Y_{X/O_2}} - \frac{r_{gs}\eta_G}{Y_{X/O_2}} \tag{5.99}$$

Species balance in the gas phase (no axial dispersion in the gas phase):

$$\frac{\partial}{\partial t}(C_{O_2,g}\varepsilon_g) + u_g \frac{\partial}{\partial z}(C_{O_2,g}\varepsilon_g) = -(k_\ell a_{g\ell})_{O_2}\left(\frac{C_{O_2,g}}{He_{O_2}} - C_{O_2,\ell}\right) - D^{eff}_{O_2,b}\left.\frac{dC_{O_2,b}}{dr}\right|_{r=r_b} \bar{a}(1-\varepsilon)(1-\eta_e) \tag{5.100}$$

The velocities at the inlet are specified in Dirichlet-type boundary conditions. At the outlet, an open boundary condition is used [124], which implies, except for pressure, zero gradients for all flow variables normal to the outflow boundary. The liquid holdup at the reactor inlet is evaluated assuming $\partial u_g/\partial z = \partial u_\ell/\partial z = 0$ and fully wetted particles and combining the momentum balance equations for the gas and liquid phases. Danckwerts boundary conditions are used for unsteady-state mass balance equations for planktonic cells and κ-size aggregates and species balance equations in gas and liquid phases [137].

The assumption of bed partial wetting entrains that the gas-phase drag has contributions due to effects located at the gas–liquid ($F_{g\ell}$) and gas–solid (F_{gs}) interfaces. Similarly, the resultant of the forces exerted on the liquid phase involves two components: (i) the drag force, $F_{\ell s}$, experienced by the liquid due to the shear stress nearby the liquid–solid boundary and (ii) the gas–liquid interfacial drag due to the slip between fluids, $F_{g\ell}$. Under trickle flow regime, the double-slit model provides satisfactory approximations for the liquid–solid, gas–solid, and gas–liquid drag forces [41, 136]. For the adaptation to the biofilm clogging context, these drag equations are recast as local functions of the local instantaneous values of the porosity and of the effective specific surface area of the bioparticle [136].

The overall effectiveness factor is defined as a weighted sum of the effectiveness factors for completely dry particle (η_g) and fully wetted particle (η_ℓ)

$$\eta_G = (1-\eta_e)\eta_g + \eta_e\eta_\ell \tag{5.101}$$

and is obtained by solving the mass balance equations for the biofilm and the simultaneous diffusion and adsorption of both phenol and oxygen within the activated carbon particles (under the assumptions given in Ref. [137]):

$$\varepsilon_b \frac{\partial C_{j,b}}{\partial t} = \frac{1}{r^2}\frac{\partial}{\partial r}\left(r^2 D^{eff}_{j,b}\frac{\partial C_{j,b}}{\partial r}\right) - \frac{\mu\rho_b}{Y_{X/j}} \quad j = O_2, S \tag{5.102}$$

$$\varepsilon_p \frac{\partial C_{S,p}}{\partial t} + \rho_p \frac{\partial q_{S,p}}{\partial t} = \frac{1}{r^2}\frac{\partial}{\partial r}\left(r^2 D^{eff}_{S,p}\frac{\partial C_{S,p}}{\partial r}\right) \quad q_s = \frac{3825.3C_S}{1+25.329C_S} \tag{5.103}$$

$$\varepsilon_p \frac{\partial C_{O_2,p}}{\partial t} = \frac{1}{r^2}\frac{\partial}{\partial r}\left(r^2 D^{eff}_{O_2,p}\frac{\partial C_{O_2,p}}{\partial r}\right) \tag{5.104}$$

The corresponding boundary and initial conditions for Equations 5.102–5.104 are given as

$$r = 0 \quad \frac{\partial C_{j,p}}{\partial r} = 0 \tag{5.105}$$

$$r = r_p \quad D^{eff}_{j,b}\left.\frac{\partial C_{j,b}}{\partial r}\right|_{r=r_p} = D^{eff}_{j,p}\left.\frac{\partial C_{j,p}}{\partial r}\right|_{r=r_p} \quad C_{j,b}\big|_{r_p} = C_{j,p}\big|_{r_p} \tag{5.106}$$

$r = r_b$—external surface area of the bioparticle is completely wetted (to calculate η_ℓ)

$$D^{eff}_{j,b}\left.\frac{\partial C_{j,w}}{\partial r}\right|_{r=r_b} = k_{\ell s,j}\left(C_{j,\ell} - C^*_{j,w}\right) \tag{5.107}$$

—external surface area of the bioparticle is completely dry (to calculate η_g)

$$C^*_{S,d} = C_{S,\ell} \quad D^{eff}_{O_2,b}\left.\frac{\partial C_{O,d}}{\partial r}\right|_{r=r_b} = k_{gs,O_2}\left(C_{O_2,g} - C^*_{O_2,d}He_{O_2}\right) \tag{5.108}$$

$$t = 0 \quad C_{j,p(b)}(r,0) = C^{in}_{j,\ell} \tag{5.109}$$

The biofilm thickness increases with the time during transient state, and consequently Equation 5.102 is a moving boundary problem. The average biofilm thickness was estimated at any time by means of the following relationship:

$$\delta_b = \frac{\varepsilon^o - \varepsilon}{\bar{a}^0} \tag{5.110}$$

Because there is no physiological change of the microbes when immobilized onto the activated carbon particles, the kinetic expression of the cell growth obtained from a suspended cell culture can be applied equally well to the immobilized system. The bacterial growth kinetics for free and captured cells is

$$r_{gf} = \mu\varepsilon_\ell c_c \tag{5.111}$$

$$r_{gs} = \mu\rho_c\sigma \tag{5.112}$$

where μ is the specific growth rate given by the Haldane-type equation developed by Tang et al. [141]:

$$\mu = \frac{\mu_{max}C_S}{K_s + C_S + C_S^2/K_i}\frac{C_{O_2}}{K_{ox} + C_{O_2}} \tag{5.113}$$

Also, the kinetic expression of cell death obtained from a suspended cell culture is applied equally to the immobilized system. Sen et al. [142] proposed an irreversible first-order reaction for death of bacterial cells:

$$r_{df} = k_{df}\varepsilon_{\ell}c_c \qquad (5.114)$$

$$r_{ds} = k_{ds}\rho_c\sigma \qquad (5.115)$$

Cell attachment rate was evaluated by coupling Equations 5.76 and 5.77. The transport of the bacterial cell to the bioparticle surface quantified by the collection efficiency was estimated with the expression developed by Rajagopalan and Tien [128], obtained by the combination of the trajectory analysis of a spherical particle in the vicinity of a spherical collector with the contribution of the Brownian diffusion.

Over the range of trickle-bed conditions fluid shear stress had no significant influence on the cell release or detachment rate. However, during reversible attachment, cells still exhibit Brownian motion and are easily removed by application of mild shear force [143]. So reversibility implies the detachment of adhering cells when conditions become unfavorable and the rate-limiting step in cell detachment is the diffusion of detached cells across the boundary layer between collector surfaces and the liquid bulk. The rate of Brownian cells and aggregate detachment was evaluated using the expression developed by Ryan and Gschwend [122] in the case of colloid detachment:

$$r^c_{rel,i} = \alpha^c_{rel,i}\bar{a}(1-\varepsilon)\eta_e d_i\sigma_i\rho_c \qquad (5.116)$$

where $\alpha^c_{rel,i}$ designates the bacterial cell detachment ($i = c$) or the κ-size aggregate detachment ($i = a, k$).

For non-Brownian cell aggregates, the rate of hydrodynamic detachment was considered to be proportional to the difference between the wall shear stress and the critical shear stress [123]:

$$r^h_{rel,a,k} = \alpha^h_{rel,a,k}\bar{a}(1-\varepsilon)\eta_e(\tau_w - \tau_{cr,a,k}) \quad \text{for } \tau_w > \tau_{cr,a,k} \qquad (5.117)$$

$$r^h_{rel,a,k} = 0 \quad \text{for } \tau_w < \tau_{cr,a,k} \qquad (5.118)$$

where critical shear stress was estimated for the condition of equilibrium between the forces acting on an attached aggregate along the tangential direction [131].

The generation of the cell aggregates is considered to take place by collision and subsequent bonding of the cells transported from the bulk liquid to the surface of the bioparticle in the liquid approaching bioparticle and by collision with the associated growth cells. So the aggregation process includes the bacterial cell autoaggregation (attachment of one species to its clonal descendants). Cells within the biofilm are individualized as in the case of IbM theory [138]. Bacterial cell aggregation rate is described by the rate at which a certain size aggregate is being formed by smaller aggregates minus the rate at which the aggregate combines to form a larger aggregate (neglecting aggregate breakup). This is given by the discrete population balance equations of von Smoluchowski [139]:

$$\frac{dn_k}{dt} = \frac{1}{2}\alpha\sum_{i+j=k}\beta_{ij}n_in_j - n_k\alpha\sum_{i=1}^{max}\beta_{ik}n_i \quad k = 2,3,\ldots,max \qquad (5.119)$$

where α is the fractional collision efficiency, β_{ij} is the rate at which cells or aggregates of volume v_i and v_j collide, n_i is the number density of cells or aggregates with volume v_i, i, j, k refer to cell or aggregate size class indices, and max refers to a maximum size class. The first summation is over the sets of sizes that, when added, produce a κ-size aggregate. The second summation reflects the loss of κ-size aggregates as they combine with all other aggregate sizes to form larger aggregates. The population balance equation for the microbial cells which collide to form aggregates is the following:

$$\frac{dn_1}{dt} = \frac{r_{dep,c}}{v_c\rho_c} + \frac{r_{gs}\eta_G}{v_c\rho_c} - \sum_{k=2}^{max}k\frac{dn_k}{dt} - \frac{r_{ds}}{v_c\rho_c} \qquad (5.120)$$

A collision between two cells, two cell aggregates, or a cell and an aggregate results in a newly formed particle, depending on the efficiency of the collision. Collision efficiency is a complex function of the surface properties, the structure of the aggregates or the diameter of the aggregates, hydrodynamic effects, and the prevailing colloidal forces [144]. The collision frequency function, β_{ij}, reflects the physical environment, such as temperature, viscosity, shear stress, and aggregation size. Brownian motion and differential settling are considered to give particle collision in the work of Iliuta and Larachi [137]. The collision frequency function due to Brownian motion is [145]

$$\beta_{ij,BR} = \frac{2\kappa_B T}{3\mu_\ell}\left(\frac{1}{r_i} + \frac{1}{r_j}\right)(r_i + r_j) \qquad (5.121)$$

Differential settling occurs when larger aggregates settle more rapidly than smaller aggregates and particles. The terminal velocity of settling particles and aggregates is assumed to follow Stokes' law. Considering differential settling rates and the rectilinear collision model, the rate constant is described by [146]

$$\beta_{ij,DS} = \frac{\pi g(\rho_a - \rho_\ell)}{72\mu_\ell}(d_i + d_j)^3|d_i - d_j| \qquad (5.122)$$

The aggregates comprised of bacterial cells can be treated using a fractal dimension [144]. Thus, the radius $r_{a,k}$ of a size class κ-aggregate can be expressed as [147]

$$r_{a,k} = r_1(k)^{\frac{1}{D_f}} \qquad (5.123)$$

where D_f is the fractal dimension, which defines the relationship between particle size and density. Fractal dimension takes values between 1 and 3. Coalescing sphere models, which include the original von Smoluchowski equation, assume that D_f is equal to 3. Extra information about two-bed membrane reactor model is given in Iliuta and Larachi [137].

5.5.2.2 Salient conclusions

From Figures 5.10 and 5.11, it is clear that the biomass is not uniformly distributed along the column and as consequence

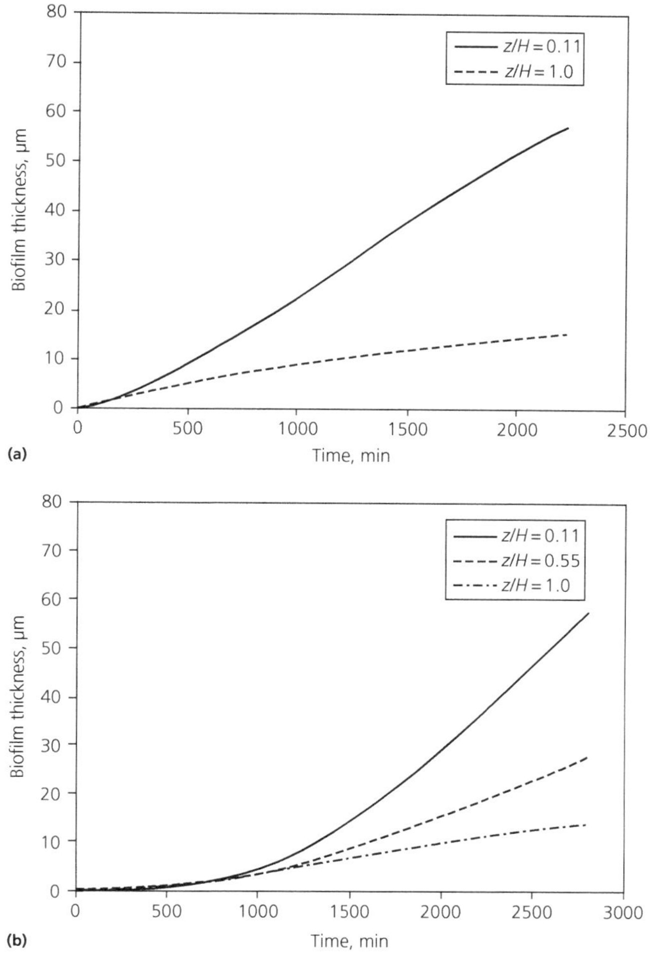

Figure 5.10 Biofilm thickness versus time at different values of inlet cell concentration (only cell detachment was considered): (a) $c_c^{in} = 1.0\,g/l$; (b) $c_c^{in} = 0.1\,g/l$ ($C_{S,\ell}^{in} = 30\,mg/l$, $C_{O_2,\ell}^{in} = 6\,mg/l$, $C_{O_2,g}^{in} = 275\,mg/l$, $\rho_b = 75000\,mg/l$, $\alpha = 0.1$, $D_f = 2.0$).

the extent of biological clogging depends upon axial distance (kinetic parameters and operating conditions are listed in Tables 5.2 and 5.3).

There is a noticeable increase in biofilm thickness in the entrance section of the bed that is coherent and also mirrored

Table 5.2 Values of kinetic parameters.

Kinetic parameter	Value
μ_{max}	$0.365\ h^{-1}$
K_s	10.948 mg/l
K_i	113.004 mg/l
K_{ox}	0.1 mg/l
Y_{X/O_2}	0.354 mg oxygen/mg phenol
$Y_{X/S}$	0.496 mg cell/mg phenol
Specific decay rate for free and captured cells	$7 \times 10^{-7}\ s^{-1}$

Table 5.3 Parameters used in simulations.

Temperature	25°C
Pressure	0.1 MPa
Reactor diameter	51 mm
Bed height	1.0 m
Diameter of support particle	3 mm
Density of support particle	$1400\ kg/m^3$
Internal porosity	0.575
External porosity	0.37
Superficial liquid velocity	0.0015 m/s
Superficial gas velocity	0.0283 m/s
Dry cell mass/cell volume	1100 g/L
Cell diameter	$1.6 \times 10^{-6}\ m$
Total dry biomass in the biofilm/wet biofilm volume	30–75 g/l
Henry's law constant	30
Hamaker constant	$1.0 \times 10^{-20}\ J$
Hydrodynamic detachment rate coefficient	$1.0 \times 10^{-8}\ kg/Ns$
Coefficient of sliding friction	$3.79 \times 10^{-6}\ m^2$

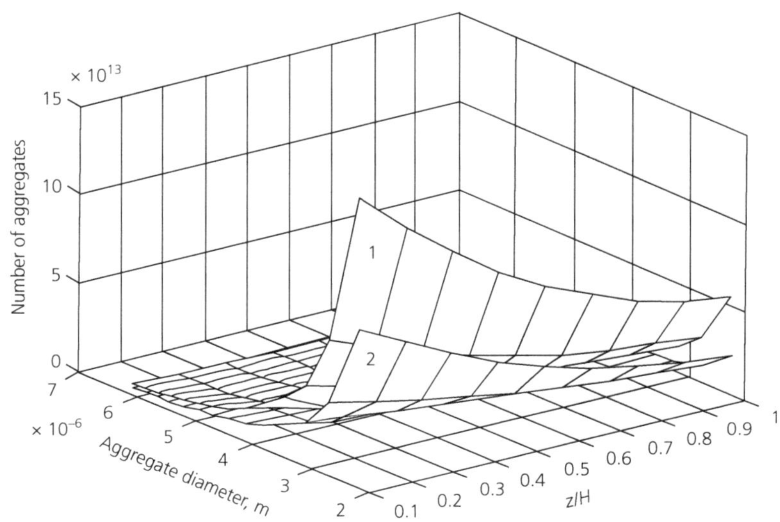

Figure 5.11 Number of aggregates versus axial distance ($C_{S,\ell}^{in} = 30$ mg/l, $C_{O_2,\ell}^{in} = 6$ mg/l, $C_{O_2,g}^{in} = 275$ mg/l, $\rho_b = 75$ g/l, $\alpha = 0.1$, $D_f = 2.0$, $t = 2000$ min, surface 1—$c_c^{in} = 1$ g/l, surface 2—$c_c^{in} = 0.1$ g/l).
(Source: Iliuta and Larachi [137]. Reproduced with permission of Elsevier.)

by the decrease in local porosity. As time advances, the biological clogging front progressively fills up the column. However, biomass accumulation is more confined in the entrance region, even if the overall effectiveness factor is lower, because the cell growth rate is higher in this region. Toward the outside of the bioreactor, the cell growth rate decreases due to the decrease in substrate concentration, and, as a result, the biomass accumulation is much lower than in the entrance section. As a consequence, the increase of the biofilm thickness is much lower. The biofilm profiles suggest that progression of biofilm accumulation exhibits sigmoid behavior. Such progression can practically be divided into three zones: (i) induction with biofilm thickness neighboring ca. 1 μm, (ii) exponential or log accumulation, and (iii) a plateau or steady-state zone.

Pressure drop across bed is probably the most straightforward criterion to sense biological clogging. Figure 5.12a shows the history of $\Delta P/\Delta P^o$ ratio (where ΔP^o is the pressure drop for the biofilm-free bed under otherwise identical conditions)

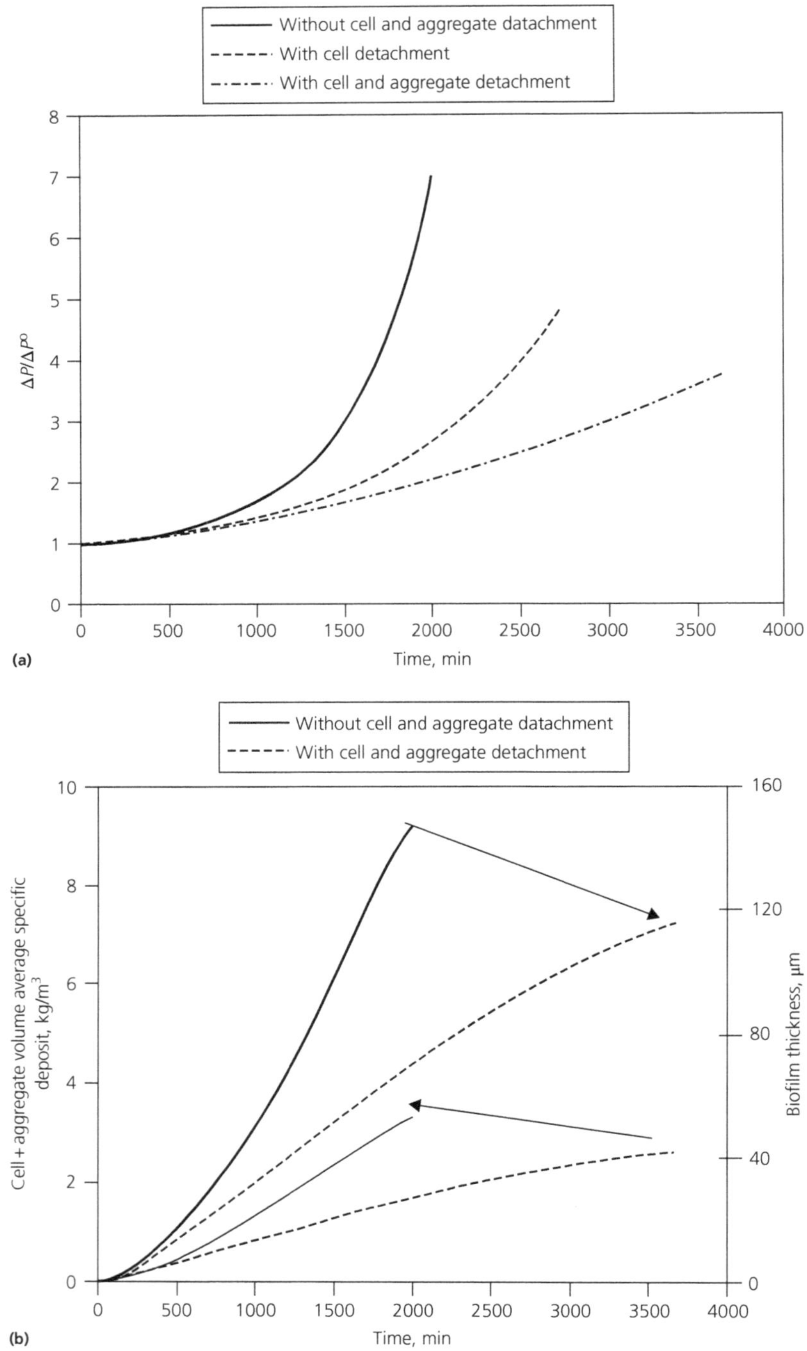

Figure 5.12 Two-phase pressure drop ratio. (a) Cell + aggregate volume average specific deposit and biofilm thickness at $z/H = 0.11$ (b) versus time in the presence or absence of biomass detachment ($C_{S,\ell}^{in} = 30\,mg/l$, $C_{O_2,\ell}^{in} = 6\,mg/l$, $C_{O_2,g}^{in} = 275\,mg/l$, $c_c^{in} = 0.1\,g/l$, $\rho_b = 30000\,mg/l$, $\alpha = 0.5$, $D_f = 1.5$).

successively with cell and aggregate detachment neglected and accounted for in the model. The cell and aggregate volume average specific deposit increases in the time yielding thicker biofilm (Figure 5.12b). As a result, the bed porosity decreases and two-phase pressure drop increases. So the major change resulting from biomass accumulation in the bed involves the reduction of bed porosity, and therefore the increase of two-phase pressure drop is the result of the decrease in the available free space. Pressure drop first increases slowly, while later on, due to increasingly severe biomass clogging, the increase becomes quasiexponential. Inclusion of the biomass detachment term reduces biomass accumulation, resulting in delayed pressure drop breakthroughs (Figure 5.12a). However, in the first period of biofilm growth, only a small number of cells and aggregates detach from the surface, indicating that the process is dominated by growth, attachment, and aggregation (Figure 5.12b).

5.5.3 Integrated aqueous-phase glycerol reforming and dimethyl ether synthesis into an allothermal dual-bed reactor

Despite the fact that aqueous-phase glycerol reforming process is more environmentally friendly than gas-phase (steam) reforming and needs less energy, there is still room for improvement. Operation at moderate temperatures of the aqueous-phase glycerol reforming process provides the opportunity to be coupled with DME synthesis process with the aim of developing a DME production integrated process where the exothermic reactions in DME synthesis unit provide the energy for the endothermic aqueous-phase glycerol reforming process and the aqueous-phase glycerol reforming produces the synthesis gas suitable for DME synthesis. This should enhance unit heat integration and thermal efficiency and also improve the economic viability of DME synthesis process by reducing costs associated with syngas production. Also, the integrated process is intended to minimize abundant glycerol by-product streams via an energy-efficient alternative for producing DME. Iliuta et al. [148] proposed a dual-bed membrane reactor for this integrated process and analyzed through simulations the new concept by means of a two-scale, nonisothermal, unsteady-state model which accounts for a detailed gas/gas–liquid dynamics whereupon DME synthesis/aqueous-phase glycerol reforming kinetics, thermodynamics, thermal effects, and variable gas flow rate due to chemical/physical contraction were accounted for.

5.5.3.1 Modeling framework

Two-bed membrane reactor for integrated process involving aqueous-phase glycerol reforming to synthesis gas coupled with DME synthesis process combines two physically separated enclosures (Figure 5.13). The first unit is a tube-in-tube fixed-bed water perm-selective membrane reactor for DME synthesis process. The reaction-side compartment (outer tube) is packed with bifunctional catalytic particles for DME synthesis, whereas the permeate-side compartment (inner tube) is an empty

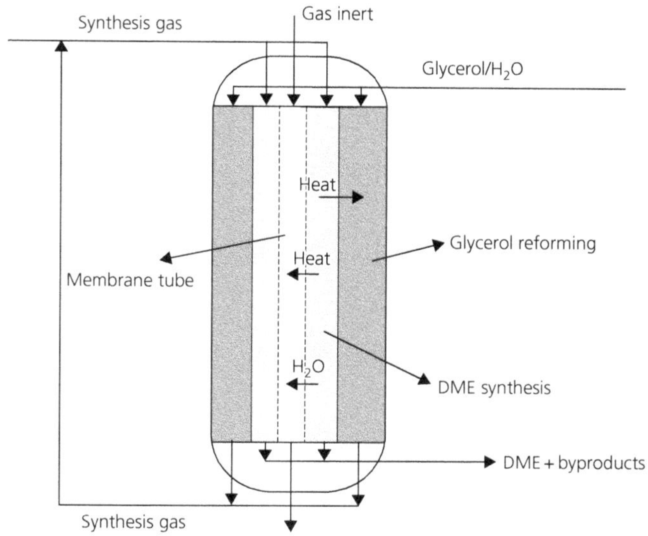

Figure 5.13 Schematic diagram of two-bed reactor system for DME production from glycerol via an integrated process involving aqueous-phase glycerol reforming coupled with DME synthesis process.

membrane tube swept by an inert gas which carries the permeating water. The wall of the inner tube is coated by a water perm-selective membrane layer; thus water permeates through the membrane layer due to a partial pressure gradient. The second unit is a gas–liquid–solid fixed-bed reactor for aqueous-phase glycerol reforming process which surrounds the fixed-bed membrane reactor for DME synthesis. The producing synthesis gas leaving the glycerol reforming reactor is directed into the fixed-bed reaction side of the DME synthesis reactor, and the heat produced by the exothermic reactions in the DME synthesis compartment is transferred to the endothermic aqueous-phase glycerol reforming process.

Membrane Fixed-Bed Reactor Model–DME Synthesis Unit. A two-scale, nonisothermal, unsteady-state model for DME synthesis membrane fixed-bed reactor takes into account the process of H_2O removal during DME synthesis [148]. The motivation for *in situ* H_2O removal during DME synthesis by means of hydrophilic membranes is to accelerate methanol production by increasing reactants' partial pressures and/or by reducing kinetic inhibition by H_2O while shifting water–gas shift reaction so as to favor CO formation. The model assumes that gas-phase backmixing in the reaction-side compartment is represented by the axial dispersion model with Danckwerts boundary conditions. The local *j*-species molar fluxes in the streamwise direction are determined by the rates of transmembrane transport and reaction in the retentate catalytic bed. Transmembrane molar flux of any *j*-species is proportional to *j*-species partial pressure difference across membrane and permeance and to membrane area [149]. Only water and hydrogen were assumed to get involved in transmembrane transits. A zeolite membrane was used assuming H_2O permeance ranging from 10^{-10} to 5×10^{-10} kmol/(sm^{2}Pa) and H_2O/H_2 permselectivity >10 at 250°C [149].

Unsteady-state mass, momentum, and enthalpy balance equations in the bulk gas phase of the reaction side are written as follows:

Species mass balance equations:

$$\frac{\partial}{\partial t}\left(\varepsilon_g^r P_{j,g}^r\right)+\frac{\partial}{\partial z}\left(\varepsilon_g^r u_g^r P_{j,g}^r\right)=D_g\varepsilon_g^r\frac{\partial^2 P_{j,g}^r}{\partial z^2}-D_{j,p}^{eff}\left.\frac{\partial P_{j,p}^r}{\partial r}\right|_{r=r_p}\bar{a}^r(1-\varepsilon)-\frac{A_m}{V_r}Q_j\left(P_j^r-P_j^p\right)RT_g^r \tag{5.124}$$

where $j = H_2O$ and H_2:

$$\frac{\partial}{\partial t}\left(\varepsilon_g^r P_{j,g}^r\right)+\frac{\partial}{\partial z}\left(\varepsilon_g^r u_g^r P_{j,g}^r\right)=D_g\varepsilon_g^r\frac{\partial^2 P_{j,g}^r}{\partial z^2}-D_{j,p}^{eff}\left.\frac{\partial P_{j,p}^r}{\partial r}\right|_{r=r_p}\bar{a}^r(1-\varepsilon) \tag{5.125}$$

where $j = CO, CO_2, CH_3OH, CH_3OCH_3$.

Overall mass balance equation:

$$\frac{\partial}{\partial t}\left(\varepsilon_g^r\rho_g^r\right)+\frac{\partial}{\partial z}\left(\varepsilon_g^r\rho_g^r u_g^r\right)=\rho_p^r(1-\varepsilon)\sum_i\sum_j \nu_{ij}\eta_i r_i M_g^r-\frac{A_m}{V_r}Q_{H_2O}\left(P_{H_2O}^r-P_{H_2O}^p\right)M_g^r-\frac{A_m}{V_r}Q_{H_2}\left(P_{H_2}^r-P_{H_2}^p\right)M_g^r \tag{5.126}$$

Momentum balance equation:

$$\frac{\partial}{\partial t}\left(\rho_g^r\varepsilon_g^r u_g^r\right)+u_g^r\frac{\partial}{\partial z}\left(\rho_g^r\varepsilon_g^r u_g^r\right)=-\varepsilon_g^r\frac{\partial P^r}{\partial z}+\varepsilon_g^r\rho_g^r g-\left[E_1\frac{\left(1-\varepsilon_g^r\right)^2}{\left(\varepsilon_g^r\right)^3 d_p^2}\mu_g^r v_{sg}^r+E_2\frac{1-\varepsilon_g^r}{\left(\varepsilon_g^r\right)^3 d_p}\rho_g^r\left(v_{sg}^r\right)^2\right]\varepsilon_g^r-\frac{A_m}{V_r}\frac{Z_r RT_g^r}{P^r}\left(y_{H_2}^r+y_{H_2O}^r\right)^{-1}\left[Q_{H_2}\left(P_{H_2}^r-P_{H_2}^p\right)+Q_{H_2O}\left(P_{H_2O}^r-P_{H_2O}^p\right)\right]\times\left[\begin{matrix}Q_{H_2}M_{H_2}\left(P_{H_2}^r-P_{H_2}^p\right)\\+Q_{H_2O}M_{H_2O}\left(P_{H_2O}^r-P_{H_2O}^p\right)\end{matrix}\right] \tag{5.127}$$

Heat balance equation:

$$\varepsilon_g^r\rho_g^r c_{p,g}^r\frac{\partial T_g^r}{\partial t}+\varepsilon_g^r\rho_g^r u_g^r c_{p,g}^r\frac{\partial T_g^r}{\partial z}=\lambda_{g,z}^{eff}\varepsilon_g^r\frac{\partial^2 T_g^r}{\partial z^2}+\sum_{i=1}^{3}(-\Delta H_{ri})r_i\eta_i\rho_p^r(1-\varepsilon)-\frac{4K_T}{D_i}\left(T_g^r-T^g\right)-\frac{4d_iK_T'}{D_i^2}\left(T_g^r-T_g^p\right)-\frac{A_m}{V_r}Q_{H_2O}\left(P_{H_2O}^r-P_{H_2O}^p\right)C_{p,H_2O}\left(T_g^r-T_g^p\right)-\frac{A_m}{V_r}Q_{H_2}\left(P_{H_2}^r-P_{H_2}^p\right)C_{p,H_2}\left(T_g^r-T_g^p\right) \tag{5.128}$$

Initial and boundary conditions for mass, momentum, and heat balance equations are as follows:

$$t=0,\ z>0\quad u_g^r=u_g^{r,in},\ P^r=P^{r,in},\ P_{j,g}^r=P_{j,g}^{r,in},\ T_g^r=T_g^{r,in} \tag{5.129}$$

$j = H_2, H_2O, CO, CO_2, CH_3OH, CH_3OCH_3$

$$t>0,\ z=0\quad u_g^r=u_g^{r,in},\ P^r=P^{r,in},\ u_g^rP_{j,g}^r\Big|_{z=0^-}=u_g^rP_{j,g}^r\Big|_{z=0^+}-D_g\left.\frac{\partial P_{j,g}^r}{\partial z}\right|_{z=0} \tag{5.130}$$

$$u_g^r\varepsilon_g^r\rho_g^r c_{p,g}^r T_g^r\Big|_{z=0^-}=u_g^r\varepsilon_g^r\rho_g^r c_{p,g}^r T_g^r\Big|_{z=0^+}-\varepsilon_g^r\lambda_{g,z}^{eff}\left.\frac{\partial T_g^r}{\partial z}\right|_{z=0} \tag{5.131}$$

$$t>0,\ z=H\quad \left.\frac{\partial P_{j,g}^r}{\partial z}\right|_{z=H}=0\quad \left.\frac{\partial T_g^r}{\partial z}\right|_{z=H}=0 \tag{5.132}$$

Unsteady-state mass, momentum, and enthalpy balance equations for permeate side are as follows:

Species mass balance equations (H_2, H_2O, inert gas):

$$\frac{\partial}{\partial t}\left(\varepsilon_g^p P_{j,g}^p\right)+\frac{\partial}{\partial z}\left(\varepsilon_g^p u_g^p P_{j,g}^p\right)=\frac{A_m}{V_p}Q_j\left(P_j^r-P_j^p\right)RT_g^p\quad j=H_2O\text{ and }H_2 \tag{5.133}$$

$$\frac{\partial}{\partial t}\left(\varepsilon_g^p P_i^p\right)+\frac{\partial}{\partial z}\left(\varepsilon_g^p u_g^p P_i^p\right)=0 \tag{5.134}$$

Overall mass balance equation:

$$\frac{\partial}{\partial t}\left(\varepsilon_g^p\rho_g^p\right)+\frac{\partial}{\partial z}\left(\varepsilon_g^p\rho_g^p u_g^p\right)=\frac{A_m}{V_p}Q_{H_2O}\left(P_{H_2O}^r-P_{H_2O}^p\right)M_g^p+\frac{A_m}{V_p}Q_{H_2}\left(P_{H_2}^r-P_{H_2}^p\right)M_g^p \tag{5.135}$$

Momentum balance equation:

$$\frac{\partial}{\partial t}\left(\rho_g^p\varepsilon_g^p u_g^p\right)+u_g^p\frac{\partial}{\partial z}\left(\rho_g^p\varepsilon_g^p u_g^p\right)=-\varepsilon_g^p\frac{\partial P^p}{\partial z}+\varepsilon_g^p\rho_g^p g+\frac{A_m}{V_p}\frac{Z_pRT_g^p}{P^p}\left(y_{H_2}^r+y_{H_2O}^r\right)^{-1}\left[Q_{H_2}\left(P_{H_2}^r-P_{H_2}^p\right)+Q_{H_2O}\left(P_{H_2O}^r-P_{H_2O}^p\right)\right]\times\left[Q_{H_2}M_{H_2}\left(P_{H_2}^r-P_{H_2}^p\right)+Q_{H_2O}M_{H_2O}\left(P_{H_2O}^r-P_{H_2O}^p\right)\right] \tag{5.136}$$

Heat balance equation:

$$\varepsilon_g^p\rho_g^p c_{p,g}^p\frac{\partial T_g^p}{\partial t}+\varepsilon_g^p\rho_g^p u_g^p c_{p,g}^p\frac{\partial T_g^p}{\partial z}=\frac{4K_T'}{d_i}\left(T_g^r-T_g^p\right)+\frac{A_m}{V_p}Q_{H_2O}\left(P_{H_2O}^r-P_{H_2O}^p\right)C_{p,H_2O}\left(T_g^r-T_g^p\right)+\frac{A_m}{V_p}Q_{H_2}\left(P_{H_2}^r-P_{H_2}^p\right)C_{p,H_2}\left(T_g^r-T_g^p\right) \tag{5.137}$$

Initial and boundary conditions for mass, momentum, and heat balance equations are the following:

$$t=0,\ z>0\quad u_g^p=u_g^{p,in},\ P^p=P^{p,in},\ P_{j,g}^p=P_{j,g}^{p,in},\ T_g^p=T_g^{p,in} \tag{5.138}$$

$j = H_2, H_2O$, inert gas

$$t>0,\ z=0\quad u_g^p=u_g^{p,in},\ P^p=P^{p,in},\ P_{j,g}^p=P_{j,g}^{p,in},\ T_g^p=T_g^{p,in} \tag{5.139}$$

The gradients of the concentration and temperature inside the catalyst particles are given by the equations describing the simultaneous mass and heat transport and reaction within the catalyst particles:

$$\varepsilon_p^r \frac{\partial P_{j,p}^r}{\partial t} = D_{j,p}^{eff} \frac{1}{r^2} \frac{\partial}{\partial r}\left(r^2 \frac{\partial P_{j,p}^r}{\partial r}\right) - \rho_p^r \sum_i \upsilon_{ij} r_i R T_p^r \tag{5.140}$$

$j = H_2$, H_2O, CO, CO_2, CH_3OH, CH_3OCH_3

$$\left(\varepsilon_p^r \rho_g^r c_{p,g}^r + \rho_p^r c_{p,s}^r\right) \frac{\partial T_p^r}{\partial t} = \lambda_p^{eff} \frac{1}{r^2} \frac{\partial}{\partial r}\left(r^2 \frac{\partial T_p^r}{\partial r}\right) + \sum_{i=1}^{3} (-\Delta H_{R,i}) r_i \rho_p^r \tag{5.141}$$

The corresponding boundary and initial conditions are as follows:

$$t > 0,\ r = r_p \quad D_{j,p}^{eff} \left.\frac{\partial P_{j,p}^r}{\partial r}\right|_{r=r_p} = k_{gs,j}\left(P_{j,g}^r - P_j^{*,r}\right) \quad \lambda_p^{eff} \left.\frac{\partial T_p^r}{\partial r}\right|_{r=r_p} = \alpha_{gs}\left(T_g^r - T_p^{*,r}\right) \tag{5.142}$$

$$t > 0,\ r = 0 \quad \frac{\partial P_{j,p}^r}{\partial r} = 0 \quad \frac{\partial T_p^r}{\partial r} = 0 \tag{5.143}$$

$$t = 0,\ r > 0 \quad P_{j,p}^r(r,0) = P_{j,g}^{r,in} \quad T_p^r(r,0) = T_g^{r,in} \tag{5.144}$$

Direct synthesis of DME from synthesis gas involves two steps, methanol synthesis, followed by *in situ* methanol dehydration (Eqs. 5.145–5.147), which in turn requires two functionally independent catalysis, that is, a methanol-forming component and a dehydration component [150]. Reactions (5.145) and (5.146) are catalyzed by a methanol synthesis catalyst and reaction (5.147) by an acidic catalyst. All these reactions are reversible and exothermic:

Methanol synthesis reaction (reaction 1):

$$CO_2 + 3H_2 \overset{r_1}{\rightleftarrows} CH_3OH + H_2O \tag{5.145}$$

Water–gas shift reaction (reaction 2):

$$CO + H_2O \overset{r_2}{\rightleftarrows} H_2 + CO_2 \tag{5.146}$$

Methanol dehydration reaction (reaction 3):

$$2CH_3OH \overset{r_3}{\rightleftarrows} CH_3OCH_3 + H_2O \tag{5.147}$$

The kinetics of DME synthesis reactions in a fixed-bed reactor with Cu–ZnO–Al_2O_3/HZSM-5 catalyst for methanol synthesis (Cu–ZnO–Al_2O_3) and for methanol dehydration (HZSM-5) is used [150]:

$$r_1 = K_1 \frac{P_{CO_2} P_{H_2}\left(1 - \frac{1}{K_{p,1}} \frac{P_{H_2O} P_{CH_3OH}}{P_{CO_2} P_{H_2}^3}\right)}{\left(1 + K_{CO_2} P_{CO_2} + K_{CO} P_{CO} + \sqrt{K_{H_2} P_{H_2}}\right)^3} \tag{5.148}$$

$$r_2 = K_2 \frac{P_{H_2O} - \frac{1}{K_{p,2}} \frac{P_{CO_2} P_{H_2}}{P_{CO}}}{\left(1 + K_{CO_2} P_{CO_2} + K_{CO} P_{CO} + \sqrt{K_{H_2} P_{H_2}}\right)} \tag{5.149}$$

$$r_3 = K_3 \left(\frac{P_{CH_3OH}^2}{P_{H_2O}} - \frac{P_{CH_3OCH_3}}{K_{p,3}}\right) \tag{5.150}$$

Trickle-Bed Reactor Model–Aqueous-Phase Glycerol Reforming Unit. A two-scale, nonisothermal, unsteady-state model describes the behavior of the gas–liquid–solid fixed-bed reactor unit for the aqueous-phase glycerol reforming [148]. A cocurrent two-phase trickle flow through a porous medium of uniform porosity and single-sized catalytic particles is considered. Two-phase flow is assumed unidirectional, and both the flowing phases are assumed as viscous Newtonian. Each flowing phase is viewed as a continuum, and the catalyst particle surface is completely covered by a liquid film and the gas flows in the remaining interstitial void. In formulating the mass and enthalpy conservation equations, the following additional assumption are made: the heat consumed by vaporization is neglected, the catalyst activity is not declining with time, and mass transfer is described by linear mathematical interrelationships. Two-phase flow model is based on the volume average form of the transport equations for multiphase systems [46]. The model equations consist of the conservation of volume, conservation of mass and momentum for the gas and liquid phases, species balance in the gas and liquid phases, and heat balance:

Conservation of volume:

$$\varepsilon_g^g + \varepsilon_\ell^g = \varepsilon \tag{5.151}$$

Continuity equations for the gas and liquid phases:

$$\frac{\partial}{\partial t}\left(\varepsilon_g^g \rho_g^g\right) + \frac{\partial}{\partial z}\left(\varepsilon_g^g \rho_g^g u_g^g\right) = \rho_p^g (1-\varepsilon) \sum_i \sum_j \nu_{ij} \eta_i r_i M_g^g \tag{5.152}$$

$$\frac{\partial}{\partial t}\left(\varepsilon_\ell^g \rho_\ell^g\right) + \frac{\partial}{\partial z}\left(\varepsilon_\ell^g \rho_\ell^g u_\ell^g\right) = 0 \tag{5.153}$$

Momentum balance equations for the gas and liquid phases:

$$\frac{\partial}{\partial t}\left(\rho_g^g \varepsilon_g^g u_g^g\right) + u_g^g \frac{\partial}{\partial z}\left(\rho_g^g \varepsilon_g^g u_g^g\right) = \varepsilon_g^g \mu_g^e \frac{\partial^2 u_g^g}{\partial z^2} - \varepsilon_g^g \frac{\partial P^g}{\partial z} + \varepsilon_g^g \rho_g^g g - F_{g\ell} \tag{5.154}$$

$$\frac{\partial}{\partial t}\left(\rho_\ell^g \varepsilon_\ell^g u_\ell^g\right) + u_\ell^g \frac{\partial}{\partial z}\left(\rho_\ell^g \varepsilon_\ell^g u_\ell^g\right) = \varepsilon_\ell^g \mu_\ell^e \frac{\partial^2 u_\ell^g}{\partial z^2} - \varepsilon_\ell^g \frac{\partial P^g}{\partial z} + \varepsilon_\ell^g \rho_\ell^g g + F_{g\ell} - F_{\ell s} \tag{5.155}$$

Species balance in the liquid phase (axial dispersion in the liquid phase):

$$\frac{\partial}{\partial t}\left(\varepsilon_\ell^g C_{j,\ell}^g\right) + \frac{\partial}{\partial z}\left(\varepsilon_\ell^g u_\ell^g C_{j,\ell}^g\right) = D_\ell \frac{\partial^2}{\partial z^2}\left(\varepsilon_\ell^g C_{j,\ell}^g\right) - D_{j,p}^{eff} \left.\frac{\partial C_{j,p}^g}{\partial r}\right|_{r=r_p} \bar{a}^g (1-\varepsilon) \tag{5.156}$$

$j = C_3H_8O_3$, H_2O

Species balance in the gas phase (no axial dispersion in the gas phase):

$$\frac{\partial}{\partial t}\left(\varepsilon_g^g P_{j,g}^g\right) + \frac{\partial}{\partial z}\left(\varepsilon_g^g u_g^g P_{j,g}^g\right) = \upsilon_j r_{ref} \eta_{ref} \rho_p^g (1-\varepsilon) R T^g \quad j = CO_2,\ H_2 \tag{5.157}$$

Pseudohomogeneous heat balance equation [151, 152]:

$$\varepsilon_\ell^g \rho_\ell^g c_{p,\ell}^g \frac{\partial T^g}{\partial t} + \varepsilon_\ell^g \rho_\ell^g u_\ell^g c_{p,\ell}^g \frac{\partial T^g}{\partial z} = \varepsilon_\ell^g \lambda_{\ell,z}^{eff} \frac{\partial^2 T^g}{\partial z^2} + (-\Delta H_r) r_{ref} \eta_{ref} \rho_p^g (1-\varepsilon) + \frac{K_T \pi D_i}{A_{ref}} \left(T_g^r - T^g \right) \quad (5.158)$$

The liquid–solid and gas–liquid drag forces are estimated using the slit analogy [132]:

$$F_{\ell s} = \left\{ E_1 \frac{(1-\varepsilon)^2 \mu_\ell^g}{d_p^2 \left(\varepsilon_\ell^g\right)^3} + E_2 \left(1 + \psi_{g\ell}\right) \frac{1-\varepsilon}{d_p \left(\varepsilon_\ell^g\right)^3} \rho_\ell^g \left| v_{s\ell}^g \right| \right\} v_{s\ell}^g \varepsilon_\ell^g \quad (5.159)$$

$$F_{g\ell} = \left\{ E_1 \frac{(1-\varepsilon)^2 \mu_g^g}{d_p^2 \left(\varepsilon_g^g\right)^3} + E_2 \left(1 + \psi_{g\ell}\right) \frac{(1-\varepsilon) \rho_g^g}{d_p \left(\varepsilon_g^g\right)^3} \left| v_{sg}^g - \varepsilon_g^g u_{g,i} \right| \right\} \left(v_{sg}^g - \varepsilon_g^g u_{g,i} \right) \varepsilon_g^g \quad (5.160)$$

where the gas interfacial velocity in the bed is

$$u_{g,i} = \frac{72}{E_1} \frac{\varepsilon_\ell^g}{\left[\bar{a}^g (1-\varepsilon)\right]^2 \mu_\ell^g} \left[\frac{1}{2} \left(-\frac{\partial P^g}{\partial z} + \rho_\ell^g g \right) \varepsilon_\ell^g + \left(-\frac{\partial P^g}{\partial z} + \rho_g^g g \right) \varepsilon_g^g \right] \quad (5.161)$$

The velocities at the inlet are specified in Dirichlet-type boundary conditions. At the outlet, an open boundary condition is used [124], which implies, except for pressure, zero gradients for all flow variables normal to the outflow boundary. The liquid holdup at the reactor inlet is evaluated assuming $\partial u_g / \partial z = \partial u_\ell / \partial z = 0$ and combining the momentum balance equations for the gas and liquid phases. Initial and boundary conditions for Equations 5.156–5.158 are as follows:

$$t=0,\ z>0 \quad P_{j,g}^g = P_{j,g}^{g,in} \quad C_{j,\ell}^g = C_{j,\ell}^{g,in} \quad T^g = T^{g,in} \quad (5.162)$$

$$t>0,\ z=0 \quad P_{j,g}^g \Big|_{z=0} = P_{j,g}^{g,in} \quad u_\ell^g \varepsilon_\ell^g C_{j,\ell}^{g,in} = u_\ell^g \varepsilon_\ell^g C_{j,\ell}^g \Big|_{z=0^+} - D_\ell \varepsilon_\ell^g \frac{\partial C_{j,\ell}^g}{\partial z} \Big|_{z=0^+} \quad (5.163)$$

$$t>0,\ z=0 \quad u_\ell^g \rho_\ell^g c_{p,\ell}^g \varepsilon_\ell^g T^{g,in} = u_\ell^g \rho_\ell^g c_{p,\ell}^g \varepsilon_\ell^g T^g \Big|_{z=0^+} - \lambda_{\ell,z}^{eff} \varepsilon_\ell^g \frac{\partial T^g}{\partial z} \quad (5.164)$$

$$t>0,\ z=\mathrm{H} \quad \frac{\partial C_{j,\ell}^g}{\partial z} \Big|_{z=\mathrm{H}} = 0 \quad \frac{\partial T^g}{\partial z} \Big|_{z=\mathrm{H}} = 0 \quad (5.165)$$

The concentration and temperature gradients in porous catalyst particles are given by the equations describing the simultaneous mass, heat transport, and reaction within the catalyst particles considering that the reactants' diffusion occurs in the liquid phase by virtue of completely internally wetted catalysts:

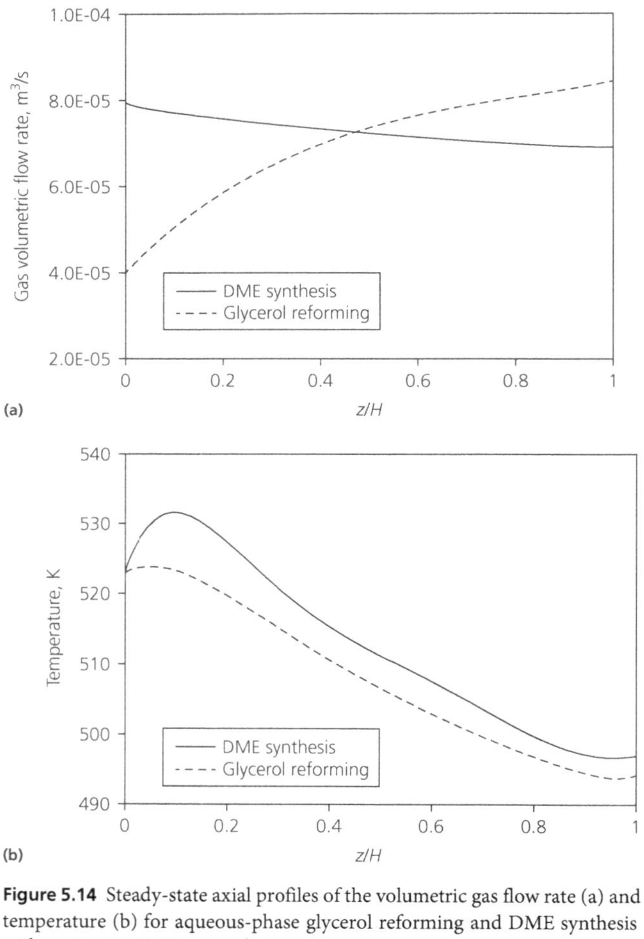

Figure 5.14 Steady-state axial profiles of the volumetric gas flow rate (a) and temperature (b) for aqueous-phase glycerol reforming and DME synthesis without *in situ* H_2O removal.

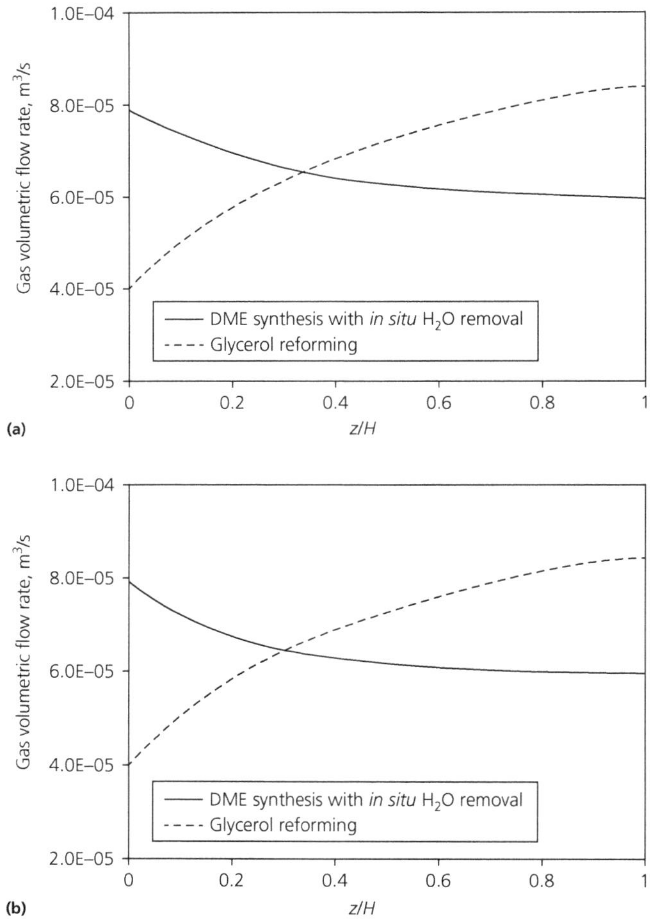

Figure 5.15 Steady-state axial profiles of the volumetric gas flow rate for aqueous-phase glycerol reforming and DME synthesis with *in situ* H_2O removal. (a) H_2O permeance $= 2.5 \times 10^{-10}\,\mathrm{kmol/m^2\,s\,Pa}$.
(b) H_2O permeance $= 5.0 \times 10^{-10}\,\mathrm{kmol/m^2\,s\,Pa}$.

$$\varepsilon_p^g \frac{\partial C_{j,p}^g}{\partial t} = \frac{1}{r^2}\frac{\partial}{\partial r}\left(r^2 D_{j,p}^{eff}\frac{\partial C_{j,p}^g}{\partial r}\right) - \upsilon_j r_{ref}\rho_p^g \quad j = C_3H_8O_3,\ H_2O \tag{5.166}$$

$$\left(\varepsilon_p^g\rho_\ell^g c_{p,\ell}^g + \rho_p^g c_{p,s}^g\right)\frac{\partial T_p^g}{\partial t} = \lambda_p^{eff}\frac{1}{r^2}\frac{\partial}{\partial r}\left(r^2\frac{\partial T_p^g}{\partial r}\right) + (-\Delta H_R) r_{ref}\rho_p^g \tag{5.167}$$

The corresponding boundary and initial conditions are given as

$$t>0,\ r=r_p \quad D_{j,p}^{eff}\left.\frac{\partial C_{j,p}^g}{\partial r}\right|_{r=r_p} = k_{\ell s,j}\left(C_{j,\ell}^g - C_j^{*,g}\right) \quad \lambda_p^{eff}\left.\frac{\partial T_p^g}{\partial r}\right|_{r=r_p} = \alpha_{\ell s}\left(T^g - T_p^{*,g}\right) \tag{5.168}$$

$$t>0,\ r=0 \quad \frac{\partial C_{j,p}^g}{\partial r} = 0 \quad \frac{\partial T_p^g}{\partial r} = 0 \tag{5.169}$$

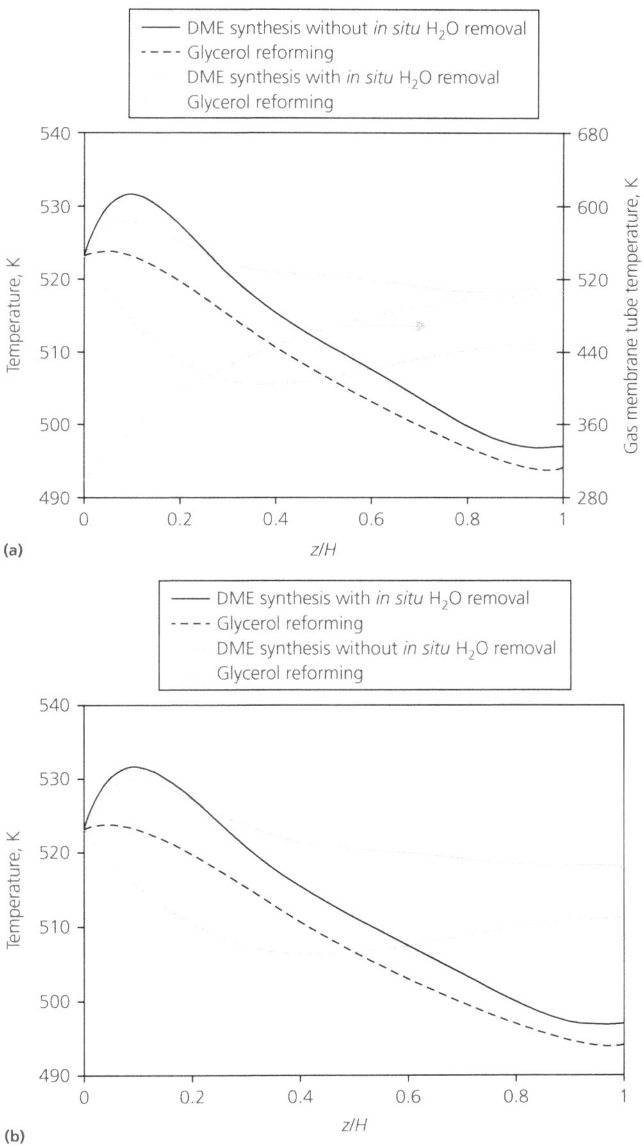

Figure 5.16 Steady-state axial profiles of the temperature for aqueous-phase glycerol reforming and DME synthesis with/without *in situ* H_2O removal. (a) H_2O permeance $= 2.5 \times 10^{-10}\,\mathrm{kmol/m^2\,s\,Pa}$. (b) H_2O permeance $= 5.0 \times 10^{-10}\,\mathrm{kmol/m^2\,s\,Pa}$.

$$t=0,\ r>0 \quad C_{j,p}^g(r,0) = C_{j,\ell}^{g,in} \quad T_p^g(r,0) = T^{g,in} \tag{5.170}$$

A first-order kinetics with respect to glycerol was adopted to describe the catalytic conversion of glycerol to hydrogen by aqueous-phase reforming on Pt/γ-alumina catalyst [148]. Extra information about two-bed membrane reactor model is given in Iliuta et al. [148].

5.5.3.2 Salient conclusions

The coupling between the aqueous-phase glycerol reforming and DME synthesis without *in situ* H_2O removal leads to synergies in the operation of these processes by avoiding the highly endothermic and exothermic steps that would result from the separate operation and eliminating the need to receive the synthesis gas necessarily in the DME synthesis process from an external source (Figure 5.14a). However, the overall energy balance is not favorable, and extra energy is required to heat the synthesis gas at the inlet in the DME synthesis process (Figure 5.14b). Moreover, the performance of DME synthesis process is not considerable because a relatively large quantity of water is produced, and the reverse water–gas shift reaction is favored.

With *in situ* H_2O removal, the exothermic reactions in DME synthesis unit make possible the endothermic aqueous-phase glycerol reforming process, and the aqueous-phase glycerol reforming process produces the synthesis gas appropriate for DME synthesis process (Figure 5.15). Simulations suggest that allothermal operation could be achieved: the overall energy balance is favorable, and the synthesis gas produced in

Table 5.4 Two-bed reactor operating conditions.

Operating conditions	Data
DME synthesis unit–fixed-bed reactor	
External reactor diameter	0.0381 m
External membrane tube diameter	0.0254 m
Fixed bed height	5.0 m
Catalyst particle density	1982 kg/m^3
Catalyst particle porosity	0.45
Porosity of fixed bed	0.39
Catalyst particle diameter	0.002 m
Inlet fixed-bed temperature	523 K
Reactor pressure	3.2 MPa
Inlet membrane inert gas temperature	298 K
Inlet fixed-bed superficial gas velocity	0.2 m/s
Permeate superficial gas velocity	0.3 m/s
Feed conditions	$H_2 = 50\%$, $CO_2 = 21.5\%$, $CO = 0.5\%$
Aqueous-phase glycerol reforming unit–trickle-bed reactor	
External reactor diameter	0.060325 m
Bed height	5.0 m
Catalyst particle density	1200 kg/m^3
Catalyst particle porosity	0.45
Porosity of fixed bed	0.37
Catalyst particle diameter	0.003 m
Inlet reactor temperature	523 K
Reactor pressure	3.2 MPa
Inlet superficial gas velocity	0.032 m/s
Inlet superficial liquid velocity	0.0028 m/s
Inlet liquid glycerol concentration	1.1 mol/l

aqueous-phase glycerol reforming unit can be heated to attain the inlet DME synthesis temperature by recovering the heat transferred to the inert gas in the membrane tube (Figure 5.16). The performance of DME synthesis process is improved: CO_2 conversion, the methanol yield, and DME selectivity are favored and the fraction of unconverted methanol is reduced.

Glycerol conversion and DME synthesis can be carried out successfully at comparable conditions (Table 5.4) in a two-bed reactor system, allowing the integrated process to be used for the production of biofuel from aqueous glycerol solutions. The integrated process minimizes the amount of waste by-products and represents an energy-efficient alternative for producing DME. Furthermore, it presents an opportunity to improve the economic viability of a green DME synthesis by reducing the costs associated with the synthesis gas production and improving the thermal efficiency of DME synthesis process.

Nomenclature

A	Helmholtz free energy function, m^2/s^2
$\bar{a}$	effective specific surface area, $m^2/m^3{}_{(bio)p}$
$a_{\alpha\beta}$	area of the $\alpha\beta$-interface per unit volume of porous medium, m^2/m^3
A_Δ	collector area loss, $A_\Delta = \frac{1}{2}\left[\frac{d_p(t)}{2}\right]^2\left[2\sqrt{3}\frac{d_f}{d_p(t)} - \sin\left(2\sqrt{3}\frac{d_f}{d_p(t)}\right)\right]$, m^2
$a_{\ell s}$	effective liquid–solid interfacial area, $m^2/m^3{}_r$, $a_{\ell s} = \bar{a}(1-\varepsilon)$
A_m	membrane tube area, m^2
A_{ref}	cross-sectional area of trickle-bed reactor for glycerol reforming, m^2
$c_{a,k}$	concentration of κ-size aggregate, $kg/m^3{}_\ell$
c_c	concentration of cells in liquid phase, $kg/m^3{}_\ell$
c_f	fine particle concentration, $kg/m^3{}_\ell$
$c_{p,\alpha}$	specific heat capacity of α-phase, J/kgK
c_{sat}	concentration of saturation, $kmol/m^3$
C_g	total concentration of the gas phase, $kmol/m^3$
$C_{j,\alpha}$	concentration of species j in α-phase, $kmol/m^3$
$C_{j,d}$	concentration of component j in the biofilm when the external surface area of the biocatalyst particle is completely dry, kg/m^3
$C_{j,w}$	concentration of component j in the biofilm when the external surface area of the biocatalyst particle is completely wetted, kg/m^3
$C_{p,j}$	heat capacity of species j, J/mol K
d_b	bioparticle diameter, m
d_c	bacterial cell diameter, m
d_e	outer diameter of membrane tube—DME synthesis, m
d_f	fine particle diameter, m
d_i	inner diameter of membrane tube—DME synthesis, m
d_p	particle diameter, m
$d_p(t)$	effective particle diameter at time t, $d_p(t) = d_p^o\sqrt[3]{1 + \frac{\sigma}{(1-\varepsilon_d)(1-\varepsilon^o)}}$, m
D_α	axial dispersion coefficient in the α-phase, m^2/s
D_{BM}	Brownian diffusion coefficient, $D_{BM} = \frac{2k_BT}{6\pi\mu_\ell d_f}$
D_i	inner diameter of the DME synthesis reactor, m
D_j^{eff}	effective diffusivity of component j, m^2/s
$\hat{e}_{\alpha\beta}^{\alpha}$	rate of mass transfer from $\alpha\beta$-interface to the α-phase, kg/m^3s
E_1, E_2	Ergun constants
$F_{g\ell}$	gas–liquid drag force, N/m^3 $F_{g\ell} = \tau_{g\ell}^{g}$
F_{gs}	gas–solid drag force, N/m^3
$F_{\ell s}$	liquid–solid drag force, N/m^3, $F_{\ell s} = \tau_{\ell s}^{\ell}$
f_s, f_v	shear and velocity slip factors
G	gas flow rate, kg/m^2s
g	gravity acceleration, m/s^2
$h_{\ell s}$	solid-to-liquid heat transfer coefficient, W/m^2K
h_w	heat transfer coefficient at the wall, W/m^2K
h_{ws}	wall-to-solid heat transfer coefficient, W/m^2K
H	bed height, m
Ha	Hamaker's constant, J
He	Henry constant
k	particle size class index, –
k_B	Boltzmann's constant, J/K
k_{df}	specific decay rates for free cells, s^{-1}
k_{ds}	specific decay rates for captured cells, s^{-1}
k_F	coefficient of sliding friction, m
k_g	gas–liquid mass transfer coefficient, m/s
$k_g a_{g\ell}$	volumetric gas-side mass transfer coefficient, s^{-1}
k_{gs}	gas–solid mass transfer coefficient, m/s
$k_{hn}K'_{H_2}$	rate constant for hydrogenation times equilibrium constant for hydrogen adsorption, s^{-1}
k_{hs}	hydrogenolysis rate constant, $kmol/kg_{cat}s$
$k_\ell a_{g\ell}$	volumetric liquid-side mass transfer coefficient, s^{-1}
$k_{\ell s}$	liquid–solid mass transfer coefficient, m/s
K_{DBT}	adsorption constant of DBT for hydrogenolysis, $m^3/kmol$
K'_{DBT}	adsorption constant of DBT for hydrogenation, $m^3/kmol$
K_{H_2}	adsorption constant of H_2 for hydrogenolysis, $m^3/kmol$
K_{H_2S}	adsorption constant of H_2S for hydrogenolysis, $m^3/kmol$
K_i	inhibition constant, kg/m^3
$K_{\ell s}$	overall mass transfer coefficient in liquid phase, m/s
K_T	overall heat transfer coefficient between DME synthesis and glycerol reforming reactors, J/m^2sK
K'_T	overall heat transfer coefficient between reaction side and permeate side in DME synthesis reactor, J/m^2sK
K_{ox}	saturation constant of oxygen, kg/m^3
K_s	saturation constant of phenol, kg/m^3
L	liquid flow rate, kg/m^2s
M_g	molecular mass of the gas mixture, kg/kmol
n_1	number density of primary bacterial cells
N_c	number of collectors in grid cell volume v, $N_c = \frac{6v}{\pi\left(d_p^o\right)^3}(1-\varepsilon^o)$

N_f number of trapped fines in grid cell volume v, $N_f = \frac{6v}{\pi d_f^3}(\varepsilon^o - \varepsilon)(1-\varepsilon_d)$
n_k number density of κ-size aggregate
∂N_f number of peripheral fines per collector, $\partial N_f = 4\beta(1-\varepsilon_d)\left[\frac{d_p(t)}{d_f}\right]^2$
P pressure, Pa
P_α pressure of α-fluid, Pa
P_c capillary pressure, Pa
Q_j membrane permeance, $kmol/m^2s$ Pa
$\dot{q}_r$ radial heat flux density, W/m^2
R ideal gas constant
r radial position within solid particle, m
r_{dep} fine (cell) attachment rate, $m^3{}_p/m^3s$ (kg/m^3s)
r_{df} decay rate for free cells, kg_{cells}/m^3s
r_{ds} decay rate for captured cells, kg_{cells}/m^3s
r_{gf} growth rate for free cells, kg_{cells}/m^3s
r_{gs} growth rate for captured cells, kg_{cells}/m^3s
r_{hn} reaction rate of DBT hydrogenation, $kmol/kg_{cat}s$
r_{hs} reaction rate of DBT hydrogenolysis, $kmol/kg_{cat}s$
r_i reaction rate, $kmol/kg_{cat}s$
r_p radius of catalyst particle, m
r_{react} reaction rate, $kmol/m^3s$
r_{rel} fine (cell) release rate, $m^3{}_p/m^3s$ (kg/m^3s)
r_{ref} glycerol reforming reaction rate, $kmol/kg_{cat}s$
s space velocity
T temperature, K
t time, s
$\underline{\hat{T}}^\alpha_{\alpha\beta}$ body supply of momentum to the α-phase from $\alpha\beta$-interface, kg/m^2s^2
u_α average interstitial velocity of α-phase, m/s
$\underline{u}_\alpha$ velocity vector for bulk α-phase, m/s
$u_{g,i}$ gas interfacial velocity in the bed, m/s
v_c volume of a cell, m^3
$v_{s\alpha}$ superficial velocity of α-phase, m/s
V_p volume of permeate side—DME synthesis, m^3
V_r volume of fixed bed side—DME synthesis, m^3
$\underline{w}_{\alpha\beta}$ velocity vector of the $\alpha\beta$-interface, m/s
X conversion
y_j mole fraction of species j in gas phase, –
Y_{X/O_2} yield coefficient, kg biomass/kg oxygen
$Y_{X/S}$ yield coefficient, kg biomass/kg phenol
z axial coordinate, m
Z compressibility factor

Greek letters

α fractional collision efficiency
α_{gs} gas-catalyst particle heat transfer coefficient, J/m^2sK
$\alpha_{\ell s}$ liquid-catalyst particle heat transfer coefficient, $J/kg_{cat}sK$
α^c_{rel} colloidal first-order release or detachment rate coefficient, $\alpha^c_{rel} = D_{BM}/\delta^2_{bl}$, s^{-1}
α^h_{rel} hydrodynamic release or detachment rate coefficient, kg/N s
β scale factor
β_{ij} collision frequency function
$\gamma_{a,k}$ fraction of the collector surface area not available for detachment of κ-size aggregate, $\gamma_{a,k} = \sin\theta_a = \tau_{cr,a,k}/\tau_w$
$\gamma_{a,k-1}$ fraction of the collector surface area not available for detachment of the $(\kappa-1)$-size aggregate, $\gamma_{a,k-1} = \sin\theta'_a = \tau_{cr,a,k-1}/\tau_w$
γ_f fraction of the collector surface area available/not available for the fine particle adhesion/detachment, $\gamma_f = \tau_{cr}/\tau_w$
$\gamma_{g\ell}$ macroscopic interfacial tension of the gas–liquid interface, N/m
$\Gamma_{\alpha\beta}$ mass of $\alpha\beta$-interface per unit area of $\alpha\beta$-interface, kg/m^2
δ separation distance between the fine particle (bacterial cell) and the collector plane, $\delta = 4\times10^{-10}$
δ_b biofilm thickness, m
δ_{bl} thickness of the boundary layer around a single spherical (bio)particle, $\delta_{bl} = 0.5d_p(2D_{BM}/u_\ell d_p)^{1/3}$
ΔH_r reaction enthalpy, J/mol
ε bed porosity
ε_d porosity of solid deposit
ε_p catalyst particle porosity
ε_α α-phase holdup
η^o single-collector efficiency
η_e external wetting efficiency
η_g effectiveness factor for a completely dry particle
η_i effectiveness factor of reaction i
η_ℓ effectiveness factor for a fully wetted particle
η_G overall effectiveness factor
λ_f filter coefficient, m^{-1}
λ flow parameter, $\lambda = \left[\frac{\rho_g}{\rho_{air}}\frac{\rho_\ell}{\rho_{water}}\right]^{0.5}$
$\lambda^{eff}_{\alpha,z}$ α-phase effective axial thermal conductivity, of α-phase, W/m K
$\lambda^{eff}_{g\ell,z}$ gas–liquid effective axial thermal conductivity, W/m K
λ^{eff}_p effective thermal conductivity of catalyst particles, W/m K
$\lambda^{eff}_{\alpha,r}$ α-phase effective radial conductivity, W/m K
$\lambda^{eff}_{g\ell,r}$ gas–liquid effective radial conductivity, W/m K
λ_{s0} bed conductivity without liquid flow, W/m K
Λ_r bed radial effective thermal conductivity, W/m K
μ_α α-phase dynamic viscosity, kg/ms
μ^e_α α-phase effective viscosity (combination of bulk and shear terms), kg/m s
ξ surface area parameter
ρ_α density of α-phase, kg/m^3
ρ_b total dry biomass in the biofilm, kg/m^3
ρ_c dry cell mass/cell volume, kg/m^3
ρ_p catalyst particle density, kg/m^3
σ volume of captured cells and aggregates in unit volume of reactor, $\sigma = \sum_{k-2}^{max}\sigma_{a,k} + \sigma_c$

$\sigma_{a,k}$ volume of κ-size aggregate in unit volume of bioreactor, m^3_{cells}/m^3_r
σ_c volume of captured cells in unit volume of bioreactor, m^3_{cells}/m^3_r
σ_f volume of captured fines in unit volume of reactor, $\sigma = (\varepsilon^o - \varepsilon)(1 - \varepsilon_d)$, m^3_f/m^3_r
σ_ℓ surface tension, N/m
τ shear stress, N/m^2
$\underline{\hat{\tau}}^\alpha_{\alpha\beta}$ nonequilibrium part of the body supply of momentum to the α-phase from $\alpha\beta$-interface, kg/m^2s^2
ψ flow parameter, $\psi = \frac{\sigma_{water}}{\sigma_\ell}\left[\frac{\mu_\ell}{\mu_{water}}\left(\frac{\rho_{water}}{\rho_\ell}\right)^2\right]^{0.33}$
$\psi_{g\ell}$ gas–liquid interaction factor

Subscripts

a,k κ-size aggregate
b biofilm
c bacterial cell
d solid deposit
f fine particle
g gas phase
hs hydrogenolysis
hn hydrogenation
i inert gas
in reactor inlet
ℓ liquid phase
p catalyst particle
ref reforming
s solid phase
w reactor wall

Superscripts

g glycerol reforming
in reactor inlet
o clean bed state
p DME synthesis— permeate side
r DME synthesis—reaction side
$*$ on solid deposit or (bio)particle surface

Abbreviations

BiPh biphenyl
CHB cyclohexylbenzene
DBT dibenzothiophene
S Phenol

References

1 Dudukovic MP, Larachi F, Mills PL. Multiphase reactors—revisited. Chem. Eng. Sci. 1999; 54: 1975.
2 Wu Y, Khadilkar MR, Al-Dahhan MH, Dudukovic MP. Comparison of upflow and downflow two-phase flow reactors with and without fines. Ind. Eng. Chem. Res. 1996; 35: 397.
3 Khadilkar RM, Wu Y, Al-Dahhan MH, Dudukovic MP, Colakyan M. Comparison of trickle-bed and upflow reactor performance at high pressure: model predictions and experimental observations. Chem. Eng. Sci. 1996; 51: 2139.
4 van Gelder KB, Borman PC, Weenink RE, Westerterp KR. Three-phase packed bed reactor with an evaporating solvent—II. Modelling of the reactor. Chem. Eng. Sci. 1990; 45: 3171.
5 Westerterp KR, Molgal EJ, van Gelder KB. Catalytic hydrogenation reactors for the fine chemicals industries. Their design and operation. Chem. Eng. Process. 1997; 36: 17.
6 Julcour C, Chaudhari RV, Le Lann JM, Wilhelm AM, Delmas H. Dynamic modeling of three-phase upflow fixed-bed reactor including pore diffusion. Chem. Eng. Process. 2002; 41: 311.
7 Chaudhari RV, Jaganathan R, Mathew SP, Julcour C, Delmas H. Hydrogenation of 1,5,9-cyclododecatriene in fixed-bed reactors: down- vs. upflow modes. AIChE J. 2002; 48: 110.
8 Stuber F, Delmas H. Partial hydrogenation in an upflow fixed bed reactor: a multistage operation for experimental optimization of selectivity. Ind. Eng. Chem. Res. 2003; 42: 6.
9 Vergel-Hernμndez CA. Les réacteurs catalytiques à lit fixe avec écoulement de gaz et de liquide. Comparaison sur le plan théorique et expérimental de la performance du réacteur dans différents sens d'écoulement. PhD Thesis, INPL, Nancy, France, 1993.
10 Trambouze P. Multiphase catalytic reactors in the oil industry, an introduction. Rev. Inst. Franç. Pétr. 1991; 46: 433.
11 Mazzarino I, Santos A, Baldi G. Catalytic oxidation of n-propanol in a multiphase upflow reactor: surface tension effects. Chem. Eng. Sci. 1994; 49: 5699.
12 Herskowitz M, Smith JM. Trickle-bed reactors: a review. AIChE J. 1983; 29: 1.
13 Satterfield CN. Trickle-bed reactors. AIChE J. 1975; 21: 209.
14 Al-Dahhan MH, Larachi F, Dudukovic MP, Laurent A. High-pressure trickle-bed reactors: a review. Ind. Eng. Chem. Res. 1997; 36: 3292.
15 Menderos FS, Ancheyta J, Chen J. Review on criteria to ensure ideal behaviors in trickle-bed reactors. Appl. Catal. A: Gen. 2009; 355: 1.
16 Charpentier JC, Favier M. Some liquid hold-ups experimental data in trickle-bed reactors for foaming and nonfoaming hydrocarbons. AIChE J. 1975; 21: 1213.
17 Talmor E. Two-phase downflow through catalyst beds, part 1: flow maps. AIChE J. 1977; 23: 868.
18 Iliuta I, Ortiz A, Larachi F, Grandjean BPA, Wild G. Hydrodynamics & mass transfer in trickle-bed reactors: an overview. Chem. Eng. Sci. 1999; 54: 5329.
19 Saroha AK, Nigam KDP. Trickle bed reactors. Rev. Chem. Eng. 1996; 12: 207.
20 Sicardi S, Hofmann H. Influence of gas velocity and packing geometry on pulsing inception in trickle-bed reactors. Chem. Eng. J. 1980; 20: 251.
21 Blok JR, Varkevisser J, Drinkenburg AAH. Transition to pulsing flow, holdup and pressure drop in packed columns with cocurrent gas–liquid downflow. Chem. Eng. Sci. 1983; 38: 687.
22 Ng KM. A model for flow regime transitions in cocurrent down-flow trickle-bed reactors. AIChE J. 1986; 32: 115.
23 Grosser K, Carbonell RG, Sundaresan S. Onset of pulsing in two-phase cocurrent down-flow through a packed bed. AIChE J. 1988; 34: 1850.
24 Holub RA, Dudukovic MP, Ramachandran PA. A phenomenological model of pressure drop, liquid holdup and flow regime transition in gas–liquid trickle flow. Chem. Eng. Sci. 1992; 47: 2343.

25 Attou A, Ferschneider G. A two-fluid hydrodynamic model for the transition between trickle and pulse flow in a cocurrent gas–liquid packed-bed reactor. Chem. Eng. Sci. 2000; 55: 491.

26 Iliuta I, Aydin B, Larachi F. Onset of pulsing in trickle beds with non-Newtonian liquids at elevated temperature & pressure—modeling & experimental verification. Chem. Eng. Sci. 2006; 61: 526.

27 Shah YT. Gas–Liquid-Solid Reactor Design. New York: McGraw-Hill, 1979.

28 Turpin JL, Huntington RL. Prediction of pressure drop for two-phase, two-component concurrent flow in packed beds. AIChE J. 1967; 13: 1196.

29 Specchia V, Sicardi S, Gianetto A. Absorption in packed towers with concurrent upward flow. AIChE J. 1974; 20: 646.

30 Mazzarino I, Sicardi S, Baldi G. Hydrodynamics and solid–liquid contacting effectiveness in an upflow multiphase reactor. Chem. Eng. J. 1987; 36: 151.

31 Rao AVR, Kumar RK, Sankarshana T, Khan A. Identification of flow regimes in concurrent gas liquid upflow through packed beds. Chem. Eng. Technol. 2011; 34: 1909.

32 Wammes WJA, Middelkamp J, Huisman WJ, de Baas CM, Westerterp KR. Hydrodynamics in a cocurrent gas–liquid trickle bed at elevated pressures, part 1: gas–liquid interfacial areas, part 2: liquid hold-up, pressure drop, flow regimes. AIChE J. 1991; 37: 1849.

33 Bensetiti Z, Larachi F, Grandjean BPA, Wild G. Liquid saturation in cocurrent upflow reactors: a state-of-the-art correlation. Chem. Eng. Sci. 1997; 52: 4239.

34 Al-Dahhan MH, Dudukovic MP. Pressure drop and liquid holdup in high pressure trickle-bed reactors. Chem. Eng. Sci. 1994; 49: 5681.

35 Larachi F, Cassanello M, Laurent A. Gas–liquid interfacial mass transfer in trickle-bed reactors at elevated pressures. Ind. Eng. Chem. Res. 1998; 37: 718.

36 Gunjal PR, Kashid MN, Ranade VV, Chaudhari RV. Hydrodynamics of trickle-bed reactors: experiments and CFS modeling. Ind. Eng. Chem. Res. 2005; 44: 6278.

37 Lopes JG, Quinta-Ferreira RM. Multiphase CFD modeling of trickle-bed reactor hydrodynamics. In Proceedings of the World Congress on Engineering and Computer Science, San Francisco, CA, USA, 2007.

38 Dudukovic MP, Mills PL. Contacting and hydrodynamics in trickle-bed reactors. In Cheremisinoff NP, ed. Encyclopedia of Fluid Mechanics. Houston, TX: Gulf Publishing Company, 1986.

39 Dudukovic MP. Catalyst effectiveness factor and contacting efficiency in trickle-bed reactors. AIChE J. 1977; 23: 6.

40 Mills PL, Dudukovic MP. Integral equation solution for the effectiveness factor of partially wetted catalysts. Ind. Eng. Chem. 1982; 21: 1.

41 Iliuta I, Larachi F, Al-Dahhan MH. Double-slit model for partially wetted trickle flow hydrodynamics. AIChE J. 2000; 46: 597.

42 Al-Dahhan MH, Dudukovic MP. Catalyst wetting efficiency in trickle-bed reactors at high pressure. Chem. Eng. Sci. 1995; 50: 2377.

43 Al-Dahhan MH, Dudukovic MP. Catalyst bed dilution for improving catalyst wetting in laboratory trickle-bed reactors. AIChE J. 1996; 42: 2594.

44 Pironti F, Mizrahi D, Acosta A, Gonzalez-Mendizibal D. Liquid–solid wetting factor in trickle-bed reactors: its determination by a physical method. Chem. Eng. Sci. 1999; 54: 3793.

45 Iliuta I, Larachi F. Hydrodynamics of power-law fluids in trickle-flow reactors: mechanistic model, experimental, verification and simulation. Chem. Eng. Sci. 2002; 57: 1931.

46 Whitaker S. The transport equations for multi-phase systems. Chem. Eng. Sci. 1973; 28: 139.

47 Dankworth DC, Kevrekidis IG, Sundaresan S. Dynamics of pulsing in trickle beds. AIChE J. 1990; 36: 605.

48 Holub RA, Dudukovic MP, Ramachandran PA. Pressure drop, liquid holdup and flow regime transition in trickle flow. AIChE J. 1993; 39: 302.

49 Sáez AE, Carbonell RG. Hydrodynamic parameters for gas–liquid cocurrent flow in packed-beds. AIChE J. 1985; 31: 52.

50 Attou A, Boyer C, Ferschneider G. Modelling of the hydrodynamics of the cocurrent gas–liquid trickle flow through a trickle-bed reactor. Chem. Eng. Sci. 1999; 54: 785.

51 Attou A, Ferschneider G. A simple model for pressure drop and liquid hold-up in packed-bed bubble reactors. Chem. Eng. Sci. 1999; 54: 139.

52 Liu S, Masliyah JH. On non-Newtonian fluid flow in ducts and porous media. Chem. Eng. Sci. 1998; 53: 1175.

53 Iliuta I, Larachi F. Hydrodynamic models for rheologically complex fluids in co- and countercurrent gas–liquid packed-bed bioreactors. Ind. Eng. Chem. Res. 2002b; 41: 2096.

54 Iliuta I, Hamidipour M, Larachi F. Non-equilibrium thermomechanical modeling of liquid drainage/imbibition in trickle beds. AIChE J. 2012; 58: 3123.

55 Hassanizadeh SM, Gray WG. Mechanics and thermodynamics of multiphase flow in porous media including interphase boundaries. Adv. Water Resour. 1990; 13: 169.

56 Achanta A, Cushman JH, Okos MR. On multicomponent, multiphase thermomechanics with interfaces. Int. J. Eng. Sci. 1994; 32: 1717.

57 Midoux N, Morsi BI, Purwasasmita M, Laurent A, Charpentier JC. Interfacial area and liquid side mass transfer coefficient in trickle bed reactors operating with organic liquids. Chem. Eng. Sci. 1984; 39: 781.

58 Iliuta I, Thyrion FC. Gas–liquid mass transfer in fixed beds with two-phase cocurrent downflow: gas/Newtonian and non-Newtonian liquid systems. Chem. Eng. Technol. 1997; 20: 538.

59 Lara-Marquez A, Larachi F, Wild G, Laurent A. Mass transfer characteristics of fixed beds with cocurrent upflow and downflow. A special reference to the effect of pressure. Chem. Eng. Sci. 1992; 47: 3485.

60 Mahajani VV, Sharma MM. Effective interfacial area and liquid side mass transfer coefficient in trickle bed reactors. Chem. Eng. Sci. 1979; 34: 1425.

61 Oyevaar MH, de la Rie T, van der Sluijs CL, Westerterp KR. Interfacial areas and gas hold-ups in bubble columns and packed bubbles columns at elevated pressures. Chem. Eng. Process. 1989; 26: 1.

62 Molga EJ, Westerterp KR. Gas–liquid interfacial area and holdup in a cocurrent upflow packed bed bubble column reactor at elevated pressures. Ind. Eng. Chem. Res. 1997; 36: 622.

63 Iliuta I, Larachi F, Grandjean BPA, Wild G. Gas–liquid interfacial mass transfer in trickle-bed reactors: state-of-the-art correlations. Chem. Eng. Sci. 1999; 54: 5633.

64 Ramachandran PA, Chaudhari RV. Three-Phase Catalytic Reactors. London: Gordon and Breach Science Publishers, 1983.

65 Gianetto A, Silveston PL. Multiphase Chemical Reactors: Theory, Design, Scale-Up. Berlin: Springer Verlag, 1986.

66 Larachi F, Cassanello M, Laurent A, Midoux N, Wild G. Gas–liquid interfacial areas in three-phase fixed bed reactors. Chem. Eng. Process. 1997; 36: 497.

67 Highfill W, Al-Dahhan M. Liquid–solid mass transfer coefficient in high pressure trickle bed reactors. Chem. Eng. Res. Des. 2001; 79: 631.
68 Sato Y, Hirose T, Takahashi F, Toda M. Performance of fixed-bed catalytic reactor with co-current gas–liquid flow. *PACHEC* 1972; Section 8, Reaction Eng. pp. 187–196.
69 Goto S, Smith JM. Trickle bed reactor performance. AIChE J. 1975; 21: 706.
70 Specchia V, Baldi G, Gianetto A. Solid–liquid mass transfer in cocurrent two phase flow through packed beds. Ind. Eng. Chem. Proc. Des. Dev. 1978; 17: 362.
71 Saterfield CN, van Eek MW, Bliss GS. Liquid–solid mass transfer in packed beds with downward concurrent gas–liquid flow. AIChE J. 1978; 24: 709.
72 Ruether JA, Yang C-H, Hayduk W. Particle mass transfer during cocurrent downward gas–liquid flow in packed beds. Ind. Eng. Chem. Proc. Des. Dev. 1980; 19: 103.
73 Saroha AK. Solid–liquid mass transfer studies in trickle bed reactors. Chem. Eng. Res. Des. 2010; 88: 744.
74 Morita BI, Smith JM. Mass transfer and contacting efficiency in a trickle bed reactor. Ind. Eng. Chem. Fundam. 1978; 17: 113.
75 Yoshikawa M, Iwai K, Goto S, Teshima H. Liquid–solid mass transfer in gas–liquid cocurrent flows through beds of small particles. J. Chem. Eng. Jpn. 1981; 14: 444.
76 Tan CS, Smith JM. A dynamic method for liquid-particle mass transfer in trickle beds. AIChE J. 1982; 28: 190.
77 Chou TS, Worley FL, Luss D. Local particle-liquid mass transfer fluctuations in mixed phase cocurrent downflow through a fixed bed in the pulsing regime. Ind. Eng. Chem. Fundam. 1979; 18: 279.
78 Rao VG, Drinkenburg AAH. Solid–liquid mass transfer in packed beds with cocurrent gas–liquid downflow. AIChE J. 1985; 31: 1059.
79 Latifi MA, Laurent A, Storck A. Liquid–solid mass transfer in a packed bed with downward cocurrent gas–liquid flow: an organic liquid phase with high Schmidt number. Chem. Eng. J. 1988; 38: 47.
80 Rode SN, Midoux N, Latifi MA, Storck A. Hydrodynamics and liquid–solid mass transfer mechanisms in packed beds operating in cocurrent gas–liquid downflow: an experimental study using electrochemical shear rate sensors. Chem. Eng. Sci. 1994; 49: 1383.
81 Goto S, Levec J, Smith JM. Mass transfer in packed beds with two-phase flow. Ind. Eng. Chem. Proc. Des. Dev. 1975; 14: 473.
82 Lakota A, Levec J. Solid–liquid mass transfer in packed beds with cocurrent downward two-phase flow. AIChE J. 1990; 36: 1444.
83 Bartelmus G. Local solid–liquid mass transfer coefficients in a three-phase fixed bed reactor. Chem. Eng. Process. 1989; 26: 111.
84 Baussaron L, Julcour-Lebigue C, Boyer C, Wilhelm AM, Delmas H. Effect of partial wetting on liquid/solid mass transfer in trickle bed reactors. Chem. Eng. Sci. 2007; 62: 7020.
85 Delaunay CB, Storck A, Laurent A, Charpentier JC. Electrochemical determination of liquid–solid mass transfer in a fixed-bed irrigated gas–liquid reactor with downward cocurrent flow. Int. Chem. Eng. 1982; 22: 244.
86 Burghardt A, Bartelmus G, Jaroszybki M, Kofodziej A. Hydrodynamics and mass transfer in a three-phase fixed-bed reactor with cocurrent gas–liquid downflow. Chem. Eng. J. 1995; 58: 83.
87 Larachi F, Alix C, Grandjean BPA, Bernis A. Nu/Sh correlation for particle-liquid heat and mass transfer coefficients in trickle beds based on Peclet similarity. Chem. Eng. Res. Des. 2003; 81: 689.
88 Mochizuki S, Matsui T. Liquid–solid mass transfer in liquid–gas upward cocurrent flow in packed beds. Chem. Eng. Sci. 1974; 29: 1328.
89 Mochizuki S. Particle mass transfer and liquid hold-up in packed beds with upward concurrent gas–liquid flow. Chem. Eng. Sci. 1981; 36: 213.
90 Mochizuki S. Empirical expressions of liquid–solid mass transfer in cocurrent gas–liquid upflow fixed beds. Chem. Eng. Sci. 1982; 37: 1422.
91 Sedahmed GH, El-Kayar AM, Farag HA, Noseir SA. Liquid–solid mass transfer in packed beds of Raschig rings with upward two-phase (gas–liquid) flow. Chem. Eng. J. 1996; 62: 61.
92 Baldi G. Heat Transfer in Gas–Liquid-Solid Reactors. Vimeiro: NATO Institute on Multiphase Chemical Reactors, 1980.
93 Weekman VW, Myers JE. Heat transfer characteristics of concurrent gas–liquid flow in packed beds. AIChE J. 1965; 11: 13.
94 Hashimoto K, Muroyama K, Fujiyoshi K, Nagata S. Effective radial thermal conductivity in cocurrent flow of a gas and liquid through a packed bed. Int. Chem. Eng. 1976; 16: 720.
95 Matsuura A, Hitaka Y, Akehata T, Shirai T. Effective radial thermal conductivity in packed beds with downward cocurrent gas–liquid flow. Heat Transf. Jpn Res. 1979; 8: 44.
96 Lamine AS, Gerth L, Le Gall H, Wild G. Heat transfer in a packed bed reactor with cocurrent downflow of a gas and a liquid. Chem. Eng. Sci. 1996; 51: 3813.
97 Mears DE. Diagnostic criteria for heat transport limitations in fixed bed reactors. J. Catal. 1971; 20: 127.
98 Larachi F, Belfares L, Iliuta I, Grandjean BPA. Heat and mass transfer in co-current gas–liquid packed beds. Analysis, recommendations and new correlations. Ind. Eng. Chem. Res. 2003; 42: 222.
99 Gutsche S, Wild G, Midoux N, Martin H. Hydrodynamik und Wärmeübergang in zweiphasig im Aufstrom durchströmten Festbettreaktoren. Chem. Ing. Tech. 1989; 61: 733.
100 Lamine AS, Colli Serrano MT, Wild G. Hydrodynamics and heat transfer in packed bed with cocurrent upflow. Chem. Eng. Sci. 1992; 47: 3493.
101 Lamine AS, Colli Serrano MT, Wild G. Influence of viscosity and foaming properties on heat transfer in packed beds with cocurrent upflow of gas and liquid. Trans. Inst. Chem. Eng. Part A 1995; 73: 280.
102 Specchia V, Baldi G. Heat transfer in trickle-bed reactors. Chem. Eng. Commun. 1979; 3: 483.
103 Sokolov VN, Yablokova MA. Thermal conductivity of a stationary granular bed with upward gas–liquid flow. J. Appl. Chem. USSR (Zh. PriM. Khim.) 1983; 56: 551.
104 Kirillov VA, Ogarkov BL. Investigation of the processes of heat and mass transfer in a three-phase fixed bed of catalyst. Int. Chem. Eng. 1980; 20: 478.
105 Marcandelli C, Wild G, Lamine AS, Bernard JR. Measurement of local particle-fluid heat transfer coefficient in trickle-bed reactors. Chem. Eng. Sci. 1999; 54: 4997.
106 Boelhouwer JG, Piepers HW, Drinkenburg AAH. Particle-liquid heat transfer in trickle-bed reactors. Chem. Eng. Sci. 2001; 56: 1181.
107 Briens CL, Del Pozo M, Trudell C, Wild G. Measurement and modeling of particle-liquid heat transfer in liquid–solid and gas–liquid-solid fluidized beds. Chem. Eng. Sci. 1999; 54: 731.
108 Boelhouwer JG. Nonsteady operation of trickle bed reactors: hydrodynamics, mass and heat transfer [PhD thesis]. Eindhoven, The Netherlands: Technische Universiteit Eindhoven, 2001.
109 Sedriks W, Kenney CN. Partial wetting in trickle-bed reactors: the reduction of crotonaldehyde over a palladium catalyst. Chem. Eng. Sci. 1973; 28: 559.

110 Hanika J, Sporka R, Ruzicka V, Krausova J. Qualitative observations of heat and mass transfer effects on the behavior of a trickle-bed reactor. Chem. Eng. Commun. 1975; 2: 19.
111 Sie ST, Krishna R. Process development and scale-up: III. Scale-up and scale-down of trickle bed processes. Rev. Chem. Eng. 1998; 14: 203.
112 Tsamatsoulis D, Papayannakos N. Axial dispersion and hold-up in a bench-scale trickle-bed reactor at operating conditions. Chem. Eng. Sci. 1994; 49: 523.
113 Gierman H. Design of laboratory hydrotreating reactors. Scaling down of trickle-flow reactors. Appl. Cat. 1988; 43: 277.
114 Kallinikos LE, Papayannakos NG. Operation of a miniscale string bed reactor in spiral form at hydrotreatment conditions. Ind. Eng. Chem. Res. 2007a; 46: 5531.
115 Kallinikos LE, Papayannakos NG. Fluid dynamic characteristics of a structured bed spiral mini-reactor. Chem. Eng. Sci. 2007b; 62: 5979.
116 Hipolito AI, Rolland M, Boyer C, de Bellefon C. Single pellet string reactor for intensification of catalyst testing in gas/liquid/solid configuration. Oil Gas Sci. Technol.—Rev. 2010; 65: 689.
117 Narayan R, Coury JR, Masliyah JH, Gray MR. Particle capture and plugging in packed-beds reactors. Ind. Eng. Chem. Res. 1997; 36: 4620.
118 Gray MR, Srinivasan N, Masliyah JH. Pressure buildup in gas–liquid flow through packed beds due to deposition of fine particles. Can. J. Chem. Eng. 2002; 80: 346.
119 Iliuta I, Larachi F, Grandjean BPA. Fines deposition dynamics in trickle flow reactors. AIChE J. 2003; 49: 485.
120 Iliuta I, Ring Z, Larachi F. Simulating simultaneous fines deposition under catalytic hydrodesulfurization in hydrotreating trickle beds—does bed plugging affect HDS performance? Chem. Eng. Sci. 2006; 61: 1317.
121 Broderick DH, Gates BC. Hydrogenolysis and hydrogenation of dibenzothiophene catalyzed by sulfided $CoO\text{-}MoO_3/\gamma\text{-}Al_2O_3$; the reaction kinetics. AIChE J. 1981; 27: 663.
122 Ryan JN, Gschwend PM. Effects of ionic strength and flow rate on colloid release: relating kinetics to intersurface potential energy. J. Colloid Interface Sci. 1994; 164: 21.
123 Khilar KC, Fogler HS. Migration of Fines in Porous Media. Dordrecht: Kluwer Academic Publishers, 1998.
124 Gresho PM, Sani RL. Incompressible Flow and the Finite Element Method. Advection–Diffusion and Isothermal Laminar Flow. Chichester: John Wiley & Sons, Inc., 1988.
125 Tien C. Granular Filtration of Aerosols and Hydrosols. Boston: Butterworths, Butterworths-Heinemann Series in Chemical Engineering, 1989.
126 Logan BE, Jewett DG, Arnold RG, Bower EJ, O'Melia CR. Clarification of clean-bed filtration models. J. Environ. Eng. 1995; 121: 869.
127 Nelson KE, Ginn TR. Colloid filtration theory and the Happel sphere-in-cell model revisited with direct numerical simulation of colloids. Langmuir 2005; 21: 2173.
128 Rajagopalan R, Tien C. Trajectory analysis of deep-bed filtration with the sphere-in-cell porous media model. AIChE J. 1976; 22: 523.
129 Tien C, Turian RM, Pendse H. Simulation of the dynamic behavior of deep bed filters. AIChE J. 1979; 25: 385.
130 Iliuta I, Larachi F. Fines deposition dynamics in packed-bed bubble reactors. Ind. Eng. Chem. Res. 2003; 42: 2441.
131 Iliuta I, Larachi F. Stretching operational life of trickle-bed filters by liquid-induced pulse flow. AIChE J. 2005; 51: 2034.
132 Iliuta I, Larachi F. Biological and physical plugging in trickle bed bioreactors for wastewater treatment. Chem. Eng. Sci. 2005; 60: 1477.
133 Ortiz-Arroyo A, Larachi F, Grandjean BPA, Roy S. Modeling and simulation of clogging in packed-bed reactors with non-aqueous media using CFD. AIChE J. 2002; 48: 1596.
134 Stephan EA, Chase GG. Development of volume-average theory for deep-bed filtration. AIChE J. 2000; 46: 1918.
135 Tien C, Payatakes AC. Advances in deep bed filtration. AIChE J. 1979; 25: 737.
136 Iliuta I, Larachi F. Biomass accumulation and clogging in trickle-bed bioreactors. AIChE J. 2004; 50: 2541.
137 Iliuta I, Larachi F. Dynamics of cells attachment, aggregation, growth and detachment in trickle-bed bioreactors. Chem. Eng. Sci. 2006; 61: 4893.
138 Kreft J-U, Picioreanu C, Wimpenny JWT, van Loosdrecht MCM. Individual-based modelling of biofilms. Microbiology 2001; 147: 2897.
139 von Smoluchowski M. Versuch einer mathematischen theorie der koagulationskinetik kolloider losungen. Z. Phys. Chem. 1917; 92: 129.
140 Iliuta I, Larachi F. Deposition and aggregation of Brownian particles in trickle-bed reactors. AIChE J. 2006; 52: 4167.
141 Tang W-T, Wisecarver K, Fan L-S. Dynamics of a draft tube gas–liquid-solid fluidized bed bioreactor for phenol degradation. Chem. Eng. Sci. 1987; 42: 2123.
142 Sen TK, Das D, Khilar KC, Suraishkumar GK. Bacterial transport in porous media: new aspects of the mathematical modeling. Colloids Surf. A 2005; 260: 53.
143 Chmielewski RAN, Frank JF. Biofilm formation and control in food processing facilities. Compr. Rev. Food Sci. Food Saf. 2003; 2: 22.
144 van Hamersveld EH, van der Lans RGJM, Caulet PJC, Luyben KChAM. Modeling brewers' yeast flocculation. Biotechnol. Bioeng. 1998; 57: 330.
145 Friedlander SJ. Smoke, Dust, and Haze. New York: Wiley Interscience, 1972.
146 Gardner KH, Theis TL, Young TC. Colloid aggregation: numerical solution and measurements. Colloids Surf. A 1998; 141: 237.
147 Kim J-W, Kramer TA. Improved models for fractal colloidal agglomeration: computationally efficient algorithms. Colloids Surf. A 2005; 253: 33.
148 Iliuta I, Iliuta MC, Fongarland P, Larachi F. Integrated aqueous glycerol reforming to dimethyl ether synthesis—a novel allothermal dual bed membrane reactor concept. Chem. Eng. J. 2012b; 187: 311
149 Rohde MP, Schaub G, Khajavi S, Jansen JC, Kapteijn F. Fischer-Tropsch synthesis with in situ H_2O removal—directions of membrane development. Microporous Mesoporous Mater. 2008; 115:126.
150 Lu W-Z, Teng L-H, Xiao W-D. Simulation and experiment study of dimethyl ether synthesis from syngas in a fluidized bed reactor. Chem. Eng. Sci. 2004; 59: 5455.
151 Carberry J. Chemical and Catalytic Reaction Engineering. New York: McGraw-Hill, 1976.
152 Froment GF, Bischoff KB. Chemical Reactor Analysis and Design. New York: John Wiley & Sons, Inc., 1990.

CHAPTER 6

Three-phase slurry reactors

Vivek V. Buwa[1], Shantanu Roy[1] and Vivek V. Ranade[2]

[1]Department of Chemical Engineering, Indian Institute of Technology-Delhi, New Delhi, India

[2]Chemical Engineering & Process Development Division, National Chemical Laboratory, Pune, India

Abstract

Many of the chemical reactors employed in various sectors of the process and chemical industries, such as petroleum refining, downstream processing and petrochemicals, bulk and specialty chemicals, and pharmaceuticals, involve reactants and products in three phases. The ubiquitous presence of three-phase reactors comes from the fact that many of the reactants are liquids under reaction conditions, to be treated with a gas like hydrogen, and in the presence of a heterogeneous catalyst that is usually a solid. This requirement leads to two broad modes of contacting one in which the catalyst is held stationary in a "fixed" or a "packed bed" and one in which the catalyst is set free to move, along with the flowing fluid phases, during their sojourn through the reactor vessel. The choice of whether the catalyst should be fixed or moving is usually dictated by the relative dominance of transport rates and intrinsic kinetics. If the reaction chemistry is intrinsically slow (when external and internal mass transfer of chemical species from the fluid phases to the porous catalyst is not rate limiting), packed bed reactor that allows very high catalyst loading is usually preferred. Packed beds, however, do not constitute the main topic of discussion in this monograph. Still, three-phase packed beds and slurry reactors are really parts of the same continuum of three-phase catalytic reactors and have been discussed comprehensively about three decades back in the classic book by P.A. Ramachandran and R.V. Chaudhari (*Three Phase Catalytic Reactors*. New York: Gordon and Breach Science Publishers; 1983). The same authors published earlier a review of three-phase slurry reactors (Chaudhari RV, Ramachandran PA. Three phase slurry reactors. *AIChE J.* 1980;26(2):177–199). This chapter builds on these authors' works and attempts to update the reader on developments in the field over the past few decades while retaining the essential reaction and reactor engineering aspects. The focus is on the design, scale-up, and operation of three-phase slurry reactor systems.

6.1 Introduction

If the reaction chemistry is intrinsically fast, external or internal mass transfer, and on occasion the heat transfer, becomes the slowest and hence the rate-controlling step. Thus, making these rates faster necessitates the choice of fine particles, which would offer a smaller characteristic length inside the particle for heat and mass to diffuse, as well as higher specific surface area for overcoming external transport limitations. Fine particles, on the other hand, also cause higher pressure drop to the flowing fluids owing to the higher external surface area per unit volume and hence higher overall skin friction. In such cases, instead of holding particles stationary as is done in packed bed reactors, it is beneficial to have freely moving catalyst particles. This will allow better heat and mass transfer characteristics without excessive pressure drop, which may occur in packed beds with fine particles. Note that since reactions are intrinsically fast, the catalyst loading in such cases is not high, which further facilitates maintenance of catalyst particles in freely moving conditions. Such reactors, in which catalyst particles are in "fluidized" state, are broadly known as three-phase slurry reactors.

A partial list of different practical three-phase reactor configurations is presented in Table 6.1. The reactors listed in the first column, involving systems in which the solid catalyst particles are held in position and the gas and liquid phases flow around them, are easier to operate. Those in which the catalyst particles move around (are "fluidized" or "slurry systems"), typically conveyed by flowing gas and liquid phases, exhibit several curious mechanical features like different flow regimes and stability issues, which make their operation challenging and design more involved. While "hydrodynamics" is important in both kinds of reactors, it is arguably more challenging in those systems that appear in the second column, that is, fluidized or slurry reactors.

In terms of industrial use, each of the aforementioned reactors and their contacting modes offer different advantages and disadvantages. Therefore, the "reactor selection" for a particular chemistry or process needs to be done after careful consideration of the operating factors and contacting options that the different reactor configurations provide. Table 6.2 provides some qualitative ratings to these factors for some of the important and industrially common reactors from the list presented in Table 6.1. Clearly, the three-phase slurry reactor types classified within the box, which are the scope of this chapter, represent a class of reactors

Multiphase Catalytic Reactors: Theory, Design, Manufacturing, and Applications, First Edition. Edited by Zeynep Ilsen Önsan and Ahmet Kerim Avci.

Table 6.1 Three-phase catalytic reactors.

Stationary solid catalyst particles (Three-phase "fixed" bed reactors)	**Freely moving solid catalyst particles (Three-phase "slurry" reactors)**
Trickle bed reactors	Slurry micro-reactor
Packed bubble column reactor	Jet loop reactor
Simulated moving bed reactor	Stirred slurry reactor
Monolith reactor	Slurry bubble column reactor
Micro-trickle bed reactor	Fluidized bed reactor
Three-phase rotating packed bed (HIGEE) reactor	Three-phase circulating fluidized bed reactor

Table 6.2 Comparison of multiphase reactors (qualitative rating: more stars mean superior performance on the pertinent metric).

Aspect	**Micro slurry reactor**	**Jet loop reactor**	**Stirred slurry reactor**	**Slurry bubble column reactor**	**Three-phase fluidized reactor**	**Packed bubble column reactor**	**Trickle bed reactor**
Ease of operation	*	**	***	****	****	*****	*****
Solid loading	*	*	**	****	****	*****	*****
Particle size	*	**	**	***	****	*****	*****
Catalyst separation	*	*	*	***	***	*****	*****
Lower catalyst attrition	**	**	**	***	***	****	*****
Heat transfer	*****	*****	****	****	****	**	*
Mass transfer	*****	*****	*****	****	****	***	***
Plug flow/lower backmixing	*****	**	*	***	***	***	****
Viscous/foaming liquid	***	****	****	****	****	****	**
Reactor volume	*	***	***	***	***	****	*****
Power per unit volume	*****	****	****	***	***	**	**
Pressure	*****	***	**	***	***	*****	*****

Scope of this chapter

that are rather well performing on different metrics and hence represent a large class of industrially applicable systems. Broad considerations leading to their choice are already discussed earlier.

As listed in Tables 6.1 and 6.2, reactors classified as "three-phase slurry reactors" broadly fall under the following categories:

1 **Three-phase slurry bubble columns**, in which the fine catalyst particles are suspended in the liquid and a gas is sparged into the vessel. The gas rises due to the buoyancy difference between the gas and the liquid–solid slurry, driving a circulation. These reactors usually exhibit vigorous mixing pattern driven by cross-sectional variation in the gas volume fraction. Typical schematic of a three-phase slurry bubble column is shown in Figure 6.1a. Particle sizes are typically less than 500 μm.

2 **Three-phase fluidized beds** usually operate with higher catalyst loads and somewhat larger catalyst particle sizes. A comparable flux of gas (and sometimes liquid, in case liquid is also in continuous throughflow) together drives fluidization of the catalytic solids. Figure 6.1b shows a schematic configuration of a three-phase fluidized bed. In general, the solid particles processed in a three-phase fluidized bed are large in size, generally belonging to the Geldart B or D category (typically, larger than 500 μm).

3 **Three-phase agitated tanks**, in which external mechanical agitation with a stirrer causes the liquid–solid suspension along with sparged gas to be vigorously mixed within the reactor vessel. Various general configurations are possible, and Figure 6.1c shows the schematic of one such layout. Particle sizes used with these reactors are typically below 500 μm. The suspension of particles and the gas distribution are maintained by the mechanical stirring and the obstructive action of the baffles positioned along the wall of the vessel.

These "three-phase slurry reactors" are employed in a wide spectrum of chemical and process engineering operations. For instance, slurry bubble columns are employed extensively in a variety of processes for hydrogenation, oxidation, chlorination, hydroformylation, and bioremediation. Slurry bubble columns are also reactors of choice for some designs of natural and syngas conversion processes to liquid fuels, such as Fischer–Tropsch (FT) reaction and liquid-phase methanol synthesis. In some of the latter applications, use is made of the fact that as a whole, bubble columns have excellent heat transfer characteristics owing to the vigorous mixing that occurs in the liquid phase. Depending on the kind (and size) of catalyst, these processes may also be conducted in three-phase fluidized beds. Other applications involving the use of three-phase fluidized beds are hydrotreating of viscous and heavy residues in petroleum refining. There are also reports on three-phase fluidized beds being used for cell cultures and biochemical processes [1].

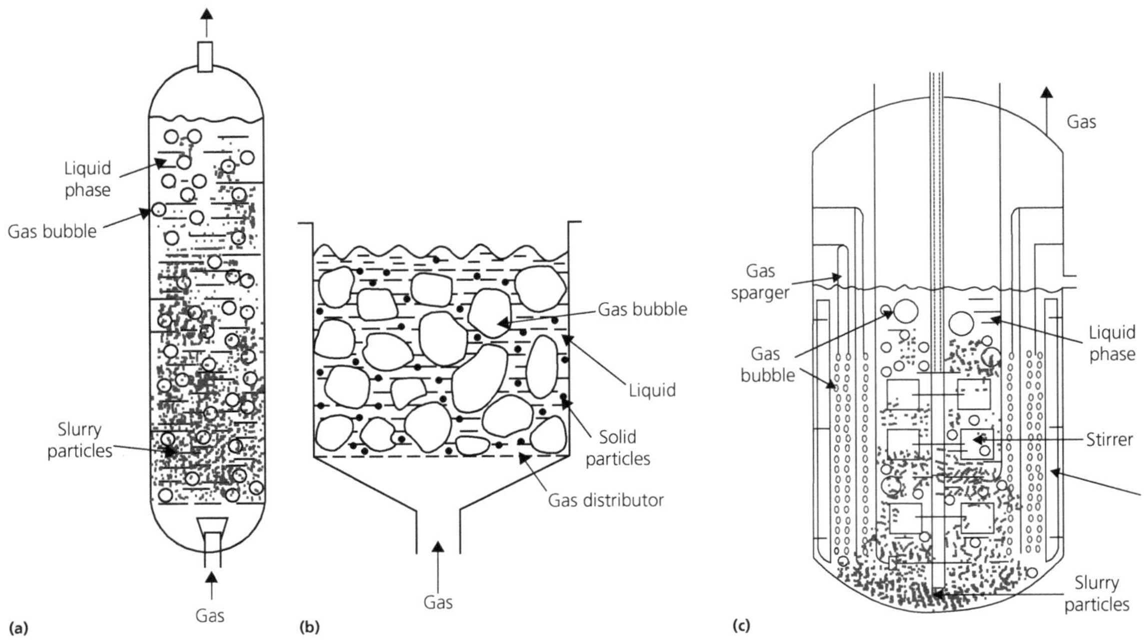

Figure 6.1 Schematic diagrams of industrial three-phase slurry reactors. (a) Slurry bubble column. (b) Three-phase fluidized bed. (c) Three-phase agitated vessel.

Three-phase stirred tanks with mechanical agitation ensure, as pointed out earlier, that all of the surface area on the catalyst particle is available for mass transport. These reactors find wide applications in catalytic hydrogenation, oxidation, ammonolysis, fermentation, and wastewater treatment. Sometimes, though not often, applications involving reacting solid particles are encountered, such as in wet metallurgy or in hydrometallurgical processes.

Table 6.3 is an illustrative list of various applications in which three-phase slurry reactors are used today and could potentially be used, detailing the system chemistry and process, catalyst types, and application sector of the economy. While this is not an exhaustive list, it is instructive to see the variety of existing and potential application areas of three-phase slurry reactors. Details about these processes may be found from the references cited in Table 6.3.

In terms of industrial use, the aforementioned three-phase slurry reactors are in themselves amenable for qualitative comparison in terms of their physical attributes and the various operating parameters. While the specifics of these attributes are determined by the process chemistry and detailed design (guidelines to which is discussed later in this chapter), Table 6.4 provides "at a glance" qualitative comparison of these attributes.

Table 6.4 shows the macroscopic attributes that may be achieved in these three-phase slurry reactors from a mechanical and configurational standpoint. For instance, in stirred slurry reactors, the action of the stirrer causes "chopping" of the bubbles emanating from the distributor, and hence the steady-state bubble size achieved in the reactor is largely determined by the breakage caused by the stirrer action. In slurry bubble columns and three-phase fluidized beds, however, fine bubbles emanate from the distributor, and as they rise, coalescence dominates and the bubbles increase in size (causing reduction in interfacial area per unit volume). In effect, the mechanics of these phenomena in the different reactor types determine the various transport parameters (such as interfacial area between gas and liquid, interfacial area between liquid and solid, effective diffusion length within the solid particle, etc.).

As discussed earlier, from profitability and volumetric productivity standpoint, we would like to minimize the transport resistances in all possible ways so that the complete activity of the catalyst for certain feed conditions and operating temperature would be realized. Prudent design of three-phase slurry reactor lies in effectively "matching" what is on offer from the mechanics of flow and the macroscopic design configuration, with the demands of the chemistry and the particle-/bubble-scale transport phenomena. In fact, the numerous applications listed in Table 6.3 are viable commercial processes for that reason. In what follows, either of these two viewpoints (i.e., the macroscopic viewpoint to generate a certain kind of phase distribution and mixing pattern, and the particle-/bubble-scale viewpoint to generate a certain concentration and temperature field that is optimal for the chemistry), are addressed to a limited degree. Our attempt has been to highlight the basic principles that are universally applicable to three-phase slurry reactor design, so that the reader may be guided to employ the ideas for developing designs specific to chosen applications.

6.2 Reactor design, scale-up methodology, and reactor selection

6.2.1 Practical aspects of reactor design and scale-up

By "reactor design," one typically envisions the evolution of reactor hardware and operating protocols which satisfy various

Table 6.3 Some illustrative applications of three-phase slurry reactors.

(A) Three-phase slurry bubble columns

S. no.	Reaction/process	Product	Catalyst	Application area	Reference
A1	Hydrogenation of *o*-nitroanisole	*o*-Anisidine	Pd/C	Dyes, fine chemicals	[2]
A2	Immobilized cell technology in synthetic and complex media	Thienamycin (β-lactam antibiotic)	*Streptomyces cattleya* enzyme	Pharmaceuticals	[3]
A3	Fischer–Tropsch synthesis for conversion of syngas ($CO + H_2$)	Liquid hydrocarbons	Fe or Co, with or without promoters (Cu, Mn, Zr, K, Na, Sc, Mo, W, Ru, Ti)	Energy and synthetic fuels	[4]
A4	Methanol synthesis for conversion of syngas ($CO + H_2$)	Methanol	—	Energy and synthetic fuels	[5]
A5	Immobilized cell action on MPI, MPII, and MPIII culture media (immobilized on Ca alginate, polyurethane sponge, active carbon, and pearlite)	Pigments (food colorants)	*Monascus purpureus*	Food industries	[6]
A6	Cell suspension cultures	Taxol (cancer treatment)	*Taxus cuspidata*	Pharmaceuticals	[7]
A7	Conversion of CO to organic acids via fermentation	Organic acids (acetic acid, butyric acid)	*Eubacterium limosum*	Energy	[8]
A8	Cell immobilization on loofa sponge for fermentation of sugar beet juice	Ethanol	*Saccharomyces cerevisiae*	Ethanol/energy	[9]
A9	Hydrodesulfurization of petroleum fractions	Sulfur-free petroleum fractions	$Co\text{-}Mo/Al_2O_3$ and $Ni\text{-}Mo/Al_2O_3$	Petroleum refining and upgradation	[10]
A10	Fischer–Tropsch (FT) process	Synthetic liquid fuel	Cobalt catalyst	Synthetic liquid fuels	[11]
A11	Aerobic biodegradation	Removal of *p*-nitrophenol (PNP)	Activated sludge	Wastewater treatment	[12]
A12	Fischer–Tropsch synthesis	Gasoline	Fe-HZSM5	Synthetic liquid fuels	[13]
A13	Absorption process	Carbon dioxide removal from flue gases	Magnesium hydroxide absorbent slurry	Greenhouse gas pollution control	[14]
A14	Adsorption of zinc from acid rock drainage	Treated surface water	Clinoptilolite adsorbent	Water pollution control	[15]

(B) Three-phase fluidized beds

S. no.	Reaction/process	Product/process	Catalyst/enzyme/ biological species	Application area	Reference
B1	Hydrocracking of unstable feedstock (shale oil)		$Ni\text{-}Mo/Al_2O_3$	Energy and synthetic fuels	[16]
B2	Photocatalytic oxidation of methyl orange	Removal of methyl orange	TiO_2 powder	Pollution prevention	[17]
B3	Catalytic wet oxidation of alkaline lignin extracted from sugarcane bagasse	Aromatic aldehydes	Pd/alumina	Fine chemicals	[18]
B4	Hydrogen production in a three-phase fluidized bed bioreactor	Hydrogen	H_2-producing sewage sludge immobilized on ethylene vinyl acetate copolymer	Fuels from waste	[19]
B5	Photooxidation of sodium lauryl sulfate	Removal of sodium lauryl sulfate (surfactant)	TiO_2/SiO_2 photocatalyst	Pollution prevention—degradation of surfactants	[20]
B6	Simultaneous adsorption and ozone decomposition of phenol	Removal of phenol from wastewater	Activated carbon granules (adsorbent) and O_3 produced *in situ* (for oxidation)	Environmental	[21]
B7	Production of enzymes	Coenzyme Q10	Gel-entrapped *Sphingomonas sp.* ZUTE03	Pharmaceuticals	[22]
B8	Extractive fermentation of L-lactic acid	Lactic acid	Immobilized *Rhizopus oryzae*	Food, pharmaceutical, leather, textile industry stock	[23]
B9	Biodegradation of diesel fuel-contaminated wastewater	Removal of diesel fuel	Biomass supported on lava rock particles	Wastewater treatment	[24]
B10	Mechanical disruption of cells	Intracellular substances such as proteins, fats, and enzymes	Baker's yeast cells (*Saccharomyces cerevisiae*)	Bioprocessing	[25]
B11	Synergetic removal of aqueous phenol	Phenol removal from wastewater by simultaneous action of ozone and activated carbon	Activated carbon adsorbent and ozone in gas phase	Wastewater treatment	[21]

(Continued)

Table 6.3 *(Continued)*

B12	CO_2 capture from combustion flue gas in a three-phase fluidized bed carbonator	Carbon dioxide removal from flue gas	$CaO/CaCO_3$ carbonator/calciner	Powder plants, cement plants, steel mills, refineries	[26]
B13	Aqueous phenol decomposition	Phenol removal from industrial wastewater	TiO_2 on silica beads, Ni or Co on carbon beads	Wastewater treatment	[27]
B14	Ethanol production in a three-phase fluidized bed bioreactor	Ethanol	Yeast	Energy, commodity chemicals	[28]
B15	Adsorption of arsenic from wastewater	Removal of arsenic	Cashew nut shale waste	Wastewater treatment	[29]

(C) Three-phase agitated vessels

S. no.	Reaction/process	Product	Catalyst	Application area	Reference
C1	Hydrogenation of glucose	Sorbitol	Raney—Ni	Pharmaceuticals	[30]
C2	Hydrogenation of *m* nitrochlorobenzenes	Chloranilines	Pt/C—sulfided	Pharmaceuticals and dyes	[31]
C3	Ethynylation of formaldehyde (Reppe process)	Butynediol	Copper acetylide	Fine chemicals	[32]
C4	Hydrogenation of 2,4-dinitrotoluene	2,4-Diaminotoluene, 4-hydroxylamino-2-nitrotoluene, 2-amine-4-nitrotoluene, 4-amino-2-nitrotoluene	Pd/alumina	Fine chemicals	[33]
C5	Hydrogenation of *o*-cresol	2-Methylcyclohexanol	Ni/SiO_2	Fine chemicals and intermediates	[34]
C6	Fischer–Tropsch synthesis	Methane to dodecane range of hydrocarbons	Iron	Synthetic fuels	[35]
C7	Hydrogenation of *m*-nitrochlorobenzene	rn-Chloroaniline	Sulfided Pt/C	Intermediates for dyes, drugs. pesticides, pharmaceuticals	[36]
C8	Partial hydrogenation of 1,5,9-cyclododecatriene	Cyclododecene	Pd/γ-alumina	Industrial intermediates	[37]
C9	Selective catalytic oxidation of glucose	Gluconic acid	Pd/Al_2O_3 catalyst	Food and pharmaceuticals	[38]
C10	Selective hydrogenation of 1,5,9-cyclododecatriene (CDT)	Cyclododecene (CDE)	Pd/Al_2O_3	Fine chemicals and intermediates	[39]
C11	Hydrogenation of *o*-aminophenol	*o*-Aminophenol (intermediate for dyes, drugs, and pesticides)	Pd/carbon	Fine chemicals	[40]
C12	Immobilized cell culture of *Streptomyces*	Kasugamycin (antibiotic)	Continuous culture of *Streptomyces kasugaensis* one Celite beads	Pharmaceuticals	[41]
C13	Hydrogenation of dinitriles	Diamines	(Raney) nickel	Fine chemicals and industrial intermediates	[42]
	Hydroxylation of iso-octenes (2,4,4-trimethyl-1-pentene (1-TMP) and 2,4,4-trimethyl-2-pentene (2-TMP))	Iso-octane (2,2,4-trimethylpentane (TMPA))	Pd/γ-alumina	Fuel additives	[43]
C14	Hydroxylation of benzene	Phenol	MFI structure titanium silica-based catalyst (TS-PQ™)	Industrial intermediates	[44]
C15	Bacterial cellulose production in an agitated culture	Bacterial cellulose	*Gluconacetobacter hansenii*, isolated from rotten apples	Clothing, pulp, and paper industry	[45]

process demands and economic interests, within the constraints set by the reaction kinetics and the multiscale transport resistances, without compromising safety, and with minimal environmental footprint. By "scale-up," we refer to the act of transforming the information from one scale of reactor to another scale (either from a laboratory-scale to pilot- and commercial-scale reactor or from commercial-scale reactor to pilot- and laboratory-scale reactor). Scale-up is needed for transforming the newly developed technology at the laboratory-scale reactor to the commercial-scale reactor or simply to interpret and extrapolate laboratory results to the commercial-scale reactor. Reverse case, known as "scale-down," is also equally important that mimics the large-scale operating process at a much smaller scale. This is used for the diagnosis of existing operating reactor using appropriate experiments on the laboratory scale.

Scale-up and scale-down may be needed for various reasons. Some of the important reasons can be:

Table 6.4 Overview of vessel designs and performance attributes of three-phase slurry reactors.

	Stirred slurry reactor	Slurry bubble column reactor	Three-phase fluidized bed reactor
Height-to-diameter ratio	1–3	1–20	1–10
Prime driver for mixing	Impeller rotation	Sparged gas	Sparged gas
Gas distributor	Ring sparger/gas inducing impeller	Bubble caps/sieve plate/branched sparger	Bubble caps/sieve plate/branched sparger
Catalyst suspension/ utilization	Critical impeller speed/nearly uniform suspension	Critical gas velocity/nearly uniform suspension but inferior than STR	Critical gas velocity/nonuniform solid suspension
Bubble size	Breakage dominated	Coalescence dominated	Coalescence dominated
Gas-phase conversion	High	Medium	Low
Liquid-phase conversion	Medium	High	High
Liquid-phase residence time distribution	Back mixed	Partially back mixed	Partially back mixed
Temperature uniformity	Excellent	Very good	Good

- Setting up of a new unit based on new or existing technology
- Modification in the existing plant or operational reactor
- Increasing the capacity of the existing reactor
- Modification in the feed or catalyst owing to various reasons
- Improvement in the product quality

Both design and scale-up exercises necessarily involve expertise from various fields ranging from thermodynamics, chemistry, and catalysis to reaction engineering, fluid dynamics, fluid mixing, and heat and mass transfer. A reactor engineer needs to combine the basic understanding of the chemistry and catalysis of interest, with the methods discussed in this book to evolve a suitable design for any reactor, and indeed the three-phase slurry system. More often than not, specific scale-up/scale-down methodologies need to be used to establish confidence in the developed design approach. Sometimes, it may be necessary to use intermediate scales between laboratory and commercial scales, especially when transformation of the information from one scale to the other is not straightforward. Appropriate scaling methodology is required depending on the specific objectives of the scaling exercise. Since three-phase slurry reactors are characterized by a rather complex hydrodynamics and multiscale transport processes, the scale-up of these systems is particularly challenging sometimes without clear-cut and fool-proof procedures.

The overall reactor design and scale-up exercise at the very basic level requires the decision with respect to the kind of reactor (reactor selection), the choice of internals and physical hardware that go into the reactor unit, and finally the decision and "recipe" of proper operating protocols. Often, these are practical decisions, which must in turn be based on established theories, firm heuristics, and operating experience. The practicing reactor engineer has to interact with chemists to understand the basic chemistry and peculiarities of the catalyst. Based on such understanding and proposed performance targets, the reactor engineer has to conceive these aforesaid suitable reactor hardware and operating protocols. A typical "wish list" of a reactor engineer for the new process can be:

- Operability within technologically feasible region
- Intrinsically safe operations
- Environmentally acceptable
- Maximum possible conversion of the feedstocks
- Maximum selectivity of reaction to the desired products
- Lowest capital and operating costs

Typical "wish list" for enhancing the performance of the existing reactor technology can be:

- More throughput per unit volume
- Improved selectivity and better quality product
- Safer operation
- Reduced energy consumption
- More environment-friendly operation

It may often turn out that some of the items in the wish list require contradictory options of hardware and operation. In such a case, a careful analysis of different items in wish list must be made to assign priorities. Operability, stability, and environmental constraints often get precedence over conversion and selectivity when such conflicting requirements arise.

Reactor design is an iterative procedure in which a preliminary design is developed based on inputs from a limited number of laboratory experiments. However, consideration of the different transport parameters and detailed exploration of kinetics aspects (such as catalyst life and deactivation) and hydrodynamic aspects (such as local volume fraction of phases, interfacial area available for mass transfer, etc.) may often lead to revision of the preliminary design either in a sequential or parallel logic. Figure 6.2 provides an overview of the procedures, though it must be said that the sequencing of these steps may depend largely on the specific application being considered. Experiments in test equipment at multiple scales, such as pilot plants, and appropriate models addressing the phenomena at various scales (such as phenomenological one-dimensional reactor models, multizone models, CFD models, etc.) are essential to develop a comprehensive reactor design toolkit.

Reactor design starts with quantification of influence of the reactant flow rate and operating temperature on the performance of the reactor (which has a direct impact on conversion of reactants and selectivity of desired products and has a secondary influence on stability). The first step in such analysis is formulating a mathematical framework to describe the rate (and mechanism) by which one chemical species is converted into

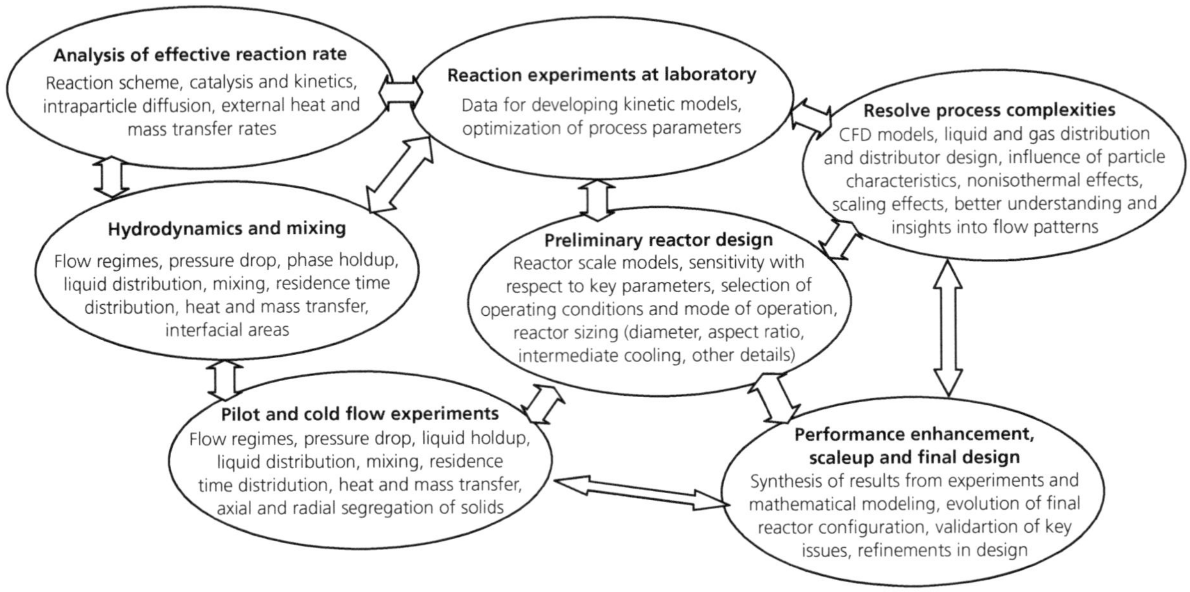

Figure 6.2 Schematic of the overall methodology for relating performance of the reactor with the hardware and the operating protocol.

another in the absence of any transport limitations (chemical kinetics). Kinetics of reactions needs to be determined from experimental measurements. Measuring the rates of chemical reactions in the laboratory is itself a specialized subject of chemical engineering. More information about laboratory reactors that are used for obtaining intrinsic kinetics can be found in textbooks like Ramachandran and Chaudhari [46], Smith [47], Levenspiel [48], and Doraiswamy and Sharma [49].

Once the intrinsic kinetics is available, the effective rate analysis can be carried out using the models discussed in the following sections. Typically, this may involve various scales. For instance, in the specific context of three-phase slurry systems, one has to develop mathematical description of transport effects at the scale of a particle in slurry and the gas–liquid bubble interface scale and possibly in clusters of particles (if relevant). At another level, one needs to be concerned about the macro-mixing of phases and conventional reaction engineering flow models involving ideal reactors and combinations thereof or more advanced phenomenological models or, more recently, CFD-based models. Clearly, these suites of models need to be all integrated, so that the impact of events at the particle or bubble scale can be adequately captured at the macroscale model. This exercise results in establishing relationships among operating conditions (flow rate, mode of operation, pressure, composition) and reactor hardware (reactor volume and reactor configuration) with reactor performance. This kind of information directly contributes to physical dimensioning and engineering of the reactor or its subparts and can hence be designated as "design" models. Subsequently, however, several "learning" models have to be developed to understand various reactor engineering issues like start-up and shut-down dynamics, multiplicity and stability of thermochemical processes occurring in the reactor, sensitivity of reactor performance with respect to mixing and residence time distributions (RTDs), and selectivity and by-product formations. This chapter concerns itself mainly with the design models; however the issue of learning models has been flagged here because of their prime importance in effective functioning of three-phase slurry reactor systems.

After establishing such a model and heuristics-based understanding and analysis of the reactor system, the next most important question facing the reactor engineer is to evaluate the consequences of the assumptions involved in such models for estimating the behavior of actual reactor. Questions like operability and stability of the flow regime need to be answered. Mixing, particle suspension and mixing, and heat and mass transfer are intimately related to flow regimes and distributors of gas and liquid. The mixing in actual reactor may significantly deviate from that assumed for the reaction engineering models. This deviation can be caused by, say, inappropriate particle suspension and mixing or by the formation of stagnant regions within the reactor. Residence Time Distribution (RTD) data and analysis are useful to determine the bounds and limiting solutions of reactor performance. It must be remembered that more than one model may fit the observed RTD data. A general philosophy is to select the simplest possible model, which adequately represents the physical phenomena occurring in the actual reactor. In recent years, the availability of CFD modeling tools has significantly enhanced the ability to predict pathological flow patterns that may exist in multiphase flow vessels and indeed in three-phase slurry systems.

It may be noted that even the small-scale hardware details like design of feed nozzles or weirs and reactor internals (cooling coils, baffles, etc.) may have dramatic influence on reactor performance. The issues of scale-up and scale-down may have to be resolved by following an iterative methodology. Knowledge of requirements of process chemistry, engineering creativity, experience, and accumulated empirical information are generally used to evolve reactor configuration and designs.

The understanding gained by development and implementation of these steps will be helpful to identify the needs for developing more sophisticated simulation models/data for establishing the desired slurry reactor design. At this stage, it is important to identify gaps between available knowledge and that required for fulfilling the "wish list." The identified gaps can then be bridged by carrying out experiments in the laboratory and/or pilot plant(s) and by developing more comprehensive fluid dynamic models. Despite the advantages, conventional chemical reaction engineering models are not directly useful for understanding the influence of reactor hardware on reactor performance. For example, the effect of distributor design on the distribution of gas and solid particles in a slurry bubble column and thereby the reactor performance will be difficult to predict without developing a detailed fluid dynamic model of the reactor or without carrying out experiments on a scale model. At this stage, Computational Fluid Dynamics (CFD)-based approach will make valuable contributions by providing the required insight, by helping to devise right kind of experiments, and by allowing screening of alternative configurations and tools for extrapolation and scale-up.

In the following sections, the reactor design for three-phase slurry reactors is addressed in the following manner. First, generic models at the particle (catalyst) scale, for gas–liquid reactions in general, are presented (Section 6.2.2). In general, these models are similar for the three kinds of three-phase slurry systems discussed in Section 6.1, the specific differences stemming from the relative sizes of solid particles and bubbles in the sparged gas. Section 6.3 will relate to further refinements of the global mixing models. These will be specific to the contactor of choice, that is, slurry bubble columns, three-phase fluidized beds, and three-phase stirred tanks. Section 6.4 relates to the estimation of transport parameters that are required for the particle scale and the mixing models. A final refinement of the reactor modeling exercise, very useful for process diagnostics and for projecting the flow patterns in new designs, is to develop detailed CFD models. This is presented in Section 6.5. Together, Sections 6.2, 6.3, 6.4, and 6.5 provide an arguably complete toolkit for the design and scale-up.

6.2.2 Transport effects at particle level

The chemical reactor is an expensive real estate, a specially designed volume or space in which various phases and streams are brought together for bringing chemical species contained therein into intimate contact so that they can react and produce a value-added product. For doing so, optimal conditions (such as temperature and pressure) are provided by this specially designed reactor volume. In continuous mode of operation, some mechanical means (such as pumping or stirring or gravity-driven flow) are used to make these streams come in contact, and this causes a spatial variation of distribution in the species concentrations. In transient systems or under dynamic operation, there are temporal variations as well. These spatial and temporal variations in concentrations of chemical species lead to sub-optimal utilization of the reactants or their conversion to the desired product. Also, the finite capacities of the reactor volume and the mechanical devices or processes that cause the phase contact and the requirements of achieving a certain processing rate often limit both the residence time of phases in the reactor vessel and the time of contact between the phases. As a whole, this act of bringing together phases within the reactor volume leads to sub-optimal utilization of incoming process streams. Thus, the capability of a reactor design in effecting a certain process depends largely on its ability to realize the "best possible" contacting pattern, provide adequate residence and contact time, and hence realize the maximum potential of the activity that is "locked in" in the catalyst. This idea has been lucidly quantified in the concept of "reactor efficiency" by Ramachandran and Chaudhari [46], which is defined as the ratio of total amount of desired product produced to the amount of desired product produced if all the catalysts were exposed to the reactant concentrations at the reactor inlet. Reactor efficiency is one of the measures that reflect the effect of flow pattern of the phases involved and also adequately respects the multistep kinetics, the multitude of timescales involved particle-scale transport, and reactor- or vessel-scale flow patterns.

The global objective of any reactor design exercise, and indeed design of three-phase slurry reactors, should be to maintain the optimal species concentrations and optimal contacting patterns between phases at all locations (or maximal region) inside the reactor. Clearly, that will lead to optimal utilization of the reactor space and the running costs, such as energy inputs, cooling requirements, etc., that are necessary for the process. Based on these facts, there must be at least two levels of design:

(I) At the level of catalyst and its contact with the gas bubbles and with continuous liquid or with liquid droplets and continuous gas

(II) At the level of reactor, including (but not limited to) the introduction of reactants streams and phases, their flow in the vessel, and their exit, as well as the energy input strategies

In addition to the aforementioned levels of design, one also has to consider the hydrodynamic peculiarities of the system, the existence of so-called flow regimes, which are of mechanical origin (fluid and particle mechanics of phases) but have an impact both at the catalyst- and reactor-level distribution of chemical species and on the operability and stability of the reactor. As briefed in Section 6.2.1, the latter consideration may outstrip the considerations of optimal reactor volume calculations in some cases.

What is proposed earlier is also in line with the three levels of reactor engineering discussed by Krishna and Sie [50] and indeed is eminently applicable in the context of three-phase slurry reactors. Naturally, the goal here is to decide on a flow pattern that optimally utilizes the catalyst. In other words, the catalyst has a certain intrinsic activity, and the contacting pattern should try and realize that activity in all parts of the reactor. Thus, level [I] design explained earlier must establish the effective performance metric at the catalyst level, which will be the major topic of discussion in this section.

Reactions that are performed in three-phase slurry systems can be generically described by a gas-phase reactant (designated as A), which dissolves in liquid and then reacts on the solid catalyst surface with a reactant B, which is in the liquid phase (for simplicity, we assume that it is practically nonvolatile and hence does not enter the gas phase). The product formed is typically in the liquid phase (which, in slurry systems, completely wets the catalyst). A typical example of such a reaction system can be the hydrogenation of unsaturated hydrocarbons, with hydrogen being in the gas phase. The catalyst, which may be native metal or metal impregnated on a porous support like alumina, is suspended in the slurry so that, after dissolving in the liquid film, the gas meets the unsaturated hydrocarbon molecules from the liquid phase on the catalyst surface (outer surface in the case of native metal catalyst, and by diffusing inside the pores of the catalyst in case the catalysts is metal impregnated on porous support). Subsequent to the reaction, the saturated hydrocarbons are released into the liquid phase. The reaction can be schematically depicted as:

$$\beta A(g) + \gamma B(l) \xrightarrow{\text{solid catalyst}} \textit{products} \tag{6.1}$$

It is possible that the actual chemistry of the reaction may be significantly more complex than what is shown in Equation 6.1, involving many species and steps. However, through appropriate lumping schemes, the kinetics may be simplified to individual gas- and liquid-phase reactants and products, with all steps eventually describable in a form presented in Equation 6.1. Thus, without any loss of generality, Equation 6.1 is completely descriptive of any general gas–liquid reaction scheme in a three-phase slurry reactor.

Naturally, for the aforementioned reaction scheme to be realized, one needs to have the presence of all the three phases (gas, liquid, and solid) in the reactor. When the stoichiometric requirement of the reaction demands a lot of liquid and less gas, the reactor of choice is the slurry bubble column (Figure 6.1a) or the three-phase stirred tank (Figure 6.1c). Sometimes, the external mass transfer limitations around catalyst particles is limiting (discussed in greater detail later), in which case the choice is clearly the stirred tank wherein mechanical stirring action is practically independent of the bubble motion-induced agitation (which is arguably limited in a typically slurry bubble column) and can be independently set by using an appropriate impeller/motor. In case the stoichiometric requirement is for higher fraction of gas and less liquid, the reactor system is the three-phase fluidized bed. The latter supports large gas-to-liquid ratios and also larger particle sizes (500 μm to a few millimeters), while the former reactors usually employ finer particles (1–200 μm range).

Ideally, purely from a reactor efficiency point of view, the ideal three-phase reactor would be a countercurrent one (shown schematically in Figure 6.3a), in which gas and liquid (with species A and B, respectively) enter from opposite directions, and where there is higher concentration of A (near gas inlet), one would have a depleted liquid stream (lower concentration of B) and vice versa at the other end. The catalyst particles would be suspended in slurry phase, and with this countercurrent trick, one would ensure relatively uniform rate on the catalyst particles no matter what their locations in the vessel. The ideal contacting flow pattern involves the countercurrent movement of gas and liquid (slurry) phases in a plug flow manner.

However, from a hydrodynamics point of view, this is very difficult to run. Primary reasons for this difficulty comes from the fact that countercurrent systems always have flooding limitations, and the window of flow rates for stable operation is relatively small. Thus, the typical choice for stable operation of slurry reactors is co-current systems (Figure 6.3b), in which the gas and liquid is arranged to flow concurrently and the

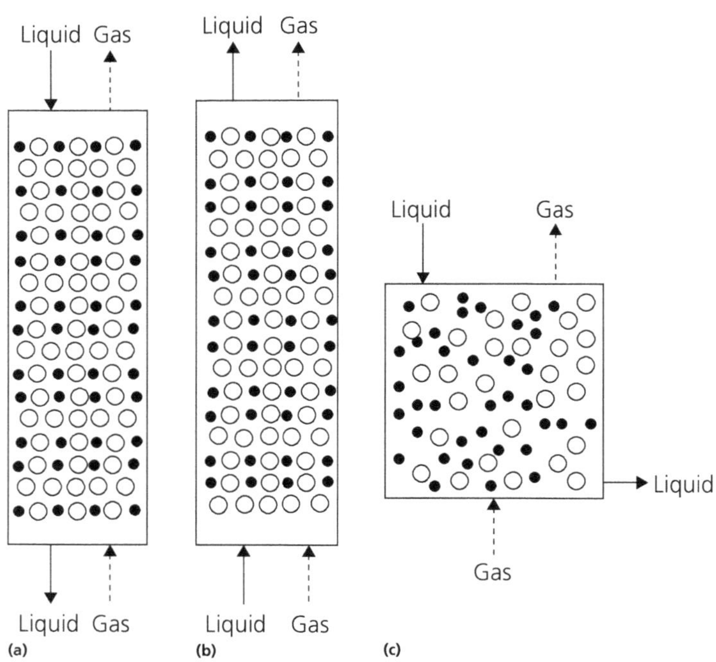

Figure 6.3 Possible ideal contacting patterns in three-phase slurry reactors. (a) Countercurrent (gas and liquid in plug flow). (b) Co-current (gas and liquid in plug flow). (c) Mixed (gas and liquid in mixed flow).

catalyst particles are suspended in slurry. The target is to have both gas and liquid in plug flow. Countercurrent flow mode leads to a somewhat less favored concentration distribution of reactants in the vessel: at the inlet plane, both gas and liquid species are at their respective highest concentrations, which progressively deplete as one moves toward the exit. At a result, the outlet zone of the reactor cannot be utilized effectively. However, in the interest of feasible and stable operation from a hydrodynamic standpoint, the co-current contacting pattern is usually the pattern of choice. Both the three-phase fluidized bed and the slurry bubble column reactor adhere to this contacting philosophy (compare Figure 6.3b with corresponding depictions in Figure 6.1a and b). However, there may be situations when either the stoichiometric requirement of relative species concentrations or the requirement of maintaining a certain gas-to-liquid flow ratio or certain transport limitations or requirements of heat addition or removal may supersede the requirements of plug flow patterns. Hence, the only feasible way (as well as ensuring stable operation) is to contact the phases to have mixed flow of gas and liquid, which is achieved in a three-phase stirred tank. Figure 6.3c shows such a contacting pattern (compare with Figure 6.1c).

No matter what the contacting pattern is, the relative configuration of a gas bubble and a catalyst particle in its immediate neighborhood is practically the same in any of the configurations shown in Figures 6.1, 6.2, and 6.3. Thus, if any of the slurry reactor systems are viewed as an assemblage of small control volumes, then each control volume will be composed of an ensemble of gas bubbles and catalyst particles separated by a small amount of liquid. For the reaction to occur, the gas-phase reactant (A) has to dissolve in the liquid (after overcoming a possible transport resistance on the gas side of the bubble), overcome a transport resistance in the liquid side of the bubble, diffuse through the bulk liquid, overcome a mass transfer resistance around the catalyst particle, and diffuse inside the particle and react at the active sites with the liquid-phase reactant (B). The reactant B would have arrived at the active sites with a similar history, except that it has to only overcome the film resistance around the particle and the internal diffusional resistance inside the catalyst particle before reaching the active site. This picture of gas–liquid–solid reaction in a slurry reactor is shown in Figure 6.4 and is extensively discussed with in standard texts on reactor engineering, such as those by Ramachandran and Chaudhari [46] and Levenspiel [48].

If reaction (6.1) is elementary, then the intrinsic kinetic rate is given by

$$-r_{A,cat} = \frac{\beta}{\gamma}(-r_{B,cat}) = kc_A c_B \tag{6.2}$$

where k is the reaction rate constant (units: $m^6_{liq}/((mol)(m^3_{cat})(s))$) expressed in terms of unit volume of catalyst. However, when the transport picture described earlier and schematically depicted in Figure 6.4 is described mathematically using a set of algebraic equations and assuming steady state within a small control volume within the reactor, the following expression for effective volumetric rate of depletion of reactant A from the liquid phase in the reactor results [46, 48]:

$$-r_{A,rxtr} = \frac{\beta}{\gamma}(-r_{B,rxtr}) = \frac{p_{Ag}}{\left[\dfrac{1}{k_{Ag}a_{g-l}} + \dfrac{H_A}{k_{Al}a_{g-l}} + \dfrac{H_A}{k_{As}a_{l-s}} + \dfrac{H_A}{k\bar{c}_B\eta_A\alpha_s}\right]} \tag{6.3}$$

Equivalently, the effective volumetric rare of depletion of reactant B from the liquid phase in the reactor would be [46, 48]

$$-r_{B,rxtr} = \frac{\gamma}{\beta}(-r_{A,rxtr}) = \frac{c_{Bl}}{\left[\dfrac{1}{k_{Bs}a_{l-s}} + \dfrac{H_A}{\left(\frac{\beta}{\alpha}k\right)\bar{c_A}\eta_B\alpha_s}\right]} \tag{6.4}$$

Equations 6.3 and 6.4 have embedded in them several parameters that are design parameters in the three-phase slurry

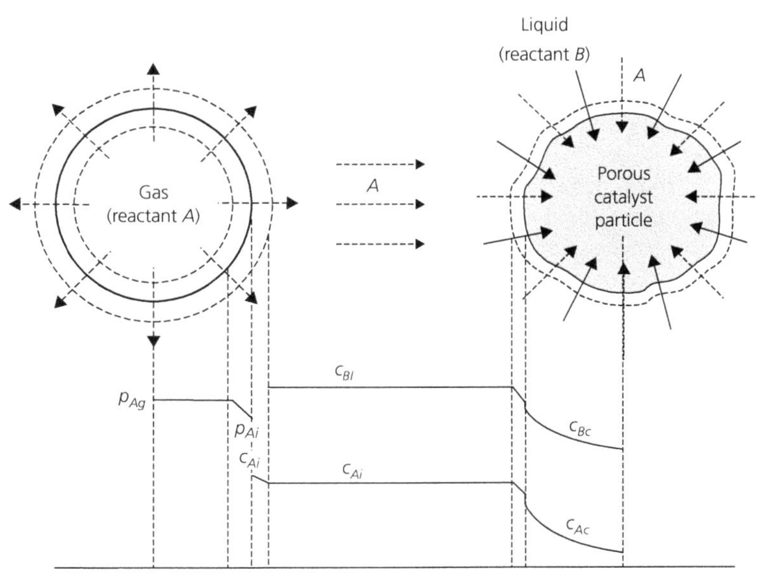

Figure 6.4 Reactant profiles in neighborhood of gas (bubble) and catalyst particle.

systems. All the independent variables in Equations 6.3 and 6.4 are measurable, such as partial pressure of gas-phase reactant A, p_{Ag}, and liquid-phase reactant B, c_{Bl}. η_A and η_B are the catalyst effectiveness factors [48] for reactants A and B, respectively, and represent the effect of mass transfer resistance inside the catalyst pellets owing to the finite porosity and tortuosity. In general, η_A and η_B would depend on the reaction orders governed by the intrinsic chemistry (for instance, for the rate form in Equation 6.2, they would be for first-order reaction for the corresponding components), but by using the generalized form reported by Hong et al. [51], it is possible to calculate these effectiveness factors for generalized n-order and Langmuir–Hinshelwood-type rate kinetics.

In Equation 6.3, the various terms in the denominator term on the right-hand side represent various controlling resistances shown in Equation 6.3. The first term in the denominator represents the mass transfer resistance to the transport of A on the gas side, the second term represents the mass transfer resistance to the transport of A on the liquid side (which has embedded the effect of finite solubility of A through Henry's law constant), the third term represents the resistance at the liquid film around the catalyst particle, and finally the fourth term represents the diffusional resistance inside the particle. For the liquid-phase reactant B (Eq. 6.4), only the last two resistances are relevant.

Even at this early stage, it is clear that since the target of any design exercise should be to maximize k_{Ag}, k_{Al}, and k_{Bl} and also a_{g-l} and a_{l-s}, the first three terms in Equation 6.3 (and first term in Equation 6.4) vanish in relation to the last term. The objective should also be to maximize η_A and η_B (and bring them as close to unity as possible), so that the entire effort toward maximization of effective reaction rate is reduced to choosing an optimal combination of a_s, a_l, and a_g. Getting an optimal combination of phase fractions depends on the mechanics of the flow and the relative flow properties like density and viscosity (in case of the fluid phases), and granular properties (in case of the solids). Maximizing mass transfer coefficients is possible by enhancing the phase slip velocities so that the film thickness (boundary layers) is reduced and the mass transfer resistances diminish. Moreover, the specific interfacial gas–liquid interfacial area may be enhanced by having fine bubbles. High catalyst effectiveness and high liquid–solid interfacial area are achieved with finer particles. Often, these methods of rate maximization put up contrary requirements and the attempt to enhance one such quantity enforces limits on the other, since the hydrodynamics effectively connects all these design variables.

In Equations 6.3 and 6.4, $\bar{c_B}$ and $\bar{c_A}$ represent the average concentrations of the reactants inside the catalyst pellets. In general, this is not known *a priori*; thus while these equations are complete in principle, they cannot be directly used for design purposes. However, several limiting forms of Equations 6.3 and 6.4 may be deduced in cases when certain of these limiting conditions hold. For example, in case the catalyst substrate is highly porous and hence internal mass transfer resistance can be ignored, then the average inside concentration of the reactants is equal to the surface concentration or $\bar{c_B} \cong c_{Bs}$ and $\bar{c_A} \cong c_{As}$. Then Equations 6.3 and 6.4 reduce to, respectively:

$$-r_{A,rxtr} = \frac{\beta}{\gamma}(-r_{B,rxtr}) = \frac{\left[p_{Ag} - H_A c_{As}\right]}{\left[\dfrac{1}{k_{Ag}a_{g-l}} + \dfrac{H_A}{k_{Al}a_{g-l}} + \dfrac{H_A}{k_{As}a_{l-s}}\right]} \tag{6.5}$$

$$-r_{B,rxtr} = \frac{\gamma}{\beta}(-r_{A,rxtr}) = k_{Bs}a_{l-s}(c_{Bl} - c_{Bs}) \tag{6.6}$$

Pore resistance may be negligible for one or both reactant species. In any case, the surface concentration will be unknown (which is really the same level of unknown as not knowing $\bar{c_B}$ or $\bar{c_A}$). However, since the intraparticle diffusion problem is eliminated, one solves simultaneously Equations 6.3 and 6.6, or Equations 6.4 and 6.5, or Equations 6.5 and 6.6. The solution can be carried out by employing an iterative procedure (initiated by guessing the unknown surface concentration), until both the individual rates of depletion (A and B) match.

Another set of simplifications can be made if one reactant is in far excess of the other. For instance, if the liquid phase involves pure reactant B and A is a sparingly soluble gas (in other words, $(p_{Ag}/H_A) << c_{Bl}$), then it may be assumed that the same concentration of B exists everywhere, whether in the liquid or inside the catalyst or on its surface. This may happen, for instance, if one wishes to do catalytic oxidation of liquid hydrocarbons, in which oxygen from air is sparingly soluble in organic phase (A), while the liquid hydrocarbon itself (B) is present in far excess. In such a situation,

$$-r_{A,rxtr} = \frac{p_{Ag}}{\left[\dfrac{1}{k_{Ag}a_{g-l}} + \dfrac{H_A}{k_{Al}a_{g-l}} + \dfrac{H_A}{k_{As}a_{l-s}} + \dfrac{H_A}{(kc_{Bl})\eta_A\alpha_s}\right]} \tag{6.7}$$

An opposite situation happens when the liquid phase is very dilute with B, while gas-phase reactant A is highly soluble (or at high pressures). A common example of such a situation is when slurry systems are employed for hydrodesulfurization of oil fractions, in which ppm levels of refractory sulfur compounds (such as dibenzothiophenes and alkyl dibenzothiophenes) may be present in fuel oils like diesel. In comparison, the gas phase is pure hydrogen, which at high pressures behaves like a fairly soluble gas in the hydrocarbon liquid. In such a case, the liquid-phase concentration of the gas-phase reactant is fairly constant throughout (and much higher than the concentration of the liquid-phase reactant B), including inside the catalyst pellets. The effective depletion rate of B (Eq. 6.4) reduces to

$$-r_{B,rxtr} = \frac{c_{Bl}}{\left[\dfrac{1}{k_{Bs}a_{l-s}} + \dfrac{H_A}{\left(\frac{\beta}{\alpha}k\frac{p_{Ag}}{H_A}\right)\eta_B\alpha_s}\right]} \tag{6.8}$$

Naturally in all cases, if the gas phase happens to be a pure species (such as in hydrogenation reactions), the gas-side film resistance also drops out and the expressions become simpler. It is also the case that for most real reactor situations, one or more limiting mass transport or reaction regimes can be identified, so that the general rate expressions 6.3 and 6.4 can be considerably simplified. Also, there may be situations in which the reaction chemistry, even after suitable lumping, is considerably more complex than Equation 6.2. In such cases, it is not trivial to obtain expressions like those listed earlier; however philosophically the same approach may be followed.

Once the effective rate forms at the particle/bubble level are established, and flow patterns as assumed in Figure 6.2 are available, one simply uses these effective rate expressions to write down the corresponding steady-state material balances for the reactor for the assumed flow patterns. Under steady-state conditions, this involves either first-order ordinary differential equations for the phases in which plug flow is assumed or simple difference equations in species concentration in phases in which completely mixed flow is assumed. The treatment in all these cases is very similar to what will be in a single-phase reactor (see, e.g., Ref. [48]), except that one has a separate differential equation balancing for each species concentration and they are coupled through the effective reaction rate term.

In reality, however, the hydrodynamics of the different phases makes the flow pattern very much more complex than what is depicted in Figure 6.3b and c. It is difficult to predict and equally challenging to measure, and hence several of the design parameters in Equations 6.3–6.8, such as the mass transfer coefficients and the interfacial areas, cannot be determined precisely. Moreover, since both quantities are coupled, phase volume fractions and the local slip velocity vary from point to point within the vessel. This, in turn, affects the overall particle-/bubble-scale reaction rates. Furthermore, in some of the flow regimes, this variation is also dynamic (i.e., varies with time, sometimes periodically, even when the overall flow system is maintained in steady state, such as with constant pumping speeds and so on). Thus, it is a challenge to estimate the governing parameters such as phase holdup interfacial areas and mass transfer coefficients. Conventionally these have been done with correlations (that are often developed in laboratory-scale vessels where the prevailing flow regime or condition is quite different from the industrial settings). In more recent studies, the attempts involve the development of phenomenological models for the flow patterns. More contemporary efforts are focused on computational fluid dynamics (CFD) tools and models, which are discussed in Section 6.5.

In may be noted that a level [I] reactor selection can be done even with the effective reaction rate expressions (Equations 6.3 and 6.4). For instance, one should always attempt to select a reactor that helps to quicken the otherwise slowest step in the effective rate. For instance, if internal diffusion within catalyst particles is the limiting step, then one has to use fine particles in a slurry bubble column. If liquid–solid mass transfer is limiting, then one may also choose fine particles but also try to enhance the slip velocity around them, which can be done by vigorous stirring and maintaining a large solid fraction in suspension in a fluidized state. This is best achieved with a slurry stirred tank. If gas phase is dilute in reactant and a large volume of gas needs to contact the solid with less liquid, then naturally the fluidized bed configuration is required. Coupled to these choices is also the issue of contacting, and the general attempt should be to keep the limiting reactant in plug flow. In general, when all other factors are equal, the effective reaction rate for the limiting reactant would be the determining factor for the reactor volume as well as contacting, and an attempt to maximize this rate globally in the reactor vessel will always be the prudent choice for design.

6.3 Reactor models for design and scale-up

A hierarchy of models, with varying degrees of complexities, is developed to predict the performance of slurry and three-phase fluidized bed reactors. Depending on the level of fluid flow information, these models are classified as idealized (mixed and plug) flow models, nonideal flow (axial dispersion model and mixing cell) models, and advanced CFD-based models. The following section provides brief description of these models and their mathematical formulation. As an example, a model reaction of catalytic conversion of synthesis gas [$CO(A) + H_2\ (B)$] to liquid fuels in the context of low-temperature FT synthesis is considered. In this reaction, the gaseous reactants $CO_{(g)}$ and $H_{2(g)}$ are first dissolved in a solvent containing suspended catalyst particles. The $CO_{(l)}$ and $H_{2(l)}$ dissolved in the liquid then react catalytically to produce liquid products.

6.3.1 Lower order models

In idealized flow models, the fluid is assumed to be completely mixed or to move in the plug flow mode. In the mixed flow model, the concentration of all reactants and products is considered to be uniform, and Equations 6.9 and 6.10, shown in Table 6.5, are solved to find conversion of reactants at the reactor outlet. In Equation 6.9, the left-hand side indicates molar rates of reactant species ($i = A, B$) entering and leaving the reactor and the right-hand side indicates the rate at which gaseous reactants ($i = A, B$) are dissolved in the liquid phase. The governing equation for the liquid phase is written in the form of balance of reactant species (A, B) in liquid phase entering and leaving the reactor. The source of these species is due to mass transfer from gas to liquid phase, and consumption of the reactants is due to the reaction taking place in the liquid phase. In the plug flow model, the fluid is assumed to move like a plug with no mixing in the direction of flow, and Equations 6.11 and 6.12 shown in Table 6.5 are solved to find changes in the concentration of reactants and products in gas and liquid phases along the length of the reactor. In aforementioned equations, k_l and a are interphase mass transfer coefficient and interfacial area per unit volume, respectively. The estimation

Table 6.5 Idealized flow and axial dispersion models (steady state).

(A) Mixed flow model

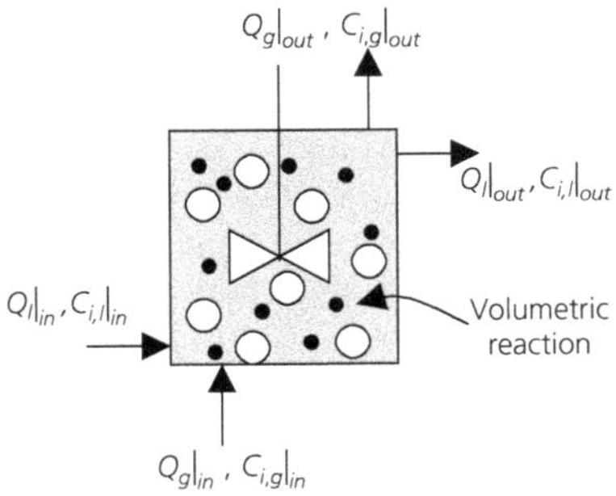

$$\left.\begin{array}{ll}\text{Gas phase: } (Q_g C_{i,g})_{in} - (Q_g C_{i,g})_{out} = -(k_l a)_i \left(\frac{(C_{i,g})_{out}}{m_i} - (C_{i,l})_{out}\right) & (6.9)\\ \text{Liquid phase: } (Q_l C_{i,l})_{in} - (Q_l C_{i,l})_{out} = +(k_l a)_i \left(\frac{(C_{i,g})_{out}}{m_i} - (C_{i,l})_{out}\right) \pm R_i & (6.10)\end{array}\right\} i = A, B$$

(B) Plug flow model

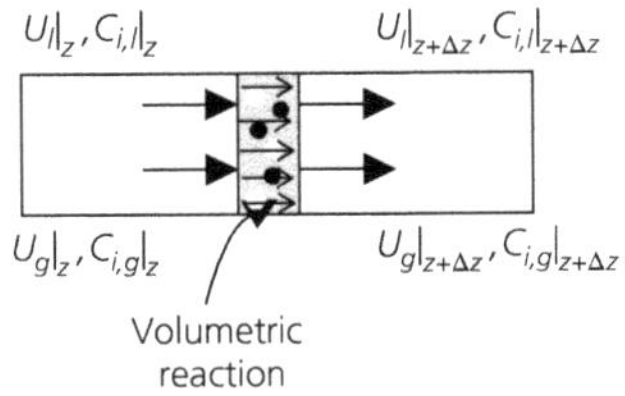

$$\left.\begin{array}{ll}\text{Gas phase: } \alpha_g U_g \frac{dC_{i,g}}{dz} = -(k_l a)_i \alpha_g \left(\frac{C_{i,g}}{m_i} - C_{i,l}\right) & (6.11)\\ \text{Liquid phase: } \alpha_l U_l \frac{dC_{i,l}}{dz} = (k_l a)_i \alpha_l \left(\frac{C_{i,g}}{m_i} - C_{i,l}\right) \pm \alpha_l R_i & (6.12)\end{array}\right\} i = A, B$$

(C) Axial dispersion model

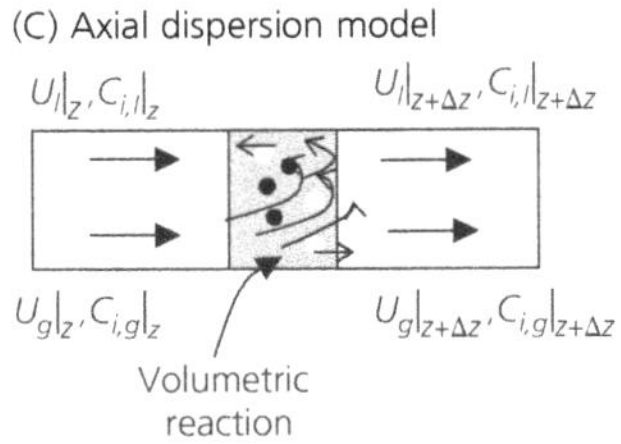

$$\left.\begin{array}{ll}\text{Gas phase: } \alpha_g U_g \frac{dC_{i,g}}{dz} = \alpha_g D_g \frac{d^2 C_{i,g}}{dz^2} - (k_l a)_i \alpha_g \left(\frac{C_{i,g}}{m_i} - C_{i,l}\right) & (6.13)\\ \text{Liquid phase: } \alpha_l U_l \frac{dC_{i,l}}{dz} = \alpha_l D_l \frac{d^2 C_{i,l}}{dz^2} + (k_l a)_i \alpha_l \left(\frac{C_{i,g}}{m_i} - C_{i,l}\right) \pm \alpha_l R_i & (6.14)\end{array}\right\} i = A, B$$

of $k_l a$ is discussed in detail in Section 6.4 with a brief review of correlations used to estimate $k_l a$.

The idealized flow approximations of fluid being completely mixed or flowing like a plug are hardly realized in the actual reactors, and often there exists concentration gradients in the reactors (deviation from the mixed flow assumption) or there exists some degree of mixing in the direction of the flow (deviation from the plug flow assumption). In order to account for such nonidealities in the flow behavior, an axial dispersion model is used. In the axial dispersion model, the axial dispersion (mixing) of gas and liquid phase is accounted through the dispersion terms D_g and D_l, respectively (see RHS of Equations 6.13 and 6.14). There exists several correlations in the literature to estimate the dispersion coefficients, and a brief review of such correlations is provided in Section 6.4. Dispersion coefficients can also be estimated by performing RTD experiments for a specific reactor geometry and operating conditions.

The idealized flow models and axial dispersion model are used extensively to predict the performance of three-phase slurry reactors. For example, mixed flow model and axial dispersion models are used to predict the reactant conversion and product distribution for conversion of synthesis gas to liquid fuels using FT synthesis [52–57], methanol synthesis in a slurry reactor [58], and wet air oxidation of sewage sludge [58].

6.3.2 Tank-in-series/mixing cell models

Though the axial dispersion model can take into account the deviations from the ideal plug flow behavior, it can also mathematically describe systems approaching complete backmixing. The tank-in-series or mixing cell models are also used to account for the gas and liquid backmixing in three-phase reactors. In these models, a reactor is divided into N parts (tanks) along the length of the reactor (see Table 6.6), and the liquid phase in each part (tank) is considered to be fully backmixed. The phase can be considered to be either in plug flow or fully back mixed. The degree of backmixing is characterized by the number of tanks, for example, $N = 1$ represents the limiting case of complete backmixing and $N \rightarrow \infty$ (~10 in practice) represents the limiting case of a plug flow behavior. A uniform distribution of catalyst particle is usually assumed. The governing equations of mixed flow model (see Table 6.5(A)) are extended for each tank to formulate governing equations for gas and liquid species (see Table 6.6). Several researchers have used mixing cell model to simulate three-phase reactors (e.g., Refs. [52, 59, 60]). Turner and Mills [52] have discussed the formulations of the axial

Table 6.6 Mixing cell model.

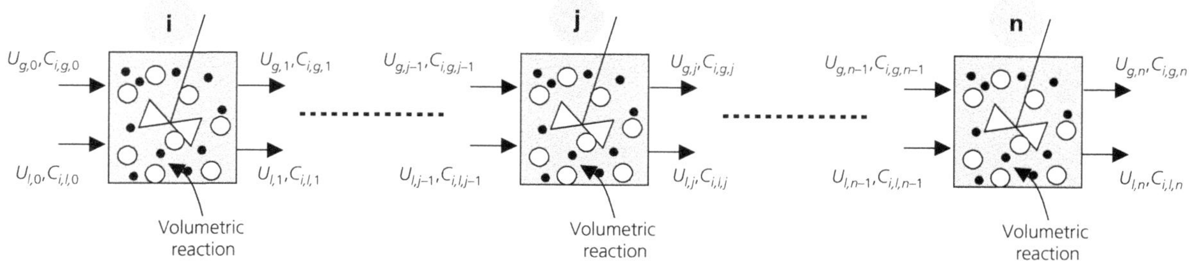

For j-th cell,

$$\left.\begin{array}{l}\text{Gas phase: } \displaystyle\int_0^{1/N} \chi_i \varphi_i \left(C'_{i,g,j} - C'_{i,l,j}\right) d\xi = \left(C'_{i,l,j} - C'_{i,l,j-1}\right) + \frac{\chi_{s,i}}{N}\left(C'_{i,l,j} - C'_{i,s,j}\right) \\ \text{Liquid phase: } C''_{i,l,j} = C''_{i,l,j-1} + \lambda_i R''_i \end{array}\right\} i = A, B \qquad \begin{array}{r}(6.15)\\(6.16)\end{array}$$

$$C'_{i,g,j} = \frac{C_{i,g,j}}{C_{i,g,0}},\ C'_{i,l,j} = \frac{m_i C_{i,l,j}}{C_{i,g,0}},\ C''_{i,l,j} = \frac{C_{i,l,j}}{C_{i,l,0}}, \qquad (6.17)$$

$$\chi_i = \frac{L k_l a}{U_g m_i},\ \varphi_i = \frac{U_g m_i}{U_l},\ \chi_{s,i} = \frac{L k_s a_p}{U_l},\ \lambda_i = \frac{L}{U_l} \qquad (6.18)$$

dispersion model and mixing cell models and their applications to simulate FT slurry reactor. Chaudhari et al. [60] have discussed the formulation of a mixing cell model and have used to simulate slurry bubble column reactor for reductive alkylation of p-phenylenediamine.

It is necessary to obtain realistic and reliable estimates of the various transport and hydrodynamic parameters used in the models described in Sections 6.2 and 6.3, as well as the kinetic parameters (estimated preferably in a transport-free environment). A full treatment of the latter is beyond the scope of this chapter, and the interested reader is referred to the extensive expositions in Ramachandran and Chaudhari [46] and Doraiswamy and Sharma [49]. The estimation of relevant transport and hydrodynamic parameters is discussed in the following section.

6.4 Estimation of transport and hydrodynamic parameters

6.4.1 Estimation of transport parameters

The main transport parameters to be estimated are the mass transfer coefficients (gas–liquid (liquid side) k_l, gas–liquid (gas side) k_g, and liquid–solid k_s)). Coupled to that is the estimation of the interfacial area per unit volume a, and often it is the combination (i.e., $k_l a$ or $k_g a$) that is estimated in a certain experimental procedure. Thermodynamic parameters, such as Henry's law constant (H) can be estimated in a simpler manner since their estimation on the flow or on any time-dependent phenomenon. Mass transfer coefficients may be evaluated in well-defined geometries with known flow fields using classical theories like film theory, penetration theory, surface renewal theory, and boundary layer theory or in laminar flow situations, by solution of the Graetz problem. In all such estimation methods, the task is to formulate the diffusion–convection problem of the species concerned and then evaluate the flux at the interface.

If better experimental data is not available in a given situation, then one may resort to make rough estimates from these theories. Consequently, accurate prediction of mass transfer coefficients, and indeed other similar transport coefficients, is frequently not possible. When the geometry and flow field is known (such as film flow over flat surface or flow in slits), often the theoretical assumptions that lead to the "knowledge" of these flow fields are not satisfied due to phenomena such as non-flat film surface because of surface instabilities or rippling, exothermicity of reactions, turbulence, etc. In other cases, the geometry or flow field may not be known at all and hence purely theoretical approach is not possible. That is why it has been conventional in reactor engineering to use groups obtained from dimensional analysis (such as Schmidt number, Reynolds number used to find correlations for Sherwood number). Unfortunately, such correlations are often reactor and process specific and depend on the local flow field. An extensive selection of such coefficients is presented by Cussler [61].

For measurement of $k_l a$, typically pure gas is used so that gas-side resistance to mass transfer is eliminated. Limiting situations can be employed in experimentation by performing the mass transfer coupled with limiting types of reactions in the homogeneous phase, such as mass transfer with no reaction (in which case the estimate is actually affected by RTD of liquid), mass transfer with slow reaction (for low Hatta numbers, the estimated $k_l a$ is independent of liquid RTD), mass transfer with instantaneous reaction (in which case also the $k_l a$ estimate works out to be independent of liquid-phase mixing and can

Table 6.7 Correlations for estimation of $k_l a$ in three-phase slurry and fluidized beds.

Nguyen et al. [62]	$k_l a = 0.39\left(1-\frac{\phi_s}{0.58}\right)U_l^{0.87}\quad \phi_s < \left(0.58-0.7U_g^{0.15}\right)$	(6.19)
Patwari et al. [63]	$k_l a = 1.68\times10^{-2}U_g^{0.36}\mu_l^{-1.3}{D_e}^{0.5}$	(6.20)
Sada et al. [64]	$k_l a = bu_g^n$ ($n = 0.86$, $b = 0.014$ for water, 0.016 for electrolyte solutions, 0.022 for electrolyte slurries) where $u_g = U_g/\left(\sigma(\rho_l-\rho_g)/\rho_l^2\right)^{0.25}$	(6.21)
Schumpe et al. [65, 66]	$k_l a = KU_g^{0.82}\mu_{eff}^{-0.39}$ ($K = 0.063$ for salt solutions and 0.042 for salt free solutions) $k_l a/D_e^{0.5} = 2988U_g^{0.44}U_l^{0.42}\mu_l^{-0.34}U_t^{0.71}$	(6.22)
Kim and Kim [67, 68]	$\frac{k_l a d_p^2}{D_e} = 0.0015\left(\frac{P_V d_p^4}{\nu^3}\right)^{0.67}\left(\frac{U_l^2\rho_{av}d_p}{W_{sLm}}\right)^{0.1} \times\left[1+0.036\left(\frac{V_f}{V_s}\right)^{1.18}-1.149\times10^{-3}\left(\frac{V_f}{V_s}\right)^{2.09}\right]$	(6.23a)
	$k_l a = 0.73U_g^{0.87}U_l^{0.45}d_p^{0.71}\left[1+0.036\left(\frac{V_f}{V_s}\right)^{1.11}-1.348\times10^{-3}\left(\frac{V_f}{V_s}\right)^{2.09}\right]$	(6.23b)
Kang et al. [69]	$k_l a = 0.256U_g^{0.56}U_l^{0.41}\mu_p^{0.52}d_p^{0.47}(1+R_b)^{1.68}$	(6.24)
Nore et al. [70]	$k_l a = 4766\left(\frac{U_l}{\alpha_l}\right)^{0.58}(U_g)^{0.59}\left(\frac{\rho_p}{\rho_l}\right)^{-4.9}$	(6.25)
Lee et al. [71]	$k_l a = 2.36\times10^{-5}U_g^{0.686}U_l^{0.469}d_p^{0.788}\sigma_l^{-1.532}\mu_l^{-0.548}$ (bubble-disintegrating regime) $k_l a = 1.10\times10^{-6}U_g^{0.940}U_l^{0.381}d_p^{0.790}\sigma_l^{-2.273}\mu_l^{-0.671}$ (bubble-coalescing or slug flow regime)	(6.26)

be evaluated from a closed form expression). Doraiswamy and Sharma [49] treat these situations extensively. Some of the widely used correlations used to estimate $k_l a$ in three-phase slurry and fluidized bed reactors are provided in Table 6.7 (see Kim and Kang [72] for a detailed review of the correlations). For estimate of $k_g a$, the typical procedure is to measure the absorption rates of very soluble gases (in the absence of reaction) or study absorption with instantaneous reaction on the gas–liquid interface. In either case, the estimate of the transport coefficient has to be free of the phase mixing. For estimate of k_s, typically one studies the dissolution of sparingly soluble solids (for instance, in an aqueous system, one would study the dissolution of naphthalene, benzoic acid, or β-naphthol).

Sometimes, interfacial area is measured independent of the mass transfer characteristics. In a low gas holdup situation, this may be done with photographic techniques, particularly with high-speed cameras. Other options include light scattering for larger size bubbles and interfacial conductance measurement of chemical absorption techniques. The latter is quite a popular method but depends on reaction used, and also physical properties of the fluids may be different in the measured and the modeled system. Also, certain constraints involving the Hatta number (*Ha*) and the enhancement factor (*E*) may not be satisfied in case the system is not an ideal measuring flow pattern (i.e., a differential reactor or a stirred system). In any case, systems like CO_2 and NaOH are used extensively for this purpose [73], as is O_2 + Na_2SO_3 solution + Co^{++} [74].

In summary, there are many known techniques to estimate the transport parameters, but most of them are application or equipment specific. Most importantly, it is a challenge to estimate them on a flow pattern-free basis. Table 6.8 shows some typical mass transfer coefficient values encountered in gas–liquid systems. These numbers are for guidance only, and the engineer interested in the design of slurry reactors is advised to refer to the various classics in this area, most of which have been cited earlier.

6.4.2 Estimation of hydrodynamic parameters

Several experimental investigations have been carried out on measurements and development of correlations to estimate various hydrodynamic parameters that are employed in the models based on idealized and nonidealized flow assumptions. While the estimation of transport parameters, for example, gas–liquid and liquid–solid mass transfer coefficients, is already discussed in Section 6.4.1, the estimation of dispersion coefficients is discussed briefly in this section. While a number of correlations exist to estimate liquid dispersion coefficients for gas–liquid two-phase systems [75–79], the studies for three-phase slurry and fluidized beds are rather limited. The liquid dispersion coefficients are usually measured through tracer experiments in which a tracer such as an electrolyte (e.g., KCl) is injected and the change in the concentration is recorded along the length of the reactor. Using these measurements, the dispersion coefficient is back calculated. Several investigators studied the effects of column diameter, superficial gas and liquid velocities, particle diameter, and liquid properties on liquid dispersion coefficients and proposed correlations. Some of these correlations are provided in Table 6.9. Similar to the liquid dispersion coefficient, the dispersion coefficients for gas and solid phases are needed. It should be noted that such correlations are valid for the range of operating and design conditions for which they are derived. Often the industrial reactors contain internal hardware, for example, baffles, cooling coils, and specific experiments are needed to measure dispersion coefficients to account for the changes in the hardware or in operating conditions. In such cases, the lower order, that is, the idealized flow, models cannot be used for *a priori* prediction of the reactor performance or for scale-up.

Table 6.8 Liquid holdup, mass transfer coefficients, and effective interfacial area in gas–liquid reactors.

Type of reactor	Liquid holdup (%)	$k_g \times 10^4$ (gmol/cm^2.s.atm)	$k_l \times 10^2$ (cm/s)	a (cm^2/cm^3 reactor)	$k_l a \times 10^2$ (s^{-1})
Packed columns					
• Countercurrent	2–25	0.03–2	0.4–2	0.1–3.5	0.04–7.0
• Co-current	2–95	0.1–3	0.4–6	0.1–17	0.04–102
Bubble column	60–98	0.5–2	1–4	0.5–6	0.6–2.4
Packed bubble columns	60–98	0.5–2	1–4	0.5–3	0.5–12
Tube reactors					
• Horizontal	5–95	0.5–4	1–10	0.5–7	0.5–70
• Coiled vertical	5–95	0.5–8	2–5	1–10	2–50
Spray columns	2–20	0.5–2	0.7–1.5	0.1–1	0.07–1.5
Mechanically agitated bubble reactors	20–95	—	0.3–4	1–20	0.3–80

Table 6.9 Correlations for liquid dispersion coefficients in three-phase slurry and three-phase fluidized beds.

El-Temtamy et al. [80]	$Pe = 0.0012 Re_p^{1.156} D_c^{-1.156}$	(6.27)
Kato et al. [81]	$Pe = Pe(BC)\left(1 + \frac{V_t}{U_L}\right)^{0.4Bo^{0.225}}$ for $U_G < 9\,\text{cm/s}$	
	$Pe(BC) = \left(\frac{\mu_l}{\mu_w}\right)^{0.07}\left[\frac{13(U_G/\sqrt{gD_T})}{1 + 6.5(U_G/\sqrt{gD_T})^{0.8}}\right]$ for $U_G = 9-25\,\text{cm/s}$	(6.28)
	$Pe = Pe(BC)$ for $U_G > 25\,\text{cm/s}$	
Kim et al. [82]	For bubble-coalescing regime:	(6.29)
	$Pe = 11.96\left(\frac{d_p}{D_C}\right)^{1.66}\left(\frac{U_l}{U_l + U_g}\right)^{1.03}$ $\begin{cases}\text{for } 0.004 \le \frac{d_p}{D_C} \le 0.024 \\ \text{and } 0.25 \le \frac{U_l}{U_l + U_g} \le 0.857\end{cases}$	
	For bubble-disintegrating regime:	
	$Pe = 20.72\left(\frac{d_p}{D_C}\right)^{1.66}\left(\frac{U_l}{U_l + U_g}\right)^{1.03}$ $\begin{cases}\text{for } 0.004 \le \frac{d_p}{D_C} \le 0.012 \\ \text{and } 0.143 \le \frac{U_l}{U_l + U_g} \le 0.857\end{cases}$	
Han et al. [83]	$Pe_z = 3.2 \times 10^{-4} Re_l Re_g^{-0.5}(1 - \alpha_S)^{2.8}$	(6.30)
	$Pe_r = 6.7 \times 10^{-3} Re_l^{1.2} Re_g^{-0.8}(1 - \alpha_S)^{0.4}$	

6.5 Advanced computational fluid dynamics (CFD)-based models

While the lower order models described in Section 6.3 are useful for the quick prediction of the overall performance of a reactor, these models often rely on simplified flow approximations and often fail to account for change in the local fluid dynamics or transport processes during the presence of internal hardware or changes in flow regimes. Moreover, these models are also based on empirical knowledge (as discussed in Section 6.4) of several parameters such as interfacial area, dispersion coefficients, and mass transfer coefficients. Some of these limitations may be avoided by using CFD models for simulations of gas–liquid–solid flows in three-phase slurry and fluidized bed.

A hierarchy of computational models is available to simulate dispersed gas–liquid–solid flows in three-phase slurry and fluidized bed reactors [84] continuum (Euler–Euler) method, discrete particle/bubble (Euler–Lagrange) method, or front tracking/capturing methods. While every method has its own advantages and limitations, the continuum method is widely used to simulate dense three-phase flows in laboratory- and industrial-scale slurry and fluidized bed reactors.

In the continuum (Euler–Euler)-type formulation, the gas, liquid, and solid phases are assumed to be continuum and the volume-averaged mass and momentum equations (see Table 6.10) are solved for each phase separately to predict the pressure, phase holdup, and phase velocity distributions. As a result of time and volume averaging, additional terms appear in the momentum conservation equations. These additional terms need closure models and such unclosed terms are highlighted in Table 6.10.

In order to close the additional Reynolds (turbulent) stresses, several different eddy viscosity-based turbulence models, in which the additional turbulent stresses are related to the mean velocity gradient as shown in Table 6.11, are used to account for the turbulence in three-phase systems. Generally, the standard k-ε turbulence model is solved only for the continuous phase or for mixture phase or for each phase. In the literature reports,

Table 6.10 Governing equations for Euler–Euler formulation.

- Mass conservation equations

Gas phase: $$\frac{\partial}{\partial t}(\rho_g\alpha_g)+\nabla\cdot(\rho_g\alpha_g\bar{U}_g)=0 \quad (6.31)$$

Solid phase: $$\frac{\partial}{\partial t}(\rho_s\alpha_s)+\nabla\cdot(\rho_s\alpha_s\bar{U}_s)=0 \quad (6.32)$$

Liquid phase: $$\frac{\partial}{\partial t}(\rho_l\alpha_l)+\nabla\cdot(\rho_l\alpha_l\bar{U}_l)=0 \quad (6.33)$$

- Momentum conservation equations

Gas phase:

$$\frac{\partial}{\partial t}(\rho_g\alpha_g\bar{U}_g)+\nabla\cdot(\rho_g\alpha_g\bar{U}_g\bar{U}_g)=-\alpha_g\nabla P_g-\nabla\cdot\bar{\bar{\tau}}_g+\alpha_g\rho_g\bar{g}-\overline{M_{gs}}-\overline{M_{gl}} \quad (6.34)$$

Solid phase:

$$\frac{\partial}{\partial t}(\rho_l\alpha_l\bar{U}_l)+\nabla\cdot(\rho_s\alpha_s\bar{U}_l\bar{U}_l)=-\alpha_l\nabla P_l-\nabla\cdot\bar{\bar{\tau}}_l+\alpha_l\rho_l\bar{g}+\overline{M_{lg}}+\overline{M_{ls}} \quad (6.35)$$

Liquid phase:

$$\frac{\partial}{\partial t}(\rho_s\alpha_s\bar{U}_s)+\nabla\cdot(\rho_s\alpha_s\bar{U}_s\bar{U}_s)=-\alpha_s\nabla P_g-\nabla P_s-\nabla\cdot\bar{\bar{\tau}}_s+\alpha_s\rho_s\bar{g}+\overline{M_{sg}}+\overline{M_{sl}} \quad (6.36)$$

Table 6.11 Turbulence closures.

Liquid phase: $$\bar{\bar{\tau}}_g=\alpha_l\mu_{eff,l}\left(\nabla\bar{U}_l+\nabla\bar{U}_l^T\right)-\frac{2}{3}\alpha_l\mu_{eff,l}\nabla\cdot\bar{U}_l\bar{\bar{I}} \quad (6.37)$$

Gas phase: $$\bar{\bar{\tau}}_g=\alpha_g\mu_{eff,g}\left(\nabla\bar{U}_g+\nabla\bar{U}_g^T\right)-\frac{2}{3}\alpha_g\mu_{eff,g}\nabla\cdot\bar{U}_g\bar{\bar{I}} \quad (6.38)$$

Solid phase: $$\bar{\bar{\tau}}_s=\alpha_s\mu_s\left(\nabla\bar{U}_s+\nabla\bar{U}_s^T\right)+\alpha_s\left(\lambda_s-\frac{2}{3}\mu_s\right)\nabla\cdot\bar{U}_s\bar{\bar{I}} \quad (6.39)$$

in which the standard k-ε model is solved only for the continuous phase, the effective viscosity of the continuous phase is obtained as $\mu_{eff,l}=\mu_{lam,l}+\mu_{tur,l}$ in which the turbulent (eddy) viscosity was estimated (with suitable assumptions about the turbulence field) as $\mu_{tur,l}=C_\mu k_l^2/\varepsilon_l$. Using $\mu_{eff,l}$, the gas-phase effective viscosity is calculated using the liquid-phase effective viscosity as $\mu_{eff,g}=\left(\rho_g/\rho_l\right)\mu_{eff,l}$. In case of a mixture turbulence model, one set of k- and ε-equations are solved using the mixture properties. In some cases, the standard k-ε model is solved for both (gas and liquid) phases and $\mu_{eff,g}$ and $\mu_{eff,l}$ are estimated separately. The estimation of solid-phase viscosities and solid-phase pressure is discussed separately later in this section.

The interphase momentum exchange between gas–liquid $\left(\overline{M_{gl}},\overline{M_{lg}}\right)$ and liquid–solid phases $\left(\overline{M_{sl}},\overline{M_{ls}}\right)$ and vice versa is accounted through various forces acting on bubbles and particles dispersed in the continuous phases, that is, drag, lift, virtual mass, and other hydrodynamic forces (Equations 6.40 and 6.41 in Table 6.12). The drag force experienced by a bubble or particle due to relative velocity is calculated using Equation 6.42. The shear in the continuous phase generates the lift force acting on bubbles or particles and is modeled as shown by Equation 6.43. In addition, the lift force can also be generated by wake generated by bubbles or due to rotation of bubbles. When a bubble or a particle accelerates, it also has to accelerate some mass of surrounding liquid, and this force is called the virtual mass force and is calculated using Equation 6.44. Often, the drag force is higher in magnitude than that of the lift and virtual mass force, and therefore the latter forces are often not

Table 6.12 Closures for interphase momentum exchange.

- Gas–liquid momentum exchange

$$\overline{M}_{gl}=-\overline{M}_{lg}=\sum\bar{F}=\bar{F}_D+\bar{F}_L+\bar{F}_{VM}+\bar{F}_{Other} \quad (6.40)$$

- Liquid–solid momentum exchange

$$\overline{M}_{sg}=-\overline{M}_{gs}=\sum\bar{F}=\bar{F}_D+\bar{F}_L+\bar{F}_{VM}+\bar{F}_{Other} \quad (6.41)$$

Drag force

$$\overline{F_D}=\frac{3}{4}\rho_l\alpha_g\alpha_l\frac{C_D}{d_B}\left|\bar{U}_g-\bar{U}_l\right|(\bar{U}_g-\bar{U}_l) \quad (6.42)$$

Lift force

$$\overline{F_L}=-C_L\rho_l\alpha_g\alpha_l(\bar{U}_l-\bar{U}_g)\times\nabla\times\bar{U}_l \quad (6.43)$$

Virtual mass force

$$\overline{F_{VM}}=-\left[\frac{D\boldsymbol{I}}{Dt}+\boldsymbol{I}\cdot\nabla\bar{U}_l\right]\text{ where }\boldsymbol{I}=C_{VM}\rho_l\frac{m_B}{\rho_B}(\bar{U}_l-\bar{U}_B) \quad (6.44)$$

considered in the simulations of gas–liquid–solid flows in three-phase slurry/fluidized bed reactors.

The motion of solid particles is modeled using the kinetic theory of granular flows (KTGF). Using the granular temperature (Θ_s) and the radial distribution function ($g_{0,ss}$), the gradient of solid phase pressure is calculated using Equation 6.45 as the sum of the kinetic and collisional parts (Table 6.13). The granular temperature is proportional to the kinetic energy of the fluctuating particles. The radial distribution function characterizes dimensionless distance between particles. When α_s is less than maximum packing limit, the solid phase is treated as "compressible." As α_s increases, the distance between the particles decreases. The solid phase is considered as "incompressible" when α_s reaches to maximum packing limit. Several different formulations of the radial distribution are available in the literature [85, 89, 90]. e_{ss} is the restitution coefficient and its value depends on the type of particles used.

The solid-phase shear viscosity (μ_s) appearing in Equation 6.39 is modeled as the sum of collisional, kinetic, and frictional viscosities (see Eq. 6.46). Following Shalala et al. [89] and Gidaspow et al. [86], the collisional and kinetic parts of the shear viscosity are modeled using Equations 6.47 and 6.48. Under dense flow conditions, solid-phase volume fraction reaches to the maximum packing limit and friction between particles leads to the generation of stress. This is accounted through the fractional viscosity term and is often modeled as shown in Equation 6.49 [87]. The bulk viscosity (λ_s) used in Equation 6.39 that accounts for the resistance of the granular particles to compression and expansion is modeled as shown in Equation 6.50 [88]. In the following part of this section, a few examples of numerical simulations of gas–liquid–solid flows in three-phase slurry and fluidized beds are given to demonstrate the applications of the aforementioned continuum CFD model.

Padial et al. [91] performed qualitative simulations of three-phase flow in a draft tube bubble column and compared the overall gas volume fraction and liquid circulation time for gas–liquid and gas–liquid–solid systems. Michele and Hempel [92] simulated flow air bubbles and PMMA particles (300 μm, 10 vol%) dispersed in water for superficial gas velocities in the range of 0.02–0.09 m/s. They compared their predictions with measured overall gas holdup and only a qualitative agreement

Table 6.13 Closures for solid phase.

• Solid pressure: $P_s = \underbrace{\alpha_s\rho_s\Theta_s}_{\text{kinetic part}} + \underbrace{2\rho_s(1+e_{ss})\alpha_s^2 g_{0,ss}\Theta_s}_{\text{collisional part}}$	(6.45)
• Viscosity for solid phase: $\mu_s = \mu_{s,col} + \mu_{s,kin} + \mu_{s,fr}$	(6.46)
Collisional viscosity [85]	
$\mu_{s,col} = \frac{4}{5}\alpha_s\rho_s d_s g_{0,ss}(1+e_{ss})\left(\frac{\Theta_s}{\pi}\right)^{1/2}$	(6.47)
Kinetic viscosity [86]	
$\mu_{s,kin} = \frac{10\rho_s d_s\sqrt{\Theta_s\pi}}{96\alpha_s(1+e_{ss})g_{0,ss}}\left[1+\frac{4}{5}g_{0,ss}\alpha_s(1+e_{ss})\right]^2$	(6.48)
Frictional viscosity [87]	
$\mu_{s,fr} = \frac{P_s \sin\phi}{2\sqrt{I_{2D}}}$	(6.49)
Bulk viscosity [88]	
$\lambda_s = \frac{4}{3}\alpha_s\rho_s d_s g_{0,ss}(1+e_{ss})\left(\frac{\Theta_s}{\pi}\right)^{1/2}$	(6.50)

could be seen. Schallenberg et al. [93] further accounted for the effect of local gas holdup on the interphase coupling forces and also included bubble-induced turbulence and reported a good agreement between the predicated overall gas holdup and radial profiles of liquid velocity. Feng et al. [94] simulated the laboratory-scale three-phase fluidization of gas–liquid nanoparticles in a rectangular column and attempted extensive validation of predicted dynamic and time-averaged flow behavior using the measurement of liquid velocity measurements performed using laser Doppler anemometer (LDA) and gas holdup using conductivity probes. The predicted transversal profiles of time-averaged axial liquid velocities and time-averaged gas holdups were in an excellent agreement with the measurements for low superficial gas velocities (0.01–0.03 m/s) and low solid loading (0–3.3%). They also reported compared time evolution of measured and predicted time series of local liquid velocity. Recently, Khopkar and Ranade [95] discussed and reviewed computational flow modeling of three-phase agitated reactors.

Wu and Gidaspow [5] performed transient 2D simulations of methanol synthesis from synthesis gas in an Air Products/DOE LaPorte three-phase slurry bubble column reactor. The geometry of the Air Products/DOE LaPorte reactor used in the simulations and the simulated instantaneous distribution of gas and solid holdup and temperature distributions are shown in Figure 6.5. Jia et al. [96, 97] applied CFD model to simulate dynamic behaviors of batch phenol biodegradation by immobilized *Candida tropicalis* in a three-phase air lift reactor, coupling the three-phase flow, interphase mass transfer, and intrinsic bioreaction, with a population balance model employed to determine the BSD in the reactor. They demonstrated that the time evolution of batch phenol biodegradation characteristics such as the oxygen, phenol, and cell concentration profiles measured in various phases could be predicted satisfactorily. Recently, Troshko and Zdravistch [98] performed Eulerian multiphase simulations of FT synthesis using a slurry bubble column reactor. They formulated a CFD model of FT reactors incorporating a population balance to predict the bubble size distribution and FT reactions (under isothermal conditions) to predict performance of FT reactor quantitatively. They simulated the effects of superficial gas velocity (0.1–0.4 m/s) and catalyst loadings of 10 and 35%. For example, the predicted instantaneous distribution of various species in gas and slurry phase at catalyst concentration of 20% is shown in Figure 6.6a and b, respectively. The experimental verification was performed by comparing measured and predicted overall gas holdup and using 1D model results to verify the syngas conversion and reactor productivity.

From the present state of literature, it can be seen that only a limited reports are available on numerical simulations gas–liquid–solid phase flows in three-phase slurry and fluidized bed reactors. Most of the simulations and verification studies are performed under very low superficial gas velocities (<10 cm/s) and low solid loadings (<10%). Though detailed experimental verification of predictive capabilities of CFD models to simulate dense three-phase flows is still lacking, significant research activity is underway to improve the predictive capabilities of CFD models and to verify these capabilities with detailed measurements of phase holdup distributions and velocity using advanced experimental techniques. Parallel to these activities, some research groups are also working on extending the CFD models to incorporate heat transport and chemical reactions (e.g., see Figure 6.6). With the advances in measurement techniques and high-performance computing, the CFD models that can be used for quantitative design and scale-up purposes will be developed. However, the available lower order/mixing cell models shall still remain indispensable for a process engineer interested in quick analysis of the three-phase reactors.

6.6 Summary and closing remarks

In this chapter, we have attempted to summarize the philosophy and procedural details for the design and scale-up of three-phase slurry reactors. First, the widespread use of three-phase slurry systems in the petroleum processing, chemicals, and process industry has been highlighted with suitable examples. The factors governing the performance of three-phase slurry

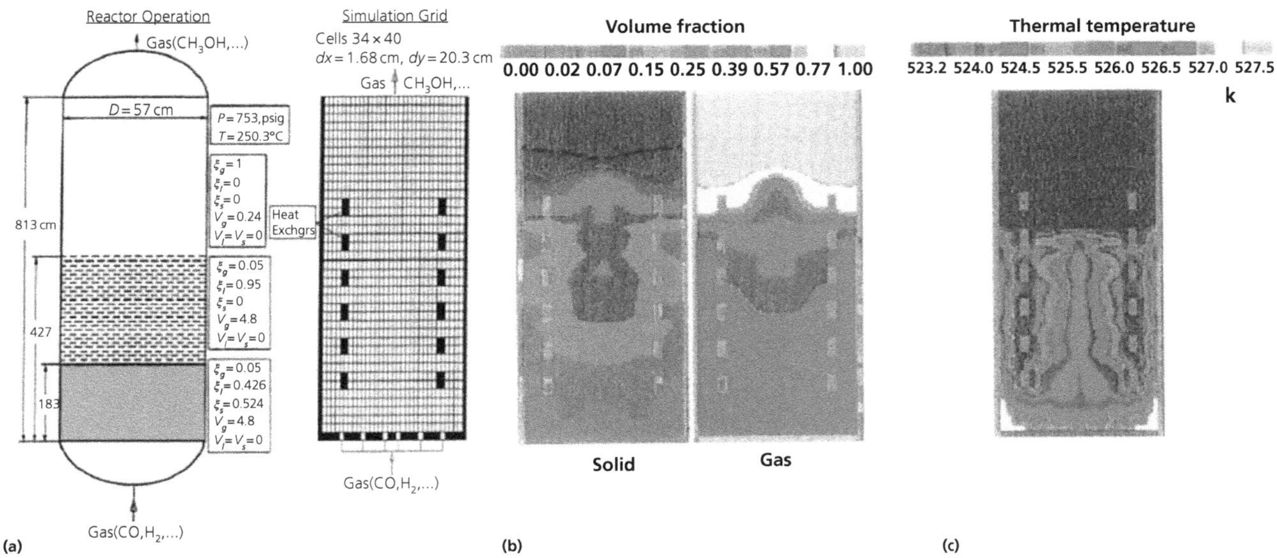

Figure 6.5 Hydrodynamic simulation of methanol synthesis in gas–liquid slurry bubble column reactors. (a) Reactor operating conditions and simulation grid. (b) Simulated instantaneous distribution of gas and solid holdup. (c) Simulated instantaneous distribution of temperature. (Source: Wu [5]. Reproduced with permission of Elsevier.)

systems as industrially relevant chemical reactors have been discussed from a first-principles approach, and how that drives the reactor selection has been highlighted. In this discussion, we have also pointed out the limitations of reactor systems in which the solid catalyst is retained as a packed bed and how those limitations are addressed by having the solid phases move freely in three-phase slurry systems. We have argued that while the characteristics of the catalyst (its chemistry and mass transfer resistance offered by the substrate and the outer surface of the catalyst) are the primary driver for choice of a process, the appropriate reactor configuration must be chosen to maximize the volumetric productivity of the reactor vessel.

Conventional reactor design, including the design of three-phase slurry systems, was based on developing models for the effective reaction rate at the particle/bubble scale (incorporating the catalyst chemistry and transport resistances at the particle scale) and then integrating this rate along the length of the reactor using a first-order differential equation (for plug flow pattern of fluids) or writing this as a discrete difference equation (for well-mixed mixing pattern). In recent times, such models have given way to the use of compartments models for three-phase reactors, several examples of which have been presented in this chapter. The compartment models involve a phenomenological way of predicting nonideality in mixing, which may be estimated using RTD experiments for the phases involved.

A further improvement in the estimation of hydrodynamic flow patterns is done with CFD models. These models are helpful in estimating the details of the flow patterns, such as velocity profiles and phase volume fraction profiles, which in turn can be directly coupled to the effective reaction rates or may be coupled using reduced order models for the hydrodynamics. Examples of this have also been presented in this chapter. It may be recognized that as of now, the uncertainties involved in CFD (and the number of parameters that are required to be estimated for the model to provide reasonable predictions) preclude it from being used as an *ab initio* design tool for three-phase slurry reactor design. However, it has already started serving its purpose for guiding the design, in terms of narrowing down the design to certain favorable flow configurations, which may be then subjected to detailed experimentation and if required, modeling, to refine it to the final design. On the other hand, CFD is also widely being accepted as a significant process diagnostics tool, for instance, in three-phase slurry reactors in which problems during operation are common. CFD provides a relatively inexpensive method for detecting and troubleshooting those problem. Certainly, one expects CFD to further improve in coming years and decades as will the validation possibilities and reliability of the models and commercial and open-source packages. This is likely to also promote a wider acceptability of CFD in the chemical reactor engineering community.

Estimation of the effective reaction rate continues to be based largely on empiricism and system-specific measurements. Several correlations reported in the literature for estimation of the transport parameters have been reported in this chapter. This is, however, only a partial list, and case-specific correlations may be either developed independently or estimated from relevant literature sources.

We have made an attempt to provide a systematic approach to understand various elements of hydrodynamics, interphase and intraparticle mass transfer, fluid phase mixing, and reaction

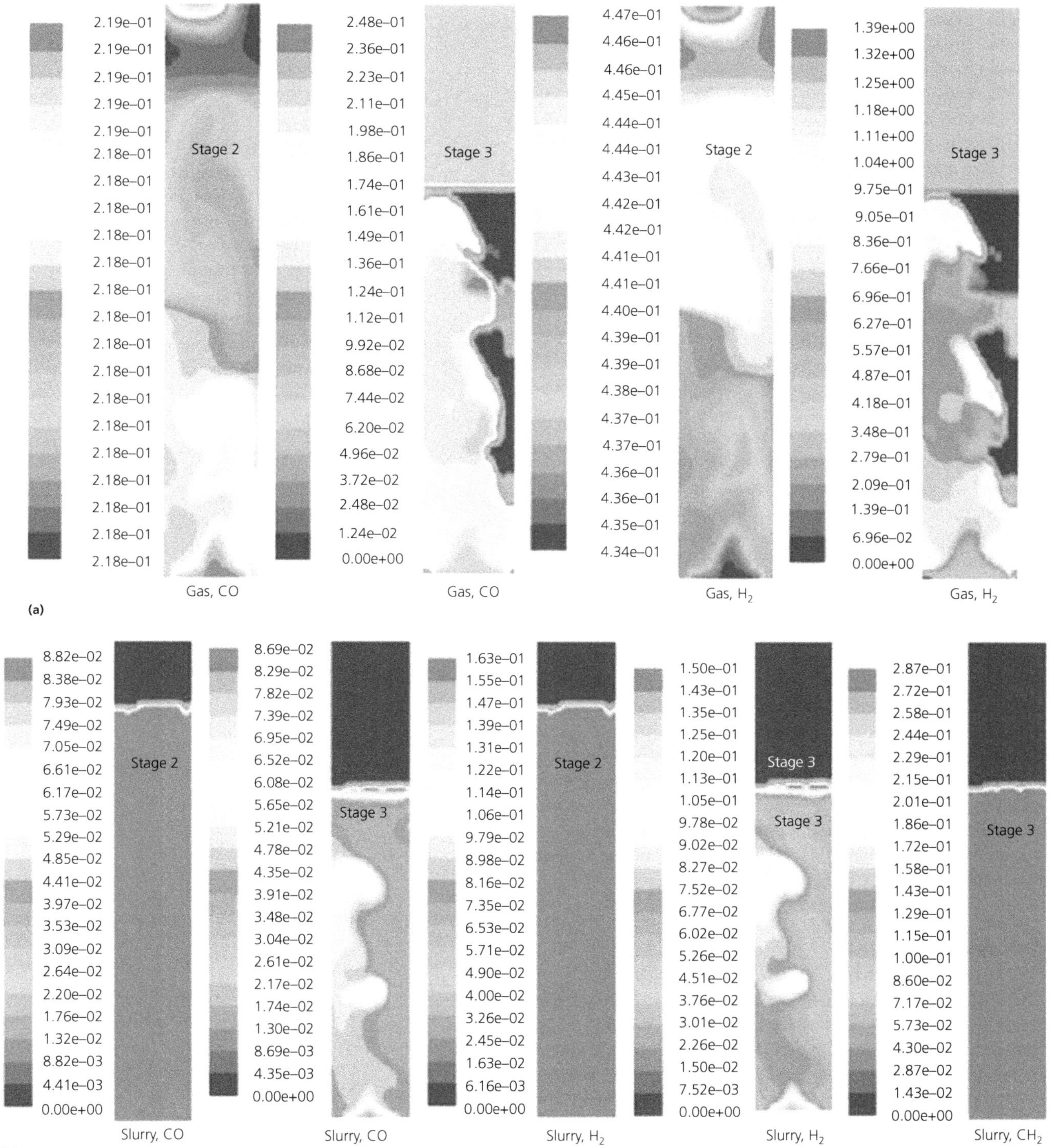

Figure 6.6 Predicted instantaneous distribution of various species in (a) gas and (b) slurry phases at catalyst concentration of 20%. (Source: Troshko and Zdravistch [98]. Reproduced with permission of Elsevier.)

kinetics relevant to three-phase slurry reactors. The discussion on key reactor engineering issues and various modeling approaches will help in selecting appropriate models and their combinations. It is important to emphasize that it is extremely important to correctly identify and define the reactor engineering objectives, analyze various key issues relevant to achieving the defined objectives, and formulate an appropriate modeling approach/tools, which are consistent with the set objectives. A diagnostic analysis of significance of various factors that may be contributing to specific process performance is helpful to simplify the models and select appropriate models for design purpose. We hope that discussion in this chapter will help the reactor engineer in making appropriate selection of modeling approach and models.

Some comments on research needs for enhancing our understanding of three-phase slurry reactors may be appropriate at this juncture. With the emergence of cheap, high-speed computing platforms and the availability of the commercial modeling tools, computational modeling needs to be harnessed to devise best possible reactor hardware. In order to fully realize the potential of such computational modeling for better reactor engineering, it is essential to focus further research on (i) the development and validation of various key physicochemical processes occurring in three-phase slurry reactors (flow regimes, distribution of gas and solid particles, coalescence and breakup of bubbles in the presence of solid particles, mixing, heat and mass transfer, and so on) and (ii) a multiscale framework that models processes occurring on different spatiotemporal scales in a coherent way. This list is merely suggestive since the complexity of reactive three-phase flows may greatly expand the list of issues on which further research is required. Accepting the limitations of knowledge of underlying physics and invoking model calibration whenever necessary are essential to combine different models for realizing optimal three-phase slurry reactor. We hope that the content of this book will be useful in engineering of three-phase slurry reactors and will stimulate further research and expand applications of these reactors.

Acknowledgments

The authors would like to acknowledge the support of IIT Delhi research scholars Deepali Chugh, Loveleen Sharma, Tejas Puneet Kaur Grewal, and Brajesh Kumar Singh for their help in the preparation of the manuscript. One of the authors (VVR) would like to acknowledge financial support from CSIR via Indus MAGIC [CSC123] project.

Nomenclature

a	interfacial area per unit volume (m^2/m^3)
C	concentration ($kmol/m^3$)
$\bar{c_B}$, $\bar{c_A}$	average concentrations of the reactants inside the catalyst pellets ($kmol/m^3$)
c_{Bs}, c_{As}	surface concentrations ($kmol/m^3$)
C_D, C_L, C_{VM}	drag, lift, and virtual mass coefficients (—), respectively
$C'_{i,g,j}$, $C'_{i,l,j}$, $C''_{i,l,j}$	concentrations defined by Equation 6.17 (—)
C_μ	constant (—)
D	dispersion coefficient (m^2/s)
d	particle/bubble diameter (m)
D, D_e	molecular diffusivity (m/s^2)
D_c, D_T	column diameter (m)
E	enhancement factor (—)
e_{ss}	coefficient of restitution
$\bar{F}_D, \bar{F}_L, \bar{F}_{VM}$	drag, lift, and virtual mass forces per unit volume, respectively (kg/m^2s^2)
$g_{0,ss}$	radial distribution function (—)
Ha	Hatta number (—)
$\bar{\bar{I}}$	unit tensor (—)
I_{2D}	second invariant of the deviatoric stress tensor
k	reaction rate constant ($m^6{}_{liq}/((mol\ B)(m^3{}_{cat})(s))$)
k_{Ag}, k_{Al}	gas- and liquid-mass transfer coefficients for A, respectively ($mol\ A/(Pa.m^2.s)$ and $m^3{}_{liq}/(m^2{}_{surface}.s)$)
k_{As}, k_{Bs}	liquid–solid mass transfer coefficients ($m^3{}_{liq}/(m^2{}_{surface}.s)$)
k_ga	volumetric gas side mass transfer coefficient ($mol/(MPa.m^3.s)$)
k_l	turbulent kinetic energy of the liquid phase per unit mass (m^2/s^2)
k_la	volumetric liquid side mass transfer coefficient (s^{-1})
$\bar{M}$	interphase momentum exchange vector (kg/m^2s^2)
m, H	Henry's constant (—)
m_B	mass of bubble (kg)
N	number of tanks (—)
P	pressure (N/m^2)
p_{Ag}	partial pressure of gas-phase reactant A (N/m^2)
P_s	solid pressure (N/m^2)
Pe	Peclet number (D_TU_l/D_l) (—)
$Pe(BC)$	Peclet number as defined in Equation 6.28 (—)
Pe_r, Pe_z	Peclet numbers in radial (D_cU_l/D_r) and axial (LU_l/D_z) directions (—)
P_v	energy dissipation rate based on total mass in beds (J/s.kg)
Q	volumetric flow rate (m^3/s)
r, R	rate of reaction ($kmol/m^3.s$)
R_b	volume ratio of floating bubble breaker to the fluidized particle (—)
Re, Re_p	Reynolds number and particle Reynolds number (—)
Re_l, Re_g	liquid-phase ($D_cU_l\rho_l/\mu_l$) and gas-phase ($D_cU_g\rho_g/\mu_g$) Reynolds numbers (—)
U	superficial velocity (m/s)

$\bar{U}$	velocity vector (m/s)
V_f/V_s	volume ratio of floating bubble breakers to solid particles (—)
W_{sLm}	adhesive energy, N/m
z	axial distance (m)

Greek letters

α	phase holdup (—)
β, γ	stoichiometric coefficients (—)
ε_l	rate of liquid-phase turbulent kinetic energy dissipation (m^2/s^3)
η_A, η_B	catalyst effectiveness factors
Θ_s	granular temperature (m^2/s^2)
λ_s	dilatational viscosity (kg/m.s)
μ	viscosity (kg/m.s)
$\mu_{s,col}, \mu_{s,kin}, \mu_{s,fr}$	collisional, kinetic, and viscosity, respectively (kg/m.s)
ν	kinematic viscosity (m^2/s)
ρ	density (kg/m^3)
σ	surface tension (N/m^2)
$\bar{\bar{\tau}}$	viscous stress tensor
ϕ	angle of internal friction (Equation 6.49) (—)
ϕ_s	solid loading (—)
$\chi_i, \varphi_i, \chi_{s,i}, \xi_i$	parameters defined by Equation 6.18 (—)

Subscripts

A, B	species *A, B*
e, eff	effective
g, G	gas phase
g–l	gas–liquid
i	species
l, L	liquid phase
lam	laminar
l–s	liquid–solid
p	particle
s, S	solid phase
tur	turbulent
u, B	bubble

References

1 Schugerl K. Three-phase-biofluidization—application of three-phase fluidization in the biotechnology—a review. Chem. Eng. Sci. 1997;52:3661–3668.

2 Chaudhari RV, Parande MG, Ramachadran PA, Brahme PH. Modelling of a batch slurry reactor for hydrogenation of *o*-nitroanisole to *o*-anisidine. Institute of Chemical Engineers Symposium. Series No. 87 (ISCRE 8); 1983. September 10–13, 1984, University of Edinburgh.

3 Arcuri EJ, Slaff G, Greasham R. Continuous production of thienamycin in immobilized cell systems. Biotechnol. Bioeng. 1986; XXVIII:842–849.

4 Maretto C, Krishna R. Modeling of a bubble column slurry reactor for Fischer–Tropsch synthesis. Catal. Today 1999;52:279–289.

5 Wu Y, Gidaspow D. Hydrodynamic simulation of methanol synthesis in gas-liquid slurry bubble column reactors. Chem. Eng. Sci.2000;55:573–587.

6 Fenice M, Federici F, Selbmann L, Petruccioli M. Repeated-batch production of pigments by immobilised *Monascus purpureus*. J. Biotechnol. 2000;80:271–276.

7 Son SH, Choi SM, Lee YH, Choi KB, Yun SR, Kim JK, Park HJ, Kwon OW, Noh EW, Seon JH, Park YG. Large-scale growth and taxane production in cell cultures of *Taxus cuspidata* (Japanese Yew) using a novel bioreactor. Plant Cell Rep. 2000;19:628–633.

8 Chang IS, Kim BH, Lovitt RW, Bang JS. Effect of CO partial pressure on cell-recycled continuous CO fermentation by *Eubacterium limosum* KIST612. Process Biochem. 2001;37:411–421.

9 Ogbonna JC, Mashima H, Tanaka H. Scale-up of fuel ethanol production from sugar beet juice using loofa sponge immobilized bioreactor Bioresour. Technol. 2001;76:1–8.

10 Matos EM, Guirardello R, Mori M, Nunhez JR. Modeling and simulation of a pseudo-three-phase slurry bubble column reactor applied to the process of petroleum hydrodesulfurization. Comput. Chem. Eng. 2009;33:1115–1122.

11 Guillen DP, Grimmett T, Gandrik AM, Antal SP. Development of a computational multiphase flow model for Fischer Tropsch synthesis in a slurry bubble column reactor. Chem. Eng. J. 2011;176–177:83–94.

12 Salehi Z, Yoshikawa H, Mineta R, Kawase Y. Aerobic biodegradation of *p*-nitrophenol by acclimated waste activated sludge in a slurry bubble column. Process Biochem. 2011;46:284–289.

13 Rahimpour MR, Jokar SM, Jamshidnejad Z. A novel slurry bubble column membrane reactor concept for Fischer–Tropsch synthesis in GTL technology. Chem. Eng. Res. Des. 2012;90:383–396.

14 Cheng L, Li T, Keener TC, Lee JY. A mass transfer model of absorption of carbon dioxide in a bubble column reactor by using magnesium hydroxide slurry. Int. J. Greenh. Gas Control 2013;17:240–249.

15 Vincenzino V, Xu W, Hebrard G, Li LY, Grace JR. Modeling of zinc adsorption onto clinoptilolite in a slurry bubble column. Chem. Eng. Sci. 2013;100:326–331.

16 Souza GLM, Afonso JC, Schmal M, Cardoso JN. Mild hydrocracking of an unstable feedstock in a three-phase fluidized bed reactor: behavior of the process and of the chemical compounds. Ind. Eng. Chem. Res. 1992;31(9):2127–2133.

17 Nam W, Kim J, Han G. Photocatalytic oxidation of methyl orange in a three-phase fluidized bed reactor. Chemosphere 2002;47:1019–1024.

18 Sales FG, Maranhão LCA, Filho NML, Abreu CAM. Experimental evaluation and continuous catalytic process for fine aldehyde production from lignin. Chem. Eng. Sci. 2007;62:5386–5391.

19 Lin CN, Wu SY, Chang JS, Chang JS. Biohydrogen production in a three-phase fluidized bed bioreactor using sewage sludge immobilized by ethylene–vinyl acetate copolymer. Bioresour. Technol. 2009;100:3298–3301.

20 Nam W, Woo K, Han G. Photooxidation of anionic surfactant (sodium lauryl sulfate) in a three-phase fluidized bed reactor using TiO_2/SiO_2 photocatalyst. J. Ind. Eng. Chem. 2009;15:348–353.

21 Charinpanitkul T, Limsuwan P, Chalotorn C, Sano N, Yamamoto T, Tongpram P, Wongsarivej P, Soottitantawat A, Tanthapanichakoon W. Synergetic removal of aqueous phenol by ozone and activated carbon within three-phase fluidized-bed reactor. J. Ind. Eng. Chem. 2010;16:91–95.

22 Qiu L, Ding H, Wang W, Kong Z, Li X, Shib Y, Zhong W. Coenzyme Q10 production by immobilized Sphingomonas sp. ZUTE03 via a conversion–extraction coupled process in a three-phase fluidized bed reactor. Enzyme Microb. Technol. 2012;50:137–142.

23 Lin J, Zhou M, Zhao X, Luo S, Lu Y. Extractive fermentation of L-lactic acid with immobilized *Rhizopus oryzae* in a three-phase fluidized bed. Chem. Eng. Process. 2007;46:369–374.

24 Lohi A, Alvarez C, Anania G, Upreti SR, Wan L. Biodegradation of diesel fuel-contaminated wastewater using a three-phase fluidized bed reactor. J. Hazard. Mater. 2008;154:105–111.

25 Charinpanitkul T, Soottitantawat A, Tanthapanichakoon W. A simple method for bakers' yeast cell disruption using a three-phase fluidized bed equipped with an agitator. Bioresour. Technol. 2008;99:8935–8939.

26 Cao C, Zhang K, He C, Zhao Y, Guo Q. Investigation into a gas-solid–solid three-phase fluidized-bed carbonator to capture CO_2 from combustion flue gas. Chem. Eng. Sci. 2011;66:375–383.

27 Mungmart M, Kijsirichareonchai U, Tonanon N, Prechanont S, Panpranot J, Yamamoto T, Eiadua A, Sano N, Tanthapanichakoon W, Charinpanitkul T. Metal catalysts impregnated on porous media for aqueous phenol decomposition within three-phase fluidized-bed reactor. J. Hazard. Mater. 2011;185:606–612.

28 Sheikhi A, Gharebagh RS, Eslami A, Sohi AH. Sequential modular simulation of ethanol production in a three-phase fluidized bed bioreactor. Biochem. Eng. J. 2012;63:95– 103.

29 Dora TK, Mohanty YK, Roy GK, Sarangi B. Adsorption studies of As(III) from wastewater with a novel adsorbent in a three-phase fluidized bed by using response surface method. J. Environ. Chem. Eng. 2013;1:150–158.

30 Brahme PH, Doraiswamy LK. Modelling of a slurry reaction. Hydrogenation of glucose on Raney nickel. Ind. Eng. Chem. Process. Des. Dev. 1976;15(1):130–137.

31 Kosak JR. Hydrogenation of haloaromatic nitro compounds. In *Catalysis in Organic Synthesis*. Ed. Jones WH. New York: Academic Press; 1980:107–117.

32 Kale SS, Chaudhari RV, Ramachandran PA. Butynediol synthesis. A kinetic study. Ind. Eng. Chem. Prod. Res. Dev. 1981;20:309–315.

33 Molga EJ, Westerterp KR. Kinetics of the hydrogenation of 2,4-dinitrotoluene over a palladium on alumina catalyst. Chem. Eng. Sci. 1988;47(7):1733–1749.

34 Hichri H, Accary A, Andrieu J. Kinetics and slurry-type reactor modelling during catalytic hydrogenation of *o*-cresol on Ni/SiO_2. Chem. Eng. Process. 1991;30:133–140.

35 Bukur DB, Nowlcki L, Lang X. Fischer-Tropsch synthesis in a stirred tank slurry reactor. Chem. Eng. Sci. 1994;49 (24A):4615–4625.

36 Rode CV, Chaudhari RV. Hydrogenation of *m*-nitrochlorobenzene to *m*-chloroaniline: reaction kinetics and modeling of a non-isothermal slurry reactor. Ind. Eng. Chem. Res. 1994;33:1645–1653.

37 Stüber F, Benaissa M, Delmas H. Partial hydrogenation of 1,5,9-cyclododecatriene in three phase catalytic reactors. Catal. Today 1995;24:95–101.

38 Nikov I, Paev K. Palladium on alumina catalyst for glucose oxidation: reaction kinetics and catalyst deactivation. Catal. Today 1995;24:41–47.

39 Benaissa M, Roux GCL, Joulia X, Chaudhari RV, Delmas H. Kinetic modeling of the hydrogenation of 1,5,9-cyclododecatriene on Pd/Al_2O_3 catalyst including isomerization. Ind. Eng. Chem. Res. 1996;35:2091–2095.

40 Choudhary VR, Sane MG, Tambe SS. Kinetics of hydrogenation of *o*-nitrophenol to *o*-aminophenol on Pd/carbon catalysts in a stirred three-phase slurry reactor. Ind. Eng. Chem. Res. 1998;37:3879–3887.

41 Kim CJ, Chang K, Chun GT, Jeong YH, Lee SJ. Continuous culture of immobilized streptomyces cells for kasugamycin production. Biotechnol.Prog. 2001;17:453–461.

42 Hoffer BW, Schoenmakers PHJ, Mooijman PRM, Hamminga GM, Berger RJ, Langeveld ADV, Moulijn JA. Mass transfer and kinetics of the three-phase hydrogenation of a dinitrile over a Raney-type nickel catalyst. Chem. Eng. Sci. 2004;59:259–269.

43 Sarkar A, Seth D, Ng FTT, Rempel GL. Kinetics of liquid-phase hydrogenation of isooctenes on a Pd/γ-alumina catalyst. AIChE J. 2006;52(3):1142–1156.

44 Chammingkwan P, Hoelderich WF, Mongkhonsi T, Kanchanawanichakul P. Hydroxylation of benzene over TS-PQ^{TM} catalyst. Appl. Catal. Gen. 2009;352:1–9.

45 Jung JY, Park JK, Chang HN. Bacterial cellulose production by *Gluconacetobacter hansenii* in an agitated culture without living non-cellulose producing cells. Enzyme Microb. Technol. 2005;37:347–354.

46 Ramachandran PA, Chaudhari RV. *Three Phase Catalytic Reactors*. New York: Gordon and Breach Science Publishers; 1983.

47 Smith JM. *Chemical Engineering Kinetics*. McGraw-Hill Chemical Engineering Series. New York: McGraw-Hill; 1970.

48 Levenspiel O. *Chemical Reaction Engineering*, 3rd Ed. New York: John Wiley & Sons, Inc.; 1999.

49 Doraiswamy LK, Sharma MM. *Heterogeneous Reactions: Analysis, Examples and Reactor Design. Vol. 2, Fluid-Fluid-Solid Reactions*. New York: John Wiley & Sons, Inc.; 1984.

50 Krishna R, Sie ST. Strategies for multiphase reactor selection. Chem. Eng. Sci. 1994;49(24A):4029–4065.

51 Hong J, Hecker WC, Fletcher TH. Predicting effectiveness factor for m-th order and langmuir rate equations in spherical coordinates. ACS Div. Fuel Chem. 1999;44(4):1011–1015.

52 Turner JR, Mills PL. Comparison of axial dispersion and mixing cell models for design and simulation of Fischer–Tropsch slurry bubble column reactors. Chem. Eng. Sci. 1990;45:2317–2324.

53 Maretto C, Krishna R. Design and optimisation of a multi-stage bubble column slurry reactor for Fischer–Tropsch synthesis. Catal. Today 2001;66:241–248.

54 de Swart JWA, Krishna R. Simulation of the transient and steady state behaviour of a bubble column slurry reactor for Fischer–Tropsch synthesis. Chem. Eng. Process. 2002;41:35–47.

55 Rados N, Al-Dahhan MH, Dudukovic MP. Modeling of the Fischer Tropsch synthesis in slurry bubble column reactors. Catal. Today 2003;79–80:211–218.

56 Ahón VR, Costa JEF, Monteagudo JEP, Fontes CE, Biscaia JEC, Lage PLC. A comprehensive mathematical model for the Fischer–Tropsch synthesis in well-mixed slurry reactors. Chem. Eng. Sci. 2005;60:677–694.

57 Wang Y, Fan W, Liu Y, Zeng Z, Hao X, Chang M, Zhang C, Xu Y, Xiang H, Li Y. Modeling of the Fischer–Tropsch synthesis in slurry bubble column reactors. Chem. Eng. Process. 2008;47:222–228.

58 Schlüter S, Steiff P, Weinspach M. Heat transfer in two- and three-phase bubble column reactors with internals. Chem. Eng. Process. 1995;34:157–172.

59 Ramachandran PA, Smith JM. Mixing-cell model for design of trickle-bed reactor. Chem. Eng. Sci. 1979;17:91–99.

60 Chaudhari AS, Rampure MR, Ranade VV, Jaganathan R, Chaudhari RV. Modeling of bubble column slurry reactor for reductive alkylation of *p*-phenylenediamine. Chem. Eng. Sci. 2007;62:7290–7304.

61 Cussler EL. *Diffusion: Mass Transfer in Fluid Systems*. Cambridge Series in Chemical Engineering. New York: Cambridge University Press; 2009.

62 Nguyen-Tien K, Patewari AN, Schumpe A, Deckwer WD. Gas-liquid mass transfer in fluidized particle beds. AIChE J. 1985;31:194–201.

63 Patwari AN. Nguyen-Tien K, Schumpe A, Deckwer WD. Three-phase fluidized beds with viscous liquid: hydrodynamics and mass transfer. Chem. Eng. Commun. 1986;40:49–65.

64 Sada E, Kumarawa H, Lee C, Fujiwara N. Gas-liquid mass transfer characteristics in a bubble column with suspended sparingly soluble fine particles. Ind. Eng. Chem. Process. Des. Dev. 1985;24:255–261.

65 Schumpe A, Saxena AK, Fang LK. Gas/liquid mass transfer in a slurry bubble column. Chem. Eng. Sci. 1987;42:1787–1796.

66 Schumpe A, Deckwer WD, Nigam KDP. Gas-liquid mass transfer in three-phase fluidized beds with viscous pseudoplastic liquids. Can. J. Chem. Eng. 1989;67:873–877.

67 Kim JO, Kim SD. Bubble breakage phenomena, phase holdups and mass transfer in three-phase fluidized beds with floating bubble breakers. Chem. Eng. Process. 1990;28:101–111.

68 Kim JO, Kim SD. Gas liquid mass transfer in a three-phase fluidized bed with floating bubble breakers. Can. J. Chem. Eng. 1990;68:368–375.

69 Kang Y, Fan UT, Min BT, Kim SD. Promotion of oxygen transfer in three-phase fluidized-bed bioreactors by floating bubble breakers. Biotechnol. Bioeng. 1991;7:580–586.

70 Nore O, Briens C, Margaritis A, Wild G. Hydrodynamics, gas-liquid mass transfer and particle-liquid heat and mass transfer in a three-phase fluidized bed for biochemical process applications. Chem. Eng. Sci. 1992;47:3573–3580.

71 Lee DH, Kim JO, Kim SD. Mass transfer characteristics and phase holdup in three phase fluidized beds. Chem. Eng. Commun. 1993;119:179–196.

72 Kim SD, Kang Y. Heat and mass transfer in three-phase fluidized-bed reactors-an overview. Chem. Eng. Sci. 1997;52:3639–3660.

73 Danckwerts PV. *Gas-Liquid Reactions*. New York: McGraw-Hill; 1970.

74 Linek V, Vacek V. Chemical engineering use of catalyzed sulfite oxidation kinetics for the determination of mass transfer characteristics of gas-liquid contactors. Chem. Eng. Sci. 1981;36:1747.

75 Deckwer WD, Serpemen Y, Ralek M, Schmidt B. Modeling the Fischer–Tropsch synthesis in the slurry phase. Ind. Eng. Chem. Process. Des. Dev. 1982;21:231–241.

76 Belfares L, Cassanello M, Grandjean BPA, Larachi F. Liquid backmixing in packed-bubble column reactors: a state-of-the-art correlation. Catal. Today 2001;64:321–332.

77 van Baten JM, Krishna R. Eulerian simulations for determination of the axial dispersion of liquid and gas phases in bubble columns operating in the churn-turbulent regime. Chem. Eng. Sci. 2001;56:503–512.

78 van Baten JM, Krishna R. Eulerian simulation strategy for scaling up a bubble column slurry reactor for Fischer–Tropsch synthesis. Ind. Eng. Chem. Res., 2004;43:4483–4493.

79 Moustiri S, Hebrard G, Thakre SS, Roustan M. A unified correlation for predicting liquid axial dispersion coefficient in bubble columns. Chem. Eng. Sci. 2001;56:1041–1047.

80 El-Temtamy SA, El-Sharnoubi YO, El-Halwagi MM. Liquid dispersion in gas—liquid fluidized beds: part I: axial dispersion. The axially dispersed plug-flow model. Chem. Eng. J. 1979;18:151–159.

81 Kato Y, Morooka S, Koyama M, Kago T, Yang S. Longitudinal dispersion coefficient of liquid in three phase fluidized bed for gas-liquid-solid systems. J. Chem. Eng. Jpn. 1985;8(4):313–317.

82 Kim SD, Kim HS, Han JH. Axial dispersion characteristics in three-phase fluidized beds. Chem. Eng. Sci. 1992;47:3419–3426.

83 Han S, Zhou J, Jin Y, Loh KC, Wang Z. Liquid dispersion in gas-liquid-solid circulating fluidized beds. Chem. Eng. J. 1998;70:9–14.

84 Ranade VV. *Computational Flow Modeling for Chemical Reactor Engineering*. London: Academic Press; 2001.

85 Ma D, Ahmadi GA. Thermodynamical formulation for dispersed multiphase turbulent flows. Int. J. Multiphase Flow 1990;16:323–351.

86 Gidaspow D, Bezburuah R, Ding J. Hydrodynamics of circulating fluidized beds, kinetic theory approach. In Fluidization VII, Proceedings of the 7th Engineering Foundation Conference on Fluidization. Ed. Potter OE, Nicklin DJ. Brisbane: Engineering Foundation; May 3–8, 1992:75–82.

87 Schaeffer DG. Instability in the incompressible granular flow. J. Differ. Equ. 1987;66:19–50.

88 Lun CKK, Savage SB, Jeffrey DJ, Chepurniy N. Kinetic theories of granular flow: inelastic particles in coquette flow and slightly inelastic particles in a general flow field. J. Fluid Mech. 1984;140: 223–256.

89 Shalala M, Rogers W, O'Brien TJ. *MFIX Documentation: Volume 1, Theory Guide*. Springfield, VA: National Technical Information Service; 1993.

90 Ibdir H, Arastoopour H. Modeling of multi-type particle flow using kinetic approach. AIChE J. 2005;51:1620–1632.

91 Padial NT, VanderHeyden WB, Rauenzahn RM, Yarbro SL. Three-dimensional simulation of a three-phase draft-tube bubble column. Chem. Eng. Sci. 2000;55:3261–3273.

92 Michele V, Hempel DC. Liquid flow and phase holdup—measurement and CFD modeling for two-and three-phase bubble columns. Chem. Eng. Sci. 2002;57:1899–1908.

93 Schallenberg J, Enss JH, Hempel DC. The important role of local dispersed phase hold-ups for the calculation of three-phase bubble columns. Chem. Eng. Sci. 2005;60:6027–6033.

94 Feng W, Wen J, Fan J, Yuan Q, Jia X, Sun Y. Local hydrodynamics of gas-liquid-nanoparticles three-phase fluidization. Chem. Eng. Sci. 2005;60:6887–6898.

95 Khopkar AR, Ranade VV. Computational flow modelling of multiphase flows in stirred vessels. In *Chemical Engineering in the Pharmaceutical Industry: R&D to Manufacturing*. Ed. Am Ende DJ. Hoboken: John Wiley & Sons, Inc. 2011:269–298.

96 Jia X, Wen J, Wang X, Feng W, Jiang Y. CFD modeling of immobilized phenol biodegradation in three-phase airlift loop reactor. Ind. Eng. Chem. Res. 2009;48:4514–4529.

97 Jia X, Wen J, Wang X, Feng W, Jiang Y. CFD modelling of phenol biodegradation by immobilized *Candida tropicalis* in a gas-liquid-solid three-phase bubble column. Chem. Eng. J. 2010;157:451–465.

98 Troshko AA, Zdravistch F. CFD modeling of slurry bubble column reactors for Fisher–Tropsch synthesis. Chem. Eng. Sci. 2009;64:892–903.

CHAPTER 7
Bioreactors

Pedro Fernandes[1,2] and Joaquim M.S. Cabral[1]
[1]Department of Bioengineering and IBB-Institute for Bioengineering and Biosciences, Instituto Superior Técnico, Universidade de Lisboa, Lisboa, Portugal
[2]Faculdade de Engenharia, Universidade Lusófona de Humanidades e Tecnologias, Lisboa, Portugal

Abstract

Bioreactors are the core of bioprocesses, as they provide the key link between the initial feedstock and the product, where chemical modifications are carried out. Vessels used as bioreactors have to be designed and operated in such a manner to accommodate the requirements of a given type of cells or enzymes to perform in conditions that maximize productivity. Therefore, bioreactor engineering is a cross-disciplinary area, where biology and engineering interact closely. Key issues for bioreactor design and operation are addressed and illustrated by some examples that highlight the relevance of biochemical process industry.

7.1 Introduction

Bioreactors are devices where biochemical reactions are carried out, aiming at the manufacture of a wide array of products, anchored in the use of microbial, mammalian, or vegetable cells and/or enzymes. When microbial cells are used, bioreactors are usually referred to as fermenters. Products formed can be very diverse, such as the traditional antibiotics, amino acids and organic acids, beer, and steroids, to biopharmaceuticals, such as monoclonal antibodies, interleukins, human growth hormone, recombinant vaccines, or human insulin, that involve either mammalian cell lines or microbial cells [1]. Bioreactors are also used: in the production of enzymes, which are themselves the catalysts in bioconversion purposes, where products such as sugars, secondary alcohols, aroma esters, or oligosaccharides are obtained [1–3]; in the production of biofuels and biopolymers, where processes combining physical and chemical pretreatment of low-cost substrates (viz., lignocellulose) combined with enzymatic saccharification and fermentation are gaining relevance as a sound strategy for sustainability [4–6]; for tissue mass culture [7] and for stem cell culture [8]; and in wastewater treatment [9], just to refer to some representative examples. Bioreactors can thus be considered the core of biologically driven production processes. Given the particular nature of biochemical reactions, bioreactors usually handle liquids, although gas and/or solid phases are often present, as in the case of aerobic fermentations or when immobilized cells/enzymes are used, either as a fixed bed or as suspended particles. This chapter aims to provide an overview on some key aspects of bioreactors, namely, their configuration, mode of operation, and model equations as well as their major advantages and limitations.

7.2 Basic concepts, configurations, and modes of operation

7.2.1 Basic concepts

Bioreactors are thus expected to operate in a large array of environments. Bioreactors currently in use in industrial environment can be roughly classified on the basis of aeration and stirring characteristics (Figure 7.1).

The design of a bioreactor is a cross-disciplinary task, involving knowledge in several complementary areas, such as reaction/growth kinetics, mass, heat and momentum transfer, mass and heat balances, and instrumentation and control. Thus, the performance of a bioreactor may be influenced by several features, such as metabolism, morphology, enzymatic activities, substrate concentration, substrate/product inhibition, mode of substrate supply, product removal, mixing efficiency, shear stress, aeration and gas–liquid mass transfer, heat transfer, and foaming. Fermentations are particularly demanding on the bioreactor, since a sterile and contained environment is mandatory along with other requirements, such as maintenance of adequate aeration and agitation, monitoring and control of pH and dissolved oxygen tension (DOT), ports for nutrient and reagents feeding, and ports for sampling and inoculation, and a frame that is scalable minimizes liquid loss and allows the growth of a large array of microorganisms. Moreover, there is a trend to tool up bench-scale bioreactors with specific probes that enable online monitoring of specific products, such as green fluorescent proteins [11] or redox potential [12]. Sterile environment is needed to prevent contamination by organisms other than those cultured. In several processes *in vitro* where enzyme reactors are used, namely, starch processing, the risk of contamination is minimized since operation is performed at relatively

Multiphase Catalytic Reactors: Theory, Design, Manufacturing, and Applications, First Edition. Edited by Zeynep Ilsen Önsan and Ahmet Kerim Avci.

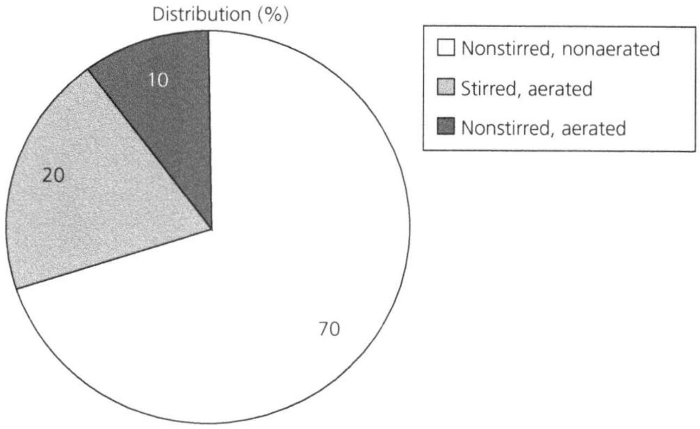

Figure 7.1 An estimate on the current use in industry of bioreactors with different basic characteristics regarding aeration and stirring.
(Source: Adapted from [10].)

high temperatures (within 50 °C to close to 100 °C). On the other hand, the design of the bioreactor must also prevent the release of microorganisms into the environment, a feature which is of particular relevance when pathogenic strains are used in the fermentation processes. Hence, different levels of containment, and accordingly of framework complexity, are required depending on the risk group of the organisms used. Particular care in the design and operation of a bioreactor is mandatory for both technological and economical successes of a given process since, depending on the characteristics of the product and its purity and concentration, the bioconversion step may account for 5–50% of the total costs. The concentration of substrates (and concomitantly of products) in the reaction media is often low, due either to inhibitory/toxic effects or to poor solubility in aqueous media, so diverse strategies have to be devised and integrated into bioreactor operation (viz., fed-batch operation, use of an auxiliary phase as substrate/product pool) not to compromise volumetric productivity [13]. Microbial cells and enzymes are both able to operate in nonconventional environments; thus, complex mass transfer and fluid dynamics scenarios occur which have to be taken into account in reactor design. Hence, the chemical compatibility of the components of the reaction media with the reactor structural materials has to be considered. Moreover, oxygen and salt titers in the reaction media as well as the need to maintain pH and temperature within relatively narrow boundaries prove critical for cell growth and/or expression of intended activity. On the other hand, the increase of microbial mass may lead to growth on the walls of the vessel and also on the probes and baffles, to flocculation, or to autolysis due to nutrient depletion in inner shells of the cell aggregates. Microbial cells and even carrier-bound biologicals are sensitive to strong shear stress which conditions mixing options.

7.2.2 Reactor configurations and modes of operation

Given the diversity of bioconversion/fermentation processes, different reactor configurations and modes of operation have been developed, so as to provide the more adequate environments for process implementation. The more common reactor configurations fall into the following groups:

- Mechanically stirred vessels: Mixing is promoted by mechanical stirring (occasionally magnetic in bench-scale vessels) using different impeller types. In addition, aeration is also employed to augment mixing. In this most commonly used type of bioreactor, the tank content is perfectly mixed and hence uniform in composition.
- Tubular reactors: Packed bed and fluidized bed reactors are included in this category. Particulate forms of enzymes/cells are bundled inside a cylindrical vessel, either in close contact with each other in a packed bed, preferably by forcing the fluid to circulate in downflow mode, or, in the form of a fluidized bed, by forcing the fluid upward, so that the bed of solid particles expands and behaves in a fluidlike manner, which favors contact between individual solid particles, hence enhancing mass transfer. The monolith reactor, where the bed of particles is replaced by a single structure, is a variation of the packed bed with minimal pressure drop.
- Bubble columns: Mixing is promoted without mechanical agitation in cylindrical vessels by introducing gas into a liquid through a sparger. One variation of bubble column reactors preferred by industry is the airlift bioreactor.
- Liquid-impelled loop reactors: In this type of reactor, liquid is forced to circulate through the action of jets or pumps.
- Membrane reactors: Using semipermeable membranes of either flat sheet or tubular type in the bioreactor allows selective separation and reaction can be combined with extraction.
- Microreactors.
- Miscellaneous (viz., drip flow, rotating film).

Illustrative representations of the most significant of these groups are given in Figure 7.2. Monoliths and microreactors are not included in this chapter since they are addressed in detail elsewhere in the book.

Bioreactors can be operated in batch, fed-batch, and continuous modes (Figure 7.3). In the batch mode of operation, all the medium components, but for gases, acid/base for pH control and antifoaming agents are added to the bioreactor at the start of the process, which is allowed to proceed for a given time. Throughout this period there is neither addition of substrates nor removal of products, only forced aeration, in the case of aerobic fermentations and addition of agents for pH and foam control. The system is in a highly dynamic unsteady state, since the concentrations of substrates and products continuously change with time. Once the intended reaction time is achieved, reactor operation is stopped and the medium is processed for product separation and purification. Although batch operation is highly flexible and the risk of contamination is relatively small, as is the risk of genetic instability of the organisms used, it is associated with a low productivity, as a result of a high downtime (unproductive time period, which is used for emptying, cleaning, sterilizing, if required, and filling the bioreactor for a new run) between consecutive batch runs. The introduction of clean-in-place (CIP) strategies and apparatus has nevertheless contributed to reduce significantly the downtime [14–16]. The introduction

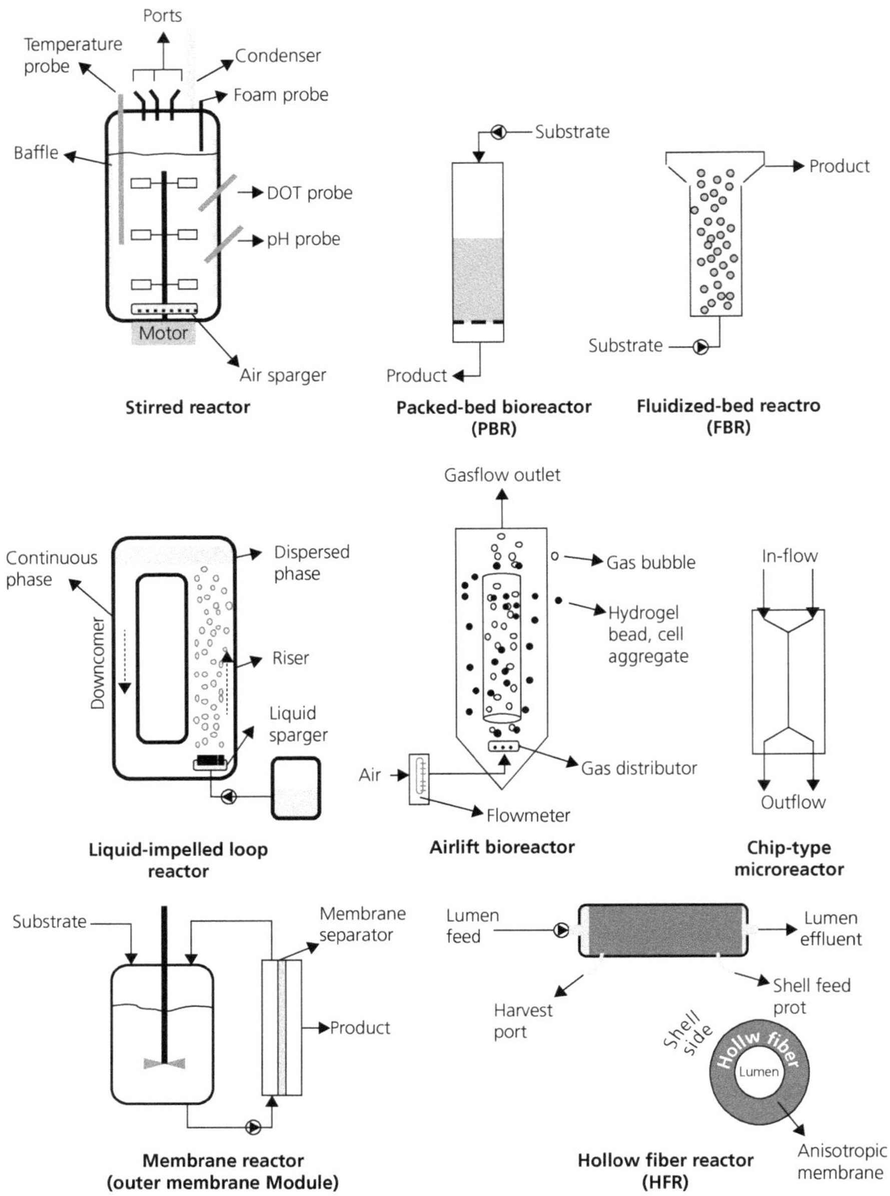

Figure 7.2 Typical configuration of commonly used bioreactors.

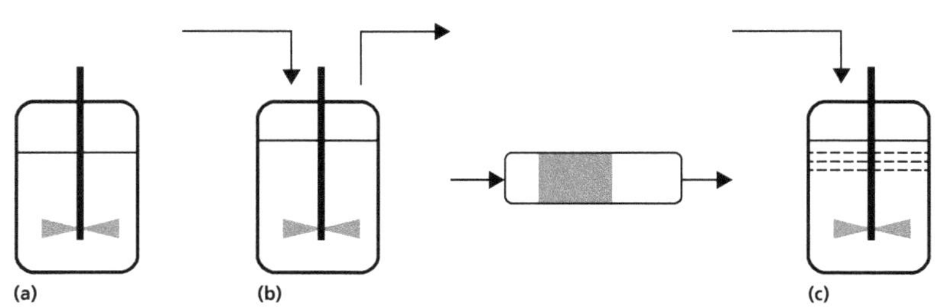

Figure 7.3 Modes of bioreactor operation: (a) batch, (b) continuous, and (c) fed batch.

of single-use bioreactors goes alongside with this trend, since they are extremely flexible, have a quick turnaround, require less infrastructures than traditional vessels, and are gaining acceptance in industrial environment, yet their use in industrial scale (and also in bench scale) is mostly in the production of biopharmaceuticals using mammalian and microbial cells [17–19].

In the continuous mode of operation, fresh medium is constantly added to the bioreactor while spent medium is simultaneously removed from the bioreactor. Since inflow and outflow rates are similar, the reaction volume inside the reactor remains constant and a steady state can be achieved, where the concentrations of all components in the medium remain constant with time. It becomes possible to establish conditions that allow for continuous production with high productivity. Still, the production of secondary metabolites under the continuous mode is not advisable, since it is difficult to maintain the low dilution rates required. Besides, contamination or strain mutation may occur under continuous mode of operation, leading to process failure [20].

When continuous operation is conducted in a perfectly mixed stirred reactor, the medium composition is homogeneous throughout the reactor, thus similar to that in the outflow. On the other hand, when reactions are carried out in an ideal tubular reactor with a plug-flow-like behavior, which displays a uniform velocity profile across the radius resulting from absence of mixing in the direction of flow (axial) but full mixing in the direction normal to the flow (radial), the concentrations of both substrates and products change along the reactor in the flow direction.

In the fed-batch mode of operation, which is an intermediate between batch and continuous, one or more substrates or nutrients (often the limiting substrate) are added to the bioreactor in a continuous manner, namely, as a concentrated solution through a dosing pump (eventually the addition can be performed intermittently). Again, after a given time period, reactor operation is stopped and the medium is directed to downstream processing for product separation and purification. Fed-batch operation may result in a variation of the volume of the reaction medium, in which case part of the working volume of the bioreactor is not used until the end of the run, and hence, the final limit of the extent of the run is the working volume of the bioreactor. Fed-batch operation does not result in a variation of the total volume if the limiting substrate can be fed as a gas or through dialysis. Moreover, a particular variant of the fed-batch mode of operation with no total volume change is a cyclic process which involves the periodic removal of most of the culture while the residual culture is diluted to the original volume using fresh medium and growth is resumed.

Fed-batch operation is particularly suitable in cases when the substrate is inhibitory/toxic or is sparingly soluble, as an approach to prevent the occurrence of by-products that are often formed in the presence of high concentrations of substrate, since only the required amount of substrate for the production of the targeted product is released. For example, controlling substrate delivery is critical when catabolic repression is to be avoided. When implemented, it allows for high cell concentrations to be obtained, with a positive impact on productivity, and thus the fed-batch strategy can be advantageously used in the production of primary metabolites, namely, gluconic acid. In the production of secondary metabolites like penicillin, which has the highest expression during the deceleration phase of growth, fed-batch mode of operation allows for periodic shifts in the growth rate, hence permitting the time span of the productive stage of a process to be extended under a controlled environment. The fed-batch mode is also useful when the process results in a considerable increase in the viscosity of the reaction medium, as observed during the production of polysaccharides such as dextran or xanthan, through the use of a water-based feed. Moreover, fed-batch operation can be used to compensate for excessive solvent (water) loss due to evaporation throughout the time course of a process. On the other hand, efficient fed-batch operation requires a detailed understanding of the mechanisms relating physiology and productivity as well as skilled personnel for designing and developing the process.

7.3 Mass balances and reactor equations

An overall mass balance for a given process can be given as

$$\frac{dM}{dt} = M_i - M_o + r_p - r_c \tag{7.1}$$

where M refers to the total mass, t to time, subscripts i and o refer to inflow and outflow, respectively, r_p to the rate of mass of products formed, and r_c to the rate of mass of substrates consumed.

In a batch reactor, where there are no inlets or outlets, and the liquid medium density can be assumed constant, the material balance for a specific substrate A is given by

$$-V\frac{dC_A}{dt} = -r_A V \tag{7.2}$$

where V is the volume of the medium (m^3), C_A is the concentration of substrate A (mol/m^3), t is the time of reaction (s), and r_A is the rate equation ($mol/m^3/s$) for substrate consumption.

7.3.1 Operation with enzymes

In enzymatic processes, the rate equation typically displays hyperbolic saturation kinetics, better known as Michaelis–Menten kinetics (7.3):

$$-r_A = \frac{V_M C_A}{K_M + C_A} \tag{7.3}$$

where V_M is the maximal reaction rate and K_M is the Michaelis constant.

Substituting Equation 7.3 into Equation 7.2 and integrating, the equation for the batch reactor Equation 7.4 is obtained:

$$-\int_{C_{A0}}^{C} \frac{dC_A}{\left(\frac{V_M C_A}{K_M + C_A}\right)} = \int_0^t dt \Leftrightarrow K_M \ \ln\left(\frac{C_{A0}}{C}\right) + (C_{A0} - C) = tV_M \tag{7.4}$$

where C_{A0} stands for the initial concentration of A.

Taking into consideration the definition of fractional conversion, $X(-)$, as

$$\frac{C_{A0} - C_A}{C_{A0}} = X \tag{7.5}$$

and substituting Equation 7.5 into Equation 7.4 leads to Equation 7.6, a more familiar equation of ideal batch reactor displaying Michaelis–Menten kinetics:

$$tV_M = C_{A0}X - K_M \ \ln(1 - X) \tag{7.6}$$

A similar approach, starting with a material balance, can be used for the characterization of bioreactors operating in the continuous mode. Thus, for a perfectly mixed reactor, or continuous stirred tank reactor (CSTR), where the term of accumulation is zero at steady state and the liquid composition is uniform, the material balance for substrate A is given by Equation 7.7:

$$FC_{A0} - FC_A = -r_A V \tag{7.7}$$

where F is the total volumetric flow rate (m^3/s). In the case of Michaelis–Menten kinetics, and considering Equations 7.5 and 7.7 can be rewritten as follows:

$$(C_{A0} - C)\frac{K_M + C}{V_M C} = \frac{V}{F} \Leftrightarrow K_M \frac{X}{1-X} + XC_{A0} = \tau V_M \tag{7.8}$$

where τ represents the residence time, the reciprocal of which, $F/V = D$, is generally known as the dilution rate.

On the other hand, for the ideal tubular reactor with a plug-flow-like profile (PFR), the material balance has to be made over a differential element of volume, $dV = Adz$, where A is the cross-sectional area of the bioreactor and dz is a differential thickness of the bioreactor (Figure 7.4). The material balance thus becomes

$$FC_A - F(C_A + dC_A) = -r_A dV \Leftrightarrow FC_A - F(C_A + dC_A) = -r_A Adz \tag{7.9}$$

Assuming that the hyperbolic Michaelis–Menten rate equation holds, Equation 7.9 becomes

$$-\int_{C_{A0}}^{C_A} dC_A = \frac{V_M C_A}{K_M + C_A}\frac{A}{F}\int_0^L dz \Leftrightarrow -\int_{C_{A0}}^{C_A} \frac{K_M + C_A}{V_M C_A} dC_A = \frac{A}{F}L$$

$$\Leftrightarrow K_M \ \ln\left(\frac{C_{A0}}{C_A}\right) + (C_{A0} - C_A) = \tau V_M \tag{7.10}$$

which can be arranged by using Equation 7.5 to give

$$\tau V_M = C_{A0}X - K_M \ \ln(1-X) \tag{7.11}$$

Equations 7.6 and 7.11 are formally identical, which suggests that the reaction time in a perfectly mixed batch reactor corresponds to the residence time in a PFR.

When a PFR and a CSTR are compared regarding conversion, in the processing of similar feed compositions, the latter requires a larger volume than the former in order to attain an equal conversion, for both Michaelis–Menten and first-order power-function kinetics, whereas the performance is unaffected by the type of the reactor in case of zero-order kinetics. On the other hand, if the substrate is inhibitory, the use of a CSTR may be favored since substrate concentration is instantaneously diluted to the concentration in the effluent, while reactions characterized by product inhibition are preferably carried out in a PFR since product concentration builds up gradually through (or along) the length of the bioreactor.

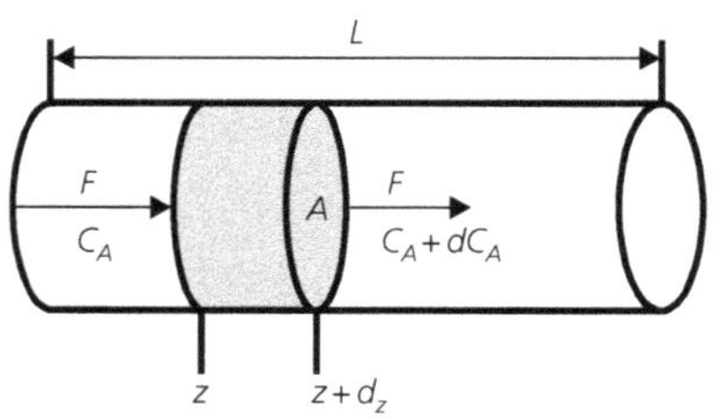

Figure 7.4 Flow through a tubular reactor of total length L, where a differential element of volume, $dV = Adz$, is highlighted.

7.3.2 Operation with living cells

When bioreactor operation relies on the use of living cells for the production of given goods or biomass, cell growth and maintenance requirements have to be considered. These processes are more complex, since they involve simultaneous substrate consumption, product formation, and cell growth, which can be schematically represented as

Cells + Substrate → More cells + Extracellular products

where product formation can occur alongside cell growth (primary metabolite), at the decline of the growth phase (secondary metabolite), or as a mixture of the both. For growth to occur new cells have to be formed; hence, the process can usually be considered as being autocatalytic Equation 7.12:

$$r_g = \frac{dC_c}{dt} = \mu C_c \tag{7.12}$$

where r_g stands for the biomass growth rate ($kg/m^3/s$, usually presented as $kg/m^3/h$), μ stands for the specific growth rate (s^{-1}, usually presented as h^{-1}), and C_c stands for cell mass concentration (kg/m^3).

Cell growth in a typical batch process can be divided into four different phases (Figure 7.5).

In the lag phase, cells adjust to a new environment. In this period, cells synthesize proteins for transporting substrates into the cell and for metabolizing substrates and for preparing all the

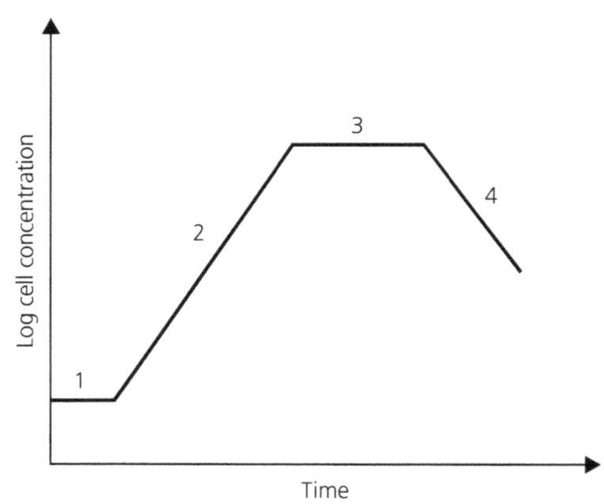

Figure 7.5 Different stages of cell growth in batch processes: (1) lag phase, (2) exponential growth phase, (3) stationary phase, and (4) death phase.

cell machinery for replication of its genetic material. The length of this phase depends on the similarity between the medium from which the inoculum is taken and the fermentation medium; the larger the similarity, the shorter is the lag phase. In the exponential phase, cells use the nutrients in the most efficient manner; thus, they replicate at the maximum rate and cell growth rate is proportional to cell concentration. In the stationary phase, there is no net growth rate, as result of the depletion of nutrients. Moreover, cell growth is also impaired due to the buildup of metabolites, such as organic acids, produced during the exponential growth phase. However, several metabolites, such as antibiotics, are synthesized during the stationary phase. The final phase, the death phase, is characterized by a marked decrease in the concentration of live cells, which can be ascribed to nutrient depletion, buildup of toxic by-products, and a harsh environment.

Several nutrients are required for cell growth to take place alongside product formation:

Cells + Carbon source + Nitrogen source + Oxygen
+ Micro-nutrients + Phosphate source + Other nutrients
→ More cells + Products

The specific growth rate, μ, can still be related to the limiting substrate, usually a carbon source such as glucose, as given in Equation 7.13. Somehow not surprisingly, growth kinetics of microorganisms are often found to follow Monod kinetics, Equation 7.13, which is formally similar to the Michaelis–Menten equation, Equation 7.3, of enzyme kinetics:

$$\mu = \frac{\mu_M C_S}{K_S + C_S} \tag{7.13}$$

where μ_M stands for the maximal specific growth rate (s^{-1}, usually given as h^{-1}), K_s is the Monod constant (kg/m^3), and C_s is the limiting substrate concentration (kg/m^3).

In order to provide a better fit for cell growth, modified expressions have been suggested. One of these takes into account substrate consumption due to maintenance requirements of the cell, expressed as μ_s (s^{-1}, usually given as h^{-1}), hence affecting the specific growth rate μ:

$$\mu = \frac{\mu_M C_S}{K_S + C_S} - \mu_S \tag{7.14}$$

Another expression considers the inhibitory effect of substrate concentration observed in some cases, where K_I is an inhibition constant:

$$\mu = \frac{\mu_M C_S}{K_S + C_S + \dfrac{C_S^2}{K_I}} \tag{7.15}$$

Inhibition as a result of product formation (viz., ethanol production) is also handled by modifying the Monod-type cell growth equation:

$$\mu = \frac{\mu_M C_S}{K_S + C_S} \frac{K_I}{K_I + C_P} \tag{7.16}$$

where C_p stands for product concentration.

Expressions other that Equation 7.13 have also been suggested to describe cell growth rate, such as Contois (7.17), Moser (7.18), and Tessier (7.19) equations:

$$\mu = \frac{\mu_M C_S}{K_{C'} C_C + C_S} \tag{7.17}$$

where $K_{C'}$ is the Contois saturation constant

$$\mu = \mu_M \left[1 - \exp\left(-\frac{C_S}{k}\right)\right] \tag{7.18}$$

$$\mu = \frac{\mu_M C_S}{1 + k C_S^{-\lambda}} \tag{7.19}$$

where k and λ are empirical constants determined by a best fit of the data.

Cell death rate, r_D, as a result of nutrient depletion, toxicity of a given medium component, or shear stress can be accounted as

$$r_D = (k_D + k_i C_{in}) C_C \tag{7.20}$$

where k_D and k_i are the respective specific death rate constants due to natural death or to toxic compounds given by C_{in}.

7.3.2.1 Stoichiometry

During the cell growth process, part of the substrate is used to build up cell components. The efficiency of this process is quantified by the yield of biomass with respect to the substrate, Y_{cs}, expressed as kg dry cells per kg substrate consumed [15]:

$$Y_{CS} = \frac{C_C - C_{C0}}{C_{S0} - C_S} \tag{7.21}$$

where C_{S0} and C_{C0} are the initial values of substrate and cell mass concentrations C_S and C_C, respectively. Plotting C_C versus C_S gives a linear correlation with $(-Y_{CS})$ as the slope. Typical values for Y_{CS} are within 0.4–0.6 (kg/kg).

Product formation can occur at different stages of cell growth. If it takes place during the growth phase, the rate of product formation, r_p (kg$_p$/m^3/h), can be given by

$$\begin{aligned} r_P &= Y_{PC} r_g = Y_{PC} \mu C_C = q_P C_C \\ &= Y_{PC} \frac{\mu_M C_S}{K_S + C_S} C_c \text{ (for Monod growth)} \end{aligned} \tag{7.22}$$

where q_p is the specific rate of product formation (kg$_p$/kgcells^3/h) and Y_{PC} is the yield of mass of product formed per mass of new cells:

$$Y_{PC} = \frac{C_P - C_{P0}}{C_C - C_{C0}} \tag{7.23}$$

Substrate may be consumed further just to comply with the maintenance requirements of the cell, the corresponding term m being calculated as the mass of substrate consumed per cell mass per time. The rate of substrate consumption for maintenance, r_{sm} (kg/m^3/h), is given by

$$r_{sm} = mC_C \tag{7.24}$$

Thus, the net rate of substrate consumption $(-r_s)$ can be formulated as

$$(-r_s) = Y_{SC}r_g + Y_{SP}r_P + mC_C \tag{7.25}$$

Moreover, when primary metabolites are produced, it is difficult to distinguish the amount of substrate used for cell growth from that used for product synthesis; hence, Y_{SC} and Y_{SP} are simply lumped into Y_{SC} and Equation 7.25 is simplified to

$$(-r_s) = Y_{SC}r_g + mC_C \tag{7.26}$$

Usually, $\mu \gg m$; hence, the maintenance term is often neglected in mass balances, which leads to further simplification of Equation 7.26 to

$$(-r_s) = Y_{SC}r_g \tag{7.27}$$

On the other hand, when product formation is independent of biomass growth, it is related solely to substrate consumption, eventually a secondary substrate, SM:

$$r_P = Y_{PSM}(-r_s) \tag{7.28}$$

where Y_{PSM} is the yield of mass of product formed per mass of substrate consumed:

$$Y_{PSM} = \frac{C_P - C_{P0}}{C_{SM0} - C_{SM}} \tag{7.29}$$

Often $C_{P0} = 0$; thus,

$$Y_{PSM}(C_{SM0} - C_{SM}) = C_P \tag{7.30}$$

The rate law for secondary metabolite formation often takes the form of the familiar Monod equation; therefore,

$$r_P = \frac{k_P C_{SM}}{K_{SM} + C_{SM}} C_C \tag{7.31}$$

where k_P is the specific rate constant for secondary metabolite formation, C_{SM} is the concentration of the substrate for secondary metabolite formation, and K_{SM} is the Monod constant for secondary metabolite formation.

Thus, the net rate of substrate consumption during the stationary phase, r_{SM}, can be given by

$$r_{SM} = mC_C + \frac{Y_{SMP}k_P C_{SM}}{K_{SM} + C_{SM}} C_C \tag{7.32}$$

In the previous discussion, extreme cases concerning the stage of product synthesis were considered, but actual production may occur in both exponential and stationary phases of growth. In order to account for this, the specific rate of product formation, q_p, is then given by the Luedeking–Piret equation:

$$q_P = \alpha\mu_g + \beta \tag{7.33}$$

and thus

$$r_P = q_P C_C \tag{7.34}$$

7.3.2.2 Mass balances

Characterization of a process occurring in a bioreactor where living cells are involved requires mass balances for cells, substrate, and extracellular product, all of them ultimately interlinked.

For stirred tank reactors, the general cell balance over the reactor volume is given by

$$V\frac{dC_C}{dt} = F_0 C_{C0} - FC_C + (r_g - r_d)V \tag{7.35}$$

Similarly, the substrate balance is given by

$$-V\frac{dC_S}{dt} = F_0 C_{S0} - FC_S - r_S V \tag{7.36}$$

and the product balance is given by

$$V\frac{dC_P}{dt} = F_0 C_{P0} - FC_P + r_P V \tag{7.37}$$

In the case of batch reactors, the cell, substrate, and product balances are simplified to

$$V\frac{dC_C}{dt} = (r_g - r_D)V \Leftrightarrow \frac{dC_C}{dt} = (r_g - r_D) \tag{7.38}$$

(which is simplified to Eq. 7.12 in the exponential growth phase):

$$-V\frac{dC_S}{dt} = -r_S V \tag{7.39}$$

$$V\frac{dC_P}{dt} = r_P V \tag{7.40}$$

In the case of a CSTR, on the other hand, all accumulation terms disappear; therefore,

$$0 = F_0 C_{C0} - FC_C + (r_g - r_D)V \tag{7.41}$$

$$0 = F_0 C_{S0} - FC_S - r_S V \tag{7.42}$$

$$0 = F_0 C_{P0} - FC_P + r_P V \tag{7.43}$$

Characterization of a process addressing the trends of substrate depletion, cell growth, and product formation involves the solution of a set of equations. Functional forms of the equations will vary, given the different representative models for growth and stages of product formation and of substrate consumption, and the solution procedure often requires a numerical approach. However, several simplifications are often made by attending to the relevance of several variables as compared to others, namely, often the maintenance coefficient is negligible.

7.3.2.3 Batch operation

In a similar manner to what was previously mentioned for enzyme reactors, the overall batch process for cell cultivation has a time span longer than solely the culture period. In addition to cell growth, batch operation of a fermenter includes development of the seed culture, cleaning, filling and sterilization of the fermenter and fermentation medium, and finally emptying the vessel for downstream processing.

In a well-mixed fermenter operated batchwise, and assuming Monod equation is valid throughout all stages of cell growth, it can still be stated that the specific growth rate μ is given by

$$\frac{dC_c}{dt} = \frac{\mu_M C_S}{K_S + C_S} C_c \tag{7.44}$$

where

$$\frac{dC_C}{dt} = \frac{dC_C}{dS}\frac{dC_S}{dt} \text{ and } \frac{dC_S}{dt} = -\left(\frac{\mu}{Y_{CS}} + \frac{r_P}{Y_{PS}}\right) C_C \tag{7.45}$$

Here Y_{CS} and Y_{PS} are assumed to be constant throughout the process and

$$C_C = C_{C0} \exp(\mu_M t) \tag{7.46}$$

as in the exponential growth phase. Combining these two equations, Equations 7.45 and 7.46, and integrating gives the time length of a batch culture for the exponential growth phase, t_b:

$$t_b = \left(\frac{1}{\mu_M}\right) \ln\left[1 + \frac{(C_{S0} - C_S)}{\left(\frac{1}{Y_{CS}} + \frac{r_P}{\mu_M Y_{PS}}\right) C_{C0}}\right] \tag{7.47}$$

If, in addition, no product is formed except cell mass itself, this equation is simplified to

$$t_b = \left(\frac{1}{\mu_M}\right) \ln\left[1 + \frac{(C_{S0} - C_S) Y_{CS}}{C_{C0}}\right] \tag{7.48}$$

7.3.2.4 Fed-batch operation

In a fed-batch culture, the volume of the liquid in the fermenter increases with time. The feed rate, which may not be constant, establishes the dilution rate D, which typically decreases with time:

$$D = \frac{F}{V} \tag{7.49}$$

and

$$F = \frac{dV}{dt} \tag{7.50}$$

Using Equation 7.12 also, the net change in cell mass with time is given by

$$\frac{d(C_C V)}{dt} = \frac{C_C dV}{dt} + \frac{V dC_C}{dt} = \mu C_C V \tag{7.51}$$

Implementing Equation 7.50,

$$\frac{dC_C}{dt} = C_C(\mu - D) \tag{7.52}$$

Therefore, when $\mu = D$, cell mass concentration remains constant. For each substrate concentration in the medium, μ is determined according to a suitable model (viz., Monod). On the other hand, C_S can be adjusted to a given μ. Typically, fed-batch operation starts in batch mode; once suitable conditions are attained (viz., cell concentration), a given feed is provided at a proper rate.

The net balance of substrate in the fed-batch fermenter is given by

$$\frac{d(VC_S)}{dt} = FC_{Si} - \left(\frac{\mu}{Y_{CS}} + \frac{r_P}{Y_{PS}}\right) C_C V \tag{7.53}$$

For a growth-associated product, where substrate is fed in such a manner as to maximize specific growth rate, and assuming negligible cell death, one obtains Equations 7.54 and 7.55:

$$C_C = \frac{\dfrac{C_{C0}\mu\exp(-\mu t)}{\left[\mu - \left(\frac{\mu}{Y_{CS}} + m + \frac{\alpha\mu}{Y_{PS}}\right)\frac{C_C}{C_S}\right]}}{\left\{\dfrac{C_{C0}}{\left[\mu - \left(\frac{\mu}{Y_{CS}} + m + \frac{\alpha\mu}{Y_{PS}}\right)\frac{C_C}{C_S}\right]} \times \left(\frac{\mu}{Y_{CS}} + m + \frac{\alpha\mu}{Y_{PS}}\right)\frac{(-1)}{C_S} \times \exp(-\mu t) - 1\right\}} \tag{7.54}$$

$$C_P = \alpha\mu / \left(\frac{\mu}{Y_{CS}} + m + \frac{\alpha\mu}{Y_{PS}}\right)\frac{(-1)}{C_S} \times \left[1 - \exp\left(\frac{\mu}{Y_{CS}} + m + \frac{\alpha\mu}{Y_{PS}}\right)\frac{F(t)}{C_S}\right] \tag{7.55}$$

For a process where the specific growth rate is maintained constant,

$$F = \frac{\mu C_{C0} V_0 \exp(\mu t)}{[(C_{S0} - C_S) Y_{CS}]} \tag{7.56}$$

$$V = V_0\left[\left(1 + \frac{C_{C0}}{C_S Y_{CS}}\exp(\mu t)\right) - \frac{C_{C0}}{C_S Y_{CS}}\right] \tag{7.57}$$

$$C_C = \mu C_{C0}\exp(\mu t) / \left[\mu\left(1 + \frac{C_{C0}}{C_S Y_{CS}}\exp(\mu t)\right) - \frac{C_{C0}}{C_S Y_{CS}}\right] \tag{7.58}$$

7.3.2.5 Continuous operation

Continuous mode of operation in well-mixed stirred tanks often starts as a fed-batch process at constant feed rate under a given substrate concentration C_{S0}, so that $D = \mu$ until an outlet valve is opened. This is performed in order to allow the effluent to be continuously recovered from the reactor at an outlet flow rate equal to the inlet flow so that the volume of the medium in the fermenter stays constant and steady state is achieved.

From the cell mass balance equation for steady state, Equation 7.38, with sterile feed and assuming negligible cell death,

$$FC_C = r_g V \Leftrightarrow D = \mu \tag{7.59}$$

that shows that the growth rate can be controlled through the dilution rate. The substrate concentration can be established by replacing μ with D in the substrate consumption equation which, in the case of Monod-type growth, results in

$$C_S = \frac{DK_S}{\mu_M - D} \tag{7.60}$$

If a single nutrient is limiting, and it is only used for cell growth, cell mass and substrate concentrations, C_C and C_S, can be related through the stoichiometry, to lead to

$$C_C = Y_{CS}\left(C_S - \frac{DK_S}{\mu_M - D}\right) \tag{7.61}$$

The dilution rate is limited by an upper boundary, a point at which all cells are washed out from the system, since D exceeds μ and the residence time is not enough to allow for cells to replicate. Hence, C_C in the reactor equals zero; thus, the dilution rate for washout, D_{wo}, is defined as

$$D_{wo} = \frac{\mu_M C_S}{C_S + K_S} \tag{7.62}$$

Another relevant parameter for continuous, stirred fermenter operation is the identification of the dilution rate which maximizes cell production. Since the volumetric cell productivity can be expressed as

$$\frac{FC_C}{V} = DC_C \tag{7.63}$$

and replacing C_C according to Equation 7.61

$$DC_C = DY_{CS}\left(C_S - \frac{DK_S}{\mu_M - D}\right) \tag{7.64}$$

The dilution rate that allows for maximum productivity, D_{prod}, is determined by differentiating the production rate, DC_C, with respect of the dilution rate and equating it to zero:

$$D_{prod} = \mu_M\left[1 - \left(\frac{K_S}{K_S + C_S}\right)^{1/2}\right] \tag{7.65}$$

As for enzyme reactors, in a CSTR, the concentrations of the different medium components in the effluent are similar to those in the reactor.

In order to maintain a high cell density inside the fermenter, cells can be recycled back to the vessel (Figure 7.6), with a recycle ratio, α, defined as

$$\alpha = \frac{F_R}{F} \tag{7.66}$$

as well as a concentration factor, β:

$$\beta = \frac{C_{CR}}{C_{C1}} \tag{7.67}$$

The net cell mass balance is

$$V\frac{dC_{C1}}{dt} = F_0 C_{C0} + (F + F_R)C_{C1} + V\mu C_{C1} \tag{7.68}$$

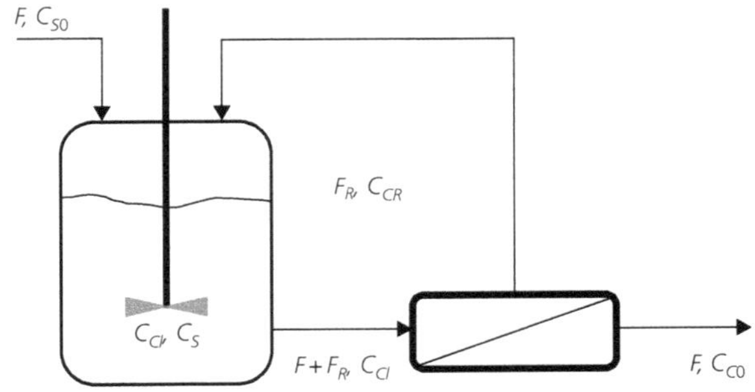

Figure 7.6 CSTR with recycle.

At steady state, assuming sterile feed and introducing α, this equation becomes

$$\mu = D[1 + \alpha(1 - \beta)] \tag{7.69}$$

which shows that cell recycle allows chemostat operation at a dilution rate higher than the specific growth rate.

From the net mass balance on the substrate,

$$V\frac{dC_S}{dt} = FC_{S0} + \alpha FC_S - V\frac{\mu C_{C1}}{Y_{CS}} - (1 + \alpha)C_S \tag{7.70}$$

Assuming steady state

$$C_{C1} = \frac{DY_{XS}(C_{S0} - C_S)}{\mu} \Leftrightarrow C_{C1} = \frac{Y_{XS}(C_{S0} - C_S)}{[1 + \alpha(1 - \beta)]} \tag{7.71}$$

For Monod-type cell growth kinetics,

$$C_S = \frac{K_S D[1 + \alpha(1 - \beta)]}{\mu_M - D[1 + \alpha(1 - \beta)]} \tag{7.72}$$

and

$$C_{C1} = \frac{Y_{CS}}{[1 + \alpha(1 - \beta)]}\left\{C_{S0} - \frac{K_S D[1 + \alpha(1 - \beta)]}{\mu_M - D[1 + \alpha(1 - \beta)]}\right\} \tag{7.73}$$

7.4 Immobilized enzymes and cells

7.4.1 Mass transfer effects

In fermentation/bioconversion processes, biomolecules can be in suspension/solution or distributed in a roughly uniform manner within solid particles or at their surface or otherwise in the form of relatively large aggregates that are not bound to a solid carrier. In the latter three cases, the enzymes/cells are immobilized by carrier-bound or carrier-free methods. Immobilization brings along increased complexity, since a substrate A must (i) migrate from the bulk of the liquid phase, where its concentration is C_{Ab} to the surface of the immobilized biomolecule by crossing the stagnant liquid film surrounding the particle where its concentration is C_{AS}, and (ii) diffuse through the pores or gel-like structure of the carrier (Figure 7.7), unless only surface immobilization is considered, which is relatively rare since this seriously limits cell/enzyme load and ultimately productivity.

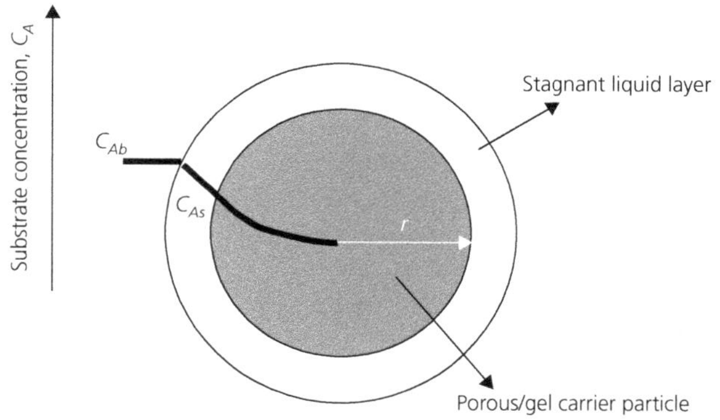

Figure 7.7 Variation of substrate concentration, C_A, from bulk liquid medium to center of support particle.

In either case, the overall reaction rate comprises both mass transfer and bioreaction kinetics, where the former is usually the rate-limiting step. It is possible to define a total effectiveness factor, η_T, which accounts for the effects of mass transfer on the reaction rate as

$$\eta_T = \frac{r_{Aobs}}{r_A} = \frac{\text{Observed rate of reaction}}{\text{Rate of reaction if } C_A = C_{Ab} \text{ everywhere in the particle}}$$

Mass transfer limitations can be broken down to two situations which are related to external and internal mass transfer resistances. Accordingly, external, η_e, and internal, η_i, effectiveness factors can be defined as

$$\eta_e = \frac{r_{As\,obs}}{r_A} = \frac{\text{Observed rate of reaction if } C_A = C_{As} \text{ everywhere in the particle}}{\text{Rate of reaction if } C_A = C_{Ab} \text{ everywhere in the particle}}$$

and

$$\eta_i = \frac{r_{A\,obs}}{r_{As}} = \frac{\text{Observed rate of reaction}}{\text{Rate of reaction if } C_A = C_{As} \text{ everywhere in the particle}}$$

where

$$\eta_T = \eta_i \eta_e \tag{7.74}$$

When external or internal mass transfer resistances are negligible, $\eta_e = 1$ or $\eta_i = 1$, respectively. If intrinsic kinetic parameters (determined while using free enzymes or cells, with no mass transfer limitations) are known, the total effectiveness factor can thus be used together with the reactor design equations as

$$\eta_{Tav} \tau V_{Mi} = C_{A0}X - K_{Mi} \ \ln(1-X) \ \text{(BSTR and PFR)} \tag{7.75}$$

$$K_M \frac{X}{1-X} + XC_{A0} = \tau\eta_T V_M \text{(CSTR)} \tag{7.76}$$

where η_{Tav} is an average total effectiveness factor which can be determined by

$$\eta_{Tav} = [C_{A0}X - K_{Mi} \ \ln(1-X)] \Big/ \left[\int_{C_{A0}}^{C} \frac{dC_A}{\eta_T C_A/(K_{Mi} + C_A)} \right] \tag{7.77}$$

The need for an average total effectiveness factor results from the variation of the total effectiveness factor with substrate concentration.

In the case of external mass transfer resistance, the rate of substrate depletion, $-r_A$, is controlled by the mass transfer across the stagnant film:

$$-r_A = k_L a(C_{Ab} - C_{As}) \tag{7.78}$$

where k_L is the liquid mass transfer coefficient, a is the surface area per unit volume of the catalyst, and C_{Ab} and C_{As} are the concentrations of the substrate in the bulk liquid phase and on the particle surface. The relevance of external mass transfer resistance can be established based on the Damköhler number (Da) that relates the maximum reaction rate and the maximum rate of mass transfer:

$$Da = \frac{-r_{A\ max}}{k_L a(C_{Ab} - C_{As})} \tag{7.79}$$

For Damköhler numbers largely exceeding unity, mass transfer is the rate-limiting step; hence, the apparent reaction rate is given by

$$-r_A = k_L a C_{Ab} \tag{7.80}$$

The internal mass transfer resistances are ascribed to the slow migration of substrate inside long, tortuous pores or layers of hydrogel. Mass transfer resistances inside the particle leads to a decrease from unity in η_i. This decrease is correlated with a dimensionless parameter, the Thiele modulus, relating the reaction rate and molecular diffusion inside the support particle. The Thiele modulus varies with the reaction mechanism and shape of the particle. However, it has been established that if a generalized Thiele modulus is used, Φ_{gen},

$$\Phi_{gen} = \frac{-r_A(C_A = C_{As})a^{-1}}{\sqrt{2D_{Aeff} \int_0^{C_{As}} (-r_A)dC_A}} \tag{7.81}$$

where D_{Aeff} is the effective diffusivity of A in the immobilization support, then η_i can be determined with a satisfactory approximation by

$$\eta_i = \frac{\tanh \Phi_{gen}}{\Phi_{gen}} \tag{7.82}$$

for different reaction kinetics and particle shapes [21].

For Michaelis–Menten-type kinetics and a spherical particle radius of R_P, Equation 7.82 becomes

$$\Phi_{gen} = \frac{V(C_A = C_{As})R_P}{3(1+\chi)\sqrt{2D_{Aeff} V(C_A = C_{As}) \left[1 - \chi \ln\left(1 + {}^1\!/_\chi\right)\right]}} \tag{7.83}$$

Other approaches for relating Thiele moduli and effectiveness factors are available in the literature [22, 23].

In the case of a packed bed reactor, a typical configuration for the use of immobilized enzymes, where mass transfer resistances are not significant and Michaelis–Menten-type kinetics is observed, the reactor design equation is simply obtained from

$$u_L \frac{dC_A}{dz} = -\frac{V_M C_A}{C_A + K_M} \tag{7.84}$$

leading to

$$L = \frac{u_L}{V_M}\left[K_M \ln\left(\frac{C_{A0}}{C_A}\right) + (C_{A0} - C_A)\right] \tag{7.85}$$

where L is the length of the packed bed and u_L is the liquid flow velocity. Naturally the reaction time, t_r, in the packed bed is determined according to

$$t_r = \frac{L}{u_L} \tag{7.86}$$

7.4.2 Deactivation effects

The performance of immobilized reactors in continuous operation can be negatively influenced by several incidents such as enzyme/cell leakage, thermal denaturation of the enzyme, disintegration of the support, or microbial contamination. These parameters can be evaluated experimentally, and approaches can thus be designed in order to counter their negative effect on bioreactor performance.

Predictive models for thermal denaturation of enzymes have been developed [24, 25], the most commonly used one being the exponential decay model:

$$-\frac{dE_a}{dt} = r_d = k_d E_a \tag{7.87}$$

where E_a stands for enzyme activity at time t, k_d is deactivation rate constant, and r_d is the rate of deactivation.

Integration of this equation leads to

$$E_a = E_{a0}\ \exp(-k_d t) \tag{7.88}$$

where E_{a0} is the enzyme activity at time 0. This equation can be used to establish the variation of V_M with time as

$$V_M = V_{M0}\ \exp(-k_d t) \tag{7.89}$$

where V_{M0} is V_M at time 0. Combining Equation 7.89 with the design equation for immobilized enzyme reactors and integration with respect to time allows the prediction of bioreactor performance when enzyme activity loss occurs due to thermal effects.

7.5 Aeration

Aeration plays a key role in aerobic fermentations, and often the rate of oxygen transfer from the gas phase sparged into the liquid phase ends up controlling the overall rate of the fermentation process, particularly when the cell biomass increases. The low solubility of oxygen in aqueous media, about 10 ppm at ambient temperature and pressure, is a major cause of such behavior. Given the small size of cells in the micron range, external mass transfer resistances to the cells can be neglected, and it may be assumed that the main resistance to mass transfer is located at the gas–liquid interface, more particularly at the surface of the gas bubbles.

Hence, the rate of oxygen transfer per unit volume of fluid, N_{O_2}, from the gas phase to the liquid phase is given by

$$N_{O_2} = k_L a\left(C^*_{O_2} - C_{O_2}\right) \tag{7.90}$$

where $C^*_{O_2}$ is the solubility of oxygen in the medium, C_{O_2} is the actual concentration of oxygen in the bulk liquid medium, and a is gas–liquid interfacial area per unit volume of fluid. The difference between the two concentrations is the driving force for mass transfer and is typically very small. The term $k_L a$, commonly named as the volumetric mass transfer coefficient, is a key parameter to measure the performance of gassed bioreactors and is often used as a reference parameter for scale-up [26]. This parameter can be experimentally determined using different methodologies [13]. Alternatively, it can be predicted using adequate empirical correlations, such as

$$k_L a = 0.026\left(\frac{P_g}{V}\right)^{0.4} u_g^{0.5} \tag{7.91}$$

where P_g is the power requirement of a gas-sparged liquid in a stirred tank and u_g is the superficial gas velocity.

For a bubble column, on the other hand, the following correlation is proposed:

$$k_L a = 0.32 u_g^{0.7} \tag{7.92}$$

Detailed information on predictive $k_L a$ correlations for various bioreactor configurations and in media with different rheologies can be found in the literature [13, 27, 28].

7.6 Mixing

Liquid mixing in stirred bioreactors can be promoted by an assorted set of impellers with different designs. When microbial cell fermentations are considered, the Rushton turbine is almost universally used. This impeller promotes a radial flow and the shear stress created does not harm cell growth. Other impellers can also be used with given specific requirements regarding rheology and suspension of solid particles. For instance, an axial flow pattern is more adequate for suspension of solid particles, which is best performed by a propeller or by pitched-blade turbines, whereas mixing of highly viscous media is best performed using gate anchors or helical impellers [11]. Mixing times in industrial-scale stirred vessels usually take less than 2 min.

Unlike stirred tanks, bubble columns and similar liquid-impelled and airlift vessels have no moving parts for mixing and hence are of simpler conception, since they are basically cylindrical vessels with a high height to diameter ratio for the gas to be bubbled. Bubble columns are less demanding on power than stirred vessels of similar volume, since the former essentially require power to feed a gas against the static head of the liquid inside the vessel; therefore, they are suitable for use when very large-scale aerobic fermentations are addressed. These bioreactors also operate with suspended solids (immobilized enzymes/cells) in which case they are often termed slurry bubble reactors. Mixing in bubble columns naturally relies on the flow of the bubbles released from the sparger. Basically, two different flow regimes are observed. At low gas flow rates, all bubbles rise with similar velocity and there is no back-mixing; this is the so-called homogeneous flow, where bubbles are homogeneously distributed and liquid mixing is poor. A different pattern typical of operating conditions, and termed as heterogeneous or turbulent flow, is observed when high gas velocities are used. In this case, chaotic circulatory patterns develop; the liquid and bubbles tend to rise at the center of the column while near the walls a downward flow of liquid is observed, which also entrains bubbles, thus leading to some back-mixing.

A condition typical of heterogeneous flow in a column of diameter ϕ is given by the following equation:

$$u_L = 0.9\left(g\phi u_g\right)^{0.33} \tag{7.93}$$

where u_L is the upward liquid velocity and g is the acceleration of gravity. Based on this correlation, the mixing time, t_m, for a column of length L can be predicted according to

$$t_m = \frac{11L\left(g\phi^{-2}u_g\right)^{-0.33}}{\phi} \tag{7.94}$$

Airlift bioreactors are used as an alternative to bubble columns; the configuration of the former permits more defined liquid flows, since upward- and downward-moving streams are physically separated. Airlift bioreactors typically allow better mixing than bubble columns, particularly if they are of the external-loop configuration, but at low liquid velocity.

7.7 Heat transfer

Heat transfer is a relevant feature of bioreactor operation, although biological reactions do not display high exo- or endothermicity typical of many chemical reactions, nor do they occur at such high temperatures. Nevertheless, the temperatures at which enzyme- or cell-catalyzed reactions (including fermentations) occur have to be controlled under strict limits. Hence, bioreactors incorporate heat transfer surfaces either as external jackets or coils, as internal coils, or as external heat exchangers. These surfaces are used to adjust the temperature of the stream fed to the reactor, to maintain the temperature of the reaction medium constant according to the endothermic/exothermic nature of the reaction involved plus the heat dissipated due to stirring during bioreactor operation, and to sterilize the medium prior to fermentation. Typically, steam and water are used as heating and cooling media, respectively.

7.8 Scale-up

The development of a bioconversion/fermentation process consists of several stages, namely, bench, pilot, and plant scale. The outcome of the bench scale establishes key conditions for the upcoming stages, and the relevant aspects of the process (viz., medium composition and concentrations of different species involved, temperature and pH of operation, conditions that optimize productivity) are expected to be reproduced as closely as possible when larger vessels are used. In order to achieve this, several aspects have to be considered. Unfortunately, it has been shown that there is not a single criterion for scale-up, yet a set of guidelines has to be considered.

The most immediate is geometric similarity, but on its own this is clearly not enough. In addition, keeping a constant k_La throughout scales is one of the most widely used approaches when aerobic fermentations are involved. Alternatively, constant volumetric power consumption has also been commonly suggested. Other alternatives proposed as criteria for scale-up include constant tip speed, similar Reynolds number, or equal mixing times [1, 21].

7.9 Bioreactors for animal cell cultures

Animal cells, particularly mammalian cells, are cultured in large scale for the production of value added goods, namely, biopharmaceuticals (interferon, monoclonal antibodies, vaccines, and the like), hence the need for a bioreactor [29]. These cells are much more fragile than microbial cells, given the absence of a cell wall, and their growth is much slower than that of microbial cells, both of which increase the cost of cell culture media and the requirements for sterility, in processes where high cell densities are envisaged, to provide the necessary product output [30, 31]. Some of these cells, such as tissue cells, further require a solid support for growth and are termed anchorage-dependent cells. Tissue engineering is actually a particularly demanding application, since each special type of tissue structure and production methodology (viz., bone, skin) has its own engineering and biological requirements, and *in vitro* conditions must emulate *in vivo* environment. Moreover, bioreactors are often required to operate under aseptic conditions for several weeks, the time gap that may be required for a functional tissue substitute to mature [32]. Supports or microcarriers used are solid particles typically with sizes of 100 μm to some mm, made of biocompatible materials such as collagen, cellulose, DEAE-dextran, gelatin, glass, or polystyrene plastic [33, 34]. Particular care has to be given to the selection of carrier used as scaffolds, since alongside the adequate affinity toward the cells targeted, mass transfer limitations related to a particular carrier have to be accounted for [35]. In contrast to these cells are anchorage-independent cells which may grow irrespective of attachment to a solid surface. Both types of cells can be cultivated in stirred reactors, albeit at low stirring rates and with an adequately designed impeller [8, 36, 37]. Alternatively, airlift bioreactors can also be used, where the shear stress promoted by the impeller is avoided, while hydrodynamic conditions may be created that allow for mixing and mass transfer requirements [38]. Hollow-fiber reactors have also been used for the growth of animal cells, where medium flows in one side of the fibers while cells are kept in the opposite side. This system has also been used to grow artificial organs [39, 40]. The wave bioreactor is of an altogether different configuration. Within the single-use approach, the wave bioreactor consists of a rocker platform to promote mixing through a wave-type motion. On top of the rocker, a single-use, sealed, cell culture bag is placed, aeration being promoted by a dedicated pump [41].

7.10 Monitoring and control of bioreactors

Monitoring and controlling the changes in different process parameters that take place inside a bioreactor is becoming ever more essential for bioprocesses to be competitive. Moreover, the knowledge gathered on the bioprocess is expected to allow the estimation of process variables throughout operation and incorporate such information in control strategies. On the whole, when monitoring and control are considered, bioconversion/fermentation processes do not differ significantly from chemical processes but for the biological nature, which may introduce some uncertainty in process dynamics; hence, variables may change unexpectedly during operation. In some cases, this variation may be counteracted, since some process variables can be monitored online, namely, pH, DOT, and CO_2, temperature can be measured online, and foam occurrence can be detected through the use of probes. Changes in these variables can be offset through the addition of concentrated alkali or acid solutions (in the case of pH), increase in stirring speed (up to a certain level) and gas flow or changes in gas composition (in the case of DOT or CO_2), addition of antifoam, and use of heat transfer equipment. Changes in pH and DOT can be related to the physiological condition of the microorganisms and to biomass concentration. Such information may ultimately be integrated in control loops, which are expected to anticipate events during operation given that data gathered act in a preventive manner. This approach requires dedicated software and hardware to process in due time the data and act according to established conditions [42, 43]. Significant technological developments have been made in order to improve online monitoring of fermentation processes. Such developments have involved mostly optics, namely, fiber optics and fluorescence, enzyme and ion-selective sensors and miniaturization, spectroscopic methods (viz., near-infrared (NIR) spectroscopy), and nuclear magnetic resonance (NMR). Online monitoring of metabolic products, intracellular metabolites, and nutrients/substrates is nevertheless far from an easy task, for it is hampered by several operational conditions. Some of the most significant limitations include the large variety of molecules, interferences resulting from the actual fermentation media (high cell concentration and biofilm formation), cost of some measuring devices, and some lack of flexibility of the latter. Online monitoring would be most welcome since it would provide a real-time, comprehensive picture of the system and thus allow for the development of more effective control systems, namely, on a time basis. Currently, monitoring of most metabolic products, intracellular metabolites, and nutrients/substrates is carried out off-line, usually requiring complex, time-consuming procedures.

Nomenclature

- a gas–liquid interfacial area per unit volume of fluid (m^{-1})
- a surface area per unit volume (m^{-1})
- A cross-sectional area of the bioreactor (m^2)
- C concentration (kg/m^3, mol/m^3)
- C_{Ab} concentration of the substrate in the bulk liquid phase (kg/m^3, mol/m^3)
- C_{As} concentration of the substrate on the particle surface (kg/m^3, mol/m^3)
- D dilution rate (s^{-1}), diameter (m)
- D_{Aeff} effective diffusivity of A (m^2/s)
- D_{prod} dilution rate for maximum productivity (s^{-1})
- D_{wo} dilution rate for washout (s^{-1})
- E_a enzyme activity at time t ($mol_{active\ enzyme}$)
- F total volumetric flow rate (m^3/s)
- g acceleration of gravity (m/s^2)
- K_c Contois saturation constant ($kg_{substrate}/kg_{biomass}$)
- k_d deactivation rate constant (s^{-1})
- k_D specific death rate constants due to natural death or to toxic compounds (s^{-1})
- k_i specific death rate constants due to toxic compounds ($m^3/s/kg_{toxic\ compound}$)
- K_I inhibition constant (kg/m^3)
- k_L liquid mass transfer coefficient (m/s)
- K_M Michaelis constant (kg/m^3, mol/m^3)
- K_s Monod constant (kg/m^3)
- M mass (kg)
- m mass of substrate consumed per cell mass per time ($kg_{substrate}/kg_{biomass}/s$)
- P_g power requirement of a gas-sparged liquid (W)
- q_P specific rate of product formation (s^{-1})
- r_A rate equation for consumption of substrate A ($mol/m^3/s$)
- r_{Aobs} observed rate of reaction ($mol/m^3/s$)
- r_c rate of mass of substrates consumed ($kg/m^3/s$)
- r_D cell death rate ($kg/m^3/s$)

r_d rate of enzyme deactivation ($mol_{active\ enzyme}/s$)
r_g biomass growth rate ($kg/m^3/s$)
r_p rate of mass of products formed ($kg/m^3/s$)
r_{sm} rate of substrate consumption for maintenance ($kg/m^3/s$)
t time (s^{-1})
u linear velocity (m/s)
V volume (m^3)
V_M maximal reaction rate ($mol/m^3/s$)
X conversion (—)
Y_{CS} biomass to substrate yield ($kg_{biomass}/kg_{substrate}$)
Y_{PC} product to biomass yield ($kg_{product}/kg_{biomass}$)
Y_{PS} product to substrate yield ($kg_{product}/kg_{substrate}$)

Greek letters

α recycle ratio (—), growth-dependent term of Luedeking–Piret equation (—)
β concentration factor (—), growth-independent term of Luedeking–Piret equation (s^{-1})
η effectiveness factor (—)
μ specific growth rate (s^{-1}, usually presented as h^{-1})
μ_M maximal specific growth rate (s^{-1}, usually presented as h^{-1})
μ_s specific rate of substrate consumption due to maintenance requirements of the cell (s^{-1}, usually presented as h^{-1})
τ residence time (s)
Φ Thiele modulus (—)
Φ_{gen} generalized Thiele modulus (—)
ϕ diameter (m)

Subscripts

0 initial
A substrate A
c biomass
e external
g gas, gaseous: inflow, internal, intrinsic
in inhibitory
L liquid
o outflow
p product
R recycle
s substrate
T total

References

1 Soetaert W, Vandamme EJ, editors. Industrial Biotechnology—Sustainable Growth and Economic Success. Weinheim: WILEY-VCH Verlag GmbH & Co. KGaA; 2010.
2 Liese A, Seelbach K, Wandrey C, editors. Industrial Biotransformations. 2nd ed. Weinheim: WILEY-VCH Verlag GmbH & Co. KGaA; 2006.
3 Meyer H-P, Eichhorn E, Hanlon S, Lütz S, Schürmann M, Wohlgemuth R, Coppolecchia R. The use of enzymes in organic synthesis and the life sciences: Perspectives from the Swiss Industrial Biocatalysis Consortium (SIBC). Catalysis Science and Technology 2013; 3: 29–40.
4 Kamm B, Gruber PR, Kamm M, editors. Biorefineries-Industrial Processes and Products: Status Quo and Future Directions. Weinheim: WILEY-VCH Verlag GmbH & Co. KGaA; 2006.
5 Gupta R, Kumar S, Gomes J, Kuhad RC. Kinetic study of batch and fed-batch enzymatic saccharification of pretreated substrate and subsequent fermentation to ethanol. Biotechnology for Biofuels 2012; 5: 16–25.
6 Suhardi VSH, Prasai B, Samaha D, Boopathy R. Combined biological and chemical pretreatment method for lignocellulosic ethanol production from energy cane. Renewable Bioresources 2013; 1: 1–5.
7 Martina Y, Vermette P. Bioreactors for tissue mass culture: Design, characterization, and recent advances. Biomaterials 2005; 26: 7481–7503.
8 Fernandes-Platzgummer A, Diogo MM, da Silva CL, Cabral JMS. Maximizing mouse embryonic stem cell production in a stirred tank reactor by controlling dissolved oxygen concentration and continuous perfusion operation. Biochemical Engineering Journal 2014; 82: 81–90.
9 Tewari PK, Singh RK, Batra VS, Balakrishnan M. Membrane bioreactor (MBR) for wastewater treatment: Filtration performance evaluation of low cost polymeric and ceramic membranes. Separation and Purification Technology 2010; 71: 200–204.
10 Najafpour GD. Biochemical Engineering and Biotechnology. Amsterdam: Elsevier Science; 2007.
11 Hisiger S, Jolicoeur M. A multiwavelength fluorescence probe: Is one probe capable for on-line monitoring of recombinant protein production and biomass activity? Journal of Biotechnology 2005; 117: 325–336.
12 Escalante-Minakata P, Ibarra-Junquera V, Rosu HC, De León-Rodríguez A, González-García R. Online monitoring of Mezcal fermentation based on redox potential measurements. Bioprocess and Biosystems Engineering 2009; 32: 47–52.
13 Kim PY, Pollard DJ, Woodley JM. Substrate supply for effective biocatalysis. Biotechnology Progress 2007; 23: 74–82.
14 Christi Y, Moo-Young M. Clean-in-place systems for industrial bioreactors: Design, validation and operation. Journal of Industrial Microbiology and Biotechnology 1994; 13: 201–207.
15 Franks JW, Seiberling DA, inventors; Electrol Specialities Company, assignee. Portable clean-in-place system for batch processing equipment. US patent 6,161,558. December 19, 2000.
16 Roy K, Undey C, Mistretta T, Naugle G, Sodhi M. Multivariate statistical monitoring as applied to clean-in-place (CIP) and steam-in-place (SIP) operations in biopharmaceutical manufacturing. Biotechnology Progress 2014; 30: 505–515.
17 Seiberling DA. Clean-in-Place for Biopharmaceutical Processes. New York: Informa Healthcare; 2007.
18 Hou JJC, Codamo J, Pilbrough W, Hughes B, Gray PP, Munro TP. New frontiers in cell line development: Challenges for biosimilars. Journal of Chemical Technology and Biotechnology 2011; 86: 895–904.
19 Eibl R, Löffelholz C, Eibl D. Disposable bioreactors for inoculum production and protein expression. In: Pörtner R, editor. Animal Cell Biotechnology—Methods in Molecular Biology, Volume 1104. New Jersey: Humana Press; 2014. p 265–284.

20 Hoskisson PA, Hobbs G. Continuous culture—making a comeback? Microbiology 2005; 151: 3153–3159.
21 Nielsen J, Villadsen J, Lidén G. Bioreaction Engineering Principles. 3rd ed. New York: Springer; 2011.
22 Doran P. Bioprocess Engineering Principles. 2nd ed. New York: Academic Press; 2013.
23 Fogler HS. Essentials of Chemical Reaction Engineering. New York: Prentice Hall; 2010.
24 Henley J, Sadana A. Deactivation theory. Biotechnology and Bioengineering 1986; 28: 1277–1285.
25 Nunes MAP, Rosa ME, Fernandes PCB, Ribeiro MHL. Operational stability of naringinase PVA lens-shaped microparticles in batch stirred reactors and mini packed bed reactors-one step closer to industry. Bioresource Technology 2014; 164: 362–370.
26 Garcia-Ochoa F, Gomez E. Bioreactor scale-up and oxygen transfer rate in microbial processes: An overview. Biotechnology Advances 2009; 27: 153–176.
27 Gupta PP, Merchant SS, Bhat AU, Gandhi AB, Bhagwat SS, Joshi JB, Jayaraman VK, Kulkarni BD. Development of correlations for overall gas hold-up, volumetric mass transfer coefficient, and effective interfacial area in bubble column reactors using hybrid genetic algorithm-support vector regression technique: Viscous Newtonian and non-Newtonian liquid. Industrial & Engineering Chemistry Research 2009; 48: 9631–9654.
28 Schmidt FR. Optimization and scale up of industrial fermentation processes. Applied Microbiology and Biotechnology 2005; 68: 425–435.
29 Wang D, Liu W, Han B, Xu R. The bioreactor: A powerful tool for large-scale culture of animal cells. Current Pharmaceutical Biotechnology 2014; 6: 397–403.
30 Meyer H-P, Schmidhalter DR. Microbial expression systems and manufacturing from a market and economic perspective. In: Agbo EC, editor. Innovations in Biotechnology (e-book). InTech; 2012. p. 211–250. Available at http://cdn.intechopen.com/pdfs-wm/28714.pdf. Accessed August 19, 2014.
31 Genzela Y, Vogela T, Buck J, Behrendt I, Ramirez DV, Schiedner G, Jordan I, Reichl U. High cell density cultivations by alternating tangential flow (ATF) perfusion for influenza A virus production using suspension cells. Vaccine 2014; 32: 2770–2781
32 Gardel LS, Serra LA, Reis RL, Gomes ME. Use of perfusion bioreactors and large animal models for long bone tissue engineering. Tissue Engineering: Part B 2014; 20: 126–146.
33 Park K-H, Pérez RA, Jin G-Z, Choi S-J, Kim H-W, Wall IB. Microcarriers designed for cell culture and tissue engineering of bone. Tissue Engineering Part B: Reviews 2013; 19: 172–190.
34 Ordovás L, Park Y, Verfaillie CM. Stem cells and liver engineering. Biotechnology Advances 2013; 31: 1094–1107.
35 Salehi-Nik N, Amoabediny G, Pouran B, Tabesh H, Shokrgozar MA, Haghighipour N, Khatibi N, Anisi F, Mottaghy K, Zandieh-Doulabi B. Engineering parameters in bioreactor's design: A critical aspect in tissue engineering. BioMed Research International 2013; 2013, Article ID 762132: 1–15.
36 Odeleye AOO, Marsh DTJ, Osborne MD, Lye GJ, Micheletti M. On the fluid dynamics of a laboratory scale single-use stirred bioreactor. Chemical Engineering Science 2014; 111: 299–312.
37 Núnez GF, Leme J, Parizotto LA, Chagas WA, Rezende AG, da Costa BLV, Monteiro DCV, Boldorini VLL, Jorge SAC, Astray RM, Pereira CA, Caricati CP, Tonso A. Influence of aeration–homogenization system in stirred tank bioreactors, dissolved oxygen concentration and pH control mode on BHK-21 cell growth and metabolism. Cytotechnology 2014; 66: 605–617.
38 Hesse F, Ebel M, Konisch N, Sterlinski R, Kessler W, Wagner R. Comparison of a production process in a membrane-aerated stirred tank and up to 1000-L airlift bioreactors using BHK-21 cells and chemically defined protein-free medium. Biotechnology Progress 2003; 19: 833–843.
39 Gramer MJ, Poeschl DM. Comparison of cell growth in T-flasks, in micro hollow fiber bioreactors, and in an industrial scale hollow fiber bioreactor system. Cytotechnology 2000; 34: 111–119.
40 Pinxteren JAM, Craeye D, inventors; ReGenesys BVBA, assignee. Expansion of stem cells in hollow fiber bioreactors. US patent application 20120308531. December 6, 2012.
41 Rodrigues ME, Costa AR, Henriques M, Azeredo J, Oliveira R. Wave characterization for mammalian cell culture: Residence time distribution. New Biotechnology 2012; 29: 402–408.
42 Venkateswarlu C. Advances in monitoring and state estimation of bioreactors. Journal of Scientific & Industrial Research 2004; 63: 491–498.
43 Julien C, Whitford W. Bioreactor monitoring, modeling, and simulation. Bioprocess International 2007; 2007: 10–17.

PART 4

Structured reactors

CHAPTER 8

Monolith reactors

João P. Lopes[1] and Alírio E. Rodrigues[2]

[1]Department of Chemical Engineering and Biotechnology, University of Cambridge, Cambridge, UK

[2]Laboratory of Separation and Reaction Engineering, Associate Laboratory LSRE/LCM, Department of Chemical Engineering, Faculty of Engineering University of Porto, Porto, Portugal

Abstract

Catalytic monoliths represent perhaps the most consolidated design embodying the class of "structured reactors." In fact, for a few decades, both terms could have been used interchangeably. The introduction of a structured environment for fluid flow and mass transfer to a catalytic surface is achieved by an arrangement of channels, whose walls are made of a material which is either incorporated with catalysts or that can act as a support for catalytic porous layers. Technologies based on this concept were found beneficial in several fields, from the synthesis of chemicals to environmental protection. In this chapter, the design principles of a monolithic washcoated channel are given, considering transport in the channel and in the washcoat, where reaction is also present. The operation of this monolith reactor can occur in several regimes, characterized by different relative magnitudes of (internal and external) mass transfer and reaction kinetics. Each regime is associated with a given physical picture and describes the system behavior in a limited range of the governing parameters. A methodology is proposed which allows the mapping of these validity regions in several operating diagrams and the analytical derivation of the boundaries that delimit each simplified description. The theoretical analysis of multiphase processes in catalytic monoliths is also briefly discussed.

8.1 Introduction

Monoliths are usually seen as single blocks with a high porosity attributed to the existence of numerous channels through which fluids flow (Figure 8.1). Each channel is sometimes referred as a "cell," and it is the repetition of several of these cells that composes an arrangement which may resemble a *honeycomb*. The channels are usually straight and parallel and are divided by a ceramic or metallic material. The skeleton that forms the structure can have catalytic properties, or a porous layer containing the catalyst can be deposited on its walls (washcoat). Monoliths have been applied in several chemical processes but are mainly known for their use in the automotive industry.

The application of these structures in catalytic processes led to the appearance of features which are so distinct from the ones found in conventional fixed beds that a new class of reactors, with its own description, was required in order to accurately characterize them. Thus, the field of *structured reactors* appeared, and nowadays apart from monoliths, foam and wall-coated microfabricated reactors are also considered under this category. Washcoated monoliths possess clear similarities with fabricated microreactors. Sometimes both terms are used to describe the same situation, but they are usually distinguished by different ranges of the characteristic dimensions, with the monoliths being associated with larger cell hydraulic diameters ($d_{ch} \sim 0.5 - 1\,\text{mm}$). Nevertheless, both technologies claim the same advantages in terms of mass/heat transfer intensification. For arrangements that are based on the channel as an elemental unit, the term "structured" is quite literal as these are spatially regular. However, the most distinctive characteristic is probably the increased control over the contact between the reactant containing fluid and the solid catalyst. This is a consequence of several particularities: (i) the monolith design can be optimized so that reactants take full advantage from the catalyst at reduced pressure drop (the catalyst activity—reactor pressure drop trade-off found in packed beds is disentangled here); (ii) there is increased freedom in the design of the channel and of the catalytic wall, which, even though coupled, are governed by different characteristic length scales (the channel diameter and the wall/washcoat thickness); (iii) it is believed that issues of flow maldistribution over the reactor cross section are less severe than those found in packed beds; (iv) the velocity profile (usually laminar) is well defined and thus the reactor behavior is easily predicted from very simple models (in contrast with the complex hydrodynamics observed in a bed of particles); and (v) these are technologies which are regarded as extremely flexible concerning possible operating modes (e.g., with respect to mass and heat transfer). Apart from the features aforementioned, these reactors have been generically associated with several advantages: (i) low pressure drops even at high throughputs, (ii) elimination of resistances to the transport of chemical species and heat, (iii) low axial dispersion and backmixing, and

Multiphase Catalytic Reactors: Theory, Design, Manufacturing, and Applications, First Edition. Edited by Zeynep Ilsen Önsan and Ahmet Kerim Avci.

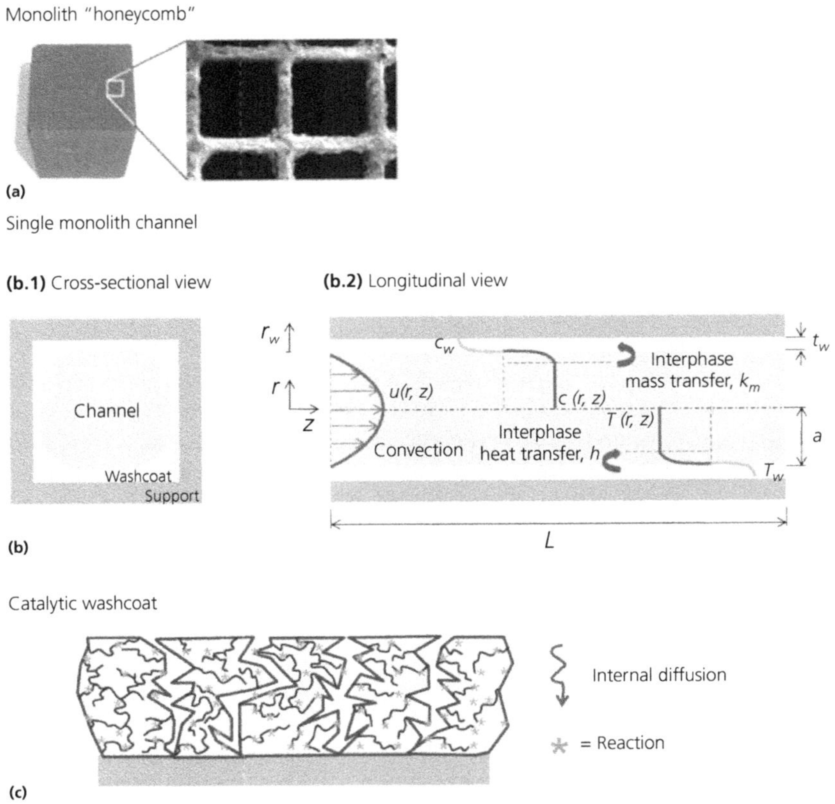

Figure 8.1 Representation of the several scales in a catalytic monolithic reactor. (a) Monolith honeycomb [1] (Source: Araki et al. [1]. Reproduced with permission of Elsevier). (b) Single washcoated monolith channel. (c) Catalytic washcoat.

(iv) large surface area per reactor volume. Thus, the highest impact on the process performance is expected for applications which (i) require short contact times, (ii) present severe mass transfer limitations (due to fast consumption of reactants) or where undesirable hot spots appear under feasible conditions in nonstructured reactors, (iii) have strict selectivity constraints, and (iv) are dominated by heterogeneous phenomena (catalytic reactions), in comparison with gas-phase homogeneous reactions.

There have been many review articles and monographs (e.g., Cybulski and Moulijn [2] and Chen et al. [3]) dedicated to this topic. In this chapter, we focus on the design and classical transport–reaction analysis of these reactors (Section 8.2). Then, it is showed how the relevant regimes of operation of a monolith can be identified in terms of ranges of dimensionless parameters, which combine the variables describing the geometry, operation, and physicochemical properties of the system. This can be done analytically as illustrated in Section 8.3. The issue of performance evaluation in isothermal monoliths is also discussed. While most part of the chapter refers to gas–solid or liquid–solid processes, Section 8.4 presents some considerations about three-phase systems.

8.1.1 Design concepts

The flexibility brought by structured reactors (and, in particular, by the catalytic monolith channel) leads to the existence of a large number of design concepts differing in configurations, materials, and operating modes. Table 8.1 shows several possibilities to classify a given monolith reactor according to these features:

(a) *Monolith channel flow*. The number of fluid phases in channel flow gives origin to completely distinct processes and required design methods. Single-phase flow applications predominate and the behavior of these reactors is better understood, owing to the relative simplicity of the hydrodynamic description. The flow in channels with these dimensions is usually laminar (Section 8.2.1), but operation in the turbulent regime or in the laminar–turbulent transition has been considered [4, 5]. In a comparison between microreactors, laminar monoliths, and turbulent monoliths, higher yield of HCN from the Andrussow process was observed in the latter. The value increased from 16% for laminar flow monoliths (realized in straight channels with diameters from 500 to 1200 μm) to 38% in turbulent flow through an irregular monolithic material with pore diameters between 500 and 800 μm [6]. Recently, applications with multiphase flow (viz., gas–liquid systems) have been studied at the laboratory scale, mainly in academic research. The key concept in this case is to take advantage from reactor structuring (by careful definition of channel dimensions and flow regime) to control the interface between one fluid phase (in the form of droplets or bubbles) and the other. The channel dimensions are typically reduced, providing a sufficiently large surface to volume ratio, improving the contact between phases and intensifying interphase transport fluxes. The implementation of gas–liquid Taylor flow (Case a.2 in

Table 8.1 Monolith reactors classified according to flow, materials, and operation features.

Classification of monolith reactor designs

(a) Phases in monolith channel flow

(a.1) Single phase (G/S or L/S processes)

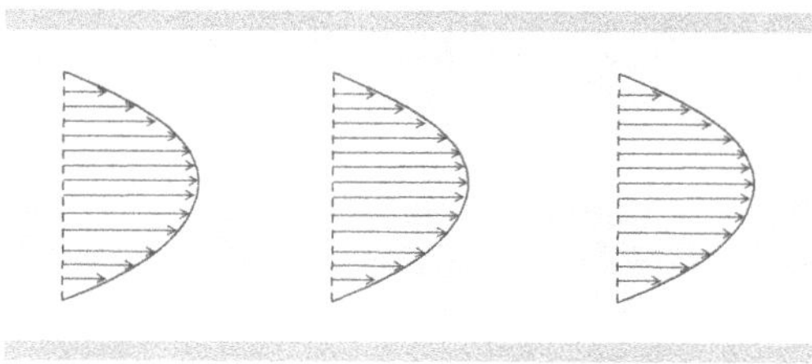

(a.2) Multiphase (G/L/S or L/L/S processes)

(b) Incorporation of catalyst

(b.1) Catalytic washcoat applied on support

(b.2) Intrinsically catalytic porous wall

(b.3) Packed monolith with inert wall

(c) Aspect ratio

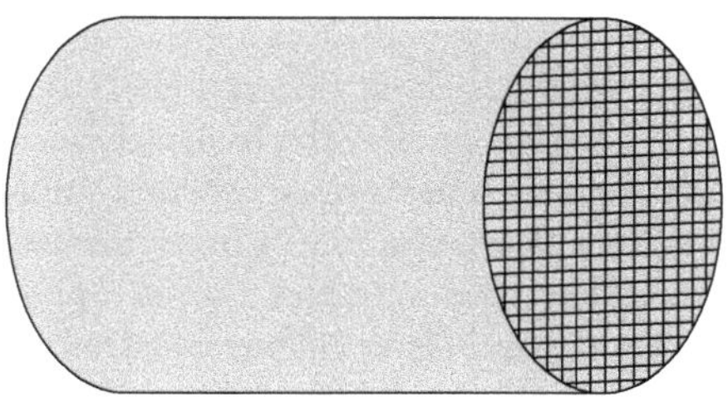

(c.1) Long channels

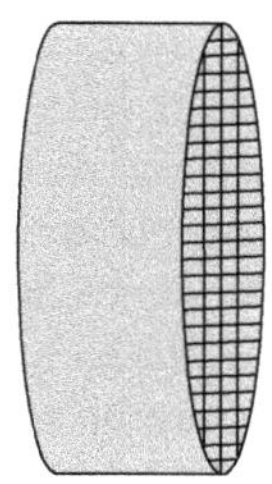

(c.2) Short channels

(d) Structure material

(d.1) Ceramic

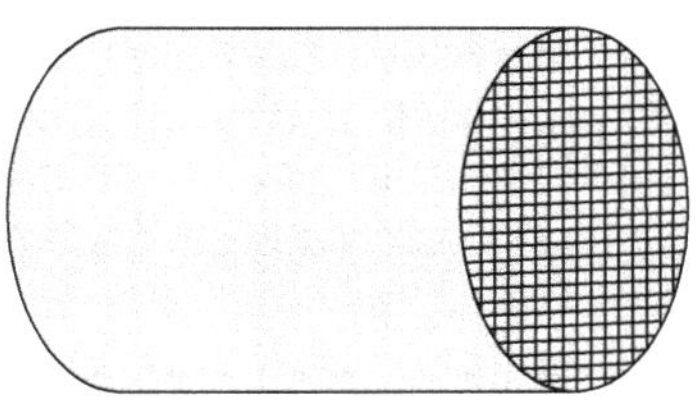

(d.2) Metallic

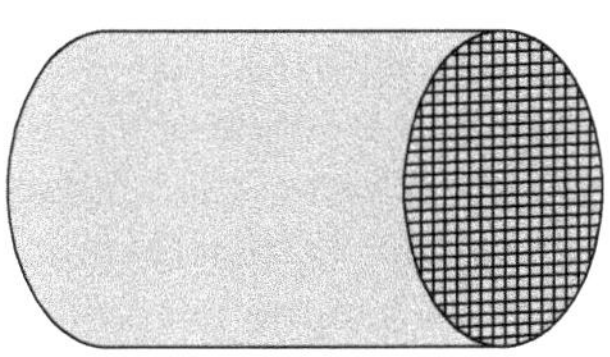

(e) Heat transfer operating modes

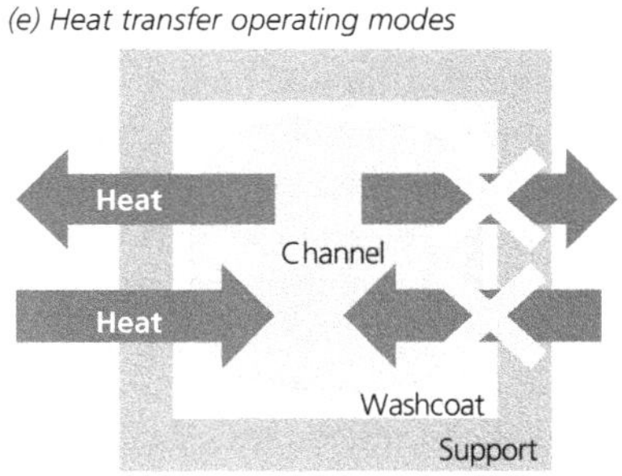

(e.1) Isothermal

(e.2) Adiabatic

Table 8.1) also leads to a fundamental change in the process description compared to steady single-phase flow, due to the transient nature of the phenomena. Other flow regimes are mentioned in Section 8.4.

(b) *Incorporation of catalyst.* There are several ways to incorporate catalyst into a monolith reactor. For heterogeneous catalytic monoliths (the subject of this chapter), this choice is critical, as it will affect the whole process synthesis. Table 8.1 shows three options:

(b.1) *Catalytic washcoat applied on support.* The intensification in mass transfer observed in microstructured reactors suggests placing the catalyst at the wall in order to take advantage from the large values of the ratio between the channel surface and volume. The design where an infinitesimal catalyst layer is deposited at the channel wall with surface area equal to the one dictated by the microchannel geometry can only be understood as an idealization. In practice, the fluid–solid interfacial area is greatly increased (10–100 times the geometrical surface) by a washcoat porous layer (e.g., γ-alumina) containing the dispersed catalyst(s), so that sufficient loading is achieved. The fluid–solid external interface area per volume may increase when decreasing channel size, but the internal surface area is determined by the washcoat porous layer. The channel cross section is usually completely available for fluid flow, even though designs have been presented where an inert packing (e.g., glass spheres) is used to promote radial mixing and approach plug flow [7].

As can be seen in Figure 8.1, a multiscale approach for the reaction–transport analysis is appropriate. Kočí et al. [8] considered three different scales (nano, micro, and macro) when modeling CO oxidation in porous Pt/γ-alumina catalyst washcoated on a monolith. The "nanolevel" consisted of a single mesoporous alumina particle. Digital reconstruction (from SEM and TEM images of the real catalyst) was used to study the agglomeration of alumina nanoparticles and subsequent deposition of individual crystalites of Pt. At the microlevel, the composition of the washcoat from the virtual packing of Pt/alumina microparticles (nanolevel model) is considered. The formulation using the effectiveness factor and average reaction rate is used. The macrolevel consists in the catalytic monolith, which was modeled assuming plug flow.

(b.2) *Intrinsically catalytic porous wall.* The monolith can be directly extruded from a mixture of support and catalyst particles. In this case, the wall itself already contains the catalyst as an integral part of its structure. In this case, the boundary conditions become more complex and channel interaction in specific arrangements has to be considered (the assumption of symmetry between different channels may not be valid).

(b.3) *Packed monolith with inert wall.* In this case, the arrangement at the reactor scale can be considered structured, even though the packing in each channel is naturally random. However, due to the small scales (bed diameter), the undesired concentration and temperature profiles that appear in macroscale reactors can be mitigated. An example of a packed monolith can be found in Kapteijn et al. [9]. Possible disadvantages include the appearance of nonuniform temperature, concentration and flow profiles (leading to increased pressure drop, channeling and favoring the appearance of hot spots), and the occurrence of clogging. The step of packing the microchannels also raises some technical issues that require additional study [10].

These cases can be seen as being associated with an increase in catalyst loading (as we move from b.1 to b.3) but also in pressure drop and internal (porous catalyst) mass transfer resistance. Matching transfer and reaction rates may be the basis not only for dimensioning the channels in these reactors but also for the selection of the appropriate design suited for each application. Therefore, the timescales for mass ($\tau_{mass\ transfer}$) and heat transfer ($\tau_{heat\ transfer}$) should be lower than the one for reaction ($\tau_{reaction}$), if the reaction is not to be mass transfer limited. Since the transport characteristic times depend on the channel diameter as $\sim d_{ch}^2$, while the time constant for a heterogeneous reaction may be proportional to $\sim d_{ch}$, there is a maximum value of the monolith channel diameter, respecting the following condition:

$$\frac{V}{S_{surf}} < \frac{k_m c_0}{R_{obs}(c_0)} \quad \text{or} \quad d_{ch} < \frac{2ShDc_0^{1-m}}{k_{surf}\eta} \tag{8.1a}$$

Alternatively, comparing with heat transfer:

$$\frac{V}{S_{surf}} < \frac{hc_0}{R_{obs}(c_0)} \quad \text{or} \quad d_{ch} < \frac{2Nu\kappa c_0^{1-m}}{k_{surf}\eta} \tag{8.1b}$$

The choice of the characteristic dimension can be nevertheless limited by other factors such as allowable pressure drop, occurrence of clogging/fouling, deviations from uniform flow distribution to individual channels, and maximization of productivity metrics based on product flow rates. For certain applications (e.g., synthesis of pharmaceuticals), larger channel diameters are also favored, since this facilitates cleaning. Beretta et al. [11] compare micropacked beds and annular microchannel reactors for intrinsic kinetic measurements at high temperature and high space velocities. The lower pressure drop and better temperature control in the coated microchannel favored the use of the structured reactor.

(c) *Aspect ratio.* The cross-sectional diameter to axial length ratio (proportional to the dimensionless number α, defined in the following) is the most distinguishing relationship between the dimensions that a monolith channel may present. Two limits are found: long channels ($\alpha \to 0$) and short channels ($\alpha = O(1)$, i.e., channel length and diameter of the same order of magnitude), represented by designs c.1 and c.2, respectively, in Table 8.1. The distinction has several implications in reactor modeling. If the aspect ratio is not negligible, axial diffusion has to be considered in the problem description. This leads to more complex solutions of the mass transfer–wall reaction problem described in Section 8.2.2. Moreover, for short monoliths, effects that would otherwise be considered only at the entrance length now gain importance over the total length of the channel. Some of these effects are developing concentration and temperature profiles (Lévêque's regime), simultaneously developing hydrodynamic flow, and stabilizing conduction effects on axial temperature gradients in the support. Kolodziej and Lojewska [12] studied the utilization of short-channel structured reactors for the combustion of volatile organic compounds (VOCs), aiming to improve mass transfer and reduce coking, in comparison with longer channels with small diameters.

Replacing a long monolith by several consecutively stacked short channels increases transfer coefficients, since the operation in each section will occur in the "inlet region." Hence, the length of the active structure is designed so as to be lower than the entrance length or at most comparable with this value. Correlations for the transfer coefficients and friction factor are distinct from the usual expressions in long channels (an apparent friction factor is used in Ref. [12], where the development of hydrodynamic boundary layers cannot be ignored). According to Ref. [12], transfer coefficients can be increased up to 10 times, reducing diffusional resistance. The physical picture which leads to an increase in the transport rates is also associated with higher pressure drop. However, even taking into account the increased flow resistance (in a criterion defined in Section 8.3.5), overall efficiencies are two to three times higher than the ones observed in long monoliths, especially if the process is limited by mass transfer. In general, short-channel structures are preferable for low to moderate (10–100) Reynolds numbers, while longer monoliths perform better regarding pressure drop for higher Reynolds numbers.

(d) *Structure material.* Ceramic or metallic materials may be used, but the former are selected more frequently. Metal structures can be found in automotive applications. The open frontal area varies between 60 and 80% in ceramic monoliths and can reach 90% in metallic ones. In the latter case, walls can be thinner [13], leading to larger channel diameters and lower pressure drop than the one observed in ceramics with comparable or greater geometric areas [14]. Thus, they are suitable for high-performance applications, where restrictions on pressure drop are severe. The thermal properties (essentially, conductivity) of the materials also need to be considered in the reactor design. Due to low thermal conduction in ceramic monoliths, adiabatic operation is approached. Metallic supports are used with the expectation of increased heat transfer (which can be one of the main drawbacks in monolith reactors). Naturally, there are also differences in the synthesis of the honeycomb and in the procedures required for monolith coating with the desired catalysts/carriers. Different pretreatments may be needed in both cases.

(e) *Heat transfer operating modes.* Regarding the heat transfer, operation may be defined by the relative magnitude of heat conduction toward and through the wall and by the ratio between heat generation and removal. These timescales decide whether the reactor operation is closer to the isothermal or adiabatic limit. Generally, it is recognized that monolith reactors have low radial heat conductivity, owing to the presence of several gas-filled channels (poorly conductive medium) across the reactor cross section. Moreover, heat transfer cannot rely on radial flow convection, which is inexistent in monoliths. These limitations should be observed mainly in honeycombs with high void fractions and thin walls with low effective conductivity.

In order to moderate hot spots in highly exothermic reactions, it is important to note that the dominant mechanism for heat removal will be by conduction through the walls (associated with a timescale given by t_w^2/κ_{wall}). Therefore, the channel walls should be designed so that due to its reduced thickness and high-conductivity construction material, no appreciable temperature gradients are registered. Hot spots are moderated in this way, and even nearly isothermal operation can be achieved when carrying the oxidation reactions of methanol, carbon monoxide, or *o*-xylene, using relatively large volume fractions of highly conductive materials in honeycombs with low channel pitch.

An intermediate scenario (finite heat transfer) is the most complete and complex description. Flytzani-Stephanopoulos et al. [15] studied heat transfer in a metal honeycomb monolith under nonadiabatic conditions in the absence of chemical reactions. The model was solved analytically to yield the solid and gas (radial and axial) temperature profiles in the monolith.

In other cases, the monolith is a multichannel configuration and heat transfer involves channels in different positions. Thus, the reactor also has heat-exchanger functionalities and different arrangements may exist (alternating layers, "checkboard" [13, 16], etc.). If through some channels process fluid circulates, while in others a heat transfer medium flows, then choosing either a concurrent or countercurrent operation is also required.

Mathematical models for monolith reactors can be classified in several manners (see, e.g., Ref. [3]), considering features such as dimensionality (one-dimensional (1D), two-dimensional (2D), or three-dimensional (3D) models) or the model multiscale nature (catalyst/washcoat–channel–reactor, washcoat–channel, etc.). Most commonly, the behavior of a single monolith channel is described, assuming it to be representative of all channels in the whole reactor. This is highly desirable since the computational cost increases with the number of cells included in the simulations. However, it should be remarked that the "single-channel analysis" may not be valid for a number of cases:

- Existence of interaction from channel-to-channel heat transfer (through the connecting support material) and from the catalytic material to the reactor wall
- Simultaneous operation of the monolith reactor as a multichannel heat exchanger
- Nonsymmetrical concentration and temperature distributions in monolithic blocks with incorporated catalysts (Case b.2 in Table 8.1 and described earlier)
- Presence of blocked or deactivated channels
- Nonuniform inlet gas distribution
- Highly permeable walls with nonzero net mass transfer between channels and unsymmetrical conditions [17] (rigorously, Cybulski and Moulijn [18] do not classify this reactor as a monolith, but certain references use this terminology)

8.1.2 Applications

The applications portfolio for monolith reactors covers the domains of environmental catalysis, production of commodities and fine chemicals, and high-added value compounds (pharmaceuticals, flavors, and fragrances) [13]:

1 *Automotive exhaust systems.* The catalytic converter system installed in automotives is a ceramic monolith honeycomb [19], containing noble metal catalysts. The treatment of the emissions involves oxidation of carbon monoxide and hydrocarbons. In three-way converters, reduction of NO_x also occurs. The channel size (or, equivalently, the cell density) is determined among other things by the particulate content of the exhaust to be treated (smaller channels can be used when dealing with clean exhausts). Nevertheless, environments with high dust can be managed without plugging. Three-way gasoline catalytic converters (TWC) simultaneously transform CO, hydrocarbons, and NO_x to CO_2, H_2O, and N_2 in the exhaust of an internal combustion engine. Cordierite monoliths with 400–600 cells per square inch (cpsi) and with wall thickness of 1.1 mm may be used, while honeycombs with 200 cpsi can be chosen if the expected amount of dry particulate (soot) is high.

2 *Reduction of emissions with VOCs.* Several substances fall into the category of VOCs, whose presence in the atmosphere may have serious environmental and public health consequences (e.g., VOCs can react with nitrogen oxides in the atmosphere to produce ozone). For this reason, the allowed concentration limits are low. Power and chemical plants can be the source of emissions, although light metal monoliths are also used in aircrafts and small vehicles engines for ozone abatement [13]. Typically, the stream that needs to be treated in these processes has several features such as high dilution of the reactants, large flow rates, and low temperatures (usually, ambient temperature) to initiate combustion without external heating [12].

3 *Removal of NO_x from stationary sources.* Several studies have been presented on selective catalytic reduction (SCR) of NO_x using monolith reactors. The decomposition of NO was studied by numerical modeling [20]. The design of the monolith honeycomb can also vary greatly from "power plant" to "chemical" applications. Lower cell densities and longer channels are used in the former [21]. Technologies for "chemicals" production are characterized by one order of magnitude higher cell density (30.6–61.2 cells/cm^2) and an order of magnitude lower channel length (around 0.1–0.3 m). Concerning the integration of catalyst in the reactor, washcoated or incorporated walls are preferred to packed beds, which lead to high pressure drops in commercial plants. Choi et al. [22] studied the catalytic reduction of NO_x by ammonia in Cu ion-exchanged mordenite (CuHM) washcoated on square-shaped cells of a cordierite honeycomb. The study involved the use of a highly active catalyst in a low-pressure drop reactor. Santos et al. [23] studied a monolithic reactor with Ti-V-W/sepiolite catalyst for removal of nitrogen oxides from power plant stack gas. Some of the typical characteristics of the process were diluted reactants in the power plant combustion gases (NO concentrations between 200 and 1200 ppm), isothermal behavior, negligible axial dispersion (axial Peclet number above 50), and fully developed laminar flow. The typical operating ranges of low dust SCR monoliths are indicated as temperatures between 300 and 380°C, space velocities (GHSV) between 2000 and 7000 h^{-1} (STP), and area velocities of 6–10 Nm/h. Mass transfer limitations have also been found important. The degree of mass transfer control, as defined by Lopes et al. [24], was above 0.6 (which indicates moderate to strong resistance to external mass transfer). Moreover, the catalyst effectiveness factor was around 0.1 (strong internal mass transfer limitations). Reconfiguration of pore structure is suggested to improve the latter, by adopting an optimum balance between surface catalytic activity and diffusivity in a bimodal pore structure with a significant percentage of macropores.

4 *Production of chemicals.* Many reactions in the gas phase are fast, are strongly exothermic, and require the use of solid catalysts. Partial oxidations fit into this category, where selectivity is also a major challenge, as strong exothermic total oxidation is undesirable. Implementation of more intense conditions is not feasible in conventional technologies due to difficulties in heat management. In this case, structured (monolith) reactors offer the possibility of enhanced control of residence time and temperature, avoiding hot spots that harm selectivity and catalyst lifetime. Gas–solid catalytic combustions [25] of fuels with large excess of air and low emission of CO, HC, or NO_x have been studied. Ethanol steam reforming was conducted in a cordierite monolith [26] with 400 cpsi and 0.9 mm of channel width and a Co_3O_4 catalyst coating. At 673 K and 2.5 s of contact time, 70% conversion was observed. Many other processes have been considered in liquid phase as well. Parikh [27] studied aromatic alkylation (toluene ethylation) over HZSM-5 washcoated honeycomb cordierite monoliths (400 cpsi). Partial oxidation of natural gas and selective oxidation of liquid hydrocarbons (isooctane and benzene) into syngas were also addressed in triangular-shaped monolith channels with α-alumina support and $LaNiPt/CeO_2-ZrO_2$ active component [28].

5 *Three-phase reactions.* The implementation of monoliths in three-phase processes is more limited compared to the cases previously described. Kreutzer [29] remarks the scarcity of commercial technologies. Nevertheless, production of hydrogen peroxide in the anthraquinone process is reported [30]. Other examples include nitroaromatic hydrogenations [31], photooxidation of cyclohexane [32], hydrogenation of 2-butyne-1,4-diol to cis-2-butene-1,4-diol and butane-1, 4-diol [33, 34], and the direct formation of hydrogen peroxide [35]. A review of multiphase flow applications in monoliths was given by Pangarkar et al. [36].

6 *Laboratory and analytical applications.* An important stage in all processes involving catalytic coatings in microchannels is the testing of the catalyst itself. Performing these tests in the monolith configuration is desirable, since conditions (temperature, flow, etc.) can be rigorously controlled and extrapolation to the production scale is based on the same geometry. Screening can be performed in packed microchannel reactors [37] and monoliths [38].

8.2 Design of wall-coated monolith channels

The analysis in this chapter is focused on flow, mass and heat transfer, and reaction in monolith channels with catalytic walls (design (b.1) in Table 8.1). Modeling of a packed inert monolith reactor (design (b.3) in Table 8.1) should follow the analysis given in Chapter 3 of this book by taking into account the issues raised by the low particle/reactor diameter ratio, which are not as relevant in large-scale packed beds. The application of the following analysis to design (b.2) in Table 8.1 is straightforward, if exact/approximate planes of symmetry can be easily identified or if negligible interaction between channels can be assumed. In this section, we mainly consider long isothermal monoliths, while nonisothermal operation is addressed in Section 8.2.4.

8.2.1 Flow in monolithic channels

The most commonly found configuration in monoliths consists in regularly arranged straight channels, with cross sections given by simple geometries (circular, square, rectangular, triangular, etc.). The actual cross-sectional area that is available deviates slightly after the washcoating procedure, especially near sharp corners, where the washcoat thickness tends to increase above the average value over the channel periphery. Therefore, triangles and rectangles with rounded corners are also geometries of interest. Flow and laminar heat transfer have already been studied in these shapes [39]; hence, mass transfer is also characterized from the analogy between transport processes.

A major simplification in the modeling of monolithic channels is to decouple fluid mechanics from the problem by assuming a well-defined velocity profile. In this respect, the laminar flow assumption is likely to be valid since the timescale for viscous diffusion over a transverse length a is much smaller than the one for convection through a channel with length L:

$$\frac{\tau_{visc,diff}}{\tau_{conv}} = \frac{a^2}{\nu}\frac{\langle u \rangle}{L} \ll 1 \quad (8.2)$$

The condition expressed by Equation 8.2 is usually fulfilled and channel flow occurs in the laminar range. This allows the problem to be treated analytically and the solutions presented in this chapter rely on this assumption. Accounting for the simultaneous development of the velocity profile compared with concentration/temperature fields requires numerical evaluation, and the importance of this effect is measured by Prandtl's number for heat transfer or by Schmidt's number for mass transfer:

$$Pr = \frac{\nu}{\kappa} \text{ and } Sc = \frac{\nu}{D} \quad (8.3)$$

When $Pr \to \infty$ and $Sc \to \infty$, the flow field develops much faster than the temperature or concentration profiles, respectively. Thus, mass transfer in a monolith channel should occur in fully developed laminar flow in liquid-phase processes. In the opposing limit ($Pr \to 0$ and $Sc \to 0$), "plug flow" can be used as an idealized inlet profile. Solutions for the mass transfer–wall reaction problem were obtained for both cases in Lopes et al. [40] as lower and upper bounds for the reactant conversion. The entrance length before which the velocity profile can be considered developed is often given by expressions of the type [41]

$$\frac{L_e}{a} \sim 0.14 Re_a \quad (8.4a)$$

$$\frac{L_e}{a} = \frac{1.20}{0.035 Re_d + 1} + 0.112 Re_d \quad (8.4b)$$

or simply, for $Re_d < 100$,

$$\frac{L_e}{d} < 6 \quad (8.4c)$$

written here for flow in a circular channel with diameter $d = 2a$ (the Reynolds number referred to diameter and radius is defined in the following). In general, the hydrodynamic entrance length corresponds to the distance from where the velocity profile $u(r,z)$ can be considered to no longer change with the axial coordinate. A criterion can be formulated by considering the deviation between the centerline velocity at distances long enough from the inlet ($u(r,\infty)$) and the one at the length L_e (to be determined) and assigning an arbitrarily low value to this difference. For example,

$$\left|\frac{u(0,L_e) - u(0,\infty)}{u(0,\infty)}\right| < 1\% \quad (8.4d)$$

The knowledge of this entrance length is important for the design of inlet sections or to define the validity of simplified models derived on the basis of fully developed laminar flow conditions. In practice, an inlet presection in the channel (with an uncoated/inert wall) can be used to allow for flow development before the fluid reaches the catalytically active region. A similarity between the entrance length of velocity and concentration/temperature profiles can be found, particularly when the wall temperature can be assumed to be uniform or severe external mass transfer control [42]. In Lopes et al. [43], the thickness of this region is discussed for the mass transfer problem with a finite wall reaction.

Pressure drop is important to quantify the process energy requirements and can even be the limiting factor in the design and performance of a monolith reactor. It is related to Darcy's friction factor (four times higher than the Fanning factor) by [41]

$$\Delta P = f \frac{L}{d_h} \frac{\rho \langle u \rangle^2}{2} \quad (8.5)$$

The (Darcy's) friction factor f in a straight channel is inversely proportional to the Reynolds number $Re_d = \langle u \rangle d_h / \nu$, according to

$$f = \frac{C_f}{Re_d} \quad (8.6)$$

Table 8.2 Typical dimensions and pressure drop in some monolith reactor processes [46].

Application	Channel length (L) (m)	$\Delta\hat{P}$(bar/m)
Respiratory protection		10^{-4} – 0.001
Automotive exhaust treatment	0.15–0.30	0.2
Cocurrent multiphase flow		0.05
Automotive catalytic reduction of NO_x	0.30–0.50	0.1
Industrial selective reduction	1	0.02–0.1

Source: Van Gulijk et al. [46]. Reproduced with permission of Elsevier.

where for fully developed velocity profile, the coefficient C_f can be found for several cross-sectional shapes [39] (it is equal to 64 in a circular channel and 96 for a planar channel). For flow in a monolith channel, the Reynolds number can take values between 10 and 1000 [44] (below the laminar–turbulent transition). In automotive exhaust gas treatment, Re_d values are between 100 and 600 [45], hence in the laminar range. Table 8.2 shows typical values for pressure drop in monolith-based processes.

In order to simplify the analysis for a given geometry, we consider a 1D model for the channel, where the transport of momentum in the cross section is described by a single (radial) coordinate, with a shape parameter S reserved to translate deviations from the common cylindrical shape ($S = 1$). The circular shape is the simplest one with a value of S different from zero and is associated with the cylindrical coordinates. Moreover, we note that even when the channels are not originally circular, the introduction of washcoat can round the channel shape significantly near sharp corners, hence approximating the geometry to that of a cylindrical channel. Another relevant situation which also transforms a noncircular channel occurs when a low BET material coating is introduced to block the macroporosity of the walls and rounds the shape of the channel, filling the poorly used regions. Pérez-Cadenas et al. [47] have shown the superior performance of these catalysts in the hydrogenation of fatty acid methyl esters compared with a square channel cordierite monolith. Better utilization and distribution of the catalytic layer as well as higher selectivity to the desired products were observed.

Other shapes are also found, where transverse transport cannot be described by a single coordinate. The detailed mass transfer problem then becomes 3D, and simplified descriptions are needed. According to the 1D model mentioned earlier, the momentum conservation equation in the axial direction (with velocity component u) for fully developed laminar flow is given by the following equation:

$$\alpha Re_a \frac{dp}{dz} = \frac{1}{r^S}\frac{d}{dr}\left(r^S \frac{du}{dr}\right) \tag{8.7}$$

The variables in Equation 8.7 were normalized as shown in Equations 8.8. The transverse position within the open channel is normalized by the characteristic scale in the transverse direction a (this can be the channel radius or the half-spacing between plates). The dimensionless axial coordinate is a fraction of the total channel length L:

$$r = \frac{\hat{r}}{a} \tag{8.8a}$$

$$z = \frac{\hat{z}}{L} \tag{8.8b}$$

$$v = \frac{u}{\langle u \rangle} \tag{8.8c}$$

$$p = \frac{\hat{p} - \hat{p}_{ref}}{\rho \langle u \rangle^2} \tag{8.8d}$$

The parameter αRe_a was given in Equation 8.2. Integrating Equation 8.7 using conditions of symmetry (at $r = 0$) and no-slip (at $r = 1$) yields

$$v(r) = \frac{\alpha Re_a}{4}\left(-\frac{dp}{dz}\right)(1 - r^2) \tag{8.9}$$

The two parameters in Equation 8.9 are the aspect ratio, α, and the Reynolds number, Re_a, defined in Equations 8.10a and 8.10b, respectively:

$$\alpha = \frac{a}{L} \tag{8.10a}$$

$$Re_a = \frac{\rho\langle u \rangle a}{\mu} = \frac{\langle u \rangle a}{\nu} \tag{8.10b}$$

According to Equation 8.8c, the average of v in the radial direction must equal $\langle v \rangle = 1$. This condition can be used to eliminate the axial pressure gradient from Equation 8.9. The fully developed parabolic velocity profile is invariant with axial position and is given by

$$v(r) = \frac{u(r)}{\langle u \rangle} = \frac{S+3}{2}(1 - r^2) \tag{8.11}$$

According to this reduced model (1D in the transverse direction), the friction factor is calculated from

$$fRe_a = -8\left.\frac{dv}{dr}\right|_{r=1} = 8(S+3) \tag{8.12}$$

The coefficient C_f in Equation 8.6 is referred to the Reynolds number based on the hydraulic diameter (d_h). Values of fRe_d for some common cross-sectional shapes are given in Table 8.3. The dimension d_h is related to the characteristic length scale a, by using the volume-to-surface area ratio according to the 1D model:

$$fRe_d = 8(S+3)\frac{d_h}{a} = 32\left(\frac{S+3}{S+1}\right) \tag{8.13}$$

In this expression, the relationship between channel volume, wall surface area (or hydraulic diameter d_h), the characteristic transverse length a, and the shape factor S is expressed as follows:

$$S = \frac{S_{surf}\,a}{V} - 1 = \frac{4a}{d_h} - 1 \tag{8.14}$$

Table 8.3 Friction factors and transfer coefficients for some common cross-sectional shapes in monolith channels [39].

Shape with characteristic dimension *a*, cross-sectional area *A*, and wet perimeter P_{surf}	$d_h = \frac{4A}{P_{surf}}$	fRe_d	$Sh'_\infty = \frac{k_m d_h}{D}$	$Sh'_0 = \frac{k_m d_h}{D}$ $Sh_{0(H_2)}^*$
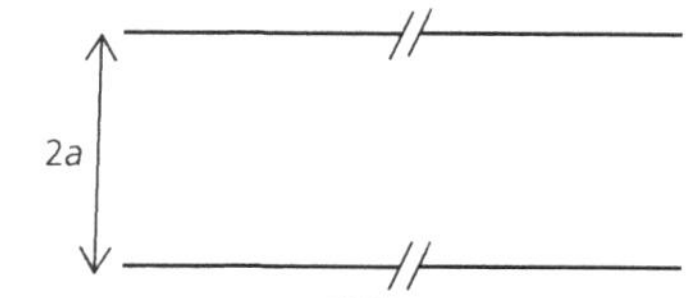	$4a$	96	7.541	8.235
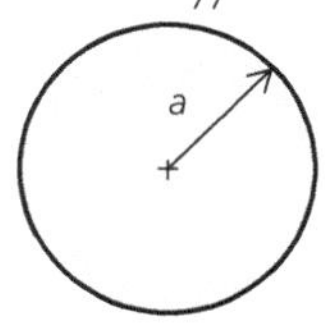	$2a$	64	3.657	4.364
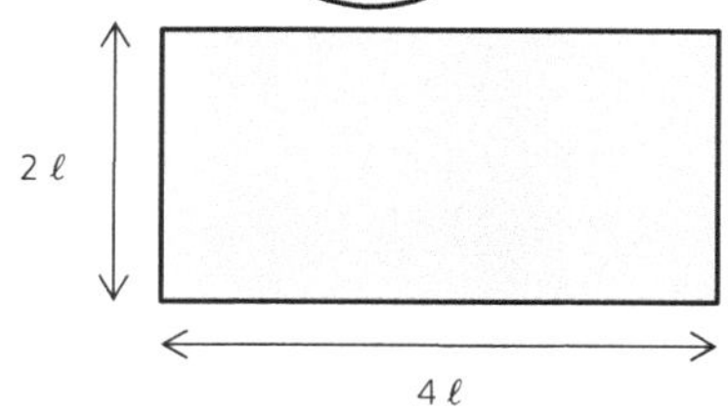	$\frac{8}{3}\ell$	62.19	3.391	4.123 3.017
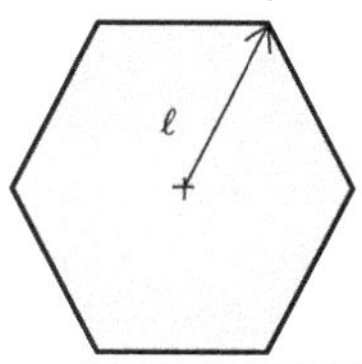	$\sqrt{3}\ell$	60.22	3.34	4.002 3.862
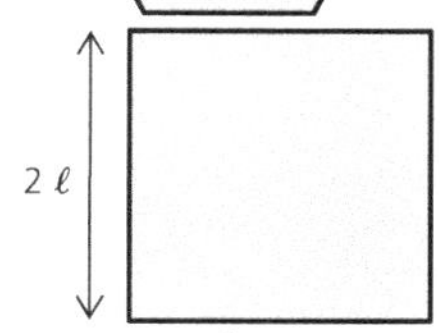	2ℓ	56.91	2.976	3.608 3.091
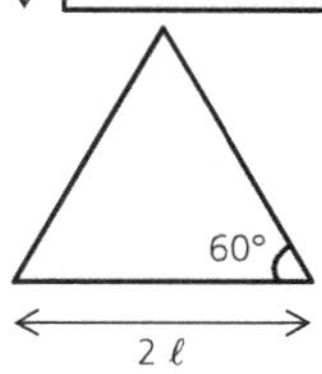	$\frac{2}{\sqrt{3}}\ell$	53.33	2.47	3.111 1.892

*$Sh_{0(H_2)}$ are the values for the peripherally uniform wall flux (H_2) boundary condition. Distinction only relevant for geometries with sharp corners.

As mentioned earlier, the dimensionless uniform velocity profile is also considered here as a limiting case of undeveloped flow (Sc or $Pr = 0$):

$$\nu(r) = \frac{u(r)}{\langle u \rangle} = 1$$

It represents a physically incorrect description for channel flow, as it does not respect the no-slip condition at the wall. However, it is useful to consider it for several reasons: (i) it yields an upper bound for the observed value of conversion; (ii) the decrease of concentration near the wall is less pronounced compared to laminar flow, yielding more strict criteria for mass transfer control (i.e., more developed flow profiles will be more mass transfer limited); and (iii) the axial velocity profile can be flattened by promoting radial mixing (e.g., using glass beads). The last motivation needs to be carefully reviewed, since its validity in packed microchannels is questionable [48].

There are a number of more complex problems regarding the fluid mechanics associated with this reactor. Some examples include the following:

- Turbulent flow in monoliths has been only considered in some cases, as discussed in Section 8.1.1.

- Internal (washcoat) flow is usually neglected, since the flow resistance that is imposed by this layer is typically large. Nevertheless, CFD simulations with hydrodynamic calculations in both domains can be performed. In this case, the Darcy–Brinkman formulation can be used to couple internal and external flows.
- Even though flow is more controlled and well-behaved inside the channels than it is in a random packed bed, flow maldistribution at the inlet may arise if a distributor is not provided. This will lead to differences in the hydrodynamics prevailing inside each channel, namely, in the residence time.
- In short-channel structures, mass/heat transfer processes have to be modeled with simultaneously hydrodynamic developing flow. For some common shapes, the "apparent" friction factor correlation applicable in these cases can be given by an expression of the form

$$f_{app}Re_d = (fRe_d)_{fd} + fxn\left(\frac{d_h Re_d}{L}\right)$$

where $fxn(d_hRe_d/L)$ is an increasing function of d_hRe_d/L, which can be described by a power-law expression, such as $fxn = (d_hRe_d/L)^y$ with exponent y equal to 0.734 for square channels, 0.5 for parallel plates, and 0.731 for equilateral triangles [12].

8.2.2 Mass transfer and wall reaction

In the case of monoliths operating in laminar flow, transverse mixing is assured only by molecular diffusion, which has a characteristic time proportional to the square of the channel radius or diameter. Thus, faster mixing implies channel diameter reduction. This also leads to an increase in pressure drop and flow rate limitation. The solution of the mass transfer equation, reflecting the balance between convection and transverse diffusion, is known as the Graetz–Nusselt problem, described in the following. The ratio between the timescales for both processes is

$$\alpha Pe_m = \frac{\tau_{diff}}{\tau_{conv}} \tag{8.15}$$

The cases where axial diffusion cannot be neglected lead to the formulation of the extended Graetz problem for which analytical solutions have been given. If axial diffusion is indeed negligible (and this should be true for long channels), the only two mechanisms at play are axial convection and transverse diffusion. From this interaction, two different regimes appear which describe the behavior for axial distances of the same order of the channel length $\hat{z} = O(L)$:

- *Graetz regime*—where the profile can be considered fully developed. In this case, both convection and transverse diffusion dominate, which translates into comparable timescales for convection and diffusion, $\alpha Pe_m = O(1)$. The solution for reactant conversion (X_{fd}) is appropriate for "long distances" and is well represented by the first term in Graetz series.
- *Lévêque regime*—when convection dominates over transverse diffusion ($\alpha Pe_m \gg 1$). Here, a concentration boundary layer is present (where the large transverse concentration gradients are restricted to and diffusion becomes important). The dimensionless thickness of the boundary layer is proportional to $(\alpha Pe_m)^{-q}$, where $q = 1/2$ for plug flow and $q = 1/3$ for laminar flow. The solution for the reactant conversion X_{dev} can be obtained by combination of variables, if the boundary condition at the wall is amenable to this treatment. In general, for a Robin boundary condition, the Laplace transform can be used, even though the resulting integral for laminar flow requires numerical evaluation.

8.2.2.1 Graetz and Lévêque analyses

The governing equations that describe the classical Graetz problem include transport in a channel by convection in the axial direction and transverse diffusion. For a channel, the mass (or heat) balance can be written as

$$\frac{1}{r^S}\frac{\partial}{\partial r}\left(r^S\frac{\partial c}{\partial r}\right) = \alpha Pe_m v(r)\frac{\partial c}{\partial z} \tag{8.16}$$

where the dimensionless independent variables are normalized according to Equation 8.8. The corresponding heat balance can be written replacing c by T (dimensionless temperature) and Pe_m by Pe_h (defined with the thermal diffusivity). Concentration and temperature are normalized according to

$$c = \frac{\hat{c}}{\hat{c}_{in}}. \tag{8.17}$$

$$T = \frac{\hat{T} - \hat{T}_{wall}}{\hat{T}_{in} - \hat{T}_{wall}} \tag{8.18}$$

The dimensionless velocity profiles are given in Section 8.2.1 and the shape factor (S) takes the values of 0 or 1, for parallel plates or circular channels, respectively. The two dimensionless parameters in Equation 8.16 are the transverse mass Peclet number

$$Pe_m = \frac{\langle u \rangle a}{D}$$

and the channel aspect ratio, already defined in Equation 8.10a. Axial diffusion is negligible up to $O(\alpha)^2$, that is, for small aspect ratio channels. Monolith channels typically present small diameter channels and relatively larger lengths ($\alpha \ll 1$). Therefore, the only parameter in Equation 8.16 is αPe_m, which compares transverse diffusion and convection at global scales, that is, evaluated with the largest transverse distance and full channel length. Therefore, it is possible to incorporate this dimensionless number in the dimensionless axial variable, defining the Graetz number:

$$Gz = \frac{\alpha Pe_m}{z}$$

This parameter can take a wide range of values. For long microchannels, the inlet Danckwerts' boundary condition is simplified to uniform inlet concentration for $Pe_m \gg \alpha$:

$$c(r,0) = 1 \tag{8.19}$$

In the transverse direction, the symmetry boundary condition is appropriate at $r = 0$. Additionally, a condition at the wall is required. In the case of a microchannel reactor where a first-order reaction is occurring in the catalytic layer, this is given by

$$\left.\frac{\partial c}{\partial r}\right|_{r=1} = -Da\, c(1,z) \tag{8.20}$$

where the Damköhler number ($Da = ak_{obs}/D$) takes the reaction timescale into account in comparison to transverse diffusion. The dimensionless parameter in Equation 8.20 includes the effectiveness factor (see Section 8.2.3), for the cases where internal diffusion inside the porous coating must be considered (hence, the surface reaction constant is the one observed in the presence of eventual mass transfer limitations). Therefore, the processes in the washcoat can be lumped into a "wall reaction boundary condition." Two particular limits of (8.19), $Da \to 0$ and $Da \to \infty$, are of importance, corresponding to Neumann (uniform surface flux) and Dirichlet (uniform surface concentration) boundary conditions, respectively. For the heat transfer problem, Equation 8.20 expresses finite wall thermal resistance [39] (R_{wall}), where the same limiting forms are found for $R_{wall} \to \infty$ and $R_{wall} \to 0$. The well-known solution [49, 50] to this problem has the following separable form:

$$c(r,z) = \sum_{n=1}^{\infty} A_n \varphi_n(r) \exp\left(\frac{-\lambda_n^2 z}{\alpha Pe_{m,\max}}\right) \tag{8.21}$$

where $\varphi_n(r)$ is an eigenfunction in the transverse coordinate, while the exponential term represents the axial dependence of the concentration profile ($Pe_{m,\max}$ being the transverse Peclet number evaluated at the maximum velocity in the channel). The 2D concentration profile is not particularly useful if in practice the measurable quantity is the mixing-cup concentration, defined by

$$\langle c \rangle(z) = \frac{\int_0^1 r^S v(r) c(r,z)\,dr}{\int_0^1 r^S v(r)\,dr} \tag{8.22}$$

The denominator in Equation 8.22 depends only on the channel geometry and flow profile (parabolic or uniform). This expression can be rewritten as

$$\langle c \rangle(z) = \sigma_C \int_0^1 r^S c(r,z)\,dr \quad \text{(plug flow)} \tag{8.23a}$$

$$\langle c \rangle(z) = \sigma_C \int_0^1 r^S (1-r^s) c(r,z)\,dr \quad \text{(laminar flow)} \tag{8.23b}$$

$$\text{with } \sigma_C = \frac{u_{max}/\langle u \rangle}{\int_0^1 r^S v(r)\,dr} = (S+1)\frac{u_{max}}{\langle u \rangle} \tag{8.23c}$$

The coefficient σ_C can be interpreted as the ratio between the flow rate given by $Q_{max} = aS_{surf}u_{max}$ (with average fluid velocity equal to u_{max} and cross-sectional area equal to the wet perimeter of the channel multiplied by the transverse length scale a) and the actual flow rate ($Q = A_{ch}\langle u \rangle$). Averaging Equation 8.21 over the channel transverse length, with the velocity profile, yields the following series solution for the mixing-cup concentration:

$$\langle c \rangle(z) = \sigma_C \sum_{n=1}^{\infty} \frac{-\varphi_n'(1)}{\lambda_n^2} A_n \exp\left(\frac{-\lambda_n^2 z}{\alpha Pe_{m,max}}\right) \tag{8.24}$$

It is possible to obtain integration constants A_n and functions of eigenvalue (λ_n^2) for several geometries and wall reaction rates. We can group them into the coefficients w_n:

$$w_n = -\varphi_n'(1) A_n \frac{\sigma_C}{\lambda_n^2} \tag{8.25}$$

The conversion of reactant at each axial position is calculated from

$$X = \frac{\hat{c}_{in} - \langle \hat{c} \rangle}{\hat{c}_{in}} = 1 - \langle c \rangle = 1 - \sum_{n=1}^{\infty} w_n \exp\left(\frac{-\lambda_n^2 z}{\alpha Pe_{m,max}}\right) \tag{8.26}$$

The solution in Equation 8.26 is inconvenient for several reasons. Each term in the series contains two coefficients (w_n and λ_n^2) which require numerical calculation. In the case of a linear wall reaction, these quantities depend on the wall kinetic parameter, and this relationship is recently obtained in a simple and explicit manner by Lopes et al. [40]. In addition, whenever this slowly convergent series is used to describe the inlet region, a large number of terms may be required so that a satisfactory result is obtained. The efficient evaluation of the terms in Graetz series has been the object of many studies. Housiadas et al. [51] presented a comparative analysis between several methods to estimate these terms, remarking the numerical issues associated with the rigorous calculation of these quantities. However, this was done for uniform wall concentration (Dirichlet boundary condition), excluding the important case of finite reaction rates.

Figure 8.2 shows the prediction of the reactant conversion from the series in Equation 8.26, when this is truncated after 1, 5, or 12 terms. It is possible to observe that for large values of the Graetz parameter (i.e., small $z/\alpha Pe_m$), the number of terms required to describe the conversion (or mixing-cup concentration) of reactant increases considerably, owing to the slower

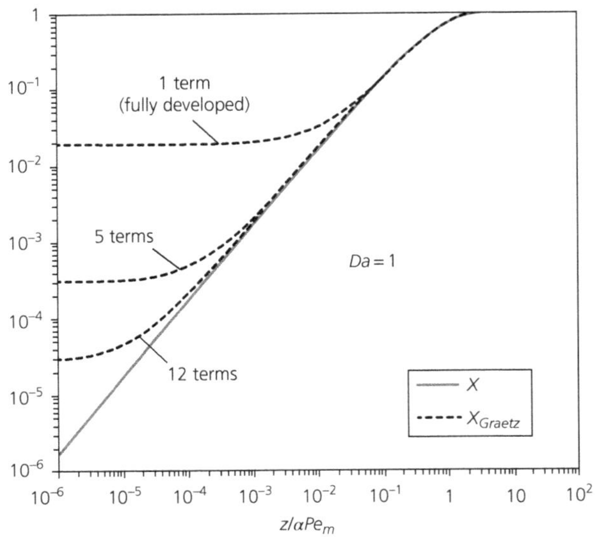

Figure 8.2 Conversion calculated according to Graetz series with different number of terms. Equation 8.26 is plotted with 1, 5, and 12 terms as X_{Graetz} for intermediate mass transfer control ($Da = 1$). The solution with only one term corresponds to the fully developed concentration profile. A uniformly valid solution for conversion X (plotted as a solid line) is given by Equation 8.37 and is in good agreement with the numerical solution.

convergence rate of the series. Since for typical values of αPe_m this situation is localized in a region near $z = 0$, mass transfer is said to occur in the "inlet regime." Lévêque [52] proposed an alternative treatment to obtain the solution to the heat transfer problem in this regime. The temperature/concentration profile presents the characteristics described in the introduction of Section 8.2.2, that is, at most part of the channel cross section $c(r,z) = 1$, while concentration variation is observed in a boundary layer near $r = 1$. To describe this latter region in a channel with parabolic velocity profile, it is convenient to stretch the transverse coordinate as

$$R = \frac{1-r}{\delta} \tag{8.27}$$

where the dimensionless thickness of the boundary layer δ is much smaller than 1 (i.e., the thickness is a small fraction of the channel length scale a). Introducing this variable transformation into Equation 8.16 leads to

$$\frac{\partial^2 c}{\partial R^2} - \frac{S\delta}{1-R\delta}\frac{\partial c}{\partial R} = 2\delta^3 \alpha Pe_{m,max} R\frac{\partial c}{\partial z} \tag{8.28}$$

where the parabolic velocity profile is normalized according to

$$v(R) = \frac{u_{max}}{\langle u\rangle}\left[1-(1-R\delta)^2\right] = 2\frac{u_{max}}{\langle u\rangle}R\delta + O\left(\delta^2\right)$$

The order of magnitude for the boundary layer length scale is obtained assuming that convective and diffusive terms are comparable in magnitude, that is, the scaling coefficients must be equal (the curvature term of diffusion of $O(\delta)$ is much smaller than 1 and can be ignored):

$$\delta \sim \frac{(S+3)^{-1/3}}{(\alpha Pe_m)^{1/3}} \tag{8.29}$$

Hence, Lévêque's problem is formulated as follows:

$$\frac{\partial^2 c}{\partial R^2} = R\frac{\partial C}{\partial z} \tag{8.30}$$

The conditions in the channel inlet and cross-sectional core must be satisfied:

$$c(z=0) = c(R \to \infty) = 1 \tag{8.31}$$

For a first-order wall reaction, the boundary condition at $R = 0$ $(r = 1)$ in Equation 8.20 is now given by

$$\left.\frac{\partial c}{\partial R}\right|_{R=0} = \delta Da\, c(R=0, z) \tag{8.32}$$

Lopes et al. [53] named the product between the Damköhler number and the dimensionless thickness of the boundary layer δDa as the *rescaled Damköhler number* Da^*. For small values of Da^*, the problem reduces Neumann (uniform flux) type. When Da^* is large, a Dirichlet-type (uniform concentration) problem arises, and in this case, Equation 8.30 can be solved by combination of variables (see, e.g., Rice and Do [54] or Basmadjian [55]). Since $\delta \ll 1$ in this regime, it is expected that much higher values of Da are required for a concentration annulment condition to apply.

In an idealized situation, where the no-slip condition near the wall is not fulfilled by the velocity profile ("plug flow"), Lévêque's analysis leads to

$$\delta \sim \frac{1}{\sqrt{\alpha Pe_m}} \tag{8.33a}$$

and

$$\frac{\partial^2 c}{\partial R^2} = \frac{\partial c}{\partial z} \tag{8.33b}$$

The following subsection provides a literature review on some of the solution techniques that have been used when dealing with the large class of Graetz and Lévêque problems in mass/heat transfer in channel flow.

8.2.2.2 Literature survey

Early independent works on the heat transfer problem in channel flow include the ones by Graetz [49], Nusselt [56], and Paneth and Herzfeld [57] as given credit by Damköhler [50]. Table 8.4 presents some of the history of the problem, with references grouped according to different solution strategies. The straightforward use of Graetz's analytical solution implies the numerical calculation of the eigenvalues and integration constants for each term of the series and for each value of the wall Damköhler number Da. Damköhler [50] reported the first three terms in the Graetz series for plug flow in a circular channel with $Da = 0.1$, 1, 10, and 100 and provided a graphical plot from

Table 8.4 Solution strategies for the Graetz–Lévêque problem with wall reaction.

Approach	Refs.
1. Graetz series: analytical solution given by Equation 8.26	
Solving the eigenproblem numerically for given values of *Da* in common cross sections with a first-order reaction occurring at the walls	Shah and London [39], Bhattacharya et al. [44], Damköhler [50], Carslaw and Jaeger [58], Bauer [59], Özisik and Sadeghipour [60]
Deriving approximate expressions valid in the asymptotic limits of an infinitely fast reaction, or for a slow reaction	Housiadas et al. [51], Sellars et al. [61], Solbrig and Gidaspow [62], Brown [63], Newman [64], Kays and Crawford [65], Bhattacharya et al. [66], Balakotaiah and West [67]
Uniformly valid analytical approximation for the first eigenvalue and weight for any value of *Da* with no fitting parameters (first-order wall reaction) (see Equations 8.34)	Lopes et al. [40]
2. Lévêque's solution	
Analytical solution with first-order wall reaction and Lévêque's original assumptions (additional numerical evaluation required for laminar flow)	Carslaw and Jaeger [58], Petersen [68], Pancharatnam and Homsy [69], Ghez [70]
Approximate solution to Lévêque's problem and first-order reaction kinetics with no numerical evaluation involved	Lopes et al. [40]
Calculating higher-order terms in Lévêque's series for finite Graetz number when $Da \rightarrow 0$ or $Da \rightarrow \infty$ (analogy with heat transfer)	Worsoe-Schmidt [71], Newman [72], Gottifredi and Flores [73], Shih et al. [74]
Computing higher-order terms in Lévêque's series for finite *Da*	Lopes et al. [40]
3. Uniformly valid solution for a first-order reaction	
Improving the convergence of Graetz series by using an asymptotic technique for series summation (see Equation 8.37)	Lopes et al. [42]
4. Nonlinear reaction kinetics	
Solving numerically the mass transfer problem with nonlinear wall reaction boundary condition	Bhattacharya et al. [44], Acrivos and Chambré [75]
Solving analytically for the special case of zero-order reactions	Sellars et al. [61], Siegel et al.[76], Compton and Unwin [77], Rosner [78]
Integral equation methods	Acrivos and Chambré [75], Chambré and Acrivos [79], Katz [80], Rosner [81], Grau et al. [82]
Explicit approximations for power-law kinetics in fully developed and developing concentration profile	Lopes et al. [40]

where the first three eigenvalues can be estimated for $0.1 < Da < 10^3$. This is perhaps one of the earliest studies (if not the first) of the Graetz problem in the context of channels with reacting walls. Subsequently, a larger number of terms in the series were considered. For example, Carslaw and Jaeger [58] presented the first six eigenvalues for plug flow in circular or planar ducts for several values of *Da*. Bhattacharya et al. [44] gave the first five eigenvalues and weights for $Da = 0.01$, 0.1, 1, 10, and 100 in channels with common cross-sectional shapes: circular, planar, square, and triangular (equilateral). They also plotted λ_1 and w_1 as a function of *Da* for plug and laminar flows ($10^{-3} < Da < 10^2$). The case where both homogeneous and heterogeneous first-order reactions occurred was dealt by Bauer [59], who calculated the first 11 eigenvalues for laminar flow in a parallel plates conduit with $Da = 0$, 0.1, 1, 10, and 100. Additional results can be obtained if the analogy between mass and heat transfer is used. In heat transfer, an equivalent problem to the one given in Section 8.2.2.1 is formulated when the channel wall thermal resistance is finite. Shah and London [39] reviewed several works addressing this problem with boundary conditions of third kind or Robin type. For example, Özisik and Sadeghipour [60] calculated the first 12 eigenvalues for laminar flow inside a circular tube for $Da = 0$, 0.1, 1, 10, and 100, as well as the quantities needed to obtain the first 12 weights w_n.

Alternatively, asymptotic approximations can be used, avoiding the numerical solution of a large number of eigenproblems. This approach can be explored in two directions: (i) large eigenvalues asymptotics (large n) and (ii) small or large Damköhler number. The first asymptotic limit has been considered for Dirichlet and Neumann boundary conditions. Sellars et al. [61] provided formulas for the calculation of high eigenvalues ($n \gg 1$) for uniform wall temperature or flux conditions. These estimates were obtained applying the Wentzel–Kramers–Brillouin (WKB) method, which constructs a composite solution for the transverse contribution to the concentration profile from three regions in this direction (one near the center at small values of r, other near the wall when r is close to 1, and one in the middle of the two latter regions). Solbrig and Gidaspow [62] attempted to provide an easier calculation formula for higher eigenvalues in the case of finite *Da*. However, the result was an implicit expression, which required numerical evaluation.

The second source of asymptotic behavior that can be explored lies in the magnitude of the wall reaction parameter *Da*. The crudest estimation for fast reactions is based on the results for Dirichlet (uniform wall temperature) condition. Brown [63] and Newman [64] performed numerical calculations for these cases. Kays and Crawford [65] also presented calculations for several cross-sectional geometries. Housiadas et al.

[51] showed estimation methods for higher integration constants and eigenfunction derivatives at the wall for Dirichlet conditions. Balakotaiah et al. [83] gave the first five eigenvalues and weights for $Da = \infty$ and circular, planar, triangular, and square geometries. The limit of low reaction rates was also considered. Bhattacharya et al. [66] presented the Taylor expansions for the first eigenvalue in the kinetic regime ($Da \longrightarrow 0$), when plug or laminar flow inside planar and circular ducts occurred.

In some sense, the two asymptotics are related since the first eigenvalues are the most important and extrapolation of the results for high n to the first couple of terms is only reasonable near the Dirichlet limit (where the first eigenvalue as a function of Da is the highest). For slow reactions, the expressions for uniform wall flux can only provide the second eigenvalue, as the first one tends to zero as Da decreases. Therefore, the following characteristics are desirable in the calculation of the Graetz eigenproblems: (i) acceptable results for the first eigenproblems, which are the most important; (ii) accuracy compared to numerical solution; (iii) validity for all values of the Damköhler number; and (iv) explicit analytical calculation formulas. The earlier approaches cannot attain these objectives simultaneously. Recently, Lopes et al. [40] presented a uniform analytical expression which has all of these advantages, providing shape and flow normalization. This will be further described in the following.

Considerably less work exists addressing Lévêque's problem for inlet channel flow with wall reaction. Carslaw and Jaeger [58] and Petersen [68] presented solutions for plug-flow conditions, using the Laplace transform. Pancharatnam and Homsy [69] used the same technique for laminar flow. The inversion of the transformed solution is given in terms of an infinite summation with coefficients given by recurrence relations (first 24 out of 50 coefficients are tabulated). Ghez [70] considered a first-order reversible reaction with the same solution method. Moreover, asymptotic expansions in the limits of fast and slow reactions were presented.

Lévêque's problem was extracted from the rescaled mass balance in Equation 8.28. As can be seen, this equation is the basis of a perturbation problem and can be decomposed into several subproblems of order $O(\delta^n)$. The concentration profile, the flux at the wall, and consequently the mixing-cup concentration (or conversion) can all be written as perturbation series on powers of the dimensionless boundary layer thickness. This series is often called as the extended Lévêque solution or Lévêque's series. Worsoe-Schmidt [71] and Newman [72] presented several terms of these series for Dirichlet and Neumann boundary conditions. Gottifredi and Flores [73] and Shih and Tsou [84] considered the same problem for heat transfer in non-Newtonian fluid flow with constant wall temperature boundary condition. Lopes et al. [40] presented approximations to the leading-order problem for all values of Da and calculated higher-order corrections for large and small values of this parameter.

For reaction kinetics other than first, the analytical work is even more scarce and numerical solution is the more common approach, if not the only one possible. Acrivos and Chambré [75] numerically solved the integral equations which yield a solution for the several reaction schemes (first-order reversible and second-order irreversible reactions and two consecutive first-order reactions). Bhattacharya et al. [44] presented the numerical results for power-law kinetics (of orders $m = 1/2$, 1, and 2) for plug flow in a short channel and also for some types of Langmuir–Hinshelwood kinetics, including parameter ranges where the solution exhibited multiplicity behavior.

Zero-order kinetics attracts special attention, due to its analytical simplicity and particular characteristics, especially when annulment of concentration at the solid surface is involved. Sellars et al. [61] and Siegel et al. [76] gave Graetz-type solutions for uniform axial heat flux, using the eigenfunction expansion method. Compton and Unwin [77] presented the Laplace's domain analytical solution of the mass transfer problem in a channel cell–crystal–electrode system under Lévêque's assumptions. Rosner [78] wrote the solutions for the wall concentration profile as $c_{wall} \sim 1 - z/z_0$ $(z < z_0)$, for several classes of boundary layer problems.

In general, integral equation methods have been found suitable to describe a class of laminar boundary layer-type flow fields past solid surfaces, where catalytic reactions with arbitrary mechanisms take place. Chambré and Acrivos [79] and Acrivos and Chambré [75] applied these techniques, which result in a nonlinear Volterra integral equation, from where the surface reaction rate can be calculated once the concentration distribution at the wall is known (measured). For a first-order reaction, the equation can be solved for certain geometries in terms of a convergent infinite series. Katz [80] considered fully developed laminar flow in a circular tube, and from an integral equation, the surface concentration was calculated from the reaction rate (which can be known from the cross-sectional average concentration). The usefulness of this approach for kinetic studies and reactor design was also discussed for arbitrary reaction rates. Rosner [81] presented the solutions obtained in a catalytic flat plate for uniform approaching velocity and power-law kinetics at the surface in the form $z = f(c_{wall})$. Grau et al. [82] transformed the original partial differential equation with arbitrary reaction kinetics into a system of integral equations using a Green's function combined with generalized Fourier expansions. The kernels of the integral operators were related to an eigenvalue problem (with homogeneous kinetic-independent boundary conditions). Expressions in slow, fast, and instantaneous reaction regimes were considered to improve the performance of the numerical solution of the resulting integral equations.

It is notorious that analytical solutions for the case of nonlinear wall reactions are substantially more difficult to obtain. Lopes et al. [40] attempted to achieve some insight on this problem by considering the limits of a slow and fast reaction. In this case, the conversion profile is perturbed from the leading-order solutions for the cases when $Da \longrightarrow 0$ and $Da \longrightarrow \infty$, respectively. The calculation of this perturbation resembles the problem of heat transfer in channel flow with a prescribed wall flux or temperature axial variation.

8.2.2.3 Calculation of reactant conversion

The approximate procedure given by Lopes et al. [40] is followed to calculate the conversion of a reactant in channel flow due to the existence of a wall reaction. This is given for fully developed and developing regimes as explained in the following.

Fully developed concentration profile. The fully developed regime asymptote is obtained by retaining just one term of the summation in Equation 8.26. This simplified solution can be frequently found in the literature [44, 54, 62, 85, 86] and is given by

$$X_{fd} = 1 - \langle c \rangle = 1 - w_1 \exp\left(\frac{-\lambda_1^2 z}{\alpha Pe_{m,max}}\right) \quad (8.34a)$$

with

$$\lambda_1^2 = \frac{Da\,\sigma_C}{1 + Da\,\sigma_C/\lambda_{1,\infty}^2} \quad (8.34b)$$

and

$$w_1 = w_{1,\infty} + \frac{w_{1,\infty}(1 - w_{1,\infty})}{\left[w_{1,\infty}^b + (1 - w_{1,\infty})^b Da^b\right]^{1/b}} \quad (8.34c)$$

The calculation of the several quantities appearing in Equations 8.34 is described as follows:

1 The first eigenvalue λ_1^2 can be calculated numerically as a function of Da for several channel cross-sectional shapes and different flow profiles (see Table 8.3 for some examples). Equation 8.34b is a correlation formulated so that the leading-order asymptotic behavior in the limits of small and large Damköhler number is respected. It was found suitable to describe both circular channel and parallel plates, with internal uniform or parabolic velocity profiles. Other cases may be described as well, by selecting appropriate values for σ_C and for the eigenvalue associated with the Dirichlet boundary condition (the latter quantity requires a single numerical evaluation, if any). Equation 8.34b represents a normalization with respect to flow profile and channel geometry, which is plotted in Figure 8.3 as a "master curve": $\lambda_1^2/\lambda_{1,\infty}^2$ versus $Da\sigma_C/\lambda_{1,\infty}^2$. Along with this curve, numerical results for different particular cases are also depicted and found to respect this dependence. As expected, the negligible deviations from this normalization occur in the intermediate range of the Damköhler number. The maximum deviation between the value of λ_1^2 predicted by Equation 8.34b and the one calculated numerically for laminar flows is 1% for parallel plates and 2% for round tube, when $Da = 2$–3. In addition to its accuracy, this expression exhibits a simple mathematical form, showing a clear dependence on the parameters, and retaining the correct asymptotic behavior. The latter feature can be verified by expanding the solution when $Da \to \infty$, from where the term of $O(1/Da)$ is correctly obtained. When expanding for low Da, the term following the leading one of $O(Da)$ is of $O(Da^2)$ and can be written in terms of the shape factor, as derived elsewhere [53].

2 According to Equation 8.34c, the dependence of the first weight on Da is expressed through the shape–flow factor σ_C and the limiting value $w_{1,\infty}$ (Dirichlet conditions). Both quantities normalize w_1 with respect to the channel shape and the flow profile (uniform or parabolic). The value of $w_{1,\infty}$ can be

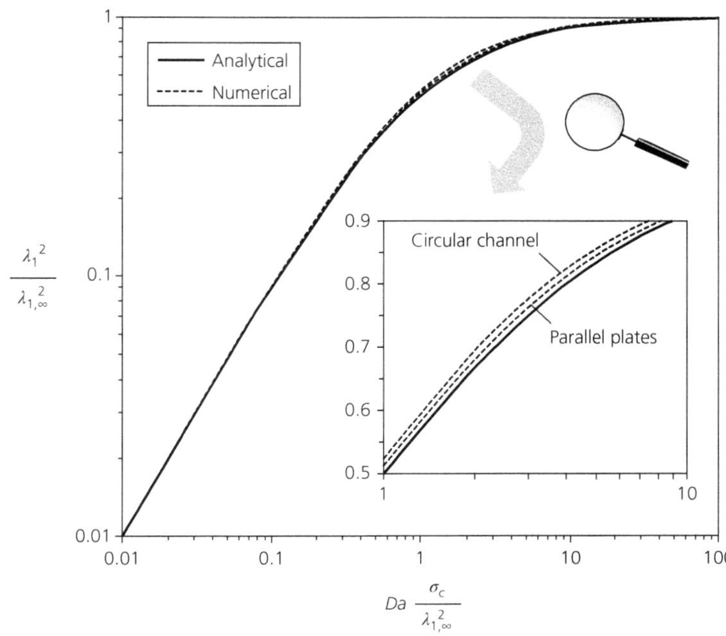

Figure 8.3 Normalized dependence of the first eigenvalue on the Damköhler number: $\lambda_1^2/\lambda_{1,\infty}^2$ versus $Da\sigma_C/\lambda_{1,\infty}^2$ curve. The analytical result given by Equation 8.34b was derived in Lopes et al. [40]. Numerical results for laminar flows are also shown (dashed lines).

calculated from the previously determined eigenvalue $\lambda_{1,\infty}^2$. The coefficient b is around 1 for most cases: plug flow between parallel plates (with 2.5% maximum relative error compared to numerical results) and for laminar flows (circular and slit ducts with less than 2% relative error). However, $b \approx 4$ is recommended when plug flow in a circular channel is studied, yielding less than 6% error. However, the numerical results are still well approximated by Equations 8.34 when b changes around these values. Thus, conversion of reactant shows low sensitivity with respect to this parameter. The maximum deviations are observed when Da varies between 1 and 10.

3 The concentration profile as estimated by Equations 8.34 is plotted in Figure 8.4, along with the numerical solution of Equation 8.16, subject to the boundary condition (8.20). The axial variation of the mixing-cup concentration (expressed in terms of the dimensionless Graetz parameter) is shown for several values of the Damköhler number (covering kinetic and mass transfer control).

As expected, when the characteristic time for diffusion is much larger than the one for convection ($\alpha Pe_m \gg 1$), higher deviations between approximate and numerical results are observed, especially for large reaction rates and curved channel geometries. For example, for $Da = 10^4$ and $\alpha Pe_m/z = 100$ in laminar flows, the relative errors are 12% for circular channels, in comparison with 6% for parallel plates.

Developing concentration profile. In conditions where the profile is developing, Lopes et al. [40] obtained for plug flow

$$X_{dev} = 1 - \langle c \rangle(z) = \frac{(S+1)z}{\alpha Pe_m} \frac{Da}{1 + \sqrt{\pi}/2\sqrt{z/\alpha Pe_m}Da} \tag{8.35a}$$

and for laminar flow

$$X_{dev} = 1 - \langle c \rangle(z) = \frac{(S+1)z}{\alpha Pe_m} \frac{Da}{1 + 0.9828Da(z/\alpha Pe_{m,max})^{1/3}} \tag{8.35b}$$

These solutions fulfill the two asymptotic limits, which in this case are observed for small ($\ll 1$) and large ($\gg 1$) values of

$$Da^* = Da\sqrt{\frac{z}{\alpha Pe_m}} \quad \text{for plug flow} \tag{8.36a}$$

or of

$$Da^* = Da\left(\frac{z}{\alpha Pe_{m,max}}\right)^{1/3} \quad \text{for laminar flow} \tag{8.36b}$$

Equation 8.35b is plotted in Figure 8.5 for a slit channel. In this case, the approximation is associated with maximum relative errors of 3, 0.1, and 0.01%, when $\alpha Pe_m/z$ equals 10, 100, and 1000. Even when $\alpha Pe_m/z = 1$, less than 5% error is observed for $Da^* < 0.5$.

The approximations given by Equations 8.35 are the solution to Lévêque's problem given in Equation 8.30 with a linear wall reaction. Since the formulation of the problem leads to a linearized velocity profile in a planar boundary layer, laminar flows (parabolic velocity profiles) in curved channels are more susceptible to present higher deviations from these results. For a fully developed flow in a round tube, the error associated with Equation 8.35b is 1.4 and 0.13% for $\alpha Pe_m/z$ equal to 100 and 1000, respectively. Lopes et al. [40] observed that these differences are visible mainly for $Da \to \infty$ and calculated corrections to account for these effects. It was shown that in the mass transfer-controlled limit,

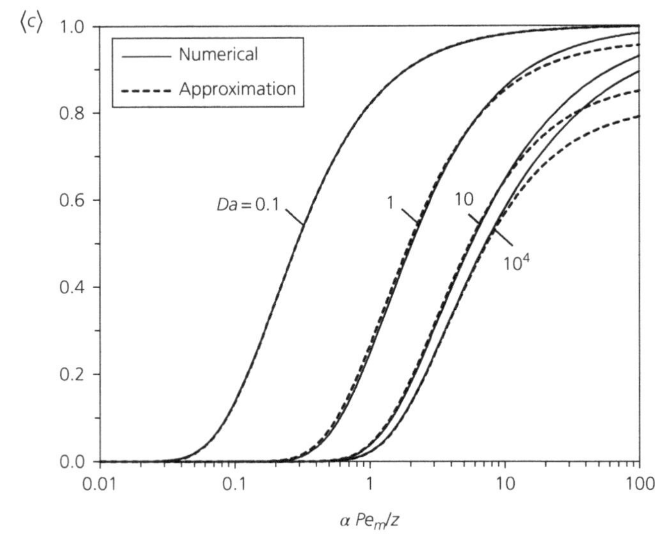

Figure 8.4 Fully developed concentration profile given by Equation 8.34. Numerical (full lines) and analytical (dashed lines) refer to laminar flow in a circular channel. The approximate solution uses eigenvalues and coefficients calculated from Equations 8.34b and 8.34c for several values of Damköhler number.

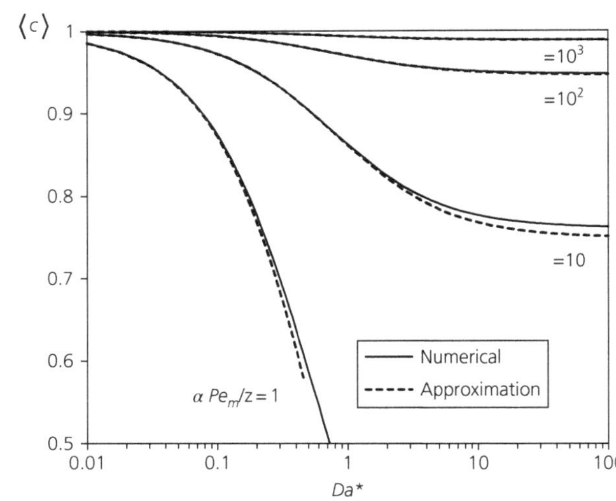

Figure 8.5 Reactant concentration (mixing cup) as a function of the rescaled Damköhler number (Da^*) given in Equation 8.36b for laminar flow between parallel plates. The dashed lines are plotted according to Equation 8.35b for several values of the Graetz parameter ($\alpha Pe_m/z$) in the ranges of Da^* leading to less than 5% relative error.

the correction for curved channels is 10 times higher than the one for nonlinear velocity profile. Both corrections are of $O[(Da^*)^2]$ under kinetically controlled conditions.

Uniformly valid solution and applicability ranges. The two previously described treatments are usually presented separately and assumed to be valid in different ranges of the dimensionless axial distance. However, it is important to remember that the same problem is being considered and that the Graetz solution given in Equation 8.21 should include the one obtained in Lévêque's regime.

This occurs since the former considers the two terms in the mass transfer problem (transverse diffusion and convection), while the latter involves a reduction where only convection is important at most part of the channel radial direction (the two terms will only need to be retained in the thin boundary layer adjacent to the channel walls). Therefore, if a sufficiently large number of terms could be conveniently retained in Graetz's eigenfunction expansion or in Lévêque's perturbation series, a uniformly valid solution for all values of the dimensionless axial distance could be achieved. This consists in the most straightforward strategy, which is however very inconvenient for a number of reasons, and inevitably breaks down in the opposite limit of the Graetz parameter range. A perhaps more interesting approach consists in the following idea: instead of approximately solving the equations in different limits, an asymptotic technique for summations can be applied to the original series for an infinite number of terms. The purpose of this strategy is to bridge the gap between limiting behaviors (finding a uniformly valid profile) and allow the derivation of criteria for the different convection–diffusion regimes. Table 8.5 compares the previously described strategies.

Further details on both Graetz's and Lévêque's extended series can be found in the references mentioned in Table 8.4, for example. The derivation of the uniformly valid solution given by Lopes et al. is described in references [42] and [43]. Briefly, the result for the conversion profile can be written as a linear combination of two contributions (from fully developed X_{fd} and developing X_{dev} profiles), weighed by two functions (Θ_{fd} and Θ_{dev}, respectively):

$$X = X_{fd}\Theta_{fd} + X_{dev}\Theta_{dev} \tag{8.37}$$

The exact form of the terms appearing in Equation 8.37 was given in Lopes et al. [42] for Dirichlet boundary condition and in Ref. [43] for the wall reaction condition (8.20). For uniform wall flux (Neumann-type problem), it is possible to apply the same technique to the infinite series from which Sherwood (or Nusselt) numbers are calculated as a function of the Graetz parameter. The structure for this solution is the same as the one shown in Equation 8.37, with X replaced by $1/Sh_0$. Regarding conversion in the mass transfer-controlled limit (nearly uniform surface concentration), three regions can be depicted in the axial profile (Figure 8.6): the two "limiting" descriptions (fully developed and developing), separated by a transition zone. The boundaries between these regions were given in Lopes et al. [42] as a function of the allowed deviations, when compared to the "exact" solution. A more complex picture is obtained in the generic case of a (possibly finite) wall reaction (Figure 8.7). When the rate of reaction compared to the one for reactant supply to the coating by diffusion is moderate or fast, then the profile development distances are close to the ones found if the Dirichlet boundary condition prevailed. It is possible to derive analytically the parameter relationship which leads to deviations between simplified model predictions and more rigorous calculations. Convenient evaluation of these deviations is given in relative (ε_{fd}) and absolute (e_{dev}) terms, for high and low conversions, respectively. Below a transition value for the Damköhler number (determined in Ref. [43]), there is a region of values for the Graetz parameter where both simplified

Table 8.5 Bridging the gap between convection and diffusion regimes.

Approach	Refs.	Comments
1. Numerical/approximate calculation of terms in Graetz series: $\langle c\rangle = \sum_{n=1}^{\infty} w_n \exp\left(\frac{-\lambda_n^2 z}{\alpha Pe_{m,\max}}\right)$, where w_n and λ_n^2 are both functions of Da	See Table 8.4 and references therein, for example, Brown [63]	• Inconvenient (involves numerical calculation, especially for finite reaction rates) • Large number of terms still required (slow convergence) • The description of the inlet behavior is limited for finite number of terms
2. Extended Lévêque solutions: $\langle c\rangle = \sum_{n=0}^{\infty} \langle c_n\rangle \delta^n$, where $\langle c_n\rangle$ is the solution of the perturbation problem at $O(\delta^n)$	Newman [64, 72], Worsoe-Schmidt [71]	• Differential nonhomogeneous problem to calculate each term $\langle c_n\rangle$ of the series • Not known for boundary conditions other than Dirichlet and Neumann • Cumbersome algebraic form • Nonuniformly valid in practice • Possibly divergent series
3. Improved convergence through asymptotic series summation	Lopes et al. [42, 43]	• Uniformly valid (respects both limits) • Maximum error in the intermediate region (acceptable) • Enables derivation of applicability criteria • Insight into the solution structure (finite Da)

descriptions yield the same result. Hence, an overlapping region for low reaction rates appears.

Nonlinear kinetics. The problem of calculating reactant conversion by a nonlinear wall reaction requires numerical solution. The most explicit approximations available in the literature were given by Lopes et al. [40] in limits regarding profile development (fully developed or developing profiles) and mass transfer control (kinetic or diffusional regimes).

8.2.2.4 Axial dispersion

For the design of a single-channel wall-coated monoliths (and other similar structured reactors), the control of the residence time is important for selectivity issues. In preliminary design methods [87] it is proposed that once the required residence time is fixed, the fluid velocity should be selected so that minimum dispersion is obtained. It is possible, however, that excessive pressure drop results from this and a trade-off must be searched.

To highlight the benefits from microstructuring in reducing dispersion, "equivalent" fixed bed and microreactor/monolith were compared in terms of a dispersion ratio (ratio of the widths of initially delta-like concentration tracers at the reactor exit) [88, 89]. In terms of the Peclet number (with D_{ax} as the axial dispersion), an expression for fixed bed reactor of the type

$$Pe = \frac{ud_p}{D_{ax}} = f(Re, Sc) \tag{8.38}$$

is considered, while for the microchannel the classical Taylor–Aris theory applies [90–92]. For $3 < ud/D < 100$, Commenge et al. [88] observed that the dispersion in a microchannel is smaller than that of a fixed bed reactor with porosity 0.4. These boundaries change for other values of porosity (the range gets broader for smaller porosities and narrower for higher porosities). The minimum dispersion is attained at $ud_{ch}/D \sim 14$ and corresponds to a 40% reduction in dispersion in a microchannel reactor compared to a fixed bed one.

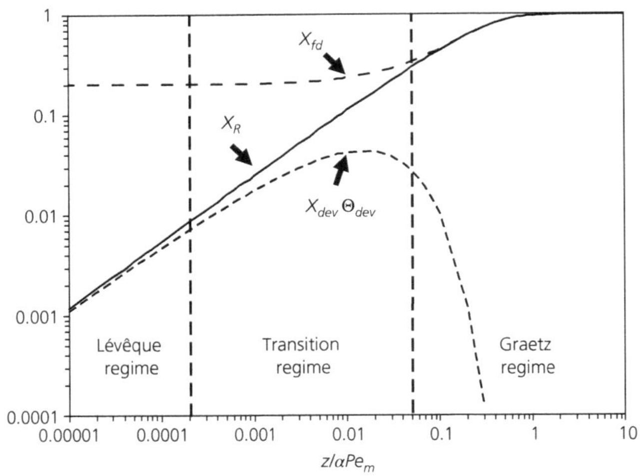

Figure 8.6 Conversion as a function of the dimensionless axial distance for laminar flow in a circular channel with instantaneous wall reaction. Three regions can be identified in different ranges of the Graetz parameter: Graetz and Lévêque descriptions, separated by a transition regime.

8.2.3 Reaction and diffusion in the catalytic washcoat

Catalytic monoliths are structured heterogeneous reactors. These two features require the consideration of the reaction–diffusion problem in the catalytic washcoat (the "internal" region), which can be designed with much more independence from the "external" domain, when compared to nonstructured reactors. In general, the operating conditions will be such that

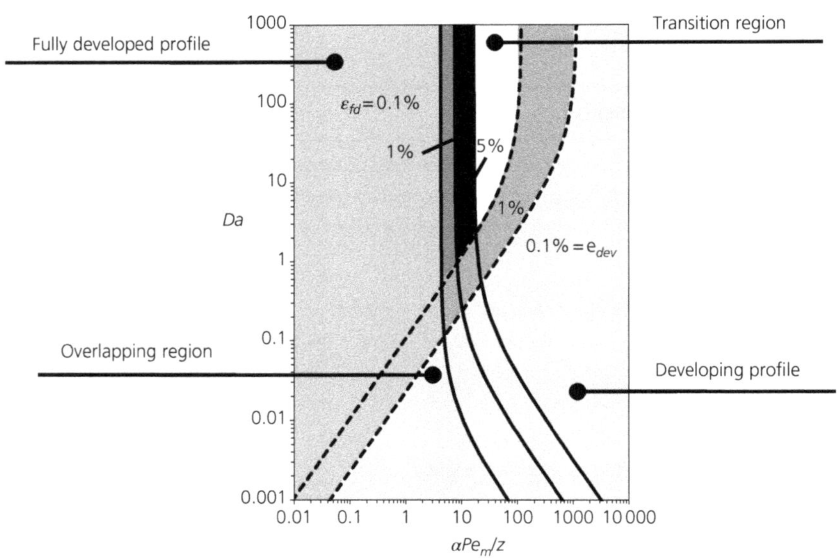

Figure 8.7 Convection–diffusion limiting cases for channel mass transfer with wall reaction in a Damköhler–Graetz parametric map ($Da - \alpha Pe_m/z$). Four regions can be depicted: fully developed profile (for $\alpha Pe_m/z$ below the prediction shown by the full lines for given values of ε_{fd}), developing profile (for Graetz numbers above the ones estimated by dashed lines for given e_{dev}), transition region (separating the two previous limits at high *Da*), and overlapping region (where similar predictions from both models are obtained at low *Da*). All boundaries were obtained analytically and the respective expressions can be found in Ref. [43].

some mass transfer resistance in the coating may be present. Therefore, the reaction rate depends on the position inside the catalyst, resulting in a difference between the averaged rate over the cross-sectional area A_w and the value at the channel–coating interface. The internal concentration profiles for each species can be determined by solving the mass balance equations, which will include the effective diffusion in the porous structure and the source/sink terms due to the occurring chemical reaction(s). The problem is identical to the one described in Chapter 3 (Section 3.3) of this book. As in the case of a catalytic nonstructured packing, the pseudo-homogeneous approach is commonly adopted, along with the formulation in terms of the effectiveness factor. The latter quantity is written as the average of the reaction rate $\hat{R}(\hat{c}_w, \hat{T}_w)$ normalized by its surface value $\hat{R}\left(\hat{c}_{surf}, \hat{T}_{surf}\right)$:

$$\eta = \frac{\iint \hat{R}(\hat{c}_w, \hat{T}_w)\, dA_w}{\hat{R}\left(\hat{c}_{surf}, \hat{T}_{surf}\right) A_w} \tag{8.39}$$

The effectiveness of a catalyst pellet is a fundamental concept in reaction engineering, which has been discussed extensively (see Section 3.3 in Chapter 3 and references therein). Its application has both qualitative and quantitative purposes. The value of the effectiveness factor provides a clear indication of the degree of utilization of the catalyst and consequently of the level of internal mass transfer resistance. In fact, this quantity was used in Lopes et al. [24] and [43] as a criterion to delimit internal regimes, which can be set as strict as desired (values close to 1 for kinetically controlled conditions and low values for mass transfer control). Numerical simulation of the reactor can also take advantage from using this quantity, since the problem in the catalyst is averaged and decoupled. The same generic observations hold for catalytic coatings. The most commonly used correlations for effectiveness factor are simplified models, which nevertheless have been found suitable to describe complex situations. Two examples are worth mentioning:

- Structural complexity—The multiscale material structure of a catalytic coating can be accounted for in more detailed terms using sophisticated experimental and modeling techniques (e.g., the work of Kočí et al. [8] involving digital reconstruction and numerical solutions at micro- and nanolevels).
- Kinetic complexity—As more reliable information on the reaction mechanisms becomes available, it seems sensible to retain information on the detailed surface chemistry when simulating the internal transport in the coating. Seyed-Reihani [93] modeled hydrogen oxidation over a Pd/PdO_x coating with a mechanism that includes 11 reversible surface reactions, 6 adsorption steps, and corresponding desorption steps.

It is curious that despite the levels of complexity involved (with different sources), the bottom line of the existing studies is still the computation of the "classical" effectiveness factor or of the average reaction rate. Whenever possible, simple effectiveness factor correlations are highly desirable. Therefore, the discussion of what we call the "internal problem" is done here in terms of this concept.

There are however some particularities when the effectiveness of a catalytic coating is considered. First, the length scale for internal diffusion (t_w) is typically much smaller than the one for external transport (a), as shown in Figure 8.1. The ratio between these characteristic distances governs the coating curvature. Second, the area available for diffusion increases as the reactant penetrates in the catalytic body and is maximum at the interface with the impermeable/symmetry boundary. The latter characteristic only requires minor adaptations, if any. The former feature can be also found in eggshell particles, which can be thought of as a catalytic coating on a (possibly, spherical) pellet. The similarity between the effectiveness of coated channels and particles was addressed in Lopes et al. [94].

In a simplified reaction–diffusion model for a pseudo-homogeneous catalytic coating (the assumptions involved can be found, e.g., in Ref. [95]), the mass balance is given by

$$\frac{1}{(1+\varepsilon r_w)^{\sigma}}\frac{\partial}{\partial r_w}\left[(1+\varepsilon r_w)^{\sigma}\frac{\partial c_w}{\partial r_w}\right] + (\alpha\varepsilon)^2\frac{\partial^2 c_w}{\partial z^2} - \phi^2 R = 0 \tag{8.40}$$

Axial diffusion is ignored based on the previous considerations regarding the typical geometric ranges for monolith reactors (see following text), and the dependence of concentration on this coordinate is only introduced by the boundary condition at the surface (concentration continuity). The dimensionless variables are defined as follows (see "Nomenclature" section at the end of this chapter):

$$c_w = \frac{\hat{c}_w}{\hat{c}_{surf}},\quad r_w = \frac{(\hat{r}_w - a)}{t_w},\quad z = \frac{\hat{z}}{L} \text{ and } R = \frac{\hat{R}(\hat{c})}{\hat{R}\left(\hat{c}_{surf}\right)} \tag{8.41}$$

It is possible to observe that a 1D model to describe diffusion in the directions normal to the axial one was adopted (similarly to that described for channel flow and mass transfer in Sections 8.2.1 and 8.2.2). Hence, a shape factor σ appears in order to translate the deviation between the actual geometry and the one modeled with reduced dimensionality. The behavior of an isothermal catalytic coating (and consequently its effectiveness) is mainly governed by two parameters (α is the previously defined aspect ratio):

$$\text{Thiele modulus } \phi^2 = \frac{t_w^2}{D_{eff}}\frac{\hat{R}\left(\hat{c}_{surf}\right)}{\hat{c}_{surf}} \tag{8.42a}$$

$$\text{Length scales ratio } \varepsilon = \frac{t_w}{a} \tag{8.42b}$$

The measurement and calculation of the effective diffusivity in a catalytic washcoat D_{eff} has been given in [96] from a 1D model for ceramic (e.g., cordierite) monoliths. A quick survey of reaction studies in monolithic catalysts (see, e.g., Ref. [97]) shows that the ratio between the timescales for internal diffusion and reaction can cover a wide range. In fact, as we discuss in Section 8.3.1, very different regimes may prevail in the catalytic coating. Thus, it is of interest to obtain the dependence of the effectiveness for all values of ϕ^2. A more distinguishing feature in monolith reactors is found when compiling the typical values for ε. This has been shown in Lopes et al. [94]. In general, the design of catalytic coatings leads to low values of this dimensionless thickness: $\varepsilon \ll 1$ or at least $\varepsilon < 1$. This magnitude statement provides an opportunity for model reduction in the internal reaction–transport problem.

Since ε is directly related with the importance of the coating curvature, the most popular strategy to account for internal diffusional limitations in the washcoat, even in detailed modeling, is to reduce the actual geometry to a slab. In this situation, the intraphase effectiveness factor is given by

$$\eta = \frac{\tanh \phi_g}{\phi_g} \tag{8.43a}$$

where the *geometric* Thiele modulus is

$$\phi_g = \frac{V_{cat}}{S_{surf}} \sqrt{\frac{k}{D_{eff}}} \tag{8.43b}$$

written here for a first-order isothermal reaction, even though a generalization regarding the form of the kinetic expression is possible for certain types of reaction rate laws (essentially, monotonic kinetics), as shown in the following. The use of the volume-to-surface area ratio as the length scale in the Thiele modulus guarantees the fulfillment of the diffusional regime asymptote, while at low reaction rates the leading-order behavior $\eta \to 1$ is obtained. Equations 8.43 represent a shape normalization which is long known for catalytic pellets [68]. It is also acknowledged that this prediction of the coating effectiveness is associated with nonnegligible errors in the intermediate region of the Thiele modulus.

While reduction of the actual geometry to a slab may be reasonable as $\varepsilon \to 0$, an improved estimate of the catalytic coating effectiveness is achieved by following the opposite reasoning (Figure 8.8). This has been proposed by Lopes et al. [94], and we describe here the main ideas associated with the procedure (full details can be found in the cited reference). We start by noting that the 1D model adopted to describe diffusion in the catalytic domain has in fact two parameters which need to be appropriately chosen so as to represent the actual geometry as closely as possible. These are the shape factor σ and the transverse characteristic length scale ℓ. In reasonably uniform geometries, ℓ is taken as the maximum distance for diffusion, which can be the coating thickness t_w. In the conventional approach described by Equations 8.43, $\ell = V_{cat}/S_{surf}$ and $\upsilon = 1$, that is, the "equivalent slab" representing the coating exhibits the same volume-to-surface area ratio (Figure 8.8). The internal concentration profile

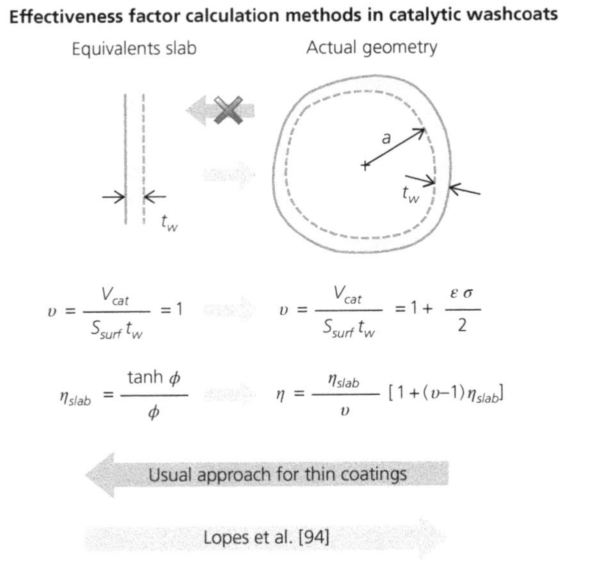

Figure 8.8 Calculation approaches for the effectiveness factor in thin catalytic coatings. The distinction between the conventional method and the one proposed in Lopes et al. [94] is illustrated. The Thiele modulus for a first-order reaction is given by $\phi = t_w\sqrt{k/D_{eff}}$. The ratio between the catalyst and channel transverse length scales is given by $\varepsilon = t_w/a$ and σ is the catalyst shape factor.

can be seen as a function of a perturbation parameter ε, which takes small (but nonzero) values:

$$c_w(r, z; \varepsilon) = c_{w(0)}(r, z) + \varepsilon c_{w(1)}(r, z) + O(\varepsilon^2) \tag{8.44}$$

By introducing Equation 8.44 into the mass balance (8.40), it is possible to extract the necessary subproblems for the calculation of $c_{w(0)}$ and $c_{w(1)}$. Then, according to Equation 8.39, the corresponding perturbation series for the effectiveness factor is obtained. Details can be found in Ref. [94], but what is relevant to highlight here is that *the effectiveness factor for a curved coating with nonnegligible thickness can be expressed in terms of the one derived for a catalytic slab with the same characteristic length.* Thus, we move from an approach which reduces the problem to a simpler formulation to one which retains all the convenient features from an approximate solution without losing information. Another important advantage is that even though the derivation of the solution is based on the assumption that ε takes low values, the final result ends up being useful even when $\varepsilon \sim 1$, which is more than enough to cover the range of geometrical features present in all practical applications.

The improved solution for the effectiveness factor was originally derived for linear reactions in [94] and is given by

$$\eta = \frac{\eta_{slab}}{\upsilon}[1 + (\upsilon - 1)\eta_{slab}] \tag{8.45}$$

where $\upsilon = V_{cat}/(S_{surf} t_w)$, and for the aforementioned kinetics,

$$\eta_{slab} = \frac{\tanh \phi}{\phi} \tag{8.46}$$

Note that the length scale in the definition of the Thiele modulus in Equation 8.46 is now t_w, as given in Equation 8.42a. The extension of this result to nonlinear kinetics and nonuniform coating geometries was also discussed in Lopes et al. [94]. Reaction rate expressions other than first order can be treated using the concept of *generalized Thiele modulus* Φ, where the previously defined parameter is modified so that its introduction in an expression such as

$$\eta_{slab} = \frac{\tanh \Phi}{\Phi} \tag{8.47}$$

yields the correct limiting behaviors at high and low reaction rates. Several possibilities to achieve this have been presented but limited to monotonic kinetics. Lopes et al. [94] proposed the following form for Φ:

$$\frac{\Phi^2}{\phi^2} = \frac{R'(1)}{1+\phi^2} + \left[2\int_0^1 R(c')dc'\right]^{-1} \frac{\phi^2}{1+\phi^2} \tag{8.48}$$

As expected, $\Phi = \phi$ when the reaction is first order. Note that $R'(1)$ is the derivative of the dimensionless reaction rate with respect to the dimensionless concentration, evaluated when the latter variable is 1 (at the surface). For power-law kinetics,

$$\frac{\Phi^2}{\phi^2} = \frac{m}{1+\phi^2} + \frac{m+1}{2}\frac{\phi^2}{1+\phi^2} \tag{8.49}$$

Equation 8.47 can be introduced into the improved solution (8.45) to predict effectiveness in catalytic coatings with nonlinear kinetics. It was shown [94] that this approach can be successfully applied for several reaction rate expressions.

The deposition of a washcoat in a monolith channel usually leads to the appearance of catalytic layers with peripherally nonuniform thickness. For example, it is possible to find washcoats deposited on square channels, which are 10 μm thick at most part of the perimeter, except near the corners where this value increases to 150 μm [44]. The nonuniformity is more severe for thicker layers and when the channel has sharp corners (due to the surface tension effects of the catalyst mixture inside the channel). An existing numerical approach to simulate these cases (presented by Papadias et al. [98, 99] and also used by Hayes et al. [100]) consists in the so-called slice method. Briefly, the method decomposes the coating cross section into several elements (the "slices"), which are treated approximately as having the geometry of a slab (for which the simplified solution (8.43) can be used). Then, the effectiveness factor of the whole coating is obtained by adding up the contributions from all "slices." Naturally, the calculation becomes more accurate as the number of slices taken increases. Mass transfer in the direction perpendicular to the channel–coating interface is accounted for, while transfer in the azimuthal coordinate (along the coating perimeter) is not considered. However, the results were found to be in good agreement with 2D numerical simulations over the washcoat cross section for a reasonable number of slices.

The contribution of the work developed in Lopes et al. [94] for this situation is shown in Table 8.6. A modification to the previously described procedure was introduced based on our improved solution for the effectiveness, given by Equation 8.45. Thus, a *curved slice method* can be formulated and is described by the following set of equations:

$$\text{Effectiveness factor of washcoat,}\quad \eta = \sum_{i=1}^{N} w_i \eta_i \tag{8.50a}$$

Table 8.6 Calculation of the effectiveness factor in uniform and nonuniform washcoats.

Washcoat shape	Uniform	Nonuniform
	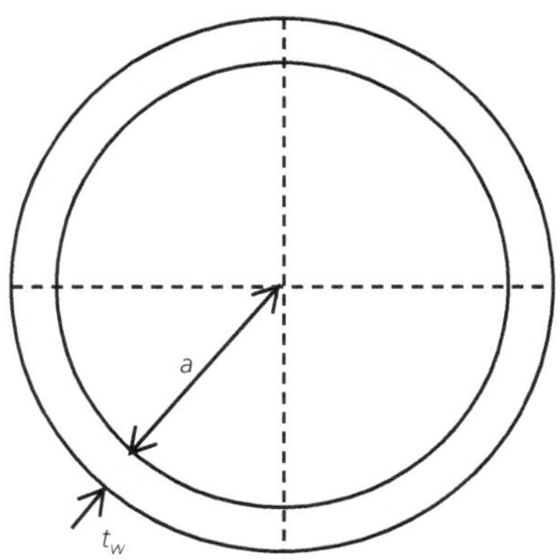	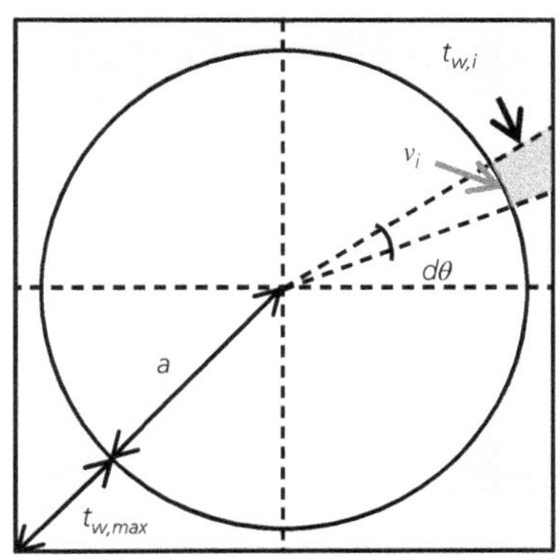
Existing approach	Exact analytical solution for hollow cylinders (Gunn [101], Wijngaarden et al. [102]) with linear kinetics or numerical solution	Numerical method with slab-shaped slices (Papadias et al. [99])
Improvement introduced by Lopes et al. [94]	Approximate solution for small ε, using the solution for the slab geometry Equation 8.45	Curved slice method Equations 8.50

Effectiveness factor of curved slice i, $\eta_i = \frac{\eta_{slab,i}}{\upsilon_i}\left[1 + (\upsilon_i - 1)\eta_{slab,i}\right]$ (8.50b)

Effectiveness factor of planar slice i, $\eta_{slab,i} = \sqrt{\frac{D_{eff}}{t_{w,i}^2 k}} \tanh\sqrt{\frac{t_{w,i}^2 k}{D_{eff}}}$ (8.50c)

Dimensionless volume to surface ratio, $\upsilon_i = \frac{V_i}{S_{surf,i} t_{w,i}}$ (8.50d)

Slice length scale, $t_{w,i} = \left\{ \begin{matrix} \text{maximum transverse distance} \\ \text{normal to the interface in } i \end{matrix} \right\}$ (8.50e)

Volume fraction of slice i, $w_i = \frac{V_i}{V}$. (8.50f)

Equation 8.50c is written for a first-order reaction, but modification for nonlinear kinetics is straightforward, as shown previously. If the geometry is reasonably defined, it is possible to calculate all the components given earlier directly. Lopes et al. [94] provide the detailed calculations for circle-in-square shape and squares with rounded corners. For the circle-in-square geometry depicted in Table 8.6, numerical results from 2D numerical simulations are available in Hayes et al. [100]. Table 8.7 compares the relative errors of several approximate methods for the calculation of η in this geometry (for several values of ϕ_g^2). It is possible to observe that (i) for the same number of slices N, the relative error from the curved slice method is always lower than the one from methods using planar slices, (ii) the reduction in the relative error is more visible in the intermediate region of the Thiele modulus, and (iii) for the same relative errors, the number of slices required can drastically decrease by adopting the curved slice concept.

Table 8.7 Accuracy of effectiveness factor calculation methods for nonuniform geometry (circle-in-square shape with a first-order reaction).

	Relative errors (%) compared to 2D numerical results in Hayes et al. [100]		
		***N* curved slices**	
ϕ_g^2	**100 planar slices**	***N* = 50**	***N* = 100**
0.01	0.52	0.29	0.27
0.09	2.37	1.20	1.07
0.49	4.18	1.91	1.61
1.94	3.07	1.07	0.69
6.37	1.67	0.43	0.05
17.82	1.40	0.74	0.38
43.82	0.97	0.65	0.33
96.94	0.63	0.49	0.21

8.2.4 Nonisothermal operation

In general, temperature gradients are present in the monolith channel and washcoat. Thus, heat transfer needs to be accounted for in the monolith reactor model, along with mass transport and species consumption/heat generation by chemical reaction(s). A 2D energy balance in channel flow can be written as in Equation 8.16. The boundary conditions at the surface include continuity of heat flux between the channel and the coating. As we have discussed in Section 8.1.1, if the transfer through the support is slow, the washcoat can be considered adiabatic, even though the case of finite heat transfer resistance can also be addressed. Results for nonisothermal effectiveness in the catalyst and for the maximum temperature expected are analogous to the ones presented in Chapter 3 for catalytic particles. However, the detailed energy balance can also be written for the washcoat, and the complete set of equations solved numerically, if required.

In this section, we formulate a 1D model with interphase mass and heat transfer coefficients. These "lumped models" [103] describe the axial variation of concentration and temperature (which are averaged over the channel cross section). The diffusion processes in the transverse directions (represented by differential terms) are replaced by a transfer term, associated with a given driving force. The use of 1D models is widespread throughout the literature on monolith reactor modeling. Chen et al. [3] reviewed some specific applications including simulation of simultaneous heat transfer in monolith catalysts [104], evaluation of the effective diffusivity in a washcoat [105], optimization of the active component distribution along an adiabatic monolith [106], and prediction of the gas exit temperature for long monoliths [107]. Limitations to their applicability have also been recognized but seem to be minimized if the appropriate correlations are chosen, as discussed in the following. Moreover, the significant reduction in computational effort associated with these models is an undeniable advantage.

In steady-state and negligible axial diffusion in the monolith channel, the governing equations can be written in terms of four dependent variables, namely, the mixing-cup and surface concentrations ($\langle c\rangle$ and c_{surf}) and the equivalent quantities for temperature ($\langle T\rangle$ and T_{surf}). The mixing-cup concentration is given by Equation 8.22, which can be defined similarly for fluid temperature as

$$\langle T\rangle(z) = \frac{\int_0^1 r^S v(r) T(r,z)\,dr}{\int_0^1 r^S v(r)\,dr} \quad (8.51)$$

A consistent derivation of the 1D model can be achieved by averaging the 2D mass and energy conservation equations (with a single transverse and axial coordinates) over the channel cross section. This results in

$$-\alpha Pe_m \frac{d\langle c\rangle}{dz} = \frac{Sh}{S+1}\left(\langle c\rangle - c_{surf}\right) \tag{8.52a}$$

$$\alpha Pe_h \frac{d\langle T\rangle}{dz} = \frac{Nu}{S+1}\left(T_{surf} - \langle T\rangle\right) \tag{8.52b}$$

The interphase coefficients for mass (k_m) and heat (h) transfer are made dimensionless into Sherwood and Nusselt numbers, both defined as

$$Sh = \frac{ak_m}{D} \tag{8.53a}$$

$$Nu = \frac{ah}{\lambda} \tag{8.53b}$$

The Graetz number for heat transfer is given by

$$\alpha Pe_h = \frac{a^2 \rho C_p}{\lambda}\frac{\langle u\rangle}{L} \tag{8.54}$$

For an adiabatic coating (no heat flux at $r_w = t_w$; see Figure 8.1), the following expressions relate mass/heat interphase transfer with the observed consumption of reactant and heat generation by chemical reaction:

$$Sh\left(\langle c\rangle - c_{surf}\right) = Da_{in}\eta R\left(c_{surf}, T_{surf}\right) \text{ or } Da_{in}\bar{\eta}R(\langle c\rangle, \langle T\rangle) \tag{8.55a}$$

$$Nu\left(T_{surf} - \langle T\rangle\right) = \beta_f Da_{in}\eta R\left(c_{surf}, T_{surf}\right) \text{ or } \beta_f Da_{in}\bar{\eta}R(\langle c\rangle, \langle T\rangle). \tag{8.55b}$$

The Damköhler number in Equations 8.55a and 8.55b refers to inlet conditions:

$$Da_{in} = \frac{a\hat{R}_{surf}\left(\hat{c}_{in}, \hat{T}_{in}\right)}{D} \tag{8.56}$$

while Prater's nonisothermal parameter is evaluated at bulk (fluid) conditions:

$$\beta_f = \frac{(-\Delta H_R)\hat{c}_{in}D}{\hat{T}_{in}\lambda} \tag{8.57}$$

The effectiveness factor (referred to surface conditions η) and the global effectiveness factor (referred to bulk conditions $\bar{\eta}$) can be described by approximations in limiting cases (see Chapter 3 for examples) and rational approximations can be derived.

In order to obtain the (mixing-cup and surface) concentration and temperature profiles, the solution of Equations 8.52 and 8.55 requires the calculation of the dimensionless transfer coefficients Sh and Nu, which are in general functions of the Graetz parameters and of the Damköhler number. Correlations for these dimensionless numbers are available in the literature (e.g., Shah and London [39]), for several conditions regarding the degree of profile development, degree of hydrodynamic flow development, and boundary condition at the wall. These features are briefly discussed in the following.

8.2.4.1 Degree of profile development

The fully developed concentration (or temperature) profile discussed in Section 8.2.2 is characterized by a uniform value of Sherwood (or Nusselt) number, that is, independent of axial distance. Young and Finlayson [108] solved the heat and mass transfer problem numerically using the orthogonal collocation technique and calculated the asymptotic (for long distances) Sherwood and Nusselt numbers for several channel geometries. In terms of the analytical development that we have been presenting, it can be calculated for laminar flows from

$$Sh_{fd,\infty} = \frac{\lambda_{1,\infty}^2}{\sigma_C} \quad \text{(uniform wall concentration condition)} \tag{8.58a}$$

$$Sh_{fd,0} = \frac{(S+5)(S+7)}{5S+17} \quad \text{(uniform wall flux condition)} \tag{8.58b}$$

If the profile is developing in a laminar flow, Lévêque's results can be used to derive expressions for Sh:

$$Sh_{dev,\infty} = \frac{6^{1/3}}{\Gamma(1/3)}\left(\frac{\alpha Pe_{m,max}}{z}\right)^{1/3} \quad \text{(uniform wall concentration condition)} \tag{8.59a}$$

$$Sh_{dev,0} = \frac{2^{1/3}}{3^{2/3}}\Gamma\left(\frac{2}{3}\right)\left(\frac{\alpha Pe_{m,max}}{z}\right)^{1/3} \quad \text{(uniform wall flux condition)} \tag{8.59b}$$

Replacing αPe_m with αPe_h yields the corresponding expressions for Nu. In this case, a dependence on the Graetz parameters is introduced, and the magnitude of the transfer coefficients increases as the inlet is approached. Lopes et al. [53] proposed the following correlation to merge both limiting behaviors regarding the profile development:

$$Sh = \left(Sh_{fd}^n + Sh_{dev}^n\right)^{1/n} \tag{8.60}$$

The coefficient n was chosen as 2 for plug flows and 4 for laminar velocity profiles, even though accurate predictions can also be obtained when n changes around these values. 1D models in nonisothermal problems have been applied by Gupta and Balakotaiah [109] and Stutz and Poulikakos [110] for studying heat and mass transfer and in the prediction of light-off/ignition [111–114]. The correlations for Sh (and Nu) reported in the literature can be classified into three categories:

1 *Theoretical*—when higher dimensional models are used, from which the coefficients can be derived analytically (as the ratio between the wall gradient and the driving force) or evaluated numerically
2 *Empirical*—obtained from fitting particular experimental results in more or less restricted ranges
3 *Combination of theoretical/empirical approaches*—where the functional dependence on the parameters may be suggested by analytical expressions but the coefficients are fitted to the particular experimentally obtained data

The works mentioned earlier are essentially of theoretical nature. We now consider some examples of correlations with partial or exclusive experimental support. Hawthorn [115] presented an expression which combines developing and fully developed behaviors:

$$Sh' = \frac{d_h k_m}{D} = 2.977\left[1 + 0.095\frac{Re_d Sc}{\varepsilon'}\left(\frac{d_h}{L}\right)\right]^{0.45} \quad (8.61)$$

The mass transfer coefficient increases from the nearly constant value for large channel lengths, according to a Lévêque-type dependence (with a different exponent, however):

$$Sh' \sim \left(\frac{ReScd_h}{L}\right)^{0.45} \sim (\alpha Pe_m)^{0.45}$$

The catalyst bed void fraction ε' is related to the hydraulic diameter and external surface area per unit volume of bed by $\varepsilon' = d_h S_{surf}/(4V_{bed})$. A correlation of the form of Equation 8.61 can be generalized for other geometries:

$$Sh' = \frac{d_h k_m}{D} = B\left[1 + CRe_d SC\left(\frac{d_h}{L}\right)\right]^{0.45} \quad (8.62)$$

The fully developed limiting value given by B depends solely on the geometry (e.g., $B = 2.976$ for square channels), while C is a surface roughness constant. The values of C are reported to be equal to 0.078 for smooth surfaces, 0.095 for automobile (washcoated cordierite) monolith catalysts, and 0.139 for extruded TiO_2-SiO_2 monolith catalysts [21].

Votruba et al. [116] presented an empirical correlation from experimental data on vaporization of water and hydrocarbons from porous monolith supports:

$$Sh' = \frac{d_h k_m}{D} = 0.705\left(Re_d\frac{d_h}{L}\right)^{0.43} Sc^{0.56} \quad (8.63)$$

In this case, a limiting value corresponding to a fully developed concentration profile is not retrieved and the values for Sh' are about three times lower than those predicted by the Hawthorn correlation. The same deviation was observed by Bennett et al. [45], who obtained data for Sh' lower than the ones given by Equation 8.61 by a factor of 10, in conditions of partial mass transfer control. The same functional form was observed, however.

Ullah et al. [117] obtained results close to the ones given by Vortruba (also without observing an asymptotic value) in CO oxidation:

$$Sh' = \frac{d_h k_m}{D} = 0.766\left(Re_d Sc\frac{d_h}{L}\right)^{0.483} \quad (8.64)$$

Uberoi and Pereira [21] studied the conversion in CO oxidation experimentally. The results were obtained under mass transfer-limited conditions, that is, in a range where no change in conversion with increasing temperature is observed. The experimental data was fitted as a function of $Re_d d_h/L$, according to the following expression:

Table 8.8 Experimental ranges used in the development of empirical mass transfer correlations.

Refs.	d_{ch} (cm)	$Re_d\frac{d_{ch}}{L}$	L (cm)
Votruba et al. [116]	0.1–1	2–400	1.2–4.0
Bennett et al. [45]	0.1	1–55	<3.8
Ullah et al. [117]	0.1	1–100	<15
Uberoi and Pereira [21]	0.345–0.605	2–30	9–30

Source: Adapted from [21].

$$Sh' = \frac{d_h k_m}{D} = 2.696\left[1 + 0.139ScRe_d\frac{d_h}{L}\right]^{0.81} \quad (8.65)$$

This correlation was applied in parametric numerical studies of SCR of NO. Its predictions are in close agreement with those from the Hawthorn correlation near the fully developed limit. Table 8.8 shows the experimental ranges in which these correlations where obtained, regarding the monolith channel geometry and flow conditions. In most cases, the values under full mass transfer control are reported. Curiously, the exact dependence of the transfer coefficients according to Lévêque's analysis is not observed but reasonably approximated. As mentioned earlier (Section 8.2.1), results for laminar flows are the most relevant for practical conditions. Nevertheless, correlations for turbulent flows have also been proposed [118]:

$$Sh' = 0.023Re_d^{0.8}Pr^{0.33} \quad (8.66)$$

The net consequence from inlet flow turbulence is to increase Sh from the value obtained for laminar flow (Holmgren and Andersson [119]).

8.2.4.2 Degree of hydrodynamic flow development

If inlet effects, such as the simultaneous development of the flow profile, are found to be important, it is known [39] that Sherwood (or Nusselt) number correlations exhibit three distinct dependences in different ranges of the Graetz parameter. For Dirichlet boundary condition, these are given by

$$Sh' = \frac{1.128(L^*)^{-1/2}}{\left[1 + (Sc/0.0468)^{2/3}\right]^{1/2}}, \quad L^* = \left(\frac{a}{d_h}\right)^2 \frac{1}{\alpha Pe_m} \le 7\times10^{-3}, \quad (8.67a)$$

$$Sh' = 1.615(L^*)^{-1/3} - 0.2, \quad 7\times10^{-3} \le L^* \le 0.03, \quad (8.67b)$$

$$Sh' = 3.657 + \frac{0.0499}{L^*}, \quad L^* > 0.03 \quad (8.67c)$$

The first branch (8.67a) accounts for hydrodynamic profile development, as given by Churchill and Ozoe [120]. The two latter ones are given by Shah and London [39], used by Kolodziej and Lojewska [12] to model short-channel structures, where advantage from these effects is taken (since transport is enhanced). The corresponding correlations for heat transfer are obtained replacing Sh, Sc, and Pe_m, by Nu, Pr, and Pe_h, respectively.

Lopes et al. [43] adapted the correlation given by Equation 8.60 to include the asymptote of developing concentration and velocity profiles. Moreover, this expression was found to be extremely convenient in the mapping of reaction–transport regimes, a topic which will be addressed in Section 8.3.

8.2.4.3 Wall boundary condition

The mass/heat wall fluxes and the concentration/temperature driving forces exhibit different dependences on the wall reaction rate, which is made dimensionless by the Damköhler number. Therefore, *Sh* is also a function of *Da*. The previously examined correlations are valid for uniform wall concentration (Sh_∞ when $Da \to \infty$) or uniform wall flux (Sh_0 when $Da \to 0$). Values of Sh_∞ are lower than those for Sh_0 for the same value of $\alpha Pe_m/z$. The difference between both values can be around 20%.

In general, values for *Sh* and *Nu* under reaction conditions are different from those observed at these limits (Hayes and Kolaczkowski [121]). Correlations which attempt to describe the variation of Sherwood (or Nusselt) numbers with the Damköhler number have been proposed. Tronconi et al. [104] and Groppi and Tronconi [122] used the result from Bräuer and Fetting [123] for mass/heat transfer correlations when modeling monolith reactors with circular, square, and triangular shape:

$$\frac{Sh' - Sh'_{H_2}}{Sh'_T - Sh'_{H_2}} = \frac{Da' Sh'}{(Da' + Sh')Sh'_T} \tag{8.68}$$

The dimensionless numbers in Equation 8.68 are referred to the channel hydraulic diameter. According to our previous notation, $Sh'_T = Sh'_\infty$. In the limit of nearly uniform wall flux ($Da \to 0$), an important distinction arises regarding the peripheral uniformity of wall concentration/temperature and mass/heat flux. This distinction is only important for geometries with corners (rectangles, triangles, etc.) and has been discussed at least since Shah and London (see pages 27 and 390 in [39]). For these geometries, the values of the transfer coefficients that should be used are the ones for the (H_2) boundary condition (zero flux along the channel perimeter), given in Table 8.3 for some common channel shapes. The same work considers the effect of rounded corners due to the introduction of a washcoat on mass transfer.

Kockmann [124] suggests using the following empirical expression:

$$\frac{1}{Sh'} = \frac{1}{Sh'_0} + \frac{Da}{Da + 1.979}\left(\frac{1}{Sh'_\infty} - \frac{1}{Sh'_0}\right) \tag{8.69}$$

In this case, the result refers to a fully developed concentration profile. Moreover, Lopes et al. [53] showed that for the purpose of regime mapping, the asymptotic values of Sh_∞ and Sh_0 could be used.

8.3 Mapping and evaluation of operating regimes

8.3.1 Diversity in the operation of a monolith reactor

The versatility of structured reactors leads to the existence of many regions in the parametric space, which require characterization. Parametric areas can be distinguished in a number of ways. Scaling analysis leads to a natural definition of "regimes," since each mechanism is associated with a timescale, and these are compared in parameters that arise when making the model dimensionless. The importance of each term is then evaluated by the magnitude of the associated parameter (since correct scaling assumes dependent variables and its derivatives of $O(1)$). In particular when seeking approximate solutions, a balance between two important effects is often looked for, leaving others ignored or restricted to nonleading-order corrections. These limiting solutions apply in a given regime which is defined a priori. In chemical reactors, the relationship between transport and reaction is often used to define limiting behaviors in the governing equations or in the boundary conditions. Mapping of all these regimes in a diagram has the benefits of systematically organizing previous results and explore new areas, which may have been left ignored in previous studies. The choice of parameters that constitute the axes of this plot also results from scaling.

If a given set of geometric and operation variables is represented by a single point in the aforementioned diagram, then one might question the possible location of such representation for typical monolith reactors. Actually, even if some geometric characteristics are shared by all designs, many possible operating regimes may prevail. We have already seen some evidences of the flexibility of monoliths, concerning the operating ranges:

- The ratio of reaction and transport timescales can cover several orders of magnitude, even when the reactions which are potential candidates to structured reactor engineering have well-defined characteristics, and monoliths can be designed to match them.
- Some authors proposed monolith-based solutions for reactions with very different characteristics ranging from kinetically controlled to transport limited.
- In many of the studies reported, several conditions such as the residence time and temperature are tested and optimized. Therefore, a change in the controlling effect may be observed in the range explored.
- In Section 8.1.2, we have reported applications with distinct purposes such as screening and chemical production applications. This naturally requires two very different sets of conditions and consequently two different regimes.

Moreover, it is interesting to obtain a comprehensive picture including as many regimes as possible, even if this is only done with the purpose of setting boundaries between them.

8.3.1.1 Intrinsic kinetic measurements

Apart from the specific applications of microreactors as devices for screening and catalyst testing, measurement of the intrinsic kinetics in a system with the advantages previously mentioned is highly desirable. Their improved mass and heat transfer properties at uniform temperature are beneficial, to avoid falsification of the observed kinetics. The following characteristics should be also mentioned:

- The monolith configuration allows easy exportation of results obtained with similar geometry and conditions between systems with different scales. The geometry for kinetic measurements may actually be the same used in chemical processing.
- Isothermal behavior may be assured due to improved thermal control in monolith configurations.
- The structured nature of the reactor allows excellent control of the catalyst layer thickness (sufficiently thin in order to guarantee the absence of internal limitations).
- External limitations are absent, since the channel diameter can be defined to be sufficiently small.
- The behavior of the reactor can be considered "ideal" ("plug flow" for $uL/D_{Aris} > 100$, with D_{Aris} given by Taylor–Aris theory [90, 92]).

Several kinetic studies in microreactors/monoliths are available in the literature over a wide range of conditions. In particular, measurements under "severe conditions" are attractive for a number of reasons. We mention some examples in the following.

McCarty [125] used an annular reactor to evaluate kinetics of methane combustion over PdO-supported catalysts. The design of the apparatus had a small gap between cylinders (0.1–0.3 mm) and a thin coating (10 µm). Using high flow rates and dilute methane and oxygen in helium, the author claims to have measured the intrinsic rate of methane oxidation up to 900°C, without contributions from gas-phase reactions. Groppi et al. [126] used the same configuration to study the same reaction over a PdO/Al_2O_3 catalyst. Beretta et al. [11] discussed the use of this reactor at high temperatures and high space velocities (GHSV) for kinetic studies of several reactions: catalytic oxidation of CO, catalytic partial oxidation of methane, and oxidative dehydrogenation of propane. In this case, a reactor with a thin catalyst layer (50 µm) and small gap (1.1 mm) was used.

The suitability of the structured annular reactor for very fast catalytic combustion reactions was confirmed, since it allowed measurement of kinetic data under conditions closer to the ones in the commercial applications (at higher GHSV and temperature). Extrapolation from lab-scale results can therefore be avoided, as changes in the reaction mechanism may occur [127]. In particular, they name the following advantages: (i) implementation of very high GHSV with negligible pressure drops (in laboratory fixed beds, ΔP amounted to nearly 1 bar at 400 cm^3 NTP/min in the combustion of methane/biogas and to 0.8 bar in highly diluted bed with GHSV ~10 times higher); (ii) well-defined geometry, flow profile, and mass transfer (which they assumed to be given by correlations for the Sherwood number at Dirichlet conditions and fully developed concentration profile); (iii) negligible internal temperature gradients; (iv) similarity between the coated catalyst tube and the monolith catalysts; (v) exclusion of homogeneous reactions (preliminary experiments with uncoated channels); and (vi) lower dilution of reacting mixture and catalyst compared to a fixed bed.

Groppi et al. [126] obtained data up to 600°C at partial conversion (due to high GHSV). The most critical phenomenon was internal diffusion and only by using a thin catalyst layer the kinetics was free from mass transfer effects ($t_w \leq 10 \mu m$ for $\eta \geq 0.9$). External mass transfer limitations played a minor role for channels with a gap between 0.2 and 0.3 mm, which should be kept as small as allowed by pressure drop. Beretta et al. [11] recognized that depending on the hydraulic diameter and on the reaction rate at the wall, operation may change from kinetic to diffusional control and that at high temperatures diffusional effects can become important. They proceeded to develop a 1D model from a Sherwood number correlation and concluded that (i) kinetic regime prevailed up to 450–500°C, (ii) decrease of conversion was thus entirely due to higher GHSV and this widened up the operating window for intermediate conversion values to very high temperatures (200 – 900°C with maximum 65% conversion in CO oxidation and methane partial oxidation), (iii) at higher temperatures CO conversion was moderated by the onset of interphase mass transfer limitations, and (iv) the estimation of intrinsic kinetic constant was in good agreement with the data from fixed bed. Annular reactors have also been considered by other groups (e.g., Yu et al. [128]).

Germani and Schuurman [129] studied the kinetics of the water–gas shift reaction over platinum/ceria/alumina catalysts washcoated on stainless steel microchannels. Despite the fact that the catalysts were very active, intrinsic kinetic measurements were performed by employing thin films and favoring plug-flow conditions in the channel. The reaction conditions cover a wide range of the operating variables. Sazonova et al. [130] analyzed the behavior of a LaNiPt catalyst supported in a honeycomb substrate for methane partial oxidation. In this case, kinetics was measured over the monolith at high temperatures and with millisecond contact times. They used a 1D model to predict the ranges where the process was kinetically controlled in the channel, since the internal resistance was found to be negligible from the low value of the Thiele modulus.

A related topic is the evaluation of internal diffusion limitations in catalyst coatings supported on microchannels. Experimentally, the two most common tests are variation of temperature and catalyst layer thickness. Both quantities are sometimes given as criteria for the chemical regime (specification of a temperature or thickness below which kinetics controls). For example, some authors [127, 131] consider internal diffusion negligible for $t_w < 50 \mu m$. Kapteijn et al. [132] concluded that washcoat layers thicker than this would lead to mass transfer control in the washcoat, when testing square channel cordierite monoliths with alumina washcoat layers of various thicknesses (20 – 110 μm) in the Fischer–Tropsch synthesis ran between 180 and 225°C. In other studies, diffusional limitations are simply excluded in the view of the thin washcoats employed [21, 117]. Walter et al. [133] considered catalyst coatings of different thickness but with same composition in microchannel reactors. For a constant reactant flow (equal hydrodynamic residence time), internal mass transfer limitation was excluded for thicknesses between 5 and 20 μm. In the selective oxidation of isoprene to citraconic anhydride, the same author excludes limitations for $t_w < 80\,\mu m$, based on an analogy with a particle in a fixed bed. Chen et al. [134] warn for the fact that internal mass transfer in a metal foam methanol microreformer must be carefully considered and that the critical thickness for the absence of diffusional effects was only 8 μm.

However, in other cases internal diffusion limitation can be significant even with very thin washcoat thicknesses [127], when temperature is high (> 700°C). This refers, for example, to catalytic combustions, which are extremely fast. Hayes et al. [135] evaluated the extent of intraphase and interphase resistances to the catalytic conversion of low concentrations of carbon monoxide in air in a tube wall reactor (coated with a platinum–alumina deposit). Above 610 K there was strong evidence of both intraphase and interphase resistances to catalytic conversion. In Sections 8.3.2, 8.3.3, and 8.3.4, we provide a systematic analysis for prediction of the extension of external and internal diffusion limitations.

8.3.1.2 Process conditions

Under most operating conditions and reactor/catalyst designs, nonnegligible sources of internal and external mass transfer limitation will exist. Actually, as it is discussed in Section 8.3.5, mass transfer-controlled conditions correspond to the monolith design associated with lower pressure drop (channel length).

Tomasic et al. [136] studied experimentally and numerically the decomposition of NO in a catalytic monolith (cordierite substrate and copper containing ZSM-5 zeolite). The behavior of the experimental monolithic reactor was found to be mainly limited by interphase mass transfer. Three models with increasing complexity were considered to describe the phenomena. A 2D heterogeneous model with distributed parameters was determined to be the most suitable mathematical model. This involved the calculation of radial and axial gradients, as well as the solution of the reaction–diffusion problem in the washcoat. The simulations revealed significant interphase mass transfer limitations. Intraphase diffusion should be also included in the mathematical model. A 1D heterogeneous model, without consideration of internal mass transfer limitations, is in good agreement with experimental results for the axial variation of concentration. The results are extremely dependent on the correlation adopted for the Sherwood number. The model is expected to be useful in the evaluation of experimental kinetic parameters and in the quantification of the effect of some design variables on the reactor performance. Kolaczkowski and Serbetcioglu [118] considered internal and external transport limitations in catalytic combustion monoliths. Raja et al. [137] compared the Navier–Stokes, boundary layer, and plug-flow model simulation results, when the problem of flow and reaction in a catalytic combustion monolith was solved. Their observations are systematized in Table 8.9. Nonnegligible mass transfer limitations tend to require more complex models.

8.3.2 Definition of operating regimes

Lopes et al. [24, 43] presented a conceptual analysis of the different operating regimes that may appear in a wall-coated monolith/microreactor. There are several particularities in the proposed methodology compared with other studies in the

Table 8.9 Comparison between models for catalytic combustion in a monolith.

Feature/model	Navier–Stokes	Boundary layer	Plug flow
Mechanisms considered and neglected	• Transport of mass, heat, and momentum • Radial and axial directions	• Radial transport to/from the walls • Negligible axial diffusion	• Radial gradients negligible • Negligible axial diffusion
Computational effort	Expensive	Intermediate	Simplest
Classification of governing equations	Elliptic PDE	Parabolic PDE	ODE for average concentration axial variation
Validity	• Less restrictive assumptions • Adequate to represent cases with low values of Re_L (where the role of axial diffusion gains importance)	• Increasingly appropriate as Re_L increases, thus for $Re_L \gg 1$	• Cases dominated by convection: $d/L \ll Re_d Sc \ll L/d$
Output comparison	• Profiles are more "spread out" due to flow-wise diffusion	• Accurate enough for most applications	• Overestimates reactant conversion when reaction is mass transfer limited

Source: Based on Ref. [137].

literature: (i) the boundaries delimiting regimes in the parametric space are derived analytically, (ii) cases of nonlinear kinetics are addressed, (iii) convective–diffusive limiting behaviors in channel flow (Figure 8.7) are superposed with the definition of reaction–transport regimes, and (iv) the interaction between channel and catalyst domains is well characterized.

As mentioned earlier, the characteristic behavior of a given reactor/operation is associated with a region in the multiparametric space. This space has as many dimensions as the number of model parameters. For typical isothermal monolith reactor designs studied in Section 8.2, we found four relevant dimensionless numbers:

1 The Graetz parameter, including the length of the channel and operating flow rate

$$\frac{\alpha Pe_m}{z} = \frac{a^2 \langle \hat{u} \rangle}{LDz} \sim \frac{Q}{LD} \tag{8.70}$$

2 The Damköhler number, defined with the channel characteristic dimension and reaction rate referred to the inlet temperature and concentration

$$Da_{in} = \frac{a\hat{R}_{surf}\left(\hat{c}_{in}, \hat{T}_{in}\right)}{D\hat{c}_{in}} \tag{8.71}$$

3 The diffusion ratio, where the diffusivities and length scales of both domains are compared

$$\Delta = \frac{D_{eff}\, a}{D t_w} \tag{8.72}$$

4 A parameter related with the catalyst geometry, namely, with the ratio between the thickness of the coating and the radius of the open channel

$$\upsilon = \frac{V_{cat}}{S_{surf}\, t_w} \tag{8.73}$$

For convenience of representation, 2D plots are preferred while the remaining dimensionless numbers are kept at fixed values. Lopes et al. [24, 43] considered several of these diagrams. In the following sections, we show some examples for linear and nonlinear kinetics. The purpose is to derive analytically the boundaries between the different regimes that appear. However, there are some preliminary questions that must be clarified, namely, (i) how should the degree of profile development be expressed quantitatively, (ii) what criterion/criteria should provide indication of the extent of mass transfer limitations, and (iii) how to classify regimes regarding not only the controlling process but also the domain to which is referred. The answer to the first question has already been mentioned in Section 8.2.2.3. Contrary to other studies in the literature, the level of profile development was not based on the agreement with the asymptotic value for the Sherwood number Sh_{fd}. Instead, we considered the deviations between actual reactant conversion (X) and the one predicted by simplified models (X_{fd} or X_{dev}):

$$\varepsilon_{fd} = \frac{|X_{fd} - X|}{X} \quad \text{(fully developed profile)} \tag{8.74a}$$

$$\varepsilon_{dev} = |X_{dev} - X| \quad \text{(developing profile)} \tag{8.74b}$$

The two remaining questions are related. The structured nature of the monolith reactor introduces additional degrees of freedom, as we have noted. Thus, it is reasonable to consider "separate" criteria for expressing mass transfer limitations in the bulk fluid flowing through the channel and in the catalytic coating. Several choices exist for these quantities. Our approach uses the degree of mass transfer control and the effectiveness as external and internal regime definers, respectively. They are defined as

Degree of mass transfer control in the channel,

$$\theta = \frac{\langle c \rangle - c_{surf}}{\langle c \rangle - c_{surf} + \hat{R}\left(c_{surf}\right)/\hat{R}\left(\hat{c}_{in}\right)} \tag{8.75a}$$

$$\text{Catalyst effectiveness factor,} \quad \eta = \frac{\iint \hat{R}\left(\hat{c}_w, \hat{T}_w\right) dA_w}{\hat{R}\left(\hat{c}_{surf}, \hat{T}_{surf}\right) A_w} \tag{8.75b}$$

The latter is a well-known quantity in the reaction–diffusion analysis in catalytic media (see Section 8.2.3) and can be written as the ratio between the average reaction rate over the washcoat cross-sectional area at a given axial position and its value at the surface. The former compares the driving force for mass transfer toward the coating, with the total potential for concentration decay (due to mass transfer and surface reaction). For a first-order reaction, θ reduces to the Carberry number (see Chapter 3), and η is a concentration ratio between the averaged value inside the washcoat and the one at the surface.

We can now suggest values for these criteria for different controlling phenomena. When the rate of reactant consumption by chemical reaction is relatively slow compared with the transport processes, $\theta \to 0$ in the channel and $\eta \to 1$ in the catalytic coating. This corresponds to *kinetically controlled conditions*. On the other hand, if *mass transfer is controlling*, $\theta \to 1$ in severely transport-limited channels and $\eta \to 0$ in a catalytic coating operating in the diffusional regime. Exact values for these criteria can be set arbitrarily by the user. The quantities are bounded between 0 and 1, but only low values or values sufficiently close to unity are sensible for regime definition, as described earlier. One of the main advantages of the proposed analytical methodology is that we can obtain the explicit relationship between the parameters which define a boundary and the referred criteria. Therefore, regime mapping based on strict values for the adopted criteria can be obtained without needing to repeat any numerical calculation.

It is now possible to clarify how the regimes can be classified regarding the domain where mass transfer or reaction processes are distinguishably controlling. In particular, the following cases can be considered:

1 Overall kinetic control and overall mass transfer control (in the sense that the same level of mass transfer resistance is present in both the channel and the catalyst)

2 Intra- and interphase mass transfer control, where the most important resistance is concentrated exclusively in one phase

These regimes will be described in more detail in the following sections. The principles of the analysis that were just presented also lead to considerations on the relative importance of internal and external transport. In Ref. [24], we have found a basis to establish whether it is external or internal phenomena which defines the overall regime. These criteria are independent of reaction rates and can be formulated as a comparison between the values for the diffusion ratio Δ and a specified value Δ^*, which governs the transition between controlling behaviors. This question is highly relevant and further information can be found in the cited Refs. [24, 43], for example, the influence of this analysis on the experimental testing of mass transfer resistances.

8.3.3 Operating diagrams for linear kinetics

The typical representation of operating regimes for a monolith channel reactor with linear kinetics is given in Figure 8.9. Three reaction–transport regimes can be identified: (i) overall mass transfer control which for the values of the diffusion ratio considered is determined by channel flow (thus, faster reactions are required to achieve the same degree of mass transfer limitation as the convective transport in the channel becomes more intense); (ii) overall kinetic control, determined by the catalyst over the shown range of the Graetz parameter, for the lower value of Δ; and (iii) intraphase (internal) mass transfer control, where nearly gradientless mass transport to a severely diffusional-limited coating occurs. The analysis of the concentration profile development is also shown in this representation. For each value of the diffusion ratio, five vertices have special meaning, as shown in Table 8.10. The coordinates for each point are given in Lopes et al. [43].

Since in microchannel reactors, the characteristic length for diffusion in the coating may be much smaller than the one for external mass transfer in the channel, the diffusion ratio (governing the relative importance of inter- and intraphase resistances) may take higher values than those expected in mass transfer through a stagnant film to a pellet in a fixed bed reactor. To highlight the role of this parameter, we construct a diffusion ratio–Damköhler number plot, shown in Figure 8.10 for linear kinetics. As mentioned previously, a transition value for Δ (which will be function of the Graetz parameter) is found to establish whether the boundary is defined by channel or by catalyst features. The value of Δ^* can be determined from analytical expressions, or for the conditions under which Figure 8.10 was constructed, it can be taken from the intersection of full and dashed lines.

Increasing the diffusion ratio results in an increase of area where the kinetic regime prevails. The increase is up to the point where transport in the channel is the controlling factor. From this point on, further expansion of this area is only obtained if the flow parameter is increased. There is higher chance of finding the regime where intraphase resistance dominates when catalyst diffusion decreases, but the upper boundary corresponding to significant resistance in the channel is delayed by moving into the developing range of the profile as the Damköhler number

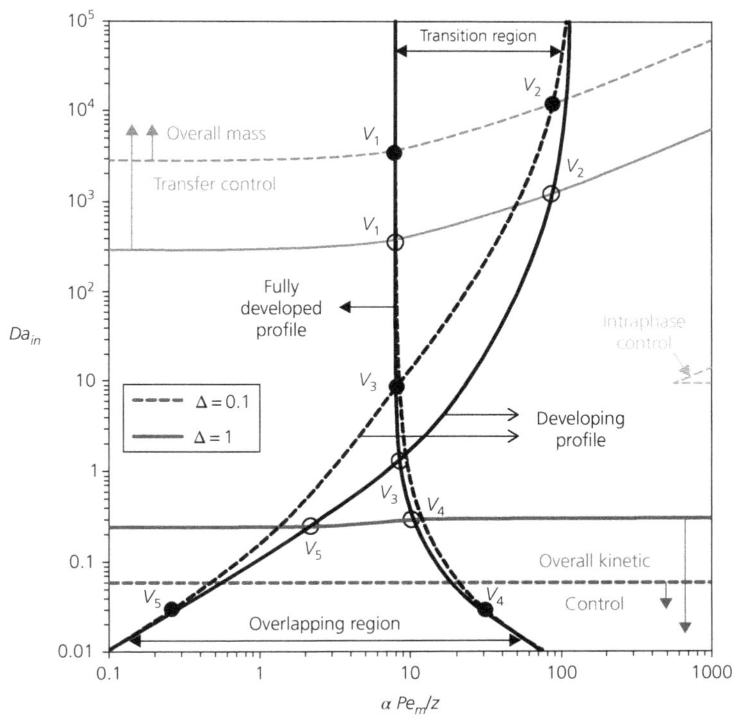

Figure 8.9 Damköhler–Graetz plot for linear kinetics with mapping of reaction–transport and convective-diffusive regimes in a washcoated monolith. Two values of the diffusion ratio are examined, and laminar flow inside a circular channel with $\upsilon = 1.05$ is considered. Regime boundaries are plotted for given values of the criteria (internal control, $\eta \leq 0.1$; external control, $\theta \geq 0.9$; no internal limitation, $\eta \geq 0.9$; no external limitation, $\theta \leq 0.1$; developed profile, $\varepsilon_{fd} \leq 1\%$; developing profile, $e_{dev} \leq 1\%$). Vertices (V) delimiting the intermediate regime are also shown.

Table 8.10 Vertices in the Damköhler–Graetz plot with reaction–transport and profile development regimes.

Vertex	Features
High conversion vertex, V_1	• For higher Da_{in} and $\alpha Pe_m/z$, the concentration profile can be considered fully developed and mass transfer controlled • Physical picture corresponds to a long microchannel with a fast wall reaction, so both reaction and transverse diffusion dominate
Hot inlet vertex, V_2	• Conditions associated with significant energetic requirements to attain mass transfer control (which increase in developing conditions) and energy dissipation rates associated with high fluid velocities • The reduced residence time in a short channel or at the inlet section penalizes conversion • It may represent the state of the inlet section of a channel attaining high conversion at its exit
Middle point, V_3	• Occurs at values of $\Delta Da_{in} \sim 1$, where convection and diffusion balance
Low conversion vertex, V_4	• Found at high Graetz parameter and low Damköhler number, where conversion is low and dominated by convection
Homogeneous microchannel vertex, V_5	• High conversions are possible in kinetic control, for sufficiently long microchannels. In this case, transport is dominated by transverse diffusion

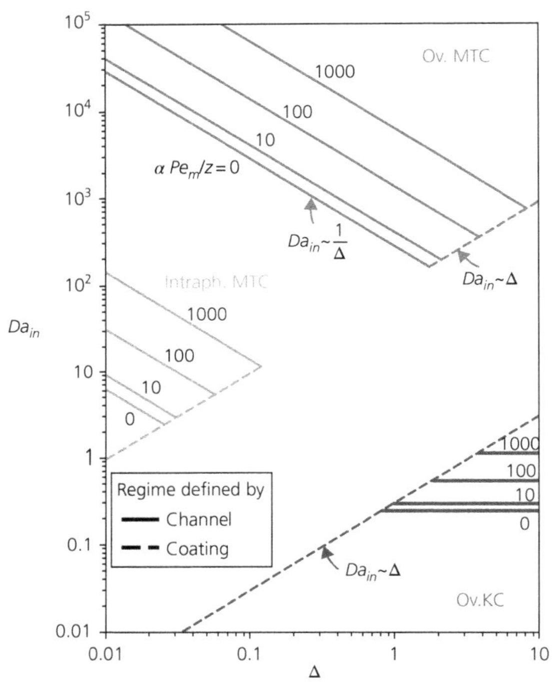

Figure 8.10 $Da_{in}-\Delta$ operating map for a circular channel with laminar flow coated with an annular catalyst layer ($\varepsilon=0.1$) and a first-order reaction. External and internal mass transfer control observed for $\theta=0.9$ and $\eta=0.1$, respectively. Negligible resistances for $\theta=0.1$ and $\eta=0.9$. Full lines associated with conditions in channel are plotted for values of the Graetz parameter. The area for overall kinetic control (Ov. KC) is depicted as well as the ones for mass transfer control: overall (Ov. MTC) and internal (Intraph. MTC).

increases. The same entry-length effect is observed for overall mass transfer control.

Due to the independence of the catalyst operating lines (associated with the effectiveness factor) with respect to the local surface distribution, the boundary for internal diffusional regime is the same for the intraphase and overall mass transfer control regimes. For a first-order kinetics, interphase mass transfer control only starts to be observed for higher diffusion ratios (above 17.3 in these conditions).

8.3.4 Influence of nonlinear reaction kinetics

One of the main advantages of the regime mapping methodology presented is the ability to deal with nonlinear kinetics. Again, the analytical derivation of all results can be found in the literature [24, 43]. Here, we focus on some illustrations of the analysis. Figure 8.11 is a more generalized version of the diagram considered in Section 8.3.3. It is plot of the Damköhler number as a function of the Graetz (flow) parameter, for a nonlinear, second-order reaction in a circular channel with laminar flow (with a thin annular washcoat, $\varepsilon=0.1$).

The parametric areas of five regimes are identified for two values of the diffusion ratio. Typical radial concentration profiles associated with each physical picture are also shown. In this case, the boundary for the kinetically controlled regime is given by the value of the Damköhler number so that the effectiveness factor is kept, for example, above 0.9. Since in this case conversion in the channel is negligible, the reactant concentration distribution at the interface with the coating is uniform and close to the inlet value.

The same cannot be said under mass transfer control, where the surface distribution of concentration and reaction rates included in the Thiele modulus also depend on the mass transfer problem in the channel. This only happens for nonlinear kinetics and gives origin the distinction between an overall regime (with low wall concentration but still allowing for strong gradients to develop in the coating) and a purely interphase resistance (where channel concentration is so low that the effectiveness factor may even approach 1).

The development of the correlation for the effectiveness factor in Section 8.2.3 also allows the accurate description of

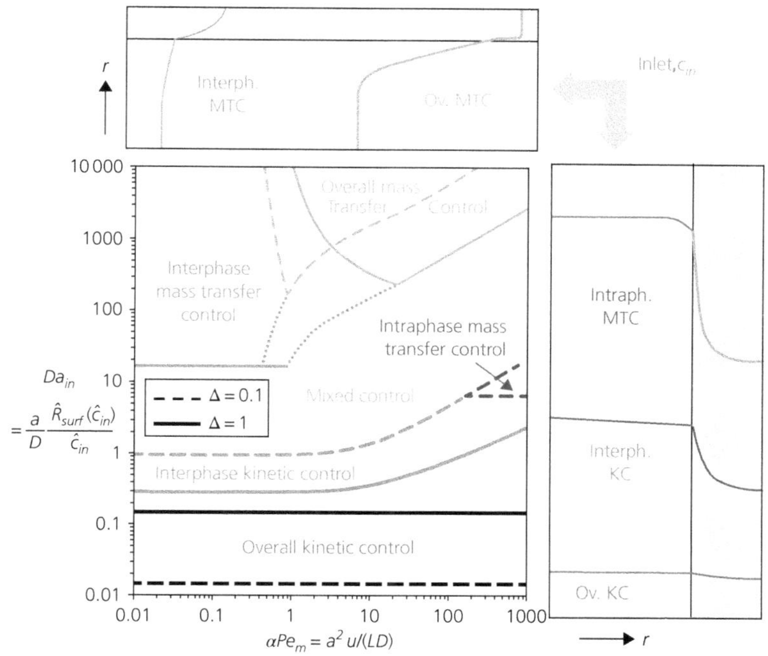

Figure 8.11 Damköhler–Graetz diagram for nonlinear kinetics. A second-order reaction occurs in the washcoat, but the remaining conditions are identical to previous representations.

interphase kinetic control, where mass transport in the channel is free from mass transfer resistance, but the coating presents a nonnegligible degree of diffusional limitation, covering the intermediate range of effectiveness factor.

Concerning the Damköhler–diffusion ratio representation when nonlinear kinetics is considered, the picture gets altered especially in the mass transfer-controlled regimes (Figure 8.12) and with orders of reaction greater than 1. Here, both boundaries of the overall transport control depend on the flow parameter. The interphase boundary is however more insensitive and defines the minimum Damköhler number for appreciable limitation (much lower than for linear kinetics). For developing profile, it only exists above a certain value of the diffusion ratio. Note that the limit where Δ tends to infinity is important in the definition of interphase mass transfer control even when $\Delta < 1$. Other consequences of kinetic nonlinearity include the following: (i) the boundaries defined by the catalyst for the overall mass transfer-controlled and intraphase mass transfer-controlled regimes do not coincide, as previously observed for linear kinetics, and (ii) for reactions with order below 1, the diffusion ratio must take very high values for transition between different behaviors to occur. These topics are discussed in more detail in Lopes et al. [43], where the effects of reaction inhibition are also addressed.

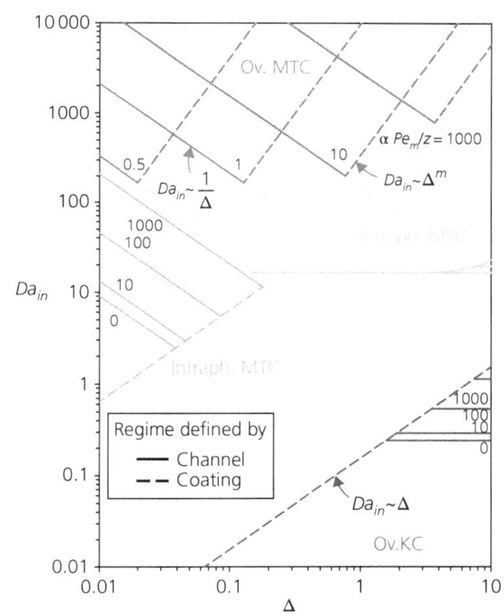

Figure 8.12 Damköhler–diffusion ratio diagram for nonlinear kinetics (second-order reaction). Same conditions from previous representations apply.

8.3.5 Performance evaluation

A topic that relates to the previous discussion is the performance evaluation of monolith designs in these well-identified regimes—in particular, the distribution of the relevant performance criteria across regime maps such as the ones presented in previous sections. Lopes et al. [43] proposed the evaluation of washcoated monolith channels in terms of the efficiencies associated with flow, catalyst, and reactant. These were measured by the following quantities:

$$\text{Dimensionless pressure drop,}\quad \frac{1}{\Delta P} = \frac{{}^{1}/_{2}\rho\langle u\rangle^{2}}{\Delta \hat{P}} = \frac{2}{(S+1)C_f Sc}\frac{\alpha Pe_m}{z} \qquad (8.76)$$

$$\text{Global effectiveness factor,}\quad \bar{\eta} = \frac{\iiint \hat{R}(\hat{c})dV_{cat}}{\hat{R}(\hat{c}_{in})V_{cat}} = \frac{\alpha Pe_{m,max}/z}{\sigma_C Da_{in}}X \tag{8.77}$$

$$\text{Reactant conversion,}\quad X = \frac{\hat{c}_{in} - \langle\hat{c}\rangle}{\hat{c}_{in}} \tag{8.78}$$

respectively. The dimensionless pressure drop in terms of the friction factor coefficient depends directly on the Graetz parameter, which with our approximate results can be expressed in terms of the desired conversion and Damköhler number. If we are constrained to attain an appreciable degree of reactant conversion, this will probably happen only under full profile development, where pressure drop is higher (Figure 8.13). Thus, the intermediate region close to full development is of interest.

On the other hand, when the pressure drop to attain a given conversion is plotted as a function of the reaction rate (see Figure 8.13), mass transfer conditions are always favored. It is known, however, that using highly active catalysts or high operating temperatures may not be possible or even desirable due to reasons such as possible catalyst deactivation, parasite reactions, and activation of homogeneous reactions. It is also hard to understand how this situation would be "optimal" from an energy efficiency point of view, especially if inlet streams require external heating. Thus, it is likely that a maximum value of the allowable reaction rate exists. The excess pressure drop to allow for not operating under full mass transfer control depends on the reciprocal of this value.

High global effectiveness is often sought with intensified microdevices but is in conflict with a strict conversion constraint, which is only possible with operation in the exactly opposite regime. The impossibility of meeting these optima due to process/energetic restrictions leads to the existence of penalties. Another problem is defined when we try to find the direction in the parametric space, along which the system would move when conversion is allowed to be slightly lower, so that costs are reduced. Changes in the design and operation were considered, and the role of the intermediate region was strongly highlighted [43, 138].

8.4 Three-phase processes

The multiphase nature of the flow inside the monolith channels influences hydrodynamics and mass transfer. We briefly review the main aspects related with modeling of monolith reactors with Taylor flow. Particular attention is devoted to the consequences from using gas–liquid flow in contact with a catalytic surface as a reactor:

1 In Taylor or slug flow, the flow pattern consists in a train of elongated bubbles separated by portions of liquid (slugs), which prevent bubble coalescence and promote recirculation, leading to improved mass transfer in the relatively more diffusion-limited liquid phase, and toward the catalytic coating in the volume occupied by slugs. The excellent mass transfer in gas–liquid (or liquid–liquid) segmented flow pattern in capillaries has been noted even in millimeter-sized channels. Kreutzer [29] showed that a correlation of the form of Equation 8.60 is suitable to describe liquid–solid mass or heat transfer in the slugs (Figure 8.14):

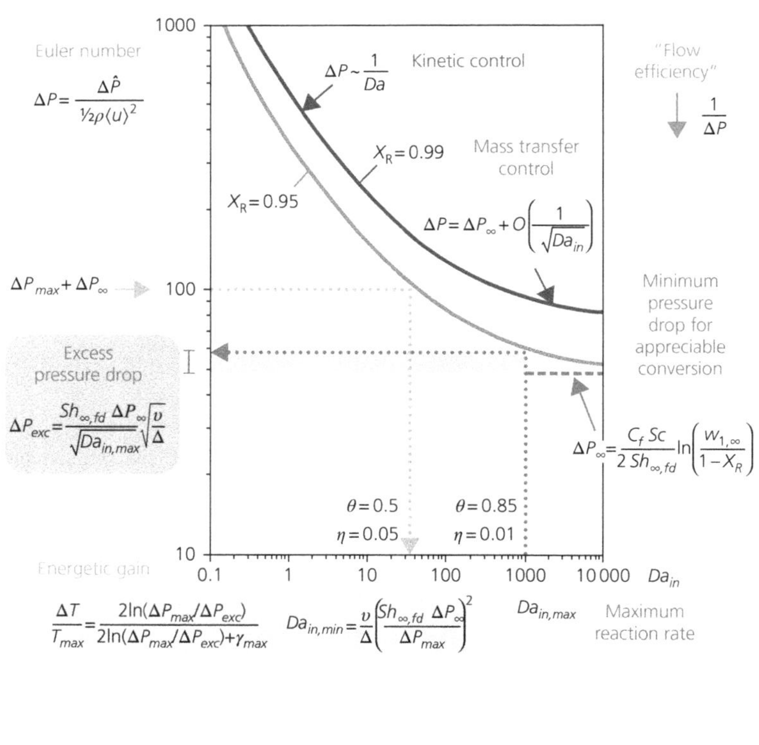

Figure 8.13 Design in the presence of a constraint on conversion. Two distinct problems are depicted: (i) increment in pressure drop due to the existence of a maximum reaction rate above which operation is not feasible and (ii) reduction in energetic requirements by increasing the allowed pressure drop.

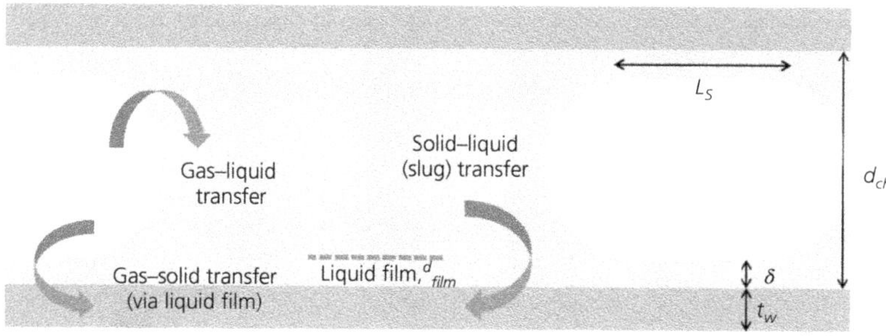

Figure 8.14 Taylor flow in washcoated monoliths. Gas–liquid–solid mass transfer routes are represented schematically. The relevant geometrical dimensions can be also depicted.

$$Sh' = \left(Sh_{fd}^n + Sh_{dev}^n\right)^{1/n} \tag{8.79}$$

where the fully developed and developing contributions are, respectively, given by

$$Sh_{fd} = 40\left(1 + 0.28\alpha_S^{4/3}\right) \tag{8.80}$$

$$Sh_{dev} = \sqrt{\frac{90 + 104\alpha_S^{4/3}}{Gz}} \tag{8.81}$$

The appropriate value for the exponent in Equation 8.79 is found to be $n = 2$, which is the same recommended by Lopes et al. [53] for single-phase systems with plug flows. Actually, the developing contribution in Equation 8.81 exhibits the same dependence on the Graetz parameter ($Gz = L/(dRe_dSc)$ with $Re_d = \rho u_{wall} d/\mu$), compared to Lévêque's solution with a uniform velocity profile, which is sensible taking into account the nonnegligible radial flow component. The slug aspect ratio (ratio between diameter and slug length) is given by

$$\alpha_S = \frac{d_{ch}}{L_S} \tag{8.82}$$

Kreutzer [29] examined values of α_S between 0.25 and 2. An order-of-magnitude increase in Sh compared to single-phase flow was observed for the same length of liquid slugs. The correlation given earlier is valid when the fraction of the channel cross section occupied by the lubricating liquid film is small (no significant film mass transfer resistance), that is, when

$$\frac{d_{film}}{d_{ch}} = \frac{0.66Ca^{2/3}}{1 + 3.335Ca^{2/3}} \ll 1 \tag{8.83}$$

where the capillary number is defined as $Ca = \mu u_{slug}/\gamma$.

2 The improvement of mass transfer is not obtained at the expense of a significant increase in pressure drop. The friction factor in Taylor flow is given by [29]

$$f = \frac{64}{Re_d}\left[1 + 0.17\alpha_S\left(\frac{Re}{Ca}\right)^{0.33}\right] \tag{8.84}$$

The increase in f is within the same order of magnitude of single-phase flow, if the same total channel volume is compared, instead of only liquid-occupied fractions.

3 A thin layer of liquid exists between the bubbles and the catalytic wall. The thickness of this film can be calculated from Ref. [139] for round channels:

$$\frac{d_{bubble}}{d_{ch}} = 0.64 + 0.36\exp\left(-3.08Ca^{0.54}\right) \tag{8.85}$$

Similar correlations are available for channels with other cross-sectional shapes. If the film thickness δ is correctly calculated by a correlation such as the one given in Equation 8.85, then the gas–solid mass transfer coefficient (through the liquid layer; see Figure 8.14) can be estimated by

$$k_{GS} = \frac{D}{\delta}$$

A film model for the transfer between the liquid slug and the catalytic surface k_{LS} can be given by the same expression. The mass transfer between bubbles and liquid slugs through the approximately spherical caps (Figure 8.13) is predicted from the penetration theory by

$$k_{GL} = \sqrt{\frac{8}{\pi}\frac{D}{d_{ch}}\left(u_{sup,L} + u_{sup,G}\right)} \tag{8.86}$$

The three mass transfer processes can be combined in an overall mass transfer coefficient:

$$k_{ov}a = k_{GS}a_{GS} + \left(\frac{1}{k_{GL}a_{GL}} + \frac{1}{k_{LS}a_{LS}}\right)^{-1} \tag{8.87}$$

The specific surface areas of the gas–solid "interface" (ignoring liquid film) a_{GS}, the bubbles caps a_{GL}, and the liquid slug–solid interface a_{LS} can be calculated from the channel diameter and liquid holdup. Kreutzer et al. [31] used a fast reaction (hydrogenation of alpha-methylstyrene) to study mass transfer in Taylor flow through monoliths. The overall mass transfer group $k_{ov}a$ was found to exceed 1 s^{-1} in a monolith with rounded corners. Improvement by one order of magnitude in the mass transfer group $k_{ov}a$ (s^{-1}) in structured monoliths was observed compared to turbulent contactors (bubble columns and agitated tanks) for the same power consumption. The same improvement holds when the walls are coated with catalyst [140].

4 A simplified isothermal reactor model, assuming uniform saturated gas-phase concentrations, is described by the following set of equations (based on the model presented in Ref. [140]):

(a) *Liquid phase*

(i) Governing equations:

$$\varepsilon_L \frac{\partial \hat{c}_{Li}}{\partial \hat{t}} + \hat{u}_{L,sup} \frac{\partial \hat{c}_{Li}}{\partial \hat{z}} = D_{ax} \frac{\partial^2 \hat{c}_{Li}}{\partial \hat{z}^2} + k_{LS} a_{LS} \left(\hat{c}_{wi}\big|_{interf} - \hat{c}_{Li} \right) \quad (8.88)$$

(ii) Boundary and initial conditions:

$$\hat{u}_{L,sup}(\hat{c}_{Li} - \hat{c}_{Li,in}) = D_{ax} \frac{\partial \hat{c}_{Li}}{\partial \hat{z}} \quad \text{at } \hat{z} = 0 \quad (8.89a)$$

$$\frac{\partial \hat{c}_{Li}}{\partial \hat{z}} = 0 \quad \text{at } \hat{z} = L \quad (8.89b)$$

$$\hat{c}_{Li}(t = 0) = \hat{c}_{i,in} \quad (8.89c)$$

(iii) Parameters:

Liquid hold-up: $$\varepsilon_L = \frac{\hat{u}_{L,sup}}{\hat{u}_{L,sup} + \hat{u}_{G,sup}} \quad (8.90a)$$

Interfacial specific area: $$a_{LS} = \frac{4\varepsilon_L}{d_{ch}} \quad (8.90b)$$

Coefficient for mass transfer in circulated slug liquid and in the film (arranged in series):

$$\frac{1}{k_{LS}} = \frac{d_{film}}{D} + \frac{d_{ch}}{Sh'D} \quad (8.90c)$$

Film thickness: d_{film}/d_{ch} given by Equation 8.83

Sherwood number in slug: Sh' given by Equation 8.80

(b) *Gas phase*

(i) Governing equations:

$$\frac{\partial(\varepsilon_G \hat{c}_{Gi})}{\partial \hat{t}} + \frac{\partial \hat{u}_{G,sup}}{\partial \hat{z}} \hat{c}_{Gi} = k_{GS} a_{GS} \left(\hat{c}_{wi}\big|_{interf} - \hat{c}^*_{Gi} \right) + k_{GS} a_{LS} \left(\hat{c}_{wi}\big|_{interf} - \hat{c}_{slug,i} \right) \quad (8.91)$$

(ii) Boundary and initial conditions:

$$\hat{c}_{Gi}(t = 0) = \hat{c}_{i,in} \quad (8.92a)$$

$$\hat{u}_{G,sup}(\hat{z} = 0) = \hat{u}_{G,in} \quad (8.92b)$$

(iii) Parameters:

Gas hold-up: $$\varepsilon_G = \frac{\hat{u}_{G,sup}}{\hat{u}_{L,sup} + \hat{u}_{G,sup}} \quad (8.93a)$$

Interfacial specific area: $$a_{GS} = \frac{4\varepsilon_G}{d_{ch}} \quad (8.93b)$$

Mass transfer coefficient from gas to solid directly (through liquid film):

$$k_{GS} = \frac{D}{\delta} \quad (8.93c)$$

Film thickness: δ can be taken as d_{film} given by Equation 8.83

Liquid concentration in circulated slug:

$$\hat{c}_{slug,i} = \hat{c}^*_{Gi} \left(1 + \frac{a_{LS}}{a_{GL}} \frac{k_{GS}}{k_{GL}} \right)^{-1} \quad (8.94)$$

Gas–liquid mass transfer coefficient: k_{GL} given by Equation 8.86

(c) *Catalytic washcoat*

(i) Governing equations:

$$\frac{\partial \hat{c}_{wi}}{\partial \hat{t}} = \frac{D_{eff,i}}{\hat{r}} \frac{\partial}{\partial \hat{r}} \left(\hat{r} \frac{\partial \hat{c}_{wi}}{\partial \hat{r}} \right) + \sum_{j=1}^{N_{RXNS}} \nu_{ij} \hat{R}_j \quad (8.95)$$

(ii) Boundary and initial conditions:

$$-D_{eff,i} \frac{\partial \hat{c}_{wi}}{\partial \hat{r}}\bigg|_{\hat{r}=a} = k_{GS} a_{GS} \left(\hat{c}_{wi}\big|_{interf} - \hat{c}^*_{Gi} \right) + k_{GS} a_{LS} \left(\hat{c}_{wi}\big|_{interf} - \hat{c}_{slug,i} \right) \quad (8.96a)$$

$$\frac{\partial \hat{c}_{wi}}{\partial \hat{r}}\bigg|_{\hat{r}=a+t_w} = 0 \quad (8.96b)$$

$$\hat{c}_{wi}(t = 0) = \hat{c}_{i,in}. \quad (8.96c)$$

These equations can be solved to obtain the distribution of the species concentrations in the different phases.

5 As in the case of single-phase flow structured reactors, the packed bed configuration can be also used in a multiphase monolithic reactor. Bauer and Haase [141] studied this design for the Pd-catalyzed hydrogenation of alpha-methylstyrene. The expected drawback of increased pressure drop was observed, with comparable space-time yields to those obtained from monoliths with empty channels. The advantages of the packed monolith reside in the possibility of using commercial catalyst pellets, with easy replacement and acceptable lifetime. In another concept, Pollington et al. [142] considered the flow of gas, liquid, and slurry (solid) through structured channels in selective glycerol oxidation.

6 The previous discussion was mainly based on Taylor flow through the channels. Other flow regimes are possible, depending on the used gas and liquid velocities. At low liquid flow rates, the most part of the channel cross section (including the core) is filled by gas, while the liquid is present as a thin flowing film wetting the walls. This is denominated of "film flow," and its application in monolith reactors has been studied by Mogalicherla and Kunzru [143] in the hydrogenation of alpha-methylstyrene on 0.5 wt% Pd/Al_2O_3.

8.5 Conclusions

The monolithic configuration is the most straightforward approach to introduce *structure* into a reactor. Apart from the obvious regular spatial arrangement, the design of the catalytic layer is rather independent from the dimensioning of the channels through which fluids flow. Several applications are tested, and commercial implementation of processes using monoliths is anticipated, with several benefits regarding energetic and material efficiency. This expectation puts pressure on a better understanding of the transport–reaction phenomena found in monolith reactors. The added complexity brought by higher degrees of freedom and flexible operating modes is counterbalanced by the reasonably simple physics involved (e.g., laminar single-phase flow). This chapter provides a review of the main aspects related with modeling of these reactors. Special attention is devoted to strategies which aim at understanding how the different dimensionless parameters governing the behavior of the system combine to predict the reactor performance, characterize its operation, and suggest its dimensioning.

A toolbox for modeling and design of monolith reactors is described and consists of four interconnected analyses. These contributions complete a *rational methodology for microstructured reactor design*:

1 **Monolith channel design**. Simplified methods for conversion prediction from a washcoated monolith channel are provided [40, 42]. The attractive asymptotic models were extended to more generic conditions. In particular, we were interested in solutions able to encompass intermediate mass transfer control and concentration profile development.
2 **Catalyst coating design**. More efficient calculation of the effectiveness factor in monolith channels is possible for nonuniform washcoat geometries and nonlinear kinetics [94]. Trade-offs in reactor performance arising from catalyst loading and washcoat thickness were also considered [43].
3 **Regime mapping and performance evaluation**. A very detailed picture of the interaction between nonlinear reactions and mass transfer at both the channel and the catalyst can be easily obtained. The analytical derivation of regime boundaries available in the literature [24, 43, 53] allows the effortless construction of parametric diagrams to evaluate operating possibilities for an envisaged application. Optimum design and operating locus can be determined for given performance measures.
4 **New windows of opportunity for perfusive structured catalysts** (not discussed in this chapter). Flow-through monoliths where the internal mass/heat transfer rates are enhanced due to the presence of convection were also considered. Non-isothermal operation [144] and design rules for maximum transport enhancement [17, 145] are among the most relevant results.

Nomenclature

A	cross-sectional channel area, m^2
a	monolith channel transverse length, m
a_{GL}	gas–liquid specific surface area, m^{-1}
a_{GS}	gas–solid specific surface area, m^{-1}
a_{LS}	liquid–solid specific surface area, m^{-1}
A_n	integration constants in Graetz eigenproblem
A_w	washcoat cross-sectional area, m^2
b	coefficient in weight dependence on Damköhler number
c	dimensionless fluid channel concentration
$\langle c \rangle$	dimensionless mixing-cup fluid concentration
Ca	capillary number, $=\mu u_{slug}/\gamma$
C_f	friction factor coefficient
$\hat{c}^*_{Gi}$	concentration in the saturated gas phase, mol/m^3
$\hat{c}_{in}$	fluid inlet concentration, mol/m^3
$\hat{c}_{Li}$	concentration in the liquid phase, mol/m^3
$\hat{c}_{Li,in}$	inlet liquid-phase concentration, mol/m^3
$\langle c_n \rangle$	term in Lévêque's perturbation series for mixing-cup concentration
C_p	fluid specific heat, J/K/kg
$\hat{c}_{slug,i}$	concentration in the circulated liquid in the slug, mol/m^3
$\hat{c}_{surf}$	concentration at the washcoat surface, mol/m^3
c_w	dimensionless concentration in the washcoat, $=\hat{c}_w/\hat{c}_{surf}$
$\hat{c}_{wi}$	concentration in the washcoat phase, mol/m^3
$\hat{c}_{wi}\vert_{interf}$	concentration at the washcoat–fluid interface, mol/m^3
Da	Damköhler number referred to inlet conditions, $a\hat{R}_{obs}(\hat{c}_{in})/(D\hat{c}_{in})$
$Da*$	rescaled Damköhler number, δDa
Da_{in}	Damköhler number referred to inlet conditions, $=\frac{a\hat{R}_{surf}(\hat{c}_{in},\hat{T}_{in})}{D}$
D_{ax}	axial dispersion, m^2/s
d_{bubble}	bubble diameter, m
D_{eff}	effective diffusivity in a catalytic washcoat, m^2/s
d_{film}	liquid film thickness
e_{dev}	absolute error of developing solution for conversion, $=\vert X_{dev}-X\vert$
f	Darcy's friction factor
f_{app}	"apparent" friction factor
GHSV	gas hourly space velocity, h^{-1}
Gz	Graetz parameter
k_{GL}	gas–liquid mass transfer coefficient, m/s
k_{GS}	gas–solid mass transfer coefficient, m/s
k_{ov}	overall mass transfer coefficient, m/s
ℓ	characteristic length scale for internal diffusion, m
L	monolith channel axial length, m
L_e	entrance length for velocity profile development, m
L_S	slug length, m

m order of reaction
Nu Nusselt number
p pressure, Pa
Pe_h heat Peclet number, $a\langle u\rangle/\kappa$
Pe_m mass Peclet number, $a\langle u\rangle/D$
Pr Prandtl's number for heat transfer, ν/κ
P_{surf} wet channel perimeter
q exponent in Lévêque's analysis
Q flow rate, m^3/s
r channel transverse dimensionless coordinate
R stretched radial coordinate (dimensionless)
R dimensionless reaction rate, $=\hat{R}(\hat{c})/\hat{R}(\hat{c}_{surf})$
$\hat{R}(\hat{c}_w,\hat{T}_w)$ reaction rate referred to the washcoat volume, $mol/(m^3s)$
Re_a Reynolds number referred to characteristic length scale a, $=\rho\langle u\rangle a/\mu$
Re_d Reynolds number referred to hydraulic diameter d_h, $=\rho\langle u\rangle d_h/\mu$
$\hat{R}_j$ rate of reaction j in the catalytic washcoat, $mol/(m^3s)$
$\hat{R}_{surf}$ reaction rate referred to the washcoat–channel area, $mol/(m^2s)$
r_w dimensionless transverse coordinate in the washcoat, $=(\hat{r}_w-a)/t_w$
R_{wall} dimensionless wall thermal resistance
S channel shape parameter
Sc Schmidt's number for mass transfer, $Sc=\nu/D$
Sh Sherwood number
T dimensionless fluid temperature
$\langle T\rangle$ dimensionless mixing-cup fluid temperature
$\hat{t}$ time, s
$\hat{T}_{in}$ inlet fluid temperature, K
T_{surf} dimensionless fluid temperature at the coating surface
$\hat{T}_{surf}$ temperature at the washcoat surface, K
$\hat{T}_{wall}$ wall temperature, K
u fluid velocity, m/s
$\langle u\rangle$ cross-sectional average fluid velocity, m/s
u_{max} maximum velocity in the channel, m/s
$u_{sup,G}$ superficial gas velocity, m/s
$u_{sup,L}$ superficial liquid velocity, m/s
X reactant conversion
z channel axial dimensionless coordinate

Greek letters

α microchannel aspect ratio
α_S slug aspect ratio
β_f Prater's nonisothermal parameter evaluated at bulk (fluid) conditions, $=(-\Delta H_R)\hat{c}_{in}D/\hat{T}_{in}\lambda$
δ dimensionless boundary layer thickness or dimensional liquid film thickness, m
Δ diffusion ratio, $=D_{eff}a/Dt_w$
Δ^* transition value for the diffusion ratio
ΔP pressure drop, Pa
ε length scale ratio, $=t_w/a$
ε_{fd} relative error of fully developed solution for conversion, $=|X_{fd}-X|/X$
ε_G gas holdup
ε_L liquid holdup
$\bar{\eta}$ global effectiveness factor referred to bulk conditions
η_{slab} effectiveness factor for slab geometry
θ degree of mass transfer control in the channel
Θ_{dev} weight function for the developing profile contribution
Θ_{fd} weight function for the fully developed profile contribution
λ fluid thermal conductivity, W/m/K
ν dimensionless fluid velocity
ν_{ij} stoichiometric coefficient of species i in reaction j
ρ density of fluid, kg/m^3
σ catalyst shape factor
σ_C flow–shape channel parameter
τ_{conv} timescale for fluid convection, s
$\tau_{visc,\ diff}$ timescale for viscous diffusion, s
υ catalyst volume-to-surface ratio normalized by coating thickness, $=V_{cat}/(S_{surf}t_w)$
ϕ^2 Thiele modulus, $=\dfrac{t_w^2}{D_{eff}}\dfrac{\hat{R}(\hat{c}_{surf})}{\hat{c}_{surf}}$
ϕ_g geometric Thiele modulus (first-order reaction), $=\dfrac{V_{cat}}{S_{surf}}\sqrt{\dfrac{k}{D_{eff}}}$
Φ generalized Thiele modulus, with respect to the kinetic expression
$\varphi_n(r)$ eigenfunction in the transverse coordinate in Graetz problem

Superscripts

$\wedge$ dimensional quantity
$'$ referred to hydraulic diameter

Subscripts

dev developing
fd fully developed
(H_2) peripherally uniform wall flux boundary condition
max maximum
min minimum
obs observed conditions (including internal mass transfer limitations)
0 Neumann conditions (uniform wall flux)
∞ Dirichlet conditions (uniform wall concentration)

References

1 Araki S, Hino N, Mori T, Hikazudani S. Autothermal reforming of biogas over a monolithic catalyst. Journal of Natural Gas Chemistry 2010;19: 477–481.

2 Cybulski A, Moulijn J. Structured Catalysts and Reactors. Boca Raton: CRC Press; 2006.

3 Chen J, Yang H, Wang N, Ring Z, Dabros T. Mathematical modeling of monolith catalysts and reactors for gas phase reactions. Applied Catalysis A: General 2008;345: 1–11.

4 Ström H, Sasic S. Heat and mass transfer in automotive catalysts—the influence of turbulent velocity fluctuations. Chemical Engineering Science 2012;83: 128–137.

5 Ström H, Sasic S, Andersson B. Effects of the turbulent-to-laminar transition in monolithic reactors for automotive pollution control. Industrial and Engineering Chemistry Research 2011;50: 3194–3205.

6 Hessel V, Ehrfeld W, Golbig K, Hofmann C, Jungwirth S, Löwe H, Richter T, Storz M, Wolf A, Wörz O, Breysse J. High temperature HCN generation in an integrated microreaction system. Microreaction Technology: 3rd International Conference on Microreaction Technology, Proceedings of IMRET 3. Berlin: Springer Verlag; 2000;152–164.

7 Redlingshöfer H, Kröcher O, Böck W, Huthmacher K, Emig G. Catalytic wall reactor as a tool for isothermal investigations in the heterogeneously catalyzed oxidation of propene to acrolein. Industrial and Engineering Chemistry Research 2002;41: 1445–1453.

8 Kočí P, Novák V, Štěpánek F, Marek M, Kubíček M. Multi-scale modelling of reaction and transport in porous catalysts. Chemical Engineering Science 2010;65: 412–419.

9 Kapteijn F, Nijhuis TA, Heiszwolf JJ, Moulijn JA. New non-traditional multiphase catalytic reactors based on monolithic structures. Catalysis Today 2001;66: 133–144.

10 van Herk D, Castaño P, Quaglia M, Kreutzer MT, Makkee M, Moulijn JA. Avoiding segregation during the loading of a catalyst-inert powder mixture in a packed micro-bed. Applied Catalysis A: General 2009;365: 110–121.

11 Beretta A, Groppi G, Majocchi L, Forzatti P. Potentialities and drawbacks of the experimental approach to the study of high T and high GHSV kinetics. Applied Catalysis A: General 1999;187: 49–60.

12 Kolodziej A, Lojewska J. Short-channel structured reactor for catalytic combustion: Design and evaluation. Chemical Engineering and Processing: Process Intensification 2007;46: 637–648.

13 Heck RM, Gulati S, Farrauto RJ. The application of monoliths for gas phase catalytic reactions. Chemical Engineering Journal 2001;82: 149–156.

14 Patcas FC, Garrido GI, Kraushaar-Czarnetzki B. CO oxidation over structured carriers: A comparison of ceramic foams, honeycombs and beads. Chemical Engineering Science 2007;62: 3984–3990.

15 Flytzani-Stephanopoulos M, Voecks GE, Charng T. Modelling of heat transfer in non-adiabatic monolith reactors and experimental comparisons of metal monoliths with packed beds. Chemical Engineering Science 1986;41: 1203–1212.

16 Moreno AM, Wilhite BA. Autothermal hydrogen generation from methanol in a ceramic microchannel network. Journal of Power Sources 2010;195: 1964–1970.

17 Lopes JP, Alves MA, Oliveira MN, Cardoso SS, Rodrigues AE. Operation of a perfusive catalytic membrane with nonlinear kinetics. Chemical Engineering Journal 2013;232: 196–212.

18 Cybulski A, Moulijn J. The present and the future of structured catalysts: An overview. In: Cybulski A, Moulijn J, editors. Structured Catalysts and Reactors, 2nd ed. Boca Raton: CRC Press; 2006.

19 Hoebink JHBJ, Marin GB. Modeling of monolithic reactors for automotive exhaust gas treatment. In: Cybulski A, Moulijn J, editors. Structured Catalysts and Reactors, 1st ed. New York: Marcel Dekker; 1998.

20 Tomasic V, Gomzi Z. Experimental and theoretical study of NO decomposition in a catalytic monolith reactor. Chemical Engineering and Processing: Process Intensification 2004;43: 765–774.

21 Uberoi M, Pereira CJ. External mass transfer coefficients for monolith catalysts. Industrial and Engineering Chemistry Research 1996;35: 113–116.

22 Choi H, Ham S-W, Nam I-S, Kim YG. Honeycomb reactor wash-coated with mordenite type zeolite catalysts for the reduction of NO_x by NH_3. Industrial and Engineering Chemistry Research 1996;35: 106–112.

23 Santos A, Bahamonde A, Schmid M, Avila P, Garcia-Ochoa F. Mass transfer influences on the design of selective catalytic reduction (SCR) monolithic reactors. Chemical Engineering and Processing: Process Intensification 1998;37: 117–124.

24 Lopes JP, Cardoso SSS, Rodrigues AE. Interplay between channel and catalyst operating regimes in wall-coated microreactors. Chemical Engineering Journal 2012;227: 42–55.

25 Groppi G, Tronconi E. Honeycomb supports with high thermal conductivity for gas/solid chemical processes. Catalysis Today 2005;105: 297–304.

26 Casanovas A, Dominguez M, Ledesma C, Lopez E, Llorca J. Catalytic walls and micro-devices for generating hydrogen by low temperature steam reforming of ethanol. Catalysis Today 2008;143: 32–37.

27 Parikh PA. Catalytic and kinetic study of toluene ethylation over ZSM-5 wash-coated honeycomb monolith. Industrial and Engineering Chemistry Research 2008;47: 1793–1797.

28 Klenov OP, Pokrovskaya SA, Chumakova NA, Pavlova SN, Sadykov VA, Noskov AS. Effect of mass transfer on the reaction rate in a monolithic catalyst with porous walls. Catalysis Today 2009;144: 258–264.

29 Kreutzer MT. Hydrodynamics of Taylor flow in capillaries and monoliths channels [dissertation]. Delft, The Netherlands: Delft University of Technology; 2003. Available from: Delft, The Netherlands.

30 Irandoust S, Andersson B, Bengtsson E, Siverstrom M. Scaling up of a monolithic catalyst reactor with two-phase flow. Industrial and Engineering Chemistry Research 1989;28: 1489–1493.

31 Kreutzer MT, Kapteijn F, Moulijn JA. Fast gas–liquid–solid reactions in monoliths: A case study of nitro-aromatic hydrogenation. Catalysis Today 2005;105: 421–428.

32 Du P, Carneiro JT, Moulijn JA, Mul G. A novel photocatalytic monolith reactor for multiphase heterogeneous photocatalysis. Applied Catalysis A: General 2008;334: 119–128.

33 Tsoligkas AN, Simmons MJH, Wood J, Frost CG. Kinetic and selectivity studies of gas–liquid reaction under Taylor flow in a circular capillary. Catalysis Today 2007;128: 36–46.

34 Natividad R, Kulkarni R, Nuithitikul K, Raymahasay S, Wood J, Winterbottom JM. Analysis of the performance of single capillary and multiple capillary (monolith) reactors for the multiphase Pd-catalyzed hydrogenation of 2-butyne-1,4-diol. Chemical Engineering Science 2004;59: 5431–5438.

35 Wang X, Nie Y, Lee JLC, Jaenicke S. Evaluation of multiphase microreactors for the direct formation of hydrogen peroxide. Applied Catalysis A: General 2007;317: 258–265.
36 Pangarkar K, Schildhauer TJ, Van Ommen JR, Nijenhuis J, Kapteijn F, Moulijn JA. Structured packings for multiphase catalytic reactors. Industrial and Engineering Chemistry Research 2008;47: 3720–3751.
37 van Herk D, Castaño P, Makkee M, Moulijn JA, Kreutzer MT. Catalyst testing in a multiple-parallel, gas–liquid, powder-packed bed microreactor. Applied Catalysis A: General 2009;365: 199–206.
38 Lucas M, Claus P. High throughput screening in monolith reactors for total oxidation reactions. Applied Catalysis A: General 2003;254: 35–43.
39 Shah RK, London AL. Laminar Flow Forced Convection in Ducts. New York: Academic Press; 1978.
40 Lopes JP, Rodrigues AE, Cardoso SS. Approximate calculation of conversion with kinetic normalization for finite reaction rates in wall-coated microchannels. AIChE Journal 2011;57: 2870–2887.
41 Bird RB, Stewart WE, Lightfoot EN. Transport Phenomena. New York: John Wiley & Sons, Inc.; 2002.
42 Lopes JP, Cardoso SSS, Rodrigues AE. Bridging the gap between Graetz's and Lévêque's analyses for mass/heat transfer in a channel with uniform concentration or flux at the wall. AIChE Journal 2012;58: 1880–1892.
43 Lopes JP, Alves MA, Oliveira MN, Cardoso SS, Rodrigues AE. Regime mapping and the role of the intermediate region in wall-coated microreactors. Chemical Engineering Science 2013;94: 166–184.
44 Bhattacharya M, Harold MP, Balakotaiah V. Shape normalization for catalytic monoliths. Chemical Engineering Science 2004;59: 3737–3766.
45 Bennett CJ, Kolaczkowski ST, Thomas WJ. Determination of heterogeneous reaction kinetics and reaction rates under mass transfer controlled conditions for a monolith reactor. Process Safety and Environmental Protection: Transactions of the Institution of Chemical Engineers, Part B 1991;69: 209–220.
46 Van Gulijk C, Linders MJG, Valdes-Solis T, Kapteijn F. Intrinsic channel maldistribution in monolithic catalyst support structures. Chemical Engineering Journal 2005;109: 89–96.
47 Pérez-Cadenas AF, Zieverink MMP, Kapteijn F, Moulijn JA. High performance monolithic catalysts for hydrogenation reactions. Catalysis Today 2005;105: 623–628.
48 Berger RJ, Kapteijn F. Coated-wall reactor modeling-criteria for neglecting radial concentration gradients. 2. Reactor tubes filled with inert particles. Industrial and Engineering Chemistry Research 2007;46: 3871–3876.
49 Graetz L. Ueber die Wärmeleitungsfähigkeit von Flüssigkeiten. Annalen der Physik und Chemie 1883;18: 79–94.
50 Damköhler G. Einfluss von diffusion, Strömung, und Wärmetransport auf die Ausbeute in Chemisch-Technischen Reaktionen. Chemical Engineering & Technology 1937;3: 359–485.
51 Housiadas C, Larrodé FE, Drossinos Y. Numerical evaluation of the Graetz series. International Journal of Heat and Mass Transfer 1999;42: 3013–3017.
52 Lévêque MA. Les lois de la transmission de chaleur par convection. Annales des Mines 1928;13: 201–299.
53 Lopes JP, Cardoso SS, Rodrigues AE. Criteria for kinetic and mass transfer control in a microchannel reactor with an isothermal first-order wall reaction. Chemical Engineering Journal 2011;176–177: 3–13.
54 Rice RG, Do DD. Applied Mathematics and Modeling for Chemical Engineers. New York: John Wiley & Sons, Inc.; 1995.
55 Basmadjian D. The Art of Modeling in Science and Engineering. Boca Raton: Chapman & Hall/CRC; 1999.
56 Nusselt W. Die Abhängigkeit der Wärmeübergangszahl von der Rohrlänge. Zeitschrift des Vereines deutscher Ingenieure 1910;54: 1154–1158.
57 Paneth F, Herzfeld KF. Free methyl and free ethyl. Zeitschrift für Elektrochemie und Angewandte Physikalische Chemie 1931;37: 577–582.
58 Carslaw HS, Jaeger JC. Conduction of Heat in Solids. New York: Oxford University Press; 1959.
59 Bauer HF. Diffusion, convection and chemical reaction in a channel. International Journal of Heat and Mass Transfer 1976;19: 479–486.
60 Özisik MN, Sadeghipour MS. Analytic solution for the eigenvalues and coefficients of the Graetz problem with third kind boundary condition. International Journal of Heat and Mass Transfer 1982;25: 736–739.
61 Sellars J, Tribus M, Klein J. Heat transfer to laminar flow in a round tube or flat conduit—the Graetz problem extended. ASME Transactions 1956;78: 441–448.
62 Solbrig CW, Gidaspow D. Convective diffusion in a parallel plate duct with one catalytic wall—laminar flow—first order reaction. The Canadian Journal of Chemical Engineering 1967;45: 35–39.
63 Brown GM. Heat or mass transfer in a fluid in laminar flow in a circular or flat conduit. AIChE Journal 1960;6: 179–183.
64 Newman J. The Graetz problem. In: Bard AJ, editor. The Fundamental Principles of Current Distribution and Mass Transport in Electrochemical Cells. New York: Dekker; 1973.
65 Kays WM, Crawford ME. Convective Heat and Mass Transfer. New York: McGraw-Hill; 1980.
66 Bhattacharya M, Harold MP, Balakotaiah V. Mass-transfer coefficients in washcoated monoliths. AIChE Journal 2004;50: 2939–2955.
67 Balakotaiah V, West DH. Shape normalization and analysis of the mass transfer controlled regime in catalytic monoliths. Chemical Engineering Science 2002;57: 1269–1286.
68 Petersen E. Chemical Reaction Analysis. Upper Saddle River: Prentice-Hall, Inc.; 1965.
69 Pancharatnam S, Homsy GM. An asymptotic solution for tubular flow reactor with catalytic wall at high Peclet numbers. Chemical Engineering Science 1972;27: 1337–1340.
70 Ghez R. Mass transport and surface reactions in Lévêque's approximation. International Journal of Heat and Mass Transfer 1978;21: 745–750.
71 Worsoe-Schmidt PM. Heat transfer in the thermal entrance region of circular tubes and annular passages with fully developed laminar flow. International Journal of Heat and Mass Transfer 1967;10: 541–551.
72 Newman J. Extension of the Leveque solution. Journal of Heat Transfer 1969;91: 177–178.
73 Gottifredi JC, Flores AF. Extended Leveque solution for heat transfer to non-Newtonian fluids in pipes and flat ducts. International Journal of Heat and Mass Transfer 1985;28: 903–908.
74 Shih YP, Huang CC, Tsay SY. Extended Leveque solution for laminar heat transfer to power-law fluids in pipes with wall slip. International Journal of Heat and Mass Transfer 1995;38: 403–408.

75 Acrivos A, Chambré PL. Laminar boundary layer flows with surface reactions. Industrial and Engineering Chemistry 1957;49: 1025–1029.
76 Siegel R, Sparrow EM, Hallman TM. Steady laminar heat transfer in a circular tube with prescribed wall heat flux. Applied Scientific Research 1958;7: 386–392.
77 Compton RG, Unwin PR. The dissolution of calcite in aqueous solution at pH < 4: Kinetics and mechanism. Philosophical Transactions of the Royal Society of London. Series A, Mathematical and Physical Sciences 1990;330: 1–45.
78 Rosner DE. Effects of convective diffusion on the apparent kinetics of zeroth order surface-catalysed chemical reactions. Chemical Engineering Science 1966;21: 223–239.
79 Chambré PL, Acrivos A. On chemical surface reactions in laminar boundary layer flows. Journal of Applied Physics 1956;27: 1322–1328.
80 Katz S. Chemical reactions catalysed on a tube wall. Chemical Engineering Science 1959;10: 202–211.
81 Rosner DE. The apparent chemical kinetics of surface reactions in external flow systems: Diffusional falsification of activation energy and reaction order. AIChE Journal 1963;9: 321–331.
82 Grau RJ, Cabrera MI, Cassano AE. The laminar flow tubular reactor with homogeneous and heterogeneous reactions. I. Integral Equations for Diverse Reaction Rate Regimes. Chemical Engineering Communications 2001;184: 229–257.
83 Balakotaiah V, Gupta N, West DH. Simplified model for analyzing catalytic reactions in short monoliths. Chemical Engineering Science 2000;55: 5367–5383.
84 Shih YP, Tsou JD. Extended Leveque solutions for heat transfer to power law fluids in laminar flow in a pipe. The Chemical Engineering Journal 1978;15: 55–62.
85 Schmidt LD. The Engineering of Chemical Reactions. New York: Oxford University Press; 1998.
86 Belfiore LA. Transport Phenomena for Chemical Reactor Design. Hoboken, NJ: John Wiley & Sons, Inc.; 2003.
87 Renken A, Kiwi-Minsker L. Chemical reactions in continuous-flow microstructured reactors. In: Kockmann N, editors. Micro Process Engineering: Fundamentals, Devices, Fabrication, and Applications. Weinheim, Germany: Wiley-VCH Verlag GmbH; 2008.
88 Commenge JM, Falk L, Corriou JP, Matlosz M. Analysis of microstructured reactor characteristics for process miniaturization and intensification. Chemical Engineering and Technology 2005;28: 446–458.
89 Hessel V, Hardt S, Löwe H. Chemical Micro Process Engineering: Fundamentals, Modelling and Reactions. Weinheim, Germany: Wiley-VCH Verlag GmbH; 2004.
90 Taylor GI. Dispersion of soluble matter in solvent flowing slowly through a tube. Proceedings of the Royal Society of London A 1953;219: 186–203.
91 Taylor GI. Conditions under which dispersion of a solute in a stream of solvent can be used to measure molecular diffusion. Proceedings of the Royal Society of London A 1954;225: 473–477.
92 Aris R. On the dispersion of a solute in a fluid flowing through a tube. Proceedings of the Royal Society of London A 1956;235: 67–77.
93 Seyed-Reihani SA, Jackson GS. Influence of thermal conditions on partial oxidation of n-butane over supported Rh catalysts. Applied Catalysis A: General 2009;353: 181–192.
94 Lopes JP, Cardoso SSS, Rodrigues AE. Effectiveness factor for thin catalytic coatings: Improved analytical approximation using perturbation techniques. Chemical Engineering Science 2012;71: 46–55.
95 Froment GF, Bischoff KB. Chemical Reactor Analysis and Design. New York: John Wiley & Sons, Inc.; 1979.
96 Zhang E, Hayes RE, Kolaczkowski ST. A new technique to measure the effective diffusivity in a catalytic monolith washcoat. Chemical Engineering Research and Design 2004;82: 481–489.
97 Lopes JP. Convection, diffusion and reaction: Bridging scales in science and engineering [dissertation]. Porto: University of Porto; 2011. Available from: Porto.
98 Papadias D, Edsberg L, Björnbom P. Simplified method of effectiveness factor calculations for irregular geometries of washcoats a general case in a 3D concentration field. Catalysis Today 2000;60: 11–20.
99 Papadias D, Edsberg L, Björnbom P. Simplified method for effectiveness factor calculations in irregular geometries of washcoats. Chemical Engineering Science 2000;55: 1447–1459.
100 Hayes RE, Liu B, Votsmeier M. Calculating effectiveness factors in non-uniform washcoat shapes. Chemical Engineering Science 2005;60: 2037–2050.
101 Gunn DJ. Diffusion and chemical reaction in catalysis and absorption. Chemical Engineering Science 1967;22: 1439–1455.
102 Wijngaarden RJ, Kronberg A, Westerterp KR. Industrial Catalysis: Optimizing Catalysts and Processes. Weinheim, Germany: Wiley-VCH Verlag GmbH; 1998.
103 Tronconi E, Forzatti P. Adequacy of lumped parameter models for SCR reactors with monolith structure. AIChE Journal 1992;38: 201–210.
104 Tronconi E, Groppi G, Boger T, Heibel A. Monolithic catalysts with "high conductivity" honeycomb supports for gas/solid exothermic reactions: Characterization of the heat-transfer properties. Chemical Engineering Science 2004;59: 4941–4949.
105 Hayes RE, Kolaczkowskib ST, Li PKC, Awdry S. Evaluating the effective diffusivity of methane in the washcoat of a honeycomb monolith. Applied Catalysis B: Environmental 2000;25: 93–104.
106 Khanaev VM, Borisova ES, Galkina LI, Noskov AS. Improvement of the catalytic monoliths efficiency for CO oxidation using non-uniform active component distribution along the monolith length. Chemical Engineering Journal 2004;102: 35–44.
107 Groppi G, Belloli A, Tronconi E, Forzatti P. A comparison of lumped and distributed models of monolith catalytic combustors. Chemical Engineering Science 1995;50: 2705–2715.
108 Young LC, Finlayson BA. Mathematical models of the monolith catalytic converter. I. Development of model and application of orthogonal collocation. AIChE Journal 1976;22: 331–343.
109 Gupta N, Balakotaiah V. Heat and mass transfer coefficients in catalytic monoliths. Chemical Engineering Science 2001;56: 4771–4786.
110 Stutz MJ, Poulikakos D. Optimum washcoat thickness of a monolith reactor for syngas production by partial oxidation of methane. Chemical Engineering Science 2008;63: 1761–1770.
111 Ramanathan K, Balakotaiah V, West DH. Light-off criterion and transient analysis of catalytic monoliths. Chemical Engineering Science 2003;58: 1381–1405.
112 Ramanathan K, Balakotaiah V, West DH. Ignition criterion for general kinetics in a catalytic monolith. AIChE Journal 2006;52: 1623–1629.

113 Ramanathan K, Gopinath A. Light-off location and front diffusion in a catalytic monolith reactor. AIChE Journal 2008;54: 1860–1873.
114 Di Benedetto A, Marra FS, Donsi F, Russo G. Transport phenomena in a catalytic monolith: Effect of the superficial reaction. AIChE Journal 2006;52: 911–923.
115 Hawthorn RD. Afterburner catalysts—effects of heat and mass transfer between gas and catalyst surface. AIChE Symposium Series 1974;70: 428–438.
116 Votruba J, Mikus O, Nguen K, Hlavacek V, Skrivanek J. Heat and mass transfer in monolithic honeycomb catalysts-II. Chemical Engineering Science 1975;30: 201–206.
117 Ullah U, Waldram SP, Bennett CJ, Truex T. Monolithic reactors: Mass transfer measurements under reacting conditions. Chemical Engineering Science 1992;47: 2413–2418.
118 Kolaczkowski ST, Serbetcioglu S. Development of combustion catalysts for monolith reactors: A consideration of transport limitations. Applied Catalysis A: General 1996;138: 199–214.
119 Holmgren A, Andersson B. Mass transfer in monolith catalysts-CO oxidation experiments and simulations. Chemical Engineering Science 1998;53: 2285–2298.
120 Churchill SW, Ozoe H. Correlations for laminar forced convection in flow over an isothermal flat plate and in developing and fully developed flow in an isothermal tube. Journal of Heat Transfer 1973;95 Ser C: 416–419.
121 Hayes RE, Kolaczkowski ST. A study of Nusselt and Sherwood numbers in a monolith reactor. Catalysis Today 1999;47: 295–303.
122 Groppi G, Tronconi E. Theoretical analysis of mass and heat transfer in monolith catalysts with triangular channels. Chemical Engineering Science 1997;52: 3521–3526.
123 Bräuer HW, Fetting F. Stofftransport bei Wandreaktion im Einlaufgebiet eines Strömungsrohres. Chemie Ingenieur Technik 1966;38: 30–35.
124 Kockmann N. Transport Phenomena in Micro Process Engineering. Berlin: Springer-Verlag; 2008.
125 McCarty JG. Kinetics of PdO combustion catalysis. Catalysis Today 1995;26: 283–293.
126 Groppi G, Ibashi W, Valentini M, Forzatti P. High-temperature combustion of CH4 over PdO/Al_2O_3: Kinetic measurements in a structured annular reactor. Chemical Engineering Science 2001;56: 831–839.
127 Hayes RE, Kolaczkowski ST. Mass and heat transfer effects in catalytic monolith reactors. Chemical Engineering Science 1994;49: 3587–3599.
128 Yu X, Tu ST, Wang Z, Qi Y. On-board production of hydrogen for fuel cells over $Cu/ZnO/Al_2O_3$ catalyst coating in a micro-channel reactor. Journal of Power Sources 2005;150: 57–66.
129 Germani G, Schuurman Y. Water-gas shift reaction kinetics over micro-structured $Pt/CeO_2/Al_2O_3$ catalysts. AIChE Journal 2006;52: 1806–1813.
130 Sazonova NN, Pavlova SN, Pokrovskaya SA, Chumakova NA, Sadykov VA. Structured reactor with a monolith catalyst fragment for kinetic studies. The case of CH_4 partial oxidation on LaNiPt-catalyst. Chemical Engineering Journal 2009;154: 17–24.
131 Oh SH, Cavendish JC. Transients of monolithic catalytic converters: Response to step changes in feedstream temperature as related to controlling automobile emissions. Industrial & Engineering Chemistry Product Research and Development 1982;21: 29–37.
132 Kapteijn F, De Deugd RM, Moulijn JA. Fischer-Tropsch synthesis using monolithic catalysts. Catalysis Today 2005;105: 350–356.
133 Walter S, Malmberg S, Schmidt B, Liauw MA. Mass transfer limitations in microchannel reactors. Catalysis Today 2005;110: 15–25.
134 Chen H, Yu H, Tang Y, Pan M, Peng F, Wang H, Yang J. Assessment and optimization of the mass-transfer limitation in a metal foam methanol microreformer. Applied Catalysis A: General 2008;337: 155–162.
135 Hayes RE, Kolaczkowski ST, Thomas WJ, Titiloye J. Intraphase diffusion and interphase mass transfer effects during the catalytic oxidation of CO in a tube wall reactor. Proceedings—Royal Society of London, A 1995;448: 321–334.
136 Tomasic V, Gomzi Z, Zrncevic S. Analysis and modeling of a monolithic reactor. Chemical Engineering and Technology 2006;29: 59–65.
137 Raja LL, Kee RJ, Deutschmann O, Warnatz J, Schmidt LD. A critical evaluation of Navier–Stokes, boundary-layer, and plug-flow models of the flow and chemistry in a catalytic-combustion monolith. Catalysis Today 2000;59: 47–60.
138 Lopes JP, Cardoso SSS, Rodrigues AE. Design of wall-coated microreactors in the intermediate reaction-transport regime. ISCRE 22 International Symposium on Chemical Reaction Engineering 2012; Maastricht, The Netherlands, September 2–5, 2012.
139 Irandoust S, Andersson B. Liquid film in Taylor flow through a capillary. Industrial and Engineering Chemistry Research 1989;28: 1684–1688.
140 Kreutzer MT, Kapteijn F, Moulijn JA. Shouldn't catalysts shape up? Structured reactors in general and gas–liquid monolith reactors in particular. Catalysis Today 2006;111: 111–118.
141 Bauer T, Haase S. Comparison of structured trickle-bed and monolithic reactors in Pd-catalyzed hydrogenation of alpha-methylstyrene. Chemical Engineering Journal 2011;169: 263–269.
142 Pollington SD, Enache DI, Landon P, Meenakshisundaram S, Dimitratos N, Wagland A, Hutchings GJ, Stitt EH. Enhanced selective glycerol oxidation in multiphase structured reactors. Catalysis Today 2009;145(1–2): 169–175.
143 Mogalicherla AK, Kunzru D. Performance of monolithic reactors in film flow. Chemical Engineering Research and Design 2010;88: 1057–1066.
144 Lopes JP, Cardoso SSS, Rodrigues AE. Convection, diffusion, and exothermic zero-order reaction in a porous catalyst slab: Scaling and perturbation analysis. AIChE Journal 2009;55: 2686–2699.
145 Lopes JP, Alves MA, Oliveira MSN, Cardoso SSS, Rodrigues AE. Internal mass transfer enhancement in flow-through catalytic membranes. Chemical Engineering Science 2013;104: 1090–1106.

CHAPTER 9

Microreactors for catalytic reactions

Evgeny Rebrov[1] and Sourav Chatterjee[2]
[1]School of Engineering, University of Warwick, Coventry, UK
[2]School of Chemistry and Chemical Engineering, Queen's University Belfast, Belfast, UK

Abstract

This chapter focuses on description of microstructured reactors for gas–solid, gas–liquid, and three-phase processes, flow regimes, mass transfer considerations for various configurations of microchannels, design criteria, and evaluation of each reactor type. Special attention is devoted to Taylor flow in microchannels, as this flow regime is the most adapted for practical engineering applications.

9.1 Introduction

Microreaction technology offers the possibility of miniaturization of conventional macroscopic reactors. The main feature of catalytic microstructured reactors (CMRs) is their high surface to volume ratio in the range of 20 000–50 000 m^2/m^3 compared to more traditional chemical reactors. The Reynolds number is typically in the range of 10–500; therefore, CMRs are operated under laminar flow regime. The small diameters of the reactor channels ensure a short radial diffusion length leading to a narrow residence time distribution (RTD). This is advantageous for consecutive processes since high selectivity to the desired intermediate can be achieved. Furthermore, microreactors provide fast mass and heat transfer. Therefore, they are often employed in fast exothermic or endothermic catalytic reactions.

The geometric surface of the microchannels can be increased for performing catalytic reactions. Porous coatings are typically applied for this purpose. The porous layer can be catalytically active or serve as a support for a catalytic active phase. Different coating techniques are developed and tested over the last years as there is a steady increase of the number of catalytic applications in pharmaceutical and fine chemical industries driven by strict environmental regulations and policies introduced during the last decade [1].

This chapter presents an overview of the fundamentals of design and operation of single-phase and multiphase catalytic microreactors. Various designs are discussed including their advantages in specific catalytic processes.

9.2 Single-phase catalytic microreactors

Single-phase microreactors have been widely used in industrial and scientific applications over the last decade. In most cases, AISI316 steel is used for fabrication of microreactors, although other metals such as aluminum, molybdenum, copper, and nickel were employed. The choice of material for a catalytic microreactor, depends on (i) microfabrication costs; (ii) the restrictions imposed by the manufacturing method, for example, technical limitations to obtain certain geometrical structures with high spatial resolution; (iii) the possibility of rapid testing of catalytic coatings or pelleted catalysts in the microchannel network with a possibility of their further optimization; (iv) the corrosion resistance of the reactor material in reaction mixture; and (v) the interaction of the reactor material with catalysts, which might cause the deactivation of the latter [2].

In several cases, operation of microreactors has shown that their expected efficiency cannot be reached either due to nonuniform distribution of reactants between different channels or due to flow maldistribution between individual microreactors working in parallel [3, 4]. The latter problem can result in substantial temperature deviations between different microreactors resulting in thermal runaway which could arise from an exothermic reaction. At the moment, there is a general agreement that both fluid flow and mass/heat transfer in single-phase microstructured reactors can be accurately described by standard theory and correlations, but scaling effects (entrance effects, conjugate heat transfer, temperature-dependent properties, nonuniform flow distribution, and singular pressure losses) have often to be accounted for in the design of individual parts of microsystems [5].

9.2.1 Residence time distribution

If the target product is an intermediate in sequential reactions, selectivity and yields are quite sensitive to the residence time. The axial dispersion in a tubular microreactor can often be described by a dispersion model. This model assumes that the RTD results from piston flow on which axial dispersion is superimposed. The latter is taken into account by means of an

Multiphase Catalytic Reactors: Theory, Design, Manufacturing, and Applications, First Edition. Edited by Zeynep Ilsen Önsan and Ahmet Kerim Avci.

effective axial dispersion coefficient, D_{ax}. The dispersion in a tubular microreactor is characterized by the Péclet number:

$$Pe = \frac{Lu}{D_{ax}} \tag{9.1}$$

The effective axial dispersion coefficient in laminar flow in cylindrical microchannels [6, 7] is given as follows:

$$D_{ax} = D_m + \frac{1}{192}\frac{U^2 d_t}{D_m} \tag{9.2}$$

By combining Equations 9.1 and 9.2, the Pe number in a microchannel can be estimated:

$$\frac{1}{Pe} = \frac{D_m}{L_t^2}\frac{L}{U} + \frac{1}{192}\frac{d_t^2}{D_m}\frac{U}{L} = \frac{\tau}{t_{D,ax}} + \frac{1}{48}\frac{t_{D,rad}}{\tau} \tag{9.3}$$

with $t_{D,ax} = L^2/D_m$ and $t_{D,rad} = d_t^2/4D_m$.

The first term corresponds to the ratio of the residence time to the characteristic axial diffusion time. The molecular diffusion coefficient is of the order 10^{-5} m^2/s for gases. Typical length of microreactors is in the order 10^{-1} m, and the residence time is typically in the range of seconds. Therefore, $\tau/t_{D,ax} \ll 1$ and the axial dispersion in microreactors is determined by the radial diffusion time. The Pe number can be estimated by

$$Pe = 48\frac{\tau}{t_{D,rad}} \cong 200\frac{D_m}{d_t^2} \tag{9.4}$$

At $Pe > 100$, the axial dispersion in a microchannel can be neglected. This condition is fulfilled if the residence time exceeds the radial diffusion time by a factor of 2. In microchannels with diameter less than 1 mm, the plug-flow behavior is achieved at the residence time of about 0.1 s. This condition was also proved experimentally [8].

9.2.2 Effect of flow maldistribution

Delsman et al. [9] proposed a method to predict the influence of flow maldistribution and manufacturing tolerances on the performance of a microreactor with a large number of parallel channels. In this method, a variable reactor parameter (e.g., diameter) is considered as a random parameter with a mean value and a standard deviation, and the microreactor is considered as a plug-flow reactor. Then, a number of relationships between the variable parameter and efficiency as compared to the ideal case are presented. The pressure drop as a function of the mean channel diameter is given by Equation 9.5:

$$\Delta p = \frac{128\mu LF}{\pi n \bar{d}^4\left(1 + 6\hat{\sigma}_d^2\right)} \tag{9.5}$$

This equation shows that a variation of the channel diameter results in a decrease in the overall pressure drop over the reactor, when the total flow rate is kept constant. A small deviation in channel diameter does not result in a large difference in the pressure drop: a standard deviation of 10% gives a pressure drop of 6% lower as compared with the ideal case.

The residence time of the fluid in a microchannel varies also between the channels. The residence time in the single channel is a function of both the fluid flow rate and the channel volume. It can be expressed as

$$\bar{\tau} = \frac{\tau_0\left(1 + 6\hat{\sigma}_d^2\right)}{1 + \hat{\sigma}_d^2} \tag{9.6}$$

Since the residence time varies between the channels, a tracer pulse at the inlet of the microreactor will be broadened similar to the case for a tubular reactor with axial dispersion. As a first approximation, the relative standard deviation in the residence time is twice the relative standard deviation in the channel diameter [9]:

$$\hat{\sigma}_\tau^2 = (2\hat{\sigma}_d)^2 \tag{9.7}$$

The influence of different process and geometrical parameters on conversion was estimated in case of an irreversible first-order catalytic reaction. The influence of temperature nonuniformity (when temperature varies between the channels) had the largest impact on conversion in a nonideal reactor as compared to non-uniform flow distribution and nonequal catalyst amount in the channels. Obtained correlations were used to estimate the influence of a variable channel diameter on the conversion in a microreactor for a heterogeneous first-order reaction. It was found that the conversion in 95% of the microchannels varied between 59 and 99% at $\hat{\sigma}_d = 0.1$ and Damköhler number of 2. Figure 9.1a shows conversion as a function of Damköhler number for an ideal microreactor and a microreactor with variations in the channel diameter ($\hat{\sigma}_d = 0.1$). It can be seen that although the conversion in individual channels can vary considerably, the effect of nonuniformity in channel diameter on the overall reactor conversion is smaller. The lower conversion in channels with a higher flow rate is partly compensated by a higher conversion in channels with a lower flow rate. Due to the nonlinear relation between the channel diameter and the flow rate, the effects do not cancel completely and a decrease in reactor conversion is observed.

The influence of temperature nonuniformity between different channels was also studied [9]. In this case the channel diameter and the amount of catalyst per channel were assumed to be equal in all channels, but the temperature was not the same in different channels. The temperature influences both the reaction rate and the molar gas flow rate (and therefore the pressure drop). The temperature nonuniformity has a much larger influence on the reactor conversion than the channel diameter. The influence of temperature deviations on the conversion is a function of the activation energy of the reaction (E_a), which is expressed via the parameter $\gamma = E_a/R\bar{T}$ assuming that the reaction rate coefficient is proportional to $\exp[-\gamma(\bar{T}/T_i)]$. Figure 9.1b shows the influence of the parameter γ on the reactor conversion. The influence of the temperature increases with an increase in the parameter γ, which means for reactions with a high activation energy and/or a low reaction temperature.

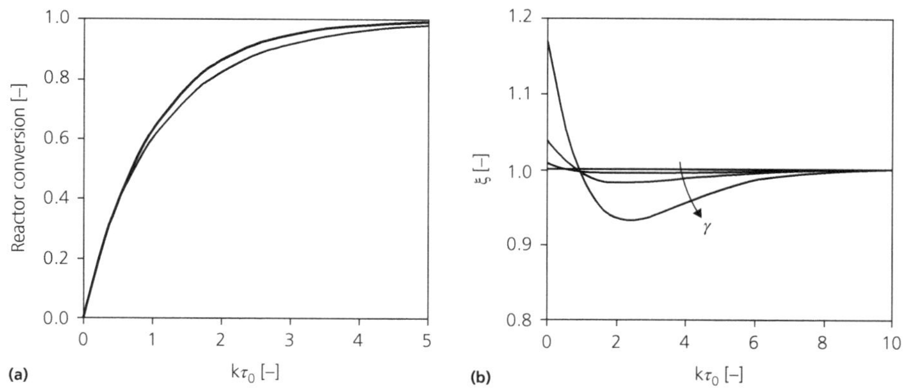

Figure 9.1 (a) Influence of channel diameter variations on the reactor conversion; ideal reactor (thick line), reactor with $\hat{\sigma}_d = 0.1$ (thin line). (b) Influence of the parameter γ on the conversion in microreactor with $\hat{\sigma}_T = 0.01$; $\gamma = 15$, 30, and 60.
(Source: Delsman et al. [9]. Reproduced with permission of John Wiley & Sons.)

9.2.3 Mass transfer

9.2.3.1 Internal mass transfer

The presence of internal mass transfer limitations depends on the reaction rate and the thickness of the porous catalytic layer and is usually expressed via the effectiveness factor, which is defined as the ratio of the observed reaction rate and the rate that would be observed in the absence of concentration gradient throughout the catalytic layer. For an isothermal layer, the maximum thickness of catalytic coating should not exceed δ_{cat} to ensure an effectiveness factor of 0.95 [10]:

$$\delta_{cat} \le b\sqrt{\frac{D_{eff}C_s}{r_{eff}}} \tag{9.8}$$

The parameter b depends on the reaction order and is reported to be equal to 0.8, 0.3, and 0.18 for a zero-, first-, and second-order reaction, respectively [10].

The Weisz modulus [11] allows for the estimation of intraparticle diffusion limitations in packed bed microchannel reactors:

$$\psi = \delta_{cat}\frac{m+1}{2}\frac{r_{eff}}{D_{eff}C_s} < 0.1 \tag{9.9}$$

This criterion was applied by Ajmera et al. [12] to the oxidation of carbon monoxide for catalyst particles incorporated into a microchannel reactor. Mass transport in the coating was not limiting the reactor performance in this case.

9.2.3.2 External mass transfer

Provided that the flow is laminar, the velocity profile develops in a microchannel from the entrance to the position where a fully developed parabolic profile is established. The length of the entrance zone (L_z) can be estimated by Equation 9.10 [13]:

$$L_z = 0.06Red_t \tag{9.10}$$

In the entrance zone, the mass transfer coefficient monotonously reduces and reaches a constant value which depends on the geometry of the channel:

$$Sh(z) = Sh_{\infty}\left(1 + 0.095\frac{d_t}{z}ReSc\right)^{0.45} \tag{9.11}$$

Table 9.1 Asymptotic Sherwood number for constant reactant concentration.

Geometry of channel cross section	Sh_{∞}
Equilateral triangle	2.47
Square	2.98
Circle	3.66
Ellipse (width/height = 2)	3.74
Parallel plates	7.54

The asymptotic Sherwood numbers (Sh_{∞}) for constant reactant concentration at the wall are listed in Table 9.1. In many cases the entrance region can be neglected and the asymptotic *Sh* number can be used for calculation of the mass transfer coefficient.

The Mears criterion is used to check for external mass transfer limitation in microchannels:

$$\frac{r_{eff}d_t}{2k_gC_{in}} < 0.15 \tag{9.12}$$

where k_g stands for mass transfer coefficient, which can be found from the Sherwood number.

9.2.4 Heat transfer

With respect to the heat transfer, the resistance of the channel wall is typically small as compared to the resistance within the porous catalytic layer. Strongly exothermic (or endothermic) reaction may cause a temperature gradient within the catalytic layer. To achieve isothermal behavior, the thickness of the catalytic layer should not exceed the value specified by Equation 9.13, which is a modified form of the Anderson criterion [14]:

$$\delta_{cat} \le 0.3\sqrt{\frac{R\lambda_{eff}T_s^2}{E_a|\Delta H_r|r_{eff}}} \tag{9.13}$$

Recent modeling studies of micro- and minichannel networks have argued the importance of solid-phase axial heat conduction [5]. The axial heat conduction in the channel wall in large pipes can indeed be neglected because the wall thickness is usually very small compared to the channel diameter. However, heat transfer in a microchannel is a combination of axial heat conduction in the solid and convection to the cooling fluid. Fedorov and Viskanta [15] investigated heat transfer in a heat sink with rectangular microchannels with a 3D numerical model for fluid flow and the heat conduction in the silicon substrate. The authors obtained a good agreement with experimental results and found rather complex heat flux patterns due to a strong coupling between convection in the fluid and conduction in the silicon substrate that can only be resolved by a detailed 3D simulation. Lee et al. [16] performed an experimental study of heat transfer in rectangular microchannels with channel widths ranging between 194 and 534 μm and with the depth of five times the width in each case using both a full 3D conjugate approach and a simplified thin wall heat transfer models. In case of either constant temperature or constant heat flux boundary conditions, the deviations of the results obtained with the 1D correlation from the 3D full conjugate analysis were 12.4 and 7.1%, respectively, demonstrating the importance of setting accurate boundary conditions in the simulations. For this reason, the authors recommend the use of numerical simulations, instead of 1D correlation, to predict the performance of microchannel reactor/heat exchangers.

9.3 Multiphase microreactors

Multiphase flows are created when two of more partially or immiscible fluids are brought in contact. Multiphase flow operation provides several mechanisms for enhancing and extending the performance of single-phase microfluidic systems. The long diffusion times and dispersion limitations often associated with single-phase laminar flow can be reduced or eliminated by adding a second, immiscible fluid stream that enhances mixing via transverse convection by inducing a recirculation motion in the liquid. The precise manipulation of immiscible fluid streams in multiphase microreactors is usually achieved by elaborate reactor designs that address fluid–fluid hydrodynamics under laminar flow conditions and fluid–solid interactions in microchannels. Such conditions enable fast heating and quenching of reaction mixture and fast mass transfer due to enhanced interfacial area under well-defined hydrodynamic conditions.

Multiphase catalytic reactors are employed in nearly 80% of industrial processes with annual global sales of about $1.5 trillion, contributing around 35% of the world's GDP [17]. Microreactors for multiphase reactions are classified based on the contact principles of gas and liquid phases: continuous-phase contacting and dispersed-phase contacting [18]. In the former type, the two phases are kept in continuous contact with each other by creating an interface. In the latter case, one fluid phase is dispersed into another fluid phase. In addition, micro trickle bed operation is reported following the path of classical chemical engineering. The study of mass and heat transfer in two-phase flow in micro trickle bed reactors still remains as a less explored area and has received growing interests in recent years due to its important applications in the synthesis of fine chemicals and pharmaceuticals.

9.3.1 Microstructured packed beds

Many microreactors use micropacked beds for gas–liquid–solid reactions [19–21]. An advantage of microstructured packed bed over catalytic coatings is the commercial availability of catalysts. The catalyst particle size is typically below 250 μm and well suited for microchannel applications. Usually a proper loading of the bed into the reactor is required to provide reliable reactor operation without a substantial pressure built up due to bed densification. In microstructured packed beds the fluid flow is rather different from larger particle systems. Often the velocity fluctuations along a streamline are observed. Therefore, the magnitude of dispersion effects depends on particle size. Another feature is that due to capillary action, microstructured packed beds are highly saturated with liquid in the liquid flow range typical of three-phase catalytic applications.

Figure 9.2 shows that the hydrodynamic behavior of multiphase flow in micropacked beds has more in common with flow in porous media than with flow in industrial packed beds.

The literature still remains very short about understanding and modeling the hydrodynamics of two-phase flow in packed bed microreactors. Thus far, few experimental studies on flow regimes (ranging from bubbling flow to pulsing flow), liquid holdup, and two-phase pressure drop have been published [19–21]. Losey et al. [23] constructed a composite silicon–glass microreactor using dry reactive ion etching (DRIE). The gas and liquid flows were premixed on-chip using high aspect ratio mixing channels. These channels delivered the reactants to 10 reactor chambers filled with catalyst particles with a diameter

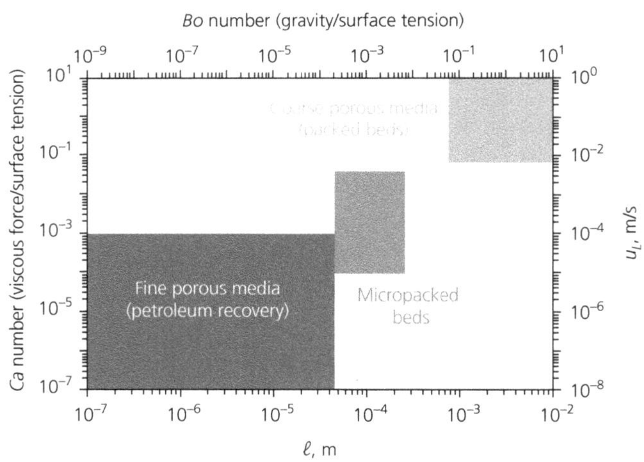

Figure 9.2 Typical ranges of force ratios in multiphase flow for packed beds, micropacked beds, and fine porous media. $Bo = (\rho_L - \rho_G)gl^2/\sigma$ is the gravity/surface tension ratio and $Ca = \mu u/\sigma$. The term u is the superficial velocity, ρ is the density, l is the characteristic hydrodynamic length, that is, wetted area over wetted perimeter, g is gravity, μ is the viscosity, and σ is the surface tension. The subscripts L and G refer to liquid and gas, respectively. (Source: Márquez et al. [22]. Reproduced with permission of American Chemical Society.)

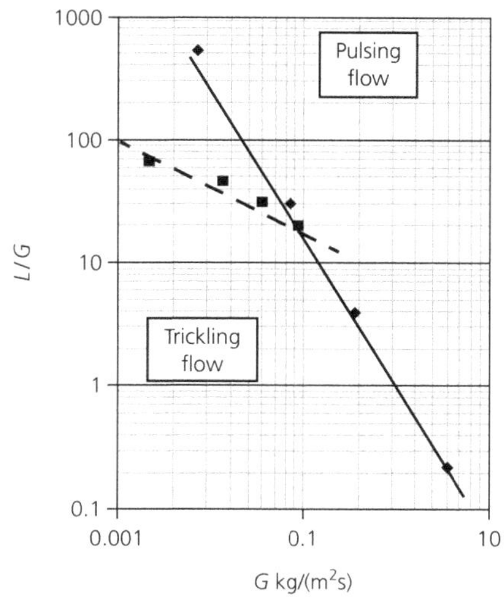

Figure 9.3 Flow transition for cyclohexene/H_2 cocurrent flow in a micro trickle bed reactor (■) versus superficial mass velocity [23]. Charpentier flow map (◆) is presented for comparison [24]. *G* is the superficial mass velocity. (Source: Losey et al. [23]. Reproduced with permission of American Chemical Society.)

of 50–75 μm. The authors observed that small bubbles were formed at the entrance at relatively low gas and liquid velocities and they were carried by the liquid stream in the packed bed. At high liquid and gas velocities, a stable interface of liquid with gas was observed. Pulsating flow regimes dominated when gas flow rate was increased at a fixed liquid flow rate. At high gas flow rate and fixed low liquid flow rate, segregation of the phases was observed which resulted in drying out of packed bed. Figure 9.3 compares the transition point from steady flow to pulsing flow with a regime map originally proposed by Charpentier and Favier [24]. It can be seen that no substantial differences were observed between these two cases.

Different flow regimes in microstructured packed bed reactor were investigated in details by van Herk et al. [19]. These authors observed pulsations and the formation of segmented flow regime at high gas flow velocities. The segmented gas–liquid flow was also reported by Tadepalli et al. [25] in a MPBR with a Pd/zeolite catalyst with particles of 45–150 μm and a length of catalyst bed of 6–8 cm.

The interaction between gas and liquid is limited at low *Re* numbers. Neither holdup nor dispersion is strongly affected by the gas flow rate, and molecular diffusion coefficient has a rather small effect on the particle Péclet number [20].

As per the guidelines available in the literature to overcome the wall effects in microstructured packed bed reactors, the maximum catalyst particle diameter that can be used in a reactor with an internal diameter of 10 mm is 0.4 mm [26, 27]. These reactors can be considered as microstructured packed bed reactors, as the flow of gas and liquid is not affected by gravity [28].

9.3.1.1 Liquid holdup

In the analysis of trickle beds, draining experiments are used to distinguish dynamic holdup from static holdup, where the static fraction is that which remains in the bed after draining has stopped. Márquez et al. [20] studied liquid holdup and dispersion in a reactor of 2 mm internal diameter filled with 0.1 mm spherical particles, for multiphase flows with hydrocarbon liquid flow rates of 10–100 μl/min and nitrogen gas flow rates of 50–1000 μl/min. Their experimental apparatus is shown in Figure 9.4.

Liquid was delivered into the packed bed with a needle. The gas feed line was connected to the feed section with a T-junction; therefore, the gas flowed through the annular area between the liquid feed line and the outer pipe. When the gas and liquid flows were stopped, the liquid entirely remained in the microstructured bed. Thus, the bed has zero dynamic holdup and only static holdup. The static holdup (ε_L) is expressed as the fraction of the space between the particles that is, on average, filled with liquid:

$$\varepsilon_L = \frac{F_L \tau}{V_p} \tag{9.14}$$

The liquid holdup does not vary significantly with the gas superficial velocity and remains always above 0.65 (or 0.28–0.37 per unit column volume) [20]. High holdup values translate into good wetting characteristics of catalysts and are desired for catalytic reactor applications. These values are considerably higher as compared to those for dynamic liquid holdup in industrial trickle bed reactors [29]:

$$\varepsilon_{L,DYN} = 3.86\varepsilon_b Re_L^{0.565} Ga_L^{-0.42}\left(\frac{a_s d_p}{\varepsilon_b}\right) \tag{9.15}$$

The use of fine particles as a diluent of trickle bed reactor packings is a common method of ensuring good liquid distribution and catalyst wetting in small-scale laboratory test reactors.

Kulkarni et al. [30] found that liquid holdup (per unit reactor volume) in a bed with 6 mm glass spheres increased from 0.05 to 0.25 upon mixing with 0.2 mm glass particles. The use of a diluent with a particle size of 0.3 mm increased liquid holdup from 0.05 to 0.20 [31]. In beds diluted with small particles, the liquid is more easily retained between the particles due to the increased surface area, leading to the increased value of the total liquid holdup. These high values are not observed in upflow beds with particles of 3 mm in diameter. Iliuta et al. [31] found static holdup in upflow packed beds of 0.03–0.08 and dynamic holdup values in the range 0.10–0.25. Bej et al. [32] found that reactor performance kept increasing for smaller particles, with optimal results obtained for dilution with 0.1 mm particles due to the high holdup values.

Márquez et al. [22] studied the step response applying various start-up procedures in their three-phase microreactor (Figure 9.4). A reproducible behavior was reported through many cycles while the different start-up procedures have little effect on the steady state that is achieved, in sharp contrast to large-scale reactors where prewetting has a remarkable impact on the hydrodynamic behavior [33, 34]. A typical start-up time

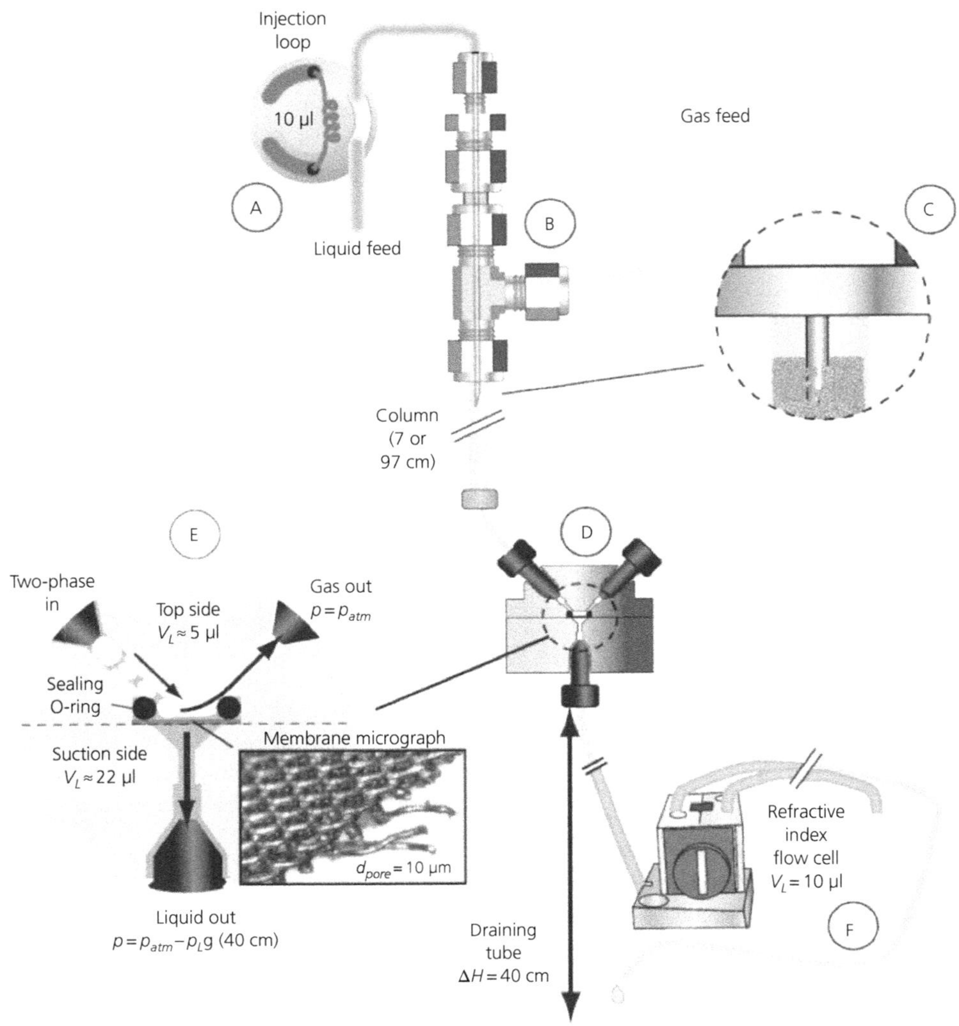

Figure 9.4 Column for two-phase dispersion measurements. To emphasize the tracer spread outside the column, most of the long column has been omitted in this graph. (A) The 10 μl injection loop; (B) a thin needle is located inside the T-connector and reducing connector; (C) the needle is inserted into the bed of particles; (D) the gas–liquid separator; (E) detail of the separator, including the relevant liquid volume; (F) a short tube connects the separator to the flow cell. (Source: Márquez et al. [20]. Reproduced with permission of John Wiley & Sons.)

is equal to four liquid residence times. This time interval was needed to achieve stable pressure drop and dispersion values over the bed [22].

9.3.2 Microchannel reactors

Microchannels reactors have channels with a diameter of 100–1000 mm and a channel length from a few centimeters to several meters. The behavior of multiphase flow in a microchannel has a substantial effect on the performance of a microreactor. In the last two decades, considerable efforts have been invested in the study of the hydrodynamics of two-phase flow in microchannels. As a result, a large body of information has become available especially concerning the flow regimes observed, ranging from bubbling flow to annular flow, and regarding the prediction of the hydrodynamic parameters for Taylor slug flow.

9.3.2.1 Hydrodynamics

The gas–liquid flow regime describes the spatial distribution of the two-phase flow in the microchannels, and it strongly influences the performance of the microreactor, particularly in terms of pressure drop, heat, and mass transfer. Different flow regimes can occur depending on the gas and liquid flow rates, the fluid physical properties (e.g., surface tension, viscosity, density), the wettability of the channel wall, the channel size, and the geometry of the cross section and inlet geometry [35]. Based on the relative importance of the surface tension over inertial forces, three overall flow regimes are identified, namely, surface tension-dominated, inertia-dominated, and transitional regimes. These three regimes consist of six main flow patterns, namely, bubbly and slug (surface tension dominated), churn and slug–annular (transitional), and dispersed and annular (inertia

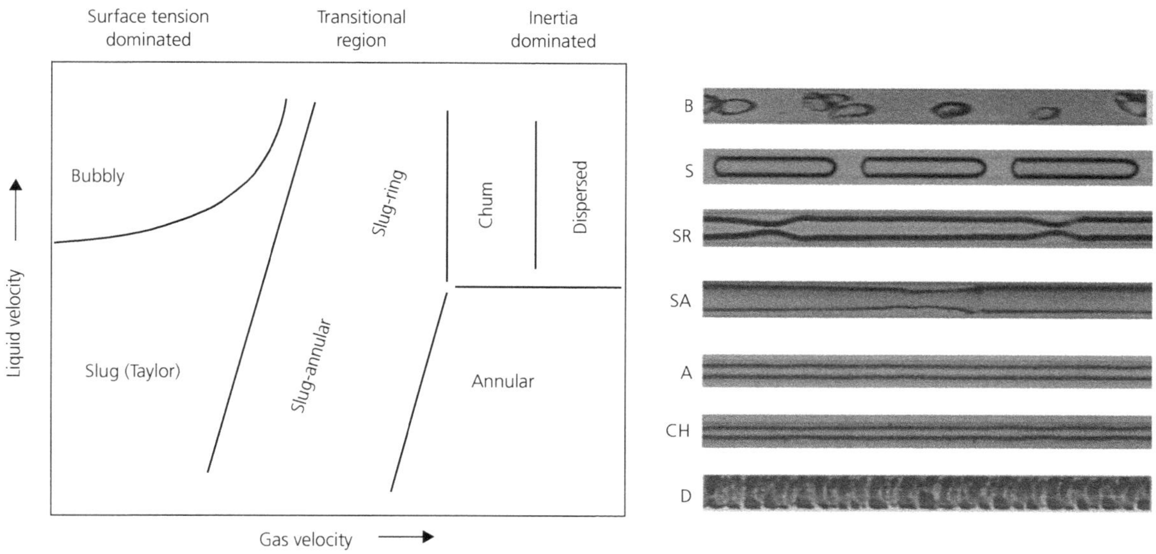

Figure 9.5 Map of two-phase flow regimes in a microchannel. (Source: Rebrov [36]. Reproduced with permission of Springer.)

dominated). Both the main regimes and the constituting flow patterns are presented in Figure 9.5.

Bubbly flow is characterized by distinct bubbles smaller or equal to the channel diameter. Zhao and Bi observed bubbles with size close to the tube diameter in an equilateral triangular channel with a hydraulic diameter of 0.886 mm which they named capillary bubble flow [37]. *Taylor flow* (also known as slug flow) is characterized by elongated bubbles with an equivalent diameter larger than that of the channel. Liquid slugs separate the gas bubbles. Depending on the channel wettability, a liquid film may form that separates the bubbles from the wall. No liquid film was observed in microchannels with contaminated walls [38]. As the gas flow rate increases in Taylor flow at relatively low U_L, the increased gas void fraction leads to the merging of Taylor bubbles to form the regime defined as *slug-annular flow* [39]. In this pattern, a gas core is surrounded by a liquid film where large-amplitude solitary waves appear and small bubbles are sometimes present in the liquid. With a decrease in the channel diameter, the size of the "rings" of the liquid constitutes a considerable proportion of the cross section of the channel. Therefore, in some studies, this flow regime in microchannels is defined as the *slug-ring* regime. With further increasing U_G, the long waves disappear and the *annular* pattern is established. From annular flow, with further increase of U_L, small liquid droplets are entrained in the gas core to form the dispersed pattern. As very high U_L and U_G are required, this pattern is commonly observed in short microchannels with a large ratio of width to depth. *Churn flow* is formed as a result of two processes. As the liquid flow rate increases, the elongated Taylor bubbles become unstable near their tail ends, leading to disruption and entrainment of gas into the liquid. In other cases, at higher gas flow rates, the liquid film that flows on the channel wall becomes very disturbed by the high inertia of the gas core and slugs and drops appear within the gas core.

The channel size, gas and liquid superficial velocities, liquid-phase surface tension, and channel wettability have the most significant effect on the position of the flow regime boundaries [35, 36]. Furthermore, different flow patterns can be observed near the inlet and outlet of a microchannel. This can be seen, for example, in Figure 9.6a where the experimental data obtained near the inlet in a 100 μm × 50 μm microchannel with a length of 20 mm. The data are compared with the flow pattern maps obtained by Hassan et al. [41], Chung and Kawaji [42], and Jones and Zuber [43]. It can be seen that the transition between slug and annular regime near the inlet is in close agreement with the results reported by Jones and Zuber [43]. The absence of the bubble and slug-annular regimes near the channel outlet (Figure 9.6b) is explained by an increase in the volume of gas bubbles as the absolute pressure decreases. In the latter case, individual gas bubbles coalesce; therefore, the bubble regime is not observed in the section located near the outlet of the capillary. However, the transition lines between the inertial regimes remain almost unchanged. Relatively poor agreement between the flow map proposed by Haverkamp et al. [40] and other experimental data is observed with larger channel sizes and fluids other than air and water.

Akbar et al. [44] suggested using Weber numbers instead of superficial fluid velocities as the two coordinates. The transition lines in Weber number coordinates predict satisfactorily the transitions in relatively large channels (Figure 9.7) [36]. Figure 9.7 shows transition lines proposed by Akbar et al. [44] and experimental data obtained by Warnier ($50 \times 50\ \mu m^2$) [45], Damianides and Westwater (i.d.:1.0 mm) [46], and Yang

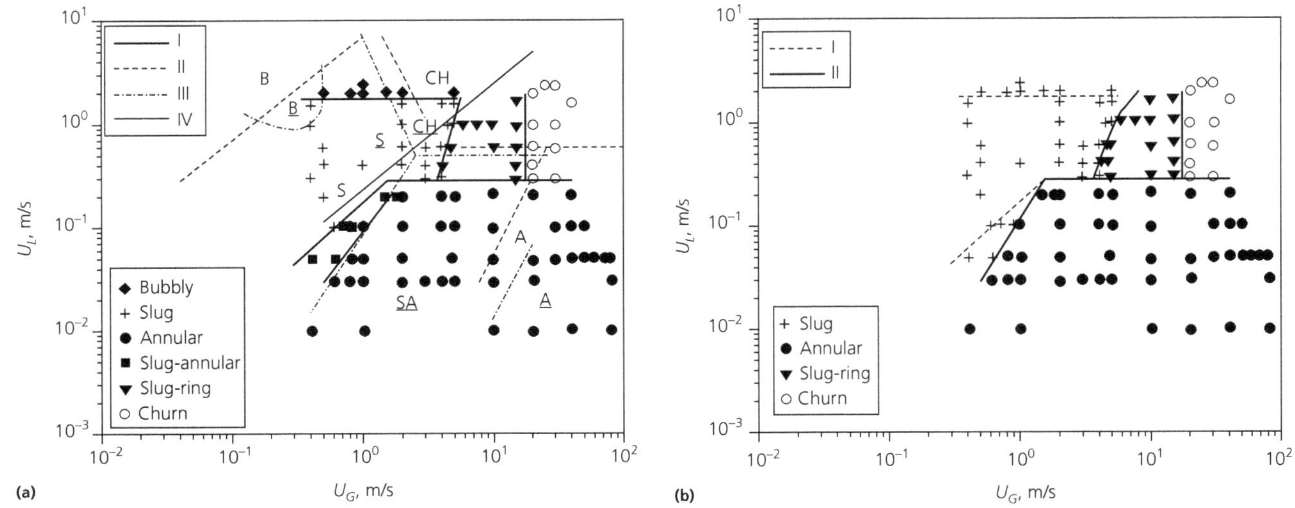

Figure 9.6 (a) Maps of two-phase flow regimes (lines I and symbols) in the capillary of the rectangular cross section with a width of 100 μm, a height of 50 μm, and a length of 20 mm directly after the mixer [40]. For comparison, the following maps of the regimes are presented: (II) Hassan et al. [41], (III) Chung and Kawaji [42], and (IV) Jones and Zuber [43]. Letters in underscore font for data Chung and Kawaji. (b) The lines of the regimes near the mixer are shown by dashed lines for comparison with the data obtained near the channel outlet (solid lines and symbols) [40].
(Source: Rebrov [36]. Reproduced with permission of Springer.)

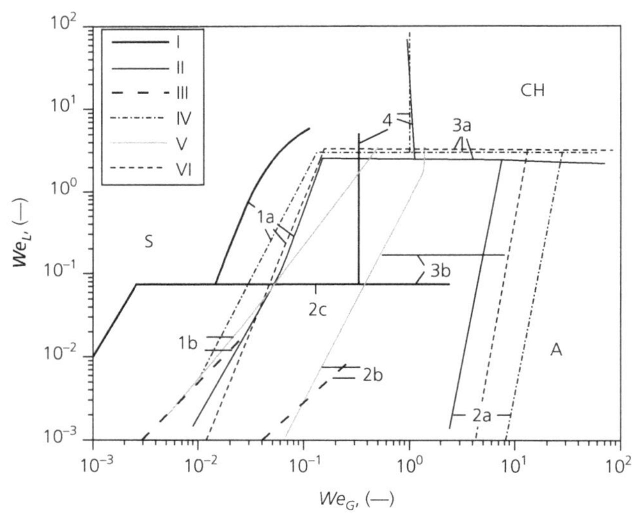

Figure 9.7 Maps of the water–nitrogen two-phase flow regimes in the microchannels: (I) 100 × 50 μm^2 [40], (II) 50 × 50 μm^2 [45], (III) 300 × 100 μm^2 [40], (IV) Ref. [46], (V) Ref. [47], and (VI) Ref. [44]. The lines represent the boundaries of the transitions between regimes: (1a), (1b) slug/slug–annular; (2a), (2b), (2c) slug–annular/annular; (3a), (3b) annular/churn; and (4) slug/churn regimes.
(Source: Rebrov [36]. Reproduced with permission of Springer.)

and Shieh (i.d.:1.0 mm) [47]. The regime transition lines fitted to all relevant data are recommended for regime transition predictions.

9.3.2.2 Bubble and slug length

The length of bubbles and slugs in Taylor flow has a direct impact on the reactor performance. The dependency of bubble and slug lengths on operating parameters in microchannels has been studied experimentally and numerically by various researchers [36, 40, 48–57]. A large number of empirical correlations for the slug length based on phase holdup and dimensionless numbers have been proposed in these studies. Haverkamp et al. [40] used two types of flow inlets, one with the phases joining perpendicularly in a cross-mixer and the other with the phases joining smoothly in a smooth mixer, and observed a profound effect of the inlet type on the air/water flow patterns formed. The transition from Taylor to churn flow was shifted to higher gas velocities in the smooth inlet. Two different bubble formation mechanisms were observed for the slug flow: squeezing (or switching) and dripping. When the pressure gradient generated by the liquid flow dominates the pressure gradient generated by the gas flow, the liquid flow leads to bubble pinch-off in the cross, Y-, and T-type mixers (Figure 9.8). If the liquid flow pressure gradient is not high enough to pinch-off the gas flow, dripping mechanism occurs, where an annular flow is created with a gas core surrounded by a liquid film. Small hydrodynamic jet instabilities lead to a breakup of the gas jet, which then forms bubbles [53].

The squeezing mechanism is observed in T-mixers at $Ca < 0.02$ when the interfacial forces dominate the shearing forces [40, 54] (Figure 9.8). Pohorecki and Kula [55] proposed an equation to estimate the bubble size for *switching* mechanism in a Y-mixer at $10 < Re < 350$ and $Ca \ll 1$:

$$\frac{L_B}{d} \approx 1 + \frac{U_G}{U_L} \tag{9.16}$$

In a T-mixer, the bubble length depends on the volume of the mixer chamber and the gas holdup. Garstecki et al. [54]

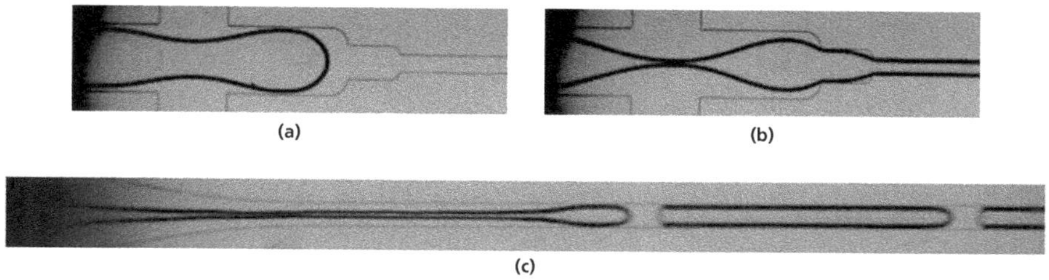

Figure 9.8 Mechanisms of the formation of gas bubbles in the microchannels: (a, b) squeezing and (c) dripping. (Source: Rebrov [36]. Reproduced with permission of Springer.)

proposed an empirical model for the bubble length in the squeezing regime in a T-mixer:

$$\frac{L_B}{w} = a_1 + a_2\frac{U_G}{U_L} \quad (9.17)$$

where $a_1 = L_B/w$ and $a_2 = w_{in}/w$. Van Steijn et al. [56, 57] improved this model by taking into account the liquid flow around the forming bubble in rectangular channels. These authors suggested correlation for parameters a_1 and a_2 in rectangular microchannels of arbitrary dimensions for $0.33 < (w_{in}/w) < 3$ and $0.1 < (h/w) < 0.5$ for $Ca < 0.01$, where h is the channel height, w is the channel width, and w_{in} is the width of the side channel. In general, the size of Taylor bubbles is independent of the fluid properties and is a function of the microchannel geometry and the gas-to-liquid flow rate ratio only.

In the dripping mechanism, the bubble size is controlled by the capillary number [58]. The bubble length follows the model developed by Garstecki et al. [54] and van Steijn et al. [56]; however, the model coefficients depend on the liquid properties [59]. The transition between the squeezing and dripping regimes happens at capillary number around 0.002–0.006 and depends on fluid properties [60, 61]. In the transition regime, the size of bubbles is controlled by the ratio of gas/liquid flow rates and the capillary number, as both the shear stresses and squeezing pressure contribute to collapse the gaseous thread and balance the surface tension force [61]:

$$\frac{\sqrt{L_B w_B}}{w} = 0.26\cdot\left(\frac{U_B}{U_L}\right)^{0.35} Ca^{-0.25} \quad (9.18)$$

where w_B, w, and L_B are in μm. Xu et al. [60] reported a similar expression for droplets generated in the transition regime:

$$\frac{L_B}{w} = 0.75\cdot\left(\frac{U_B}{U_L}\right)^{0.33} Ca^{-0.2} \quad (9.19)$$

Therefore, when predicting the bubble size for process engineering applications, the fluid properties should be taken into account in inertia-dominated flow conditions.

9.3.2.3 Liquid film thickness

The thickness of the liquid film surrounding a Taylor gas bubble is an important parameter for describing the hydrodynamics of gas–liquid Taylor flow. It has been the subject of study since Fairbrother and Stubbs [62] first showed that a single bubble in a capillary moves faster as compared to the two-phase velocity. Bretherton [63] expanded on this work and analytically derived an expression for the liquid film thickness in channels with a circular cross section for $Ca_B < 0.01$ and negligible inertial effects:

$$\frac{\delta_L}{d} = 0.67\cdot Ca_B^{2/3} \quad (9.20)$$

Aussillous and Quéré [64] have applied a scaling analysis to Bretherton's result for the liquid film thickness and suggested an empirical equation for the liquid film thickness:

$$\frac{\delta_L}{d} = \frac{0.67\cdot Ca_B^{2/3}}{1 + 3.34\cdot Ca_B^{2/3}} \quad (9.21)$$

In cylindrical capillaries where the effects of gravity can be neglected, the liquid film around the bubble has a constant thickness, which increases with the capillary number. Under inertia-dominated flow conditions, the liquid film thickness decreases and then increases with increasing Reynolds number [65, 66]. Han and Shikazono [66] studied hydrodynamics of Taylor flow in circular tubes with different diameters of 0.3, 0.5, 0.7, 1.0, and 1.3 mm, and they proposed empirical correlations for the dimensionless film thickness for $Re < 2000$:

$$\frac{\delta_L}{d} = \frac{0.67\cdot Ca_B^{2/3}}{1 + 3.13\cdot Ca_B^{2/3} + 0.504\cdot Ca_B^{0.672} Re^{0.589} - 0.352\cdot We^{0.629}} \quad (9.22)$$

As the Re number approaches 1, Equation 9.22 follows Bretherton's theory. In a rectangular channel, the film thickness at the lateral walls differs from that above and below the bubble [67, 68], and these, in turn, differ from the diagonal film thickness in the channel corners. At small capillary numbers, the liquid film formed on the channel center becomes very thin. However, as capillary number increases, the interface shape becomes axisymmetric. As the Reynolds number increases, transition from nonaxisymmetric to axisymmetric flow pattern occurs. Based on their experimental work in capillaries of 0.3, 0.5, and 1.0 mm, Han and Shikazono [69] provided a correlation which calculates the film thickness in microchannels with a square cross section with an accuracy within ±5%:

$$\frac{\delta_{CORN}}{d} = 0.122 + \frac{1.22\cdot Ca_B^{2/3}}{1 + 7.28\cdot Ca_B^{2/3} - 0.255\cdot We^{0.215}}, \quad (9.23a)$$

$Re < 2000$, in the channel corner

$$\begin{cases} \frac{\delta_{CENT}}{d} = 0 & \frac{\delta_{CORN}}{d} < 0.22 \\ & \qquad Re < 2000, \text{ in the channel center} \\ \frac{\delta_{CENT}}{d} = \frac{\delta_{CORN}}{d} & \frac{\delta_{CORN}}{d} \geq 0.22 \end{cases} \tag{9.23b}$$

9.3.2.4 Pressure drop

9.3.2.4.1 Taylor flow

Consider a gas–liquid Taylor flow moving through a channel with a cross-sectional area A (Figure 9.9).

The gas bubbles have a velocity u_B and a length L_B and occupy a fraction of the cross-sectional area of the channel A_B/A. The flow is divided into unit cells consisting of one gas bubble, its surrounding liquid film, and one liquid slug and the unit cell length is $L_B + L_S$. Continuity requires that the overall average velocity through any cross section of the channel perpendicular to the direction of flow is equal to the sum of the superficial gas u_G and liquid u_L velocities over the channel cross section. The flow through plane A_1 consists of the gas bubble moving with velocity u_B through a cross section A_B and the liquid film moving at an average velocity u_F through a cross section $A - A_B$ (Eq. 9.24). The flow through plane A_2 consists solely of liquid, which occupies the whole cross section of the channel A, and the average velocity of the liquid in the slug is therefore equal to $u_G + u_L$:

$$\frac{A_B}{A}u_B + \left(1 - \frac{A_B}{A}\right)u_F = u_G + u_L \tag{9.24}$$

An additional liquid volume near the nose and tail parts of a bubble can be written as the bubble cross-sectional area A_B times a length δ. In case of a bubble with a circular cross section and hemispherical bubble caps, δ is equal to 1/3 of the bubble diameter d_B. The superficial gas velocity U_G is then given by

$$u_G = \frac{A_B f_B}{A}(L_B - \delta) \tag{9.25}$$

The bubble formation frequency f_B is equal to the number of unit cells passing a certain location in the channel per unit of time, which is the reciprocal of the time it takes for a unit cell to travel a distance equal to its own length:

$$f_B = \frac{u_B}{L_G + L_L} \tag{9.26}$$

Equation 9.24 can be rewritten as

$$u_L = \frac{A_B}{A}f_B(L_S + \delta) + \left(1 - \frac{A_B}{A}\right)u_F \tag{9.27}$$

The frictional pressure loss in one liquid slug is given by

$$\Delta p_L = \frac{16}{Re_{GL}}\frac{4}{d}\left(\frac{\rho_L(U_G + U_L)^2}{2}\right)(L_L + \delta) = \frac{32\mu_L(U_G + U_L)}{d^2}(L_L + \delta) \tag{9.28}$$

Bretherton's result for the pressure drop over a single Taylor gas bubble is modified to account for a nonnegligible liquid film thickness [70]:

$$\Delta p_G = \frac{7.16 \cdot \sigma \cdot (3 \cdot Ca_B)^{2/3}}{\left(1 + 3.34 \cdot Ca_B^{2/3}\right)d} \tag{9.29}$$

The pressure drop over a unit cell is then given by the sum of the frictional pressure drop of the liquid flowing in the slug (Δp_L) and the pressure drop over the Taylor gas bubble (Δp_G). The length of a unit cell is $L_B + L_S$ and thus the number of unit cells per unit length of channel is $1/(L_B + L_S)$. The pressure drop over a unit length of channel is therefore given by

$$-\left(\frac{dp}{dz}\right)_c = \frac{\Delta p_L + \Delta p_G}{L_B + L_S} = \frac{1}{L_B + L_S}\left(\frac{32\mu_L(u_G + u_L)}{d^2}(L_L + \delta) + \frac{7.16 \cdot \sigma \cdot (3 \cdot Ca_B)^{2/3}}{\left(1 + 3.34 \cdot Ca_B^{2/3}\right)d}\right) \tag{9.30}$$

It can be shown from the mass balance-based model that $(L_S + \delta)/(L_B + L_S) = U_L/(U_G + U_L)$ [70]. Equation 9.24 for a stagnant liquid film ($u_F = 0$) can be rewritten as

$$\frac{\sigma}{\mu_L(u_G + u_L)} = \frac{A}{A_B}\frac{1}{Ca_B} \tag{9.31}$$

By substituting Equation 9.31 into Equation 9.30, the following expression for the frictional pressure drop over a unit length of channel is obtained [71]:

$$-\left(\frac{dp}{dz}\right)_c = \frac{32\mu_L u_L}{d^2}\left(1 + \frac{7.16 \cdot 3^{2/3} d}{32}\frac{A}{A_B(L_S + \delta)}\frac{1}{\left(Ca_B^{1/3} + 3.34 \cdot Ca_B\right)}\right) \tag{9.32}$$

Using Equation 9.27 for a stagnant liquid film, the term $A/(A_B \times (L_S + \delta))$ can be rewritten as f_B/u_L. This term is not a function of bubble velocity and thus is not dependent on the pressure. Calculating the pressure profile from the model requires an additional step since the pressure drop depends

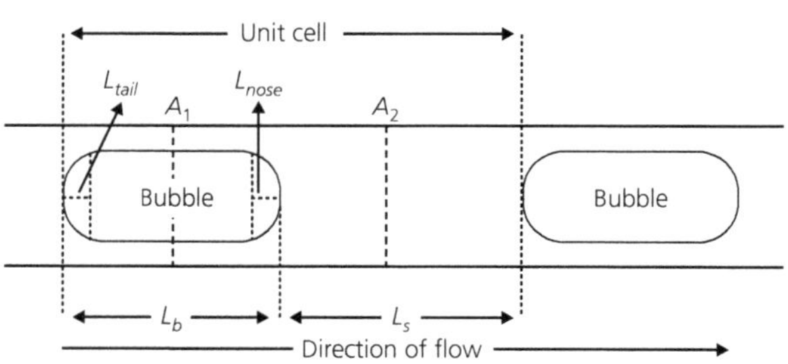

Figure 9.9 Schematic of Taylor flow showing the definitions of the unit cell, gas bubble length L_b and the liquid slug length L_s. The lengths of the nose L_{nose} and tail L_{tail} sections of the gas bubble are also indicated. Mass transfer contributions in the Taylor flow are indicated with the numbers: (1) bubble to wall through film, (2) bubble to slug, and (3) slug to wall.

Table 9.2 Two-phase laminar–laminar frictional pressure drop correlations.

C-factor correlation	Remarks	Reference
$C = 5$	d = 1.49–25.83 mm, adiabatic, water, oils, hydrocarbons	[73]
$C = 21 \cdot [1 - e^{-319d}]$ for rectangular channel, d [m]	d = 1.05–4.08 mm, adiabatic, air–water	[74]
$C = 21 \cdot [1 - e^{-333d}]$ for circular channel, d [m]	d = 1.05–4.08 mm, adiabatic, air–water	[74]
$C = 6.833 \times 10^{-8} \lambda^{-1.317} Re^{0.557} Ca^{0.719}$, $\lambda = \frac{\mu_L^2}{\rho_L \sigma d}$	d = 0.78–6.67 mm, adiabatic, air–water	[75]
$C = 7.599 \times 10^{-3} \lambda^{-0.631} Re_L^{-0.008} Ca^{0.005}$	d = 0.15, 0.22, 0.53 mm, adiabatic, air–water	[76]
$C = 0.227 \cdot Re^{0.452} X^{-0.32} N_{conf}^{-0.82}$	d = 0.244, 0.430, 0.792 mm adiabatic, air–water	[77]
$N_{conf} = \frac{1}{d}\sqrt{\frac{\sigma}{g(\rho_L - \rho_G)}}$		[78]

on the bubble velocity u_B rather than on U_G. The bubble velocity is related to the superficial gas velocity and the cross-sectional area of the gas bubble A_B as described in Equation 9.24 and thus depends on the liquid film thickness. In turn, the liquid film thickness depends on the bubble velocity as described by Aussillous and Quéré (Eq. 9.21). Therefore, the bubble velocity and the liquid film thickness are calculated by an iteration of Equations 9.21 and 9.24 for every axial position in the channel while calculating the pressure profile. The model describes the experimental data within an accuracy of ±4% in a glass channel with a circular cross section and an inner diameter of 250 μm for $41 < Re_{GL} < 159$ and $Ca_{GL} < 0.01$.

9.3.2.4.2 *Annular flow*

Lockhart and Martinelli proposed that the pressure drop in two-phase flow can be related to the equivalent pressure drop in single-phase flow using a two-phase parameter [72]. For the gas phase, the single-phase pressure drop is given by

$$-\left(\frac{dp}{dz}\right)_G = \frac{32\mu_G u_G}{d^2} \tag{9.33}$$

For the liquid phase, the single-phase pressure drop is given by

$$-\left(\frac{dp}{dz}\right)_L = \frac{32\mu_L u_L}{d^2} \tag{9.34}$$

The Lockhart–Martinelli parameter χ is given by

$$\chi^2 = \frac{(dp/dz)_L}{(dp/dz)_G} \tag{9.35}$$

Chisholm suggested the following equation for the two-phase multiplier parameter [73]:

$$\phi_L^2 = 1 + \frac{C}{\chi} + \frac{1}{\chi^2} \tag{9.36}$$

The two-phase pressure drop is calculated by

$$-\left(\frac{dp}{dz}\right)_c = \phi_L^2 \left(\frac{dp}{dz}\right)_L \tag{9.37}$$

The main uncertainty related with the Lockhart–Martinelli model is the determination of the C-factor. Different C-factor correlations used for laminar gas–liquid flow are listed in Table 9.2.

9.3.2.4.3 *Liquid–liquid microreactors*

Salim et al. [79] studied two-phase oil–water flows and pressure drop in horizontal microchannels. The pressure drop measurements were interpreted by using the homogeneous and Lockhart–Martinelli models. The two-phase pressure drop was correlated to the single-phase pressure drop of each phase over the whole length of the capillary:

$$\left(\frac{\Delta p}{L}\right)_{TP} = \left(\frac{\Delta p}{L}\right)_c + \eta \varepsilon_d \left(\frac{\Delta p}{L}\right)_d \tag{9.38}$$

where $(\Delta p/L)_{TP}$ is the two-phase pressure drop, $(\Delta p/L)_c$ and $(\Delta p/L)_d$ are the continuous and dispersed single-phase pressure drops, respectively, ε_d is the dispersed-phase volume fraction, and η is a parameter which depends on the wettability of the capillary wall. The empirical parameter η was determined from the experimental results, with values of 0.67 and 0.80 for the quartz and glass microchannels, respectively. The main drawback of this approach is the absence of the surface tension and slug length influence on the pressure drop. Therefore, this model underestimates the experimental data obtained by other researchers.

Jovanovic et al. [80] proposed a pressure drop model which account for frictional pressure losses in the discrete and continuous phases (Δp_{FRIC}) and the interfacial pressure drop (Δp_I):

$$\Delta p_{SLUGFLOW} = \Delta p_{FRIC} + \Delta p_I = \Delta p_{FRIC,C} + \Delta p_{FRIC,d} + \Delta p_I \tag{9.39}$$

The frictional pressure drop in both phases was calculated from the Hagen–Poiseuille equation for laminar flow in a tube. The interfacial pressure drop is described by the Bretherton's solution for the pressure drop over a single bubble in a capillary (Eq. 9.40) [63]:

$$\Delta p_I = C_1 (3Ca)^{0.67} \frac{\sigma}{d} \tag{9.40}$$

where the constant C_1, accounting for the influence of the interface curvature, equals 7.16. The theory of Bretherton is in good agreement with experimental data for $Ca < 5 \times 10^{-3}$ and $We \ll 1$.

The film velocity was found to be of negligible influence on the pressure drop. The pressure drop model (Eqs. 9.39 and 9.40) was in good agreement with the experimental data, with a mean relative error less than 5%. The value of the curvature parameter of 7.16 provides a good agreement with the experimental results in the 248 μm capillary, while in the 498 μm capillary a value of 3.48 should be used to fit the experimental data due to the asymmetrical distribution of the liquid film [80]. With increasing velocity two effects occur, namely, increase of the film thickness and deformation of the front and back meniscuses. The increase of the film thickness squeezes the slug cap, thus deforming the curvature. As a result, the curvature parameter decreases at higher We numbers, which should be accounted for by modifying the curvature parameter values as a function of the slug velocity.

9.3.2.5 Mass transfer

An important advantage of microreactors for gas–liquid reactions is the significant intensification of the mass transfer processes. The characteristic values governing gas–liquid mass transfer in a variety of conventional equipment and microreactors are listed in Table 9.3 from Yue et al. [81].

It can be seen that the volumetric mass transfer coefficient, k_La, is from one to two orders of magnitude larger in microreactors compared with that in conventional process equipment. However, it is mainly due to a huge increase in the specific interfacial area in microreactors. The absolute value of mass transfer coefficient is of the same order of magnitude as that obtained in static mixers and in tubular reactors.

The most interesting regime of the gas–liquid flow in a microreactor is Taylor flow. In this regime, a number of contributions to the mass transfer processes may be distinguished as demonstrated in Figure 9.9 and listed as (1) bubble to wall through film, (2) bubble to slug, and (3) slug to wall. In case of a reaction at the wall, all these contributions may occur simultaneously [82]. In case of a *nonreactive wall*, only bubble-to-film and bubble-to-slug mass transport occurs [83–85]. Van Baten and Krishna [86] proposed that the gas-to-liquid mass transfer process may be described as

$$k_La = k_{L,C}a_C + k_{L,F}a_F \tag{9.41}$$

The overall gas–liquid interfacial area is composed of the area of bubble caps (a_C) and the area of the liquid film (a_F) [64, 87]. The contribution of the liquid film area may exceed 90%. Since the film thickness is usually 1–2 μm, it is fast saturated with the absorbed component and thus may cease to contribute actively to the mass transfer process. Pohorecki suggested a criterion that enables to identify the film contribution to the mass transfer [87]:

$$\frac{Ld}{u_B\delta_L^2} \ll 1 \tag{9.42}$$

The criterion should be met in order to avoid saturation of the liquid film covering the channel walls.

Vandu et al. [88] investigated the absorption of oxygen from an oxygen–nitrogen mixture in water in microchannels of square and circular cross sections (diameter 1, 2, or 3 mm). These authors suggested an empirical correlation for k_La as a function of the gas and liquid superficial velocities, the unit cell length, and the diffusivity in the liquid phase (D_m):

$$k_La = \frac{4.5}{d}\sqrt{\frac{D_m u_G}{L_B + L_S}} \tag{9.43}$$

More recently, Yue et al. [81] measured the rate of pure CO_2 absorption into a CO_3^{2-}/HCO_3^- solution in a 1 mm × 0.5 mm microchannel. They proposed a correlation for k_La as follows:

$$Sh_La \cdot d = 0.084 Re_G^{0.213} Re_L^{0.937} Sc_L^{0.5} \tag{9.44}$$

In a subsequent study, these authors investigated physical absorption of oxygen from air into water in air–water Taylor flow in a square microchannel with hydraulic diameters of 400 μm.

They obtained an empirical correlation for the estimation of k_La in this microchannel:

$$k_La = \frac{2}{d}\sqrt{\frac{DU_G}{L_B + L_S}}\left(\frac{L_B}{L_B + L_S}\right)^{0.3} \tag{9.45}$$

Table 9.3 Comparison of mass transfer parameters in different gas–liquid contactors [81].

Type of contactor	$k_L \times 10^5$ (m/s)	A (m²/m³)	$k_La \times 10^2$ (s⁻¹)
Bubble columns	10–40	50–600	0.5–24
Couette–Taylor flow reactor	9–20	200–1200	3–21
Impinging jet absorbers	29–66	90–2050	2.5–122
Packed columns, cocurrent	4–60	10–1700	0.04–102
Packed columns, countercurrent	4–20	10–350	0.04–7
Spray column	12–19	75–170	1.5–2.2
Static mixers	100–450	100–1000	10–250
Stirred tank	0.3–80	100–2000	3–40
Tube reactors, horizontal and coiled	10–100	50–700	0.5–70
Tube reactors, vertical	20–50	100–2000	2–100
Gas–liquid microchannel contactor	40–160	3400–9000	30–2100

Source: Yue et al. [81]. Reproduced with permission of Elsevier.

The standard deviation was 10.8%, in the range of $0.4\ m/s < u_G < 2\ m/s$, $1.4 < L_B/d < 6.3$, and $1 < L_S/d < 3.2$. It is seen from Equation 9.38 that the k_La increases when the bubble velocity increases or the liquid slug length is reduced.

Sobieszuk et al. [89] measured the CO_2 absorption rate from a CO_2/N_2 mixture into a CO_3^{2-}/HCO_3^- solution in a 0.4 mm microchannel. In this study, the interfacial areas of the film and caps were determined independently by image analysis. The shape of gas bubbles was defined as the cylinder and two hemispheres:

$$a_C = \frac{4}{L_B + L_S} \tag{9.46}$$

$$a_F = \frac{4(L_B - d)}{(L_B + L_S)d} \tag{9.47}$$

The experimental values of $k_{L,F}$ and $k_{L,C}$ were very close to each other. The authors suggested an expression for the Sherwood number:

$$Sh = \frac{k_L d}{D} = 0.1 \cdot Re_{GL}^{1.12} Sc^{0.05} \tag{9.48}$$

where k_L means both $k_{L,C}$ and $k_{L,F}$ and the Reynolds number is calculated based on the two-phase superficial velocity. A good agreement with the data of Yue et al. [90] was obtained at low liquid velocities; however, their values were lower than those calculated from the Vandu et al. [88] correlation.

Van Baten and Krishna [91] simulated liquid-to-wall mass transfer for Taylor flow in circular capillaries of 1.5, 2.0, and 3.0 mm diameter. In their analysis, the wall mass transfer process consisted of two separate contributions: (i) wall–slug contribution of the regions in contact with the liquid slug and (ii) wall–film contribution of the region in contact with the liquid film surrounding the bubble. A correlation for total liquid-to-wall mass transfer coefficient was proposed:

$$Sh_W = \frac{1}{2 \cdot Gz_t^{\alpha}\left(\dfrac{Gz_S}{\varepsilon_G}\right)^{0.15}}, \quad \alpha = 0.61 \cdot Gz_S^{0.025}, \quad Gz_S = \frac{L_S D_L}{d^2 u_B}. \tag{9.49}$$

This correlation is valid for $Gz_t < 0.01$ and it is in reasonably good agreement with earlier findings by Berčič and Pintar [92]. However, further research on liquid-to-wall mass transfer would be desirable.

9.4 Conclusions and outlook

Design rules for microreactors have already been formulated and many aspects are generally well understood. Heat and mass transfer in single-phase fluid flow in microchannels has been extensively studied during the last 15 years, and it was demonstrated that classical engineering correlations for calculation of heat and mass transfer coefficients can be used. For example, for fully developed laminar flow, the Nusselt number is a constant whose value depends only on cross-sectional geometry and boundary conditions. However, when a microreactor is made of stainless steel or other materials with high thermal conductivity, heat transfer in a microchannel is a combination of axial heat conduction in the solid and convection to the cooling fluid. The temperature distribution in a single-phase flow in a microreactor can be found with a high accuracy by using various commercially available numerical codes.

More recent studies have focused on multiphase flows in microchannels. The flow regime maps with the transition lines in Weber number coordinates allow to predict an order of magnitude of gas and liquid velocities for a required flow regime. Simple models for a reliable prediction of bubble and liquid slug lengths in Taylor flow have been developed. At low capillary numbers, the slug length of both gas and liquid depends solely on the gas and liquid flow rates and channel dimensions in the mixer. At higher *Ca* numbers, when inertia forces cannot be neglected, the effects of fluid properties should be taken into account when predicting the bubble and liquid slug length. Due to very strong capillary forces, multiphase packed bed microreactors have very high liquid holdup values and, therefore, more stable hydrodynamic conditions when compared with conventional trickle bed reactors.

The liquid film thickness surrounding the Taylor bubble in a microchannel can be precisely predicted with a number of correlations derived from Bretherton's theory. In the closed microchannels, the pressure drop in Taylor flow can be correctly estimated by the Warnier model and in the annular regime by the Lockhart–Martinelli model. A number of correlations for the *C*-factor are available for the Lockhart–Martinelli model.

The liquid-side volumetric mass transfer coefficient in microreactors is one to two orders of magnitude larger as compared to conventional three-phase reactors. This is mainly due to higher specific interfacial area in microreactors. The values of the liquid-side and gas-side mass transfer coefficients in falling film microreactors (FFMRs) are in the ranges of k_L from 1×10^{-6} to 1×10^{-5} m/s and k_G from 10^{-3} to 10^{-2} m/s, respectively.

Many microreactors are at the research level but a few are already commercially available. Microwave-assisted flow processing in microreactors (MAFP) [93] or radiofrequency-heated flow reactors [94, 95] are promising alternatives for conventionally heated, multistep production of fine chemicals in batch reactors. Realization of MAFP at an industrial scale requires a proper design of multitubular reactors integrated with microwave heating [96].

Process intensification is a novel design approach which aims at reduction of equipment size by several orders of magnitude leading to substantial savings in capital cost, improvement of intrinsic safety, and reduction of environmental impact. Conventional nonadiabatic reactors can be intensified utilizing designs based on compact micro heat exchangers, where at least one of the streams is replaced by a reacting mixture. The concept of integrated reactors/heat exchangers was first proposed by Eigenberger and coworkers for coupling of methane steam reforming with methane combustion in mesoscaled reactors made from ceramic monoliths and steel foils [97, 98]. More recently, the implementation of a declining temperature gradient into water gas shift (WGS) microreactors/heat exchangers by counterflow cooling has been realized for various systems up to 2 kW FC power range [99, 100]. The WGS reaction is an exothermic equilibrium reaction and the equilibrium conversion decreases with increasing temperature. Therefore, a significantly better performance was achieved by operating at a relatively high temperature, thereby exploiting the reaction kinetics when the gas composition is far from equilibrium, and then lowering the temperature as thermodynamics begin to limit the CO conversion [101].

Considerable stocks of unsymmetrical dimethylhydrazine (UDMH), a component of high-energy propellant for liquid-fueled rockets, have been accumulated in several countries by the present time [102]. UDMH production capacity changes from year to year, and actual production and sales data are carefully guarded company secrets. UDMH is a highly toxic compound, close to war toxic agents according to its effect on biological objects. UDMH was completely oxidized in a series of catalytic microreactors at a near-stoichiometric ratio with oxygen at relatively low temperatures [103]. The UDMH micro fuel processor consists of a vaporizer and four catalytic reactors: two microreactors for UDMH oxidation followed by a DeNOx reactor for selective catalytic reduction of nitrogen oxides in the presence of oxygen and then a catalytic afterburner for oxidation of added ammonia and carbon monoxide. UDMH is oxidized over a Ir/Al_2O_3 catalyst in the main catalytic microreactor/heat exchanger. A supported Cu/magnesium spinel catalyst is used in the second UDMH burner to decrease the formation of nitrogen oxides as the oxygen excess increases from 1.2 to ca. 3 with increasing the UDMH conversion. The energy produced in the UDMH oxidation is utilized to run the whole system without external heat supply. Here microreactors are superior to conventional technology because of the very high heat and mass transfer rates that can be achieved. Furthermore, even if a microreactor fails, the small quantity of chemicals released accidentally could be easily contained. Two-stage UDMH oxidation in microreactors suppresses considerably the formation of "thermal" or "fuel" nitrogen oxides. Among others, this is achieved by the use of highly active catalytic coatings, having rather high thermal conductivity and so providing an efficient removal of reaction heat from the reaction zone.

Gas–liquid–solid reactions are an important class of reactions in chemical and pharmaceutical industry. The production processes in this type of industry, however, are often directly based on the stoichiometric chemical synthesis and are far from optimized from a process technology point of view. The FFMRs utilize a multitude of thin liquid films that move by gravitational force, providing a typical liquid residence time of few seconds up to 1 min. The FFMRs have been widely applied for different chemical reactions, such as photochemical chlorination [104], sulfonation [105], and hydrogenation [106] that are usually limited by the mass transfer within the liquid phase. Catalytic octanal oxidation with oxygen was demonstrated at 100°C in a FFMR with varying reaction plates bearing different in-channel mixing structures [107]. Using finned reactor plates, the liquid-side mass transfer rate in catalytic oxidation of octanal with oxygen was enhanced by a factor of 1.2 compared to that on a standard plate with parallel channels.

Still many practical problems and issues associated with the application of microchemical systems as chemicals production devices exist, such as plugging and fouling of microchannels, cost and availability of microreactor modules, lack of reliable catalyst coating techniques, supply and removal of reactants and products, use of reliable sensors, transducers and actuators for process monitoring and control, and, last but not least, production capacity and scale-up. However, a tremendous, worldwide effort is being made by many academic, governmental, and industrial research groups that have made significant commitments to push the technology forward, while also successful commercial applications of micro and miniature devices have already been realized. It would not be long before one will see small chemical plants instead of large and potentially dangerous reactor tanks.

Nomenclature

a	area of bubble caps, m^2
A	channel cross-sectional area, m^2
a_1, a_2	parameters in Equation 9.17, —
A_B	bubble cross-sectional area, m^2
a_s	specific surface area of packing, m^2/m^3 (Eq. 9.15)
b	parameter in Equation 9.8, —
$Bo = \frac{(\rho_L - \rho_G) g l^2}{\sigma}$	Bond number, —
C	parameter in Equation 9.36, —
C_1	curvature parameter in Equation 9.40, —
$Ca = \frac{\mu u}{\sigma}$	capillary number, —
$Ca_B = \frac{\mu u_B}{\sigma}$	capillary number based on the liquid properties and the gas bubble velocity, —
$Ca_{GL} = \frac{\mu u_{GL}}{\sigma}$	capillary number based on the liquid properties and the sum of the superficial gas and liquid velocities, —
C_{in}	inlet concentration, mol/m^3
C_s	limiting reactant concentration near the surface, mol/m^3
D_{ax}	axial dispersion coefficient, m^2/s
D_{eff}	effective diffusivity, m^2/s
D_L	liquid-phase diffusivity, m^2/s
D_m	diffusion coefficient, m^2/s
d, d_t	channel diameter, m
$\bar{d}$	mean channel diameter
d_p	particle diameter, m

E_a	activation energy of the reaction, J/mol
f	bubble formation frequency, s^{-1}
F	volumetric flow rate, m^3/s
g	acceleration due to gravity, m/s^2
G	mass flux, $kg/(m^2{\cdot}s)$
$Ga = \frac{\rho^2 g d_p^3}{\mu^2}$	Galileo number, —
$Gz_S = \frac{L_S D_L}{d_t^2 u_B}$	Graetz number based on slug length, —
$Gz_t = \frac{L_t(1-\varepsilon_G)D_L}{d_t^2 u_B}$	Graetz number based on tube length, —
h	channel height, m
ΔH_r	reaction enthalpy, J/mol
k	kinetic constant, 1/s
k_g	mass transfer coefficient (gas–solid), m/s
$k_L a$	mass transfer coefficient (gas–liquid), m/s
L	channel or slug length, m
l	characteristic length, m
L_{nose}	length of the bubble nose part, m
L_{tail}	length of the bubble tail part, m
L_z	length of the reactor entrance zone, m
m	reaction order, —
n	number of channels, —
$Pe = \frac{Lu}{D_{ax}}$	Péclet number, —
Δp	pressure drop, Pa
R	ideal gas constant, 8.3145 J/mol·K
$\mathrm{Re} = \frac{\rho u d}{\mu}$	Reynolds number, —
$\mathrm{Re}_{GL} = \frac{\rho u_{GL} d}{\mu}$	Reynolds number based on the sum of the superficial gas and liquid velocities, —
$\mathrm{Re}_L = \frac{\rho u_L d}{\mu}$	Reynolds number based on the superficial liquid velocity, —
r_{eff}	measured (effective) reaction rate, $mol/m^3{\cdot}s$
$Sc = \frac{\mu\rho}{D_m}$	Schmidt number, —
$Sh = \frac{k_L d}{D_m}$	Sherwood number, —
$\bar{T}$	average temperature, K
$t_{D,ax} = \frac{L^2}{D_m}$	characteristic axial diffusion time, s
$t_{D,rad} = \frac{d_t^2}{4D_m}$	characteristic radial diffusion time, s
T_i	local temperature (at location *i*), K
T_s	surface temperature, K
u	superficial velocity, m/s
u_B	bubble velocity, m/s
u_F	liquid film average velocity, m/s
u_{GL}	the sum of the superficial gas and liquid velocities, m/s
u_L	liquid velocity, m/s
V	volume, m^3
V_p	volume between particles in a reactor, m^3
w	channel width, m
w_B	width of the bubble in Equation 9.18, μm
$We = \frac{U^2 d\rho}{\sigma}$	Weber number, —
w_{in}	width of the side channel in a mixer, m
z	axial coordinate, m

Greek letters

$\gamma = E_a/R\bar{T}$	temperature nonuniformity parameter, —
δ	additional length near the nose and tail parts of a bubble, m
δ_{cat}	thickness of catalytic coating, m
δ_L	liquid film thickness, m
ε	holdup, —
ε_b	overall porosity of the packed bed, —
η	wettability parameter in Equation 9.38, —
λ_{eff}	effective thermal conductivity of catalytic layer, W/m·K
μ	fluid viscosity, Pa·s
ξ	ratio of conversion in a nonisothermal microreactor with temperature nonuniformity parameter γ to that in an isothermal microreactor, —
ρ	density, kg/m^3
σ	surface tension, N/m
$\hat{\sigma}_d$	variation in the channel diameter, m
$\hat{\sigma}_d = \frac{\sigma_d}{d}$	relative variation in the channel diameter, —
$\bar{\tau} = \frac{\tau_0\left(1+6\hat{\sigma}_d^2\right)}{1+\hat{\sigma}_d^2}$	mean residence time in a nonideal microreactor, s
τ_0	mean residence time in an ideal microreactor (without variation in the channel diameter), s
ϕ_L	two-phase multiplier, —
χ	Lockhart–Martinelli parameter, —
$\psi = \delta_{cat}\frac{m+1}{2}\frac{r_{eff}}{D_{eff}C_s}$	Weisz modulus, —

Subscripts

B	bubble
C	bubble cap
c	continuous phase
$CENT$	in the channel center
$CORN$	in the channel corner
d	dispersed phase
F	film
G	gas
GL	two-phase (gas–liquid)

i	channel number
I	interface
L	liquid
s	slug
t	tube
TP	two-phase (liquid–liquid)
w	liquid-to-wall

References

1 Farina V, Reeves JT, Senanayake CH, Song JJ. Asymmetric synthesis of active pharmaceutical ingredients. Chem. Rev. 2006; 106(7): 2734–2793.

2 Rebrov EV. Sol–gel synthesis of zeolitic coatings and their application in catalytic microstructured reactors. Catal. Ind. 2009; 1(1): 322–347.

3 Rebrov EV, Ismagilov IZ, Ekatpure RP, de Croon MHJM, Schouten JC. Header design for flow equalization in microstructured reactors. AIChE J. 2007; 53(1): 28–38.

4 Rebrov EV, Ekatpure RP, de Croon MHJM, Schouten JC. Design of a thick-walled screen for flow equalization in microstructured reactors. J. Micromech. Microeng. 2007; 17(3): 633–641.

5 Rebrov EV, Schouten JC, de Croon MHJM. Single-phase fluid flow distribution and heat transfer in microstructured reactors. Chem. Eng. Sci. 2011; 66(7): 1374–1393.

6 Taylor G. Dispersion of soluble matter in solvent flowing slowly through a tube. Proc. R. Soc. A 1953; 219(1137): 186–203.

7 Aris R. On the dispersion of a solute in a fluid flowing through a tube. Proc. R. Soc. A 1956; 235: 67–77.

8 Rouge A. Periodic operation of a microreactor for heterogeneously catalyzed reactions: The dehydration of isopropanol. Ph.D. thesis. EPFL, Lausanne, Switzerland, 2001.

9 Delsman ER, De Croon MHJM, Elzinga GD, Cobden PD, Kramer GJ, Schouten JC. The influence of differences between microchannels on micro reactor performance. Chem. Eng. Technol. 2005; 28(3): 367–375.

10 Mears DE. Tests for transport limitations in experimental catalytic reactors. Ind. Eng. Chem. Process. Des. Dev. 1971; 10(4): 541–547.

11 Levenspiel O. Chemical Reaction Engineering. New York: John Wiley & Sons, Inc.; 1999.

12 Ajmera SK, Delattre C, Schmidt MA, Jensen KF. Microfabricated differential reactor for heterogeneous gas phase catalyst testing. J. Catal. 2002; 209(2): 401–412.

13 Hoebink JHBJ, Marin GB. Modeling of monolithic reactors for automotive exhaust gas treatment, in: Structured Catalysts and Reactors, (Eds.) Cybulki A, Moulijn JA. New York: Marcel Dekker; 1998.

14 Anderson JB. A criterion for isothermal behavior of a catalyst pellet. Chem. Eng. Sci. 1963; 18(2): 147–148.

15 Fedorov AG, Viskanta R. Three-dimensional conjugate heat transfer in the microchannel heat sink for electronic packaging. Int. J. Heat Mass Transf. 2000; 43(3): 399–415.

16 Lee PS, Garimella SV, Liu D. Investigation of heat transfer in rectangular microchannels. Int. J. Heat Mass Transf. 2005; 48(9): 1688–1704.

17 Ma Z, Zaera F. Heterogeneous catalysis by metals, in: Heterogeneous Catalysis by Metals. Encyclopedia of Inorganic Chemistry 2nd ed., (Ed.) King RB. Chichester: John Wiley & Sons, Ltd; 2005.

18 Hessel V, Angeli P, Gavriilidis A, Lowe H. Gas–liquid and gas–liquid-solid microstructured reactors: Contacting principles and applications. Ind. Eng. Chem. Res. 2005; 44(25): 9750–9769.

19 Van Herk D, Kreutzer MT, Makkee M, Moulijn JA. Scaling down trickle bed reactors. Catal. Today 2005; 106(1–4): 227–232.

20 Márquez N, Castaño P, Makkee M, Moulijn JA, Kreutzer MT. Dispersion and holdup in multiphase packed bed microreactors. Chem. Eng. Technol. 2008; 31(8): 1130–1139.

21 Márquez NMML. Hydrodynamics of multi-phase packed bed micro-reactors. Ph.D. thesis. Delft University of Technology, Ridderprint, Ridderkerk, the Netherlands, 2010.

22 Márquez N, Castaño P, Moulijn JA, Makkee M, Kreutzer MT. Transient behavior and stability in miniaturized multiphase packed bed reactors. Ind. Eng. Chem. Res. 2010; 49(3): 1033–1040.

23 Losey MW, Schmidt MA, Jensen KF. Microfabricated multiphase packed-bed reactors: Characterization of mass transfer and reactions. Ind. Eng. Chem. Res. 2001; 40(12): 2555–2562.

24 Charpentier JC, Favier M. Some liquid holdup experimental data in trickle-bed reactors for foaming and nonfoaming hydrocarbons. AIChE J. 1975; 21(6): 1213–1218.

25 Tadepalli S, Halder R, Lawal A. Catalytic hydrogenation of *o*-nitroanisole in a microreactor: Reactor performance and kinetic studies. Chem. Eng. Sci. 2007; 62(10): 2663–2678.

26 Zimmerman SP, Ng KM. Liquid distribution in trickling flow trickle-bed reactors. Chem. Eng. Sci. 1986; 41(4): 861–866.

27 Chu CF, Ng KM. Flow in packed tubes with a small tube to particle diameter ratio. AIChE J. 1989; 35(1): 148–158.

28 Chander A, Kundu A, Bej SK, Dalai AK, Vohra DK. Hydrodynamic characteristics of cocurrent upflow and downflow of gas and liquid in a fixed bed reactor. Fuel 2001; 80(8): 1043–1053.

29 Colombo AJ, Baldi G, Sicardi S. Solid–liquid contacting effectiveness in trickle bed reactors. Chem. Eng. Sci. 1976; 31(12): 1101–1108.

30 Kulkarni RR, Wood J, Winterbottom JM, Stitt EH. Effect of fines and porous catalyst on hydrodynamics of trickle bed reactors. Ind. Eng. Chem. Res. 2005; 44(25): 9497–9501.

31 Iliuta I, Thyrion FC, Muntean O, Giot M. Residence time distribution of the liquid in gas–liquid cocurrent upflow fixed-bed reactors. Chem. Eng. Sci. 1996; 51(20): 4579–4593.

32 Bej SK, Dabral RP, Gupta PC, Mittal KK, Sen GS, Kapoor VK, Dalai AK. Studies on the performance of a microscale trickle bed reactor using different sizes of diluent. Energy Fuels 2000; 14(3): 701–705.

33 Gunjal PR, Kashid MN, Ranade VV, Chaudhari RV. Hydrodynamics of trickle-bed reactors: Experiments and CFD modeling. Ind. Eng. Chem. Res. 2005; 44(16): 6278–6294.

34 Saroha AK, Nandi I. Pressure drop hysteresis in trickle bed reactors. Chem. Eng. Sci. 2008; 63(12): 3114–3119.

35 Shao N, Gavriilidis A, Angeli P. Flow regimes for adiabatic gas–liquid flow in microchannels. Chem. Eng. Sci. 2009; 64(11): 2749–2761.

36 Rebrov EV. Two-phase flow regimes in microchannels. Theor. Found. Chem. Eng. 2010; 44(4): 355–367.

37 Zhao TS, Bi QC. Co-current air–water two-phase flow patterns in vertical triangular microchannels. Int. J. Multiphase Flow 2001; 27 (5): 765–782.

38 Serizawa A, Feng Z, Kawara Z. Two-phase flow in microchannels. Exp. Thermal Fluid Sci. 2002; 26(6–7): 703–714.

39 Triplett KA, Ghiaasiaan SM, Abdel-Khalik SI, Sadowski DL. Gas–liquid two-phase flow in microchannels, part I: Two-phase flow patterns. Int. J. Multiphase Flow 1999; 25(3): 377–394.

40 Haverkamp V, Hessel V, Löwe H, Menges G, Warnier MJF, Rebrov EV, de Croon MHJM, Schouten JC, Liauw M. Hydrodynamics and mixer-induced bubble formation in microbubble columns with single and multiple channels. Chem. Eng. Technol. 2006; 29(9): 1015–1026.
41 Hassan I, Vaillancourt M, Pehlivan K. Two-phase flow regime transitions in microchannels: A comparative experimental study. Microscale Thermophys. Eng. 2005; 9(2): 165–182.
42 Chung PMY, Kawaji M. The effect of channel diameter on adiabatic two-phase flow characteristics in microchannels. Int. J. Multiphase Flow 2004; 30(7–8): 735–761.
43 Jones OC Jr, Zuber N. Slug-annular transition with particular reference to narrow rectangular ducts, in: Two-Phase Momentum, Heat and Mass Transfer in Chemical, Process and Energy Engineering Systems, (Eds.) Durst F, Tsiklauri GV, Afgan NH. Washington, DC: Hemisphere; 1979, vol. 1.
44 Akbar MK, Plummer DA, Ghiaasiaan SM. On gas–liquid two-phase flow regimes in microchannels. Int. J. Multiphase Flow 2003; 29(5): 855–865.
45 Warnier M. Taylor flow hydrodynamics in gas–liquid-solid micro reactors. Ph.D. thesis. Eindhoven University of Technology, Eindhoven, the Netherlands, 2009.
46 Damianides CA, Westwater JW. Two-Phase flow patterns in a compact heat exchanger and in small tubes. Proceedings of the 2nd UK National Conference on Heat Transfer, Glasgow: Mechanical Engineering Publications; September 14–16, 1988. p 1257.
47 Yang CY, Shieh CC. Flow pattern of air–water and two-phase R-134a in small circular tubes. Int. J. Multiphase Flow 2001; 27(7): 1163–1177.
48 Thulasidas TC, Abraham MA, Cerro RL. Bubble-train flow in capillaries of circular and square cross section. Chem. Eng. Sci. 1995; 50 (2): 183–199.
49 Laborie S, Cabassud C, Durand-Bourlier L, Lainé JM. Characterisation of gas–liquid two-phase flow inside capillaries. Chem. Eng. Sci. 1999; 54(23): 5723–5735.
50 Liu H, Vandu CO, Krishna R. Hydrodynamics of Taylor flow in vertical capillaries: Flow regimes, bubble rise velocity, liquid slug length, and pressure drop. Ind. Eng. Chem. Res. 2005; 44(14): 4884–4897.
51 Qian D, Lawal A. Numerical study on gas and liquid slugs for Taylor flow in a T-junction microchannel. Chem. Eng. Sci. 2006; 61(23): 7609–7625.
52 Abiev RSh. Method for calculating the void fraction and relative length of bubbles under slug flow conditions in capillaries. Theor. Found. Chem. Eng. 2010; 44(1): 86–101.
53 Gordillo JM, Ganan-Calvo AM, Perez-Saborid M. Monodisperse microbubbling: Absolute instabilities in coflowing gas–liquid jets. Phys. Fluids 2001; 13(12): 3839–3842.
54 Garstecki P, Fuerstman MJ, Stone HA, Whitesides GM. Formation of droplets and bubbles in a microfluidic T-junction—scaling and mechanism of break-up. Lab Chip 2006; 6(3): 437–446.
55 Pohorecki R, Kula K. A simple mechanism of bubble and slug formation in Taylor flow in microchannels. Chem. Eng. Res. Des. 2008; 86(9a): 997–1001.
56 Van Steijn V, Kreutzer MT, Kleijn CR. μ-PIV study of the formation of segmented flow in microfluidic T-junctions. Chem. Eng. Sci. 2007; 62(24): 7505–7514.
57 Van Steijn V, Kleijn CR, Kreutzer MT. Predictive model for the size of bubbles and droplets created in microfluidic T-junctions. Lab Chip 2010; 10(19): 2513–2518.
58 De Menech M, Garstecki P, Jousse F, Stone HA. Transition from squeezing to dripping in a microfluidic T-shaped junction. J. Fluid Mech. 2008; 595: 141–161.
59 Abadie T, Aubin J, Legendre D, Xuereb C. Hydrodynamics of gas–liquid Taylor flow in rectangular microchannels. Microfluid. Nanofluid. 2012; 12(1–4): 355–369.
60 Xu JH, Li SW, Tan J, Luo GS. Correlations of droplet formation in T-junction microfluidic devices: From squeezing to dripping. Microfluid. Nanofluid. 2008; 5(6): 711–717.
61 Fu T, Ma Y, Funfschilling D, Zhu C, Li HZ. Squeezing-to-dripping transition for bubble formation in a microfluidic T-junction. Chem. Eng. Sci. 2010; 65(12): 3739–3748.
62 Fairbrother F, Stubbs AE. Studies in electro-endosmosis, part VI: The Bubble-tube method of measurement. J. Chem. Soc. 1935; 1: 527–529.
63 Bretherton FP. The motion of long bubbles in tubes. J. Fluid Mech. 1961; 10(2): 166–188.
64 Aussillous P, Quéré D. Quick deposition of a fluid on the wall of a tube. Phys. Fluids 2000; 12(10): 2367–2371.
65 Heil M. Finite Reynolds number effects in the Bretherton problem. Phys. Fluids 2001; 13(9): 2517–2521.
66 Han Y, Shikazono N. Measurement of the liquid film thickness in micro tube slug flow. Int. J. Heat Fluid Flow 2009; 30(5): 842–853.
67 Hazel AL, Heil M. The steady propagation of a semi-infinite bubble into a tube of elliptical or rectangular cross-section. J. Fluid Mech. 2002; 470: 91–114.
68 Kuzmin A, Januszewski M, Eskin D, Mostowfi F, Derksen JJ. Three-dimensional binary-liquid lattice Boltzmann simulation of microchannels with rectangular cross sections. Chem. Eng. J. 2011; 178: 306–316.
69 Han Y, Shikazono N. Measurement of liquid film thickness in micro square channel. Int. J. Multiphase Flow 2009; 35(10): 896–903.
70 Warnier MJF, Rebrov EV, De Croon MHJM, Hessel V, Schouten JC. Gas hold-up and liquid film thickness in Taylor flow in rectangular micro channels. Chem. Eng. J. 2008; 135S(1): 153–158.
71 Warnier MJF, De Croon MHJM, Rebrov EV, Schouten JC. Pressure drop of gas–liquid Taylor flow in round micro capillaries for low to intermediate Reynolds numbers. Microfluid. Nanofluid. 2010; 8(1): 33–45.
72 Lockhart RW, Martinelli RC. Proposed correction of data for isothermal two-phase component flow in pipes. Chem. Eng. Prog. 1949; 45(1): 39–48.
73 Chisholm D. A theoretical basis for the Lockhart–Martinelli correlation for two-phase flow. Int. J. Heat Mass Transf. 1967; 10(12): 1767–1778.
74 Mishima K, Hibiki T. Some characteristics of air-water two-phase flow in small diameter vertical tubes. Int. J. Multiphase Flow 1996; 22(4): 703–712.
75 Ju Lee H, Yong Lee S. Pressure drop correlations for two-phase flow within horizontal rectangular channels with small heights. Int. J. Multiphase Flow 2001; 27(5): 783–796.
76 Sairson S, Wongwises S. The effects of channel diameter on flow pattern, void fraction and pressure drop of two-phase air-water flow in circular micro-channels. Exp. Thermal Fluid Sci. 2010; 34(4): 454–462.
77 Hwang YW, Kim MS. The pressure drop in microtubes and the correlation development. Int. J. Heat Mass Transf. 2006; 49(11–12): 1804–1812.

78 Sun L, Mishima K. Evaluation analysis of prediction methods for two-phase flow pressure drop in mini-channels. Int. J. Multiphase Flow 2009; 35(1): 47–54.
79 Salim A, Fourar M, Pironon J, Sausse J. Oil–water two-phase flow in microchannels: Flow patterns and pressure drop measurements. Can. J. Chem. Eng. 2008; 86(6): 978–988.
80 Jovanovic J, Rebrov EV, Nijhuis TA, Hessel V, Schouten JC. Liquid-liquid slug flow: Hydrodynamics and pressure drop. Chem. Eng. Sci. 2011; 66(1): 42–54.
81 Yue J, Chen G, Yuan Q, Luo L, Gonthier Y. Hydrodynamics and mass transfer characteristics in gas–liquid flow through a rectangular microchannel. Chem. Eng. Sci. 2007; 62(7): 2096–2108.
82 Haase S, Bauer T. New method for simultaneous measurement of hydrodynamics and reaction rates in a mini-channel with Taylor flow. Chem. Eng. J. 2011; 176–177: 65–74.
83 Sun R, Cubaud T. Dissolution of carbon dioxide bubbles and microfluidic multiphase flows. Lab Chip 2011; 11(17): 2924–2928.
84 Tan J, Lu YC, Xu JH, Luo GS. Mass transfer performance of gas–liquid segmented flow in microchannels. Chem. Eng. J. 2012; 181–182: 229–235.
85 Eskin D, Mostowfi F. A model of a bubble train flow accompanied with mass transfer through a long microchannel. Int. J. Heat Fluid Flow 2012; 33(1): 147–155.
86 Van Baten JM, Krishna R. CFD simulations of mass transfer from Taylor bubbles rising in circular capillaries. Chem. Eng. Sci. 2004; 59(12): 2535–2545.
87 Pohorecki R. Effectiveness of interfacial area for mass transfer in two-phase flow in microreactors. Chem. Eng. Sci. 2007; 62(22): 6495–6498.
88 Vandu CO, Liu H, Krishna R. Mass transfer from Taylor bubbles rising in single capillaries. Chem. Eng. Sci. 2005; 60(22): 6430–6437.
89 Sobieszuk P, Cygański P, Pohorecki R. Volumetric liquid side mass transfer coefficient in a gas–liquid microreactor. Chem. Process Eng. (Inz. Chem. Procesowa) 2008; 29(3): 651–661.
90 Yue J, Luo L, Gonthier Y, Chen G, Yuan Q. An experimental study of air-water Taylor flow and mass transfer inside square microchannels. Chem. Eng. Sci. 2009; 64(16): 3697–3708.
91 Van Baten JM, Krishna R. CFD simulations of wall mass transfer for Taylor flow in circular capillaries. Chem. Eng. Sci. 2005; 60(4): 1117–1126.
92 Berčič G, Pintar A. The role of gas bubbles and liquid slug lengths on mass transport in the Taylor flow through capillaries. Chem. Eng. Sci. 1997; 52(21–22): 3709–3719.
93 Rebrov EV. Microwave assisted organic synthesis in microstructured reactors. Russ. J. Gen. Chem. 2012; 82(12): 2060–2069.
94 Houlding T, Rebrov EV. Application of alternative energy forms in catalytic reactor engineering. Green Processes. Synth. 2012; 1(1): 19–31.
95 Houlding T, Tchabanenko K, Rahman MdT, Rebrov EV. Direct amide formation using radiofrequency heating. Org. Biomol. Chem. 2013; 11(25): 4171–4177.
96 Patil NG, Benaskar F, Rebrov EV, Meuldijk J, Hulshof LA, Hessel V, Schouten JC. Continuous multi-tubular milli-reactor with a Cu thin film for microwave assisted fine-chemicals synthesis. Ind. Eng. Chem. Res. 2012; 51(44): 14344–14354.
97 Frauhammer J, Friedrich G, Kolios G, Klingel T, Eigenberger G, von Hippel L, Arntz D. Flow distribution concepts for new type monolithic co- or countercurrent reactors. Chem. Eng. Technol. 1999; 22(12): 1012–1016.
98 Frauhammer J, Eigenberger G, von Hippel L, Arntz D. A new reactor concept for endothermic high-temperature reactions. Chem. Eng. Sci. 1999; 54(15–16): 3661–3670.
99 Kolb G, Schurer J, Tiemann D, Wichert M, Zapf R, Hessel V, Lowe H. Fuel processing in integrated micro-structured heat-exchanger reactors. J. Power Sources 2007; 171(1): 198–204.
100 Dubrovskiy AR, Rebrov EV, Kuznetsov SA, Schouten JC. A microstructured reactor/heat-exchanger for the water gas shift reaction operated in the 533–673 K range. Catal. Today 2009; 147(S1): S198–S203.
101 Rebrov EV. Advances in low temperature water-gas shift technology: Modern catalysts and improved reactor concept, in: Clean Hydrocarbon Fuel Conversion Technology, (Ed.) Rashid Khan M. Woodhead Publishing, Cambridge, UK, 2011, pp. 387–412, ISBN 978-1-84569-727-3.
102 Schmidt EW. Hydrazine and Its Derivatives: Preparation, Properties, Applications. New York: Wiley-Interscience; 2001, vol. 1.
103 Ismagilov IZ, Michurin EM, Sukhova OB, Tsykoza LT, Matus EV, Kerzhentsev MA, Ismagilov ZR, Zagoruiko AN, Rebrov EV, de Croon MHJM, Schouten JC. Oxidation of organic compounds in a microstructured catalytic reactor. Chem. Eng. J. 2008; 135(1): 57–65.
104 Ehrich H, Linke D, Morgenschweis K, Baerns M, Jahnisch K. Application of microstructured reactor technology for the photochemical chlorination of alkylaromatics. Chimia Int. J. Chem. 2002; 56(11): 647–653.
105 Muller A, Cominos V, Hessel V, Horn B, Schurer J, Ziogas A, Jahnisch K, Hillmann V, Großer V, Jam KA, Bazzanella A, Rinke G, Kraut M. Fluidic bus system for chemical process engineering in the laboratory and for small-scale production. Chem. Eng. J. 2005; 107(1–3): 205–214.
106 Yeong KK, Gavriilidis A, Zapf R, Hessel V. Catalyst preparation and deactivation issues for nitrobenzene hydrogenation in a microstructured falling film reactor. Catal. Today 2003; 81(4): 641–651.
107 Rebrov EV, Duisters T, Lob P, Meuldijk J, Hessel V. Enhancement of the liquid-side mass transfer in a falling film catalytic microreactor by in-channel mixing structures. Ind. Eng. Chem. Res. 2012; 51(26): 8719–8725.

PART 5

Essential tools of reactor modeling and design

CHAPTER 10

Experimental methods for the determination of parameters

Rebecca R. Fushimi[1], John T. Gleaves[2] and Gregory S. Yablonsky[3]
[1]Materials Science & Engineering Department, Idaho National Laboratory, Idaho Falls, ID, USA
[2]Department of Energy, Environmental and Chemical Engineering, Washington University, St. Louis, MO, USA
[3]Parks College of Engineering, Aviation and Technology, Saint Louis University, St. Louis, MO, USA

Abstract

The determination of kinetic parameters is a key element in catalyst research and multiphase catalytic process development. In this chapter, the experimental methods used to determine kinetic parameters, the nature of the parameters, and the areas of application and advantages of the different experimental approaches are discussed. Emphasis is placed on transient techniques and methods that provide intrinsic kinetic parameters, that is, parameters that can be directly linked to the surface composition/structure of a heterogeneous catalyst.

10.1 Introduction

Reactions in multiphase processes occur at an interface. The chemical properties of the interface give rise to kinetic processes such as adsorption and surface reaction, and kinetic parameters are characteristics of these processes. Kinetic parameters provide a quantitative link between the rate of a reaction, the temperature, and the chemical properties of the interface. The degree to which experimentally determined kinetic parameters can be related to the composition and structure of the interface depends on the interface and the experimental approach. There are three main experimental approaches, experiments in flow reactors, surface science experiments, and a newer approach, the Temporal Analysis of Products (TAP) approach which combines elements of the both flow reactors and surface science techniques.

In flow reactors, reaction conditions more closely resemble that of the larger-scale process and the objectives are typically to measure steady-state kinetic properties for a specific gas-phase composition, temperature, flow rate, etc. In addition to being a function of its preparation and pretreatment, the steady-state kinetic parameters of a technical catalyst are a function of its reaction history and the reaction conditions during the steady-state experiment. Changes in reaction conditions can change the composition and structure of a sample, which in turn changes its kinetic properties. Physical characterization of a sample can be performed before and after kinetic characterization using a number of different spectroscopic techniques. Generally, structure/properties relationships are difficult to establish using the steady-state approach. Transient flow experiments such as steady-state isotopic transient kinetic analysis (SSITKA) can provide information about surface adspecies and lifetimes. A pulsed experiment in a *microcatalytic* reactor configuration allows one to observe how a material changes in its approach to a steady operating state.

With the surface science approach reactions occur at a vacuum–solid interface on an atomically ordered surface maintained at ultrahigh-vacuum (UHV) conditions. The objective is to accurately define the structure and composition of the surface, measure its chemical or kinetic properties, and establish the relationship between structure and properties. Kinetic parameters determined under UHV conditions using molecular beam scattering (MBS) and thermal desorption experiments on an atomically ordered surface can be directly related to elementary processes and can be used to construct detailed models that describe how different atomic or molecular species interact with the surface. In recent years, studies have moved beyond single crystals, and a number of groups have fabricated model catalysts with greater complexity to more closely represent the important features of technical materials [1, 2].

With the TAP approach reactions also occur at a vacuum–solid interface. However, unlike the surface science approach, the solid surface in a TAP experiment is generally a disordered "technical" material such as industrial catalyst particles that are generally studied in flow reactors. The objective of a TAP experiment is to measure nonsteady-state kinetic properties and track how changes in reaction conditions and sample composition alter these properties. Incrementally increasing or decreasing the amount of a single component while simultaneously making kinetic measurements charts the correspondence between a component

Multiphase Catalytic Reactors: Theory, Design, Manufacturing, and Applications, First Edition. Edited by Zeynep Ilsen Önsan and Ahmet Kerim Avci.

concentration and the kinetic properties of the sample. This procedure provides a method of identifying the active components in a complex multicomponent catalyst and determining how different components influence individual steps in a multistep process.

In this work our presentation of well-known techniques will be brief since they are thoroughly documented in the literature. Toward the end of this chapter, the TAP approach is introduced, which is not widely known in the industry and incorporates many of the advantages from more commonly used techniques such as flow reactors, transient operation, and MBS.

10.2 Consideration of kinetic objectives

The experimental options for obtaining kinetic information on catalytic systems range in complexity and cost depending on the nature of information sought after. If the goal is to optimize an existing process and simply capture the activity, selectivity, and catalyst life of an industrial sample, then, in general, the steady-state differential flow reactor is the simplest and the most cost-effective choice. Here one might capture a rate expression to optimize process conditions for process scale-up. For the most part, technical catalysts are developed using a trial-and-error approach. Technical catalysts are routinely screened for activity and selectivity in microreactor libraries, and the performance feedback guides the catalyst development. When a catalyst meets specific performance criteria, then steady-state kinetic parameters are used to construct transport–kinetic models needed for the design of commercial-scale processes.

The properties of technical catalytic materials are formed kinetically under unsteady-state conditions as they are the product of an evolutionary process that emerges as a result of a complex interaction of physical and chemical conditions. Controlling factors include the preparation procedure, starting materials, reaction conditions, chemical and thermal pretreatments, etc. Exposure to a reactant mixture while a material is being tested can also alter the composition and structure of a catalyst [3]. Even the adsorption of simple molecules can induce structural and compositional changes that can alter the kinetic properties.

While flow reactors operate at conditions that closely mirror the operating environment witnessed in practice, these reactors can typically only offer a global kinetic description and lack the details of elementary reaction steps and mechanism that reveal how materials operate on a more fundamental level. If the goal is catalyst development, then one typically needs more detailed rate expressions that describe elementary reaction steps for the development of a mechanism. The goal is to understand not only more than just the properties of the catalyst but also how it functions. In order to understand such details, transient experiments will provide the most insight into the elementary steps as well as secondary processes (e.g., surface/bulk diffusion) that make up the complex catalytic system.

10.3 Criteria for collecting kinetic data

In general, a number of criteria can be identified that the ideal kinetic measurement might embody:

1 The experiment must be isothermal and uniform in chemical composition (i.e., present perfect mixing). Temperature gradients may be minimized by intensive heat exchange, dilution of the active material, or rapid recirculation of fluid phase. The reaction zone may be minimized in order to render the reactor "gradientless," such as in the cases of differential plug flow reactor (Section 10.4.1.1) or thin-zone TAP reactor (TZTR) (Section 10.6.1.2).
2 The technique must provide the ability to separate transport effects, such as pore diffusional resistance, and film mass and heat transfer, from kinetic effects. Transport should be well defined and the measurement should be conducted in a regime where surface reaction is controlling.
3 The kinetic measurement should be made in the absence of deactivating effects (e.g., poisoning, coking, sintering). The catalyst should not change during the measurement. It is an advantage, however, if the catalyst can be incrementally changed and tested to observe how the system evolves. The evolution of a material from one kinetic state to another can reveal processes (e.g., surface-to-bulk diffusion) that represent the interdependence of multiple parts of a complex system.
4 Kinetic data should be collected over a wide range of temperature and reactant concentrations for comparison to other studies and translation to a working environment.

10.4 Experimental methods

In this work, the most commonly used bench-scale fixed-bed flow reactors, the nature of the kinetic data that they provide, and their advantages and limitations will first be highlighted. Such techniques (e.g., using fixed-bed tubular reactors and basket-type mixed gas reactors at steady state) are well documented and hence will not be discussed in detail [4–8]. The same reactors operated in a transient mode (e.g., SSITKA (Section 10.4.2.1) or pulsed tracer flow) offer additional detail of catalyst storage capacity and information about number of working sites. We wish to compare the nature of kinetic data provided by reactors operating in a convective-flow mode (steady-state or transient) with kinetic data obtained from low-pressure surface science-type experiments. The microkinetic approach is briefly discussed, followed by a relatively new technique known as TAP, which complements the microkinetic approach encompassing features from both convective-flow reactors and surface science techniques. The TAP approach provides a new method for kinetic analysis which specifically allows one to study how the catalyst system *evolves* from one state to another.

10.4.1 Steady-state flow experiments

Bench- and laboratory-scale flow reactors provide an inexpensive and simple means to collect kinetic data for a catalytic process and hence are nearly a ubiquitous entity in catalyst development. The tubular plug flow reactor is most commonly used in the differential configuration for extracting kinetic data, while the integral configuration is useful for closely mimicking process conditions. Fixed-bed mixed gas recycle reactors such as the Berty reactor and spinning basket-type reactors similar to the Carberry design are commercially available and simple to use. Other reactor types such as stirred catalyst or fluid bed are available but may be more suited for specific applications. The use of these devices is well documented and discussed in numerous books [4, 6, 8]; thus the focus of the chapter will be on the nature of kinetic data from flow-type experiments in relationship to other techniques that are available.

Steady-state flow experiments are more often used for optimizing process conditions when the catalyst has been identified, for process scale-up, or to validate models rather than for catalyst development. One advantage is the ability to collect data over a wide range of conditions that closely mimic the larger-scale process. These experiments offer information about catalyst performance at a specific set of operating conditions (temperature, pressure, gas composition, space velocity, etc.), but do not give a detailed information about the fundamental processes that work together to comprise the catalytic system. Steady-state kinetic dependencies are usually related to the rate-limiting step(s) of a complex catalytic process. The complexity inherent in catalytic systems can be described as *emergent*, that is, the function of the system as a whole is distinct from the sum of its parts. In order to characterize the individual steps of a process, transient techniques are required.

For any given catalyst/reaction system a number of different reactor types can be selected to collect rate/concentration data. The choice is based on the process conditions, the thermodynamics of the process, the material properties of the catalyst, and the type of kinetic data that is desired.

10.4.1.1 Flow reactor configurations

Any tubular reactor can be operated either in integral or in differential mode. The fixed-bed integral reactor operates at high conversions (generally at high temperatures and low space velocities). As a result there is a large variation of the gas composition and rate along the length of the reactor, even when operated isothermally. Rates cannot be determined directly with an integral reactor. Such reactors are generally not useful for obtaining kinetic data that can be easily translated to other systems. This configuration is however useful for measuring overall conversion and catalyst life under realistic operating conditions. If the process operates at high conversions, then kinetic data directly from an integral reactor configuration might be desirable since this most closely resembles the process.

The same fixed-bed reactor operated at low conversions (generally low T and high space velocity) becomes a differential reactor where concentration and temperature gradients will be small. For practical matters, the catalyst particles must be small enough and the gas flow sufficiently high such that intra- and interparticle transport effects do not influence the observed kinetics. This can be determined by measuring the rates for a number of particle sizes. Ideal kinetic experiments will be conducted in a "gradientless" reactor that is isothermal and uniform in chemical composition. The differential operating mode is set up so that the rate can be considered constant at all points within the reactor. When this is the case, the tube becomes a continuous stirred-tank reactor (CSTR), and the analysis is greatly simplified.

Another experimental strategy aims to ensure perfect mixing by recirculating the reacting fluid at a very high rate. As a result the concentration of the circulating fluid becomes equal to that of the effluent stream. This strategy allows one to use a packed bed with behavior close to ideal mixed flow. In recycle reactors, small deviations from perfect mixing, for example, recycle flow rates that are too low, as well as heat and mass transport resistances can dramatically introduce error into the kinetic data. The Berty internal recycle reactor [9, 10] adopts a different configuration where the catalyst is fixed in a basket and an impeller rapidly recirculates the reacting gas. This configuration makes it possible to control the flow rate through the catalyst bed with greater accuracy, and as a result one can establish flow regimes with negligible heat and mass transfer resistances.

The Carberry reactor [11] presents similar experimental strategy to the Berty reactor but one where the catalyst bed is moved instead; the catalyst is trapped in a basket that spins through the fluid phase. The interparticle resistances can be reduced by spinning the basket at high speed and intraparticle diffusion effects minimized by using small catalyst particles. The basket-type reactor presents good fluid–solid mixing which helps to minimize thermal and concentration gradients. The internal recycle and basket-type reactors, like other differential reactors, allow one to measure the rate directly from concentration. These reactors are particularly useful for the study of deactivation and catalyst lifetime.

10.4.1.2 Kinetic analysis from flow reactors

Using flow reactors under steady-state conditions, we can easily collect data for process optimization, record activity, and selectivity and study catalyst life and deactivation processes. If we know the contacting pattern in the reactor, then we can explore the kinetics from the reactor performance equation. All of the flow reactors described previously present data for the average reactor concentration versus time. Generally the activity, selectivity, and stability are presented as a function of different process variables such as temperature, pressure, and space velocity. From conversion we can calculate the rate from the performance equation of the reactor, for example, for a CSTR:

$$\frac{V}{F_{A_0}} = \frac{X_A}{-r_A}, \tag{10.1}$$

for an ideal plug flow reactor:

$$\frac{V}{F_{A_0}} = \int_0^{X_{A_f}} \frac{dX_A}{-r_A}, \tag{10.2}$$

for recycle reactor:

$$\frac{V}{F_{A_0}} = (R+1)\int_0^{X_{A_f}} \frac{dX_A}{-r_A}, \tag{10.3}$$

where, $-r_A$ is rate of disappearance of species A, V is the reactor volume or catalyst volume, F_{A_0} is the initial molar flow rate of reactant A, X_A is the conversion of A, R is the recycle ratio or volume of gas returned divided by the volume of gas leaving the reactor.

The rate of reaction or production is the amount (moles or mass) of chemical species that is converted or formed per unit time per unit volume. The reaction rate is a function of both temperature and reactant concentrations:

$$-r_A = k(T)f(X_i) \tag{10.4}$$

The reaction rate constant, k, indicates the rate of the chemical reaction:

$$k = Ae^{(-E_a/RT)} \tag{10.5}$$

where A is the preexponential factor, E_a is the activation energy, R is the gas constant (8.314 J/mol K), and T is the temperature, K.

From the reactor data the Arrhenius dependence of $\ln(r_A)$ versus $1/T$ may be presented to give an *apparent* activation representing the overall catalytic reaction including many of the elementary reaction steps that precede the rate-determining step. This is the most basic analysis and it should be noted that changes in the rate-determining step, heats of adsorption, or equilibrium constants will give variations in the temperature dependence. Indeed the temperature dependence of the equilibrium constants is part of a complex Langmuir–Hinshelwood rate expression.

Since the concentration of reactant at the catalyst surface is highly temperature dependent, the apparent activation energy may not be an accurate representation of the true activation energy and should be used with caution to represent the catalytic activity. If transport resistances are not well accounted (for pore diffusion or film mass transfer), then the observed activation energy will be much lower than the true activation energy of a chemically controlled process since it will include the temperature dependence of the transport processes.

The turnover frequency (TOF) can be used as another measure of catalytic activity. It is simply the number of times an overall catalytic reaction occurs per catalytic site per unit time at a fixed set of reaction conditions (temperature, concentration, etc.):

$$\text{TOF} = \frac{\text{Molecules of product}}{\text{Number of active sites} \times \text{time}} = \frac{1}{S}\frac{dn}{dt} \tag{10.6}$$

In general it is difficult to determine the number of active sites (S) and thus it's often represented as the total surface area of the exposed catalyst, which will represent the lower limit of the TOF.

Even with well-controlled flow reaction conditions, often only the apparent activation energy and TOF can be determined from flow experiments. Apparent activation energies and TOFs are useful for screening samples performing the same chemistry at identical operating conditions. Transient flow experiments such as SSITKA (Section 10.4.2.1) and MBS experiments (Section 10.4.3.1) offer different routes for experimental measurement of the TOF for comparison. If indeed the flow experiment is performed in a chemically controlled regime, then from the rate data one may calculate rate constants from elementary reactions and plot $\ln k$ versus $1/T$ to determine the true activation energy. In order to determine the rate expression or rate model from the experimental data, one can use several methods: integral analysis (section "Integral Analysis of Experimental Flow Data"), differential analysis (section "Differential Analysis of Experimental Flow Data"), the method of initial rates [12, 13], the method of half-lives [12, 13], or microkinetic analysis [14]. Integral and differential analysis techniques are the most commonly used ones and will only be briefly presented here since the reader can refer to most chemical engineering textbooks.

10.4.1.2.1 Integral analysis of experimental flow data

For a packed-bed reactor at a fixed concentration and temperature, the mass of the catalyst, W, or the initial molar flow rate, F_{A_0}, should be varied to get a range of W/F_{A_0} at different values of conversion. Next a rate expression is tested using the performance equation for the reactor with the candidate kinetic expression inserted for the rate. A plot is then constructed with W/F_{A_0} data plotted against the integral $\int_0^x dX/(-r_A)$ (determined numerically) and tested for linearity. A judge of linearity is used to validate if the rate equation can be used to represent the experimental data. This method works well only for simple kinetic expressions (e.g., first-order irreversible); the integral forms of more complex rate expressions can become cumbersome. Since this method averages the rate, it cannot be relied upon when changes in reaction order occur with changes in concentration or temperature. Also, since there are an infinite number of rate expressions, a linear fit does not guarantee the correctness of the expression and, in general, several rate expressions may present an adequate fit. This analysis method requires some additional information of the reaction mechanism if the proper rate expression is to be determined.

10.4.1.2.2 Differential analysis of experimental flow data

For a packed-bed reactor at a fixed concentration and temperature, the mass of the catalyst, W, or the initial molar flow rate, F_{A_0}, should be varied to get a range of W/F_{A_0} at different values of conversion. The conversion versus W/F_{A_0} is plotted, and the data is fitted to a polynomial equation to give the curve of best fit. The curve of best fit is then differentiated and the derivative is determined at various intervals. The derivative represents the reaction rate

$$\text{Slope} = dX = (-r_A)\left[d\left(\frac{W}{F_{A_0}}\right)\right] \tag{10.7}$$

The reaction order is then determined from a log plot of the rate versus concentration. For example, for an irreversible reaction described by a power law $-r_A = kC^a$, a plot of $\ln(-r_A)$ obtained as derivatives from experimental data versus $\ln(C)$ will be linear with a slope a.

The differential method allows one to directly evaluate the rate at incremental values over the data set and can indicate if changes in reaction kinetics occur with changes in concentration and temperature. In reality, the experimental data is typically noisy, and the process of smoothing and differentiating may skew the results.

10.4.1.3 *Operando* spectroscopy

Kinetic information obtained from flow reactors can be greatly enhanced by observing conversion or reaction rate together with spectroscopic data on the catalyst bed. Such *operando* techniques can use methods such as UV/Vis, Raman, NMR, IR, XRD, etc. to present important information about surface intermediates [15, 16]. The chemical engineering aspects of the reactor must be carefully considered as spectroscopic cells often cannot be considered "kinetically appropriate" reactors [17]. For example, channeling, bypassing, nonisothermality, and deviations from plug flow can lead to unreliable kinetic data. Reactor design can be made simpler when using Raman and UV/Vis measurements where it is possible to transmit/collect light via a fiber-optic probe inserted into the reactor.

10.4.2 Transient flow experiments

Steady-state flow experiments are straightforward and useful for screening multiple samples for comparison of basic kinetic quantities but will typically reflect only the rate-limiting step(s) of a complex mechanism. Under steady-state conditions the elementary steps proceeding in series take place at the same rate. When a transient is imposed on the system, either the concentrations of intermediates will change according to the intrinsic reaction mechanism, or the changes in the rate constants of some elementary steps associated with side reactions may occur. The simplest transients to impose experimentally are concentration and temperature, and the responding change in the system results from the elementary steps of the process. More detailed information can be obtained by looking at the response of the system to a transient input. In this section we will present the more commonly used transient flow experiments, namely, SSITKA, step transient, pulsed transient, and microcalorimetry.

10.4.2.1 SSITKA

SSITKA experiments can be performed in plug flow or mixed flow reactors. This approach was proposed by Happel et al. [18] and further developed by Bennett [19], Biloen [20], and Shannon and Goodwin [21]. In these experiments a step change or pulsed input is induced in the isotopic label of one reactant in the reactant flow. The total concentration of labeled plus nonlabeled reactants, adsorbates, and products is maintained at steady state under isothermal and isobaric conditions. The reactor effluent species are then monitored versus time. The mean surface residence time and abundance of adsorbed surface intermediates leading to the product are determined by the incorporation of the isotopic elements of the reactant into the product species. These values are determined independent of the assumption of a kinetic model. For overall reactions with first-order kinetic dependence, the apparent kinetic constant may be directly determined. For reversible reactions this value will be underestimated. It is possible to obtain surface coverage of intermediates from separate measurements as well as estimates of the number of "working" active sites.

Product readsorption at reactive sites can lead to substantial contributions to the transient response, lowering the measured activity and reaction rate. Product readsorption at nonreactive sites will also inflate the measurement of surface intermediates leading to the observed product and overestimate the mean surface residence time. Effects of product readsorption can be addressed by decreasing the bed length or increasing the space velocity.

10.4.2.2 Step transients and pulse response in flow reactors

In a step transient experiment, two separate feeds are established and diverted to either the reactor or a vent line by switching a four-way valve. Here the feeds should be set up to contain equal total molar flow rates of inert and reaction mixture. Equal pressures at the outlets must be established by compensating for the reactor pressure drop using a needle valve at the vent exit. The four-way valve enables fast switching of one feed from the reactor to the vent, and vice versa, to create a step transient in the reactant concentrations. For highly exo- or endothermic reactions, a step change in reactant concentration may result in a significant change in catalyst temperature, making it difficult to maintain isothermal conditions. Experimentally, an ideal step change can be difficult to attain due to the switching time of the four-way valve and inertia in the sample lines. In addition, the time resolution of product sampling in flow reactors is generally too low to generate mechanistic detail.

A step transient experiment induces a change from one steady state to another, while a pulse transient experiment starts and ends with the same stationary state of the catalyst. Kobayashi and Kobayashi [22] proposed a transient response method based on a differential plug flow reactor. Müller and Hofman [23] have tested dynamic methods in packed-bed reactors including step transients and the studies of time derivatives. Tobin, Kokes, and Emmet [24, 25] presented a reactor where a microquantity of reactant is pulsed into an inert stream passed over a fixed bed of catalyst, the "microcatalytic" reactor. This mode of operation promises finite conversion with infinitesimal heat release/abstraction ensuring isothermal operation. When the reactor effluent is analyzed in a chromatographic column, the time scale of the experiment is extended which can make mechanism development more difficult. The authors indicate that the pulsing technique can be misleading if compared to reaction data obtained under steady-state conditions. For example, initial pulsing over materials could indicate initially lower/higher activity that would be achieved under steady-state conditions recorded in a flow reactor. A comprehensive review of micropulse techniques in small catalytic packed beds is presented by Christoffell [26].

The flow reactor offers kinetic data at only one particular set of conditions, that is, the gas composition, temperature, and flow rate that will yield a particular surface coverage of reacting species. The transient reactor, however, offers far more insight and allows one to study how materials are kinetically formed, that is, one can observe the evolution of the material as it progresses toward a steady-state. For example, one may observe the system as the surface coverage changes where the highest oxidation state may not yield the desired performance. A catalyst may perform best when it is slightly reduced which would require numerous steady-state observations. A transient experiment allows one to directly observe how the kinetics responds to changes in surface coverage. While transient kinetic data may not directly reflect operations at steady state, they offer far more detail for the complex link between properties such as surface coverage and kinetic performance.

10.4.2.3 Microcalorimetry

Microcalorimetry provides a kinetic measurement at thermodynamic equilibrium under isothermal conditions. The adsorption setup is based on the work of Spiewak and Dumesic [27]. In this experiment, small doses of adsorbing gas are expanded into a calorimeter which measures the resulting heat flow. The amount of adsorbed molecules is measured volumetrically. Measurements are very time consuming, and typically several days are needed for sample pretreatment, degassing, and reaching thermal equilibrium before an automatic dosing sequence begins. Calorimetric data and pressure data are collected simultaneously to present the differential heat of adsorption, Q, versus coverage and adsorption isotherms (coverage vs. pressure). Integration of the heat flow for each pulse yields the heat evolved.

10.4.3 Surface science experiments

The strategy of the surface science approach is to collect structural and kinetic data on well-defined model surfaces, develop structure–activity relationships, and compare the kinetic properties with those of technical surfaces containing the same components. In principle, a model surface is comparable to a technical surface if both surfaces exhibit the same kinetic properties. When properties are comparable, the operation of the model surface may provide insight into how the technical surface works and information to guide the development of new and/or improved catalysts.

Simple metal single crystals prepared under well-controlled conditions are extensively characterized with respect to their geometric and electronic properties [28, 29]. With techniques such as MBS, one can describe details of reaction mechanisms and obtain reaction probabilities and rate constants for elementary reaction steps, activation barriers for surface processes, and adsorption quantities [30–33]. Kinetic data from single crystal experiments may be vastly different from that of technical catalytic materials as the activity and selectivity may be greatly altered by changes in coordination number, morphology, and the interaction of the metal and support. In order to understand this interaction as well as the role of particle size and morphology more clearly, surface science experiments are extended to well-defined model systems for highly dispersed metal catalysts supported on oxides, for example, palladium metal mounted on single crystal oxide such as Al_2O_3, SiO_2, ZrO_2, and TiO_2 [28, 34–36]. Experiments using metals mounted on single crystal oxides allow one to capture the electronic properties, adsorption/desorption behavior, and catalytic activity of the technical system. Surface science techniques, however, find difficulty in describing real catalytic materials that are often complex multicomponent, multiscalar metal oxides operating at high pressures, known as the "materials and pressure gap," a term first coined by Bonzel [37].

In recent years, studies moved beyond single crystals and a number of groups have fabricated model catalysts with greater complexity to more closely represent the important features of technical materials. For example, metal–support interactions are modified by depositing metal atoms or nanoparticles on metal oxide single crystals [35, 38, 39].

10.4.3.1 Molecular beam scattering

Surface-catalyzed reactions begin with collisions at the vacuum–surface interface, which result in adsorption of one or more of the reactants. Reactant species may scatter elastically or inelastically back into the vacuum phase, become momentarily trapped, or remain adsorbed on the surface long enough to undergo reaction. Adsorption results from the formation of a specific bond between the reactant species and the surface, and an adsorbed species can have a surface lifetime ranging from microseconds to hours. The adsorption rate is a key characteristic in all surface-catalyzed reactions and may be the controlling factor in determining the rate of the overall reaction. The most fundamental way to represent the rate of adsorption in a gas–solid reaction is in terms of the sticking coefficient:

$$\frac{d\sigma_s}{dt} = \nu_{coll} \cdot S_0 \cdot e^{(-E_{ads}/kt)} \cdot f(\sigma) \qquad (10.8)$$

where σ_s is the number of adspecies per unit surface area, ν_{coll} is the collision frequency of reactant species with the surface, S_0 is the initial sticking coefficient at vanishing coverage, E_{ads} is the activation energy for adsorption, and $f(\sigma)$ accounts for the loss of adsorption sites due to accumulation of adspecies.

The initial sticking coefficient S_0 ($0 \geq S_0 \leq 1$) is a key parameter in the rate equation that depends strongly on the surface composition and structure of the solid surface and the reactant species. The sticking coefficient for a well-defined surface can be determined in an MBS experiment, which is conceptually depicted in Figure 10.1.

In an MBS experiment a pulsed or chopped beam of atoms or molecules is directed at a target surface (e.g., a specific plane of metal single crystal) typically mounted on a substrate heater in a

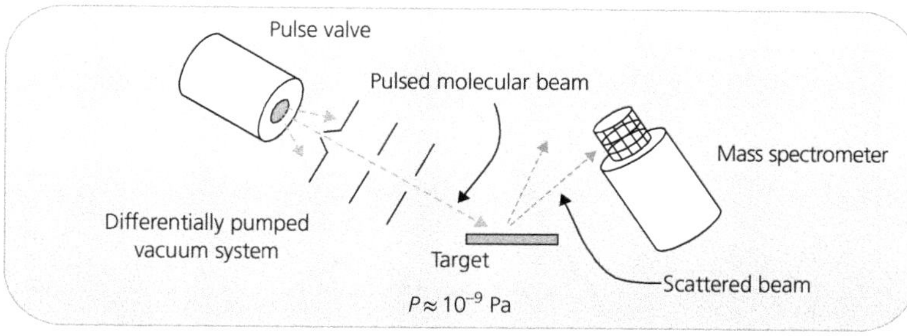

Figure 10.1 Conceptual diagram of a molecular beam scattering (MBS) experiment comprised of pulsed beam source, sample target, mass spectrometer detector, and a differentially pumped ultrahigh-vacuum system.

vacuum chamber. The molecular beam simulates the reactants, the gas pressure, and the gas temperature, variables which can all be varied independently. The scattered particles are analyzed using a mass spectrometer. A variety of MBS setups are described in the literature, and they generally fall into two different classes: (i) systems designed to study gas–surface dynamics and (ii) systems designed to study the kinetics and mechanism of surface reactions [31, 40–43]. A typical MBS apparatus is comprised of a differentially pumped beam generating system with modulated source and collimating apertures, which is connected to a sample chamber maintained at UHV ($\approx 10^{-11}$ torr). In addition to the target, the sample chamber contains the mass spectrometer, surface analysis techniques (e.g., low-energy electron diffraction Auger electron spectroscopy (LEED-AES)), and an ion gun for cleaning the target between experiments. The target and mass spectrometer are mounted on an axes perpendicular to beam axis. The mass spectrometer can be rotated about the target in the scattering plane.

The mass spectrometer measures the scattered particle flux, which is a function of the incident particle flux and its angular and energy distributions, the scattered particle mass and its angular and energy distributions, and the target composition, temperature, and cleanliness. The initial condition of the target is crucial since the sticking coefficient and the scattering distribution are sensitive to surface contaminants. Great care is taken in preparing, cleaning and characterizing the target surface before it is exposed to the beam to insure a well-defined surface structure that is free of contamination. Amplitude modulation of the beam and temperature control of the crystal surface can provide a means to measure the lifetimes of surface processes.

At the start of a scattering experiment, when the beam is switched on, coverage on the target is zero. After molecules strike the target, those that scatter are sampled by the mass spectrometer, and those that remain on the surface contribute to the coverage. Molecular beam experiments are performed under single collision conditions so that reactant molecules collide once with the target surface but not with each other. Rotating the spectrometer about the target, the angular scattering distribution can be determined as a function of the target temperature and coverage. The observed distributions contain dynamic information, which describes the process of energy exchange between gas and surface atoms, as well as kinetic and mechanistic information. Tilting the target relative to the beam, the scattering distribution can be determined as a function of the beam angle of incidence. Scattered molecules from a collimated beam may also exhibit diffraction effects, which provide information on surface structure and bond lengths [44–46]. The absolute sticking coefficient can be determined by measuring the scattered signal as a function of time, starting with an initially clean surface.

From MBS experiments one can determine the reaction probability, R_p, which is expressed as follows:

$$R_p = \frac{\text{Rate of formation of product}}{\text{Rate of reactant incident on catalyst}} \tag{10.9}$$

This quantity can be directly compared with the reaction probability determined from flow experiments (Section 10.4.1.2) where

$$R_p = \frac{\text{TOF}}{\text{Flux of reactant upon the catalyst}}. \tag{10.10}$$

Figure 10.2 shows a plot of the CO flux scattered from an initially clean Pd(111) surface versus time. The plot is from a molecular beam study by Engel [47] in which He, CO, and O_2 beams were scattered off a Pd target. For CO on Pd(111) the initial sticking coefficient, S_0, is very close to unity (0.96), shows no dependence on temperatures less than 650 K, and is independent of the angle of incidence. The initial sticking coefficient for O_2 on Pd(111) is independent of the incident angle of the O_2 beam but exhibits a pronounced temperature dependence (see Figure 10.3) and maximum value of 0.5.

The sticking coefficients for CO on Rh(110) [48] and (111) [49], Pd(110) [50], and Pt(111) [51] range in value between 0.4 and 0.9. The coefficients vary with the substrate temperature and the beam energy. For example, MBS studies by Jones et al. [52] measured S_0 for CO and O_2 on Pd(110) and obtained values of 0.5 for CO and 0.4 for O_2 at a substrate temperature of 300 K. The sticking coefficient for CO decreases with temperature and is 0.3 at 472 K. CO is molecularly adsorbed on Pd(110) and has a desorption peak at 470 K. As a result there is less net sticking at temperatures near this value. S_0 for CO on Pt(111) varies inversely with the energy of the incident beam and is 0.88 at a beam energy of $E_i \approx 6$ kJ/mol and 0.2 at a beam energy of ≈ 170 kJ/mol [51]. The sticking probability for CO on Ni(111) and Ni(100) decreases with increasing energy, falling from a

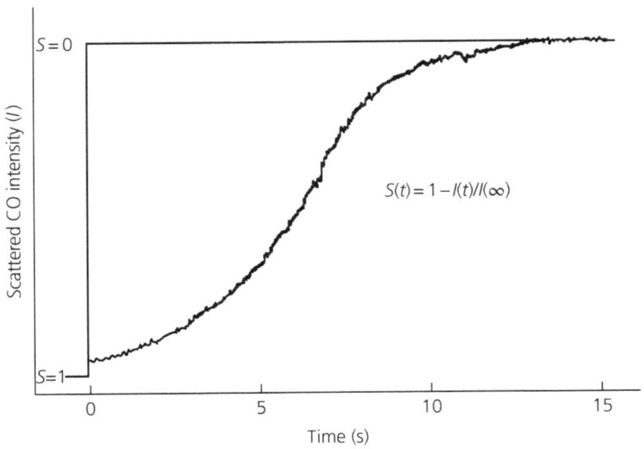

Figure 10.2 Plot showing the scattered CO intensity as a function of time from an initially clean Pd(111) surface. The initial sticking coefficient can be determined from the mass spectrometer signal $I(t)$ using the relationship $S(t) = 1 - I(t)/I(\infty)$.
(Source: Engel [47]. Reproduced with permission of AIP Publishing LLC.)

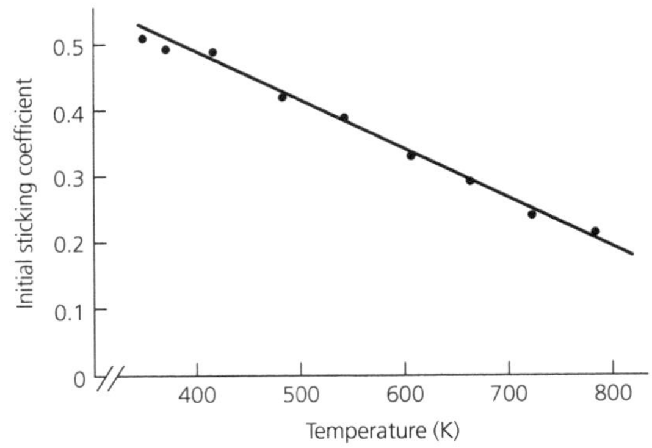

Figure 10.3 S_0 for O_2 on Pd (111) as a function of substrate temperature. (Source: Engel [47]. Reproduced with permission of AIP Publishing LLC.)

value $S_0 \approx 1.0$ to 0.5 when the beam energy (E_i) increases from $E_i \leq 5$ to 50 kJ/mol. The initial sticking coefficient for dissociative chemisorption of N_2 on W(110) [53] is a strong function of the beam energy, ranging from a value of $S_0 \approx 2.5 \times 10^{-3}$ to 0.4 when $E_i = 30$ kJ/mol is raised to 100 kJ/mol.

The present data allows construction of a scheme that relates the rate of adsorption of reactant, intermediate, and product molecules over different surfaces with different surface coverages. In essence one captures the fundamental modes of energy exchange between the surface and the gas. This reveals information about the dynamics of gas–surface interactions that are intrinsic to the material.

10.4.3.2 Temperature-programmed desorption

Temperature-programmed methods including temperature-programmed desorption (TPD) and temperature-programmed reaction (TPRx) can be applied to both model catalysts and practical materials. The temperature at which a species desorbs from the surface is an indication of its bond strength with the surface. Temperature-programmed experiments are used to determine binding energies and binding states of adsorbed molecules; to measure surface acidity, surface area, surface coverage, and the dispersion of metals; to determine valence states of metal atoms; and to study adsorption/desorption mechanisms and kinetics [54–57].

Generally, a species of interest is preadsorbed at low temperature and the sample is then heated at a constant rate, while desorbing species are monitored using mass spectrometry. TPD can be performed in either flow reactors or under vacuum conditions. For flow-based TPD studies involving industrial catalysts, care must be taken with the experimental conditions to minimize the effects of pore diffusional and film mass transfer limitations. This can be addressed by using low heating rates and a differential catalyst bed. Readsorption of species can dramatically influence kinetic results which can be minimized when experiments are performed in a vacuum. As the temperature increases, the rate constant for desorption also increases but more rapidly following the exponential Arrhenius dependence:

$$E = RT_m \left[\ln\left(\frac{\nu T_m}{\beta} \right) - 3.64 \right] \tag{10.11}$$

where E = activation energy of desorption, T_m = desorption peak maximum, β = heating rate, and ν = Arrhenius preexponential factor.

The desorption reaction order and Arrhenius parameters along with the saturation coverage for adsorption can be determined by varying the initial adsorbate coverage and the heating rate. For example, first-order desorption kinetics are indicated when TPD peaks do not shift as a function of adsorbate coverage. The area under the TPD peak is an indicator of the surface coverage of molecules, and the binding energies (heats of adsorption as well) can be determined as a function of coverage. Adsorption-induced surface reconstruction and heterogeneity of sites can lead to much more complicated desorption traces. Adsorbate-induced surface restructuring is well-known from LEED studies of crystal surfaces [58–61]. For example, metal atoms are known to relocate around the adsorption site to optimize the adsorbate–metal chemical bond [62, 63]; the exothermic adsorption process is compensated with the weakening of metal–metal bonds. Two well-known examples include the microfaceting of Pt(100) under the influence of CO or the creation of oxide surface phases on Ni(111) when exposed to O_2 [64].

In temperature-programmed oxidation (TPO), temperature-programmed reduction (TPR), and TPRx, the sample is exposed to a reactant gas while it is being heated. For example, examination of the desorption spectroscopy of previously adsorbed surface carbon species in a flow of hydrogen is useful for understanding the amounts and types of deactivating coke that may have formed during reaction. Different surface carbon species such as hydrocarbon fragments (e.g., CH_2 or CH_3), atomic carbon, polymeric carbon, aromatic clusters, amorphous graphitic carbon, or crystalline graphite can be identified according to peak temperatures during TPR with hydrogen. Such information is useful for identifying intermediate species and carbon species that contribute to deactivation in both hydrogenation and reforming reactions.

10.5 Microkinetic approach to kinetic analysis

In comparison to kinetic measurements collected from flow reactors, the microkinetic approach provides a different research emphasis that is centered more on the development of the catalytic material than on the development of a process. The goal of a microkinetic analysis is to estimate independently the rates of elementary steps and surface coverages by integrating data obtained from multiple sources such as surface science studies (e.g., sticking coefficients), thermodynamic data (e.g., equilibrium constants), spectroscopic characterization of surface species, and the calculation of bond energies and activation energies based on electronic structure theory. While the catalytic reaction consists of numerous elementary steps, most can be adequately described by considering only kinetically significant steps. To this end the microkinetic approach consolidates experimental data and theoretical observations pertaining to elementary steps to develop more complex models that can be substantiated with global observations from steady-state reactors. The microkinetic analysis presented in 1993 by Dumesic et al. [14] is defined as the examination of catalytic reactions in terms of elementary steps and their relation with each other during a catalytic cycle. The approach is best presented by reviewing its application in examples reported in the literature [65–68].

One caveat to the microkinetic approach that should be considered is that the surface structure of a material can often change with time and/or gas composition. For example, surface science studies of Rh(111) have indicated that CO and NO gas adsorption properties are highly dependent on the reactant gas concentration. The initial sticking coefficients of NO and CO are reported to vary by as much as an order of magnitude depending on the ratio of NO/CO in the gas phase [69]. Boreskov et al. [70, 71] were among the first to indicate that the chemical composition and structure of the catalyst change depending on the reaction mixture and temperature. *In situ* Raman spectroscopy [72], sum-frequency generation (SFG) [61], and other techniques [73] have established that the structure and composition of a catalyst change when exposed to a reactant gas [74–77]. Surface analysis before and after high-pressure reaction revealed that the surface structure and composition are dramatically changed by the reaction mixture [61, 62, 78].

As a result, one can't generally assume a catalyst has a uniform, nonchanging surface during a kinetic measurement and for comparison with other kinetic measurements. *Dynamic* microkinetic modeling should be used to take into account the changes in gas composition as this can show a significant effect on surface morphology, reaction rate, and mechanism. Changes in observed kinetics can result not only from changes in the surface but also from changes in bulk compositions. For example, absorption of hydrogen into palladium bulk has been demonstrated to manifest new catalytic properties different from the pure material [61, 62, 78]. Microkinetic models are more effective if the rates of morphological changes can be integrated. Material properties can change during a kinetic experiment, and if one monitors catalyst composition directly, it is difficult to relate observed kinetics to material properties. In the next section the TAP approach to kinetic analysis is presented. One advantage of this approach is the ability to incrementally manipulate the surface composition in conjunction with the kinetic measurement.

10.6 TAP approach to kinetic analysis

The TAP approach to kinetic analysis integrates concepts from all of the experimental strategies described hereto now. Similar to flow experiments, TAP uses real materials, often catalysts taken directly from industrial reactors. With flow experiments, however, transport and kinetics display a complex interdependence that cannot be easily decoupled. Often the strategy is to minimize the contribution of diffusion by operating at high space velocities. The phenomenon of diffusion can be suppressed but never eliminated. TAP takes a different approach by eliminating convective flow (small amounts of reactant molecules are injected under vacuum conditions) and uses well-defined transport in the Knudsen diffusion regime as a standard. In terms of uniformity in gas and catalyst compositions, the TAP technique can be considered like a CSTR. TAP is a pulsed transient technique which enables one to separate the introduction of two reactants and monitor product species with millisecond time resolution. Such information is key to understanding the reaction mechanism.

The microkinetic approach integrates a variety of different data sources to estimate rates and surface coverages. Difficulties may arise when the surface structure changes with time and/or gas composition. With TAP the surface coverage can be directly manipulated, and dynamic models of surface coverage/kinetic relationships can be investigated.

Similar to the surface science approach, TAP is useful for determining rate constants of elementary reaction steps, activation barriers for surface processes, and adsorption quantities. Surface science techniques utilize well-defined model systems and are useful for understanding the substituent parts of a complex system. The TAP technique uses real materials with complexity intact which makes it possible to study how the system works together to present *emergent* properties (properties that are more than just the sum of parts), for example, how does surface composition influence surface-to-bulk transport which then manifest an effect on the global kinetics?

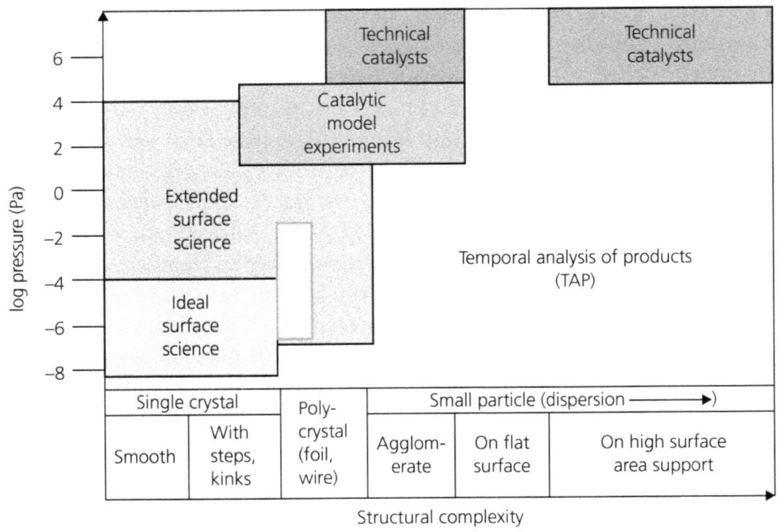

Figure 10.4 Adaptation of the "pressure/materials" correlation proposed by Bonzel [37] indicating pressure regimes and types of materials found in various types of catalyst characterization experiments. (Source: Bonzel [37]. Reproduced with permission of Elsevier.)

As initially described by Bonzel [37], the difference in experimental conditions and materials creates a "pressure/materials" gap that prevents direct comparison of data from the surface science and technical catalyst regimes. Figure 10.4 is an adaptation of the original diagram proposed by Bonzel that shows the pressure regimes and types of materials found in surface science and technical catalyst experiments. The region in which the TAP experiment operates is shown in Figure 10.4, illustrating the overlap between the surface science, technical catalyst, and TAP regimes.

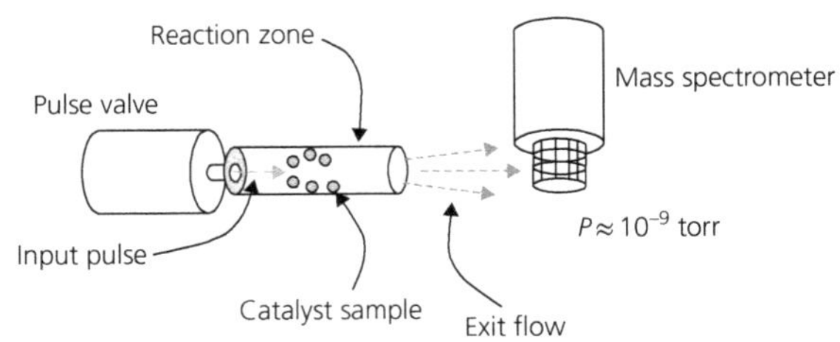

Figure 10.5 Key components of a TAP experiment.

10.6.1 TAP experiment design

Early TAP designs were aimed at retaining the time-dependent features of a molecular beam experiment, minimizing gas-phase interactions, and providing a way to extract intrinsic kinetic information from reactions on bulk catalysts. The basic elements of a TAP experiment, depicted in Figure 10.5, have some characteristics common to MBS experiment (recall Figure 10.1) and others common to conventional microreactor experiments. The key components are a "reaction zone" or microreactor, a fast-pulse gas feed system, a mass spectrometer detector, and a high-throughput UHV system. The reaction zone (Figure 10.5), which holds the catalyst sample, is a temperature-controlled cylindrical tube, which is typically made of stainless steel, inconel, or quartz. One end of the tube receives input from the feed system, and the opposite end is open to vacuum. During an experiment the reaction zone and catalyst sample are held under high-vacuum conditions. The reaction zone can hold a practical catalyst, similar to what one might study in conventional microreactor experiments, as well as single particles, single crystals, or model catalysts, which are studied in MBS experiments.

MBS experiments obtain essential kinetic and mechanistic information by modulating the reactant flux and measuring the shift in arrival times at the detector between scattered reactant and product molecules. The difference in arrival times can be used to determine reaction sequences, surface lifetimes of adspecies, and rates of surface reactions. TAP pulse response experiments [79–81] extract kinetic information in a similar fashion. Injection of a narrow gas pulse into the reaction zone initiates an experiment. The gas molecules travel through the reaction zone where they encounter the catalyst and can react to form product molecules. Molecules that exit the reaction zone are monitored by the mass spectrometer positioned at the outlet. The observed characteristic feature in a TAP experiment is the time-dependent gas flow, $F(t)$ [mol/s] or [molecules/s], that escapes from the exit of the microreactor. It is important to keep in mind that the "flow" in a TAP experiment is Knudsen diffusion, a random walk, such that a molecule's flow is not influenced by other molecules in the reactor. By comparing the exit flow of an adsorbing/reacting molecule with that of an inert molecule, one can easily separate transport effects from kinetic effects by comparing the integral characteristics of the flow/time dependency. Kinetic and mechanistic information is obtained by analyzing these time-weighted dependencies, and the inert gas transport is used as a "measuring stick" against which one can calculate the rates of chemical transformations.

In the TAP reactor system the reactor can be isolated from the vacuum system via a slide valve. When the slide valve is closed, the microreactor can be operated as a continuous plug flow-type reactor at atmospheric or higher pressures. This allows one to rapidly switch the catalyst from one kinetic regime to another (for catalyst pretreatment or conditioning) without exposing the sample to atmosphere. In the "high-pressure" mode the bulk of the reactor effluent exits through an external vent. A small leak into the mass spectrometer can be introduced via a needle valve to monitor the reaction. The slide valve assembly makes it possible to perform plug flow, step transient, and SSITKA experiments at higher pressures and TPD, TPR, and TAP pulse response experiments at low pressure.

10.6.1.1 Surface coverage

In both TAP and MBS experiments, the sample is maintained under vacuum conditions. Prior to performing a TAP experiment, it is common to heat the catalyst sample and monitor the desorption spectrum. Standard TPD and TPR experiments are easily performed. During the heating process, adspecies (e.g., water, CO, CO_2, etc.), which cover the surface at ambient pressures, vacate the surface leaving the coverage to more closely resemble that of an MBS target. Conversely, pulsing a molecule into the reaction zone can increase the molecule's coverage so that it resembles the coverage at ambient pressures. By pulsing different molecules into the reaction zone in an alternating sequence, the surface coverage of two or more species can be adjusted. As a result, coverage in a TAP experiment can be manipulated to resemble coverage in an MBS experiment or coverage in a conventional microreactor experiment.

10.6.1.2 Pulse size

The input pulse in a TAP experiment typically contains only 10 nmol of reactant, and the local pressure in the reaction zone may reach ca. 10^{-3} torr during a pulse. In a packed reactor, the mean free path is ca. 4000 μm, which is significantly larger than the space between particles. As a result, in a packed-bed reactor, molecules collide with particles, but seldom with one another.

At sufficiently small pulse intensities, a one-pulse TAP experiment can be considered a "state-defining" experiment, since the number of molecules in a reactant pulse is typically much smaller (10^2–10^5 times smaller) than the number of surface atoms in the catalyst sample being probed [82]. As a result, the reactant pulse does not significantly perturb the catalyst surface.

10.6.1.2.1 Kinetic analysis of TAP pulse response data

There are three basic concepts that distinguish TAP from other kinetic experiments and form the basis for extracting kinetic information from TAP pulse response data:

1 *Well-defined Knudsen diffusion is a "measuring stick" for measuring chemical reaction rates and extracting kinetic parameters.* Since Temkin's [83, 84] and Denbigh's [85] times, a common approach for extracting kinetic information is to measure the rate of chemical reaction using the rate of mass transport as a "measuring stick." In traditional steady-state experiments with perfect mixing, convectional transport is the "measuring stick," and diffusional transport is neglected.

   In TAP Knudsen pulse response experiments, there is no convective flow, and Knudsen diffusion is the only gas transport process. In the absence of reaction, the gas exit flow from the microreactor is described by a standard diffusion curve. In the "reaction–diffusion" case, the exit flow response curve is different from the standard diffusion curve, and this difference is attributed to kinetic phenomena. Comparison of the "reaction–diffusion" data to the "transport only" data (the "measuring stick") allows the TAP technique to separate kinetic effects from transport effects.

2 *During a single-pulse experiment the solid catalyst changes insignificantly, and a controlled change occurs in a multipulse experiment.* When the number of reactant gas molecules in a single pulse is significantly smaller than the number of catalytically active sites on the catalyst, the catalytic system remains in the same state after the measurement. A long series of small pulse experiments can induce a change in the catalyst state, and this change is characterized by the amount of consumed/released gaseous substances.

3 *The solid catalyst surface composition will remain uniform if the active zone is a sufficiently small fraction of the total bed length.* When a heterogeneous catalyst is exposed to a reactant gas, the composition and kinetic properties of the catalyst can change as a result of reaction with the gas. In a packed-bed reactor, the gas flow can cause the change to occur nonuniformly. The inlet of the bed will see the highest reactant concentration and will change by the largest amount. In a nonsteady-state experiment the bed composition can also change in time. An important and unique feature of a TAP pulse response experiment is that the catalyst composition remains essentially uniform when the thickness of the catalyst zone is small compared to the total length of the packed bed. In practice a small amount of catalyst can be packed in a "thin zone" between two beds of inert particle (Section 10.6.1.2), or in some cases a single particle can be used [86]. The advantage and properties of a thin zone reactor have been discussed in detail in the literature [87–89]. Using a thin zone reactor, a catalyst sample can be characterized "state by state" to determine how its catalytic properties incrementally evolve as a result of reaction. This process has been demonstrated using the selective oxidation of hydrocarbons over a transition metal oxide catalyst (e.g., vanadyl pyrophosphate (VPO)) [82]. Here, an oxidized VPO sample was exposed to a series of hydrocarbon pulses, and the change in kinetic properties was determined as a function of the oxidation degree.

10.6.1.3 Model-free analysis of experimental data

Ideally, kinetic information should be presented without the imposition of a kinetic expression or reactor characteristics, which is defined as a "model-free" analysis. When plug flow reactor data is analyzed using the differential method (section "Differential Analysis of Experimental Flow Data"), the rate is calculated directly from the experimental data without assumption of a kinetic model. Temkin and Denbigh applied a model-free approach to the analysis of steady-state kinetics over 50 years ago [85]. In their approach, the rate of chemical substance transformation is equal to the difference between the inlet and outlet molar flow rates divided by the catalyst volume or surface area. In CSTR steady-state experiments, no assumptions regarding the type of kinetic dependence, the reaction mechanism, or the corresponding model are needed to determine the rate of chemical transformation.

Model-free kinetic analysis of nonsteady-state reactions is a recent development that began with the thin zone microreactor configuration [82, 88, 89]. A model-free kinetic method known as the "Y-procedure" has been used to extract the nonsteady-state rate of chemical transformation from reaction–diffusion data with no assumptions regarding the kinetic model; the reader is referred to [90] for more details describing this procedure.

10.6.2 TAP experimental results

Excellent summary lists of publications where the TAP technique is applied to different catalytic systems can be found in [79, 80]. These lists include the many types of catalytic materials that are studied including supported metals, mixed metal oxides, zeolites, metal particles, metals deposited on screens, catalytic monoliths, and nanoparticles or atoms deposited on microparticles, single crystals, and other model catalysts.

10.6.2.1 Bridging the pressure gap

With the TAP reactor system, a slide valve at the reactor exit allows a catalyst sample to be rapidly cycled between vacuum and high pressure without exposing it to the atmosphere. In the high-pressure position, the reactor effluent flows through the slide valve and an external vent where it can be analyzed using a GC. A small portion of the effluent can be diverted to the mass spectrometer chamber through an adjustable needle valve, and the mass spectrum of the reaction products can be collected in real time. After running atmospheric pressure experiments, the reactor can be evacuated, and the slide valve can be moved to the vacuum position.

During the switch from high pressure to vacuum, the reactor effluent can be monitored, and desorbing adspecies left on the surface during pressure experiments can be measured. TAP pulse response experiments are performed after the reactor reaches vacuum and desorption of adspecies stops. Switching back and forth between high-pressure and vacuum operations typically takes less than 30 s. Recently, results were reported comparing atmospheric pressure and vacuum pulse response experiments on the oxidation of CO [86]. The catalyst sample was a single 400 μm diameter polycrystalline platinum (Pt) particle, which was placed in a microreactor bed with ca. 100 000 inert quartz particles with diameters between 210 and 250 μm.

Figure 10.6 presents a scaled drawing of the reactor configuration. The particle occupies less than 0.3% of the cross-sectional area of the microreactor, so the reaction zone can be considered a point source. Gas concentration or temperature gradients across the catalyst zone can be assumed to be negligible since the zone is a single particle.

Vacuum pulse response experiments are performed using a "pump-probe" format illustrated in Figure 10.7. Oxygen/Ar and CO/Ar mixtures are injected from different pulse valves in an alternating sequence, and the CO_2 response is measured during each pulse. Argon is used as an internal standard. CO_2 does not appear on the first oxygen pulse since no CO is present in the reactor. CO_2 appears on the first CO pulse and all subsequent oxygen pulses.

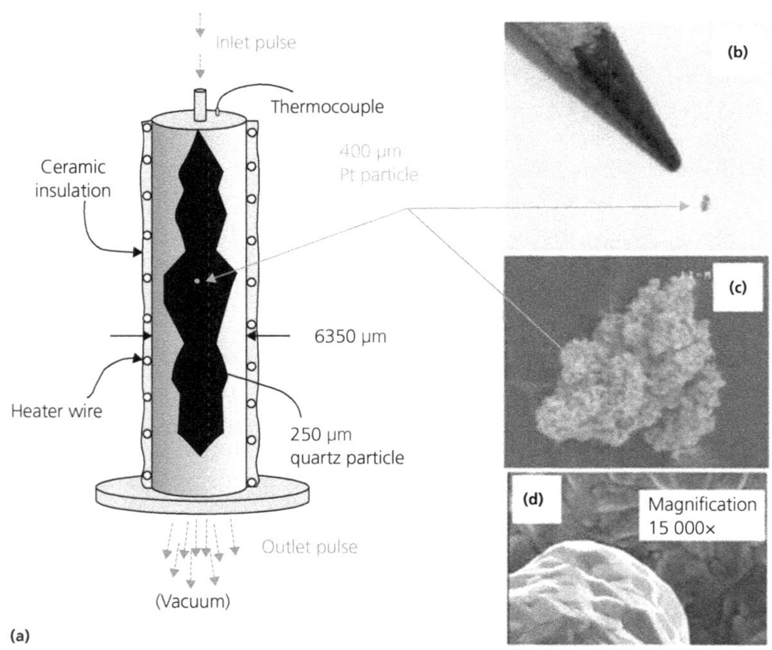

Figure 10.6 (a) Schematic of TAP single particle microreactor configuration. The 400 μm diameter Pt particle is packed within a bed of inert quartz particles with diameters between 210 and 250 μm. (b) Image comparing a 400 μm Pt particle to a pencil point. (c) SEM image showing the complex surface structure of a polycrystalline Pt particle. (d) Higher magnification (15 000×) of the particle shown in (c), which shows the surface is nonporous [86]. (Source: Zheng et al. [86]. Reproduced with permission of Elsevier.)

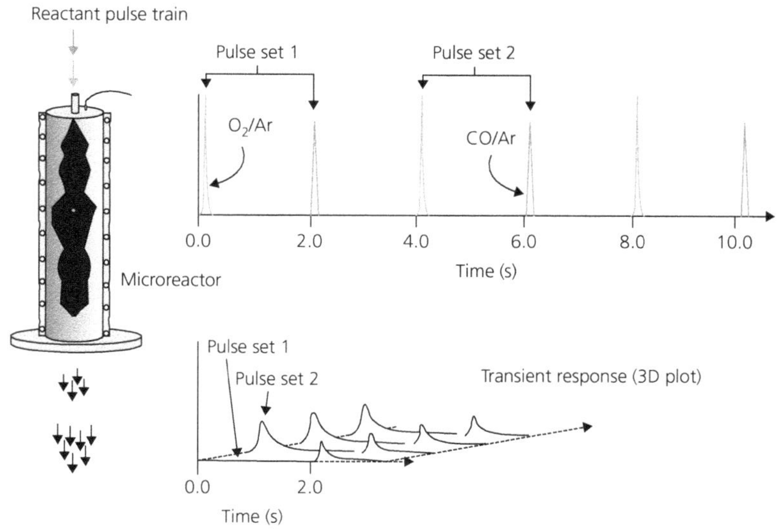

Figure 10.7 Illustration of a TAP pump-probe experiment in which O_2/Ar and CO_2/Ar are pulsed in an alternating sequence and the CO_2 transient response is measured during each pulse.
(Source: Zheng et al. [86]. Reproduced with permission of Elsevier.)

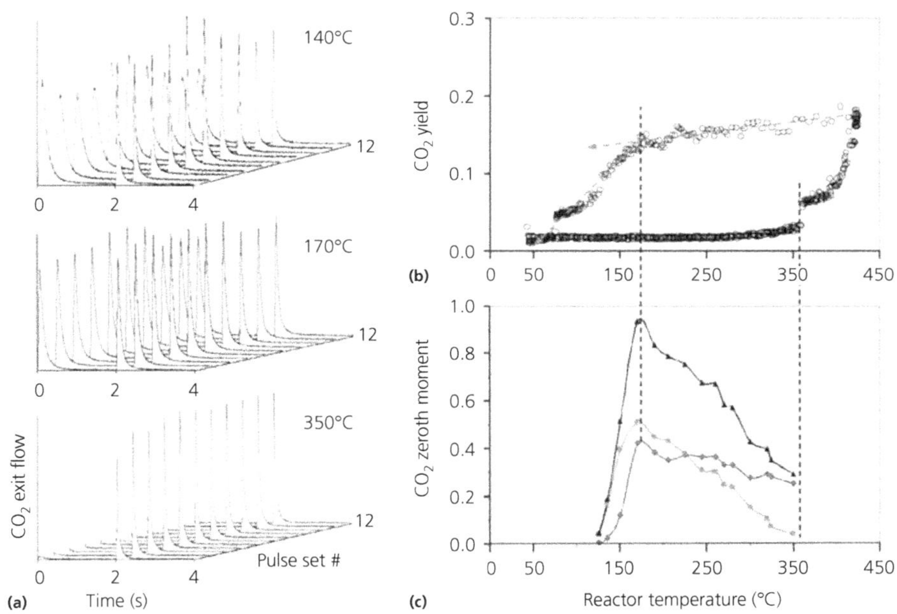

Figure 10.8 Comparison of CO_2 produced during TAP vacuum pump-probe experiments and atmospheric flow experiments for CO oxidation over single Pt particle with the same composition of reactants. (a) A typical set of pump-probe CO_2 responses ($m/e = 44$) for reaction at 140, 170, and 350°C. There is a shift in the amount of CO_2 produced during both CO and oxygen pulses as temperature increases. (b) CO_2 production observed from atmospheric flow experiment. The CO_2 produced while increasing reactor temperature is less than the CO_2 produced during reactor temperature decrease as shown by the counterclockwise hysteresis loop. (c) CO_2 production observed from vacuum pump-probe experiment. The black line represents the total CO_2 yield. The circle and diamond points represent the CO_2 yield on the oxygen pulse and CO pulse, respectively.
(Source: Zheng et al. [86]. Reproduced with permission of Elsevier.)

Figure 10.8(a) displays pump-probe data sets obtained at 140, 170, and 350°C. Figure 10.8(c) shows the plots of the amount of CO_2 produced during each pulse and the total amount for each pump-probe cycle at a series of temperatures. The amount of CO_2 produced is determined by measuring the area (zeroth moment) under the transient response curve [91]. The zeroth moment is normalized relative to the amount of CO per pulse to determine the CO_2 yield. The absolute amounts of O_2, CO, and CO_2 are determined by comparing the response obtained in a reaction with one obtained using a standard blend. At 170°C the yield for the individual O_2 and CO pulses reaches a maximum, making the total yield equal to 95% during one pump-probe cycle, indicating that at least 95% of CO molecules pulsed into the reactor must strike the particle. Above 170°C the CO_2 yield decreases more rapidly for the O_2 pulse than the CO pulse.

Atmospheric flow experiments are performed after closing the slide valve. The particle bed is first exposed to a hydrogen flow (20 cc/min diluted in Ar, H_2/Ar = 1) at 350°C for 1 h. More details can be found in Zheng et al. [86]. The temperature dependence of CO_2 production is obtained by heating or cooling the reactor at a constant rate while maintaining an input flow of

50 cc/min. A small amount of the reactor effluent is diverted into the mass spectrometer chamber, and its mass spectrum is continuously monitored.

Both TAP vacuum and atmospheric flow data exhibit a turning point in CO_2 production, indicating a transition from reaction controlled by one adsorbed species to one controlled by a different adsorbed species at 170°C. The upper branch in the CO_2 curve of the atmospheric flow experiment corresponds to an O_2-covered surface, and the lower branch corresponds to a CO-covered Pt surface (Figure 10.8(b)). From the pulse response data, in the region of the CO_2 maximum, the areas under the CO_2 response curves (CO_2 yield) corresponding to the O_2 and CO pulses are approximately the same, indicating nearly equal coverages of O_2 and CO on the Pt particle (Figure 10.8(c)). *The correspondence in "turning points" indicates that the coverage in vacuum and atmospheric pressure experiments is approximately the same and intrinsic kinetic data obtained in vacuum experiments can be used to describe kinetic behavior in the atmospheric pressure domain.*

Taking the conversion of CO or CO_2 yield at the "turning point" from vacuum pulse response and atmospheric flow data, the apparent kinetic rate constant can be calculated. In combination with the gas residence time (τ) in the catalyst zone, the apparent kinetic rate constant is given by the following expression:

$$k_{apparent} = \frac{X}{(1-X)\tau} \tag{10.12}$$

Using an approximate conversion of 90% in the vacuum pulse response experiment, the apparent kinetic rate constant is calculated to be 9000 s^{-1}. In the atmospheric flow experiment involving 20% conversion, the apparent kinetic rate constant is calculated to be 9280 s^{-1}. The two apparent kinetic rate constants differ by approximately 3%. Although there may be some error involved in experimentation, values of the apparent kinetic rate constants are in the same order of magnitude. *The ability to relate data both qualitatively and quantitatively in the atmospheric pressure domain to the data obtained in vacuum pulse response experiments using a single Pt particle is a significant step toward bridging the pressure gap.*

10.6.2.2 Tracking the evolution of catalytic properties

Structure–activity correlations can be established using a surface science strategy and measuring reactions on a well-defined model surface. Another approach is based on the application of *in situ* spectroscopic techniques, which try to measure surface adspecies or identify surface structures that change during reaction. Surface spectroscopy techniques can also be used to characterize catalyst samples before and after reaction. In this case a change in catalyst performance can often be associated with a change in surface composition or structure. Transient kinetic experiments can also be used to indirectly measure changes in the composition and structure while simultaneously measuring changes in kinetic properties. This latter approach is the one adopted in TAP reactor studies and is illustrated by the following examples.

10.6.2.2.1 C_4 oxidation over VPO catalysts

Catalysts based on vanadium oxides are used extensively in selective oxidation processes. The selective oxidation of butane and other C_4 molecules over VPO-based catalysts to maleic anhydride and intermediate compounds (e.g., furan, butadiene, butene) is strongly influenced by the feed conditions, especially the oxygen-to-hydrocarbon ratio. A combination of atmospheric pressure and TAP pulse response experiments determined that "reactor-equilibrated" VPO adsorbs oxygen at elevated oxygen pressures to form a more active–selective catalyst [92]. If the oxygen-treated catalyst is heated in vacuum, the adsorbed oxygen desorbs leaving a less active–selective catalyst. Figure 10.9(a) shows three oxygen uptakes at 430, 450, and 460°C and at an oxygen pressure of 800 torr and the oxygen desorption spectrum when an oxygen-18 treated VPO sample is heated in vacuum. The ^{16}O to ^{18}O ratio, shown in Figure 10.9(b), indicates that oxygen is only taken up in the first few monolayers of the VPO surface.

Oxygen-treated VPO catalysts are determined to be more active and selective, provided the oxidation did not lead to less active $VOPO_4$ phases [93]. The oxidation of butane and other C_4 compounds is investigated in TAP pulse response experiments by pulsing a C_4/Ar mixture over VPO and measuring the pulse response curves of the reactants and products [82]. Figure 10.10 shows typical response curves and an Arrhenius plot for *n*-butane oxidation over an oxygen-treated catalyst. The change in kinetic parameters as the surface oxygen concentration is altered is shown in Figure 10.11, which plots the activation energy and apparent equilibrium constant for different products at different stages in the reduction of VPO

10.6.2.2.2 Surface lifetimes of reactive species

The surface lifetime of an adspecies under reaction conditions is a function of the rate of reaction, the rate of desorption, and the rate at which the adspecies diffuses into the catalyst bulk. The adsorption/desorption characteristics of a species can be determined in TAP pulse response experiments by comparing the exit flow curve of the species with the standard diffusion curve (SDC) [81]. If the curve falls inside the SDC, then the species is irreversibly adsorbed. The surface concentration of an active species can also decrease if the species diffuses into the catalyst bulk or is depleted by reaction with some other surface species. The reactive lifetime of an adspecies can be measured in TAP pump-probe experiments by changing the pump-probe interval [94].

For example, in recent pump-probe experiments by Zheng [95], the reactive lifetime of oxygen on a Pt particle was measured in a series of pump-probe experiments using two separate reactant mixtures of O_2/Ar (70/30 ratio) and CO/Ar (70/30 ratio), which were injected from separate pulse valves. The mixtures were pulsed in an alternating sequence into a microreactor containing a single 400 μm Pt particle packed in a bed of inert quartz

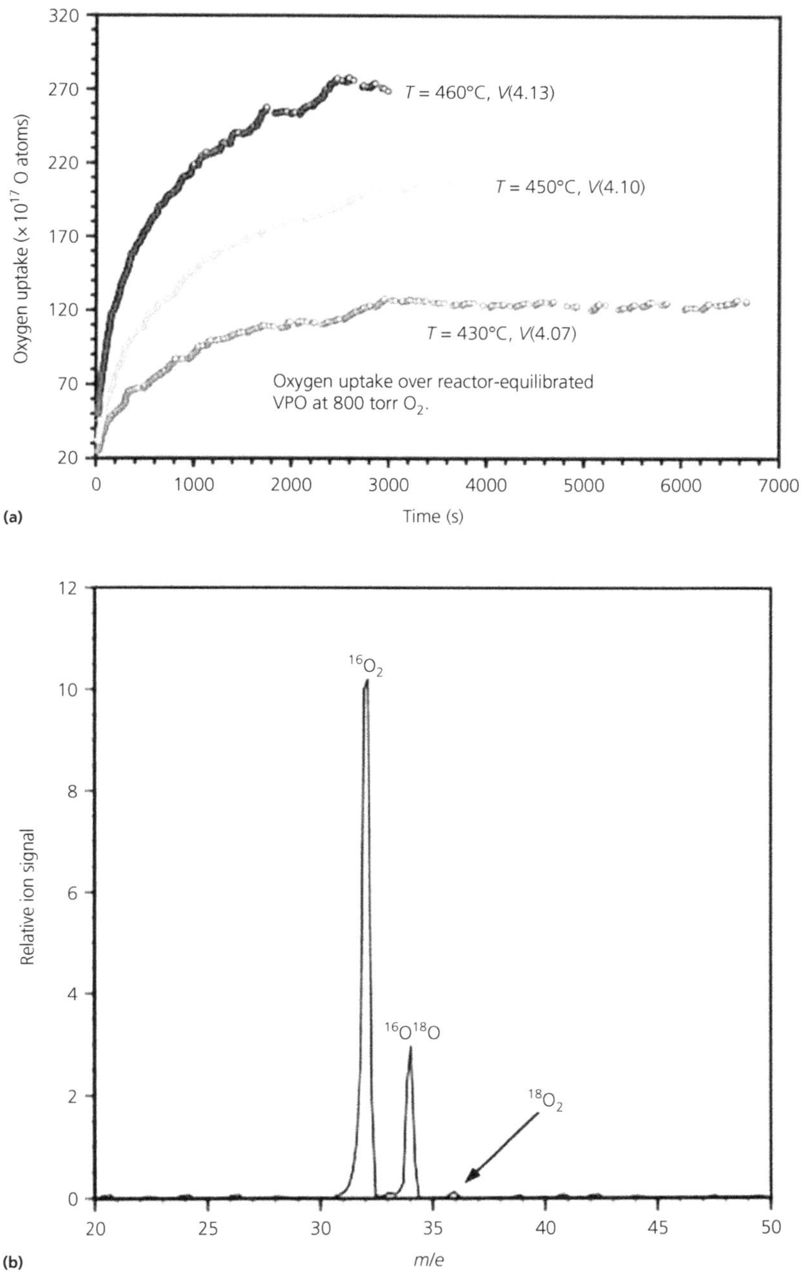

Figure 10.9 (a) Oxygen uptake over reactor-equilibrated VPO at 800 torr O_2. (Source: Schuurman and Gleaves [92]. Reproduced with permission of Elsevier.) (b) O_2 desorption spectrum under vacuum from $^{18}O_2$-treated $(VO)_2P_2O_7$.

particles. The interval separating the oxygen pulses and the CO pulses was varied between 1 and 9 s. In all cases the pump-probe cycle time was 10 s. Thus when the interval between the oxygen and CO pulse is 5 s, the interval between the CO and following oxygen pulse is also 5 s. Figure 10.12 shows the CO_2 production for different pump-probe intervals at 150 and 350°C.

CO_2 production on the oxygen pulse at 350°C is significantly lower than production at 150°C. It is independent of the pump-probe interval. CO_2 production on the CO pulse at 150°C is also independent of the pump-probe interval. At 350°C CO_2 decreases with the pump-probe interval. The drop in CO_2 production can be attributed to a decrease in the amount of active oxygen.

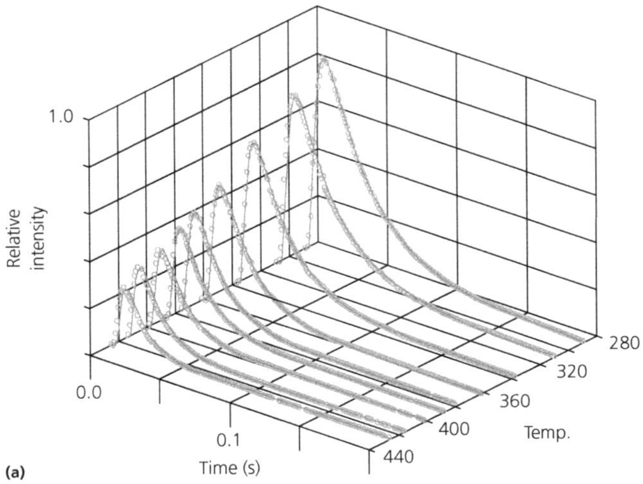

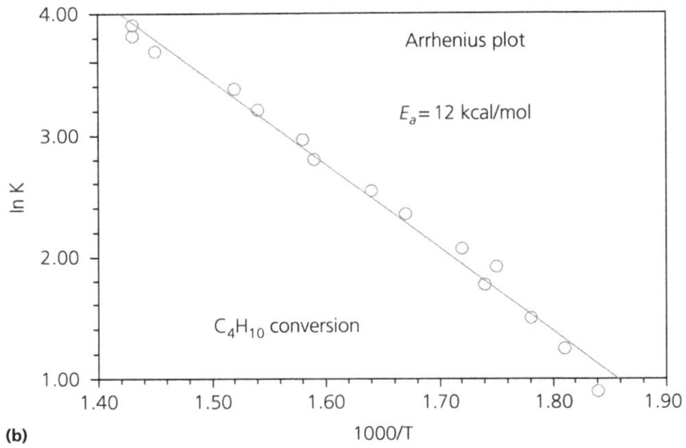

Figure 10.10 (a) n-Butane pulse response curves over oxygen-treated VPO at various temperatures. Each curve is obtained for the same initial VPO oxidation state. (b) Arrhenius plot obtained from the temperature dependence of the n-butane conversion giving an activation energy of 12 kcal/mol.
(Source (b): Schuurman and Gleaves [92]. Reproduced with permission of Elsevier.)

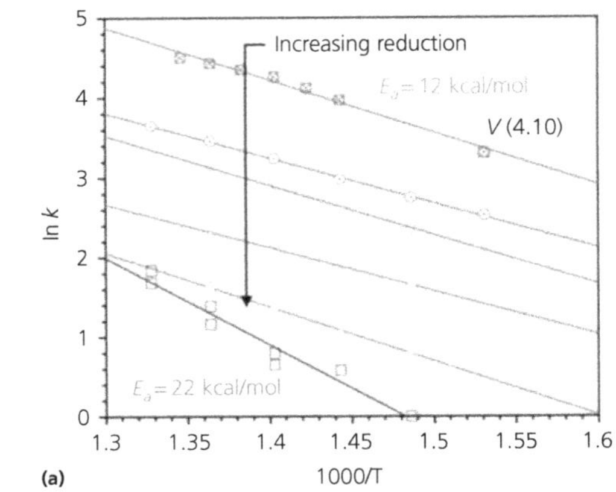

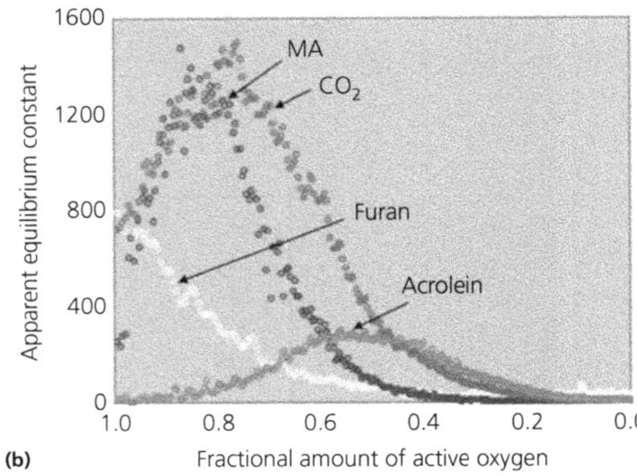

Figure 10.11 (a) Arrhenius plots for n-butane oxidation over a single VPO sample reduced by a long series of n-butane pulses. As the surface is reduced, the activation energy increases.
(Source: Schuurman and Gleaves [92]. Reproduced with permission of Elsevier.)
(b) Apparent equilibrium constant for various products as a function of VPO oxidation state.)
(Source: Shekhtman et al. [82]. Reproduced with permission of Elsevier.)

The rate of the drop in active oxygen can be calculated from the zeroth moments of the CO_2 and is plotted in Figure 10.13.

10.6.2.3 Probe molecules

The TAP pulse response methodology allows one to test a material using a variety of different probe molecules. Most often the reactant is used, but in many cases using a reaction intermediate, a product, or a base (ammonia or pyridine) can be very insightful. For example, when comparing samples pretreated with different amounts of oxygen or varying surface metal concentration, it can be very insightful to compare how CO, a double bond, or ammonia might interact with these different surfaces. Even though these molecules are not part of the process of interest, they can be useful to characterize and compare materials with subtly different preparations. For example, one can compare the capacity to store CO and the adsorption/desorption strength of ethylene or measure the acidity of similar materials. Even if the reaction kinetics of an industrial process are not being measured, one can still gain insight into how the material functions on a more fundamental level.

10.7 Conclusions

The determination of kinetic parameters is an essential step in developing a catalytic process. Parameters determined in laboratory-scale steady-state reactors are necessary to formulate models for scale-up to pilot plant and process-scale reactors. Kinetic parameters also provide insight into the fundamental processes that occur during a catalytic reaction and form the basis for creating microkinetic models that describe the individual steps (e.g., adsorption, surface reaction, and desorption) of a complex reaction.

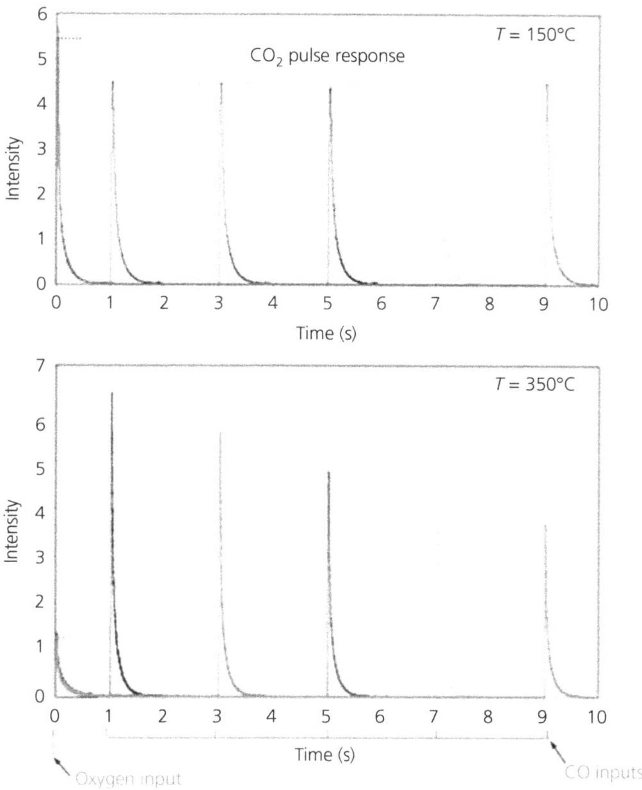

Figure 10.12 Pump-probe data showing CO_2 production as a function of temperature and pump-probe interval. At 150°C the CO_2 production is essentially independent of the pump-probe interval out 9 s separation. At 350°C the CO_2 production drops as the pump-probe interval increases, indicating a drop in the active oxygen concentration with time. (Source: Reproduced with kind permission of Zheng [95].)

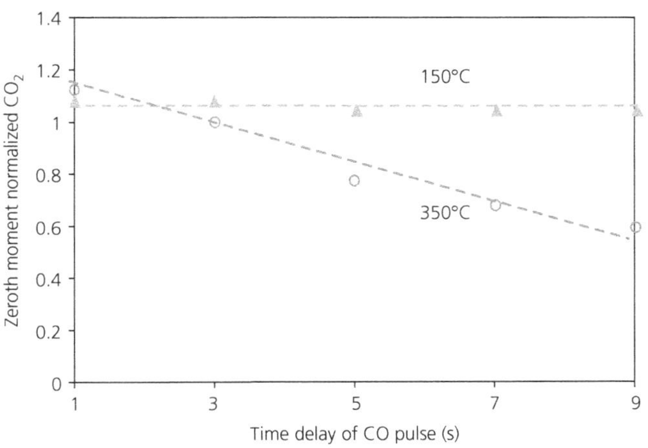

Figure 10.13 Normalized CO_2 production on the CO pulse calculated from the zeroth moment of the pulse response curve. CO_2 production is constant at 150°C and drops to ca. 0.5 times its value in 9 s at 350°C. (Source: Reproduced with kind permission of Zheng [95].)

When intrinsic parameters can be determined, as in MBS or TAP experiments, it becomes possible to establish relationships between catalyst structure or composition and catalytic properties. Such information provides a basis for developing a fundamental understanding of how catalytic materials operate and can be used to validate quantum-based models of the active site.

References

1 F. Gao and D. W. Goodman, Model catalysts: simulating the complexities of heterogeneous catalysts, Annual Review of Physical Chemistry, vol. 63, pp. 265–286, 2012.

2 M. Sterrer and H.-J. Freund, Towards realistic surface science models of heterogeneous catalysts: influence of support hydroxylation and catalyst preparation method, Catalysis Letters, vol. 143, pp. 375–385, 2013.

3 D. Teschner, Z. Révay, J. Borsodi, M. Hävecker, A. Knop-Gericke, R. Schlögl, D. Milroy, S. D. Jackson, D. Torres, and P. Sautet, Understanding palladium hydrogenation catalysts: when the nature of the reactive molecule controls the nature of the catalyst active phase, Angewandte Chemie, International Edition, vol. 47, pp. 9274–9278, 2008.

4 C. H. Bartholomew and R. J. Farrauto, Fundamentals of Industrial Catalytic Processes, Second Edition. Hoboken, NJ: John Wiley & Sons, Inc., 2006.

5 R. J. Wijngaarden, A. Kronberg, and K. R. Westerterp, Industrial Catalysis: Optimizing Catalysts and Processes. Weinheim: Wiley-VCH Verlag GmbH, 1998.

6 J. M. Berty, Experiments in Catalytic Reaction Engineering. Amsterdam: Elsevier, 1999.

7 E. G. P. Z. Christoffel, Laboratory Studies of Heterogeneous Catalytic Processes. Amsterdam: Elsevier, 1989.

8 D. Murzin, Engineering Catalysis. Berlin: De Gruyter, 2013.

9 J. Berty, Testing commercial catalysts in recycle reactors, Catalysis Reviews: Science and Engineering, vol. 20, pp. 75–96, 1979.

10 J. M. Berty, Ethylene oxide synthesis, B. Leach, Ed., Applied Industrial Catalysis, Elsevier, 1983, pp. 41–67.

11 J. Carberry, Designing laboratory catalytic reactors, Industrial and Engineering Chemistry, vol. 56, pp. 39–46, 1964.

12 O. Levenspiel, Chemical reaction engineering, Industrial & Engineering Chemistry Research, vol. 38, pp. 4140–4143, 1999.

13 H. S. Fogler, Elements of Chemical Reaction Engineering. Englewood Cliffs, NJ: Prentice-Hall, 1999.

14 J. A. Dumesic, D. F. Rudd, L. M. Aparicio, J. E. Rekoske, and A. A. Treviño, The Microkinetics of Heterogeneous Catalysis. Washington, DC: American Chemical Society, 1993.

15 S. J. Tinnemans, J. G. Mesu, K. Kervinen, T. Visser, T. A. Nijhuis, A. M. Beale, D. E. Keller, Ad M.J. van der Eerden, and B. M. Weckhuysen, Combining *operando* techniques in one spectroscopic-reaction cell: new opportunities for elucidating the active site and related reaction mechanism in catalysis, Catalysis Today, vol. 113, pp. 3–15, 2006.

16 T. X. Nijhuis, S. J. Tinnemans, T. Visser, and B. M. Weckhuysen, Operando spectroscopic investigation of supported metal oxide catalysts by combined time-resolved UV-VIS/Raman/on-line mass spectrometry, Physical Chemistry Chemical Physics, vol. 5, pp. 4361–4365, 2003.

17 F. C. Meunier, The design and testing of kinetically-appropriate operando spectroscopic cells for investigating heterogeneous catalytic reactions, Chemical Society Reviews, vol. 39, pp. 4602–4614, 2010.

18 J. Happel, H. Cheh, M. Otarod, S. Ozawa, A. Severdia, T. Yoshida, and V. Fthenakis, Multiple isotope tracing of methanation over nickel catalyst: II. Deuteromethanes tracing, Journal of Catalysis, vol. 75, pp. 314–328, 1982.

19 C. O. Bennett, Understanding Heterogeneous Catalysis Through the Transient Method, vol. 178. Washington, DC: American Chemical Society, 1982.

20 P. Biloen, Transient kinetic methods, Journal of Molecular Catalysis, vol. 21, pp. 17–24, 1993.

21 S. L. Shannon and J. G. Goodwin Jr, Characterization of catalytic surfaces by isotopic-transient kinetics during steady-state reaction, Chemical Reviews, vol. 95, pp. 677–695, 1995.

22 H. Kobayashi and M. Kobayashi, Transient response method in heterogeneous catalysis, Catalysis Reviews: Science and Engineering, vol. 10, pp. 139–176, 1974.

23 E. Müller and H. Hofmann, Dynamic modelling of heterogeneous catalytic reactions—I. Theoretical considerations, Chemical Engineering Science, vol. 42, pp. 1695–1704, 1987.

24 W. K. Hall and P. Emmett, An improved microcatalytic technique, Journal of the American Chemical Society, vol. 79, pp. 2091–2093, 1957.

25 R. J. Kokes, H. Tobin Jr, and P. Emmett, New microcatalytic-chromatographic technique for studying catalytic reactions, Journal of the American Chemical Society, vol. 77, pp. 5860–5862, 1955.

26 E. G. Christoffel, Laboratory reactors and heterogeneous catalytic processes, Catalysis Reviews: Science and Engineering, vol. 24, pp. 159–232, 1982.

27 B. Spiewak and J. Dumesic, Microcalorimetric measurements of differential heats of adsorption on reactive catalyst surfaces, Thermochimica Acta, vol. 290, pp. 43–53, 1997.

28 M. Baumer and H.-J. Freund, Metal deposits on well-ordered oxide films, Progress in Surface Science, vol. 61, pp. 127–198, 1999.

29 M. Bäumer, J. Libuda, and H. Freund, Metal deposits on thin well ordered oxide films: Morphology, adsorption and reactivity, In Lambert, R. M., Pacchioni, G. Chemisorption and Reactivity on Supported Clusters and Thin Films: Towards an Understanding of Microscopic Processes in Catalysis. Boston: Kluwer Academic Publishers, NATO ASI Series E Applied Sciences-Advanced Study Institute, vol. 331, pp. 61–104, 1997.

30 M. Asscher and G. Somorjai, Atomic and Molecular Beam Methods, vol. 1. Oxford: Oxford University Press, 1988.

31 J. Barker and D. Auerbach, Gas—surface interactions and dynamics; Thermal energy atomic and molecular beam studies, Surface Science Reports, vol. 4, pp. 1–99, 1984.

32 C. Rettner, D. Auerbach, J. Tully, and A. Kleyn, Chemical dynamics at the gas-surface interface, The Journal of Physical Chemistry, vol. 100, pp. 13021–13033, 1996.

33 M. P. D'Evelyn and R. J. Madix, Reactive scattering from solid surfaces, Surface Science Reports, vol. 3, pp. 413–495, 1983.

34 S. Shaikhutdinov, M. Heemeier, J. Hoffmann, I. Meusel, B. Richter, M. Bäumer, H. Kuhlenbeck, J. Libuda, H.-J. Freund, R. Oldman, S.D. Jackson, C. Konvicka, M. Schmid, and P. Varga, Interaction of oxygen with palladium deposited on a thin alumina film, Surface Science, vol. 501, pp. 270–281, 2002.

35 M. Frank and M. Bäumer, From atoms to crystallites: adsorption on oxide-supported metal particles, Physical Chemistry Chemical Physics, vol. 2, pp. 3723–3737, 2000.

36 W.-J. Shen, M. Okumura, Y. Matsumura, and M. Haruta, The influence of the support on the activity and selectivity of Pd in CO hydrogenation, Applied Catalysis A: General, vol. 213, pp. 225–232, 2001.

37 H. Bonzel, The role of surface science experiments in understanding heterogeneous catalysis, Surface Science, vol. 68, pp. 236–258, 1977.

38 B. R. Cuenya, Synthesis and catalytic properties of metal nanoparticles: size, shape, support, composition, and oxidation state effects, Thin Solid Films, vol. 518, pp. 3127–3150, 2010.

39 M. Haruta, Catalysis of gold nanoparticles deposited on metal oxides, Cattech, vol. 6, pp. 102–115, 2002.

40 G. A. Somorjai and J. Y. Park, Concepts, instruments, and model systems that enabled the rapid evolution of surface science, Surface Science, vol. 603, pp. 1293–1300, 2009.

41 M. J. Cardillo, Gas-surface interactions studied with molecular beam techniques, Annual Review of Physical Chemistry, vol. 32, pp. 331–357, 1981.

42 D. Padowitz, K. Peterlinz, and S. J. Sibener, New modulated molecular beam scattering methods for probing nonlinear and coverage-dependent reaction kinetics at surfaces, Langmuir, vol. 7, pp. 2566–2573, 1991.

43 R. David, K. Kern, P. Zeppenfeld, and G. Comsa, High-resolution He-scattering apparatus for gas–surface interaction studies, Review of Scientific Instruments, vol. 57, pp. 2771–2779, 1986.

44 J. Toennies, Scattering of molecular beams from surfaces, Applied Physics, vol. 3, pp. 91–114, 1974.

45 A. Lahee, J. Manson, J. Toennies, and C. Wöll, Helium atom differential cross sections for scattering from single adsorbed CO molecules on a Pt (111) surface, The Journal of Chemical Physics, vol. 86, pp. 7194–7203, 1987.

46 M. Van Hove, R. Koestner, P. C. Stair, J. Biberian, L. L. Kesmodel, I. Bartoš, and G.A. Somorjai, The surface reconstructions of the (100) crystal faces of iridium, platinum and gold: I. Experimental observations and possible structural models, Surface Science, vol. 103, pp. 189–217, 1981.

47 T. Engel, A molecular beam investigation of He, CO, and O_2 scattering from Pd (111), The Journal of Chemical Physics, vol. 69, pp. 373–385, 2008.

48 M. Bowker, Q. Guo, and R. Joyner, CO and O_2 adsorption on Rh(110), Surface Science, vol. 253, pp. 33–43, 1991.

49 P. A. Thiel, E. D. Williams, J. T. Yates Jr, and W. H. Weinberg, The chemisorption of Co on Rh(111), Surface Science, vol. 84, pp. 54–64, 1979.

50 I. Z. Jones, R. A. Bennett, and M. Bowker, CO adsorption on Pd(110), Surface Science, vol. 402–404, pp. 595–598, 1998.

51 J. Harris and A. Luntz, Sticking and scattering in the molecular chemisorption regime: CO on Pt (111), The Journal of Chemical Physics, vol. 91, pp. 6421–6428, 1989.

52 I. Z. Jones, R. A. Bennett, and M. Bowker, CO oxidation on Pd(110): a high-resolution XPS and molecular beam study, Surface Science, vol. 439, pp. 235–248, 1999.

53 H. Pfnür, C. Rettner, J. Lee, R. Madix, and D. Auerbach, Dynamics of the activated dissociative chemisorption of N_2 on W (110): a molecular beam study, The Journal of Chemical Physics, vol. 85, pp. 7452–7466, 1986.

54 J. L. Falconer and J. A. Schwarz, Temperature-programmed desorption and reaction: applications to supported catalysts, Catalysis Reviews: Science and Engineering, vol. 25, pp. 141–227, 1983.
55 A. De Jong and J. Niemantsverdriet, Thermal desorption analysis: comparative test of ten commonly applied procedures, Surface Science, vol. 233, pp. 355–365, 1990.
56 J. S. Rieck and A. T. Bell, Influence of adsorption and mass transfer effects on temperature-programmed desorption from porous catalysts, Journal of Catalysis, vol. 85, pp. 143–153, 1984.
57 R. Cvetanovic and Y. Amenomiya, Application of a temperature-programmed desorption technique to catalyst studies, Advances in Catalysis, vol. 17, pp. 103–149, 1967.
58 S. Haq and F. M. Leibsle, Acetate on Cu(110): evidence for long-range intermolecular interactions and molecular-induced restructuring of a metal surface, Surface Science, vol. 355, pp. L345–L349, 1996.
59 U. Starke, M. Van Hove, and G. Somorjai, Adsorbate-induced relaxations of close-packed fcc and hcp metal surfaces, Progress in Surface Science, vol. 46, pp. 305–319, 1994.
60 H. Shih, F. Jona, D. Jepsen, and P. Marcus, Metal-surface reconstruction induced by adsorbate: Fe (110) p (2× 2)-S, Physical Review Letters, vol. 46, p. 731–734, 1981.
61 G. A. Somorjai and G. Rupprechter, Molecular studies of catalytic reactions on crystal surfaces at high pressures and high temperatures by infrared-visible sum frequency generation (SFG) surface vibrational spectroscopy, The Journal of Physical Chemistry B, vol. 103, pp. 1623–1638, 1999.
62 G. Somorjai, The surface science of heterogeneous catalysis, Surface Science, vol. 299, pp. 849–866, 1994.
63 G. A. Somorjai, Modern surface science and surface technologies: an introduction, Chemical Reviews, vol. 96, pp. 1223–1236, 1996.
64 D. Murzin and T. Salmi, Catalytic Kinetics. Amsterdam: Elsevier, 2005.
65 D. Chen, R. Lødeng, A. Anundskås, O. Olsvik, and A. Holmen, Deactivation during carbon dioxide reforming of methane over Ni catalyst: microkinetic analysis, Chemical Engineering Science, vol. 56, pp. 1371–1379, 2001.
66 S. Dahl, J. Sehested, C. Jacobsen, E. Törnqvist, and I. Chorkendorff, Surface science based microkinetic analysis of ammonia synthesis over ruthenium catalysts, Journal of Catalysis, vol. 192, pp. 391–399, 2000.
67 M. Maestri, D. G. Vlachos, A. Beretta, G. Groppi, and E. Tronconi, Steam and dry reforming of methane on Rh: microkinetic analysis and hierarchy of kinetic models, Journal of Catalysis, vol. 259, pp. 211–222, 2008.
68 C. Ovesen, B. Clausen, B. Hammershøi, G. Steffensen, T. Askgaard, I. Chorkendorff, J.K. Nørskov, P.B. Rasmussen, P. Stoltze, and P. Taylor, A microkinetic analysis of the water–gas shift reaction under industrial conditions, Journal of Catalysis, vol. 158, pp. 170–180, 1996.
69 C. S. Gopinath and F. Zaera, Transient kinetics during the isothermal reduction of NO by CO on Rh (111) as studied with effusive collimated molecular beams, The Journal of Physical Chemistry B, vol. 104, pp. 3194–3203, 2000.
70 G. Boreskov and Y. S. Matros, Unsteady-state performance of heterogeneous catalytic reactions, Catalysis Reviews: Science and Engineering, vol. 25, pp. 551–590, 1983.
71 G. K. Boreskov, Heterogeneous Catalysis. New York: Nova Publishers, 2003.
72 H. Knözinger and G. Mestl, Laser Raman spectroscopy—a powerful tool for in situ studies of catalytic materials, Topics in Catalysis, vol. 8, pp. 45–55, 1999.
73 G. Thomas and G. Somorjai, Eds., Topics in Catalysis, Special Issue on In-Situ Characterization of Catalysts. vol. 8 (1–2) New York: Springer; 1999.
74 D. Wang and M. A. Barteau, Oxidation kinetics of partially reduced vanadyl pyrophosphate catalyst, Applied Catalysis A: General, vol. 223, pp. 205–214, 2002.
75 G. Centi, F. Trifiro, J. R. Ebner, and V. M. Franchetti, Mechanistic aspects of maleic anhydride synthesis from C_4 hydrocarbons over phosphorus vanadium oxide, Chemical Reviews, vol. 88, pp. 55–80, 1988.
76 J. Haber, Molecular mechanism of heterogeneous oxidation—organic and solid state chemists' views, Studies in Surface Science and Catalysis, vol. 110, pp. 1–17, 1997.
77 C. O. Bennett, Experiments and processes in the transient regime for heterogeneous catalysis, Advances in Catalysis, vol. 44, pp. 329–416, 1999.
78 F. Ribeiro, C. Gerken, G. Rupprechter, G. Somorjai, C. Kellner, G. Coulston, L.E. Manzer and L. Abrams, Structure insensitivity and effect of sulfur in the reaction of hydrodechlorination of 1, 1-dichlorotetrafluoroethane (CF_3–$CFCl_2$) over Pd catalysts, Journal of Catalysis, vol. 176, pp. 352–357, 1998.
79 J. T. Gleaves, G. Yablonsky, X. Zheng, R. Fushimi, and P. L. Mills, Temporal analysis of products (TAP)—recent advances in technology for kinetic analysis of multi-component catalysts, Journal of Molecular Catalysis A: Chemical, vol. 315, pp. 108–134, 2010.
80 J. Perez-Ramirez and E. V. Kondratenko, Evolution, achievements, and perspectives of the TAP technique, Catalysis Today, vol. 121, pp. 160–169, 2007.
81 J. T. Gleaves, G. S. Yablonskii, P. Phanawadee, and Y. Schuurman, TAP-2: an interrogative kinetics approach, Applied Catalysis A: General, vol. 160, pp. 55–88, 1997.
82 S. O. Shekhtman, G. S. Yablonsky, J. T. Gleaves, and R. Fushimi, "State defining" experiment in chemical kinetics—primary characterization of catalyst activity in a TAP experiment, Chemical Engineering Science, vol. 58, pp. 4843–4859, 2003.
83 M. Temkin, The kinetics of some industrial heterogeneous catalytic reactions, Advances in Catalysis, vol. 28, pp. 173–291, 1979.
84 M. Temkin, S. Kiperman, and L. Luk'yanova, Flow-recycling method for studies of kinetics of heterogeneous catalytic reactions, Doklady Akademii USSR (Proceedings of the USSR Academy of Sciences), vol. 74, pp. 763–767, 1950.
85 K. Denbigh, The thermodynamics of the steady state, American Journal of Physics, vol. 20, pp. 385–385, 1952.
86 X. Zheng, J. T. Gleaves, G. S. Yablonsky, T. Brownscombe, A. Gaffney, M. Clark, and S. Han, Needle in a haystack catalysis, Applied Catalysis A: General, vol. 341, pp. 86–92, 2008.
87 S. O. Shekhtman and G. S. Yablonsky, Thin-zone TAP reactor versus differential PFR: analysis of concentration nonuniformity for gas-solid systems, Industrial & Engineering Chemistry Research, vol. 44, pp. 6518–6522, 2005.
88 S. O. Shekhtman, G. S. Yablonsky, S. Chen, and J. T. Gleaves, Thin-zone TAP-reactor - theory and application, Chemical Engineering Science, vol. 54, pp. 4371–4378, 1999.
89 S. O. Shekhtman, G. S. Yablonsky, J. T. Gleaves, and R. R. Fushimi, Thin-zone TAP reactor as a basis of "state-by-state transient screening", Chemical Engineering Science, vol. 59, pp. 5493–5500, 2004.

90 G. S. Yablonsky, D. Constales, S. O. Shekhtman, and J. T. Gleaves, The Y-procedure: how to extract the chemical transformation rate from reaction–diffusion data with no assumption on the kinetic model, Chemical Engineering Science, vol. 62, pp. 6754–6767, 2007.
91 G. S. Yablonskii, S. O. Shekhtman, S. Chen, and J. T. Gleaves, Moment-based analysis of transient response catalytic studies (TAP experiment), Industrial & Engineering Chemistry Research, vol. 37, pp. 2193–2202, 1998.
92 Y. Schuurman and J. T. Gleaves, A comparison of steady-state and unsteady-state reaction kinetics of n-butane oxidation over VPO catalysts using a TAP-2 reactor system, Catalysis Today, vol. 33, pp. 25–37, 1997.
93 Y. Schuurman and J. T. Gleaves, "Alkane oxidation over oxidized VPO catalysts," Proceedings of the 210th ACS National Meeting, Chicago, IL. August 20–24, 1995, pp. COLL–036.
94 X. Zheng, G. M. Veith, E. Redekop, C. S. Lo, G. S. Yablonsky, and J. T. Gleaves, Oxygen and CO adsorption on Au/SiO_2 catalysts prepared by magnetron sputtering: the role of oxygen storage, Industrial & Engineering Chemistry Research, vol. 49, pp. 10428–10437, 2010.
95 X. Zheng, Getting to the point: Bridging the gap between simple and complex catalytic systems using Temporal Analysis of Products (TAP), Ph.D. Dissertation, Washington University in St. Louis, 2009.

CHAPTER 11

Numerical solution techniques

Ahmet Kerim Avci[1] and Seda Keskin[2]

[1]Department of Chemical Engineering, Boğaziçi University, Istanbul, Turkey

[2]Department of Chemical and Biological Engineering, Koc University, Istanbul, Turkey

Abstract

The present chapter provides an overview of several numerical techniques that can be used to solve model equations of ordinary and partial differential type, both of which are frequently encountered in multiphase catalytic reactor analysis and design. Brief theories of the ordinary differential equation solution methods are provided. The techniques and software involved in the numerical solution of partial differential equation sets, which allow accurate prediction of nonreactive and reactive transport phenomena in conventional and nonconventional geometries, are explained briefly. The chapter is concluded with two case studies that demonstrate the application of numerical solution techniques in modeling and simulation of hydrocarbon-to-hydrogen conversions in catalytic packed-bed and heat-exchange integrated microchannel reactors.

11.1 Techniques for the numerical solution of ordinary differential equations

Consider the following ordinary differential equation (ODE) and the initial condition:

$$\frac{dy}{dt} = f(y) \qquad y(0) = y_0 \tag{11.1}$$

Equation 11.1 resembles an initial value problem, which describes the structure of the models used for simulation and design of various reactors such as steady-running tubular packed beds with negligible radial gradients. In simulation problems, in which boundaries of integration domain (i.e., length or the volume of the reactor) and the inlet conditions are known, the objective is to solve the model equations for calculating the changes in the reactor operating parameters (e.g., temperature, molar flow rate of the species, pressure) with respect to the reactor volume or length. The model equations can also be used for handling design problems in which the aim is to estimate the boundaries of integration domain that allows the realization of the design target (e.g., specific values for parameters such as reactant conversion, product selectivity, or yield). In this case, model equations of ordinary differential type that are subject to the specified initial conditions are solved within the integration domain whose upper bound is selected on an arbitrary basis. This selection is continued on a trial-and-error basis, and the model is solved for each trial value of integration domain (e.g., reactor volume or length) until the pertinent design objective is satisfied.

Numerical solution of the ODEs describing initial value problems is possible by explicit and implicit techniques, which are described in the Sections 11.1.1 and 11.1.2, respectively. It is worth noting that the techniques are formulated for the solution of a single equation (Equation 11.1), but they can be used for solving multiple ODEs as well. Theoretical background of these methods as well as their stabilities are described elsewhere [1, 2] and will not be discussed here.

11.1.1 Explicit techniques

Explicit techniques calculate the state of a system at a later time by using information from the state of the system at the present time. These methods are also known as step-by-step methods, which start from the initial condition given in Equation 11.1 and continue stepwise by computing approximate values of the solution at points that increase by a fixed number, h, which is known as the step size. Some of the explicit techniques used to obtain numerical solutions of Equation 11.1 are summarized as follows.

11.1.1.1 Euler's and Runge–Kutta methods

Euler's method is first order and executes the integration through the following formulation:

$$y^{n+1} = y^n + h \cdot f(y^n) \tag{11.2}$$

A more accurate technique used for numerical integration is the classical fourth-order Runge–Kutta (RK) method. In this

Multiphase Catalytic Reactors: Theory, Design, Manufacturing, and Applications, First Edition. Edited by Zeynep Ilsen Önsan and Ahmet Kerim Avci.

technique, whose formulation is outlined in Equations 11.3–11.7, the quantities k_1, k_2, k_3, and k_4 are computed at each integration step to calculate y^{n+1}:

$$y^{n+1} = y^n + \frac{1}{6}(k_1 + 2k_2 + 2k_3 + k_4) \tag{11.3}$$

$$k_1 = h \cdot f(t^n, y^n) \tag{11.4}$$

$$k_2 = h \cdot f\left(t^n + \frac{1}{2}h, y^n + \frac{1}{2}k_1\right) \tag{11.5}$$

$$k_3 = h \cdot f\left(t^n + \frac{1}{2}h, y^n + \frac{1}{2}k_2\right) \tag{11.6}$$

$$k_4 = h \cdot f(t^n + h, y^n + k_3) \tag{11.7}$$

An alternative fourth-order formulation is the Runge–Kutta–Gill method:

$$y^{n+1} = y^n + \frac{1}{6}(k_1 + k_4) + \frac{1}{3}(b \cdot k_2 + d \cdot k_3) \tag{11.8}$$

$$k_1 = h \cdot f(t^n, y^n) \tag{11.4}$$

$$k_2 = h \cdot f\left(t^n + \frac{1}{2}h, y^n + \frac{1}{2}k_1\right) \tag{11.5}$$

$$k_3 = h \cdot f\left(t^n + \frac{1}{2}h, y^n + a \cdot k_1 + b \cdot k_2\right) \tag{11.9}$$

$$k_4 = h \cdot f(t^n + h, y^n + c.k_2 + d.k_3) \tag{11.10}$$

$$a = \frac{\sqrt{2}-1}{2},\; b = \frac{2-\sqrt{2}}{2},\; c = -\frac{\sqrt{2}}{2},\; d = 1 + \frac{\sqrt{2}}{2}$$

Runge–Kutta–Fehlberg (RKF) is another method involving fourth- and fifth-order integration formulations (Equations 11.11 and 11.12, respectively) whose difference $(y^{n+1} - z^{n+1})$ can be used for step size control:

$$z^{n+1} = y^n + \frac{25}{216}k_1 + \frac{1408}{2565}k_3 + \frac{2197}{4104}k_4 - \frac{1}{5}k_5 \tag{11.11}$$

$$y^{n+1} = y^n + \frac{16}{135}k_1 + \frac{6656}{12825}k_3 + \frac{28561}{56430}k_4 - \frac{9}{50}k_5 + \frac{2}{55}k_6 \tag{11.12}$$

These formulations require evaluation of six functions per step:

$$k_1 = h \cdot f(t^n, y^n) \tag{11.13}$$

$$k_2 = h \cdot f\left(t^n + \frac{1}{4}h, y^n + \frac{1}{4}k_1\right) \tag{11.14}$$

$$k_3 = h \cdot f\left(t^n + \frac{3}{8}h, y^n + \frac{3}{32}k_1 + \frac{9}{32}k_2\right) \tag{11.15}$$

$$k_4 = h \cdot f\left(t^n + \frac{12}{13}h, y^n + \frac{1932}{2197}k_1 - \frac{7200}{2197}k_2 + \frac{7296}{2197}k_3\right) \tag{11.16}$$

$$k_5 = h \cdot f\left(t^n + h, y^n + \frac{439}{216}k_1 - 8k_2 + \frac{3680}{513}k_3 - \frac{845}{4104}k_4\right) \tag{11.17}$$

$$k_6 = h \cdot f\left(t^n + \frac{1}{2}h, y^n - \frac{8}{27}k_1 + 2k_2 - \frac{3544}{2565}k_3 + \frac{1859}{4104}k_4 - \frac{11}{40}k_5\right) \tag{11.18}$$

11.1.1.2 Adams–Bashforth methods

Euler's and RK methods are also known as *one-step techniques* which use function values only in a single step, that is, in the preceding step. However, in the *multistep techniques*, evaluation of each step requires function values from more than one of the preceding steps. The benefit of the multistep techniques is the use of additional information to obtain more accurate solutions. The Adams–Bashforth methods for explicit solution of Equation 11.1 are multistep in nature and are given in second and fourth orders in Equations 11.19 and 11.20, respectively, as follows:

$$y^{n+1} = y^n + \frac{h}{2} \cdot \left[3f(y^n) - f(y^{n-1})\right] \tag{11.19}$$

$$y^{n+1} = y^n + \frac{h}{24} \cdot \left[55f(y^n) - 59f(y^{n-1}) + 37f(y^{n-2}) - 9f(y^{n-3})\right] \tag{11.20}$$

11.1.2 Implicit techniques

Implicit integration techniques involve various interpolation formulas that include the y^{n+1} term. Some of the techniques are described in the following sections.

11.1.2.1 Backward Euler method

The method is first order and is formulated as follows:

$$y^{n+1} = y^n + h \cdot f(y^{n+1}) \tag{11.21}$$

11.1.2.2 Trapezoidal rule

The method is second order that is given with the following equation:

$$y^{n+1} = y^n + \frac{h}{2} \cdot \left[f(y^n) + f(y^{n+1})\right] \tag{11.22}$$

11.1.2.3 Gear's backward difference formulas

The method involves the use of the following formulas that vary with the orders:

First order: $y^{n+1} = y^n + h \cdot f(y^{n+1})$

Second order:

$$y^{n+1} = \frac{4}{3}y^n - \frac{1}{3}y^{n-1} + \frac{2}{3}h \cdot f(y^{n+1}) \tag{11.23}$$

Third order:

$$y^{n+1} = \frac{18}{11}y^n - \frac{9}{11}y^{n-1} + \frac{2}{11}y^{n-2} + \frac{6}{11}h \cdot f(y^{n+1}) \tag{11.24}$$

Fourth order:

$$y^{n+1} = \frac{48}{25}y^n - \frac{36}{25}y^{n-1} + \frac{16}{25}y^{n-2} - \frac{3}{25}y^{n-3} + \frac{12}{25}h \cdot f\left(y^{n+1}\right) \tag{11.25}$$

Fifth order:

$$y^{n+1} = \frac{300}{137}y^n - \frac{300}{137}y^{n-1} + \frac{200}{137}y^{n-2} - \frac{75}{137}y^{n-3} + \frac{12}{137}y^{n-4} + \frac{60}{137}h \cdot f\left(y^{n+1}\right) \tag{11.26}$$

The main advantage of the implicit techniques is their stability for any given value of the step size. This advantage, however, requires the solution of a set of nonlinear equations via an iterative approach. For this purpose, methods such as successive substitution or Newton–Raphson can be used [1].

Gear's formulas are particularly useful for stiff problems, which are frequently encountered in chemical reaction engineering problems. In stiff problems, parameters such as temperature and molar flow rate of a species change rapidly in a small spatial scale. The sharp temperature rise as a result of hydrocarbon oxidation (i.e., the light-off phenomenon) in a packed-bed tubular reactor is an example of a stiff problem [3]. Here, significant temperature change occurs in a very small portion of reactor tube length, and the solution of the differential equation describing the spatial temperature change requires special attention. In such cases, numerical solution via the explicit RK techniques is possible by using very small step sizes. Alternatively, implicit integration techniques such as Gear's method, which offers a higher chance of convergence, can be employed.

Most of the explicit and implicit algorithms introduced previously form the basis of numerical integration packages that are used in commercial suites such as Polymath [4], MATLAB [4, 5], and Mathematica [6]. Single or multiple nonstiff ODEs of initial value type can be solved in MATLAB by using either "ode45" or "ode113" functions that are based on RK and Adams–Bashforth techniques, respectively [5]. For the solution of ODEs describing stiff problems, "ode15s" function that involves numerical differentiation formulas (e.g., Gear's method) or "ode23t" function that is based on the trapezoidal rule can be used [5]. In addition to commercial software, some open sources such as NIST (http://gams.nist.gov) and Netlib (http://www.netlib.org) offer downloadable ODE packages that can be used in the solution of stiff and nonstiff problems.

In some problems of chemical reaction engineering, set of ODEs needs to be coupled with nonlinear algebraic set of equations. An example of such a case is the use of heterogeneous packed-bed reactor models in which product and temperature distributions in bulk fluid and in catalyst phases need to be solved simultaneously. Temperature and concentration changes in the bulk fluid are described by differential mole and energy balances forming the set of ODEs, while the transport of mass and energy from bulk fluid to the external catalyst surface is mathematically described by algebraic set of equations. Simultaneous solution of the so-called differential-algebraic equation (DAE) set requires coupling of the ODE and algebraic equation solvers, the latter which are not discussed here, but can be found in detail elsewhere [1]. Description of a DAE set and its solution in the context of a one-dimensional (1D) heterogeneous packed-bed reactor model for autothermal conversion of methane to hydrogen is available in the literature [7]. It is also worth noting that packages such as DASSL and DAEPACK are also available for the solution of coupled DAE sets.

11.2 Techniques for the numerical solution of partial differential equations

ODEs describe the change of a dependent variable (y in Equation 11.1) with respect to a single independent variable (t in Equation 11.1). If the number of independent variables increases, the mathematical expression becomes a partial differential equation (PDE). Two-dimensional (2D) steady-state tubular reactor models (which quantify the axial and radial changes of reaction parameters (e.g., temperature, species concentrations)) and 1D unsteady-state models (which describe temporal and spatial changes of reaction parameters) are examples of cases involving set of PDEs. Single PDEs can be solved by analytical techniques such as separation of variables. Analytical techniques will not be discussed here, as most of the chemical reactor models involve change of more than one dependent variable in multiple spatial directions either at steady-state or at transient conditions. Solution of set of PDEs requires numerical techniques such as finite difference method (FDM) [8]. In FDM, the solution domain is divided into structured grids (such as quadrilateral cells), and the partial derivatives are approximated by finite difference expressions based on Taylor series expansions. The derivative of the function at a node (one of the connection points of the grids) can then be written in terms of the function values at the neighboring nodes. As a result of this formulation, which can be done explicitly or implicitly, a set of linear algebraic equations, with the unknowns being the values of dependent variables at each node, is obtained. In explicit formulation, function values at node $n + 1$ can be evaluated directly if the values at node n and at lower nodal points are known. In implicit formulation, however, values at all nodes have to be evaluated altogether by simultaneous solution of the algebraic equations. Even though implicit formulation is more expensive than the explicit one in terms of computational complexity, the likelihood of convergence and stability is higher in the former one.

An alternative technique is the method of orthogonal collocation. The method involves several number of spatial collocation points at which the partial derivatives are formulated in terms of ODEs. In other words, the method allows the discretization of PDEs into an ODE set over the collocation points, with the number of ODEs being determined by the number points used. The resulting ODE set can then be solved by one of the methods outlined in Section 11.1. Using higher number of collocation points improves the spatial discretization but increases the cost of ODE solution, which is less of an issue with the help of fast computers. Details of orthogonal collocation technique can be found elsewhere [1, 8–11].

FDM and orthogonal collocation are well-known techniques, but they may become inefficient in solving the model equations involving stiffness, which is common in chemical reaction engineering problems. The use of structured grids in FDM may fail in parsing of irregular modeling domains and in predicting phenomena involving sharp gradients. These disadvantages of FDM can be dampened by the use of unstructured grids (such as triangular cells), which are more suitable for discretizing irregular flow paths with varying intensity to handle stiffness. Unstructured grids are less compatible with FDM but are used in finite element and finite volume methods (FEM and FVM, respectively), the techniques which are more robust than FDM in solving set of PDEs. FEM and FVM, which are explained briefly in Sections 11.3.2 and 11.3.3, respectively, are also known as computational fluid dynamics (CFD) techniques.

11.3 Computational fluid dynamics techniques

CFD is the general term used for the numerical analysis of systems involving fluid flow, heat, and mass transfer and associated phenomena such as flow in porous media and chemical reactions [12]. Analysis and engineering of multiphase catalytic reactors are among the many application areas of the CFD analysis such that all the physical and reactive phenomena are interrelated to each other. Wide availability of high-performance computing tools and commercial software has led to a notable increase in the number of studies that handle catalytic reaction engineering problems using CFD techniques. The advantages of CFD over experimental testing systems and pilot plants can be summarized as follows:

- Faster and cheaper delivery of new designs by reducing dependence on experimentation
- Ability to investigate systems when controlled experimentation is difficult, expensive, or impossible to perform (e.g., chemical reactors near explosive limits)
- Production of large number of parametric results with high level of detail at no additional cost

11.3.1 Methodology of computational fluid dynamics

Numerical algorithms that are developed to solve fluid flow problems form the basis of the CFD codes [12], which discretize the problem on a given grid that represents the model geometry, and then solve the resulting system of equations in order to obtain the nodal values of the field variables. Most of the commercial CFD suites are able to examine, display, and visualize the results. A complete CFD code involves three main elements, namely, a preprocessor, a solver, and a postprocessor [12]. The following tasks are carried out in the preprocessing step:

- Definition of the computational domain, the geometry to be simulated
- Generation of a grid consisting of smaller, nonoverlapping subdomains called *elements* or *cells*
- Association of the physical and/or chemical phenomena with each group of subdomains and specification of the boundary conditions
- Definition of fluid properties and incorporation of constitutive and user-defined functions (UDFs) (e.g., rate expressions of chemical reactions)

Solution to a flow problem is defined at nodes inside each element (cell). In general, it can be said that the larger the number of elements, the better the solution, since the accuracy of a CFD solution is governed by the number of elements in the grid. Moreover, some field variables go through rapid variations in certain regions in the domain, for example, as in the case of fast exothermic oxidation reactions. This necessitates the refining of the grid, or increasing the number of elements in that region. Refining continues until the solution is guaranteed to be grid-free. It is worth noting that refining the grid increases the computational cost associated with the CFD solution of the problems [12].

The solver, second element of a CFD package, transforms the equations of transport phenomena into a system of algebraic equations via discretization and solves them by an iterative approach. These steps can be carried out either by the FEM or by the FVM that are explained in the following sections. The following balance equation in an arbitrarily chosen element or cell can be written for a general flow variable ϕ, for example, a velocity component or enthalpy [12]:

$$\begin{bmatrix}\text{Rate of change}\\ \text{of } \phi \text{ in the}\\ \text{control}\\ \text{volume with}\\ \text{respect to time}\end{bmatrix} = \begin{bmatrix}\text{Net rate of}\\ \text{increase of}\\ \phi \text{ due to}\\ \text{convection into}\\ \text{the control}\\ \text{volume}\end{bmatrix} + \begin{bmatrix}\text{Net rate of}\\ \text{increase of}\\ \phi \text{ due to}\\ \text{diffusion into}\\ \text{the control}\\ \text{volume}\end{bmatrix} + \begin{bmatrix}\text{Net rate of}\\ \text{generation of}\\ \phi \text{ inside the}\\ \text{control}\\ \text{volume}\end{bmatrix} \tag{11.27}$$

Outcomes of the solvers are analyzed by using the postprocessors, the collection of tools that, in general, allows detailed visualization of the computational domain and the grid architecture, editing and refining the 2D/three-dimensional (3D) plots of the distributions of dependent variables (e.g., velocity, temperature, species concentration) along the computational domain and displaying time-variable results in animated forms.

11.3.2 Finite element method

One of the methods that is incorporated into the CFD packages is the FEM. The method is based on the integral formulation of the governing PDEs, and, when compared with the FDM, it poses less stringent requirements for regularity and offers higher chance of obtaining a numerical solution [13, 14]. FEM typically involves the following steps [14]:

(a) Discretization of the solution domain into a grid structure that is composed of elements such as triangles and quadrilaterals. The elements are generated by the connection of nodes. A description of elements and nodes are presented in Figure 11.1.

(b) Determination of the interpolation (or shape) function, which is used to describe the change of a field variable (e.g., velocity, temperature) over a particular element.

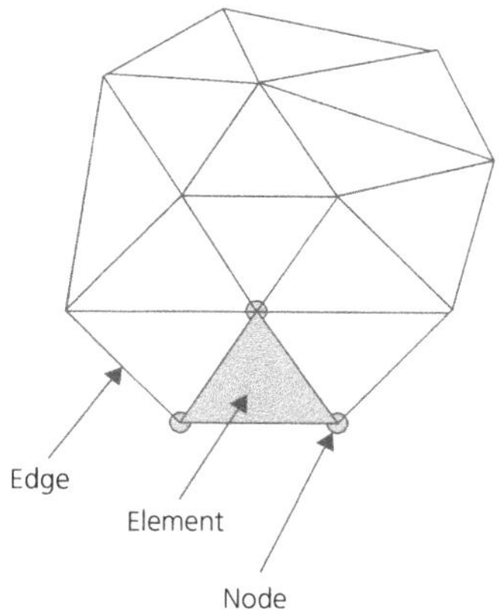

Figure 11.1 Structure of a finite element grid.
(Source: Lewis et al. [14]. Reproduced with permission of John Wiley & Sons, Inc.)

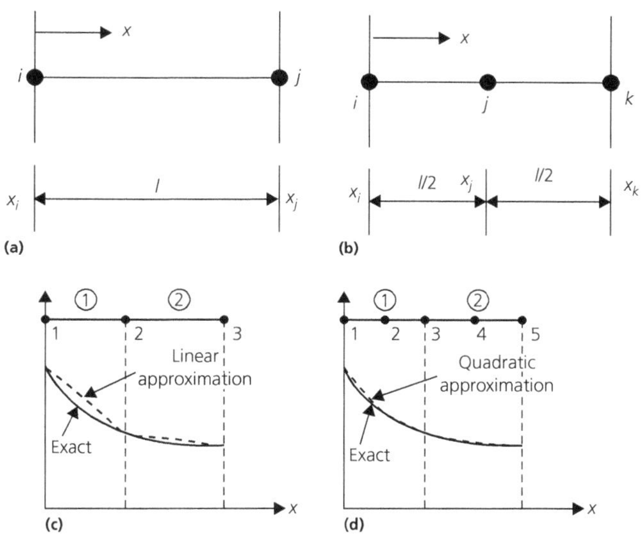

Figure 11.2 One-dimensional finite elements. (a) A linear element, (b) a quadratic element, and (c) linear and (d) quadratic variation of field variable ϕ over an element. Black circles represent the nodes, while the white, numbered circles represent the particular element.
(Source: Lewis et al. [14]. Reproduced with permission of John Wiley & Sons, Inc.)

(c) Formulation of the element equations whose unknowns are the values of the field variables at the nodal points.
(d) Combination of the element equations into a set of algebraic equations to represent the change of the field variable in the complete solution domain.
(e) Solution of the set of algebraic equations to compute values of the field variable at the nodal points.

The shape (or interpolation) functions mentioned in the previous algorithm are mostly considered to be polynomial type due to their ease of integration and differentiation. Another advantage of the polynomial shape functions is the possibility of obtaining a more accurate approximation of the exact solution simply by increasing the order of the polynomial. Comparison of the linear and quadratic approximations of the change of a field variable in an element is presented in Figure 11.2.

Fundamentals and theory of the FEM are covered extensively in the literature and will not be covered in this chapter. However, the reader is directed to Refs. [8] and [13] for a concise explanation of FEM theory and to Ref. [14] for in-depth description and formulation of shape functions, elements, and applications of FEM to various heat transfer problems.

COMSOL Multiphysics is one of the commercial computing software that uses FEM technique to solve continuum equations. The solvers used by COMSOL Multiphysics for the solution of the matrices assembled in the FEM break down the linear (or nonlinear) problems into one or several linear systems of equations by approximating the pertinent problem with a linearized one [15]. The so-called linear system solvers available in COMSOL Multiphysics and their key features are presented in Table 11.1.

The direct solvers listed in Table 11.1 solve the assembled set of equations by the Gaussian elimination method [1, 8] and deliver solutions often faster than the iterative solvers for 1D and 2D problems. Direct solvers can be used for 3D models if the degrees of freedom is less than ~10^5. Iterative solvers, on the other hand, are used in models with degrees of freedom above ~10^5 and in the solution of 3D problems, for which the memory requirements of the direct solvers are excessive. The readers are directed to Ref. [15] for further details of the solvers listed in Table 11.1.

Table 11.1 Linear system solvers available in COMSOL Multiphysics.

Direct solvers	
UMFPACK	• Highly efficient direct solver for one- and two-dimensional problems • Used for nonsymmetric systems
SPOOLES	• Efficient direct solver for one- and two-dimensional problems • Used for both symmetric and nonsymmetric systems • Uses less memory than UMFPACK
PARDISO	• Highly efficient direct solver for one- and two-dimensional problems • Used for symmetric and nonsymmetric systems • Often uses less memory than UMFPACK
Cholesky (TAUCS)	• Efficient direct solver for one- and two-dimensional problems • Used for symmetric positive-definite systems
Iterative solvers	
GMRES	• Used for nonsymmetric systems
FMGRES	• Used for nonsymmetric systems • Able to handle more general preconditioners • Requires more memory than GMRES
Conjugate gradients	• Used for symmetric positive-definite systems
Geometric multigrid	• Used for elliptic or parabolic systems

Source: Adapted from [15].

11.3.3 Finite volume method

The FVM is similar to FEM in the sense that it is based on the integral formulation of the PDEs describing the system. The solution steps involved in FVM can be summarized as follows:

(a) Discretization of the solution domain into control volumes (cells)
(b) Integration of the equations on the individual control volumes in order to reduce them to algebraic equations for the dependent (unknown) variables
(c) Linearization of the discretized equations and solution of the linear system(s) of equations in order to update the values of the variables until convergence criteria are met

The solution strategy is somewhat varied by the last step since the approach used to linearize and solve the discretized equations varies with the solver type. The two commonly employed solvers in the FVM are *pressure-based* and *density-based* solvers [12, 16]. In both methods the velocity field is obtained from the momentum equations. In the *density-based* approach, the continuity equation is used to obtain the density field, while the pressure field is determined from an equation of state. On the other hand, the *pressure-based* solver extracts the pressure field by solving the pressure or pressure correction equation, which is obtained by manipulation of the momentum and continuity equations [16]. Implementation of the pressure-based solver via the so-called Semi-Implicit Method for Pressure-Linked Equations (SIMPLE) algorithm [12] is explained later. Details of the density-based solver are extensively covered elsewhere [16] and will not be discussed here.

Application of the pressure-based solver via the SIMPLE algorithm can be explained in the context of transport equations that quantify hydrodynamics, mass, and heat transfer inside a single porous catalyst-coated channel of a monolith reactor shown in Figure 11.3 [17]. The relevant transport equations are given in their generalized form as follows:

Continuity equation:

$$\frac{\partial \rho_f}{\partial t} + \nabla \cdot \left(\rho_f \vec{v}\right) = 0 \tag{11.28}$$

Momentum conservation:

$$\frac{\partial}{\partial t}\left(\rho_f \vec{v}\right) + \nabla \cdot \left(\rho_f \vec{v}\vec{v}\right) = \vec{F} - \nabla p + \nabla \cdot \tau + \rho_f \vec{g} \tag{11.29}$$

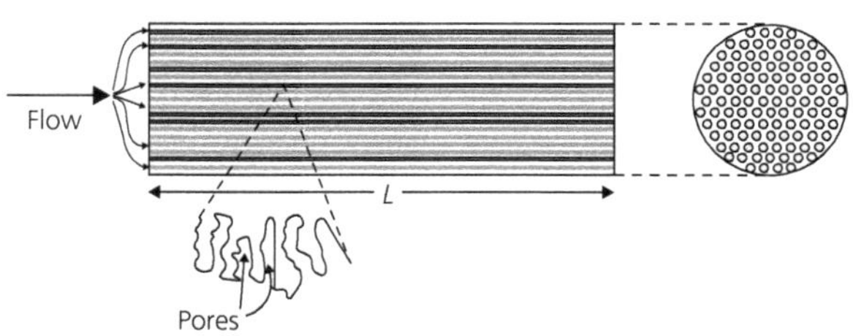

Figure 11.3 Schematic presentation of a monolith reactor with porous catalyst-coated channels
(Source: Onsan and Avci [17]. Reproduced with permission of Elsevier.)

Species mass conservation:

$$\frac{\partial}{\partial t}\left(\rho_f Y_m\right) + \nabla \cdot \left(\rho_f \vec{v} Y_m\right) = -\nabla \cdot \vec{J}_m + M_m R_m^v + M_m R_m^s A_s \tag{11.30}$$

Energy equation:

$$\varepsilon \frac{\partial}{\partial t}\left(\rho_f c_{pf} T\right) + (1-\varepsilon)\frac{\partial}{\partial t}\left(\rho_s c_{ps} T\right) + \nabla \cdot \left(\rho_f c_{pf} \vec{v} T\right) = \nabla \cdot \left(k_{eff} \nabla T - \sum_m h_m \vec{J}_m + \left(\tau \vec{v}\right)\right) + S_h \tag{11.31}$$

The transport equations given earlier are all functions of the velocity field $\vec{v}$, which is an unknown in the CFD calculations. The momentum conservation equation that gives the velocity field includes the pressure gradient term. If pressure gradient is known, the momentum equation can readily be discretized as has been done for any other scalar. If the flow is compressible, the continuity and energy equations may be used as the transport equations for density and temperature, respectively, which allow for the calculation of pressure through an equation of state. However, if the flow is incompressible, then the density is a constant and, by definition, not linked to pressure. In this case pressure–velocity coupling introduces a constraint in the solution of the flow field: if the correct pressure field is applied in the momentum equation, the resulting velocity field should satisfy the continuity equation. By adopting an iterative solution strategy such as the SIMPLE algorithm [18, 19], the pressure–velocity coupling can be resolved. In this algorithm the mass flux $(\rho_f \vec{v})$ terms are evaluated from so-called guessed velocity components. Furthermore, a guessed pressure field is used to solve the momentum equation, and a pressure correction equation, deduced from the continuity equation, is solved to obtain a pressure correction field, which is in turn used to update the velocity and pressure fields. The SIMPLE algorithm can be summarized as follows [12, 16, 18, 19]:

(a) Set the boundary conditions and the initial guesses $(p^*, \vec{v}^*, \phi^*)$.
(b) Compute the velocity and pressure gradients.
(c) Calculate coefficients of the equations and solve the discretized momentum equation to compute the intermediate velocity field $(\vec{v}^*)$.
(d) Solve the pressure correction equation, and update velocity and pressure and any other scalars $(p, \vec{v}, \phi)$.
(e) Check for convergence; if not converged, store the computed values and go back to step (b).

The pressure-based solver and the SIMPLE algorithm, as well as the density-based solvers, are implemented into ANSYS Fluent, another CFD package used in many applications [16]. While it is successful in simulating many continuum problems, some additional steps needs to carried out for simulation of the catalytic reactors in ANSYS Fluent. If the catalytic phenomena are described by mechanistic rate laws, these expressions have to be introduced to the software in the form of UDFs, which are collection of computer codes needed to be prepared separately. However, empirical rate expressions can be defined into ANSYS Fluent directly. Alternatively, software such as CHEMKIN can be directly integrated to solve reactive flow problems. In some suites such as COMSOL Multiphysics (Section 11.3.2), it is possible to define the mechanistic kinetic information directly into the software interface without requiring the need for coding of UDFs. Apart from ANSYS Fluent and COMSOL Multiphysics, other commercial suites such as OpenFOAM, CFD-ACE+, and STAR-CD, can be used to model and simulate catalytic reactors by the CFD approach.

11.4 Case studies

11.4.1 Indirect partial oxidation of methane in a catalytic tubular reactor

11.4.1.1 Problem description

Hydrogen production is conventionally carried out by steam reforming (SR) of natural gas over Ni-based catalysts. Since most of the natural gas is composed of methane, SR can be described by the following reaction:

$$\text{SR}: CH_4 + H_2O = CO + 3H_2 \quad \Delta H^{o}_{298} = 206.2\,\text{kJ/mol} \tag{11.32}$$

The reaction is carried out in packed-bed tubular reactors which are heated externally due to the strong endothermic nature of the reaction. External energy demand can be minimized by coupling SR with catalytic total oxidation (TOX) of part of the fuel fed to the reactor over a Pt-based catalyst:

$$\text{TOX}: CH_4 + 2O_2 = CO_2 + 2H_2O \quad \Delta H^{o}_{298} = -802.3\,\text{kJ/mol} \tag{11.33}$$

Combination of SR and TOX is known as indirect partial oxidation (IPOX). Due to the fast nature of TOX, it occurs earlier than SR and elevates the reactor temperature via exothermic heat release. In IPOX, oxygen is always used as the limiting reactant, and once it is completely consumed, the remaining methane is converted to hydrogen via Reaction 11.32. While SR and TOX are the main reactions, product distribution is strongly affected by the water–gas shift (WGS):

$$\text{WGS}: CO + H_2O = CO_2 + H_2 \quad \Delta H^{o}_{298} = -41.2\,\text{kJ/mol} \tag{11.34}$$

11.4.1.2 Reactor model and numerical solution

In IPOX it is essential to operate the reactor in adiabatic mode as the exothermal heat has to be kept within the reactor for use in the reforming step. As the adiabatic operation hinders the radial heat transfer, gradients in this direction can be considered as negligible compared to the ones in the axial (z)-direction. Assuming that the convective effects are dominant over the diffusive ones in the z-direction, the spatial changes in the molar flow rate of the species and temperature at steady state can be described by the 1D packed-bed tubular reactor model [20]:

$$\frac{dF_m}{dz} = A_c \rho_{bed} R'_m \tag{11.35}$$

$$\frac{dT}{dz} = \frac{A_c \rho_{bed} \sum_{e=1}^{3} (-\Delta H_e)(-r_e)}{\sum_{m=1}^{7} F_m c_{pm}} \tag{11.36}$$

$$\text{At } z = 0,\; F_m = F_m^{in};\; T = T^{in} \tag{11.37}$$

As summarized earlier, IPOX can be represented by three reactions. The presence of seven components (CH_4, H_2O, CO, CO_2, H_2, O_2, and N_2) is considered in the model. The catalyst is considered to be a bimetallic Pt/Ni/Al_2O_3, which involves the metals active for TOX (Pt) and SR (Ni). It is worth noting that the model is formulated by assuming the catalytic reactor as a pseudohomogeneous entity and neglects internal (intraparticle) and external (particle-to-fluid) mass and heat transfer resistances. The rate expressions needed to describe the coexistence of TOX, SR, and WGS reactions and physical parameters needed for the solution of equations are obtained from the literature [3, 21]. The differential mole and energy balances, Equations 11.35 and 11.36, respectively, together with the initial conditions given in Equation 11.37 form a system of ODEs that are solved using a nonstiff, low-order ODE solver, the "ode23" function of the MATLAB software.

Equations 11.35–11.37 describe the steady-state variations of species molar flow rates and temperature along the reactor. When time is considered as an independent variable in addition to the reactor length to account for the temporal variations, the reactor model can be expressed in terms of a set of PDEs as follows [20]:

$$\varepsilon \frac{\partial c_m}{\partial t} = -\frac{1}{A_c}\frac{\partial F_m}{\partial z} + \rho_{bed} R'_m \tag{11.38}$$

$$\varepsilon \sum_{m=1}^{7} (c_m c_{pm}) \frac{\partial T}{\partial t} = -\frac{\sum_{m=1}^{7} (F_m c_{pm})}{A_c}\frac{\partial T}{\partial z} + \rho_{bed} \sum_{e=1}^{3} (-\Delta H_e)(-r_e) \tag{11.39}$$

$$\text{At } t = 0, F_m = F_{m0};\; T = T_0 \tag{11.40}$$

$$\text{At } z = 0,\; F_m = F_m^{in}; T = T^{in} \tag{11.41}$$

The species concentration term in Equations 11.38 and 11.39 can be expressed in terms of species molar flow rate through the following relationship:

$$F_m = c_m \upsilon \tag{11.42}$$

Assuming that the gaseous reaction mixture can be considered to behave ideally, the volume change during the reaction can be evaluated by the following equation:

$$\upsilon = \upsilon^{in} \left(\frac{F_{Total}}{F_{Total}^{in}}\right)\left(\frac{T}{T^{in}}\right) \tag{11.43}$$

Time dependence of υ is implicitly accounted by the dependence of F_T and T via Equations 11.38 and 11.39, respectively.

Spatial dependence of υ can be expressed by the following differential equation which is coupled to Equations 11.38 and 11.39 at each time step:

$$\frac{d\upsilon}{dz} = \frac{\upsilon^{in}}{F^{in}_{Total}T^{in}}\left(T\frac{dF_{Total}}{dz} + F_{Total}\frac{dT}{dz}\right) \tag{11.44}$$

The system of equations and the initial conditions (Equations 11.38–11.44) are solved by using the FDM. If the 1D domain is divided into I intervals with $I+1$ nodal points and the temporal and spatial derivatives are approximated by central differences, the following explicit scheme is obtained:

$$\frac{(c_m)_i^{n+1} - \frac{1}{2}\left[(c_m)_{i+1}^n - (c_m)_{i-1}^n\right]}{\Delta t} = -\frac{1}{\varepsilon A_c}\frac{(F_m)_{i+1}^n - (F_m)_{i-1}^n}{2\Delta z} + \frac{\rho_{bed}(R'_m)_i^n}{\varepsilon} \tag{11.45}$$

$$\frac{T_i^{n+1} - \frac{1}{2}\left[T_{i+1}^n - T_{i-1}^n\right]}{\Delta t} = \frac{\sum_{m=1}^{7}(F_m c_{pm})_i^n}{\varepsilon\sum_{m=1}^{7}(c_m c_{pm})_i^n A_c}\frac{T_{i+1}^n - T_{i-1}^n}{2\Delta z} + \frac{\rho_{bed}\sum_{e=1}^{3}(-\Delta H_e)(-r_e)}{\varepsilon\sum_{m=1}^{7}(c_m c_{pm})_i^n} \tag{11.46}$$

$e = 1, 2, 3;\ m = 1, 2, \ldots 7;\ i = 1, 2, \ldots I$

At each time step, values of the variables are calculated at every i, and these are used as input for the next time stepping. This scheme of discretization is known as Lax–Friedrichs finite difference scheme, which is first order accurate [22, 23]. In order to ensure stability during time stepping, the variables at time n are approximated as the average of their values at $(i-1)$th and $(i+1)$th nodes instead of simple forward differencing, that is, $(c_m)_i^{n+1} - (c_m)_i^n$ or $T_i^{n+1} - T_i^n$. According to the Courant–Friedrichs–Levy (CFL) criterion [22, 23], the system retains its stability as long as the following condition is satisfied:

$$\left(\frac{1}{\varepsilon A_c}\right)\frac{\Delta t}{\Delta z} \le 1 \tag{11.47}$$

At the end of the spatial domain, which is quantified by the boundary conditions given in Equation 11.41, a hypothetical nodal point $(I+1)$ must be included. Therefore, an artificial boundary condition, which is of Neumann type that specifies the derivative of the function, is required:

$$\text{At } z = L,\ \frac{\partial f}{\partial z} = 0 \tag{11.48}$$

11.4.1.3 Results and key findings

Output of the numerical solution is given in terms of temperature profiles shown in Figure 11.4 that demonstrate dependence of temperature on the inlet conditions. The results show that maximum value of the temperature increases with decrease in the molar inlet methane-to-oxygen (CH_4/O_2) ratio from 2.24 to 1.89. This outcome is expected due to the increased extent of Reaction 11.33 that releases higher amount of exothermal heat to raise the temperature. Increasing the amount of steam (i.e., changing molar inlet steam-to-methane ratio (H_2O/CH_4) from 1.17 to 2.34) while keeping the amount of inlet oxygen constant leads to an expected decrease in the maximum temperature due to the pronounced effect of endothermic SR that acts as a heat sink.

Outcomes of the solution of the unsteady-state (transient) reactor model by the finite difference technique outlined earlier are given in Figures 11.5, 11.6, and 11.7 in terms of temperature profiles obtained at different combinations of molar inlet CH_4/O_2 and H_2O/CH_4 ratios. It can be observed that the reactor operation reaches to its steady state between 60 and 120 s, depending on the feed combination. Comparison of the solutions of the steady-state (Figure 11.4) and unsteady-state models (Figures 11.5 and 11.6) shows that the "steady-state" profiles are close to each other with minor differences that possibly come from the

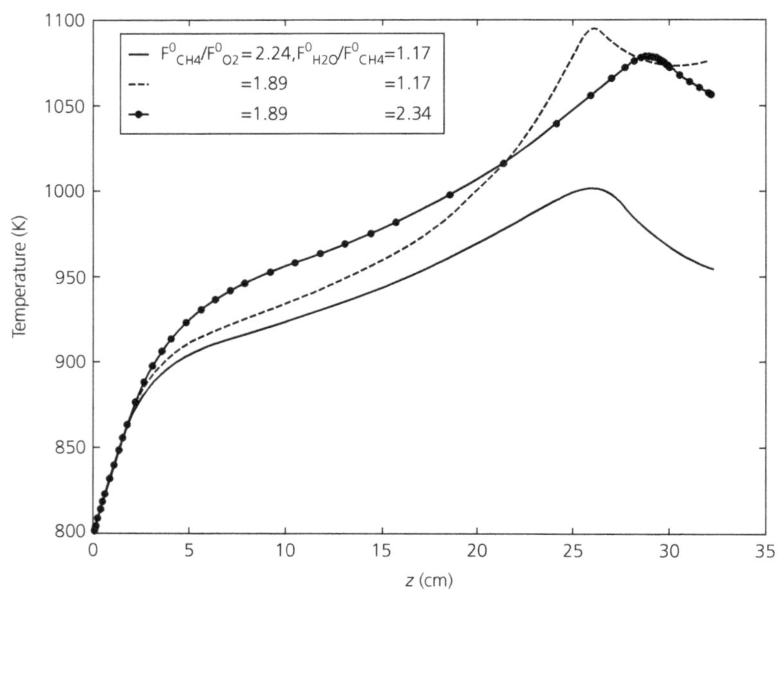

Figure 11.4 Steady-state reactor temperature profiles obtained at different values of molar CH_4/O_2 and H_2O/CH_4 ratios at the inlet.
(Source: Karakaya et al. [20]. Reproduced with permission of Elsevier.)

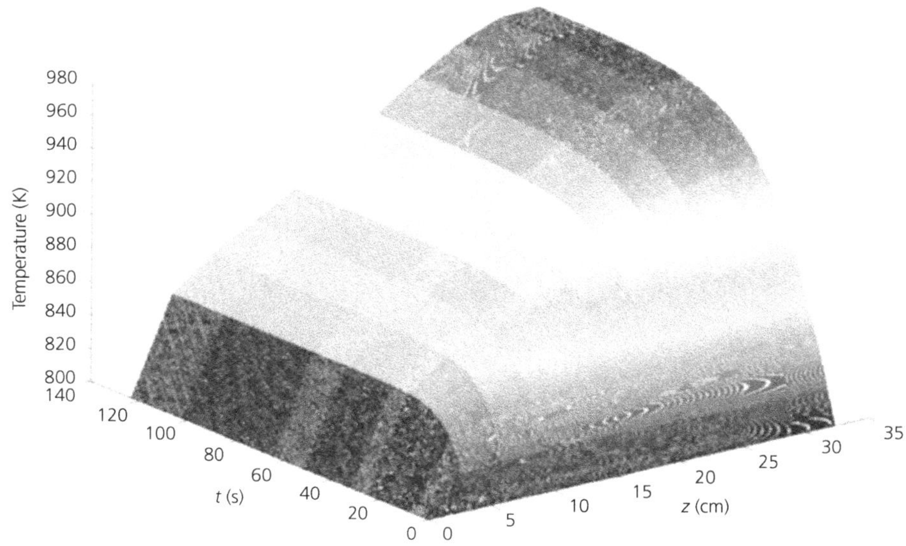

Figure 11.5 Transient reactor temperature profile obtained at $CH_4/O_2 = 2.24$ and $H_2O/CH_4 = 1.17$.
(Source: Karakaya et al. [20]. Reproduced with permission of Elsevier.)

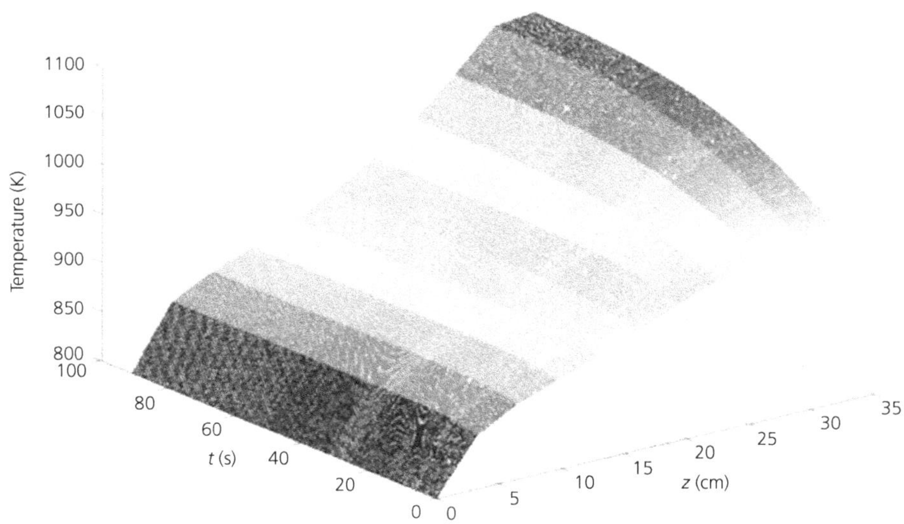

Figure 11.6 Transient reactor temperature profile obtained upon changing CH_4/O_2 from 2.24 to 1.89 ($H_2O/CH_4 = 1.17$).
(Source: Karakaya et al. [20]. Reproduced with permission of Elsevier.)

approximations involved in the formulation of the finite difference scheme. Responses of the transient model against the changes in the feed conditions (Figures 11.5, 11.6, and 11.7) are similar to those of the steady-state model explained earlier.

11.4.2 Hydrocarbon steam reforming in spatially segregated microchannel reactors

11.4.2.1 Problem description

Apart from coupling the exothermic and endothermic reactions in the same volume, these reactions can be physically decoupled to run in different volumes which are separated by a solid medium allowing conductive heat exchange between the reactions. This configuration can be realized in microchannel reactors, the units known with heat and mass transfer rates that are much higher than those involved in the conventional units. The reader is referred to Chapter 9 of this book for a detailed description and analysis of microchannel reactors.

A microchannel reactor configuration, in which catalytic endothermic (hydrocarbon SR) and exothermic (hydrocarbon combustion) reactions can be coupled, is shown in Figure 11.8 [24]. The reactor is composed of parallel groups of endothermic and exothermic channels which are separated by thin solid walls. The reactive flows are considered to be co-current. Each channel is square shaped, and the inner walls of the channels are wash-coated with a porous supported metal catalyst specific for the reaction type. Washcoat thickness is assumed to be uniform

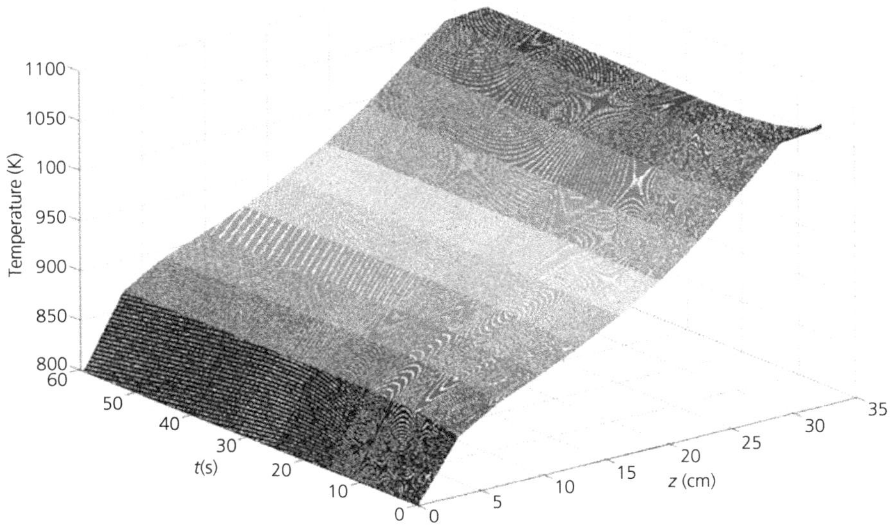

Figure 11.7 Transient reactor temperature profile obtained upon changing H_2O/CH_4 from 1.17 to 1.56 ($CH_4/O_2 = 1.89$).
(Source: Karakaya et al. [20]. Reproduced with permission of Elsevier.)

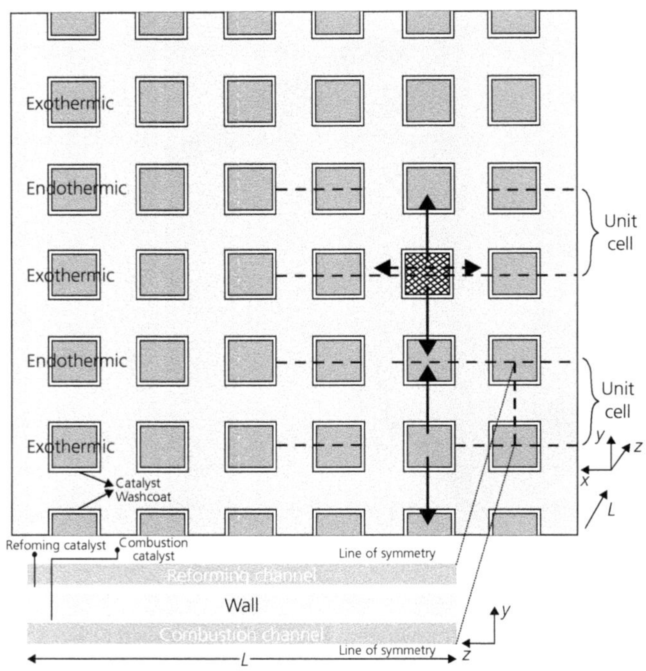

Figure 11.8 Frontal view of the heat-exchange integrated microchannel reactor concept and the two-dimensional unit cell (shown at the lower inset).
(Source: Karakaya and Avci [24]. Reproduced with permission of Elsevier.)

on all sides and along the channels. It is also assumed that (i) the same reaction takes place in a horizontal array of channels, each of which has the same inlet conditions, and (ii) the outermost surfaces of the reactor block is well insulated, and the heat loss to the surroundings become negligible.

11.4.2.2 Reactor model and numerical solution

Modeling of the reactor domain shown in Figure 11.8 requires simultaneous solution of continuity and conservation (of momentum, energy, and species mass) equations in the fluid and catalytic washcoat phases in every channel as well as of the heat flow between each channel in three dimensions. These sets of equations and the necessary boundary conditions, presented in Tables 11.2 and 11.3, respectively, can be simplified as a result of the following assumptions arising from the particular positioning of the channels:

- Considering the shaded exothermic combustion channel in Figure 11.8, it can be understood that the released heat is transferred in all directions. However, due to the fact that identical reactions occur in x-axis, temperature gradients occurring in x-direction become negligible compared to the ones occurring in y-axis. Moreover, heat flow out of the shaded channel in x-direction (shown by the dashed arrows) is counterbalanced by the heat flow from the neighboring exothermic channels. As a result, heat transfer in x-direction, that is, between the channels of the same horizontal group, becomes negligible in the whole reactor domain.
- The alternating and symmetrical arrangement of the channels allows the use of the “unit cell” as the representative geometry of the whole reactor block. Since the vertical heat fluxes (shown by the solid arrows in y-direction) that enter to/leave from an endothermic/exothermic channel group cancel each other out, the unit cell becomes adiabatic.

Based on the simplifications summarized earlier, 3D modeling of the multichannel reactor block can be simplified to 2D modeling of the unit cell which is composed of half of the exothermic and endothermic channels involving the catalytic washcoats and the thermally conductive wall. Validity of this simplification is checked by solving the 3D model equations

Table 11.2 Model equations used to simulate heat-exchange integrated microchannel reactor operation shown in Figure 11.8.

Fluid phase:

Equation of continuity

$$\frac{\partial v_{xb}}{\partial x_b}+\frac{\partial v_{yb}}{\partial y_b}+\frac{\partial v_{zb}}{\partial z_b}=0$$

Equation of motion

$$\rho_{fb}\left(v_{xb}\frac{\partial v_{xb}}{\partial x_b}+v_{yb}\frac{\partial v_{xb}}{\partial y_b}+v_{zb}\frac{\partial v_{xb}}{\partial z_b}\right)=-\frac{\partial p_b}{\partial x_b}+\mu_b\left[\frac{\partial^2 v_{xb}}{\partial x_b^2}+\frac{\partial^2 v_{xb}}{\partial y_b^2}+\frac{\partial^2 v_{xb}}{\partial z_b^2}\right]$$

$$\rho_{fb}\left(v_{xb}\frac{\partial v_{yb}}{\partial x_b}+v_{yb}\frac{\partial v_{yb}}{\partial y_b}+v_{zb}\frac{\partial v_{yb}}{\partial z_b}\right)=-\frac{\partial p_b}{\partial y_b}+\mu_b\left[\frac{\partial^2 v_{yb}}{\partial x_b^2}+\frac{\partial^2 v_{yb}}{\partial y_b^2}+\frac{\partial^2 v_{yb}}{\partial z_b^2}\right]$$

$$\rho_{fb}\left(v_{xb}\frac{\partial v_{zb}}{\partial x_b}+v_{yb}\frac{\partial v_{zb}}{\partial y_b}+v_{zb}\frac{\partial v_{zb}}{\partial z_b}\right)=-\frac{\partial p_b}{\partial z_b}+\mu_b\left[\frac{\partial^2 v_{zb}}{\partial x_b^2}+\frac{\partial^2 v_{zb}}{\partial y_b^2}+\frac{\partial^2 v_{zb}}{\partial z_b^2}\right]$$

Equation of species continuity

$$v_{xb}\frac{\partial c_{mb}}{\partial x_b}+v_{yb}\frac{\partial c_{mb}}{\partial y_b}+v_{zb}\frac{\partial c_{mb}}{\partial z_b}=D_{AB,b}\left[\frac{\partial^2 c_{mb}}{\partial x_b^2}+\frac{\partial^2 c_{mb}}{\partial y_b^2}+\frac{\partial^2 c_{mb}}{\partial z_b^2}\right]$$

Equation of energy

$$\rho_{fb}c_{pf,b}\left(v_{xb}\frac{\partial T_b}{\partial x_b}+v_{yb}\frac{\partial T_b}{\partial y_b}+v_{zb}\frac{\partial T_b}{\partial z_b}\right)=\lambda_{fb}\left[\frac{\partial^2 T_b}{\partial x_b^2}+\frac{\partial^2 T_b}{\partial y_b^2}+\frac{\partial^2 T_b}{\partial z_b^2}\right]$$

Catalytic washcoat phase:

Equation of continuity

$$\frac{\partial v_{xb}}{\partial x_b}+\frac{\partial v_{yb}}{\partial y_b}+\frac{\partial v_{zb}}{\partial z_b}=0$$

Equation of motion

$$\left(\frac{\mu_b}{\kappa_b}\right)v_{xb}=-\frac{\partial p_b}{\partial x_b}+\left(\frac{\mu_b}{\varepsilon_b}\right)\left[\frac{\partial^2 v_{xb}}{\partial x_b^2}+\frac{\partial^2 v_{xb}}{\partial y_b^2}+\frac{\partial^2 v_{xb}}{\partial z_b^2}\right]$$

$$\left(\frac{\mu_b}{\kappa_b}\right)v_{yb}=-\frac{\partial p_b}{\partial y_b}+\left(\frac{\mu_b}{\varepsilon_b}\right)\left[\frac{\partial^2 v_{yb}}{\partial x_b^2}+\frac{\partial^2 v_{yb}}{\partial y_b^2}+\frac{\partial^2 v_{yb}}{\partial z_b^2}\right]$$

$$\left(\frac{\mu_b}{\kappa_b}\right)v_{zb}=-\frac{\partial p_b}{\partial z_b}+\left(\frac{\mu_b}{\varepsilon_b}\right)\left[\frac{\partial^2 v_{zb}}{\partial x_b^2}+\frac{\partial^2 v_{zb}}{\partial y_b^2}+\frac{\partial^2 v_{zb}}{\partial z_b^2}\right]$$

Equation of species continuity

$$v_{xb}\frac{\partial c_{mb}}{\partial x_b}+v_{yb}\frac{\partial c_{mb}}{\partial y_b}+v_{zb}\frac{\partial c_{mb}}{\partial z_b}=D_{AB,eff,b}\left[\frac{\partial^2 c_{mb}}{\partial x_b^2}+\frac{\partial^2 c_{mb}}{\partial y_b^2}+\frac{\partial^2 c_{mb}}{\partial z_b^2}\right]-\rho_{bed,b}R'_{mb}(c_{mb},T_b)$$

Equation of energy

$$\rho_{bed,b}c_{ps,b}\left(v_{xb}\frac{\partial T_b}{\partial x_b}+v_{yb}\frac{\partial T_b}{\partial y_b}+v_{zb}\frac{\partial T_b}{\partial z_b}\right)=\lambda_{eff,b}\left[\frac{\partial^2 T_b}{\partial x_b^2}+\frac{\partial^2 T_b}{\partial y_b^2}+\frac{\partial^2 T_b}{\partial z_b^2}\right]+\rho_{bed,b}\sum_{e=1}^{N_{rxn}}(-\Delta H_e)(-r_e(c_{mb},T_b))$$

Solid wall phase:

Equation of energy

$$\lambda_w\left[\frac{\partial^2 T_w}{\partial x_w^2}+\frac{\partial^2 T_w}{\partial y_w^2}+\frac{\partial^2 T_w}{\partial z_w^2}\right]=0$$

Source: Karakaya and Avci [25]. Reproduced with permission of Elsevier. Subscript *b* in the equations defines the quantities in channel *b*.

Table 11.3 Boundary conditions related to the mathematical model given in Table 11.2.

Channel entrance: $z_b=0,\ \forall x_b, y_b$

$u_b=u_b^{in};\ c_{mb}=c_{mb}^{in};\ T_b=T_b^{in}$

Along the fluid–wall interfaces (internal wall boundaries)

$\vec{n}\cdot\vec{v}_j=0$

$\vec{n}\cdot(-D_{AB}\nabla c_{mb}+\vec{v}_b c_{mb})=0$

$\vec{n}\cdot(-\lambda_w\nabla T_w)=\vec{n}\cdot(-\lambda_f\nabla T_b+\vec{v}_b\rho_{fb}c_{pf,b}T_b)$

Channel exit: $z_b=L,\ \forall x_b, y_b$

$p_b=p_b^{out}$

$\vec{n}\cdot(-D_{AB}\nabla c_{mb})=0$

$\vec{n}\cdot(-\lambda_f\nabla T_b)=0$

External wall boundaries

$\vec{n}\cdot(-\lambda_w\nabla T_w)=0$

Source: Karakaya and Avci [25]. Reproduced with permission of Elsevier. Subscript *b* in the equations defines the quantities in channel *b*.

for a nine-channel microchannel reactor (see Section 11.4.2.3) and by comparing the outcomes with those obtained from modeling of the 2D unit cell. The study is carried out by using *i*-octane SR (Reaction 11.49) and methane combustion (Reaction 11.33) as the representative endothermic and exothermic reactions, respectively [25]:

$$i\text{-}C_8H_{18}+8H_2O=8CO+17H_2\quad \Delta H_{298}^{o}=1274.8\ \text{kJ/mol} \tag{11.49}$$

$$CH_4+2O_2=CO_2+2H_2O\quad \Delta H_{298}^{o}=-802.3\ \text{kJ/mol} \tag{11.33}$$

The rate laws describing *i*-octane SR over a Ni/Al_2O_3 catalyst and methane TOX over a Pt/Al_2O_3 catalyst, the necessary physical properties, and structural and operating parameters of the reactor needed to execute the solution of the equations given in Table 11.2 together with the pertinent boundary conditions (Table 11.3) are available in the literature [25].

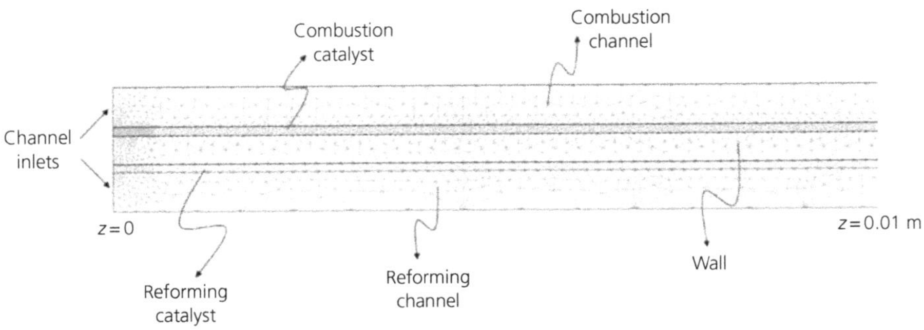

Figure 11.9 The first one-tenth of the computational grid used in the 2D simulation of the microchannel reactor for coupling catalytic methane combustion and *i*-octane steam reforming.

The model equations presented in Table 11.2, subject to the boundary conditions given in Table 11.3, are solved using the FEM (see Section 11.3.2) on COMSOL Multiphysics platform. The simplifying assumptions stated previously allow the variations along the x-axis to be neglected and take the changes on y- and z-directions into account within the domain of the 2D unit cell shown in Figure 11.8. A representative grid structure needed for the solution of 2D model is presented in Figure 11.9. The unstructured grids consist of 2D quadratic triangular elements, with the number of elements ranging between 22 674 and 40 251. At regions near the channel inlets and in the subdomain where combustion is modeled (i.e., the combustion catalyst layer), the grid is refined so that sharp changes in concentration and temperature can be resolved. It is worth noting that the outcomes of the numerical solution can vary with the type of meshing and the number of grid points. In the present case, 33 980 triangular elements are found to give a mesh-independent solution. UMFPACK solver (Table 11.1) is employed for the solution of the linear system of equations obtained after discretization.

Inclusion of the variations in x-axis leads to the formulation of the reactor model in three dimensions. FEM-based solution of the 3D model is carried out by meshing with tetrahedral elements. Mesh independency of the mathematical solution is investigated by increasing the number of elements progressively from 25 800 to 84 923 and is guaranteed when a minimum of 70 388 elements are used [25]. The cost of the 3D model is significantly higher than that of the 2D one. The former requires 32 h of CPU time for convergence, whereas the 2D model can be solved in 2–4 min on a HP z800 workstation equipped with 8×2.67 GHz Xeon^{TM} processors and 32 GB memory [24].

11.4.2.3 Results and key findings

Solution of the 3D model for a nine-channel microchannel reactor is shown in Figure 11.10, which shows the temperature profiles in the x- and z-axes at five different, evenly spaced locations in y-axis. The profiles clearly show that the variations in x-axis are almost negligible. Comparison of these outcomes with those of the 2D model, presented in Figure 11.11, indicates that the differences between the average channel temperatures and the exit temperatures are limited to only 4 K and 2 K, respectively. These findings verify the validity of the simplifying assumptions made

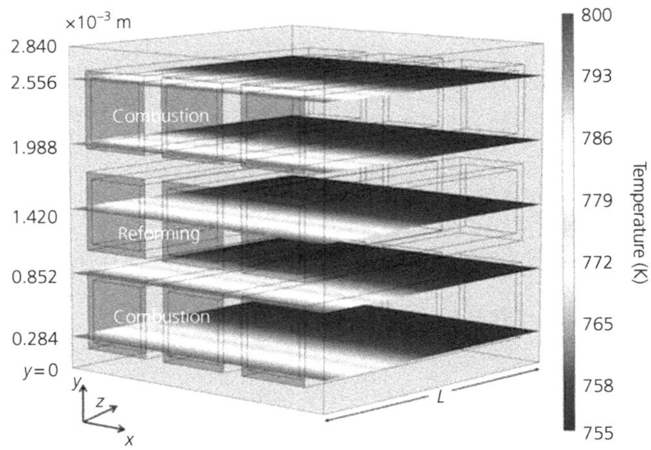

Figure 11.10 Temperature profiles obtained from the solution of 3D model (side length of the microchannel = 5.6×10^{-4} m, thickness of the wall separating the channels = 2×10^{-4} m, reactor material: AISI-steel). (Source: Karakaya and Avci [25]. Reproduced with permission of Elsevier.)

earlier and show that a lower-cost 2D model can be used to simulate the heat-exchange integrated microchannel operation. The model can also be used to design the reactor by investigating the effects of structural parameters (thickness of the wall separating the channels, material of construction of the reactor, side length of the square-shaped microchannel) on temperature and product distributions. An example of such an investigation is presented in Figures 11.11 and 11.12. Figure 11.11 shows the effect thickness of the stainless steel made separating wall on distribution of average bulk temperature along the SR channel. For this purpose the 2D model is solved for the wall thickness values that are increased from 1×10^{-4} to 4×10^{-4} m by increments of 1×10^{-4} m. The results show that as the wall thickness is increased, temperature decrease is observed to be sharper at the channel inlet, but the rate of decrease then becomes smaller along the remainder of the channel. This finding is linked with the evolution of the axial (z-direction) heat flux through the steel wall-SR washcoat interface, whose details are provided in the literature [25].

Figure 11.12 presents the dependence of i-octane conversion and H_2 yield on the structural parameters of the microchannel reactor. The outcomes, obtained by the solution of the 2D model, show that both H_2 yields and hydrocarbon conversion increase

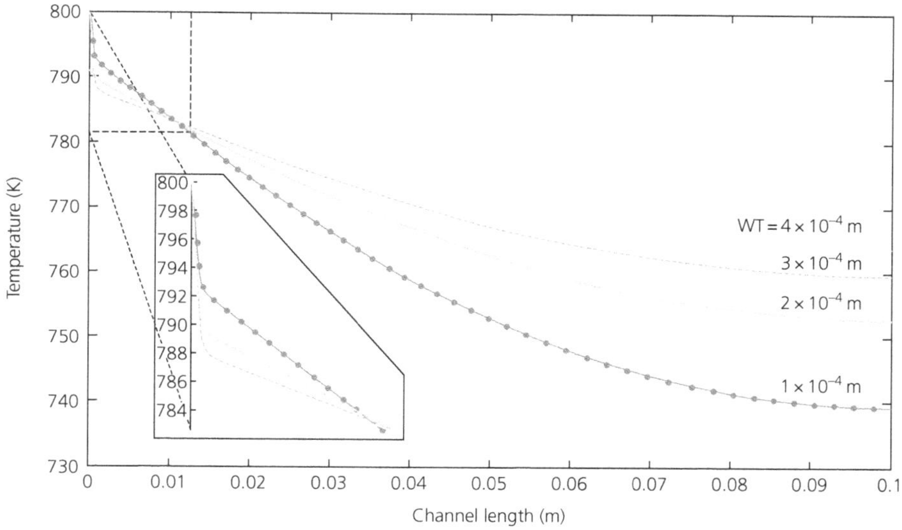

Figure 11.11 Profiles of average bulk temperatures along the steam reforming channel obtained at different wall thickness values by the solution of the 2D model.
(Source: Karakaya and Avci [25]. Reproduced with permission of Elsevier.)

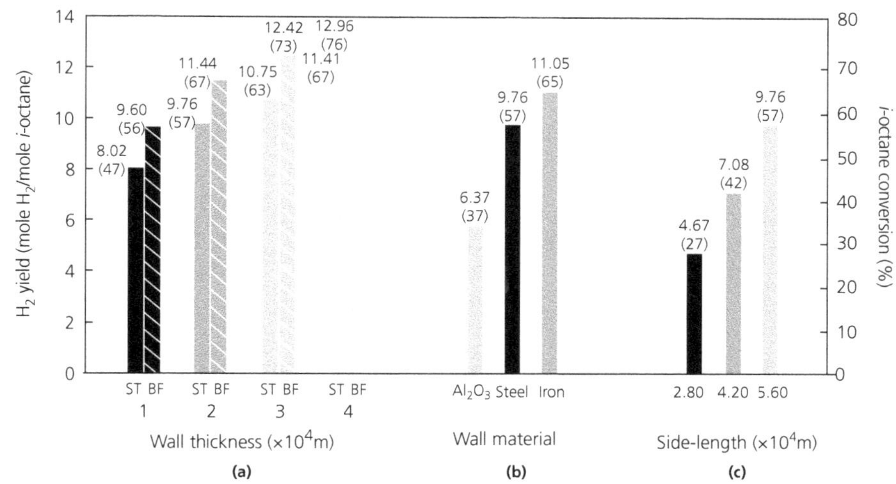

Figure 11.12 Hydrogen yields and *i*-octane conversions (in parentheses) obtained for different (a) wall thickness values, (b) wall materials, and (c) channel side lengths by the solution of the 2D model.
(Source: Karakaya and Avci [25]. Reproduced with permission of Elsevier.)

when channels with higher side lengths are separated by thicker walls made from materials of construction with higher thermal conductivities. These changes are shown to lead to the presence of higher temperatures within the reforming channel and faster reaction rates due to improved heat transfer from the combustion channel. In addition to these structural parameters, the texture of the channel wall is found to affect product distribution. This result is obtained by simulating two channel wall textures, straight-through (ST) and microbaffled (BF), both defined in Figure 11.13 [26], at identical conditions. It is observed that the use of microbaffles improves *i*-octane conversion and H_2 yield by favoring static mixing between the fluid and catalyst layer and by slightly increasing the contact time of the fluid and the catalyst in both channels. Comparison of the grid structures needed to explore the differences between the ST and BF configurations (Figure 11.13) via the 2D model is shown in Figure 11.14.

The study and the outcomes provide insight into the importance of modeling and simulation tools in the design and analysis of catalytic microchannel reactors that can operate with improved conversion and yield figures. It is worth noting that this conclusion is not limited to microchannel reactors and is also valid for other catalytic reactor types.

11.5 Summary

Mathematical models of multiphase catalytic reactors involve a set of conservation equations that describe the transport of momentum, heat, and mass in a specified volume mathematically bounded with several conditions. Incorporation of the catalytic chemistry and reaction rates into the model equations, all of which need to be solved simultaneously, increases the nonlinearity of the problem. At this point, analytical techniques for solving the model equations become insufficient, and one has to employ a numerical solution method suitable to the nature of the equations. In this chapter, numerical techniques for the

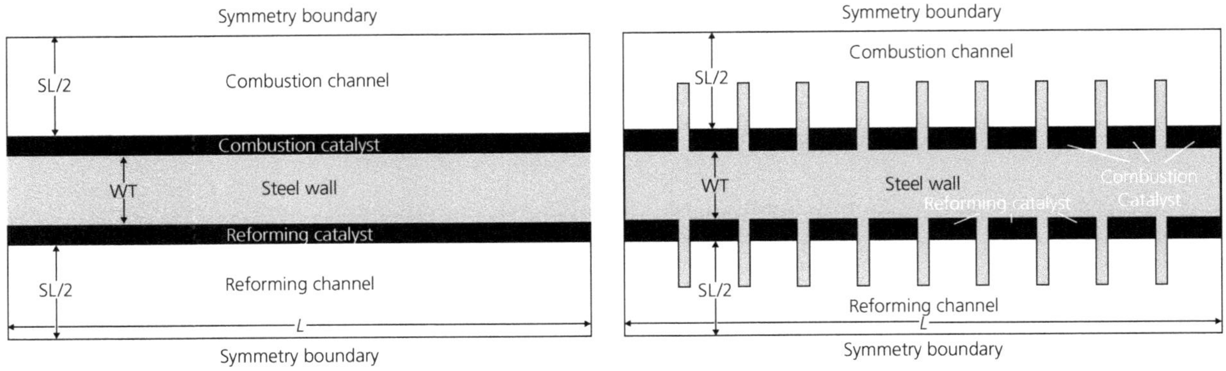

Figure 11.13 Two-dimensional unit cell with (a) straight-through (ST) and (b) microbaffled (BF) channel wall configurations. (Source: Karakaya and Avci [26]. Reproduced with permission of Springer.)

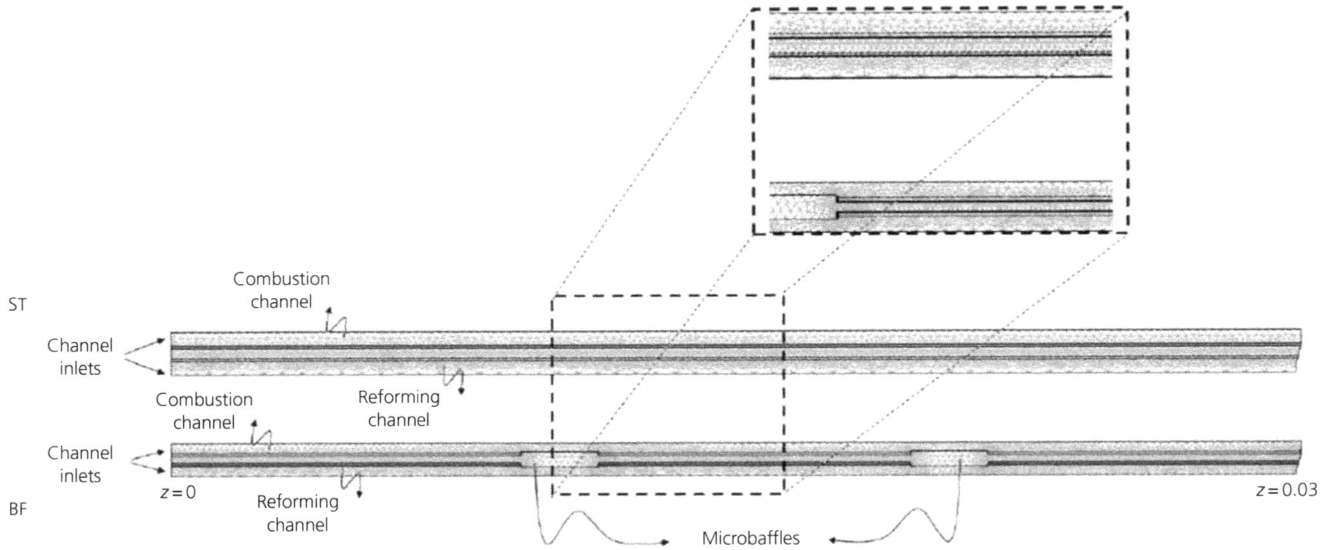

Figure 11.14 The first one-third of the computational grid used in the 2D simulation of the microchannel reactors with ST and BF wall configurations for coupling catalytic methane combustion and *i*-octane steam reforming.

solution of ODEs and PDEs, which are used in understanding reactive transport of mass and heat in most of the reactor types, are outlined. A brief overview of the CFD techniques is given, as their use in detailed analysis and design of reactor operations is increasing with the availability of technical opportunities for fast computing. The content of the chapter is enriched with two case studies in which the problem description, mathematical model, accompanying numerical solution methods, and the descriptive results are presented in detail.

Nomenclature

A_c	cross-sectional area, m^2
A_s	surface area-to-volume ratio of the solid catalyst, m^{-1}
c_m	concentration of species *m*, mol/m^3
c_{pf}	heat capacity of the fluid mixture, kJ/kg/K
c_{pm}	heat capacity of species *m*, kJ/kg/K
c_{ps}	heat capacity of the solid catalyst, kJ/kg/K
D_{AB}	diffusivity of component A into stagnant B, m^2/s
$D_{AB,eff}$	effective species diffusivity in the porous catalyst, m^2/s
$\vec{F}$	momentum source/sink term, $kg/m^2/s^2$
F_m	molar flow rate of species *m*, mol/s
F_{Total}	total molar flow rate, mol/s
$\vec{g}$	gravitational acceleration term, m/s^2
h	step size
h_m	enthalpy of species *m*, kJ/kg
$\vec{J}_m$	diffusive mass flux of species *m*, $kg/m^2/s$
k_{eff}	effective thermal conductivity, W/m/K
L	microchannel length, upper limit of the spatial coordinate *x*, m
M_m	molecular weight of species m, kg/mol

$\vec{n}$ normal unit vector
p total pressure, Pa
$-r_e$ rate of reaction e, mol/kg_{cat}/s
R'_m consumption/production rate of species m per unit mass of catalyst, mol/kg_{cat}/s
R^s_m consumption/production rate of species m in the catalyst surface reactions, mol/m^2/s
R^v_m consumption/production rate of species m in the gas phase reactions, mol/m^3/s
S_h heat source/sink term due to chemical reaction, W/m^3
T temperature, K
t time, s
u linear fluid velocity, m/s
$\vec{v}$ fluid velocity vector, m/s
v_x, v_y, v_z components of the fluid velocity, m/s
x, y, z spatial coordinates, m
y^n value of the dependent variable y at the nth nodal point
Y_m mass fraction of species m

Greek letters

ΔH^o_{298} reaction enthalpy at 298 K, kJ/mol
Δt time step, s
Δz grid size, m
ε void fraction of the porous catalyst
κ permeability of the porous catalyst, m^2
λ_{eff} effective thermal conductivity in the porous catalyst, W/m/K
λ_f fluid thermal conductivity, W/m/K
λ_w thermal conductivity of the wall separating the microchannels, W/m/K
μ fluid viscosity, kg/m/s
ρ_{bed} bulk density of the catalyst bed, kg_{cat}/m^3
ρ_f fluid density, kg/m^3
τ stress tensor due to flow, kg/m^2/s^2
υ volumetric flow rate, m^3/s
ϕ value scalar quantity at the cell center

Subscripts/superscripts

0 initial value of a dependent variable
b channel index
bed catalyst bed
e reaction index
eff effective
f fluid
i, j, k spatial nodal point indices
in reactor inlet
m species index
n temporal nodal point index
out reactor exit
s solid catalyst
w solid wall separating the microchannels

References

1 Finlayson BA. Nonlinear analysis in chemical engineering. Seattle: Ravenna Park Publishing, Inc.; 2003.
2 Kreyzig E. Advanced engineering mathematics. 9th ed. Hoboken (NJ): John Wiley & Sons, Inc.; 2006.
3 Ma L, Trimm DL, Jiang C. The design and testing of an autothermal reactor for the conversion of the light hydrocarbons to hydrogen I. The kinetics of the catalytic oxidation of light hydrocarbons. Applied Catalysis A: General 1996;138: 275–283.
4 Cutlip MB, Shacham M. Problem solving in chemical and biochemical engineering with POLYMATH, Excel and MATLAB. 2nd ed. Upper Saddle River (NJ): Prentice Hall; 2008.
5 The MathWorks Inc. 2015. Ordinary differential equations. Available at http://www.mathworks.com/help/matlab/math/ordinary-differential-equations.html (accessed January 12, 2016).
6 Wolfram Research, Inc. Mathematica, Version 10.0. Champaign (IL); 2014.
7 Avci AK, Trimm DL, Onsan ZI. Heterogeneous reactor modeling for simulation of catalytic oxidation and steam reforming of methane. Chemical Engineering Science 2001;56:641–649.
8 Finlayson BA, Biegler LT, Grossmann IE. Mathematics in chemical engineering. In: Ullmann's modeling and simulation. Weinheim: Wiley-VCH; 2007. p 3–149.
9 Villadsen J, Stewart WE. Solution of boundary-value problems by orthogonal collocation. Chemical Engineering Science 1967;22:1483–1501.
10 Finlayson BA. The method of weighted residuals and variational principles. New York: Academic Press; 1972.
11 Rice RG, Do DD. Applied mathematics and modeling for chemical engineers. 2nd ed. Hoboken (NJ): John Wiley & Sons, Inc.; 2012.
12 Versteeg HK, Malalasekera W. An introduction to computational fluid dynamics: the finite volume method. 2nd ed. New York: Pearson Education Limited; 2007.
13 Peiro J, Sherwin S. Finite difference, finite element and finite volume methods for partial differential equations. In: Yip S, editor. Handbook of materials modeling. Berlin: Springer; 2005. p 2415–2446.
14 Lewis RW, Nithiarasu P, Seetharamu KN. Fundamentals of the finite element method for heat and fluid flow. Hoboken (NJ): John Wiley & Sons, Inc.; 2004.
15 COMSOL AB. COMSOL Multiphysics User's Guide, Version 3.4. COMSOL AB, Stockholm; 2007.
16 ANSYS Inc. ANSYS FLUENT 14.0 User's Guide. SAS IP, Inc., Canonsburg (PA); 2011.
17 Onsan ZI, Avci AK. Reactor design for fuel processing. In: Shekhawat D, Spivey JJ, Berry DA, editors. Fuel cells: technologies for fuel processing. Amsterdam: Elsevier Science; 2011. p 451–516.
18 Patankar SV, Spalding DB. A calculation procedure for heat, mass and momentum transfer in three-dimensional parabolic flows. International Journal of Heat and Mass Transfer 1972; 15:1787–1806.

19 Patankar SV. Numerical heat transfer and fluid flow. New York: Hemisphere; 1980.
20 Karakaya M, Avci AK, Aksoylu AE, Onsan ZI. Steady state and dynamic modeling of indirect partial oxidation of methane in a wall-coated microchannel. Catalysis Today 2009;139: 312–321.
21 Numaguchi T, Kikuchi K. Intrinsic kinetics and design simulation in a complex reaction network: steam-methane reforming. Chemical Engineering Science 1988;43:2295–2301.
22 Strikwerda JC. Finite difference schemes and partial differential equations. Belmont (CA): Wadsworth & Brooks/Cole; 1989.
23 LeVeque RJ. Finite volume methods for hyperbolic problems. Cambridge: Cambridge University Press; 2002.
24 Karakaya M, Avci AK. Comparison of compact reformer configurations for on-board fuel processing. International Journal of Hydrogen Energy 2010;35:2305–2316.
25 Karakaya M, Avci AK. Microchannel reactor modeling for combustion driven reforming of iso-octane. International Journal of Hydrogen Energy 2011;36:6569–6577.
26 Karakaya M, Avci AK. Simulation of on-board fuel conversion in catalytic microchannel reactor-heat exchanger systems. Topics in Catalysis 2009;52:2112–2116.

PART 6

Industrial applications of multiphase reactors

CHAPTER 12

Reactor approaches for Fischer–Tropsch synthesis

Gary Jacobs and Burtron H. Davis
Center for Applied Energy Research, University of Kentucky, Lexington, KY, USA

Abstract

This chapter begins by considering the first reactors ever used for FTS, which were fixed-bed reactors operated at atmospheric pressure, and examines historical developments through World War II and into the 1950s. In the time period between 1950 and 1985, three types of reactors were developed for commercial use, including fixed-bed, fluid-bed, and slurry-phase reactors, with significant advances being made by Shell on fixed-bed technology and by Sasol on fixed-bed, fluid-bed, and slurry-phase reactor technology. From 1985 to the present, the three reactor technologies have continue to be used the most extensively for commercial operations with, for example, both fixed-bed (Pearl GTL) and slurry-phase (Oryx GTL) reactors being installed most recently in Qatar. In addition, significant strides have been made to develop small-scale reactor technology that may prove to be useful for converting remote gas, as well as biomass. These include micro-channel reactors (e.g., Velocys), modular and mobile small-scale GTL plants (e.g., Compact GTL), and small channel reactors (e.g., Chart Energy & Chemicals, Inc.). In academia, researchers have taken advantage of the liquid-like extraction property and gas-like transport property of supercritical media for FTS to prevent wax holdup and achieve high productivity.

12.1 Introduction

The reactor is an essential component of the conduct of catalytic reactions. However, the reactor cannot improve the performance of a catalyst; it can only decrease its performance. Badger [1] outlined how early reactor engineering developments occur. He indicates that initial developments usually utilize makeshift reactors. Only later will a sudden advance in engineering lead to an improvement in equipment that will offer a great advance in unit operations. As understanding of the advance develops, varieties of the new advance are introduced. Further along, these varied developments are narrowed into one or a limited number of reactors. The industry then settles on these few reactors until another advance introduces new and radical changes. This was certainly the case for reactors for the Fischer–Tropsch (FT) synthesis.

There are two general routes to convert coal and other solid fuels to transportation fuels: direct liquefaction and indirect liquefaction. Direct liquefaction converts the solid to transportation fuels by heating to a high temperature (ca. 450°C) and pressure (160 atm). These are severe conditions that produce a highly aromatic product that still contains significant heteroatoms that cause environmental problems when combusted. The FT indirect liquefaction involves two steps; the first step involves the conversion of the solid to synthesis gas (syngas) and then the conversion of the syngas to transportation fuels. The syngas is amenable to cleaning to remove the elements that are of environmental concern (sulfur, nitrogen, CO_2, etc.). Thus, the fuel products from indirect liquefaction are of interest because they are virtually free of environmentally harmful heteroatoms and because their paraffinic nature results in a diesel fuel with a very high cetane number. During the energy crisis of the 1970s, the direct coal liquefaction process was the one that received most attention, since it produced a high octane gasoline that could meet the less stringent environmental regulations. However, today it is the indirect liquefaction pathway that is receiving the most attention, because of its fuel quality and environmental benefits.

The conversion of syngas to transportation fuels is a highly exothermic reaction so that both the catalyst and the reactor must be designed to handle the heat of reaction. In addition, the reaction may involve three phases (solid, liquid, and gas), and this complicates reactor design. For the high-temperature synthesis conditions, operation involves only solid- and gas-phase reactants and products. Because the reaction follows a simple polymerization mechanism in which the monomer is some $C_1(O_xH_y)$ unit, it has not been possible to design a catalyst that produces a narrow range of products. Thus, if the catalyst has a high termination probability relative to the chain growth probability (a low alpha catalyst), the products will contain a high fraction of gaseous (C_1–C_4) products and these are not desirable products today. Thus, today one prefers a catalyst with a low termination probability relative to the chain growth

Multiphase Catalytic Reactors: Theory, Design, Manufacturing, and Applications, First Edition. Edited by Zeynep Ilsen Önsan and Ahmet Kerim Avci.

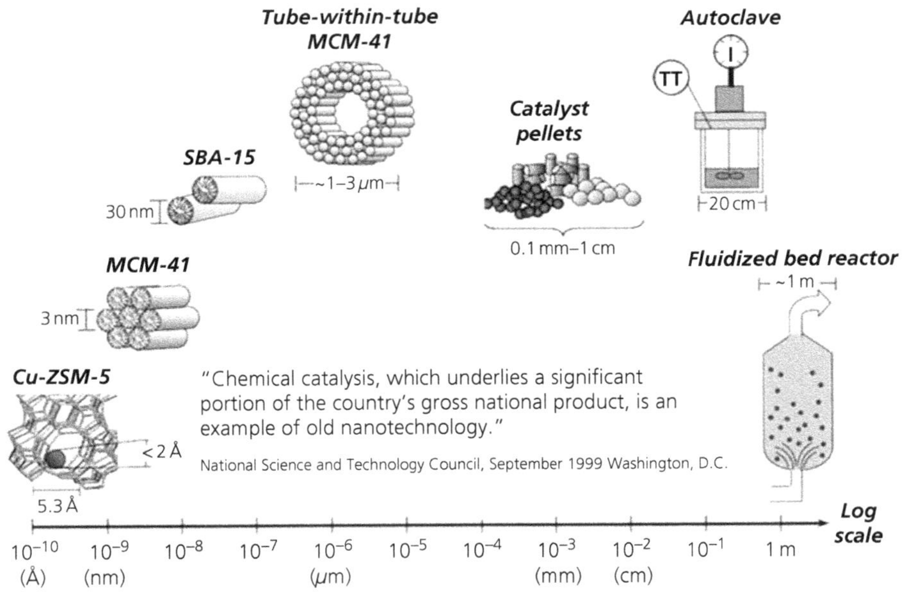

Figure 12.1 Scale of catalytic materials and reactors [2].
(Source: Davis [2]. Reproduced with permission of Springer.)

probability, so that the reactor must be operated with three phases (solid, liquid, and gas) being present. This means that today's reactor for FT synthesis is either a fixed-bed or a slurry reactor, and both of these are used at the commercial scale today. These two reactor types each have many advantages and problems associated with them. While significant advances in our understanding of these two reactor types have been made during the past 15 years, much still remains to be developed. The objective of this chapter is to outline many of the types of reactors that have been utilized for FT synthesis and review current understanding of the science and engineering principles involved with their operation.

Catalyst particles that are placed in the reactors are really small chemical reactors. As shown in Figure 12.1 [2], the catalyst particle can provide the smallest scale to be considered and is at the angstrom level. At the other extreme, we can visualize a chemical reactor that is on the order of 10 plus meters in diameter and more than 30 m in height. The catalyst particle may impart both physical and chemical effects upon the reaction. For example, the porosity of the catalyst particle may impose diffusional limitations for both the reaction rate and the product selectivity. Early scientists and engineers were developing catalytic materials with large surface areas and extensive porosity [3, 4]; however, an accurate assessment of these improvements had to wait for analytical methods to measure them. Three advances started our ability to assess these physical properties: the development of a reliable method to measure the surface area [5], the extent of the area of supported metals [6], and an assessment of the porosity of the catalyst support or the catalytic material on its performance [7, 8]. A variety of these small catalyst pellets had been designed, usually to handle problems encountered for mass and/or heat transfer [9] (Figure 12.2). While the design of the catalyst particle—the smallest reactor—is an important factor in the process, it will not be covered in any detail in this chapter, which will focus on the impact of the laboratory scale and larger reactors.

12.2 Reactors to 1950

The initial work by Fischer and Tropsch was carried out in simple fixed-bed reactors and in many of the studies they were operated at atmospheric pressure. For their commercial operations, German scientists and engineers settled on the fixed-bed reactor with sufficient tubes for a single unit to produce about 15 bbl/day of product. For a large plant, they would then utilize a sufficient number of these 15 bbl/day units to attain the design production rate for the plant. The early German work in the 1925–1945 period focused on the fixed-bed reactor, but they conducted research on a variety of reactor types and on other catalysts. The German work fits well to the model described by Badger [1] since a basic plant design was settled upon and then individuals or small groups explored a variety of reactor types.

The status of the reactors used in pilot and commercial scale by the end of World War II was summarized by Tramm [10]. These reactor types are shown in Figure 12.3.

Figure 12.3a illustrates an atmospheric pressure fixed-bed reactor used in German FT plants. The catalyst is located between vertical cooling plates, interconnected by horizontal cooling-water pipes. The heat of the reaction is led away from the catalyst by boiling water inside the pipes. The large-scale plants in Germany, France, and Japan were usually operated with these atmospheric pressure reactors.

Figure 12.3b shows a middle-pressure reactor in which the catalyst is located in vertical tubes. Some 2000 such tubes were connected into bundles and immersed into boiling water. This reactor type was also utilized in many large installations, and it was found that they could be operated satisfactorily.

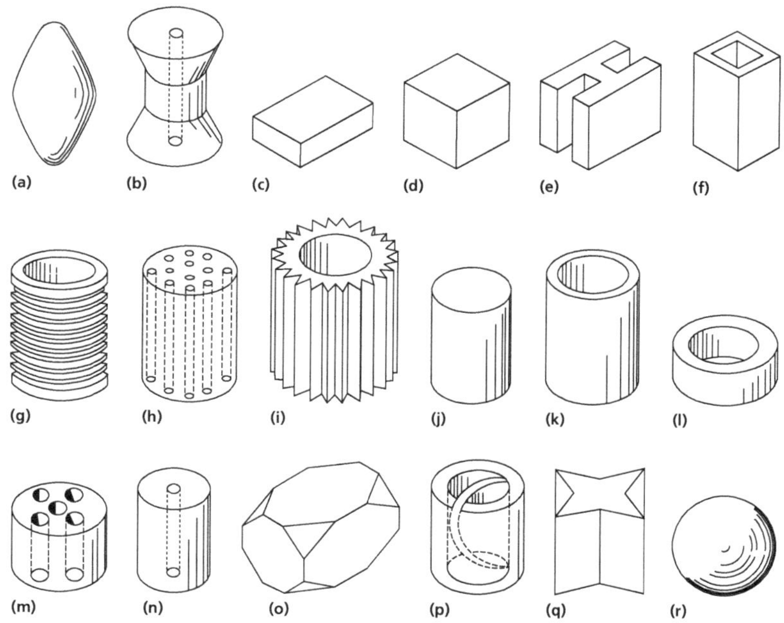

Figure 12.2 Schematic of some possible forms of catalyst pellets, the smallest chemical reactor [9].

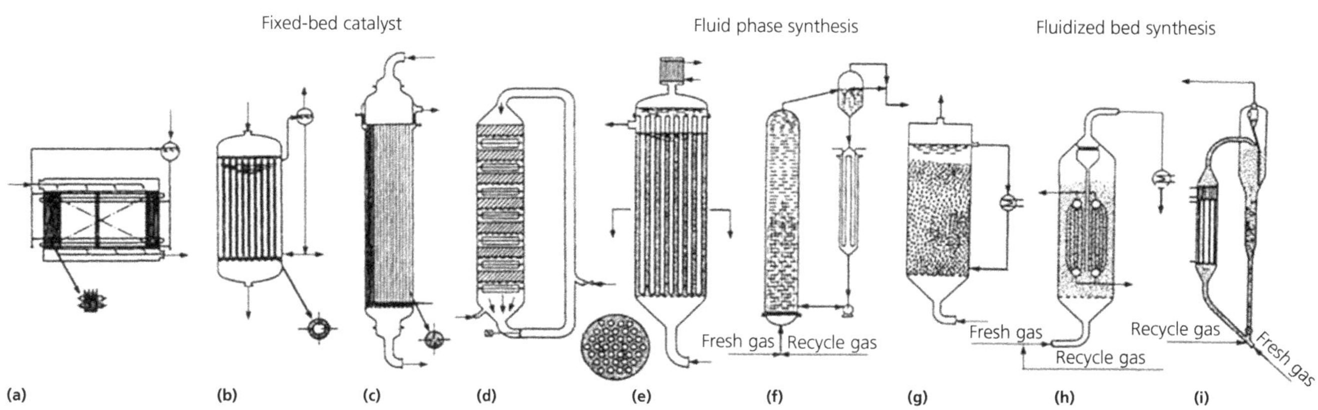

Figure 12.3 Schematic of German reactors used in the pre-1955 period according to Tramm [10]. (Source: Tramm [10]. Reproduced with permission of Oil Gas Publisher.)

Figure 12.3c provides a schematic of a higher-performance medium-pressure reactor that was developed shortly after World War II by the Ruhrchemie/Lurgi Development Company. They differ from those in Figure 12.3b primarily in their greater height and the use of larger diameter catalyst tubes. A number of these reactors were used in Ruhrchemie plants for several years in full-scale operation. Synthesis in these reactors was preferentially conducted with gas recycling. These reactors were utilized in the initial Sasol plants, and this type of reactor is still being utilized at the Sasol I plant today.

Figure 12.3d represents reactors with catalyst layers bounded with cooling surfaces. Only a part of the synthesis gas is converted in each layer of catalyst, and the heat that is produced is partly or completely removed by the cooling system located below that catalyst layer. Gas was recycled in this reactor also.

Figure 12.3e illustrates a liquid-phase slurry reactor with cooling coils located inside the reactor. In this reactor, the catalyst is no longer fixed in place but is suspended in the start-up oil initially and finally in the FT products. Synthesis gas bubbles rise through the suspension. Heat is conducted from the reactor through pipes located inside the reactor that are filled with water boiling under pressure.

Figure 12.3f represents a liquid-phase reactor with cooling located outside the reactor. Here the catalyst slurry (called sludge in much of the German literature of that period) was carried by forced circulation by pumps through the reactor and the cooler; details of this reactor are described in Ref. [11].

Figure 12.3g is a liquid-phase system with a granulated catalyst that is in motion. Just as in the two previous reactor types, the catalyst is suspended in oil inside the reaction space. Gas and oil are introduced at such a rate that the catalyst particles are kept in slight motion to prevent caking but are not carried over by the gas stream. As a result, the catalyst remains in the reactor while the oil is recycled through a cooling system located above the reactor. This design was developed so that the catalyst should not plug the cooling system.

Figure 12.3h describes a fixed fluid catalyst bed reactor. The system is characterized by a lively motion of the catalyst induced by interactions with the ascending gas flow. The heat is removed using a cooling system filled with water or other media that is located in the fluidized catalyst bed. The 300 000 ton/year (ca. 8000 bbl/day) Brownsville, Texas, plant used this design, and, while problems were initially encountered, the unit was being successfully operated until high natural gas prices forced closure of the plant.

Figure 12.3i represents a recycle fluid catalyst bed reactor. This design is similar to that developed during World War II to carry out catalytic cracking. The gas velocity is sufficiently high that the powdered catalyst is carried overhead with the unconverted syngas and reaction products. The catalyst that is carried over settles in the parallel vessel and the reaction products are withdrawn. Heat is removed from the parallel vessel for catalyst settling and holding vessels.

In addition to the commercial reactors, the Germans tested at the lab or pilot scale many types of reactors. These reactors are included in the reviews mentioned later in the text and will not be described, except for work with the slurry bubble column reactor and the foam reactor.

The German gas circulation process was essentially the one that was developed by HRI following World War II, as described later in the text. The German work was directed toward producing good quality gasoline and an iron catalyst was utilized in this work [11]. After completing work at the laboratory scale, the reactor was scaled up to contain a 4 m^3 catalyst volume with the catalyst bed making up the bulk of the available space in the reactor. The German operators noted marked disturbances of the flow in the catalyst bed, a problem that was to later trouble the HRI workers. The German operators overcame this problem by adding baffle plates and dividing the reactor space into three zones. The German operators concluded that the design of the reactor they were using had to be changed and suggested the following design. They proposed alternating catalyst and cooling elements into, for example, seven layers as shown in the Figure 12.4 [11]. The war ended before the new concept could be implemented and was not taken up in the following years. A summary of the major reactor technologies utilized in Germany during the 1935–1955 period follows [12].

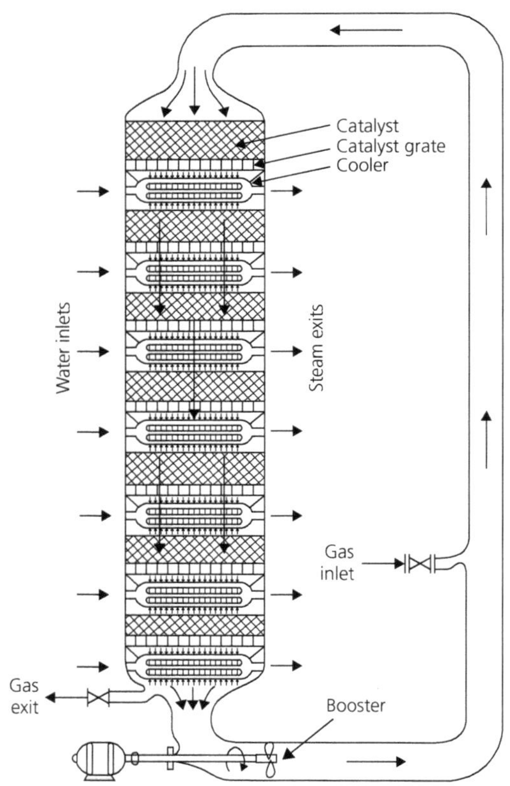

Figure 12.4 Schematic of the slurry reactor with multilayer beds [11].

The Ruhrchemie normal-pressure synthesis utilized a fixed-bed reactor where the catalyst was located between plates and internal water pipes were used for cooling. For the medium-pressure synthesis, Ruhrchemie used a fixed-bed reactor with the catalyst either filling the tube or located in the gap between concentric tubes with cooling by water/steam located in the spaces exterior to the out tubes. The IG Farben "Michael process" used a fixed-bed reactor with external gas recycle for cooling. The ARGE reactor uses a multitubular fixed-bed reactor with water/steam located within the reactor shell for cooling; external gas recycle may also be used for cooling. The Lurgi staged reactor used segmented fixed-bed reactors with a distributed feed of fresh gas along the reactor plus external gas recycle for cooling. A fixed-bed reactor was used for the BASF Duftschmid process together with external liquid recycle for cooling, whereas the BASF foam process used the same cooling with a reactor in which the reactor contained a suspension of fine catalyst particles dispersed on the liquid film of the gas bubbles. The Rheinpreussen–Koppers synthesis utilized a slurry reactor with a suspension of a fine catalyst in a liquid; both water/steam circulation within cooling coils in the reactor and external liquid recycle were used for cooling.

Bubble column reactors were considered for the FT synthesis during this period. In one version, a reactor with a relatively small cross-sectional area to height was described [13]. The reactor was claimed to have a height from 10 to 40 ft (3–12 m) and a diameter from about 1 to 6 inches (2.5–15 cm). Presumably, this patent is the reason why Exxon later made claims for a reactor with a diameter larger than 6 inches. Another bubble column reactor was described that had an expansion at the top to facilitate retaining the catalyst in the reactor [14].

12.3 1950–1985 period

Three types of reactors were utilized for the FT synthesis during this period: fixed-bed, fluidized-bed, and slurry bubble columns (Figure 12.5) [2, 15, 16]. The fluidized-bed reactor (Figure 12.5c and d) must be a two-phase system, solid and gas, for FTS; if this is not the case, liquid will lead to solid agglomeration and loss of the fluid phase. The slurry-phase reactor (Figure 12.5a) will

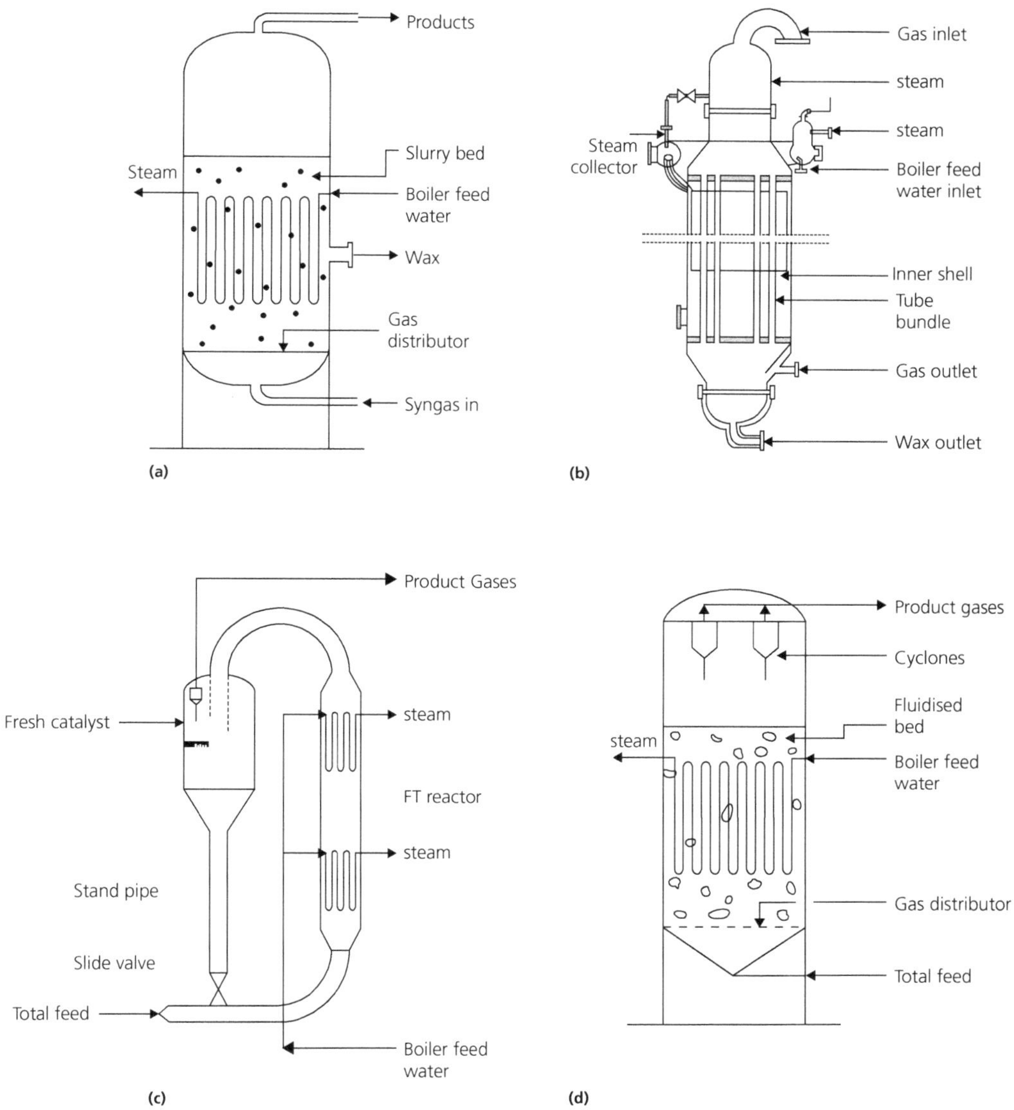

Figure 12.5 Possible reactors for Fischer–Tropsch synthesis. (a) Slurry bubble column reactor, (b) multitubular trickle bed reactor, (c) circulating fluidized-bed reactor, and (d) fixed fluidized-bed reactor [15].
(Source: Steynberg et al. [15].)

always be a three-phase system for FTS. Biardi and Baldi [17] briefly reviewed three-phase catalytic reactors and concluded that trickle bed and slurry reactors are the most important. Another reviewed the advantages and prospects for FTS in fixed-bed and slurry reactors but could not settle upon either as the preferred one [18]. Since Shell utilizes the fixed-bed and Sasol the slurry reactor in Qatar, this is also the situation today.

The Brownsville, Texas, plant (Carthage Hydrocol, a consortium led by Texaco) utilized a fixed fluidized-bed reactor designed by Hydrocarbon Research, Inc. (HRI). The reactor was 17 ft (5.2 m) in diameter and 17 ft tall. Its construction and movement to the plant site were at the forefront of engineering, manufacturing, and transportation facilities available at the time. The Brownsville plant encountered and overcame many operational problems. Many of the problems encountered were due to using new technologies for syngas production as well as for the FT synthesis. However, in the FT reactor, many of the problems were due to the scale-up factor. The initial design was based on reactor results from a 2 inch (5 cm) laboratory reactor and later from a 7 inch (17.8 cm) pilot plant. The use of a scale-up factor of 100 was overly optimistic, and a number of unexpected flow problems were encountered at the Brownsville plant. Among these was the formation of large gas bubbles that caused both operational problems and a conversion that was much lower than anticipated, based on lab data. Just as the operational problems were being overcome, the price of natural gas, the plant's feedstock, dramatically increased and this forced the closure of the plant.

Following World War II, Sasol was established to produce fuel in South Africa, a country without petroleum reserves but an abundant supply of coal. The company chose to develop the circulating fluid-bed reactors; this type of reactor had been successfully developed and scaled up several times during World War II in the United States for the catalytic cracking process. While initially encountering operational problems, Sasol solved these to the point where its operation was reliable. Two of the major problems they encountered were: (i) the formation of "large cannon-sized balls" of catalyst/liquid product that impeded gas flow through the smaller tubing and (ii) the growth of carbon fibers that broke the catalyst particle or changed the density so much that they clogged the filters and impeded the gas flow.

Sasol also operated fixed-bed reactors but on a smaller scale than the circulating fluid-bed reactors. The fixed-bed reactor utilizes a 2000 bundle of small diameter (5 cm) 12 m (40 ft) tubes enclosed in a shell; the cooling fluid, usually steam, is contained within the shell and on the exterior of the tube. The catalyst must be carefully loaded into the tubular reactor so that each tube has essentially the same pressure drop to prevent uneven flow through the tubes. Today, Sasol uses an iron catalyst in the ARGE tubular reactors to produce about 8000 bbl/day of high molecular weight products. Because of the exothermicity of the FT reaction, there will be a hot spot unless special attention is paid to the operation. This can be accomplished by a number of operations including dilution with inert gas, recycle of unconverted reactants, and limiting the conversion. However, the analysis of the catalyst from their commercial-sized plant indicated that temperature gradients and gas composition along the catalyst bed impacted the chemical state of the catalyst (Figure 12.6) [19].

The development of the low-temperature liquid-phase process began in 1938 by Kölbel and Ackermann [20]. It was interrupted in 1944 by the events of World War II but was resumed in 1951. In 1953 it led to the start-up of a pilot plant (Rheinpreussen–Koppers) with a production rate of 11.5 tons (ca. 75 bbl/day) of hydrocarbons per day that was operated under a cooperative agreement among Kölbel at the Technische Universität, Berlin-Charlottenburg, Rheinpreussen AG for Mining and Chemical, Hamburg, and Heinrich Koppers GmbH, Essen. The primary goal was the production of fuels as well as products for further chemical processing [21–24]. In spite of questionable mass balance data that were reported for this unit, Kölbel and his coworkers developed much of the theoretical groundwork needed for developing an understanding of the slurry reactor [25].

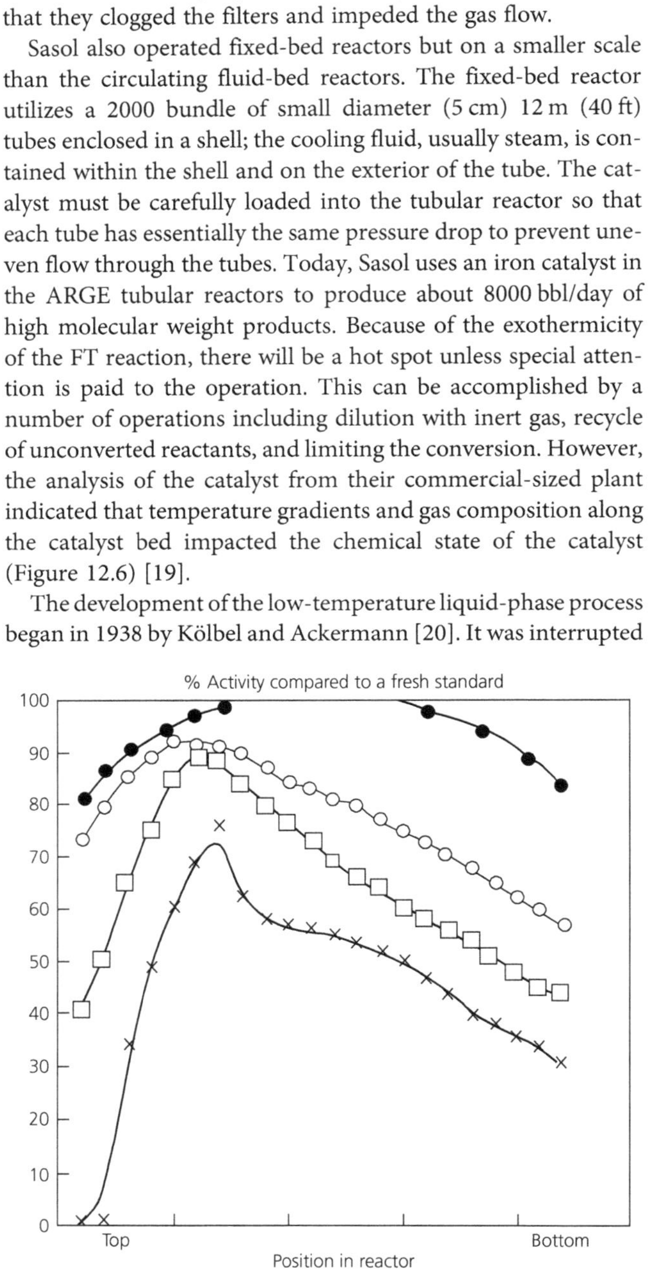

Figure 12.6 Relative activity profile of an precipitated iron catalyst after different periods of time on stream. Times are in relative units (●: 1 tu; ○, 50 tu; □, 270 tu; x, 1000 tu (tu = time units)).
(Source: Duvenhage and Coville [19]. Reproduced with permission of Elsevier.)

Further investigations with the same objectives were conducted during the war by BASF and after the war by the US Bureau of Mines, the Fuel Research Station (London), the Indian Institute of Technology, and the University of Tokyo. Minor variations of the reactor design utilized by Kölbel and coworkers were developed, but the overall design was the same. With the exception of the BASF process modified by the Bureau of Mines, development work did not exceed pilot-plant scale. The Bureau of Mines assembled a 75 bbl/day plant in Louisiana, Missouri (including German researchers from Operation Paperclip), but the development of the Middle East oil fields removed the fear of a shortage of petroleum and the US government terminated funding of this project soon after the election of President Eisenhower in 1952.

With the "oil crisis" of the 1970s, new activities in the use of the liquid-phase process for the production of primary chemical products have become apparent. Interest at that time was being shown mainly in short-chain C_2 to C_4 olefins, the medium-chain C_5 to C_{14} olefin fraction, and oxygen-containing products (aldehydes) [26].

Some of the work at the Pittsburgh Energy Technology Center (PETC) utilized tube wall reactors for both methanation and the FT synthesis [27–29]. These reactors had the catalyst adhering as a thin layer on the wall of the tube. Troubles were encountered in attaching and in long-term retention of these catalysts. They utilized high temperatures to attach the catalyst so that this limited the catalytic productivity. While a promising idea, this approach was abandoned.

During this period, Sasol was the only organization that maintained operational status. Being very isolated during much of this period, they did not have to worry about others utilizing their process, and their international patents are very limited during this period. It is surprising that the company allowed as much information about their process and its operation to be made available in public meetings and the open literature as they did.

12.4 1985 to present

12.4.1 Fixed-bed reactors

The fixed-bed reactor is highly desirable since it offers the highest density of catalyst/reactor volume that is attainable and, if the temperature can be made uniform, the potential for the highest productivity/reactor volume. This has not been accomplished to date so that there is much room for improvement in the commercial operation. In addition, there is a problem with diffusion and associated methane, which is made as the particle size of the catalyst increases. It would appear that the fixed-bed reactor is where the greatest advantage of the "rim-loaded" catalyst would be obtained. A "rim-loaded" catalyst is one where the catalytic component is located on the exterior of the catalyst particle. It is therefore not surprising that there has been a lot of patent activity by both Shell and Exxon in this area. Shell indicates that they have made significant progress in making improvements in the operation of the fixed-bed reactor and that some of these were included in the Bintulu plant when it was restarted in 2000. It has been implied that Shell does not utilize a rim-loaded catalyst in their commercial operation although they have reported at least two improvements that led to increases in production for their Bintulu plant due to improvements in the catalyst (i.e., the smallest reactor).

In spite of the widespread use of fixed-bed reactors, much remains to be done to define the dynamics of the reactor [30]. Most of these reactors are operated in the concurrent mode at which the gas and the liquid both flow from the top to the bottom. A number of flow regimes have been distinguished for packed columns in downflow operation [31, 32]. Based on the Reynolds number for liquid and gas flows, the flow regimes include (i) trickle flow, (ii) pulsed flow, (iii) dispersed bubble flow, (iv) wavy flow, and (v) spray flow (Figure 12.7) [17]. In general, countercurrent flows lead to much larger pressure drops across the bed, and this would be the case for FTS. Thus, the countercurrent flow mode is not used in today's plants.

The fixed-bed reactor has been scaled to the largest size that is likely to be used. The shipment of the outer vessel limits the number of tubes that can be placed in a single bundle of tubes. Likewise, when the diameter of the bundle is fixed, the number of tubes becomes limited since adequate space must be left

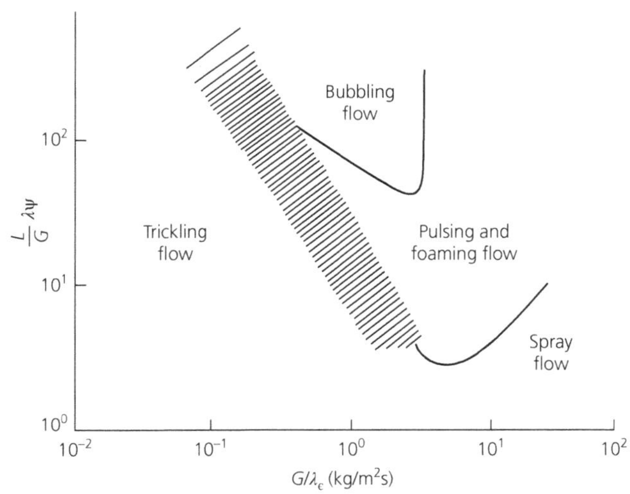

Figure 12.7 Flow regime maps for trickle-bed reactors [17]. (Source: Biardi and Baldic [17]. Reproduced with permission of Elsevier.)

around each tube for circulation of the cooling material. This limits one to about 3000 tubes or less for each reactor. In addition, catalyst loading becomes a more complex problem as the number of tubes is increased since each tube must have essentially the same pressure drop across them in order to have the same flow of reactants and products through each tube. Liquid holdup has a significant impact in the fixed-bed reactor but this has been rarely explained. Recently, de Klerk has provided a detailed description of this effect during the FT synthesis [33]. Shell has experience with the slurry reactor since runs were made at the La Porte facility with their catalyst [34]. However, for their 140 000 bbl/day plant in Qatar, they chose to use the fixed-bed reactor. This clearly demonstrates the viability of the fixed-bed reactor for the FT synthesis at the commercial scale.

Flow through a packed bed reactor was investigated during the FT work at the Bureau of Mines [35]. The studies showed the extraordinary effect that the voids in the bed and the degree of roughness of the solid had on the pressure drop along the reactor. They developed equations to account for these. Al-Dahhan et al. [36] developed a model that is able to predict the pressure drop together with the liquid holdup of a fixed-bed reactor at operating conditions appropriate for industrial processes. A theoretical tortuosity model was developed for fixed beds of identical particles. It was found to be proportional to a packing structure factor that was given in terms of bed voidage and particle shape. In another study, the use of high-pressure trickle-bed reactors has been authoritatively reviewed with respect to flow regime transitions, pressure drop, liquid holdup, gas–liquid interfacial area and mass-transfer coefficient, catalyst wetting efficiency, catalyst dilution with fines, and evaluation of trickle-bed models for liquid-limited and gas-limited reactions [37]. While this chapter does not deal with FT synthesis directly, the issues encountered for this reaction are covered. The issues of process development and scale-up and scale-down issues have been addressed by Sie and Krishna [38]. Considerations that went into the selection by Shell to commercialize the fixed-bed reactor have been described [39]. It was reported that the main factor in choosing the multitubular trickle-bed reactor for the Bintulu plant was timing. At that time, Shell personnel decided that the attractive cost and the acceptable risk would allow them to proceed to construction of the plant. They expected that further increases in reactor capacity for later plants should be possible by reactor improvements and advances in catalyst design. These decisions, based on the 140 000 bbl/day Pearl GTL plant in Qatar (i.e., a joint venture between Shell and Qatar Petroleum), proved to be correct. One of the great advantages of the fixed-bed reactor is that one only has to utilize one tube in the laboratory setting in order to evaluate the performance at the commercial scale.

Nontraditional types of fixed-bed reactors are beginning to be considered. A gas-lift reactor with monolith type catalyst packing has been described for FT synthesis [40]. It is reported that this reactor has been tested at the laboratory scale with a cobalt

Cell density (cells/in^2)	25	25	50
Shape	Square	Square finned	Square
Void fraction (%)	66	73	68
Surface/volume ratio (m^2/m^2)	640	1040	920
Hydraulic diameter (mm)	4.14	2.81	2.96
Number of full channels	44	44	88

Figure 12.8 Examples of monolith reactors illustrating cell size and shape possibilities [41].
(Source: Heibel et al. [41]. Reproduced with permission of Elsevier.)

catalyst and that it showed advantages over slurry and fixed-bed reactors so that it permitted an increase in the efficiency of FTS units as a whole. The influence of channel geometry (Figure 12.8) [41] on the hydrodynamics and mass transfer in monolith film flow reactors has been described, and the liquid distribution and the flooding boundaries were determined by experiments. It was found that flooding limits were in the range of other commercial structured packings and operation could be conducted using industrially relevant conditions. Larger channel sizes and lower surface tension expanded the window of operation but the viscosity seemed to have a minor effect. Gas–liquid mass transfer was a strong function of the surface to volume ratio that was defined by the channel dimensions.

Reactive distillation tray columns are receiving much attention for some applications [42]. This technique has been applied for the FT synthesis, but it has not been widely accepted as a useful approach [43].

Microchannel reactors are compact FT reactors that have channels with diameters in the micrometer range. The small-sized channels dissipate heat more quickly than conventional reactors with larger channel diameters in the 20–30 mm (i.e., inch) range so that more active catalysts can be used. As a result, microchannel reactors exhibit conversions in the range of 70% per pass, and this limit is introduced by the sensitivity of the catalyst to water vapor. Microchannel reactors are designed for economical production on a small scale. A single microchannel reactor block might be so small that it will produce up to 50 barrels (bbls) of liquid fuel per day. However, it can be scaled to produce up to 500 bbl/day.

At least three companies are active in developing the microchannel reactors for FT synthesis. Velocys' developments are based on initial research conducted at Battelle labs and has invested more than $300 million in developing the technology over the past 15 years [44]. They report that they are at the cusp of commercialization, having completed one demonstration plant with two more scheduled, and they have been recently selected to provide reactors for a plant in the 1000–2000 bbl/day scale.

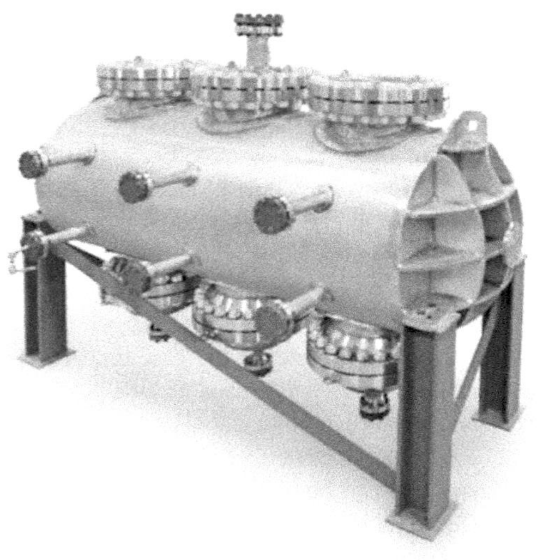

Figure 12.9 The Velocys commercial-scale Fischer–Tropsch reactor.
(Source: LeViness et al. [44]. Reproduced with permission of Velocys, Inc.)

The scale of their reactor core is indicated in Figure 12.9. They indicate that their productivity is much greater than that obtained by two other organizations that use conventional reactors, as illustrated in Figures 12.10 and 12.11 [44, 45].

Compact GTL is another company that is actively developing small-scale GTL plants [46]. They completed a test with Petrobras in 2012 of a modular system that is suitable for deploying at remote locations in order to utilize stranded natural gas. In 2012, they completed testing of a 20 bbl/day pilot plant at the Petrobras location. With Petrobras they have completed a conceptual engineering study for a 2000 bbl/day plant that will be installed on a floating vessel.

Chart Energy & Chemicals, Inc. is a manufacturing company that is interested in developing their facility to produce small channel reactors. Two materials of construction, stainless steel and aluminum, have been considered [47]. Aluminum has the advantage that it is easier and faster to use to manufacture larger units than can be done with stainless steel. This will permit them, using existing manufacturing facilities, to construct a single unit that could produce about 500 bbl/day, whereas today's single unit produced from stainless steel only produces about 20–50 bbl/day. Aluminum has the disadvantage that it has an upper operating temperature that is much lower than stainless steel. The isothermal operation of this reactor is illustrated by the results in Figure 12.12 [48]. The products obtained using the aluminum fixed-bed compact heat exchange reactor are illustrated in Figure 12.13 [48] and are comparable to those obtained with the same catalyst in a liquid slurry-phase reactor with the same catalyst. In another study, a multichannel microstructured reactor packed bed reactor was utilized with a high-activity cobalt–alumina catalyst [49]. It was found that the conversion and selectivity of the catalyst obtained using the microstructured catalyst were very similar to that obtained with the same diluted catalyst in a fixed-bed reactor.

It appears that the above small channel reactors utilize a packed bed catalyst design. Another approach is provided in which a monolithic-structured catalyst bed is utilized (Figure 12.14) [50]. In this configuration, the catalyst is attached to the wall of the reactor. The authors reported very surprising results with a cobalt catalyst since even at nearly 100% CO conversion and obtained less than 10% methane and very low CO_2 (Figure 12.15). These two products show increases to the range of 20% or higher as the conversion approaches 90% in a slurry reactor where the temperature remains constant the same as for the small channel reactor.

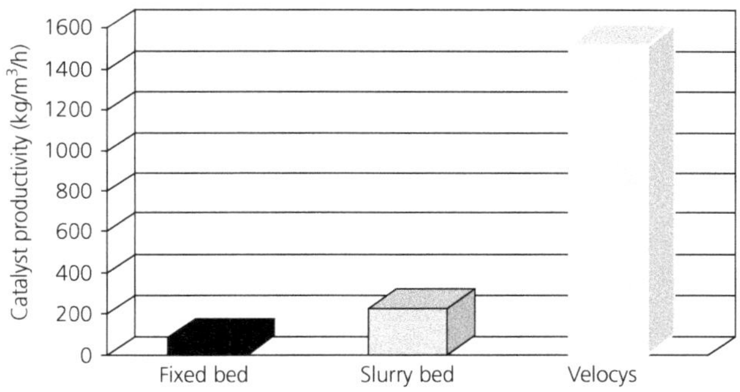

Figure 12.10 Comparison of catalyst productivity for Shell, Sasol, and Velocys Oxford Catalysts Group (OCG) reactors (courtesy of Velocys). (Source: LeViness et al. [44]. Reproduced with permission of Velocys, Inc.)

UOP [51] made a comparison of the entrained bed reactor, designed by Kellogg [52] and operated commercially by Sasol, a tube-wall reactor developed by US DOE [53], a slurry reactor, designed by Kölbel [25] and operated on a semicommercial scale at Rheinpreussen–Koppers, and the ebullating bed reactor, developed by the US Bureau of Mines [54] and utilized by ChemSystems, Inc. [55]. It was concluded that, since the FT reactor represented only a small part of the total cost of an indirect coal liquefaction plant, the value of the products was of primary importance in the selection of a particular reactor [51]. These authors also mentioned that in spite of uncertainties in the yield structures that have been reported for some reactor systems, the differences in comparing the above reactor systems are large enough to indicate clear advantages for the slurry reactor. The slurry reactor has the lowest investment and catalyst costs. It was expected to give the highest gasoline yields and thermal efficiencies [51]. While the slurry reactor appears to be the one currently being considered by many organizations, it appears that it will be operated to maximize wax yield, which will then be cracked to diesel fuel with the goal of minimizing the gasoline yields. Changes in perception and in product demand occur rapidly in the fuel industry.

Guettel and Turek [56] compared four reactor types, namely, fixed bed, slurry bubble column, monolith loop, and microstructured. A one-dimensional (no axial mixing) modeling approach was used to make the comparisons. One of their conclusions was

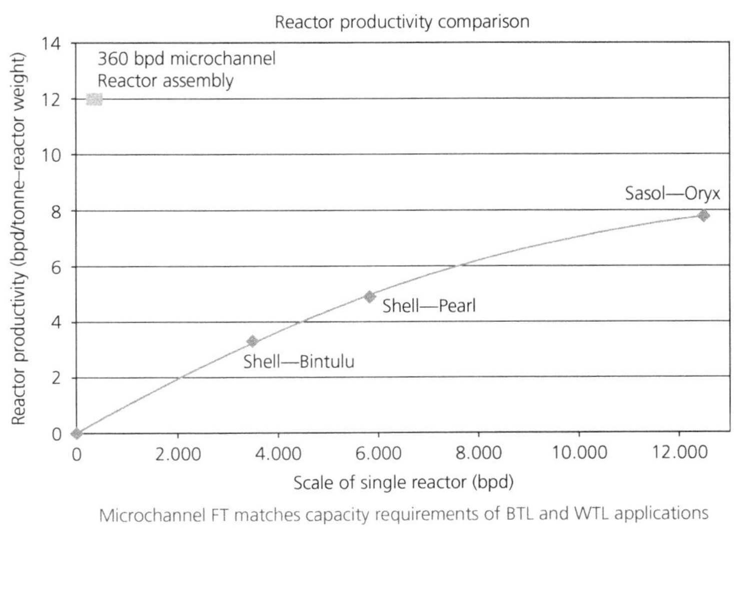

Figure 12.11 Productivity of commercial Velocys microchannel, fixed-bed, and slurry reactors compared to that of the reactor scale (courtesy of Velocys).
(Source: LeViness et al. [44]. Reproduced with permission of Velocys, Inc.)

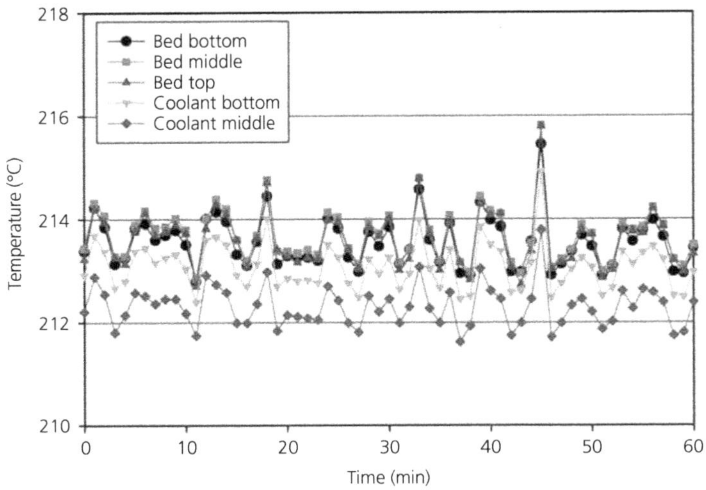

Figure 12.12 Profile of reactor temperature at the top, middle, and bottom of a small channel reactor during Fischer-Tropsch synthesis [48].
(Source: Jia et al. [48]. Reproduced with permission of John Wiley & Sons.)

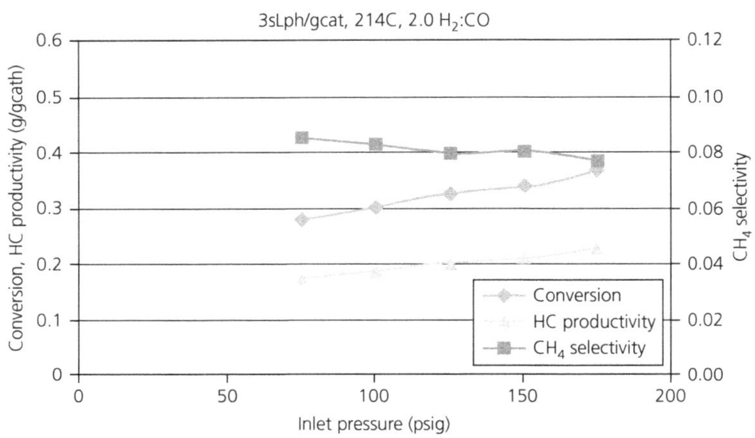

Figure 12.13 CO conversion, hydrocarbon productivity, and methane selectivity for Fischer–Tropsch synthesis at 214°C with a Pt-Co-alumina catalyst [48].
(Source: Jia et al. [48]. Reproduced with permission of John Wiley & Sons.)

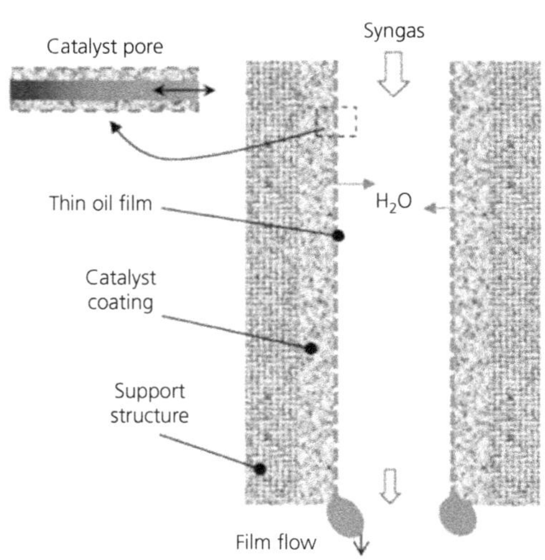

Figure 12.14 Model for structured bed of catalyst in a straight flow channel reactor [50].
(Source: Liu et al. [50]. Reproduced with permission of John Wiley & Sons.)

that a monolith loop reactor could approach the productivity of the slurry bubble column reactor and at the same time had the advantages of a fixed-bed reactor. They also concluded that microstructured reactors could achieve very high efficiencies but would suffer from their low productivity per reactor volume. The research team [57] later found that high catalyst and reactor volume-specific productivities could be obtained using milli-structured fixed-bed reactors having channel widths as high as 1.5 and 3 mm, with catalyst particle sizes of 100, 200, and 350 μm providing high effectiveness factors and suitable pressure drops.

12.4.2 Fluidized-bed reactors

Circulating fluidized-bed reactors have been operated successfully at Sasol for many years. The major disadvantages for them are the need to operate at high temperature to obtain sufficient productivity with the lower surface area catalysts that are needed to obtain attrition resistance, the low alpha operation where only vapor-phase products are produced as is required to eliminate catalyst agglomeration, the energy required for catalyst

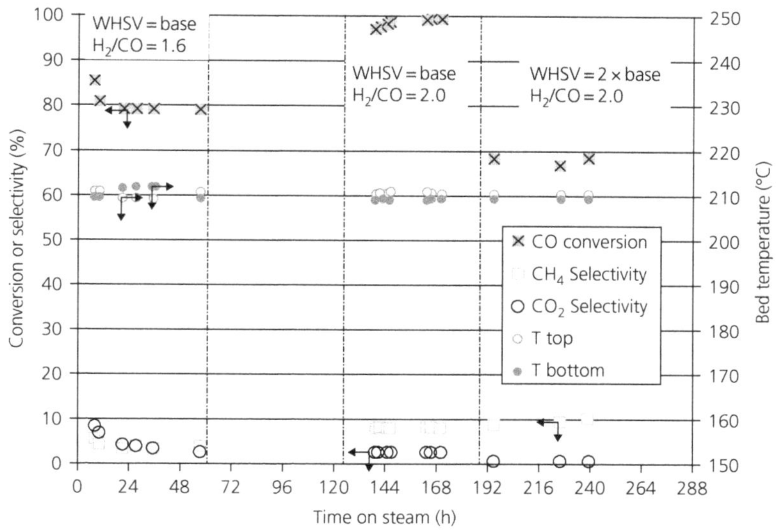

Figure 12.15 Steady-state performance of a structured catalyst bed (25 bar) [50].
(Source: Liu et al. [50]. Reproduced with permission of John Wiley & Sons.)

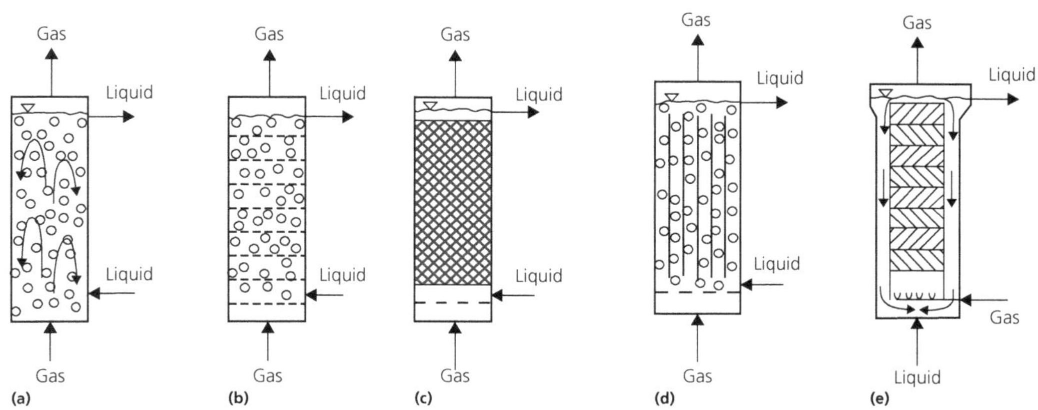

Figure 12.16 Illustration of possible types of slurry bubble column reactors. (a) Simple bubble column, (b) cascade bubble column with sieve trays, (c) packed bubble column, (d) multishaft bubble column, and (e) bubble column with static mixers [61].
(Source: Reproduced with kind permission of Lee [61].)

circulation, and the pressure drop through the catalyst bed. Many of the operational disadvantages have been overcome by the development of the fixed fluidized-bed reactor [58]. The Sasol fixed fluidized-bed reactors have now replaced all of the circulating bed reactors. Analogies between fluidized multiple phase reactors and bubble column reactors have been made [59, 60]. They show that the hydrodynamic behavior is both qualitatively and quantitatively similar in nature and that much cross-fertilization of concepts is applicable in scaling up.

12.4.3 Slurry bubble column reactors

The major attention, both industrial and academic, during the past 15 years, has been given to the slurry bubble column reactors and the literature on this subject in general, and specifically for the FT synthesis, has exploded. Several options exist today for the slurry reactors; some of these are illustrated in Figure 12.16 [61]. The simple bubble column reactor (type A) is most commonly encountered and has been applied for the FT synthesis. The multishaft bubble column reactor (type D) has also been utilized for the FT synthesis or at least one of the shafts of 6 inch diameter has been utilized by Exxon. Several laboratory-scale bubble column reactors with static mixers have been evaluated for the FT synthesis at the laboratory scale.

The slurry-phase methanol process has been developed and commercialized [62–64]. Deckwer has published a major treatise on bubble column reactors [65]. Saxena [66] reviewed the use of bubble column reactors for FTS. A review of more recent studies has been presented [67]. The reader is referred to these for the description of reactor details. In the following, only brief accounts will be given for some of the major topics.

Two major types of slurry bubble column reactors are illustrated in Figure 12.17, one from Exxon [68] and the other from Sasol [69]. The reactor containing cooling coils immersed in the slurry is the one that is the most frequently encountered. The concept of one type of reactor patented by Exxon [68] has many similarities to the ARGE fixed-bed reactor except the fixed

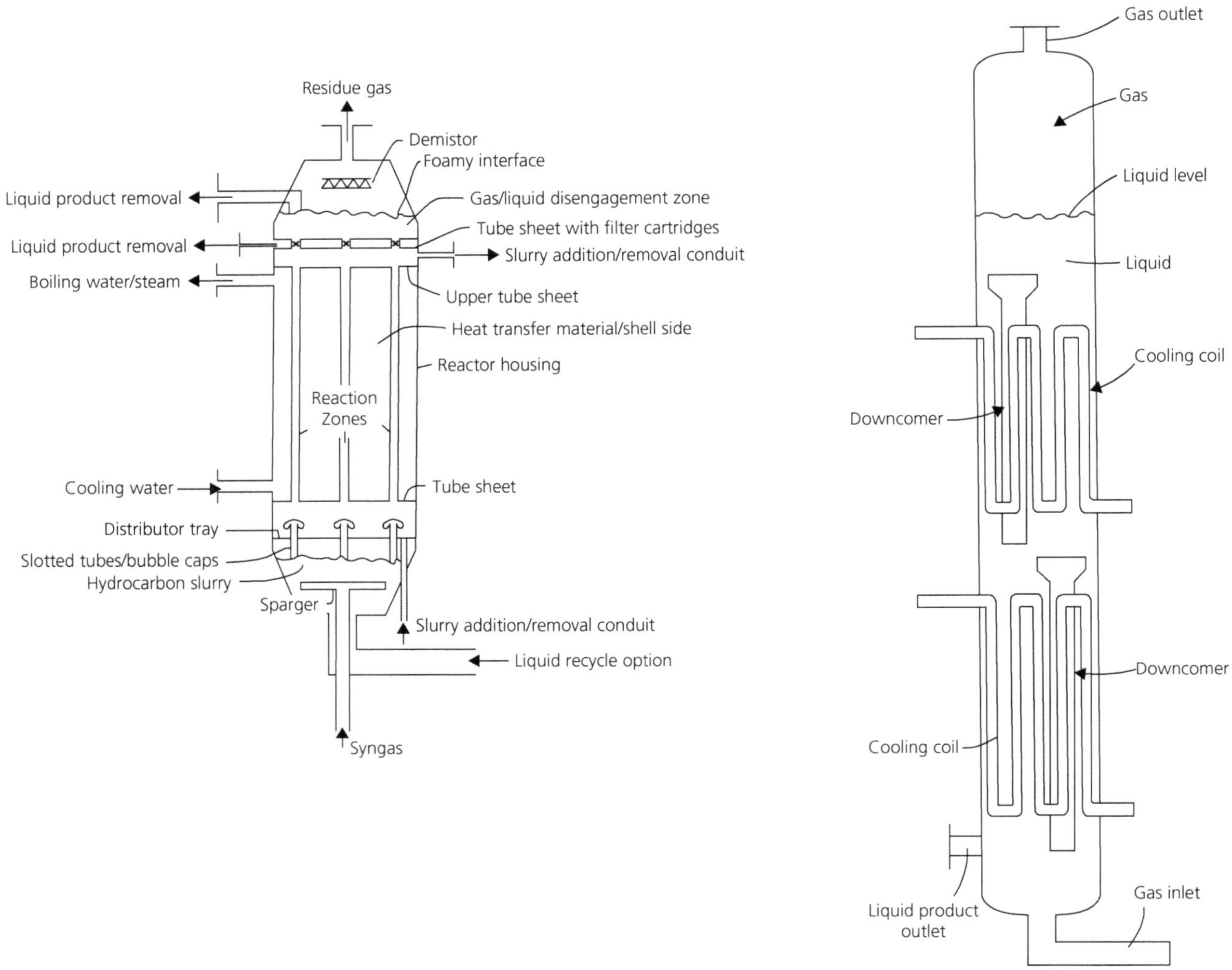

Figure 12.17 Schematic of an Exxon slurry reactor (left) [68]. Schematic of Sasol slurry bubble column reactor (right) [69].

catalyst bed has been replaced with a slurry catalyst. Both utilize the "tube-in-a-shell" reactor concept.

As with the three-phase slurry reactor in other applications, the distributor plate is usually treated as a proprietary topic. Kölbel contends that when the bubble column reactor is operated in the proper gas flow range, the bubbles within the reactor will be in dynamic equilibrium with the formation of larger bubbles and their spontaneous breakup. Some of the concepts for the gas sparger are illustrated in Figure 12.18. Most patents for the application of bubble column reactors for the FT synthesis would include multiple tube types of spargers. In practice, the tube spargers would be covered with a bubble cap to prevent solids from entering the tube, as shown in Figure 12.18.

For modeling purposes, the bubbles within the slurry are usually considered to have a spherical shape, their size distribution taken as some average size, and their breakup and reformation are neglected. However, it was shown that the bubbles are clearly not spherical but are as presented in Figure 12.19 [70]. The influence of elevated pressure and temperature on the shape and rise velocity of a single bubble in a liquid–solid suspension has been investigated in a unique experimental setup [71]. For very low flows (e.g., gas velocity of 0.01 m/s), it is possible to have a reasonably uniform distribution of small (<5 mm) bubbles that rise vertically [72] and can be modeled as plug flow. However, to operate in the small bubble, plug flow regime DeSwart et al. [72] calculated that it would require seventeen 11 m diameter reactors to produce the same amount of product as could be produced in four 7.5 m diameter reactors that operate in the churn turbulent slurry regime.

It is found that the bubble rise velocity decreases with increasing pressure and with decreasing temperature. This is mainly due to the changes in the density of the gas and the liquid velocity. Not only do the bubbles have a shape like a comet with a tail of smaller bubbles, they also do not rise straight to the vertical but may exhibit a corkscrew path to the top of the reactor, leaving regions rich in bubbles and regions with a lower concentration (Figure 12.20) [73]. The bubble shape is frequently modeled with variable aspect ratios as cross sections of ellipsoids [74].

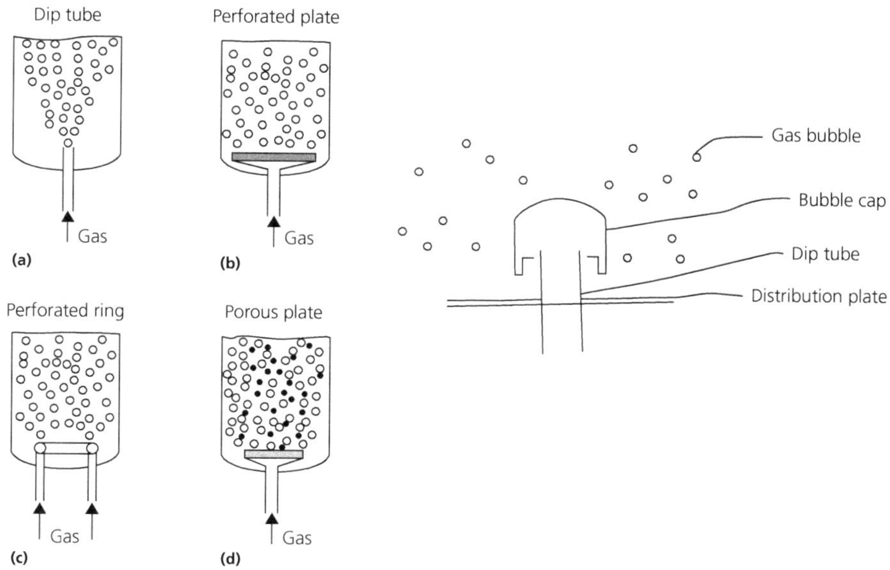

Figure 12.18 Illustration of some concepts for the sparger for the bubble column reactor. Left, from [61]: (a) dip tube, (b) perforated plate, (c) perforated ring, and (d) porous plate. (Source: Reprinted from Lee [61], © 2003, with permission from the author.). Right: closeup view of a sparger from an Exxon patent [68].

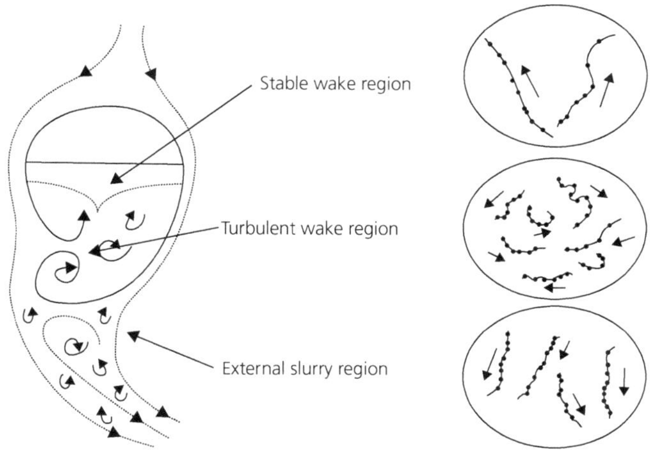

Figure 12.19 Typical particle trajectories within three different flow regions around a rising bubble in a slurry bubble column reactor [70]. (Source: Luewisuthichat et al. [70]. Reproduced with permission of the Society of Chemical Engineers, Japan.)

The US DOE had a major effort to understand the many variables affecting the performance of a bubble column reactor. Dudukovic and Toseland [75] outlined the cooperative study by Air Products and Chemicals (APC), Ohio State University (OSU), Sandia National Laboratory (SNL), and Washington University in St. Louis (WU). The efforts of this group have developed valuable unique experimental techniques for the measurement of gas holdup, velocity, and eddy diffusivities in bubble columns. They have obtained data that allows improved insight in churn-turbulent flow and have assessed the impact of various effects (internals, solid concentration, high gas velocity, pressure, etc.). General ideal flow pattern-based models do not reflect bubble column reality; to date the models are based on a combination where some parameters are evaluated from first principles and some from the database.

It has been observed that the time-dependent flow behavior in a cylindrical column is chaotic and not predictable. However, some periodic structures may be detected in the velocity time series in the axial and tangential directions that may lead to some kind of periodicity [76]. It has been reported that the transition from the homogeneous to the heterogeneous flow regime in bubble columns could be quantitatively found with high accuracy by analyzing the chaotic characteristics of the pressure fluctuation signal. A distinctive feature of the pressure signal from bubble columns is that it is composed of two different parts: a low-frequency part resulting from the motion of the large bubbles and a high-frequency part resulting from all other processes (coalescence, collapse, breakup, etc.) [77]. Delnoij [78, 79] conducted an extensive study and concluded that their computed flow structures resembled the experimentally observed patterns

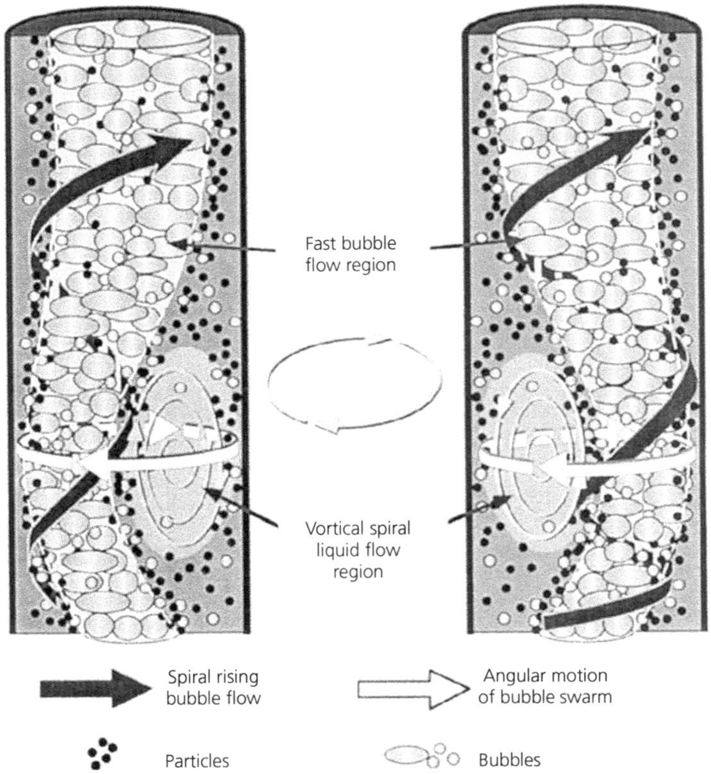

Figure 12.20 Model of the flow scheme in the slurry bubble column reactor [73].
(Source: Ohkawa et al. [73]. Reproduced with permission of Elsevier.)

although the time-dependent behavior predicted by the model differed from that observed in their bubble column.

The extensive theoretical and experimental data for bubble column reactors have been applied to situations that are applicable to FTS. Inga and Morsi [80, 81] have constructed a 0.3 m diameter, 2.8 m tall slurry bubble column reactor and have operated it in the churn-turbulent regime with catalyst concentrations up to 50 wt.% and pressures to 8 bars. The impact of gas velocity, pressure, and catalyst concentration on gas holdup (ε_G) has been determined, and it was found that under given operating conditions, the gas bubbles could be classified into "small" and "large" bubbles [80, 81]. The "large" bubbles showed plug flow behavior, but the "small" bubbles recirculated with the liquid. The gas holdup increased with superficial gas velocity and pressure and decreased with catalyst concentration. At high catalyst loading, coalescence of gas bubbles increased, reducing the number of small bubbles and the gas interfacial area. The gas–liquid mass transfer coefficient (k_La) values followed the trends of the gas–liquid interfacial area [80, 81].

Van der Laan [82] reported attempts to model FT in a bubble column reactor. His model exhibited well-mixed liquid and two gas bubble regimes: small bubbles that were well mixed and large bubbles that showed plug flow behavior (Figure 12.21). Van der Laan [82] also provided a summary of bubble column reactor models that others have utilized (Tables 12.1 and 12.2). He concluded that the FT slurry bubble column reactor is reaction controlled due to the low activity of the iron catalyst and the volumetric mass transfer coefficient of the large bubbles due to frequent bubble coalescence and breakup. Recent models involve the application of multicomponent reaction engineering models for FTS in bubble column reactors. Examples of these and other models can be found in the literature [72, 82, 96–105].

The internals of the bubble column reactor may have a dramatic impact on the flow patterns of the bubbles and the liquid. Companies have not divulged details about the internals to date. Some details of the US DOE pilot plant (22.5 inch; 0.57 m diameter) have been published [106]. In this report the dimensions of the cooling tubes, their location, and their number are provided. These cooling coils occupied about 10% of the total volume of their commercial reactors' slurry volume. The gas holdup and bubble characteristics as well as their radial profiles were determined in a column that was about the size of the US DOE reactor [107–109]. Dense internals were found to increase the overall gas holdup and to alter the radial gas profile at various superficial gas velocities. The tube bundle in the column increased the liquid recirculation and eliminated the rise of bubbles in the wall region of the column. These results indicate that further studies of bubble column hydrodynamics are directed toward larger scale units equipped with heat exchange tubes.

The fate of bubbles in the slurry reactor is a complex issue since they undergo formation and breakup [110, 111]. Single-bubble and two-bubble models have been evaluated [112]. Two models for the churn-turbulent flow regime were developed, and a comparison indicated that increasing reactor

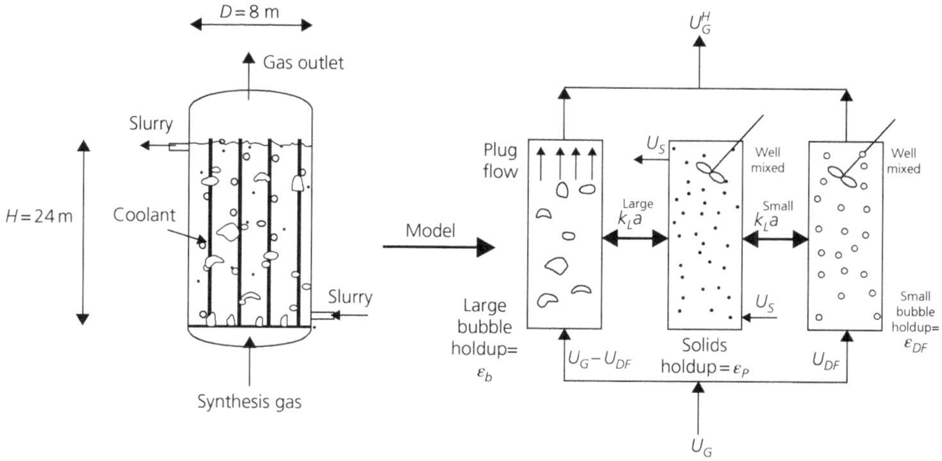

Figure 12.21 Hydrodynamic model of a slurry bubble column reactor in the heterogeneous flow regime [82]. (Source: van der Laan [82]. Reproduced with permission of Elsevier.)

Table 12.1 Comparison of reaction engineering models for the Fischer–Tropsch synthesis in slurry bubble column reactor.

Reference	Gas phase	Liquid phase	Catalyst distribution	Energy balance	Components	FT	WGS Table 12.2
Calderbank et al. [83]	PF	PF	Uniform	Isothermal	H_2	1	—
Satterfield and Huff [84]	PF	PM	Uniform	Isothermal	H_2	1	—
Deckwer et al. [85]	PF	PM	Uniform	Isothermal	H_2	1	—
Deckwer et al. [86]	AD	AD	Non-uniform	Non-isothermal	H_2	1	—
Bukur [87]	PF	PF,PM	Uniform	Isothermal	H_2	1	
Stern et al. [88]	PF	PM	Uniform	Isothermal	H_2, CO, CO_2, H_2O	1	—
Kuo [89]	PF	PM,PF,AD	Non-uniform	Isothermal	C_5H_{10}	1	—
Kuo [89]	PF	PF	Non-uniform	Isothermal	H_2	2	2
Stenger and Satterfield [90]	AD	AD	Non-uniform	Isothermal	H_2, CO, CO_2, H_2O, C_5H_{10}	1	1
Prakash and Bendale [91]	AD	AD	Non-uniform	Isothermal	H_2, CO, CO_2, H_2O, C_{1-3}	2	2
Prakash [92]	AD	AD	Non-uniform	Isothermal	H_2, CO, CO_2, H_2O, C_{1-3}	2	2
De Swart [93][a]	AD	AD	Non-uniform	Non-isothermal	H_2	1	—
De Swart [93][b]	PF	PM	Uniform	Isothermal	H_2	1	—
Mills et al. [94]	AD	AD	Non-uniform	Non-isothermal	H_2	1	—
Inga and Morsi [95]	PF	MC	Uniform	Isothermal	H_2, CO, H_2O	1,3	2
Krishna, et al. [96][b]	PF	PM	Uniform	Isothermal	H_2, CO	3	—
Van der Laan model[b]	PF	PM	Uniform	Isothermal	H_2, CO, CO_2, H_2O, N_2, C_{1-100}		

AD, axial dispersion; MC, mixing cells; PF, plug flow; PM, perfectly mixed.
[a] Heterogeneous flow regime: large bubbles: PF, small bubbles and liquid: AD.
[b] Heterogeneous flow regime: large bubbles: PF, small bubbles and liquid: PM.
Source: van der Laan [82]. Reproduced with permission of Elsevier.

Table 12.2 Kinetic models for Table 12.1.

FT kinetic expressions	WGS kinetic expressions
1 $kC_{H_2,L}$	1 $k(C_{CO,L}C_{H_2O,L} - C_{CO_2,L}C_{H_2,L}/K_P)$
2 $\dfrac{kC_{H_2,L}C_{CO,L}}{C_{CO,L} + aC_{H_2O}}$	2 $\dfrac{k(C_{CO,L}C_{H_2O,L} - C_{CO_2,L}C_{H_2,L}/K_P)}{P_{CO} + aP_{H_2O}}$
3 $\dfrac{kC_{H_2,L}C_{CO,L}}{(1 + aC_{CO,L})^2}$	

Source: van der Laan [82]. Reproduced with permission of Elsevier.

diameter and reducing catalyst concentration enhanced the difference between the two models. However, because the bubbles were continuously undergoing formation and breakup, the distinction between the two models was found to be small. In the same study, a numerical model was developed to predict the bubble size distribution in turbulent bubbly flows [112]. A square bubble column operated at a superficial gas velocity of 2 cm/s was chosen as a simulation base to evaluate the parameters. "Daughter" bubbles are produced from the breakup of a "parent" bubble in a bubble column reactor. The critical Weber number, the daughter size distribution, and the superficial gas velocity were varied, and it was found that changes both in the Weber number and the daughter size distribution had a significant impact on the overall bubble size distribution, but a different shape did not have a significant impact. Krishna and van Baten [113] illustrated the relation of gas holdup and the superficial gas velocity for an air–water system (Figure 12.22). Operating in the churn-turbulent regime, the hydrodynamic and mass transfer characteristics of four gases (H_2, CO, N_2, and CH_4) were obtained for a liquid hexane mixture using a 0.3 m diameter and 2.8 m tall bubble column. The gas holdup and mass transfer coefficient values for all four gases increased with superficial velocity and pressure increases. However, the values decreased with increasing catalyst concentrations. The hydrodynamics (i.e., gas holdup, Sauter mean bubble diameter)

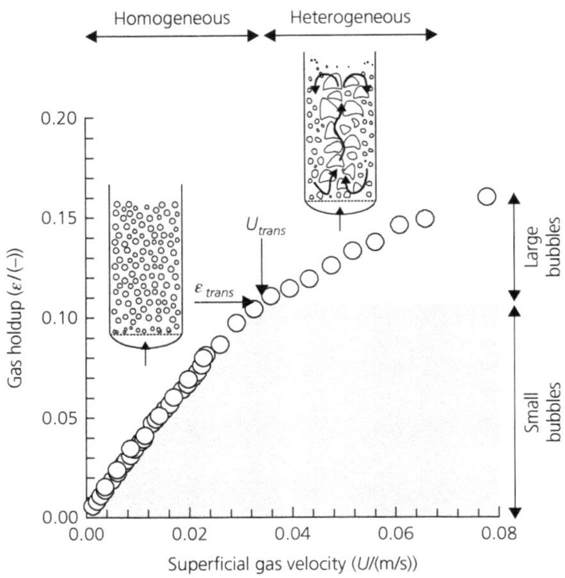

Figure 12.22 Experimental data for gas holdup in a 0.1 m diameter bubble column operating with air–water system spanning both the homogeneous and heterogeneous flow regimes [113].
(Source: Krishna and van Baten [113]. Reproduced with permission of Elsevier.)

and the overall volumetric liquid-side mass transfer coefficients were measured in the same-sized column using three liquids: liquid paraffin mixture, light FT cut, and heavy FT cut [114–116]. It was found that for nitrogen, the gas holdup, the overall k_La, and the population of small bubbles decreased as the molecular weight of the liquid was increased. Two novel empirical correlations for predicting gas holdup and the overall k_La for these gases in the FT cuts were proposed.

12.4.4 Structured packings

The use of structured packings has been investigated during recent years, and some examples of the packing used are shown in Figure 12.23 [117], and packing for a 0.1 m column is shown in Figure 12.24 [118]. Results demonstrate that the application of structured packings in gas–liquid–solid reactors offers good opportunities for the removal or supply of energy. The application of a structured catalyst in an FT reactor has revealed clear advantages of the improved radial heat transport properties, allowing the application of more active catalysts and the use of larger reactor tube diameters. It gives an outlook for process intensification in fixed-bed FTS reactors. It is thus concluded that, although there are still challenges to be overcome, structured packings have great potential for application in multiphase fixed-bed reactors [119].

Several structured packing elements have been compared with regard to catalyst holdup, heat transfer performance, and pressure drop [120]. The results indicate that using catalyst coating gave lower pressure drop than packed beds but had a much lower catalyst inventory per reactor volume. On the other hand, a particle-packed structure exploited the advantages of structured flow while not sacrificing much in catalyst holdup compared to a randomly packed bed. This alternative retained the favorable pressure drop characteristics so that smaller particle-sized catalysts could be used. Despite their lower

Figure 12.23 Examples of structured packing: (a) open cross flow structure (OCFS), (b) closed cross flow structure (CCFS), (c) knitted wire, and (d) aluminum foam [117].
(Source: Pangarkar et al. [117]. Reproduced with permission of Elsevier.)

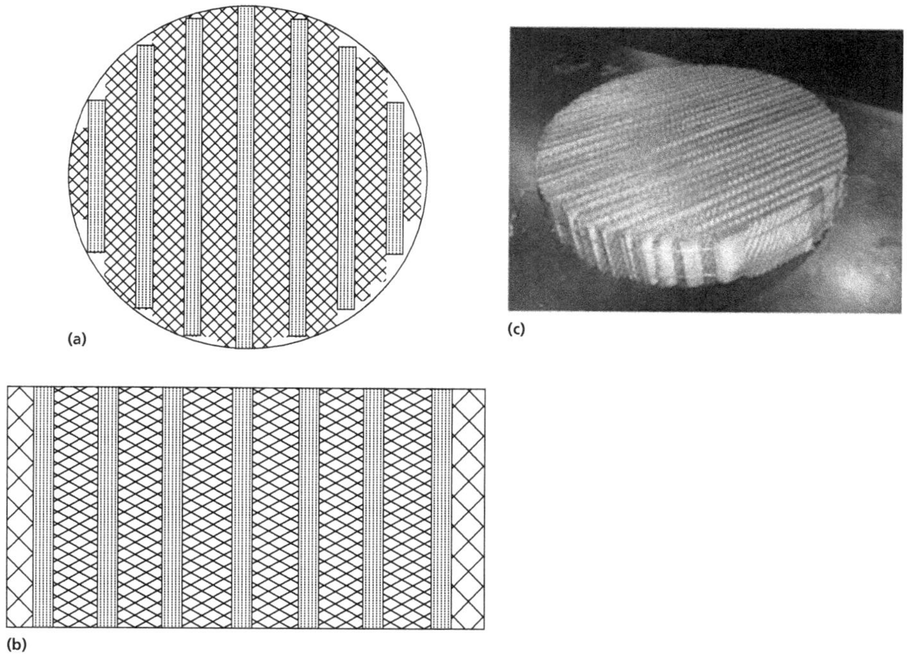

Figure 12.24 Schematic representation of structured catalytic packing made by BUCT: (a) cross section, (b) side view, and (c) actual photography [118].
(Source: Lei et al. [118]. Reproduced with permission of Elsevier.)

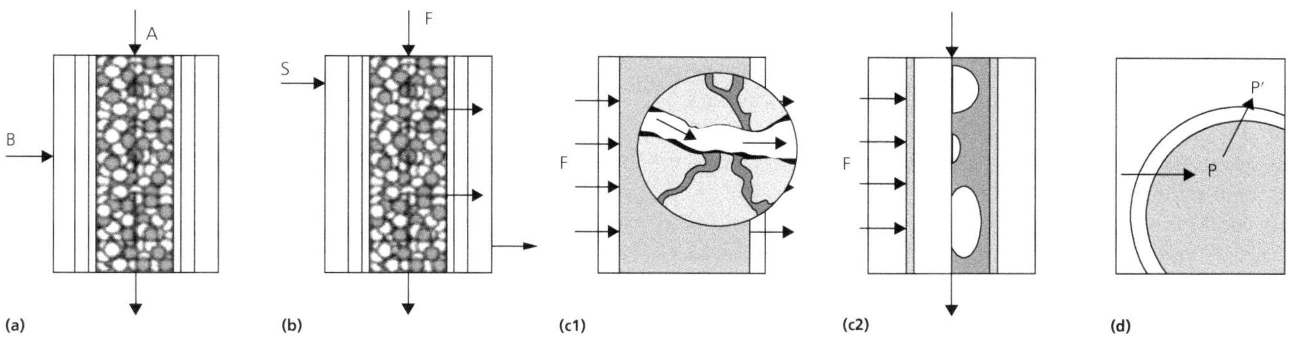

Figure 12.25 Membrane reactors for FT synthesis from the literature: (a) distributed feeding of reactants *A* and *B*, (b) *in situ* water removal by selective membrane (F, feed; S, sweep), (c1) plug-through contactor membrane (PCM) with wide transport pores, (c2) forced-through flow membrane contactor, product and heat removal by circulated liquid product, (d) zeolite encapsulated FT catalyst, *P′*, modified product [123].
(Source: Rohde et al. [123]. Reproduced with permission of Elsevier.)

catalyst loading than the randomly packed bed, they could have 25% higher C_5+ productivity per tube volume. This increased productivity was a result of the increased catalyst effectiveness that could be combined with better temperature control [121].

The application of permeable composite monolith membranes for the FT synthesis has been tested [122]. An overview of concepts associated with this reactor type has been presented (Figure 12.25) [123]. Novel uses of this concept have been advanced, and some experimental results have demonstrated the ability to operate at high CO conversion with metal FT catalysts by removal of the water produced during the synthesis [124] and the encapsulation of an FT catalyst by a zeolite membrane layer to effect upgrading reactions in the FT reactor [125]. The potential of this technique merits further studies to evaluate the ability to scale to a commercial level.

Compositional modulation has been practiced for the FT synthesis in catalytic reactors [126]. It was found that the cyclic feeding of synthesis gas (CO/H_2) had an influence on the selectivity of the FT products. In the early studies, only low conversions could be utilized due to the exothermic nature of the reaction. It was concluded that for an iron catalyst, the methane selectivity increased with periodic operation as did the molar ratio of alkene/alkane. Higher conversion studies were conducted in a CSTR, and it was found that periodic operation had an influence on the selectivity of the products from the FT synthesis using an iron catalyst [127]. First, there was a decrease in the alpha value for synthesis with increasing period. In addition, the alkane/alkene ratio increased with an increase in the period. There was a change in the CO_2 production but this could be attributed to the change in CO conversion and not the periodic operation of the reactor. As found in earlier studies, the fraction of methane in the hydrocarbon products increased with increasing length of the periodic operation, which did not bring any desirable effects of activity and selectivity of an iron catalyst.

12.4.5 Operation at supercritical conditions (SCF)

At the high end of the pressure range, researchers have reported significant benefits by operating FT synthesis with supercritical media at supercritical pressure, including enhanced wax extraction, catalyst lifetime extension through carbon removal, and improvements in selectivity [128–132]. As previously described, FTS can be operated by the traditional gas-phase route or by more advanced liquid-phase methods, and either approach offers advantages as well as drawbacks. Gas-phase FTS (e.g., carried out in a fixed or fluidized-bed reactor) produces higher yields of product, because the catalyst is concentrated, on average, in a smaller reactor volume. However, heat removal is a significant problem, as high initial rates and poor heat removal capacities result in hot spots, which cause sintering and alter selectivities toward lighter products. Moreover, heavy waxes deposit within catalyst pores, which further exacerbates catalyst deactivation. With liquid-phase FTS, there is more uniformity in temperature throughout the reactor through better heat control of conducting FTS in the liquid phase; this is a direct consequence of the higher heat removal capacities of the liquid. Deactivation rates are further lowered because liquid media helps dissolve wax products in catalyst pores. The main drawback is that the liquid is a diffusional resistance to the transport of gas-phase reactants to active sites, which, in turn, tends to decrease of the reaction rate relative to gas-phase FTS. One can combine the high transport features of the gas with the extractive and heat transfer properties of the liquid by conducting FT in supercritical media. Fan and Fujimoto [128] have established criteria for choosing the supercritical fluid, as follows: (i) the critical temperature and pressure should be slightly lower than the typical reaction temperature and pressure, (ii) the solvent should be one that does not poison the catalyst and should be stable under reaction conditions, and (iii) the solvent should have a high affinity for aliphatic hydrocarbons to extract the wax from the catalyst surface and reactor.

A density versus pressure diagram of the supercritical fluid mixture of 55% hexane and 45% pentane at 220°C is shown in Figure 12.26. A pressure of 5.45 MPa gives a high liquid-like density and was used in experiments carried out over a 25% Co/Al_2O_3 catalyst using a fixed-bed supercritical reactor (Figure 12.27) [131]. By tuning the supercritical fluid pressure, Figure 12.28 demonstrates that deactivation by pore plugging by heavy wax can be avoided. Moreover, because the products are extracted rapidly from the reactor, a greater fraction of primary products, that is, olefins, were obtained in the product distribution as compared to conventional CSTR (i.e., slurry phase) and fixed-bed (non-supercritical conditions) reactors (Figure 12.29). The state-of-the-art of supercritical FTS technology was recently reviewed [132]. Despite the numerous benefits outline earlier, obvious drawbacks are costs associated with operating at higher pressure (e.g., compression costs at the gasifier), as well as separation and recycling of the supercritical fluid.

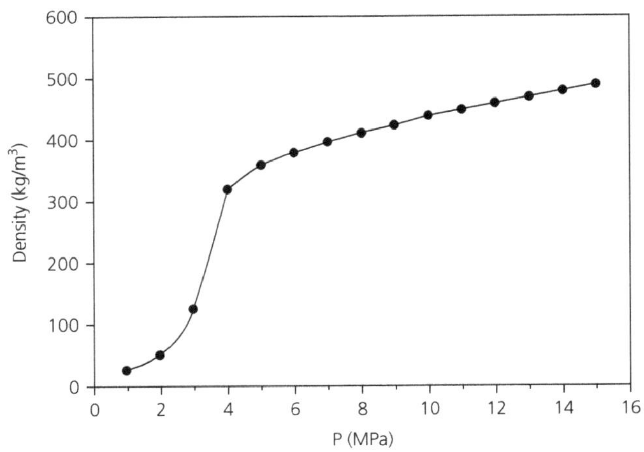

Figure 12.26 Density versus pressure of the mixture of 55% hexane and 45% pentane at 220°C [131].
(Source: Jacobs et al. [131]. Reproduced with permission of Elsevier.)

12.5 The future?

The commercial plants that are operating or are in the shakedown stage are given in Table 12.3. The Sasol and Shell plants in Qatar have been scaled to the largest-sized reactor for operating in both the fixed and slurry bubble column reactors, which appears possible today. As has been the case since 1950, the construction, including the heat treatment, and the transportation available have set the limit for the size of reactor that can be utilized. Today's reactors are at the limit imposed by today's manufacturing and transportation facilities. To scale to a larger size would represent a first time event and would induce a very large cost to build the facilities for the construction and transportation of larger reactors. Nevertheless, it is anticipated that the output of the current Qatar reactors will be increased by a factor of 1.5–2.0 because of catalyst improvements. For example, Sasol envisions being able to increase production from 17 500 bbl/day that is now produced in Qatar to 30 000 bbl/day by the mid-2020s due to catalyst improvements while maintaining the same reactor size as is now used [133]. Thus, at the commercial scale, it appears that one must return to consider improvements in the smallest reactor, the catalyst [134]. In another study, it was concluded that a productivity increase of 20% was possible for the slurry bubble column and 40% for the fixed-bed reactors through reactor structuring. For the slurry reactor, they found that decreasing the liquid axial dispersion by a factor of 4 would greatly enhance the C_{5+} production. Likewise, improvements in heat transfer and diffusion effects in the fixed-bed reactor were considered to be possible.

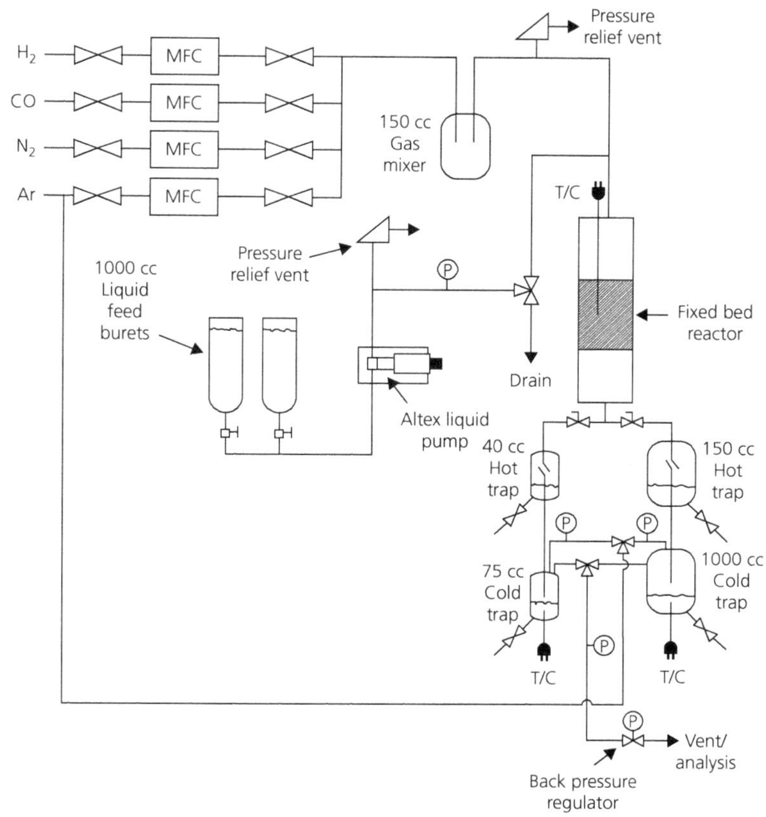

Figure 12.27 Schematic of a laboratory-scale supercritical fixed-bed reactor [131]. T/C: thermocouple.
(Source: Jacobs et al. [131]. Reproduced with permission of Elsevier.)

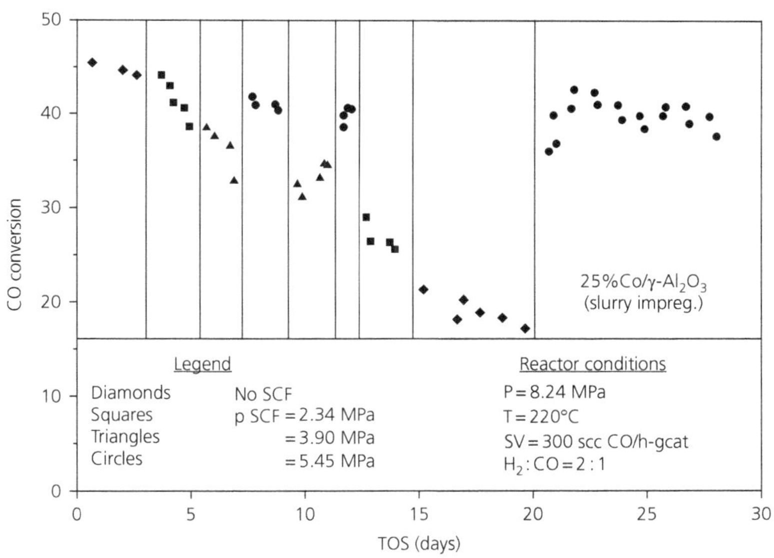

Figure 12.28 CO conversion versus time on stream on a 25% Co/γ-Al_2O_3 slurry-phase impregnation catalyst in a fixed-bed reactor with varying partial pressure of supercritical fluid (SCF).
(Source: Jacobs et al. [131]. Reproduced with permission of Elsevier.)

The FT industry has been driven since 1950 by the "economy of scale" philosophy. This philosophy has advanced the petroleum refining industry during this period, and the commonality of the two processes has influenced developers in the FT industry. The influence of this philosophy can be seen in Figure 12.30, which was taken from [135], where the cost per barrel of production was shown as being nearly linearly related to the size of the plant. Based on this philosophy, one should not consider building a small FT plant.

If one is to build a plant based upon renewables or on remote natural gas associated with oil production, one must consider a small (1000–5000 bbl/day) plant. The availability of renewables

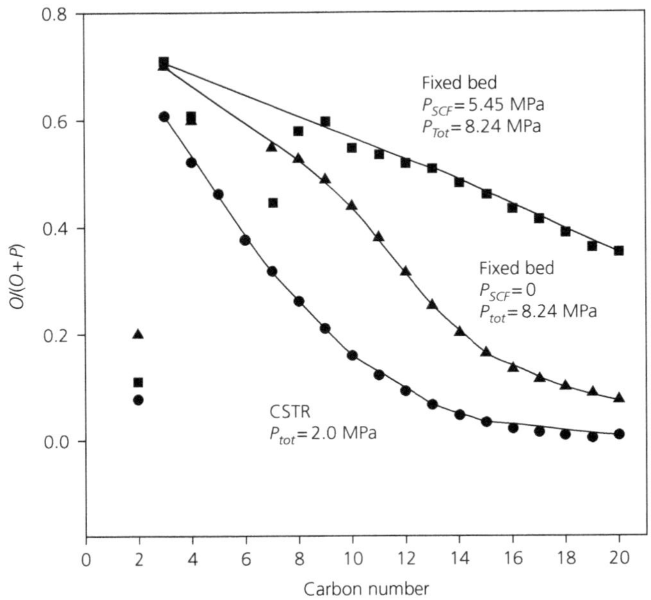

Figure 12.29 Olefin selectivity as an function of carbon number for a variety of reactors.
(Source: Jacobs et al. [131]. Reproduced with permission of Elsevier.)

Table 12.3 Existing operating Fischer–Tropsch plants.

Company	Site	Capacity bpd	Raw material	Start-up date
Sasol	Sasolburg	14 000[a]	Natural gas[b]	1955
Sasol	Secunda	80 000	Coal	1980
Sasol	Secunda	80 000	Coal	1984
PetroSA	Mossel Bay	30 000	Natural gas	1992
Shell	Bintulu	14 500	Natural gas	1993
Qatar/Sasol	Qatar	34 000	Natural gas	2006
Shell	Qatar	140 000[c]	Natural gas	2013
Chevron	Escravos	33 000	Natural gas	2014

[a] Expanding to nearly double this capacity.
[b] Switched from coal to natural gas about 2010.
[c] Total to be on-line at end of 2014.

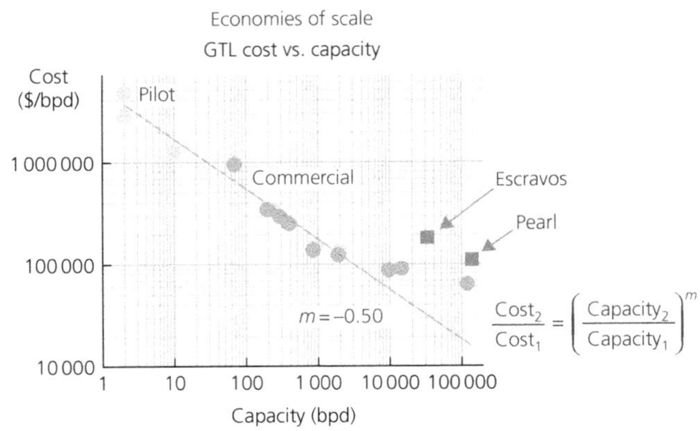

Figure 12.30 Cost of plant versus the plant capacity [135]. (Source: Reproduced with kind permission of Tijm [135].)

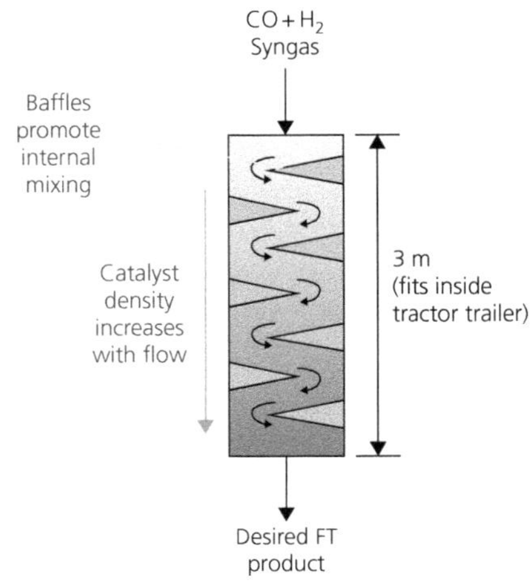

Figure 12.31 Schematic of a FT reactor suitable for a mobile biorefinery [136].
(Source: Reprinted from Pratt [136], Copyright (2012), with permission from Sandia National Laboratories.)

is limited because of high transportation costs compared to that of coal or natural gas. This means that one must think small or abandon improving the environmental impact of producing oil in some locations or of using a renewable carbon source. A number of approaches have been advanced to overcome the view that one needs the "economy of scale" to proceed with the development of an FT plant. For example, a new concept has been developed for the reactor used for liquid biofuel production [136]. Pratt, in his study, found three main lessons from his modeling of traditional reactors: the need for a shorter length, greater mixing, and variable active catalyst density. Based on his work, he designed a new type of reactor (Figure 12.31). His reactor is 3 m tall and is short enough for mobile installations. Baffles inside the reactor promote more radial mixing of the reactants, while the use of a variable catalyst density along the length of the reactor serves to eliminate, or greatly reduce, the hot spot. At the other extreme, the slurry reactor is considered to be applicable for the small- and medium-scale plant that is needed for processing associated petroleum gas. The successful operation of a 1000 bbl/day plant can be handled at this scale; the only uncertainty is their economics.

The microchannel (small channel) reactor is ideally suited for small-scale operations. As of this date, it has not moved beyond the pilot plant stage. However, Velocys has recently announced an agreement for sale of a 175 bbl/day reactor, together with their catalyst, to a customer in the Commonwealth of Independent States. They also indicate that they have been selected to provide the reactor for a 2800 bbl/day plant located in Ohio that will utilize natural gas from shale deposits in that region.

It would appear that further increases in the size of the FT reactor must await the construction of larger manufacturing and transportation capabilities. The cost of one, or even two, of these reactors does not appear to be sufficient to support the development of this equipment specifically for the FT industry. Thus, for the near future, the maximum benefits of economy of scale for reactor size are achieved. It appears that environmental concerns will cause the industry to begin a new cycle with reactor sizes approaching the small size utilized by the Germans in the 1930s. However, these reactors will be much more productive than those utilized in the 1930s.

Fürnsinn [137] has discussed at length means of overcoming the perception of the impossibility of utilizing small-scale reactors for FT synthesis. He concluded that it was possible to achieve both high yields and high overall efficiency at the small scale. He reached the conclusion that for atmospheric steam generation, one needs to utilize atmospheric pressure FT synthesis.

Even if the reactor construction and transportation limit has been reached, there remains much work to be done to provide scientific and engineering understanding of the operation of the current reactors. At the other reactor and pressure scale, much work needs to be accomplished to make the smaller-scale reactors competitive at this scale.

References

1 Badger WL. Prospects for engineering developments. Ind. Eng. Chem. 1944;36:397–399.

2 Davis BH. Fischer-Tropsch synthesis: overview of reactor development and future potentialities. Top. Catal. 2005;32:143–168.

3 Davis BH. Development of the science of catalysis. In: Ertl G, Knözinger H, Weitkamp J, editors. Handbook of Heterogeneous Catalysis. New York: Wiley-VCH; 1997. p 13–35.

4 Davis BH, Sing KSW. Historical aspects. In: Schüth F, Sing KSW, Weitkamp J, editors. Handbook of Porous Solids. New York: Wiley-VCH; 2002. p 1–23.

5 Brunauer S, Emmett PH, Teller E. Adsorption of gases in multimolecular layers. J. Am. Chem. Soc. 1938;60:309–319.

6 Brunauer S, Emmett PH. Chemisorptions of gases on iron synthetic-ammonia catalysts. J. Am. Chem. Soc. 1940;62:1732–1746.

7 Thiele EW. Relation between catalytic activity and size of particle. Ind. Eng. Chem. 1939;31:916–920.

8 Barrett EP, Joyner LG, Halenda PP. The determination of the pore volume and area distributions in porous substances. I. Computations from nitrogen isotherms. J. Am. Chem. Soc. 1951;73:373–380.

9 Foster AL, inventor; Phillips Petroleum Co., assignee. Catalyst preparation. US Patent 2,408,164. 1946.

10 Tramm H. Technique of carbon monoxide hydrogenation. Erdöl. Kohle 1952;5:10–17.

11 Zorm H, Faragher WF. The CO–H_2 Synthesis at I.G. Farben A.G. (1949) Field Information Agency (F.I.A.T.). Final Report 1,267 (PB 97,368); 1949. Joint Intelligence Objectives Agency, Washington, DC—(see www.fischer-tropsch.org).

12 Guettel R, Kunz U, Turek T. Reactors for Fischer-Tropsch synthesis. Chem. Eng. Technol. 2008;31:746–754.

13 Moore FJ, inventor; Texas Co., assignee. Method of effecting catalytic reactions. US Patent 2,440,109. 1948.

14 Atwell HV, inventor; Texas Co., assignee. Method of effecting catalytic conversions. US Patent 2,438,029. 1948.

15 Steynberg AP, Dry ME, Davis BH, Breman BB. Fischer-Tropsch reactors. In: Steynberg AP, Dry ME, editors. Studies in Surface Science and Catalysis. Amsterdam, The Netherlands: Elsevier B.V.; 2004. Vol. 152. p 64–195.

16 Davis BH. Overview of reactors for liquid phase Fischer-Tropsch synthesis. Catal. Today 2002;71:249–300.

17 Biardi G, Baldi G. Three-phase catalytic reactors. Catal. Today 1999;52:223–234.

18 Slivinskii EV, Kliger GA, Kuz'min AE, Abramova AV, Shuikin AN. Fischer-Tropsch synthesis in a fixed-bed reactor in the slurry process: advantages and prospects. Kinet. Catal. 1999; 40:338–344.

19 Duvenhage DJ, Coville NJ. Deactivation of a precipitated iron Fischer-Tropsch catalyst—a pilot plant study. Appl. Catal. A Gen. 2006;298:211–216.

20 Kölbel H, Ackermann P, inventors; Steinkohlen-Bergwerk "Rheinpreussen", assignee. Hydrocarbons from gaseous mixtures of carbon monoxide, hydrogen, and carbon dioxide. German patent DE 728217. November 23, 1942.

21 Kölbel H, Ralek M. The Fischer-Tropsch Synthesis in the Liquid Phase. Chemical Engineering Institute, Technical University, Berlin, FRG. Translation No. UCRL-Trans. 1568, University of California, 1979.

22 Poutsma ML. Assessment of Advanced Process Concepts for Liquefaction of Low H_2/CO Ratio Synthesis Gas Based on the Kölbel Slurry Reactor and the Mobil-Gasoline Process. ORNL Report ORNL-5635; 1980. National Technical Information Service, US Dept. of Commerce, Springfield.

23 Deckwer WD. A Comparison of Different Fischer-Tropsch Processes. Hannover, FRG: University of Hannover; 1974.

24 Zimmer J, Elstner M, Falkenain G. Generation of Primarily Petrochemical Raw and Basic Products by the Fischer-Tropsch Synthesis—Interim Report I. Union Rheinisch Braunkohlen Kroftstoff A.B. report to U.S. Energy Research and Development Admin. Translation of German Report BMFT 316-7291 ET 34, October 1974, issued May 1977. OCLC number: 4139619. Washington, DC; 1977.

25 Kölbel H, Ralek M. The FT synthesis in the liquid phase. Catal. Rev. Sci. Eng. 1980;21:225–274.

26 Büssemeier B, Frohning CD, Cornils B. Lower olefins via Fischer-Tropsch. Hydrocarb. Proc. 1976;55:105–112.

27 Pennline HW, Schehl RR, Haynes WP. Operation of a tube wall methanation reactor. Ind. Eng. Chem. Process Des. Dev. 1979;18:156–162.

28 Haynes WP, Baird MJ, Schehl RR, Zarochak MF. Fischer-Tropsch studies in a bench-scale tube wall reactor using magnetite, Raney iron and taconite catalysts. ACS Division of Petroleum Chemistry Preprints, Anaheim, CA, March 12–17, 1978.

29 Zarochak MF, Pennline HW, Schehl RR. Fischer-Tropsch Studies with Catalyst-Sprayed Tube Wall Reactors. DOE/PETC/TR-82/14 (DE82020084); 1982. Public National Technical Reports Library (NTRL), National Technical Information Service (NTIS), US Dept. of Commerce, Alexandria, VA.

30 Andrigo P, Bagatin R, Pagani G. Fixed bed reactors. Catal. Today 1999;52:197–221.

31 Ramachandran PA, Chaudhari RV. Three-Phase Catalytic Reactors. New York: Gordon and Breach; 1983.

32 Fukushima S, Kusaka K. Interfacial area and boundary of hydrodynamic flow region in packed column with concurrent downward flow. J. Chem. Eng. Jpn. 1977;10:461–467.
33 de Klerk A. Liquid holdup in packed beds at low mass flux. AIChE J. 2004;49:1597–1600.
34 Bhatt BL. Liquid Phase Fischer-Tropsch (III & IV) Demonstration in the LaPorte Alternative Fuels Development Unit, Topical Report DE-FC22-95PC93052-39 and ON: DE00014026; 1999. Federal Energy Technology Center (FETC), Morgantown, WV and Pittsburgh, PA. Work site: LaPorte, TX.
35 Leva M, Weintraub M, Grummer M, Pollchik M, Storch HH. Fluid flow through packed and fluidized systems, Bulletin 504. Washington, DC: U.S. Bureau of Mines, U.S. Government Printing Office; 1951.
36 Al-Dahhan MH, Khadilkar MR, Wu Y, Duduković MP. Prediction of pressure drop and liquid holdup in high-pressure trickle-bed reactors. Ind. Eng. Chem. Res. 1998;37:793–798.
37 Al-Dahhan MH, Larachi F, Dudukovic MP, Laurent A. High-pressure trickle-bed reactors: a review. Ind. Eng. Chem. Res. 1997;36:3292–3314.
38 Sie ST, Krishna R. Process development and scale-up: III. Scale-up and scale-down of trickle bed processes. Rev. Chem. Eng. 1998;14:203–252.
39 Sie ST. Process development and scale-up: IV. Case history of the development of a Fischer-Tropsch synthesis process. Rev. Chem. Eng. 1998;14:109–157.
40 Mesheryakov VD, Kirillov VA, Kuzin NA. A multifunctional reactor with a regular catalyst packing for Fischer-Tropsch synthesis. Chem. Eng. Sci. 1999:54:1565–1570.
41 Heibel AK, Heiszwolf JJ, Kapteijn F, Moulijn JA. Influence of channel geometry on hydrodynamics and mass transfer in the monolith film flow reactor. Catal. Today 2001;69:153–163.
42 Hiwale RS, Bhate NV, Mahajan YS, Mahajani SM. Industrial applications of reactive distillation: recent trends. Int. J. Chem. React. Eng. 2004;2:1–54.
43 Srinivas S, Mahajani SM, Malik RK. Reactive distillation for Fischer-Tropsch synthesis: simulation-based design methodology using Aspen Plus. Ind. Eng. Chem. Res. 2010;49:9673–9692.
44 LeViness S, Schubert P, McDaniel J, Dritz T. Velocys Fischer-Tropsch Technology. Economical smaller scale GTL enabled by microchannel reactor technology and superactive catalyst. Syngas Convention – Fuels and Chemicals from Synthesis Gas: State of the Art, Cape Town, South Africa, April 1–4, 2012.
45 Atkinson D. Fischer-Tropsch reactors for biofuels production: new technology needed! Biofuels Bioprod. Biorefin. 2010;1:12–16.
46 Baxter I. Small-scale GTL: back on the agenda, World Gas, 2012, Petroleum-Economist, pp 84–85.
47 Vallee S, Gadalla H, Jia Z. Comparison of materials of construction for compact heat exchange reactors. AIChE Spring Meeting, San Antonio, TX, April 28–May 2, 2013.
48 Jia Z, Gadalla H, Vallee S, Davis BH, Sparks DE. Pilot plant test for Fischer-Tropsch synthesis using a fixed bed compact heat exchange reactor, 2011. AIChE Meeting, Minneapolis, MN, October 16–21, 2011.
49 Myrstad R, Eri S, Pfeifer P, Rytter E, Holmen A. Fischer-Tropsch synthesis in a microstructured reactor. Catal. Today, 2009;147S: S301–S304.
50 Liu W, Wang Y, Wilcox W, Li S. A compact and high throughput reactor of monolithic-structured catalyst bed for conversion of syngas to liquid fuels. AIChE J. 2012;58:2820–2829.
51 Riekena ML, Vickers AG, Haun EC, Koltz RC. Comparison of Fischer-Tropsch reactors. Chem. Eng. Prog. April 1982;78(4):86–90.
52 Kellogg P. Synthol Feasibility Study for Standard Oil Company (Indiana)—Base Case. J-5109, Revised Version, October, 1977. Pullman Kellogg, Division of Pullman Inc., Houston, TX.
53 Haynes WP, Baird MJ, Schehl RR, Zarowchak MF. Fischer-Tropsch studies in a bench-scale tube-wall reactor using magnetite Raney iron, and taconite catalyst. ACS Meeting, Anaheim, CA, March 12–17, 1978.
54 Benson HE, Field JH, Bienstock D, Nagel RR, Brunn LW, Hawk CO, Crowell JH, Storch HH. Development of the Fischer-Tropsch Oil-Recycle Process. U.S. Bureau of Mines Bulletin; 1957. Vol. 568. p 72.
55 Blum DB, Sherwin MB, Frank ME. Liquid-phase methanation of high concentration CO synthesis gas. ACS Adv. Chem. Ser. 1974;146:149–159.
56 Guettel R, Turek T. Comparison of different reactor types for low temperature Fischer-Tropsch synthesis: a simulation study. Chem. Eng. Sci. 2009;64:955–964.
57 Knochen J, Guettel R, Knobloch C, Turek T. Fischer-Tropsch synthesis in milli-structured fixed-bed reactors: experimental study and scale-up considerations. Chem. Eng. Process. Process Intensif. 2010;49:958–964.
58 Jager B, Dry ME, Singles T. Experience with a new type of reactor for Fischer-Tropsch synthesis. Catal. Lett. 1990;7:293–302.
59 Krishna R, van Baten JM, Ellenberger J. Scale effects in fluidized multiphase reactors. Powder Technol. 1998;100:137–146.
60 Krishna R, Ellenberger J. A unified approach to the scale-up of "fluidized" multiphase reactors. Chem. Eng. Res. Des. 1995;73:217–222.
61 Lee QF. 2003. Bubble column reactors, Department of Chemical and Biomolecular Engineering, The University of British Columbia. Available at www.chmltech.com/reactors/bubblecolumn.pps.
62 Report prepared by Air Products and Chem., Inc., Allentown, PA and Eastman Chem. Co., Kingsport TN. Design and Fabrication of the First Commercial Scale Liquid Phase Methanol (LPMEOHTM) Reactor. Report prepared for the Air Products Liquid Phase Conversion Co., L.P. and US DOE PETC, Pittsburgh, PA under Cooperative Agreement No. DE-FC22-92PC90543; 1998.
63 Wu Y, Gidaspow D. Hydrodynamic simulation of methanol synthesis in gas-liquid slurry bubble column reactors. Chem. Eng. Sci. 2000;55:573–587.
64 Shollenberger KA, O'Hern TJ. Characterization of Slurry-Phase Flow in the LaPorte Alternative Fuels Development Unit (AFDU) Using Differential Pressure Measurements. SAND97-0530, Sandia National Labs; 1997.
65 Deckwer W-D. Reaktionstechnik in Blasensaeulen. Otto Salle Verlag GmbH & Co, Frankfurt am Main Verlag Sauerlaender AG, Aarau, Switzerland; W.-D. Deckwer, Bubble Column Reactors (translated by Valeri Cottrell), New York: John Wiley and Sons; 1992.
66 Saxena SC. Bubble column reactors and Fischer-Tropsch synthesis. Catal. Rev. Sci. Eng. 1995;37:227–309.
67 Al-Dahhan MH, Fan L-S, Dudukovic MP, Toseland B. Advanced Diagnostic Techniques for Three-phase Slurry Bubble Column Reactors (SBCR). Final Technical Report, DE-FG-26-99FT40594; August 2003. Washington University in St. Louis, St. Louis, MO.
68 Koros RM, inventor; Exxon Research & Engineering Co., assignee. Bubble column, tube slurry process and apparatus. US Patent 5,384,336. January 24, 1995.

69 Vogel AP, Steynberg AP, van Berge PJ, inventors; Sasol, assignee. Slurry-bed Fischer-Tropsch reactor with in-situ filtration zone for separation of catalyst from product waxes. Internat. Patent WO 0045948. August 10, 2000.

70 Luewisuthichat W, Tsutsumi A, Yoshida K. Deterministic chaos analysis of particle dynamics in three-phase systems. J. Chem. Eng. Jpn. 1996;29:675–682.

71 Luo X, Zhang J, Tsuchiya K, Fan L-S. On the rise velocity of bubbles in liquid-solid suspensions at elevated pressure and temperature. Chem. Eng. Sci. 1997;52:3693–3699.

72 DeSwart JWA, Krishna R, Sie ST. Selection, design and scale up of the Fischer-Tropsch reactor. Natural Gas Conversion IV. Stud. Surf. Sci. Catal. 1997;107:213–218.

73 Ohkawa WM, Kawata N, Uchida S. Cross-sectional distributions of gas and solid holdups in slurry bubble column investigated by ultrasonic computed tomography. Chem. Eng. Sci. 1999;54:4711–4728.

74 Raymond F, Rosant J-M. A numerical and experimental study of the terminal velocity and shape of bubbles in viscous liquids. Chem. Eng. Sci. 2000;55:943–955.

75 Dudukovic MP, Toseland BA. DOE Meeting. Cincinnati, OH; September 22, 1999.

76 Becker S, De Bie H, Sweeney J. Dynamic flow behavior in bubble columns. Chem. Eng. Sci. 1999;54:4929–4935.

77 Letzel HM, Schouten JC, Krishna R, van den Bleek CM. Characterization of regimes and regime transitions in bubble columns by chaos analysis of pressure signals. Chem. Eng. Sci. 1997;52:4447–4459.

78 Delnoij E. Fluid dynamics of gas-liquid bubble columns: a theoretical and experimental study [Ph.D. Thesis]. Enschede, The Netherlands: Twente University; 1999.

79 Delnoij E, Westerweel J, Deen NG, Kuipers JAM, van Swaaij WPM. Ensemble correlation PIV applied to bubble plumes rising in a bubble column. Chem. Eng. Sci. 1999;54:5159–5171.

80 Inga JR. Scaleup and scaledown of slurry reactors: a new methodology [Ph.D. thesis]. Pittsburgh (PA): University of Pittsburgh; 1997.

81 Inga JR, Morsi BI. Effect of catalyst loading on gas/liquid mass transfer in a slurry reactor: a statistical approach. Proceedings of the 13th International Pittsburgh Coal Conference, Pittsburgh, PA. Vol. 2, September 3–7, 1996. pp. 924–929.

82 van der Laan G. Kinetics, selectivity and scale up of the Fischer-Tropsch synthesis [Ph.D. Thesis]. Groningen, The Netherlands: University of Groningen; 1999.

83 Calderbank PH, Evans F, Farley R, Jepson G, Poll A. Rate processes in the catalyst-slurry Fischer-Tropsch reaction. Proc. Symp. Catal. Pract. (Inst. Chem. Eng.: London) 1963;66–74.

84 Satterfield CN, Huff GA. Effects of mass transfer on Fischer-Tropsch synthesis in slurry reactors. Chem. Eng. Sci. 1980; 35:195–202.

85 Deckwer W-D, Serpemen Y, Ralek M, Schmidt B. On the relevance of mass transfer limitations in the Fischer-Tropsch slurry process. Chem. Eng. Sci. 1981;36:765–771.

86 Deckwer W-D, Serpemen Y, Ralek M, Schmidt B. Modeling the Fischer-Tropsch synthesis in the slurry phase. Ind. Eng. Chem. Proc. Des. Dev. 1982;21:231–241.

87 Bukur DB. Models for Fischer-Tropsch reaction in slurry bubble column reactors. Chem. Eng. Sci. 1983;38:441–446.

88 Stern D, Bell AT, Heinemann H. Effects of mass transfer on the performance of slurry reactors used for Fischer-Tropsch synthesis. Chem. Eng. Sci. 1983;38:597–605.

89 Kuo JCW. Slurry Fischer-Tropsch/Mobil Two Stage Process of Converting Synthetic Gas to High Octane Gasoline. Final Report DOE-PC-3022-10; 1985. Mobil Research and Development Corp., Paulsboro, NJ.

90 Stenger HG, Satterfield CN. Effects of sulfur poisoning of a reduced fused magnetite catalyst in the Fischer-Tropsch synthesis. Ind. Eng. Chem. Proc. Des. Dev. 1985;24:415–420.

91 Prakash A, Bendale PG. Design of Slurry Reactor for Indirect Liquefaction Applications. DoE Final Report, DE-AC22-89PC89870; 1991. U.S. DOE, Pittsburgh Energy Technology Center, Pittsburgh, PA.

92 Prakash A. On the effects of syngas composition and water-gas-shift reaction rate on FT synthesis over iron based catalyst in a slurry reactor. Chem. Eng. Commun. 1993;128:143–158.

93 De Swart JWA. Scale-up of a Fischer-Tropsch reactor [Ph.D. Thesis]. Amsterdam, The Netherlands: University of Amsterdam; 1996.

94 Mills PL, Turner JR, Ramachandran PA, Dudukovic MP. The Fischer-Tropsch synthesis in slurry bubble column reactors: analysis of reactor performance using the axial dispersion model. In: Nigan KDP, Schumpe A, editors. Three-phase Sparged Reactors. Topics in Chemical Engineering 8. Amsterdam: Gordon & Breach; 1996. p 339–386.

95 Inga JR, Morsi BI. A novel approach for the assessment of the rate-limiting step in Fischer-Tropsch slurry process. Energy Fuels 1996;10:566–572.

96 Krishna R, Maretto C. Scale up of a bubble column slurry reactor for Fischer-Tropsch synthesis, Natural Gas Conversion V, (A. Parmaliana et al., Eds). Stud. Surf. Sci. Catal. 1998;119:197–202.

97 Sie ST, Krishna R. Fundamentals and selection of advanced Fischer-Tropsch reactors. Appl. Catal A Gen. 1999;186:55–70.

98 Maretto C, Krishna R. Modeling of a bubble column slurry reactor for Fischer-Tropsch synthesis. Catal. Today 1999; 52:279–289.

99 Krishna R. A scale-up strategy for a commercial scale bubble column slurry reactor for Fischer-Tropsch synthesis. Oil Gas Sci. Technol. Rev. 2000;55:359–393.

100 Krishna R, Sie ST. Design and scale-up of the Fischer-Tropsch bubble column slurry reactor. Fuel Sci. Technol. 2000;64:73–105.

101 Schanke D, Wagner M, Taylor P. Scale-up and Demonstration of Fischer-Tropsch technology, Proceedings of the 1st Annual Gas Processing Symposium (Alfadala H, Rex Reklaits GV, El-Halwagi MM, Editors) Elsevier, pp 1–8; January 10–12, 2009, Doha, Qatar.

102 Neathery JK, Davis BH. FT catalyst performance: comparison between pilot-scale SBCR and CSTR systems. Catal. Today 2003;84:3–8.

103 Schweitzer JM, Viguié JC. Reactor modeling of a slurry bubble column for Fischer-Tropsch synthesis. Oil Gas Sci. Technol. Rev. 2009;64:63–77.

104 Rados N, Al-Dahhan MH, Duduković MP. Modeling of the Fischer-Tropsch synthesis in slurry bubble column reactors. Catal. Today 2003;79–80:211–218.

105 Rados N, Al-Dahhan MH, Duduković MP. Dynamic modeling of slurry bubble column reactors. Ind. Eng. Chem. Res. 2005;44:6086–6094.

106 Bhatt BL, Tijm PJA. Slurry phase Fischer-Tropsch synthesis process development. Fifteenth Annual Pittsburgh Coal Conference Proceedings, Pittsburgh, PA; September 14–18, 1998.

107 Youssef AA, Hamed ME, Grimes JT, Al-Dahhan MH, Duduković MP. Hydrodynamics of pilot-scale bubble columns: effect of internals. Ind. Eng. Chem. Res. 2013;52:43–55.
108 Youssef AA, Al-Dahhan MH, Duduković MP. Bubble columns with internals: a review, Int. J. Chem. React. Eng. 2013;11:1–5.
109 Youssef A. Fluid dynamics and scale-up of bubble columns with internals [Ph.D. Thesis]. St. Louis (MO): Washington University; 2010.
110 Chen P, Sanyal J. Duduković MP. Chem. Eng. Sci. 2005; 60:1085–1101.
111 Shaikh A, Al-Dahhan MH. A review on flow regime transition in bubble columns. Int. J. Chem. React. Eng. 2007;5:R1.
112 Ghasemi S, Sohrabi M, Rahmani M. A comparison between two kinds of hydrodynamic models in bubble column slurry reactor during Fischer-Tropsch synthesis: single-bubble and two-bubble class. Chem. Eng. Res. Des. 2009;87:1582–1588.
113 Krishna R, van Baten JM. Mass transfer in bubble columns. Catal. Today 2003;79–80:67–75.
114 Lemoine R, Morsi BI. Hydrodynamic and mass transfer parameters in agitated reactors. Part II: gas-holdup, Sauter mean bubble diameters, volumetric mass transfer coefficients, gas-liquid interfacial areas, and liquid-side mass transfer coefficients. Int. J. Chem. React. Eng. 2005;3:A20.
115 Sehabiague L, Morsi BI. Hydrodynamic and mass transfer characteristics in a large-scale slurry bubble column reactor for gas mixtures in actual Fischer-Tropsch cuts. Int. J. Chem. React. Eng. 2013;11:1–20.
116 Sehabiague L. Modeling, scaleup and optimization of slurry bubble column reactors for Fischer-Tropsch synthesis [Ph.D. Thesis]. University of Pittsburgh; 2012.
117 Pangarkar K, Schildhauer TJ, Ruud van Ommen J, Nijenhuis J, Moulijn JA, Kapteijn F. Experimental and numerical comparison of structured packings with a randomly packed bed reactor for Fischer-Tropsch synthesis. Catal. Today 2009;147S:S2–S9.
118 Lei Z, Yang Y, Li Q, Chen B. Catalytic distillation for the synthesis of tert-butyl alcohol with structured catalytic packing. Catal. Today 2009;147S:S352–S356.
119 Pangarkar KV. Solving the heat transport issue in multiphase fixed bed reactors [dissertation]. Delft, The Netherlands: Delft University of Technology; 2010.
120 Vervloet D, Kapteijn F, Nijenhuis J, Ruud van Ommen J. Process intensification of tubular reactors: considerations on catalyst hold-up structured packings. Catal. Today 2013;216:111–116.
121 Pangarkar K, Schildhauer TJ, Ruud van Ommen J, Nijenhuis J, Moulijn JA, Kapteijn F. Heat transport in structured packings with co-current downflow of gas and liquid. Chem. Eng. Sci. 2010; 65:420–426.
122 Khassin AA, Sipatrov AG, Yurieva TM, Chermashentseva GK, Rudina NA, Parmon VN. Performance of a catalytic membrane reactor for Fischer-Tropsch synthesis. Catal. Today 2005;105:362–366.
123 Rohde MP, Unruh D, Schaub G. Membrane application in Fischer-Tropsch synthesis reactors—overview of concepts. Catal. Today 2005;106:143–148.
124 Espinoza RL, Santamaria JM, Menendez MA, Coronas J, Irusta S. (1999) Production of Primary Hydrocarbons via the Fischer-Tropsch Reaction with Removal of Water via a Semi-permeable Membrane, WO 9964380; 1999. Sasol Technology (Proprietary) Limited, Johannesburg.
125 He J, Yoneyama Y, Xu B, Nishiyama N, Tsubaki N. Designing a capsule catalyst and its application for direct synthesis of middle isoparaffins. Langmuir 2005;21:1699–1702.
126 Silveston PL. Compositional Modulation of Catalytic Reactors. Amsterdam, The Netherlands: Gordon and Breach Science Publisher, pp. 464–497; 1988.
127 van Vuuren MJ, Davis BH. Fischer-Tropsch synthesis: compositional modulation study using an iron catalyst. In: Davis BH, Occelli ML, editors. Fischer-Tropsch Synthesis, Catalysts and Catalysis. Amsterdam, The Netherlands: Elsevier; 2007. p 201–215.
128 Fan L, Fujimoto K. Fischer-Tropsch synthesis in supercritical fluid: characteristics and application. Appl. Catal. A Gen. 1999; 186:343–354.
129 Bochniak DJ, Subramaniam B. Fischer-Tropsch synthesis in near-critical n-hexane: pressure-tuning effects. AIChE J. 2004; 44:1889–1896.
130 Lang X, Akgerman A, Bukur DB. Steady state Fischer-Tropsch synthesis in supercritical propane. Ind. Eng. Chem. Res. 1995; 34:72–77.
131 Jacobs G, Chaudhari K, Sparks DE, Zhang Y, Shi B, Spicer R, Das TK, Li J, Davis BH. Fischer–Tropsch synthesis: supercritical conversion using a Co/Al_2O_3 catalyst in a fixed bed reactor. Fuel 2003;82:1251–1260.
132 Elbashir NO, Bukur DB, Durham E, Roberts CB. Advancement of Fischer-Tropsch synthesis via utilization of supercritical fluid reaction media. AIChE J. 2010;56:997–1015.
133 Armstrong J. 2011. The commercialization progress of Sasol coal to oil, Tianjin, China. Available at http://www.upcholding.com/admin/eWebEditor/UploadFile/2011530143336985.pdf.
134 Pretorius T. GTL technology and commercialization. Oil & Gas Africa. March 16, 2010.
135 Tijm PJA. Gas to Liquids, Fischer-Tropsch, Advanced Energy Technology, Future's Pathway, February 2010.
136 Pratt JW. A Fischer-Tropsch Synthesis Reactor Model Framework for Liquid Biofuels Production. Sandia Report SAND2012-7848; September, 2012. Sandia National Laboratories, Albuquerque, NM and Livermore, CA.
137 Fürnsinn S. Outwitting the Dilemma of Scale: Cost and Energy Efficient Scale-Down of the Fischer-Tropsch Fuel Production from Biomass. Saarbrücken, Germany: VDM Verlag Dr. Müller; 2009.

CHAPTER 13

Hydrotreating of oil fractions

Jorge Ancheyta[1], Anton Alvarez-Majmutov[1] and Carolina Leyva[2]

[1]Instituto Mexicano del Petróleo, Management of Products for the Transformation of Crude Oil, Mexico City, Mexico

[2]Centro de Investigación en Ciencia Aplicada y Tecnología Avanzada, Unidad Legaria, Instituto Politécnico Nacional, Mexico City, Mexico

Abstract

Catalytic hydrotreating represents a fundamental process in modern petroleum refining operations. It allows removing hydrocarbon contaminants, such as sulfur, nitrogen, oxygen, and metals, saturating aromatic rings and olefins, and breaking high molecular weight molecules into lighter compounds. Due to its flexibility, the process can be employed to upgrade a variety of petroleum streams, ranging from naphtha to vacuum residues, or even full-range crude oils. Conventional hydrotreating is typically used as a pretreatment step to provide suitable quality feeds for conversion processes, such as reforming, catalytic cracking, and hydrocracking, and also as a finishing step to produce transportation fuels that meet ecological standards. Over the years, hydrotreating has also been gaining acceptance in the primary upgrading of heavy and extra-heavy crude oils to produce the so-called synthetic crudes. Among all the available hydrotreating reactor technologies, fixed-bed reactors are the most frequently used in the petroleum industry. In such systems, both gas and liquid flow cocurrently down through a catalyst bed in trickle-flow regime. Modeling and simulation of hydrotreating becomes a challenging task due to the complex interaction of numerous physical and chemical processes: vapor–liquid equilibrium, gas–liquid and liquid–solid mass transfer of reactants and products, diffusion inside the catalyst particle, vast reaction networks, and catalyst deactivation. The purpose of this chapter is to provide a comprehensive overview of the hydrotreating process. It covers general aspects such as process chemistry, reaction kinetics, and thermodynamics, but it also presents more process-oriented topics including process variables, reactor characteristics, and reactor modeling and simulation.

13.1 Introduction

Catalytic hydrotreating (HDT) is a mature technology used in the petroleum refining industry for the upgrading of hydrocarbon streams for the last 60 years. For conventional distillate HDT, the main purpose of the process is to remove impurities such as heteroatoms (sulfur, nitrogen, and oxygen) and saturate aromatic compounds, whereas in the case of heavy oils and residues, it also comprises the elimination of metals (nickel and vanadium) and conversion of asphaltene molecules. Its major applications in current refinery operations can be grouped in the following categories: (i) feed pretreatment for conversion processes such as catalytic reforming, catalytic cracking, and hydrocracking (HCR) and (ii) post-HDT of distillates. In the first case, generally the objective is to reduce the amount of sulfur, basic nitrogen compounds, metals, and polynuclear aromatics (PNA), which act as deactivation agents in acid-catalyzed processes. The second group is mainly the finishing step to produce transportation fuels that meet ecological standards (e.g., ultralow sulfur gasoline and diesel).

There are numerous HDT processes for handling all kinds of refinery streams and for each specific objective [1]. They differ in reactor technology, catalyst type, operating conditions, and process configuration. Among all the reactor technologies, fixed-bed reactors (FBRs) are still the most widely used in HDT operations due to their flexibility and relative simplicity [2]. Other types of reactors such as moving-bed reactor (MBRs) and ebullated-bed reactors (EBRs) are also available specifically for upgrading the heaviest fractions, and their main feature is their capacity for replacing spent catalyst without interrupting the operation. The selection between each type of technology is dictated by catalyst deactivation, which depends on the nature of the feedstock [3].

HDT is carried out under a wide range of operating conditions. The severity of the process is adjusted depending on the properties of the feed and required product composition. The main process variables are pressure, temperature, hydrogen-to-oil (H_2/oil) ratio, and space velocity. Each variable affects the process in different ways; therefore, the set of operating conditions must be carefully adjusted to achieve an efficient operation.

The reactions that take place during the HDT process are hydrodesulfurization (HDS), hydrodenitrogenation (HDN), hydrodearomatization (HDA), hydrodeoxygenation (HDO),

Multiphase Catalytic Reactors: Theory, Design, Manufacturing, and Applications, First Edition. Edited by Zeynep Ilsen Önsan and Ahmet Kerim Avci.

and hydrodemetallization (HDM). Another type of reaction in which high molecular weight compounds are broken down into lighter molecules is HCR. In the case of HCR of asphaltenes, the process is often referred to as hydrodeasphaltenization (HDAs). The chemistry of these reactions can be visualized as a hydrogen exchange process where hydrogenolysis and hydrogenation mechanisms consume the externally supplied hydrogen in order to replace heteroatoms and stabilize unsaturated products. However, in reality, HDT chemistry is far more complicated due to the intrinsic complexity of oil composition.

For proper design and simulation of HDT reactors, kinetic and reactor modeling are aspects that need to be deeply studied; however, this is not a trivial task due to the numerous physical and chemical processes that occur simultaneously in the reactor: phase equilibrium, mass transfer of reactants and products between the gas–liquid–solid phases, diffusion inside the catalyst particle, a complex reaction network, and catalyst deactivation. Ideally, the contribution of all these events must be coupled into a robust reactor performance model. The level of sophistication of the model is generally defined based upon the pursued objectives and prediction capability [4].

The main objective of this chapter is to serve as an updated overview of the HDT process. The chapter covers general aspects such as chemistry, reaction kinetics, and thermodynamics, but it also presents more process-oriented topics including process variables, reactor characteristics, and reactor modeling and simulation.

13.2 The HDT process

13.2.1 Overview

Catalytic HDT is a fundamental refining process for the upgrading of a wide variety of streams, ranging from straight-run naphtha to vacuum residue (VR) or even heavy and extra-heavy crude oils. The HDT process is commonly employed for reducing the contents of hydrocarbon impurities such as sulfur, nitrogen, oxygen, and aromatics by the so-called hydrogen addition route. When it is applied to the processing of heavy feeds, it also has the virtue of bringing down the concentration of metals (Ni and V) and simultaneously increasing distillate yield at the expense of the heaviest fractions such as VR and asphaltenes [5]. The process has gained significant relevance in the industry due to the growing demand for transportation fuels and the strict environmental legislations.

From a practical standpoint, HDT is a process in which the hydrogen/carbon ratio of the feed is increased in the presence of a catalyst in a hydrogen-rich atmosphere [6]. HDS, HDN, HDO, and HDM reactions are considered to follow the hydrogenolysis mechanism, where the carbon–heteroatom bond is cleaved by a hydrogen molecule. Olefin and aromatic molecules are directly hydrogenated without undergoing any bond cleavage. HCR and HDAs reactions, on the other hand, belong to the acid-catalyzed cracking chemistry in combination with hydrogenation.

The typical configuration of an HDT unit is presented in Figure 13.1. The sequence of operations begins with the preparation of the feedstock. The hydrogen recycle stream is mixed with the feed oil, and afterward the mixture is heated to the required reaction temperature in the feed heater. The gas–liquid mixture is subjected to HDT conditions in a single reactor or a series of reactors, typically operating in downflow fixed-bed mode. Depending on the amount of heat release, the reactor can have multiple beds separated by quench zones in order to inject cold hydrogen streams, typically from the recycle stream. In addition, it is preferable to use adequate reactor internals for distributing and mixing reactants in the interbed zones, as well as at the top of the reactor. The reactor effluent is sent to a high-pressure separator (HPS) where the liquid products are recovered from the gases. The liquid hydrocarbon stream from the HPS passes through a stripping unit in order to remove the remaining dissolved hydrogen sulfide and ammonia. The gas

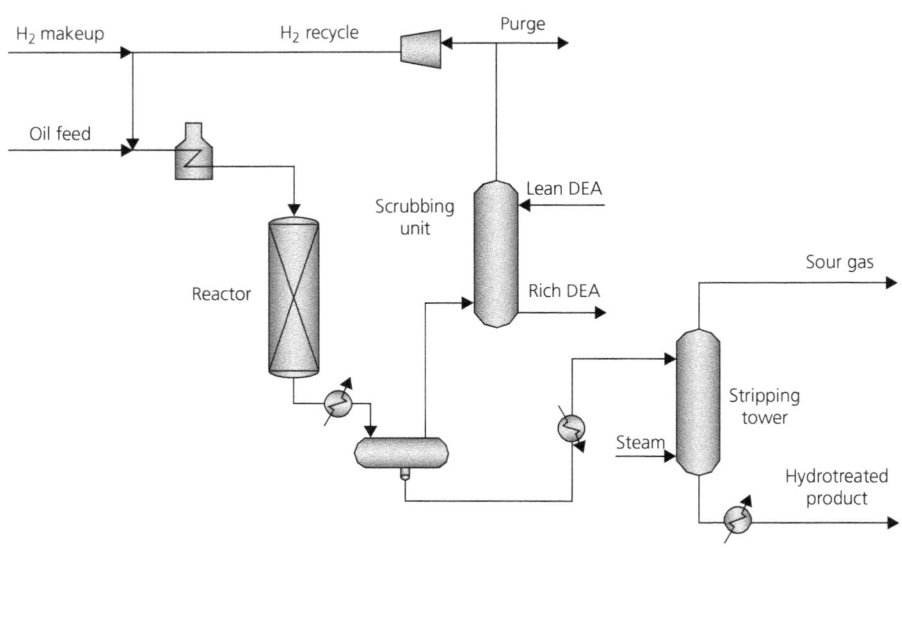

Figure 13.1 Flow diagram of a typical hydrotreating unit.

stream is scrubbed with diethylamine (DEA) in order to remove the excess of hydrogen sulfide and ammonia generated by chemical reaction. The resulting high-purity hydrogen stream is recompressed and recycled to the reaction system. Depending on the extent of HCR, the hydrotreated product can be fractionated into various distillate products.

The HDT catalyst is typically a CoMo/NiMo alumina-supported catalyst, the composition and textural properties of which vary according to the different purposes. The severity of the operating conditions depends on the feed type and the final product quality. In general, the process is carried out at high pressures and temperatures; typical industrial units operate at pressures of 2–20 MPa, temperatures of 320–440°C, H_2/oil ratios of 350–1800 Nm^3/m^3, and liquid hourly space velocities (LHSVs) between 0.2 and 8 h^{-1} [7]. Concerning reactor technology, there are currently four main kinds of HDT reactors that can be classified according to the type of catalytic bed: FBR, MBR, EBR, and slurry-phase reactor (SPR) [2]. This classification is illustrated in Figure 13.2.

FBRs are the most widely used technology for all types of HDT operations due to their relative simplicity, flexibility, and ease of operation. MBRs and EBRs, on the other hand, are designed for the upgrading of heaviest feeds [6]. Initially, FBRs were employed for processing light feeds (naphtha and gas oil), but eventually they were adapted for tougher feeds such as vacuum gas oil (VGO) and residues. Their main disadvantage for upgrading heavy feeds is the drastically reduced length of run owing to the rapid catalyst deactivation caused by metals accumulation and coking [3]. Layered catalyst systems were introduced to make better use of the catalyst inventory in order to extend significantly the length of run [8]. The main feature of this technology is the front-end HDM macroporous catalyst that allows the disaggregation of asphaltene molecules for metal removal, so that the downstream HDS/HCR catalysts can operate with low metal and coke precursor content hydrocarbons.

MBRs and EBRs overcome the catalyst cycle life limitations of FBRs. Such technologies allow for replacing spent catalyst without interrupting operation; therefore, they are the best option for handling the most problematic feeds with high contents of metals and asphaltenes. MBRs combine plug-flow operation with the periodic replacement, typically once or twice a week, of portions of spent catalyst during time-on-stream. The application of these reactors is specifically in front-end demetallization to protect subsequent FBRs for HDS and HCR [3]. EBRs constitute the most advanced hydroprocessing technology, specifically designed for upgrading extra-heavy feeds directly, without any kind of pretreatment [1]. The continuous catalyst replacement feature allows the use of conventional high-activity HDS/HCR catalysts. The operation is very flexible, conversion is very efficient (up to 90 vol%), and products have low levels of sulfur, metals, and nitrogen [9]. Nevertheless, ebullated-bed technologies suffer from considerable sediment formation and high catalyst consumption.

13.2.2 Role in petroleum refining

In the first step of petroleum refining, crude oil is fed to the atmospheric fractionation tower to obtain straight-run distillates such as naphtha, kerosene, and gas oil. Such raw products cannot be used directly as transportation fuels due to several technical limitations such as high amounts of impurities (sulfur, nitrogen, and aromatics) and low octane and cetane numbers in the cases of gasoline and diesel fuels, respectively. These streams are converted into valuable products through a variety of refining processes. In this context, HDT operations play a major role at several stages in a refinery. According to a recent worldwide refining survey [10], HDT has in fact the largest processing capacity among all refining operations (45 431 300 bpcd).

The typical configuration of a representative petroleum refinery is shown in Figure 13.3. In this type of configuration, the diverse purposes of HDT can fall into two categories: (i) feed pretreatment for conversion processes such as isomerization, catalytic reforming, catalytic cracking, and HCR and (ii) postprocessing of distillates. The first category has the primary objective of providing upgraded feedstock with less catalyst deactivating agents for acid-catalyzed processes. For instance, sulfur removal from straight-run naphtha is essential

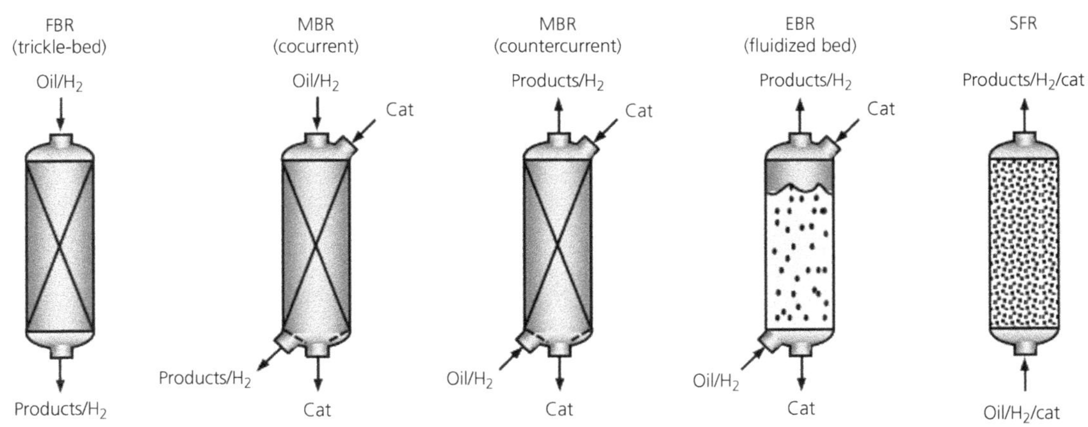

Figure 13.2 Hydrotreating reactor technologies.

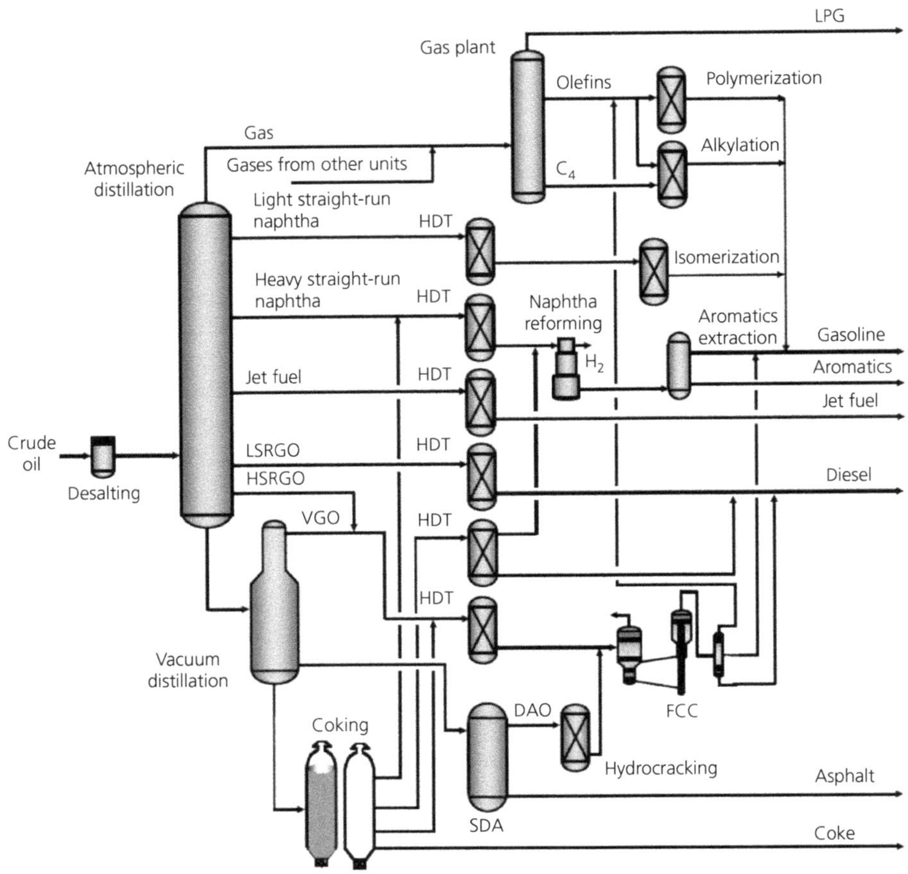

Figure 13.3 Process scheme of a typical petroleum refinery.

to prevent poisoning of the noble metal catalyst in the reforming unit. In the case of VGO, previous HDT is required for removing nitrogen, especially basic nitrogen, sulfur, and metal compounds, as well as for saturating aromatics, in order to produce a better and more crackable feed for the fluid catalytic cracking (FCC) or HCR processes. Basic nitrogen compounds have a temporary poisoning effect on acid catalysts; PNA compounds favor coke formation, whereas metals deactivate zeolitic catalysts and promote dehydrogenation reactions, which increase selectivity toward dry gas [11–13]. The second category is a finishing step to produce clean fuels such as ultralow sulfur gasoline and diesel. This application of HDT has gained much attention due to the stringent environmental legislations. Low sulfur, nitrogen, and aromatic contents in automotive fuels are becoming mandatory in order to reduce the emissions of sulfur oxides (SO_x), nitrogen oxides (NO_x), and volatile organic matter (VOM) during their combustion.

Besides the aforementioned conventional applications, there are many other HDT technologies that have emerged in the changing environment of the oil industry. The majority of these are focused on the upgrading of the heaviest feedstocks. The atmospheric residue/vacuum residue desulfurization processes (RDS/VRDS), which have been developed to meet a variety of objectives such as preparing feed for FCC, residue FCC, coker, and residue HCR, are cases in point [14, 15]. Ebullated-bed residue HCR with simultaneous desulfurization and demetallization is also an example belonging to this category. Recently, HDT has been extended to the primary upgrading of heavy and extra-heavy crude oils (before the atmospheric distillation tower) to produce the so-called synthetic oils [16]. In comparison to traditional thermal upgrading processes, HDT certainly offers much better selectivity to liquid yield, much lower coke formation, and substantially cleaner products [17]. Therefore, it is expected that this particular group of technologies will be a defining step in future refineries due to the increasing production and supply of heavy crudes.

13.2.3 World outlook and the situation of Mexico

It is no secret that the worldwide energy consumption is growing year by year. Petroleum currently represents 33% of the total world energy supply [18] and by far is the most commonly used source of liquid fuels. It is expected that this scenario will continue for the next 50 years [19]. Figure 13.4 shows how the world consumption of oil refining products has been increasing during the last 20 years. In particular, the last decade has witnessed a 12.2% growth in petroleum product usage. This behavior is a result of the fast-developing demand for automotive and aviation fuels, particularly in developing countries such as China, Russia, and Latin America [20].

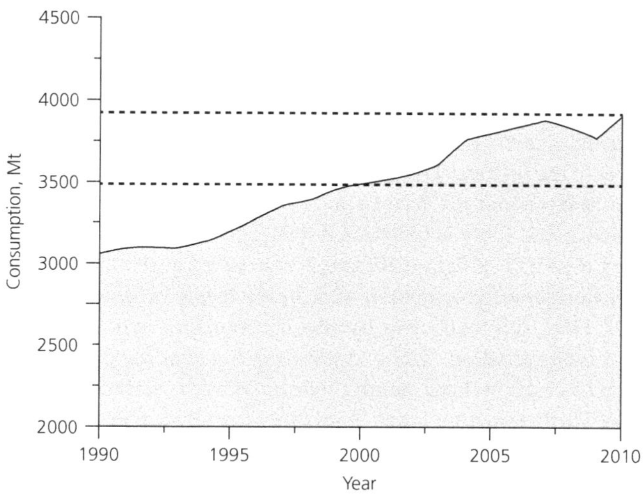

Figure 13.4 World consumption of oil refined products from 1990 to 2010. (Source: Reprinted from Enerdata [20], © 2011, with permission from Enerdata Publications http://www.enerdata.net/enerdatauk/press-and-publication/publications/.)

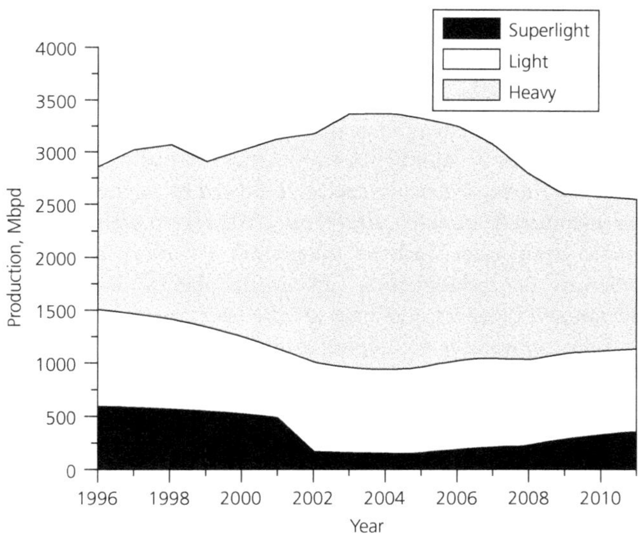

Figure 13.5 Crude oil production in Mexico. (Source: Adapted from [21].)

In the meanwhile, the worldwide stock of crude oils has been changing scenario toward heavy and extra-heavy oils, which means that the average quality of crude oil has deteriorated. The case of Mexico is a representative example of this trend, where heavy crudes represent more than 50% of the total production (Figure 13.5) [21]. Additionally, the Mexican heavy oil stock was composed of heavy Maya (21°API) up until recently, but now this is shifting toward even heavier crudes of 10–16°API. Consequently, the supply of light crudes to the refineries is being gradually substituted by lower-quality crudes. The refineries that have been conventionally designed to process light crude oils now face drastic changes in feed composition, that is, feeds with higher content of sulfur, metals, and asphaltenes.

Table 13.1 Properties of various Mexican crude oils (obtained at IMP analytical lab).

	Maya	Ku-Maloob-Zaap	Isthmus
API gravity	21.3	12.9	33.2
S, wt%	3.57	5.19	1.80
N, wppm	3200	4771	1446
Ni, wppm	53	83	20
V, wppm	298	501	79
Insolubles in C_7, wt%	11.32	17.03	3.06
AR (343°C +) yield, vol%	56.4	73.88	45.31

The complex nature of heavy crudes is the reason why refining of these feeds is so difficult. Table 13.1 shows the properties of two types of heavy crude oils available in Mexico (Maya and Ku-Maloob-Zaap) and one light oil (Isthmus) which was included to stress out the difference. In comparison to light crudes, heavy crudes are characterized by much higher content of impurities such as sulfur, nitrogen, metals (Ni and V), and asphaltenes. Also, the amount of material boiling above 343°C, termed atmospheric residue (AR), is quite significant (>50 vol%), which indicates that these types of oils have a poor yield of distillable fractions. In this regard, it is evident that heavy crudes are composed mainly of residue, and therefore, the residual fraction has a strong influence on the overall oil properties since it concentrates most of the impurities found in crude oil. Consequently, most of the research effort must be focused on this fraction, because it will be the main raw material for obtaining valuable liquid products in the near future.

There are several technical problems that are encountered during the processing of heavy oils and residues [5]: (i) permanent catalyst deactivation due to metal deposition in catalytic cracking and HCR processes; (ii) temporary poisoning of acid catalysts caused by the presence of basic nitrogen in the feed; (iii) higher coke formation and lower liquid product yield, as a result of high Conradson carbon and asphaltene contents; (iv) higher sulfur levels in products; and (v) fouling in downstream equipment. Additionally, even before refining, heavy crudes impose serious difficulties in their transportation through pipelines due to their extremely poor flow properties and stability/compatibility problems when mixing with other crudes.

Summing up, the current trends of oil production and fuel market impose an enormous challenge for the refining industry. There are three main critical factors that refiners must address to keep up with the situation: (i) higher production capacity to satisfy the growing demand for fuels, (ii) higher quality of the final product, and (iii) much lower quality of the feedstock. Certainly, all this evidence pushes the refining industry to focus on primary upgrading in order to handle higher amounts of heavier feedstock. HDT in this sense will play a central role because it offers many advantages for the processing of heavy feeds.

13.3 Fundamentals of HDT

13.3.1 Chemistry

HDT chemistry is the basis for understanding the transformations that the hydrocarbon goes through at process conditions and offers the basic guidelines for process modeling. The chemistry of HDT reactions can be simply visualized as a hydrogen transfer process [22]. In this generalization, hydrogen is supplied from an external source to replace heteroatoms and reduce the molecular weight of the original hydrocarbon mixture by means of various hydrogenation and hydrogenolysis mechanisms. Nevertheless, the events that take place at the molecular level are far more intricate.

The main problem in elucidating the details of HDT reactions arises from the complex composition of commercial feedstocks. Depending on the boiling range, oil fractions may contain from a few hundred to several thousands of different components. Strictly speaking, this implies a huge network of series and competing parallel reactions and a wide distribution of reactivities due to the high degree of polydispersity of hydrocarbon mixtures. On top of this, collecting information on the process at the molecular level is still an issue because of the analytical limitations in obtaining the detailed composition of petroleum streams, especially of the heaviest fractions.

HDT reactions are classified according to the impurity that is removed from the hydrocarbon (sulfur, nitrogen, oxygen, metals, and asphaltenes). In the cases of HDS, HDN, HDO, and HDM reactions, carbon–heteroatom bond energies can be used for explaining some of the differences in reactivity toward preferential reactions. However, this does not account for the various configurational effects of complex three-dimensional structures typically found in petroleum nor does it consider the unknown interactions between reactants and products in such a vast reaction network [23]. The latter suggests that the best way to understand HDT chemistry is through model compounds and simple mixtures. As pointed out by Girgis and Gates [24], HDT chemistry studies should focus on the most refractory compounds (i.e., those difficult to react), as they are the most relevant for process modeling and catalyst development.

The nature of the chemical transformations that take place during catalytic HDT varies significantly according to the objectives of the process. The complexity of the process increases with the heaviness of the feed because the compounds become less reactive and more sophisticated in structure. This is the reason why the HDT of light distillates is conducted at relatively lower reaction severity compared with heavy feeds. In general, catalytic HDT can follow two parallel mechanisms: hydrogenolysis and hydrogenation. The first pathway involves the direct scission of a carbon–heteroatom single bond by the action of hydrogen. The heteroatom is any atom other than hydrogen or carbon present in petroleum, such as sulfur, nitrogen, oxygen, and metals. In hydrogenation, hydrogen is added to the molecule without cleaving bonds. In many cases, prior hydrogenation is necessary for the hydrogenolysis mechanism to proceed. The main hydrogenolysis and hydrogenation reactions in catalytic HDT are briefly described in the following.

13.3.1.1 HDS

During HDS, sulfur is extracted from hydrocarbons and released as hydrogen sulfide (H_2S). Reactivity of S-components can vary greatly depending on the structure of the molecule. In general, HDS reactivity increases according to the hydrocarbon type: paraffins > naphthenes > aromatics. Mercaptans and sulfides are the most reactive species, followed by naphthenic and six-membered aromatic structures [25]. Five-membered aromatic compounds such as thiophenes are more refractory, whereas benzothiophenes, dibenzothiophenes, and alkyl-substituted dibenzothiophenes are the least reactive species.

Sulfur removal is accomplished directly by a hydrogenolysis mechanism or indirectly by prior hydrogenation. Figure 13.6 shows the two possible pathways for dibenzothiophene HDS. In the direct route, the S-atom is eliminated and replaced by hydrogen. The other mechanism requires saturation of one

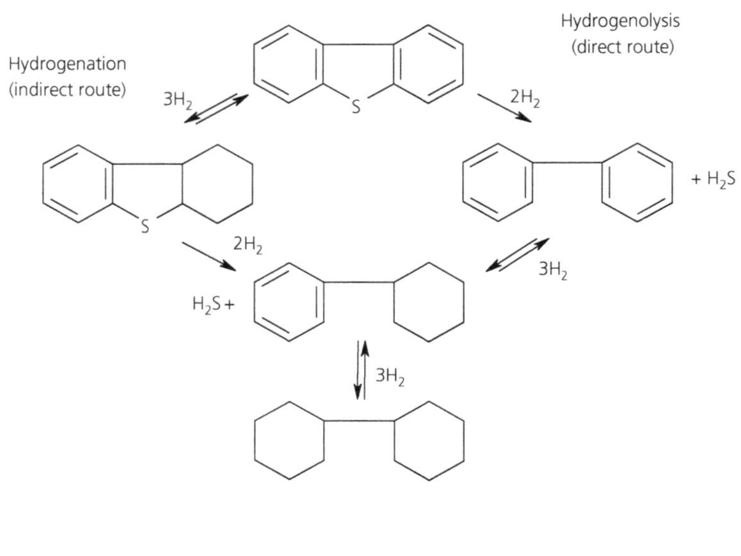

Figure 13.6 Reaction pathways of dibenzothiophene HDS.

aromatic ring before sulfur removal. The sulfur-free products may undergo further hydrogenation until the molecule is completely saturated.

It is evident that the direct route would be the preferred pathway because it is more efficient in terms of hydrogen consumption; nevertheless, it is strongly affected by the structure of the molecule. Aromatic ring condensation or alkyl group substitution close to the S-atom makes the molecule sterically hindered and consequently less reactive. The indirect pathway is apparently less influenced by structural features. Hydrogenation of the aromatic ring reduces the energy of the S—C bonds, allowing facile bond scission. It is believed that the saturated structure reduces the impediments for the S-atom to access the active site of the catalyst [25]. However, this pathway is thermodynamically restricted because hydrogenation is equilibrium limited at low pressures and high temperatures [26]. Another aspect affecting HDS is the strong inhibition by adsorption of reaction products such as H_2S and NH_3 and organic nitrogen compounds. Therefore, the predominant reaction mechanism will depend on the nature of the feedstock and process conditions.

13.3.1.2 HDN

Nitrogen in petroleum is commonly present in aromatic rings. These species are categorized as basic N-compounds (e.g., quinoline, acridine, and pyridine) and nonbasic N-compounds (e.g., indole, carbazole, and pyrrole). In commercial practice, basic N-compounds are more relevant because nonbasic compounds are rapidly hydrogenated into basic compounds [27]. It is well accepted that nitrogen extraction by hydrogenolysis requires complete prehydrogenation of the aromatic rings in order to weaken the strong aromatic C—N bond [24]. In general, HDN is much more difficult than HDS, and even than HDM and HDAs in the case of residue HDT, and consequently requires more severe conditions. Figure 13.7 presents the reaction mechanism for the HDN of quinoline, proposed by Gioia and Lee [28]. The preferred path (fast route) proceeds through the saturation of both rings before cleaving the C—N bond. The other route is more direct and hydrogen efficient, but it is generally neglected because it is very slow [29]. The end products in this case are a saturated hydrocarbon and ammonia (NH_3).

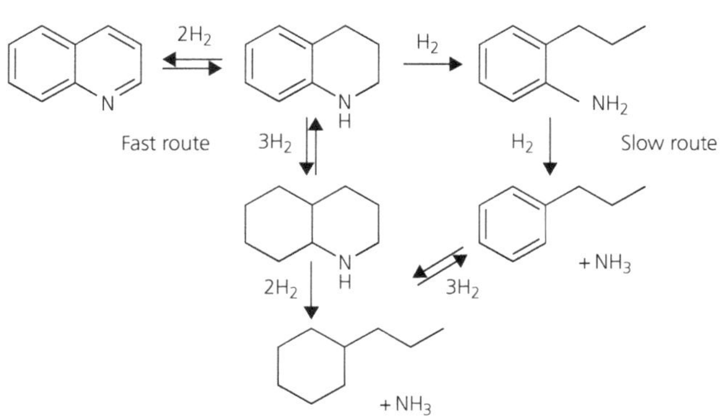

Figure 13.7 Reaction scheme of quinoline HDN.

Besides the thermodynamic limitations of HDN, the reactivity of N-compounds is strongly influenced by their molecular structure, which interferes with the absorption of the N-atom on the active site [30]. Basic N-compounds are also the strongest inhibitors of hydrogenation sites and consequently of all reactions that may follow this pathway (e.g., HDS). Therefore, removing N-inhibitors is an essential strategy to achieve deep desulfurization of diesel fuels.

13.3.1.3 HDO

HDO reactions have received less attention than HDS and HDN because organic oxygen compounds are present in low concentrations in petroleum. HDO consists of the removal of the O-atom from hydrocarbons and its conversion to water (H_2O). O-compounds are found as phenols, naphthol, furan, and their derivatives [31]. As in HDS and HDN, the reactivity of O-compounds decreases with molecular weight. It has been reported that HDO proceeds preferably through partial saturation of the aromatic rings rather than through direct hydrogenolysis [24].

13.3.1.4 HDM

Nickel and vanadium are found in petroleum generally in the form of metalloporphyrins. These entities are concentrated in the heaviest fractions, particularly in the asphaltenic fraction. During HDM, Ni and V are converted into their respective metal sulfides and deposited on the catalyst surface producing irreversible deactivation. Janssens et al. [32] proposed a reaction scheme for HDM of metalloporphyrins in which the molecule undergoes hydrogenation of the pyrrole subunits, followed by Ni—N or V—N bond cleavage by hydrogenolysis, and ending with the fragmentation of the porphyrinic structure, as shown in Figure 13.8. Many experimental observations indicate that generally the V-atom is easier to remove than Ni. This difference in reactivity has been attributed to the fact that V-porphyrins tend to concentrate in the periphery of the asphaltene molecule, whereas Ni-porphyrins are hidden somewhere inside the core and therefore require previous disaggregation and HCR of the molecule [33]. In addition, it has been proposed that the perpendicular oxygen atom linked to V-porphyrin structures provides enhanced adsorption to the catalyst surface [34].

13.3.1.5 Saturation reactions

Hydrogenation reactions can be grouped as the saturation of olefins and HDA. Hydrogenation of olefins to produce their saturated homologues is essentially irreversible. Saturation of aromatics to naphthenes is reversible under the commercial operating conditions. Figure 13.9 illustrates the hydrogenation reactions of naphthalene to tetrahydronaphthalene and butenylbenzene to *n*-butylbenzene. Aromatics are found in petroleum fractions as monoaromatics, diaromatics, triaromatics, and PNA. Saturation of monoaromatics is the most difficult due the stability of the benzene molecule and its homologues. Hydrogenation of multiaromatics proceeds sequentially ring

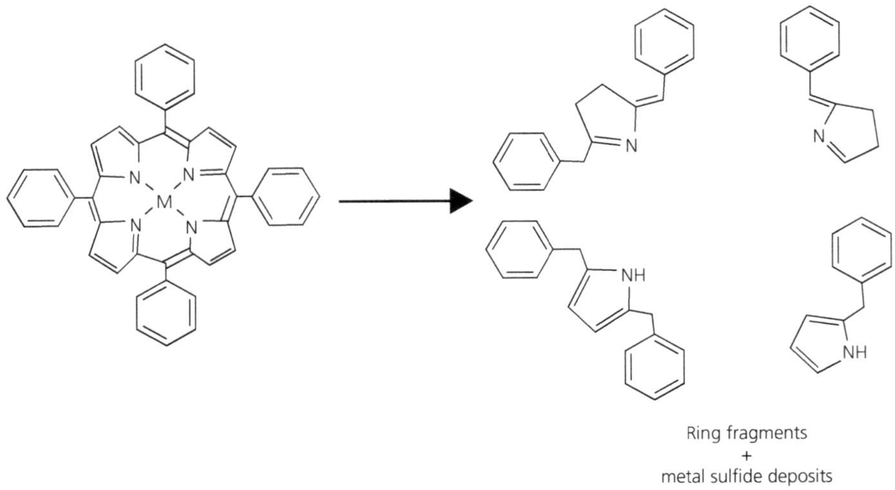

Figure 13.8 Simplified representation of the HDM of metalloporphyrins.

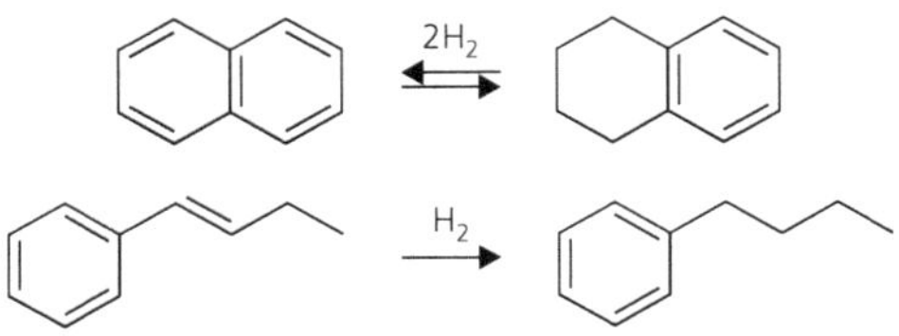

Figure 13.9 Examples of saturation reactions.

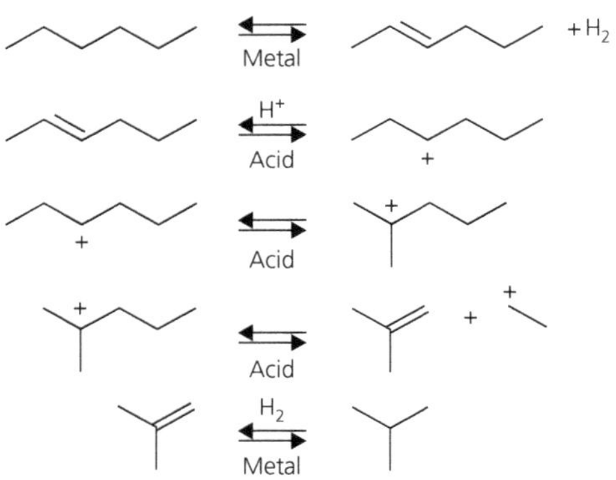

Figure 13.10 Reaction mechanism of paraffin hydrocracking.

by ring, where equilibrium favors the saturation of the first ring [35]. Hydrogenation also plays a defining role in HDS and HDN reactions, as discussed earlier.

13.3.1.6 HCR

The purpose of HCR reactions is to break C—C bonds in order to reduce the molecular weight of the feedstock. The chemistry resembles that of catalytic cracking but with cocurrent hydrogenation. HCR catalysts are designed to have an acid-based cracking function and a metal-based hydrogenation/dehydrogenation function. Figure 13.10 illustrates the dual-site HCR mechanism of an *n*-paraffin [36]. Initially, the paraffin molecule is chemisorbed on a metal site and dehydrogenated into the corresponding olefin. The olefin then migrates to a Brønsted acid site to be protonated into a carbenium ion. The paraffinic carbenium ion can undergo a rearrangement, such as a methyl shift, to form a more stable carbenium ion. This ion is cracked through a β-scission to produce an olefin and a smaller carbenium ion, and the resulting olefin can undergo further cracking, or it can be saturated on a metal site.

One relevant aspect of HCR is that the feed must have low nitrogen and PNA concentrations to avoid catalyst poisoning and coke formation. Organic N-bases are known to strongly inhibit the cracking functionality of the catalyst. Additionally, coke growth must be controlled by keeping hydrogen partial pressure at sufficiently high levels in order to stabilize olefinic products.

13.3.1.7 HDAs

Asphaltenes are a solubility class of compounds found in the residual fractions of petroleum. When asphaltenes are subjected to hydroprocessing conditions, they undergo a series of complex chemical transformations. The reaction mechanism depends on the operating conditions and the structure of the molecule. For instance, HDAs is very sensitive to reaction temperature, which has been attributed to a change in reaction selectivity from hydrogenation dominated at low temperature to HCR dominated at high temperature [37]. Asphaltene conversion is closely linked to the elimination of sulfur, nitrogen, and metals since many heteroatom-bearing molecules are fused into the asphaltenic structure [38]. Recently, it has been proposed that large asphaltene aggregates evolve into smaller low molecular weight aggregates [39]. Depending on the hydroconversion level, the first step involves the dissociation of asphaltene aggregates, followed by the dealkylation of small side chains, and ends with the cracking of alkylic bridges connecting large polyaromatic cores. Under high conversion conditions (>400°C), the remaining asphaltene molecules are composed of large polyaromatic cores, which are responsible for the formation of sediments. At high temperatures, it has been proven that hydrothermal noncatalytic cracking reactions also contribute to the conversion of asphaltenes [40].

13.3.2 Reaction kinetics

Petroleum feedstocks are complex hydrocarbon mixtures characterized by their intricate structural nature and composition at the molecular level. For process modeling purposes, this translates into a gigantic reaction network.

Kinetic models of refining processes have evolved historically attempting to predict the vast product distribution or at least part of it. Up to now, most kinetic models are based on the lumping approach for practical reasons. Lump-based kinetics can be easily derived from conventional analytical information. However, the advances in analytical methods, which currently allow obtaining relevant information on the molecular structure of complex hydrocarbons, and the increase in computational capacity have enabled the development of detailed molecule-based models [41].

In the lumping approach, the feedstock and products are represented in terms of chemical lumps or pseudocomponents, which are defined by standard analyses. The chemical lumps are generally delimited depending on the modeling purpose. For example, the traditional lumping scheme for catalytic cracking and HCR is formulated based on the so-called partition lumping [42], in which the feedstock is partitioned by boiling point or solubility into a finite number of pseudocomponents. The lumps are composed of groups of molecules with relatively similar physical and chemical properties. Typically, the pseudocomponents are defined according to the boiling range of petroleum fractions of practical interest (e.g., naphtha (IBP: 204°C), middle distillates (204–343°C), VGO (343–538°C), and VR (538°C+)). Another form of partition lumping is by hydrocarbon class (paraffins, isoparaffins, naphthenes, and aromatics) and sometimes even including sublumps defined by the carbon number (C_5 paraffin, C_6 paraffin, and so on), as in the catalytic reforming of naphtha. The resulting chemical lumps are connected through a relatively simple reaction scheme.

Another type of lumping is when the kinetic equations for a group of certain chemical species are formulated in terms of an overall lump. These lumps are usually defined from routine petroleum analyses such as total sulfur, nitrogen, aromatics, metals (Ni + V), and sometimes solubility classes (e.g., asphaltenes). In this case, lumping is more drastic because each aggregate generally contains a polydisperse sample of molecules. A more elaborate representation can be obtained by combining partition and overall lumping [43]: first, a set of pseudocomponents is defined by the fractionation method and then subdivided by their elemental analyses (sulfur, nitrogen, etc.). In this way it is possible to predict the evolution of HCR products as well as their respective atomic composition.

Fundamental kinetic models offer a more detailed description of the process based on the actual chemistry. This modeling approach incorporates the transformation of the feed at the molecular level. These models are developed on fundamental knowledge of reaction mechanisms and on a detailed description of the feedstock. In HDT, only a few models of this nature have so far been reported in the literature [44–46]. They focus on the HDS, HDN, and HDA of relatively light feeds such as naphtha, gas oil, and light cycle oil, for which sufficient molecular structure information is available. The formulation of these models makes it possible to track the conversion of all observable species but does not describe the process at the mechanistic level. Due to the huge number of reactions, these models employ diverse techniques for reducing the number of rate parameters. HCR is one step ahead in this sense, because it has already been modeled at the mechanistic level. Froment and coworkers introduced the "single-event" concept by which several acid-catalyzed processes can be represented based on the elementary steps of carbenium ion chemistry [47]. This approach has been applied to the kinetic modeling of commercial heavy feedstocks such as VGO [48].

The lumping and the molecule-based modeling approaches are clearly extreme opposites. The differences between them are quite evident. It is definitely desirable to have the tools for predicting in detail the complex product distribution of petroleum refining processes. However, this is a really challenging task due to the enormous requirements in computational capacity and analytical techniques to measure the molecular attributes of complex feedstocks. Molecular modeling is therefore restricted to relatively light feedstocks. Extending this approach to heavy oils and residues is still an issue and will continue to be a subject of research effort. Furthermore, molecule-based models generally are not appropriate for simulation of complex hydrodynamics with computational fluid dynamic (CFD) packages and plant monitoring and control due to the excessive computer capacity demand [42]. The situation is clear: the higher the number of reactants considered in the model, the larger the size of the reaction network and the number of rate parameters that need to be estimated and consequently the more experimental information that is required.

On the other hand, lumping is the most used kinetic modeling method nowadays because it simplifies the problem drastically. In fact, lump-based models have been used for several decades in the modeling of hydrocarbon conversion processes. Many commercial units are still being designed based on this approach. Among other areas in which lumps are extensively employed are catalyst screening, real-time monitoring, process control, basic process studies, and dynamic modeling. However, the problem with lumping is that reaction kinetics are not based on the chemistry of the process, and hence there is a substantial loss of information in each multicomponent aggregate. No matter the number of pseudocomponents, the fundamental drawback of lumped kinetics is the dependency of the rate parameters on the composition of the feedstock. Therefore, this type of model is very specific and lacks prediction and extrapolation capability. Every time the feedstock or even the process configuration changes, the model must be refitted on the basis of a new experimental program. In spite of these limitations, the complexity of real feedstocks, especially the heaviest ones, suggests that lump-based models will continue to be used for the study of HDT and HCR kinetics or at least until the advances in analytical methods allow direct measuring of the most complex molecules such as asphaltenes.

Most kinetic studies with model compounds agree that HDT reactions follow simple first-order kinetics. Nevertheless, real feedstocks usually produce higher-order kinetics or even fractional kinetics. A generalized form of the typical rate expression of a given HDT reaction is as follows:

$$r_j = k_j C_i^n C_{H_2}^m \tag{13.1}$$

where k_j is the Arrhenius-type rate coefficient of reaction j; C_i and C_{H_2} are the concentrations of chemical lump i and hydrogen, respectively; n is the order of reaction with respect to lump i; and m is the reaction order of hydrogen. This form of model equation considers explicitly the effect of hydrogen concentration on the overall rate of reaction. This basic expression has been used for all types of HDT reactions and for a variety of feedstocks.

It is also possible to split the chemical lump i into reactive and refractory compounds to account for the difference in reactivity when the feed includes a broad distribution of chemical species. The resulting expression comprises a set of two competing first-order reactions:

$$r_j = \gamma k_{j,1} C_j + (1-\gamma) k_{j,2} C_j \tag{13.2}$$

where parameter γ represents the fraction of reactive species and consequently $(1-\gamma)$ is the fraction of nonreactive species and $k_{j,1}$ and $k_{j,2}$ are the rate coefficients of each fraction. It can be noticed that hydrogen concentration is implicit in this case, but the expression can be modified to account explicitly for this variable.

In other cases, it is possible to consider the inhibitory effect of certain compounds such as H_2S, NH_3, organic N-compounds, or even asphaltenes, by competitive adsorption. By factoring out the adsorption term from the lumped rate coefficient, the following Langmuir–Hinshelwood expression can be obtained:

$$r_j = k_j \frac{C_i^n C_{H_2}^m}{\left(1 + \sum K_k C_k\right)^2} \tag{13.3}$$

where K_k is the adsorption coefficient of species k and C_k is the concentration of adsorbing species k. It is generally assumed that adsorption occurs on two sites, hence the exponent 2 in the denominator.

Besides the three basic forms presented previously, the literature offers a long list of rate expressions, which are basically combinations and variations of Equations 13.1–13.3. Depending on the available information about the reaction pathway, simplified reaction networks have been proposed to represent, for instance, reversible hydrogenation of aromatic rings, sequential hydrogenation of multiaromatic rings, and parallel and series HCR reactions.

The effect of grouping a large spectrum of different components into a single lump is mostly reflected in the order of reaction n. In fact, reaction orders between 0.5 and 2.0 are a common observation for HDT kinetics. Reaction orders higher than 1.0 are commonly attributed to the wide disparity of molecular reactivities in the lump. Figure 13.11 shows the first-order rate coefficients for the HDS of benzothiophene, dibenzothiophene, and its alkyl-substituted derivates [49]. The difference in reactivity between each class is around one order of magnitude, which makes clear that the structural effect has a significant impact on the reactivity of the molecule. Therefore, a lumped mixture can be visualized as a set of competing first-order reactions of reactive and refractory species. The second-order reaction is given then by the persistence of the most refractory species in the reaction system [50].

Fractional-order kinetics are commonly observed for heavy oil and residue HDT [51]. Certainly, fractional reaction orders are more difficult to explain. A convincing explanation to this observation can be found in the work of Gray et al. [52]. From a mathematical analysis of a simple network of two competing parallel reactions (direct hydrogenolysis of thiophene and hydrogenation of thiophene followed by direct desulfurization), it was concluded that grouping these two pathways into a single lump expression can literally give any reaction order if there is partial conversion in one path, which is the case in reversible hydrogenation. This applies directly to heavy fractions because any chemical lump generally groups series and parallel reactions (hydrogenation, hydrogenolysis, and hydrothermal removal).

The development and formulation of lump-based models are heavily based on the kinetic experiments. A basic set of

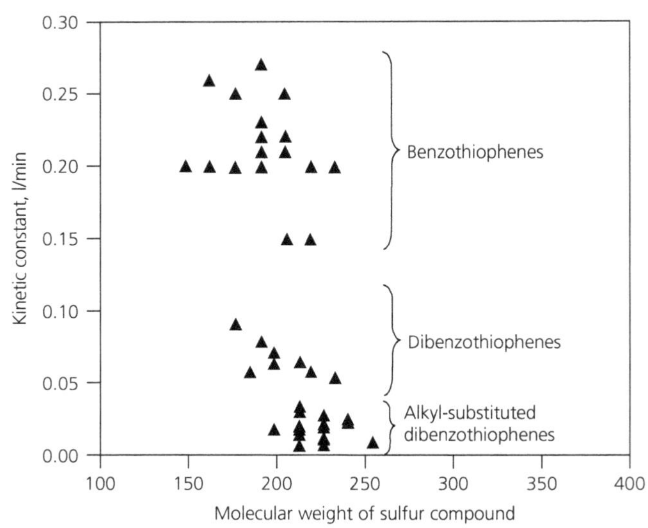

Figure 13.11 First-order rate coefficient values for the HDS of various sulfur compounds.
(Source: Ancheyta [49]. Reproduced with permission of John Wiley & Sons, Inc.)

experiments with variable residence time and temperature allows determining only the global order of reaction and activation energy. The reaction order of hydrogen is obtained by studying the effect of varying total pressure, keeping other variables constant. When considering inhibitory effects (e.g., H_2S or NH_3), the adsorption coefficients are evaluated by varying the initial concentration of those species, again keeping other parameters constant. Ideally, the experimental conditions must ensure the absence of internal mass-transfer effects (small particle size) to obtain intrinsic rate parameters. For scale-up and process development purposes, kinetic tests are generally performed using commercial-size catalysts in trickle-flow systems to emulate industrial conditions [53]. The result is an apparent kinetic model that incorporates implicitly the effect of intraparticle mass transfer. Nevertheless, these models generally are not readily scalable due to the hydrodynamic effects at different reactor scales and therefore require a correction parameter of some sort, such as catalyst wetting efficiency.

13.3.3 Thermodynamics

Knowledge of the thermodynamics of HDT reactions is necessary for predicting chemical equilibrium. Detailed thermochemical data on individual reactions of complex feedstocks are generally unavailable [24]. The literature offers some generalizations about each type of reaction made from theoretical calculations and few experimental studies with model compounds. For modeling proposes, the equilibrium constants are usually approximated using group contribution methods.

The standard heat and equilibrium constants of selected HDT reactions are listed in Table 13.2. The positive values of the equilibrium constants of HDS and HDN indicate that these reactions are essentially irreversible in the practical operating temperature range and can proceed to completion if hydrogen is present in stoichiometric quantity. On the contrary, hydrogenation of aromatics is reversible under typical HDT conditions, and consequently, complete conversions are not possible because equilibrium shifts toward dehydrogenation at high temperatures. In general, equilibrium constants tend to decrease with reaction temperature, which is consistent with the exothermic nature of HDT reactions. In the case of HDS, the constants become much less than unity only at temperatures considerably higher than those required in practice (>425°C) [35]. Table 13.2 also shows that the enthalpies of reaction vary significantly from one reaction to another. Usually for each class of reaction, the amount of heat release depends on the extent of stoichiometric hydrogen consumption.

Hydrogenation reactions in hydrocarbon processing comprise saturation during alkane cracking and ring opening, HDA, and saturation of olefins [54]. The first type is related to the hydrogenolysis of σ C—C bond in a naphthene ring or an alkylic side chain. This is the least exothermic reaction of all hydrogenations because σ C—C bond scission absorbs high amounts of energy. Olefin saturation is a fast reaction, highly exothermic (105–120 kJ/mol of H_2), and irreversible under practical conditions. Hydrogenation of an aromatic ring is more difficult than olefin saturation due to the stabilization of the aromatic structure by mesomeric effect; it is also less exothermic than olefin hydrogenation since the saturation of conjugate π C—C bonds of aromatic rings into σ bonds absorbs more energy than that of nonconjugated π C—C bonds as in olefins. Although the heat release of aromatic hydrogenation increases proportionally with hydrogen consumption, the amount of heat per mole of hydrogen is fairly constant (58–70 kJ/mol of H_2).

Hydrogenation of PNA is carried out successively ring by ring to the saturated end products, as discussed earlier. It has been

Table 13.2 Standard enthalpies and equilibrium constants of representative HDT reactions.

		$\log_{10} K_{eq}$		
	$\Delta H^{\circ a}$	200°C	300°C	400°C
HDS				
$C_3H_7-SH + H_2 \Leftrightarrow C_3H_8 + H_2S$	−57	6.92	5.87	5.15
Thiophene + $3H_2 \Leftrightarrow nC_4H_{10} + H_2S$	−262	14.13	9.33	6.04
Benzothiophene + $H_2 \Leftrightarrow$ ethylbenzene + H_2S	−203	16.65	12.85	10.20
Dibenzothiophene + $2H_2 \Leftrightarrow$ biphenyl + H_2S	−148	15.23	12.50	10.61
HDN				
Indole + $3H_2 \Leftrightarrow$ ethylbenzene + NH_3	−49	—	7.8	5.0
Carbazole + $2H_2 \Leftrightarrow$ biphenyl + NH_3	−126	—	6.8	5.1
Pyridine + $5H_2 \Leftrightarrow$ *n*-pentane + NH_3	−362	—	8.9	4.4
Quinoline + $4H_2 \Leftrightarrow$ propylbenzene + NH_3	−272	—	7.0	3.3
HDA				
Naphthalene + $2H_2 \Leftrightarrow$ tetralin	−140	1.26	−1.13	−2.80
Tetralin + $3H_2 \Leftrightarrow$ trans−decalin	−193	0.74	−2.95	−5.56
Cyclohexylbenzene + $3H_2 \Leftrightarrow$ cyclohexylhexane	−295	2.47	−1.86	−4.91
Phenanthrene + $4H_2 \Leftrightarrow$ octahydrophenanthrene	−251	1.16	−3.64	−7.12

[a] Standard enthalpy of reaction in kilojoule per mole of organic reactant.
Source: Ancheyta [49]. Reproduced with permission of John Wiley & Sons, Inc.

determined that equilibrium favors the saturation of the first ring, whereas the final ring is the most difficult one due its resonance stability [35]. Hydrogenation of aromatics also participates in the HDS, HDN, and even HDM of complex feedstocks. When heteroatoms are fused into aromatic structures, the reaction generally proceeds through the hydrogenation route, in which at least one aromatic ring adjacent to the heteroatom-containing ring is first hydrogenated before heteroatom–C bond cleavage. The conversion through this route might be restrained at high temperatures and low hydrogen partial pressures due to a lower equilibrium concentration of partially hydrogenated intermediates. Therefore, the overall rates of HDS, HDN, and HDM reactions may be affected by the unfavorable thermodynamics of aromatic hydrogenation.

The reversibility of aromatic hydrogenation reactions has several implications in industrial practice. The maximum aromatic conversion is strongly subjected to operating condition constraints, particularly reaction temperature, space velocity, hydrogen partial pressure, and catalyst type, as well as the type and amount of aromatic compounds in the feed. Increasing the reaction temperature to achieve deeper HDS and HDN reduces equilibrium conversion of HDA. This is particularly important in gas oil HDS, for instance, as it is desirable to reduce simultaneously sulfur, nitrogen, and aromatic contents in order to meet fuel specifications. In FCC feed pretreatment, it is also necessary to reduce the concentration of these impurities in order to prevent coke formation on acidic catalysts and at the same time to obtain clean products. Therefore, it is necessary to tailor reaction conditions to balance hydrogenation equilibrium with the extent of HDS and other reactions. A common solution is to operate at higher hydrogen partial pressures to increase the equilibrium concentration of saturated products.

Except for naphtha and kerosene HDT, most of these operations are considered to be highly exothermic as a result of the contribution of all the reactions to the total heat release. The temperature increase down through the catalyst bed depends primarily on the concentration of aromatics and sulfur compounds and the extent of their respective reactions. It has been reported that the temperature rise in some hydrotreaters can exceed 100°C [36].

The availability of heat release data is crucial for safe reactor design. It is possible to make good approximations of the global heat release from each individual reaction using group contribution methods, but this would require developing a complex molecule-based kinetic model. The literature generally reports information on individual HDT reactions, typically HDS, HDN, and HDA, with model compounds. Some selected works report global heats of reaction of specific processes which can be used judiciously for reactor modeling. The overall heat of reaction is a fitting parameter originated from several heat balances of similar HDT processes and which accounts for the contribution of all reactions (HDS, hydrogenation, etc.) [55]. These parameters generally require fine-tuning to accurately reproduce industrial data of any specific process. Their values are commonly reported as the amount of heat generated with respect to a certain reactant (removed sulfur, cracked hydrocarbon, consumed hydrogen, etc.). For instance, Mohanty et al. [56] reported a value of −42 MJ/kmol of consumed H_2 for VGO HCR, whereas Shah and Paraskos [57] reported the following values for various processes:

Residue HDS: −7820 kJ/kg sulfur
Gas oil HCR: −582 kJ/kg converted oil
Shale oil HDN: −8147 kJ/kg nitrogen

13.3.4 Catalysts

Typical commercial HDT catalysts are composed of active metals such as Mo or W in their oxidic state, with promoters such as Co or Ni, supported on γ-alumina (γ-Al_2O_3). γ-Alumina is the preferred support material due to its flexibility for tailoring textural properties, possibility to achieve a high dispersion of active metals, high mechanical resistance, and low cost [25]. The acidity level of γ-alumina is relatively low, which reduces the extent of cracking reactions and coke formation. The active phase is formed by distributing the active metals in the pores of the catalyst support by impregnation. Typically the active phase consists of Mo precursor (15–20 wt%) and either Ni or Co promoter (1–5 wt%), depending on the application. The main function of the promoter is to improve substantially the activity and selectivity of the active metal.

HCR catalysts are characterized by their dual functionality. The cracking function is promoted by the highly acidic support, whereas the active metal phase is responsible for the hydrogenation/dehydrogenation function. The typical support is made of amorphous silica alumina or crystalline zeolites (X and Y zeolites), the latter being the most acidic [36]. The hydrogenation function can be catalyzed by noble metals such as Pd and Pt or by metal sulfides of NiMo and NiW. Noble metals exhibit the best hydrogenation activity; however, they are very expensive and are easily poisoned by hydrogen sulfide. The metallic phase is responsible for both initiating the cracking mechanism by dehydrogenation of saturate hydrocarbons into the corresponding olefin intermediates and for stabilizing the end cracking products by hydrogenation [58]. The metal functionality must be balanced with the cracking functionality in order to avoid excessive coke formation from unstable products.

The appropriate selection of hydroprocessing catalysts depends on a careful inspection of the chemical properties of the feedstock and the expected products. For conventional distillate HDT (naphtha, kerosene, and gas oil), the specific surface area and the composition of the active phase are the most relevant features. CoMo-based catalysts are the traditional HDS catalysts used in the industry. NiMo catalysts are better for saturation reactions and HDN due to their higher hydrogenation power. Therefore, Ni-promoted catalysts are preferred over CoMo catalysts when the chemistry of the process proceeds through the hydrogenation route (i.e., gas oil feeds containing aromatic S- and N-compounds). NiW catalysts exhibit remarkable hydrogenation properties, but they are rarely used in commercial practice due to their elevated cost. Commercial VGO hydrocrackers generally include front beds loaded with highly active NiMo catalysts to protect the HCR catalyst from poisoning by N- and S-species and coke precursors.

The case of heavy feeds such as heavy oils and residues is the most sophisticated in terms of catalyst design. Besides the basic hydrogenation and HCR functions, the catalyst must be resistant to catalyst deactivation caused by metal and coke deposits. In this context, the textural properties (porosity) of the catalyst play a major role to ensure an acceptable length of run. Support acidity must be adjusted to achieve moderate HCR, which is necessary for avoiding excessive coking and sediment formation. Usually it is preferred to use a combination of selective catalysts of different characteristics to accomplish the best process performance. A typical graded-bed system comprises a front-end macroporous HDM catalyst, a mid-end balanced HDM/HDS catalyst, and a tail-end highly active HDS/HDN/HCR catalyst. The front-end HDM catalyst is perhaps the most relevant aspect of the catalyst system. Its purpose is to disaggregate asphaltene molecules to allow Ni and V removal, so that the downstream catalysts can operate with partially upgraded feed. Since intraparticle diffusion of large molecules such as asphaltenes is a rate-determining step, pore size in this case becomes an important factor [2]. If the pore size is too narrow, the catalyst will experience rapid pore mouth plugging, leading to early shutdown. The capacity of the HDM catalyst to store high amounts of metals from the feedstock is what determines the overall catalyst stability and consequently the length of run. The middle section comprises a more active catalyst that provides partial HDS and some additional metal removal. The tail-end catalyst resembles typical highly active HDT catalysts for HDS/HDN and HCR. In general, in graded-bed systems catalyst pore size decreases toward the tail end, whereas activity tends to increase in that same direction.

Besides the composition and textural properties, the shape and size of HDT catalysts are other characteristics that must be carefully selected according to the type of feed and reactor technology for optimal performance. Conventional catalyst shapes such as spheres and pellets are well suited for distillate HDT. For heavy feeds, such shapes are inadequate because large molecules do not have access to the interior of the particle as a result of diffusional limitations. Therefore, it is necessary to shorten the diffusion path by reducing particle size. However, decreasing particle size will often lead to a high pressure drop in FBRs. A practical solution to this problem is the use of other types of shape such as tri- and tetralobules, which design is characterized by an extended surface area to reduce significantly the diffusion path.

Finally, the overall catalyst performance during commercial operation is evaluated on the basis of the product quality, process selectivity, initial activity, and stability during time-on-stream [58]. The last two elements can be categorized as operational criteria since they determine the operating temperature program during the operation cycle. The initial activity establishes the start-up temperature to meet product specifications, whereas catalyst stability is more relevant in the mid- and end-of-run, as it determines the rate of temperature increase to keep the product within those specifications. When processing heavy feeds, catalyst stability is of particular importance for achieving acceptable cycle life due to the rapid catalyst deactivation caused by metals and coke.

13.4 Process aspects of HDT

13.4.1 Process variables

An accurate selection of the set of operating conditions ensures the best process performance. The main process variables (temperature, pressure, space velocity, and H_2/oil ratio) are adjusted according to the specific HDT application. Table 13.3 shows typical operating conditions of various processes [59, 60]. Most of these processes are generally carried out in fixed-bed units, with the exception of ebullated-bed residue HCR. Naturally, the severity of the process increases with the heaviness of the feedstock. Distillate HDT is carried out at relatively mild conditions compared to residue HDT. HCR processes require more severe conditions than HDT and are much more demanding in terms of hydrogen supply. A brief discussion on the effect of these variables is presented later.

13.4.1.1 Reaction temperature

Reaction temperature is certainly the most influential process variable. The extent and selectivity of HDT and HCR reactions

Table 13.3 Typical process conditions of various hydrotreating and hydrocracking processes.

Process	*T*, °C	P_{H_2}, MPa	LHSV, h^{-1}	H_2/oil, Nm^3/m^3	H_2 cons, Nm^3/m^3
Hydrotreating					
Naphtha	320	1–2	3–8	60	2–10
Kerosene	330	2–3	2–5	80	5–15
Gas oil	340	2.5–4	1.5–4	140	20–40
VGO	360	5–9	1–2	210	50–80
Residue	370–410	8–13	0.2–0.5	>525	100–175
Hydrocracking					
VGO	380–430	9–20	0.5–1.5	1000–2000	150–300
Residue	400–440	12–21	0.1–0.5	1000–2000	150–300

Source: Adapted from [59, 60].

are very sensitive to this process parameter because the rate coefficients increase exponentially with reaction temperature. Therefore, reaction temperature is matched to the chemistry of the process in order to achieve the desired selectivity. Light distillate HDT is performed in a temperature range (320–340°C) that is sufficient for almost complete HDS and HDN (Table 13.3). Above these temperature levels, there could be degradation of light hydrocarbons as a result of thermal cracking and unfavorable equilibrium for hydrogenation of aromatics in the case of gas oil HDT. Residues and VGO are more refractory in nature, and therefore, HDT requires higher temperatures, in which case it is important to operate at higher hydrogen partial pressures to improve the equilibrium concentration of saturated rings. HCR processes are carried out at the highest temperatures because C—C bond scission requires more energy than HDT reactions.

Process temperature has several side effects that must be carefully evaluated. Even a small excess in temperature leads to a loss in selectivity and higher catalyst deactivation. Consequently, there is a specific reaction temperature limit for each case. Increasing reaction temperature inevitably enhances coke growth as a result of a higher rate of condensation of unstabilized cracking products. In residue HDT, high temperatures increase the extent of HDM, but this at the same time accelerates irreversible catalyst deactivation by metal deposits. Temperatures above 410°C promote thermal cracking of valuable hydrocarbon constituents to generate considerable amounts of low molecular weight liquids and gases. Additionally, severe cracking of residues may lead to the formation of sediments, which have the tendency to produce fouling in all types of equipment.

In industrial fixed-bed units, reactor temperature increases as the reacting stream moves down through the catalyst bed. For this reason, temperature control is a major concern of hydroprocessing operations. Usually, the total heat release is limited to smaller and safer portions by dividing the total catalyst volume into several beds for injecting quench fluids in between [7]. Naphtha and kerosene HDT typically require only a single bed because the heat release is relatively small, but in the case of heavier feeds, single-bed reactors would be unpractical due to the excessive temperature rise. In those cases, the reactor is configured to obtain a more favorable temperature distribution. Figure 13.12 illustrates a three-bed reactor with two quenches.

The number of beds required in an HDT reactor varies according to the total amount of heat release. Bed depth is established by the allowable upper temperature limit. Ideally, the bed distribution must ensure equal temperature differences (delta-*T*s) in every bed (therefore equal average temperature in all beds) so as to improve the usage of the total catalyst inventory [58]. Generally, the allowed delta-*T* per bed varies somewhere between 15 and 30°C depending on the technical specifications of the process. It is typical to find multibed designs of increasing length because the rate of heat release decreases progressively from top to bottom. In most cases, gas oil hydrotreaters have two beds, whereas HCR units can have from four to six consecutive beds [36].

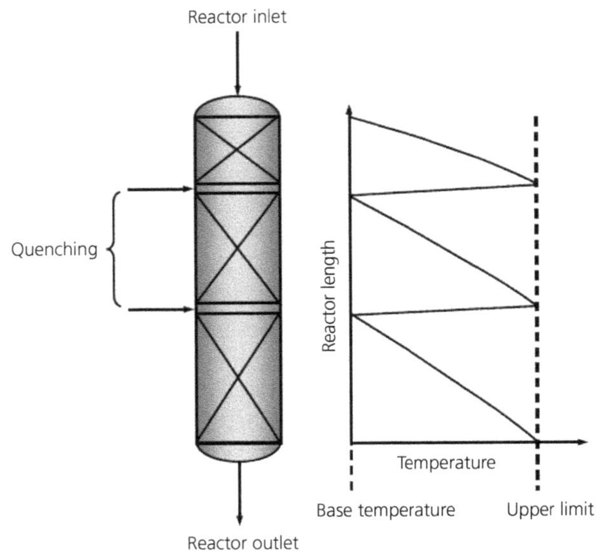

Figure 13.12 Multibed hydrotreating reactor with quenching.

The temperature profile of the reactor is also influenced by catalyst deactivation. During operation, the loss of catalyst activity is counterbalanced by periodically increasing reactor temperature, which progressively displaces the temperature profile upward. The cycle is terminated when the upper temperature level reaches the metallurgical limit of the construction material of the reactor. If axial temperature is not properly distributed, early shutdown is likely to happen, especially when the deactivation process is too fast as in residue HDT. Therefore, in such cases it is desirable to have the lowest possible bed delta-*T*s in order to delay the time to reach the maximum allowable limit. This implies more catalyst beds and consequently a larger reactor vessel with additional quench zone hardware.

13.4.1.2 Hydrogen partial pressure

Hydroprocessing units are typically operated in a pressure range of 1–30 MPa. High pressures are necessary for reducing coke formation on the catalyst particles, increasing the availability of hydrogen in the liquid phase, increasing conversion, improving heat transfer, and handling higher volumes of gas [59]. Hydrogen partial pressure (P_{H_2}) is simply the product of multiplying the total pressure by the hydrogen purity (mol%) of the gas recycle stream. As in the case of reaction temperature, the required hydrogen partial pressure increases with the heaviness of the feed and the desired conversion level. Hydrocrackers are operated at the most elevated hydrogen partial pressures to attain the highest conversions.

From a practical point of view, the main function of hydrogen partial pressure is to improve process performance by promoting hydrogenation reactions. Only a small fraction of gaseous hydrogen is dissolved in the hydrocarbon and eventually is available for reacting, which is why increasing total pressure is so important. In general, hydrogen partial pressure favors hydrogenation equilibrium, which in return increases the saturation of aromatics and improves substantially the HDS and

HDN of refractory compounds that follow the prehydrogenation route. In HCR units, high hydrogen partial pressure is necessary for saturating cracking products into more stable components and to keep coke growth under control. The consequence of operating below the preestablished hydrogen partial pressure is rapid catalyst deactivation by coke deposits.

Although the higher the hydrogen partial pressure the better the process performance, this process variable has some practical implications. Pressure is restricted by the design specifications of the plant equipment. Therefore, it is essential to maintain hydrogen recycle purity as high as possible to close the gap between hydrogen partial pressure and design pressure. The other important factor is that high-pressure equipment is very expensive because of the required wall thickness to resist such pressure levels.

13.4.1.3 H_2/oil ratio and gas recycle

H_2/oil ratio is a standard measure of the volume of hydrogen circulating through the reaction system with respect to the volume of liquid feed. It is defined by the following relationship:

$$H_2/\text{oil} = \frac{\text{Total } H_2 \text{ to the reactor } \text{Nm}^3/\text{h}}{\text{Total feed to the reactor } \text{m}^3/\text{h}} [=] \frac{\text{Nm}^3}{\text{m}^3} \qquad (13.4)$$

Gas recycling is required to generate a sufficiently rich hydrogen environment in the reactor. This ensures keeping hydrogen partial pressure at appropriate levels and increases the availability of hydrogen in contact with the catalyst and hydrocarbon molecules. The recycle rate is established on the basis of process objectives and economic aspects. Similar to hydrogen partial pressure, a recycle rate below the design value will induce coke formation and reduce the extent of conversion. This is particularly important in HCR in order to extend the life cycle of the catalyst, which is why it is common to operate these units at very high H_2/oil ratios (1000–2000 Nm^3/m^3). The downside of recycling higher gas volumes is the higher input of heating and cooling and especially the necessity for a larger gas compressor.

One relevant aspect of gas recycling is its effect on the gas–liquid equilibrium in the reactor [61]. It is typical for most HDT units to operate with partially vaporized hydrocarbon feed. This effect alters gas composition and reaction rates. Increasing the H_2/oil ratio can be useful for concentrating the heaviest and most refractory compounds (e.g., dibenzothiophenes in gas oil feeds) in the liquid phase, providing them more contact time with the catalyst. However, special care must be taken with excess recycle rates because some of the species in the vaporized fraction may not have access to the active sites of the catalyst particle.

Figure 13.13 illustrates the recycle gas circuit throughout a hydroprocessing unit. First, the reaction product is sent to an HPS where the liquid products are recovered from the gas mixture. Then, the gas stream is sent to a scrubbing unit in order to remove hydrogen sulfide and ammonia by contacting with an amine. The resulting hydrogen-rich stream (~80–85 mol% H_2) is recompressed in the recycle compressor and fed to the top of the reactor and/or used for interbed quenching. Part of the gas stream coming from the scrubbing tower (10–15%) is purged to the fuel gas system or to a hydrogen purification unit (e.g., pressure swing adsorption (PSA)). The product from the hydrogen purification unit is combined with the makeup hydrogen (~96–99.9 mol% H_2) and sent to the reactor along with the recycle stream from the compressor. Makeup

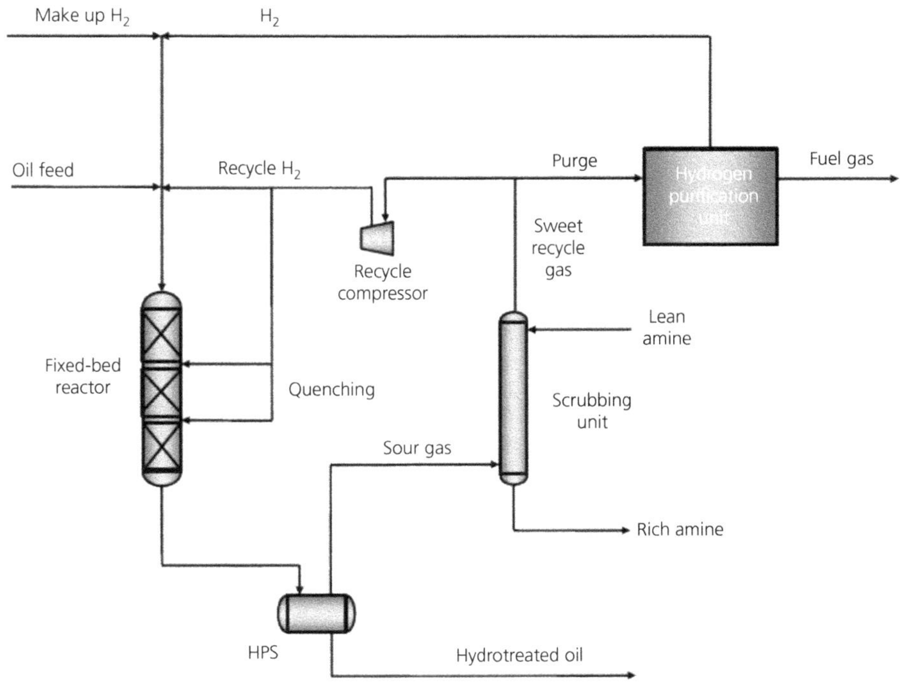

Figure 13.13 Hydrogen circuit in a hydroprocessing unit.

hydrogen is intended to compensate for the chemically consumed hydrogen and for losses through the gas purge, scrubbing unit bottom, and dissolved hydrogen in the hydrocarbon stream.

13.4.1.4 Space velocity

Space velocity is a ratio between the hydrocarbon feed rate and the amount of catalyst loaded in the reactor. It indicates how many reactor volumes of feed (considering only the volume occupied by the catalyst) can be processed in a unit of time. Space velocity can be established on a volume basis (LHSV) or on a weight basis (weight hourly space velocity (WHSV)):

$$\text{LHSV} = \frac{\text{Total volumetric liquid feed rate to the reactor m}^3/\text{h}}{\text{Total catalyst volume m}^3} [=] \text{h}^{-1} \quad (13.5)$$

$$\text{WHSV} = \frac{\text{Total liquid mass feed rate to the reactor kg/h}}{\text{Total catalyst weight kg}} [=] \text{h}^{-1} \quad (13.6)$$

The severity of the process increases inversely with LHSV. A low value of LHSV indicates that less amount of feed is being processed per hour (i.e., more contact time with the catalyst inventory). Usually, distillate HDT is carried out at higher LHSVs (>1.0) than residue HDT and HCR in general (<1.0). LHSV is usually a preestablished design parameter that determines the amount of catalyst and therefore reactor capacity for a required production rate.

13.4.2 Reactors for hydroprocessing

The currently available hydroprocessing reactor technologies differ mainly in the type of catalyst bed. According to this notion, the types of reactor fall into the following categories: FBR, MBR, EBR, and SPR. FBRs can be considered as the most well-established reactor technology for hydroprocessing, whereas the other three are more complex and specific for certain applications. The major selection criterion between each type of technology is the rate of catalyst deactivation, which essentially depends on the amount of metals and asphaltenes in the feed [3]. MBRs, EBRs, and SPRs are appropriate for the heaviest feedstocks, because they allow for replacement of spent catalyst without interrupting continuous operation. FBRs can also be used for heavy feed processing but only when the expected length of run is within economically acceptable limits (typically above 6 months). The main features of each technology are discussed as follows.

13.4.2.1 FBRs

Most HDT and HCR commercial units employ FBRs. Historically, FBRs were meant for processing naphtha, kerosene, and gas oil, but they were gradually modified to handle tougher feeds such as VGO and AR/VR. They are the preferred choice of refiners due to their relative simplicity, flexibility, and ease of operation.

Hydroprocessing reactors are generally three-phase (gas–liquid–solid) reaction systems. The gas phase is composed mainly of hydrogen, gaseous reaction products, and partially vaporized hydrocarbons; the hydrocarbon feed is the liquid phase, whereas the catalyst bed is the solid phase. The only exception is naphtha HDT which exhibits two phases (gas–solid) as a result of complete vaporization of hydrocarbons. The coexistence of these three phases puts hydroprocessing FBRs into a special category often referred to as trickle-bed reactors (TBRs). A TBR is a system where a cocurrent gas–liquid flow carrying both reactants and products moves downward through a fixed bed of catalyst particles [62]. Figure 13.14 shows a representation of the trickle-flow regime. It is considered that in this regime the liquid is the disperse phase and the gas is the continuous phase. The liquid flow tends to form a thin film over the catalyst particle and the gas flows separately, filling the remaining empty space of the catalyst bed. In general, trickle flow is achieved at relatively low gas and liquid velocities due to a low interaction between these two phases.

In trickle-flow regime, reactants and products must be transported between the three phases. Figure 13.15 provides a schematic illustration of the mass-transfer phenomena occurring in a TBR, in accordance with the two-film theory [63]. Hydrogen, being the main reactant in the gas phase, must diffuse from the bulk gas to the gas–liquid interface and then cross it into the main body of liquid. It is common to assume that the resistance in the gas film can be neglected and that the concentration of the gaseous species at the gas–liquid interface is in equilibrium with their partial pressure in the gas phase. The gaseous reactants, organic compounds, and hydrocarbons in the liquid phase travel to the catalyst particle across the liquid–solid interface in order

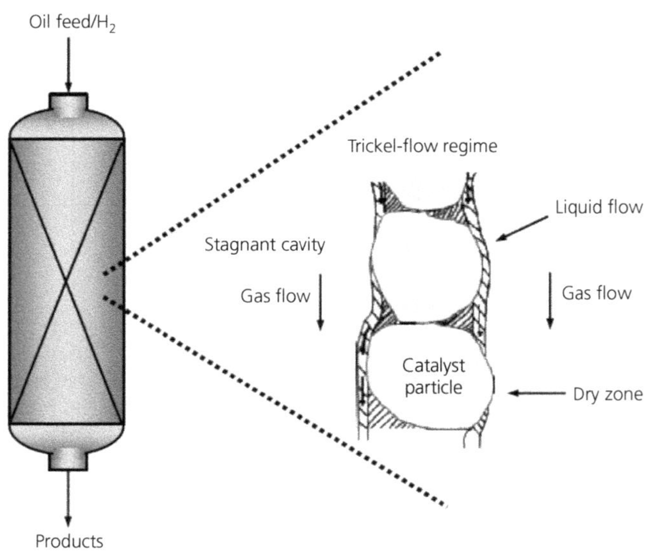

Figure 13.14 Representation of the trickle-flow regime in a hydroprocessing reactor.
(Source: Mederos et al. [4]. Reproduced with permission of Taylor & Francis LLC.)

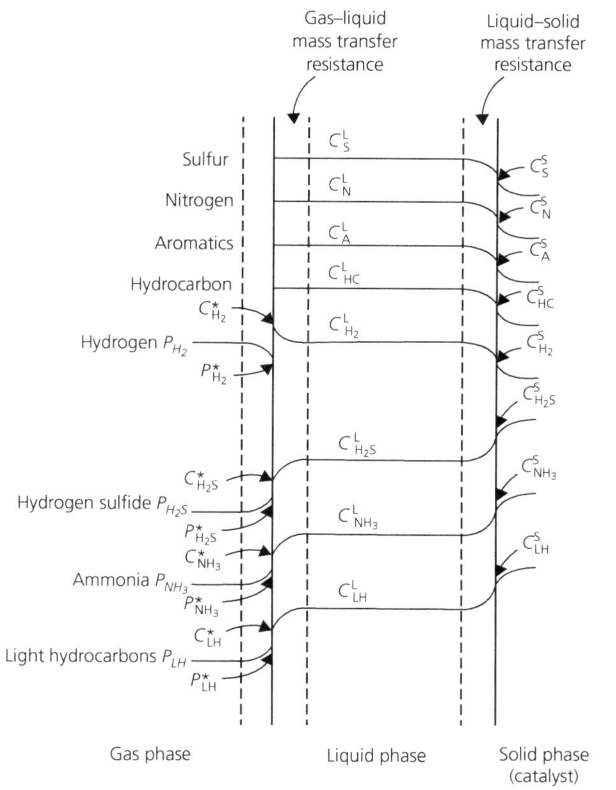

Figure 13.15 Concentration profiles in a TBR reactor.
(Source: Korsten and Hoffmann [63]. Reproduced with permission of John Wiley & Sons, Inc.)

to react with each other. The hydrocarbon products return to the liquid phase, while the gases generated by chemical reaction, such as H_2S, NH_3, and light hydrocarbons, are transported back to the gas phase in the same fashion.

One of the main advantages of TBRs is that liquid flow nearly approaches to plug flow, and therefore, TBRs exceed in performance other three-phase reactors such as EBRs or SPRs. They also exhibit a higher ratio of catalyst loading capacity per liquid volume. In practical terms, TBRs are very simple in construction, require low investment, and are the most flexible with respect to the demanded throughput and reaction severity for different conversion levels.

The drawbacks of TBRs include the presence of diffusional limitations inside the catalyst due to the particle size. This is because in commercial practice, the reactor must be loaded with relatively large catalyst particles to avoid an excessive pressure drop. However, the main disadvantage of this type of reactor is certainly the loss of catalyst activity over time. In addition, TBRs are exposed to fouling-related problems because of the presence of solids in the feedstock (iron scale, salts, coke fines, etc.) and reaction products (coke plugs and metal deposits) [64]. This means that the operation must be interrupted after a certain period of time for replacing the catalyst, which procedure takes around 1 month and requires complete dismantling of the reactor. The length of each operation cycle is determined by pressure drop buildup or catalyst deactivation, depending on the type of process. Most units typically operate during 2 years; however, cycle lengths can be as short as 6–12 months as in residue HDT or as long as 5 years [36]. It is also common that TBRs suffer from flow maldistribution of the reactants across the catalyst bed due to the poor performance of the reactor internal hardware [65]. This produces overuse of certain parts of the catalyst bed which may induce hot spot formation, whereas the rest of the bed becomes underused leading to overall reactor underperformance and shorter cycle lengths.

Quenching in FBRs. Industrial fixed-bed hydroprocessing reactors operate in the adiabatic mode and therefore may require quenching depending on the extent of heat release. Thermal stability is essential for safe operation, meeting product specifications, and achieving acceptable catalyst cycle lengths. The traditional method for keeping reaction temperature under control in fixed-bed hydroprocessing is by mixing the hot process fluids from the preceding bed with quench gas. Although historically hydrogen has been the quench fluid of choice, the use of liquids has been also reported [66]. Quenching takes place in the interbed sections of the reactor vessel, also termed quench zones, which are basically mixing chambers where the heated process stream is mixed with the cooling medium.

Hydrogen quenching is typical of most HDT and HCR processes with multiple catalytic beds. Quench streams are withdrawn from the gas recycle stream and injected into the interbed sections of the reactor, as presented in Figure 13.13. Hydrogen quenching has the advantage of replenishing some of the chemically consumed hydrogen in the preceding beds; enriching the gas phase with hydrogen, which helps keeping clean the catalyst from coke formation; and simultaneously diluting the concentration of reaction inhibitors such as H_2S and NH_3 [66]. The availability of quench gas primarily depends on the rate of recycle gas, which, of course, is governed by economic considerations. Usually, HCR units can have five or more gas injection points, as they are designed to operate at high H_2/oil ratios (~2000 Nm^3/m^3). At lower recycle volumes, the maximum rate of quench gas that can be withdrawn from the recycle stream is also restricted by the design H_2/oil ratio in the reactor, because every quench stream reduces the amount of recycle gas fed through the top of the reactor.

In contrast to hydrogen quenching, liquid quenching-based processes are not so common [66]. The use of liquids may become attractive because of their higher heat capacity and lower compression costs. The processes with liquid quenching can be classified in two general categories: (i) multiple feeding and (ii) product recycling. Multiple feeding, also known as split-feed hydroprocessing, involves splitting the feed into a number of fractions and introducing them selectively at different positions of the reactor length (Figure 13.16a). Generally, the heaviest fraction is fed at the top of the reactor for full contact with the total catalyst volume, whereas the lighter fractions are introduced as quench streams into downstream beds where they are provided with treatment in admixture with the heavy fraction. The advantage of this technique is that the feeding and catalyst bed arrangement improves process selectivity. Recent patent literature describes several multiple feed processes for upgrading light and middle distillates, FCC naphtha, VGO, and Fischer–Tropsch products [66]. The other type of processes is based on recycling portions of the liquid effluent for quenching and also providing a second-pass opportunity to unreacted species (Figure 13.16b).

Liquid quenching has the disadvantage of affecting reactor performance by increasing the LHSV (i.e., decreasing reaction severity), decreasing H_2 partial pressure if vaporization takes place, diluting the concentration of reactants, and adding new and more reactive species to the mixture. Such an effect must be compensated by designing larger reactor vessels for adding extra catalyst volume or by increasing reaction temperature. The first option evidently will increase the investment, whereas the second one may reduce the length of run and change the specification of the reactor construction material.

Reactor internals. One vital aspect for FBR performance is the internal hardware design [65]. Reactor internals are responsible for efficient catalyst utilization and process operation by means of uniform volumetric and thermal reactant distribution across the catalyst bed and for quenching performance and fouling protection [67]. Inappropriate reactor internal designs cause poor catalyst utilization due to maldistribution of reactants and deficient quenching performance.

A general design of hydroprocessing reactor internals is shown in Figure 13.17. Reactor internals may be located at the inlet, between catalyst beds, and at the outlet. The inlet hardware consists of a distributor tray along with fouling abatement trays and/or top-bed grading materials. For highly exothermic processes, multibed reactors with interstage quench boxes are employed to limit the heat release. Quench boxes comprise a fluid collection tray, a quench fluid injection device, a chamber for mixing the cooling medium with the hot reactants, and a redistribution tray. Finally, the hardware at the outlet provides fluid collection and catalyst retention. Figure 13.17 also shows the axial and radial delta-T profiles. The axial delta-T represents the typical temperature rise observed in hydroprocessing units. The drop in axial temperature is caused by the quenching/mixing feature of the interbed section. The radial delta-T reflects the performance of reactor internals. A good performance is characterized by narrow radial temperature differences after distribution and quenching, whereas a gradual widening in radial delta-T is an evidence of flow maldistribution. Such a negative effect has a cumulative character when the distributor trays and quench boxes are not working adequately.

In terms of catalyst utilization, the most relevant reactor hardware is the distributor tray as it is responsible for the liquid distribution across the catalyst bed. In general, traditional distributor designs such as sieve trays, chimney trays, and bubble cap trays are known for their poor performance, whereas state-of-the-art distributors facilitate complete irrigation of the catalyst bed (e.g., Shell's HD tray, Topsøe Vapor-Lift tray, Exxon's Spider Vortex technologies, Akzo Nobel's Duplex tray, and Fluor's Swirl Cap tray) [65].

Figure 13.18 presents a comparison of several design features of distributor trays. Tray spacing refers to the distance between the centers of two drip points (Figure 13.18a). With narrower tray spacing, uniform liquid distribution is achieved closer to the top of the catalyst bed because there are more liquid source points. On the contrary, wide tray spacing reduces catalyst utilization, requiring more bed depth for correcting liquid distribution by radial dispersion. Figure 13.18b compares a bubble cap tray against a Topsøe Vapor-Lift tray, in terms of tray spacing. Evidently, the Topsøe tray has a higher density

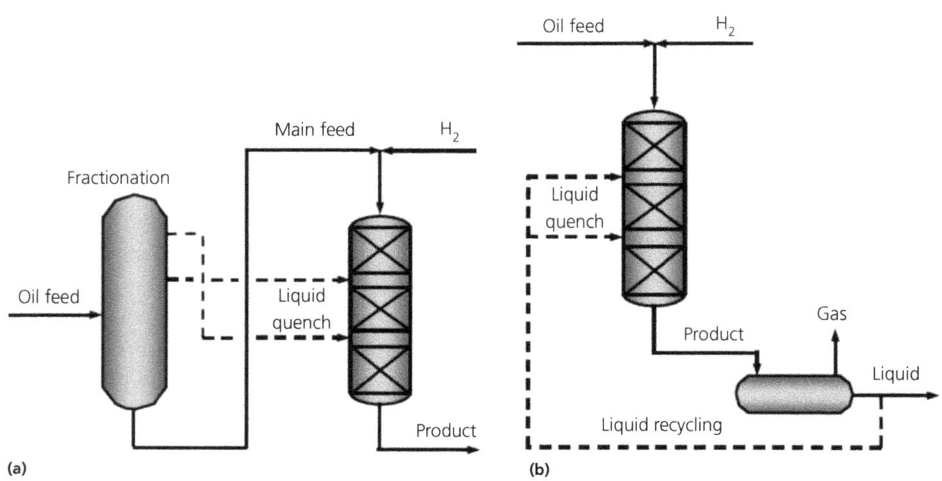

Figure 13.16 General representation of liquid quenching-based processes. (a) Multiple feeding. (b) Product recycling. (Source: Alvarez et al. [66]. Reproduced with permission of American Chemical Society.)

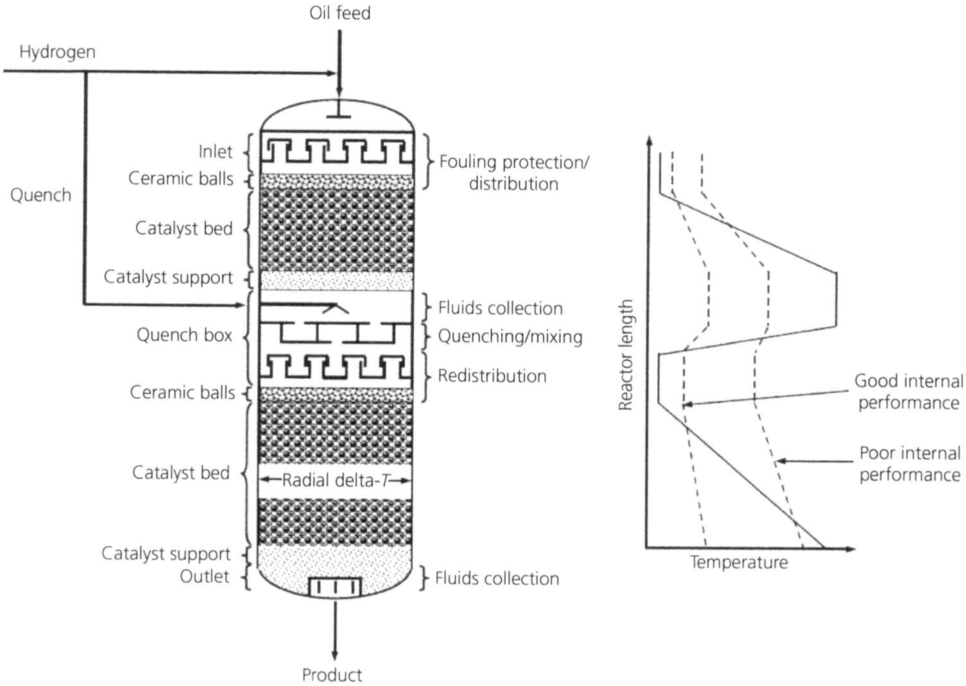

Figure 13.17 Hydroprocessing reactor internals. (—) Axial delta-*T*. (---) Radial delta-*T*.
(Source: Alvarez et al. [65]. Reproduced with permission of American Chemical Society.)

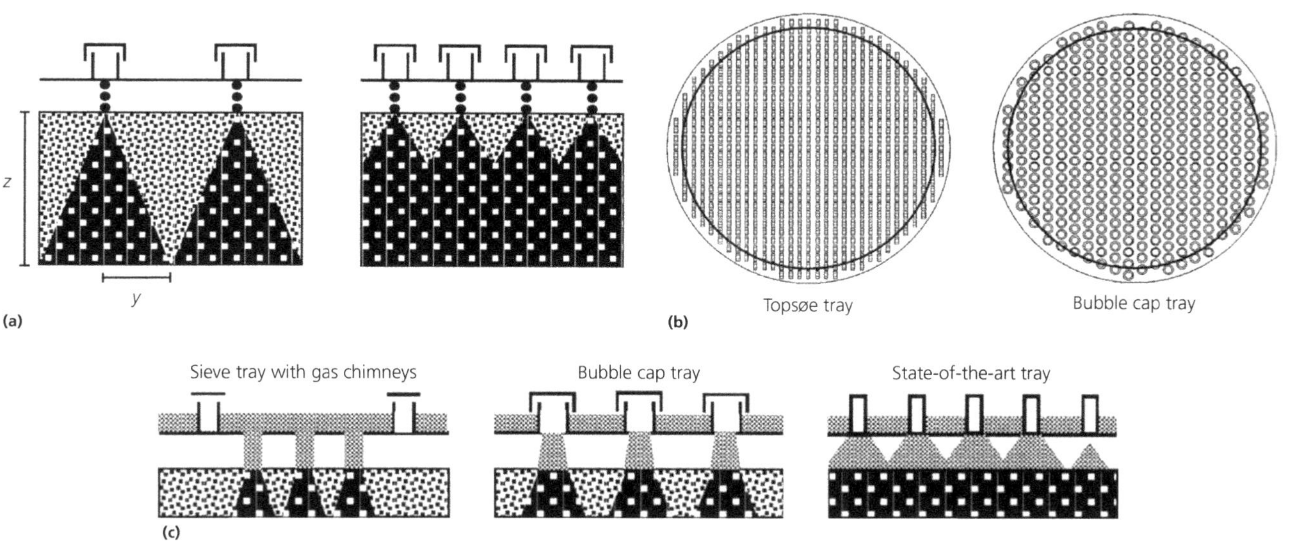

Figure 13.18 Distributor tray design parameters. (a) Effect of tray spacing on liquid distribution, (b) comparison of tray spacing, and (c) discharge pattern of several tray designs.
(Source: Alvarez et al. [65]. Reproduced with permission of American Chemical Society.)

of distribution points and therefore provides much better liquid distribution. The conventional tray suffers from wide tray spacing owing to the size of the bubble caps. This defect also promotes the presence of dead zones (with no liquid source) near the reactor wall leaving a great amount of unused catalyst vulnerable to hot spot formation. Perhaps the most important design feature is the liquid discharge pattern, as it determines the percent of wetted catalyst across the top of the bed. Conventional trays produce disc-type discharge pattern, wetting only the surface right beneath the discharge point (10–30% of the bed surface), as shown in Figure 13.18c. Modern trays, on the other hand, form a wide spray-type discharge pattern covering the entire catalyst bed surface. Such a discharge pattern operates under the gas-assisted principle, which takes advantage of the high gas velocity to drag the liquid held on the tray forming a highly dispersed liquid phase.

The quench box is the chamber where the hot reactants are mixed with the quench fluid. These internals must provide the following functions: (i) injecting the cooling medium, (ii) mixing with the hot reactants from the previous bed, and (iii) redistributing the liquid and gaseous reactants across the following catalyst bed. Generally, quench boxes must have a quench fluid injection device and a certain internal layout to provide effective gas–liquid mixing. Typically, the quench box has an arrangement of vanes and baffles that creates passageways and constrictions to produce a turbulent swirling motion of the fluids enhancing gas–liquid contacting. Shell's UFQ, ExxonMobil's Spider Vortex quench zone, Chevron Lummus' Nautilus, and ISOMIX reactor internals are some examples of the newest quench hardware technologies [65].

13.4.2.2 MBRs

MBR technologies overcome the problem of shutting down the operation every time the catalyst is completely deactivated. The main feature of MBRs is that they combine plug-flow mode operation of FBRs with the possibility to replace portions of spent catalyst during time-on-stream [2]. Therefore, they are well suited for handling tough feeds rich in metals and asphaltenes.

The term moving bed arises from the mode in which the spent catalyst is replaced. The catalyst bed is displaced periodically downward by gravitational forces. The fresh catalyst enters at the top of the reactor, and the deactivated catalyst leaves the reactor through the bottom. Liquid flow can be supplied either cocurrently or countercurrently with respect to the movement of the bed. The rate of deactivation determines how frequently the catalyst is replaced. Commonly, catalyst replacement is a batch operation and is done once or twice a week [68].

In comparison to FBRs, MBRs offer a much more favorable catalyst activity distribution along the reactor [3]. The periodical addition of fresh catalyst in MBRs increases the overall HDM and HDAs performance. Contrary to FBRs, the substantial amount of metals and coke deposits on the catalyst particles is removed through the bottom of the reactor during operation. This feature of MBRs allows for operating at higher pressures (200 MPa) and temperatures (400–430°C) than those in typical FBR units [2]. Thus, MBRs are more tolerant to metals and other contaminants than the FBR, even with the same type of catalyst and under more severe conditions. However, the catalysts used in MBRs should have improved mechanical properties in order to resist severe grinding and abrasion effects during replacement.

The most representative commercial developments using moving-bed technology are the bunker-flow reactor of the HYCON process developed by Shell and Chevron's onstream catalyst replacement (OCR) reactor [6]. The bunker reactor operates in the cocurrent downflow mode, whereas in the OCR reactor the hydrocarbon is fed in the upflow mode (countercurrent). Figure 13.19 shows a diagram of the OCR reactor. The countercurrent mode is apparently more efficient in terms of catalyst consumption, because the fresh feed is initially processed with spent catalyst at the bottom of the reactor, and then the partially demetallized feedstock is passed through the fresh catalyst at the top of the reactor. These technologies are used specifically in the front-end HDM stage in residual oil upgrading, which also serves as protection for subsequent FBRs for HDS and HCR [3].

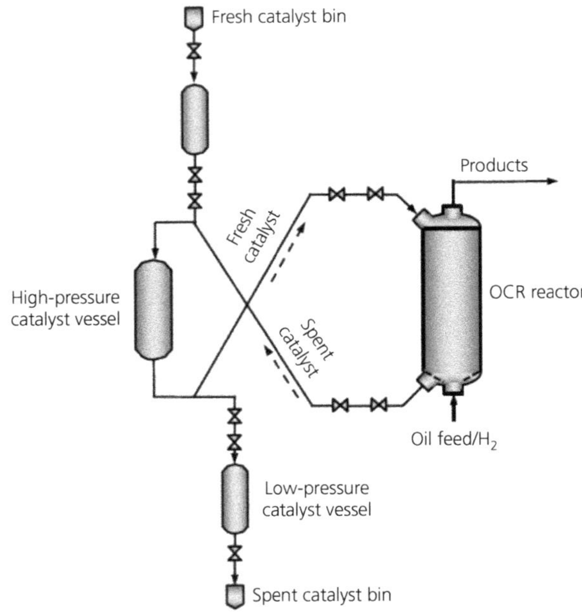

Figure 13.19 Onstream catalyst replacement reactor.
(Source: Ancheyta [6]. Reproduced with permission of Taylor & Francis LLC.)

13.4.2.3 EBRs

EBRs are perhaps the most sophisticated hydroprocessing technologies. They are specifically suited for the upgrading of heavy and extra-heavy feeds, directly without any pretreatment [1]. Spent catalyst is replaced continuously in EBRs, and the selection of the feedstock or conversion level is not restricted by catalyst deactivation. This feature allows for using conventional high-activity HDS and HCR catalysts instead of the front-end HDM catalysts used in FBR HDT. Residue HCR and HDM/HDS are the two major applications of EBR processes. Examples of commercial EBR technologies include the H-Oil process (Figure 13.20), licensed by Axens, and the LC-FINING process, licensed by Chevron Lummus. Both technologies have very similar characteristics in terms of process parameters and reactor design, but they differ in some mechanical details [9].

EBR hydroprocessing is a three-phase system in which the recycle gas bubbles up through the hydrocarbon mixture and the catalyst particles, creating a turbulent suspension [69]. The oil feed and hydrogen are delivered to the EBR in upflow

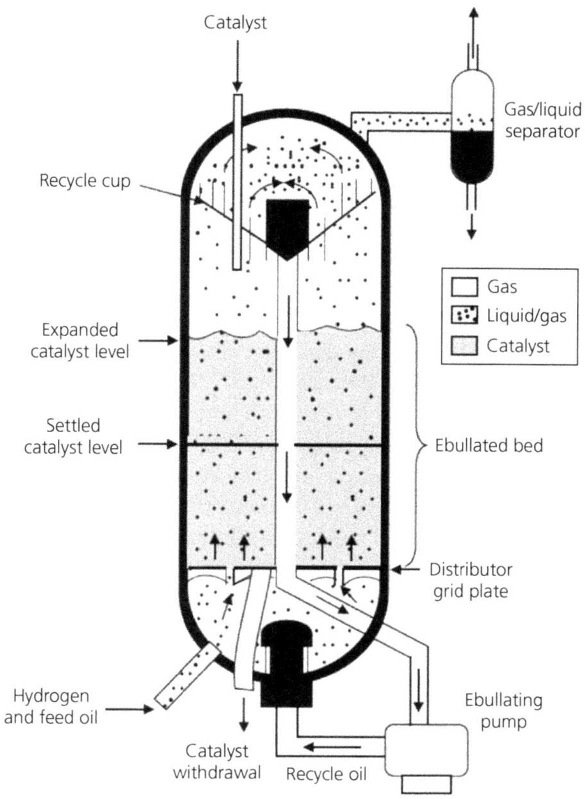

Figure 13.20 H-Oil ebullated-bed reactor.

mode passing through an expanded bed of catalyst particles. At the top of the reactor, the process fluids are separated from the catalyst and sent to a gas–liquid separator. Most of the catalyst particles are returned to the reactor. The oil is recycled to the bottom of the EBR and mixed with the fresh feed. The ebullating pump is responsible for keeping the liquid in circulation inside the reactor. EBRs can be assumed to behave almost as continuous stirred-tank reactors (CSTR) due to their large recycle rates. Therefore, the reaction conditions are nearly isothermal. Catalyst replacement in EBRs is done intermittently, where fresh catalyst is supplied through the top of the reactor and the spent catalyst is removed from the bottom of the reactor.

EBRs exhibit many advantages for processing heavy oils. EBRs are very flexible in operation, conversion can be as high as 90 vol%, and the end products have low levels of sulfur, metals, and nitrogen. The ebullated bed allows free movement of solids, which minimizes fouling and, consequently, the pressure drop [2]. Particle size is not restricted by pressure drop, and therefore smaller particles can be employed, reducing diffusional limitations significantly and making the catalyst less susceptible to pore mouth plugging by metal deposits. Operation in EBRs is almost isothermal, which increases product selectivity, improves heat transfer, minimizes the risk of local overheating, and reduces coke yield. Unlike FBRs, the overall catalyst activity remains constant during the whole cycle and so does the product quality.

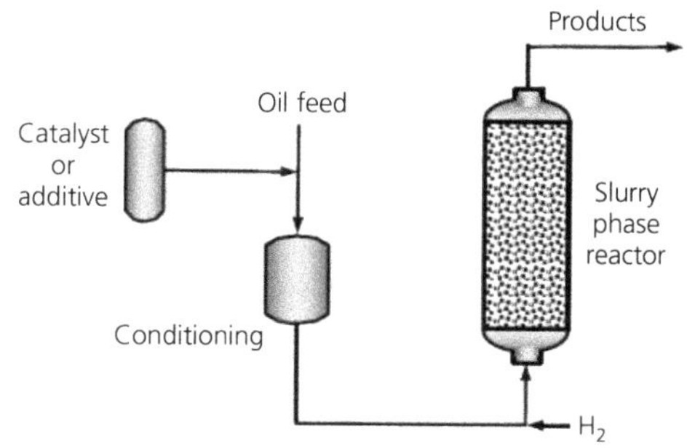

Figure 13.21 Slurry-phase hydroprocessing reactor. (Source: Ancheyta [6]. Reproduced with permission of Taylor & Francis LLC.)

One of the problems of EBRs is that they suffer from excessive catalyst consumption. The back-mixed character of EBRs is kinetically less favorable compared with the plug-flow regime. The catalyst must have improved mechanical strength, because replacement conditions in an EBR promote serious attrition and erosion of the catalyst particles. EBRs require a larger volume with respect to FBRs due to their small ratio of catalyst per liquid volume. Sediment formation is a major concern as a result of the high conversion levels (>50 vol%). Scale-up and design of EBRs is particularly more difficult due to the complex hydrodynamics [9].

13.4.2.4 SPRs

SPRs are catalogued as unconventional technology for hydroprocessing. Unlike typical hydroprocessing, the main characteristic of SPR technologies is the use of low-cost disposable catalysts aiming at reducing the cost of the catalyst inventory. The major application of SPR processes is in the primary upgrading of heavy oils and bitumen. The process is typically carried out at severe reaction temperature (>450°C) and pressure (14–21 MPa) [36] in order to operate in the thermal cracking regime. The oil feed is added with finely powdered additives in low concentrations, typically carbon or iron salts, to control coke formation. It has been reported that there are several slurry-phase hydroconversion technologies currently at the demonstration scale (Veba Combi-Cracking, CANMET HCR process, UOP's Aurabon, and Intevep's HDH Cracking) [2].

A simplified diagram of an SPR is shown in Figure 13.21. First, the catalyst powder is added to the feed and is subsequently brought to the reactor temperature. The heated slurry is then mixed with hydrogen, and both are fed through the bottom of the reactor which is basically an empty vessel. The liquid–solid suspension behaves as a homogenous phase due to the small catalyst particle size, whereas the gas phase bubbles up through this suspension. The SPR behaves like a plug-flow reactor because the liquid and catalyst particles flow cocurrently [23]. The spent catalyst leaves the SPR in admixture with the liquid product stream and eventually concentrates in the unconverted fraction in a nonhazardous form.

In general, slurry-phase hydroconversion can be advantageous for the upgrading of heaviest feedstocks due to the remarkably high levels of conversion (>90 vol%) achieved as well as the low costs associated with the catalyst stock and the simple design of the reactor vessel. The slurry phase is characterized by improved mass transfer and is thermally more stable. The main drawback, however, is the extremely poor quality of the unconverted fraction with very high contents of sulfur and metals.

13.4.3 Catalyst activation in commercial hydrotreaters

HDT catalysts are commercially produced with the metallic phase in its oxidic form (e.g., CoO/MoO_3). These metal oxides must be transformed into sulfides in order to obtain a more activate state of the metallic phase. This activation process, also termed presulfiding, is achieved by the action of a sulfiding agent. In current practice, there are three main sulfiding techniques: (i) with a regular sulfur-containing hydrocarbon feedstock (typically straight-run gas oil), (ii) with a H_2/H_2S mixture in the gas phase, or (iii) with a spiked feedstock in which the hydrocarbon is added with a spiking agent [70].

The sulfiding process is traditionally carried out *in situ*; however, *ex situ* sulfiding has also shown a lot of promise in terms of improved catalyst activity [71]. The essence of this process is passing the sulfiding stream through the catalyst bed in a hydrogen-rich atmosphere so that the metal oxides can be converted into sulfides. It is widely recognized that liquid-phase sulfiding is more efficient and rapid compared to gas-phase sulfiding. This is because the hydrocarbon flow assists in irrigating the catalyst bed, and thus a more uniform sulfide distribution is achieved. It also serves as a heat absorption medium for controlling the exothermality of sulfiding reactions. In the case of spiked feedstocks, the feed oil is added with a spiking agent such as carbon disulfide (CS_2), dimethyl sulfide (DMS), dimethyl disulfide (DMDS), or ethyl mercaptan (EM), to name a few. Among all of them, DMDS has proven to be the most effective both at laboratory- and commercial-scale sulfiding.

The sulfiding procedure is carried out at the decomposition temperature of the sulfiding agent. Activation with spiking agents generally requires much lower temperatures (~160–260°C) than with nonspiked oil (~300–350°C) [70]. It is important to keep an adequate sulfur concentration in the sulfiding feedstock in order to control the heat release of sulfiding reactions. A sulfur content of 0.5–2.0 wt% in the feedstock is typically recommended, whereas sulfur concentration must be below 1 wt% in spiked feeds.

Figure 13.22 shows the details of the liquid-phase sulfiding procedure for a NiMo HDS catalyst [71]. Before starting the sulfiding itself, the catalyst is dried with hydrogen flow at low temperature in order to remove traces of water that can be potentially harmful for the catalyst at higher temperatures, and after that, the catalyst bed is completely soaked in oil to prevent the presence of dry areas. Once the catalyst bed is prepared for sulfiding, the oil feed is switched to the sulfiding feed (e.g., straight-run gas oil spiked with DMDS), and the temperature is increased to 260°C. At this first stage of the sulfiding process, the temperature level must be high enough to decompose the spiking agent, but it should not exceed 260–315°C in order to avoid coking and metal oxide reduction before sulfiding is complete. This first step is also known as initiation. After this stage, the temperature is increased to a second level (290–350°C). Experimental results have indicated that temperature levels close to the upper limit are more effective for achieving a higher catalyst activity. This is basically a finishing step that ensures the maximum sulfur uptake. It should be mentioned that sulfiding reactions liberate substantial amounts of H_2S, which serves as an indicator of the state of the activation process. The appearance of considerable amounts of H_2S in the gas outlet indicates that the process has reached a breakthrough point, where the catalyst is being saturated with sulfur [36].

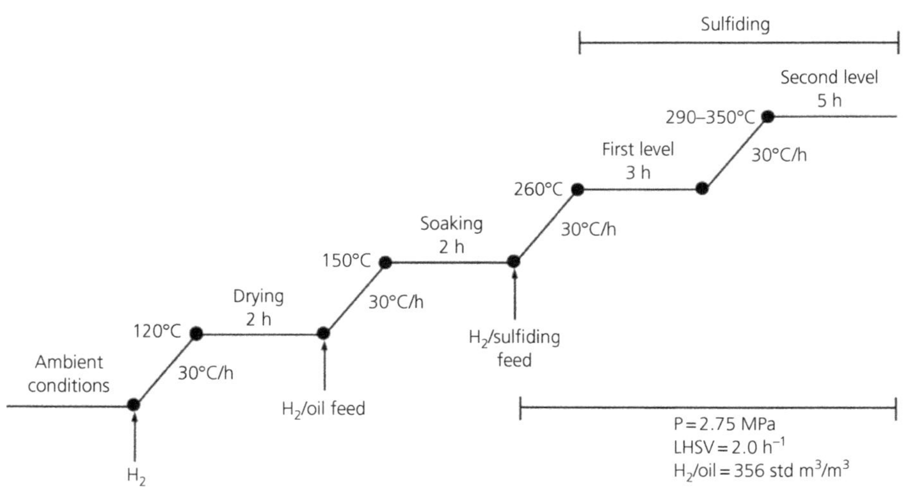

Figure 13.22 Sulfiding conditions of an HDS catalyst.
(Source: Marroquín et al. [71]. Reproduced with permission of Elsevier.)

13.5 Reactor modeling and simulation

This final section of the chapter focuses on the modeling and simulation of catalytic hydroprocessing. Modeling gathers all aspects of the process reviewed earlier into a computational tool intended for the design and optimization of commercial units. In this section, the construction of a hydroprocessing reactor model and its application at several stages of process design and simulation are illustrated through a practical example. The case study is derived from a series of recent publications on the modeling of a heavy oil upgrading process developed by the Mexican Institute of Petroleum (IMP) [51, 72, 73]. The modeled process is appropriate for the purposes of this chapter as it couples the most relevant phenomena that occur during hydroprocessing.

13.5.1 Process description

The IMP process is a catalytic hydroprocessing technology for the primary upgrading of heavy and extra-heavy crude oils to meet pipeline specifications and to provide better quality feedstock to refineries [16]. The upgraded oils, also known as "synthetic oils," exhibit similar properties to those of intermediate crude oils (22–30°API) but have lower sulfur and other impurity contents. The process is characterized by an arrangement of FBRs in series loaded with a graded-bed catalyst system consisting of selective catalysts for HDT and hydroconversion and in combination with low-pressure operating conditions to minimize sludge formation.

The basic process scheme of the IMP technology is presented in Figure 13.23. The initial step involves the splitting of a full-range heavy crude oil into a light fraction and a heavy fraction (typically an AR). The heavy fraction is subjected to hydroprocessing conditions in the first FBR where substantial metal and asphaltene removal is achieved, and at least a portion of the sulfur and nitrogen content is eliminated. The partially converted products from this stage enter a second FBR for achieving substantial HDS and HDN as well as a moderate level of HCR. The reactor effluent is sent to an HPS where the liquid products are recovered from the gases. The liquid stream from the HPS undergoes additional stripping in order to remove the remaining dissolved hydrogen sulfide. The gas mixture from the HPS is fed to the scrubbing unit in order to remove hydrogen sulfide and ammonia, and the resulting high-purity hydrogen stream is recompressed and recycled to the reaction system. Finally, the liquid stream is mixed with the light fraction to obtain the upgraded oil.

13.5.2 Summary of experiments

The reactor model of the heavy oil upgrading process was developed and fitted on the basis of two sets of detailed bench-scale experiments. The first experimental program was designed to collect reaction kinetic data at various operating conditions, whereas the second program served to obtain information about catalyst deactivation. A description of each set of experiments is given later.

13.5.2.1 Kinetic tests

The set of kinetic experiments was conducted in a bench-scale HDT unit. The plant consisted of two FBRs in series (~500 cm^3 of catalyst in each reactor) which were loaded with a triple layer of catalyst extrudates: a front-end HDM catalyst, a mid-end catalyst with balanced HDM/HDS activity, and a tail-end highly active catalyst for HDS and HCR. Catalysts were activated *in situ* by sulfiding with spiked straight-run gas oil (1.46 wt% sulfur, also containing 1.0 wt% dimethyl disulfide). The feedstock was an AR (343°C+) from a Mexican heavy crude oil (13°API).

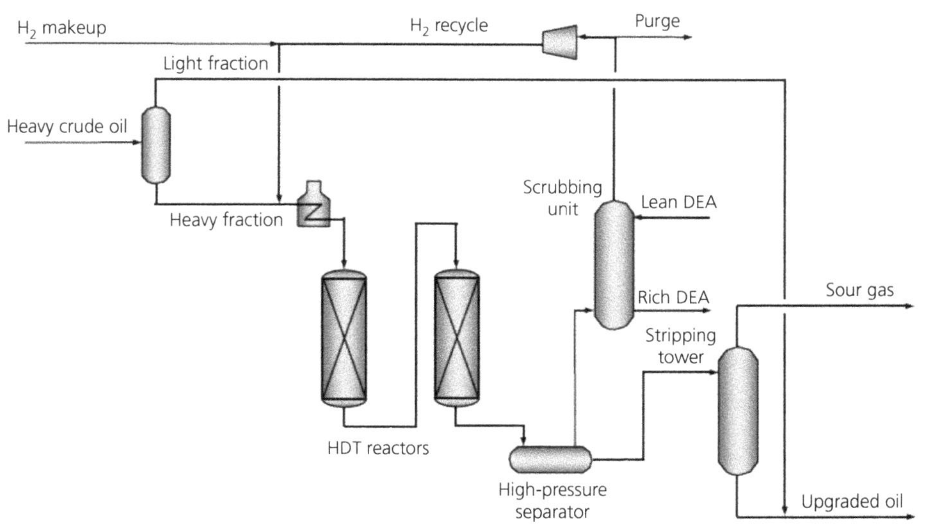

Figure 13.23 IMP heavy oil upgrading process. (Source: Adapted from [16].)

In this case, the residue feed represented the heavy fraction of the process diagram of the IMP technology (Figure 13.23). Physical and chemical properties of the feed (AR1) are presented in Table 13.4.

The details of the entire experimental program are shown in Figure 13.24. In order to perform the tests under steady-state catalyst activity, the fresh catalyst system was first stabilized at low temperature (360°C) by processing the feed during 100 h. After the initial deactivation period, the experiments were conducted consecutively by varying LHSV (0.25–1.0 h^{-1}) and temperature (380–420°C). The pressure and the H_2/oil ratio were kept constant during the entire run at 9.81 MPa and 891 std m^3/m^3 ("std" refers to standard gas volume at 0°C and 100 kPa), respectively. Hydrogen makeup was supplied between reactors at a rate of 151 std m^3/m^3 to compensate for hydrogen consumption in the first reactor. For each test, operating conditions were adjusted during a period of 5 h, and after that, the process was left to stabilize for a period of 5–12 h (depending on the flow rate) to obtain representative products. Under stable operation, two consecutive mass balances were performed by collecting the product for 3 h. The tests were carried out in less than 250 h of operation so as to minimize the influence of catalyst deactivation on the kinetic analysis. After the experiments, catalyst activity was monitored with a checkback test using the conditions of the first experiment (first and checkback experiments are marked with circles in Figure 13.24).

Table 13.4 Properties of the feedstocks.

	AR1	AR2
Density at 15.6°C, g/cm^3	1.0326	1.0475
API gravity	5.4	3.2
Kinematic viscosity at 121°C, mm^2/s	1637.9	1832.4
S, wt%	5.74	6.21
Metals, wppm		
Ni	102	117
V	620	578
Ni + V	722	695.6
Insolubles in C_7, wt%	21.77	25.10
ASTM D-1160 distillation		
IBP, °C	296	380.3
5 vol%, °C	372	415.4
10 vol%, °C	401	447
15 vol%, °C	438	474
20 vol%, °C	475	504
30 vol%, °C	521	551
40 vol%, °C	541	—
vol% recovery at 538°C	35.25	26.90

13.5.2.2 Catalyst aging test

The stability of the layered catalyst system was evaluated under process conditions in a 5.2 month-long test. The study was carried out in another bench-scale plant with two FBRs in series (~900 cm^3 of catalyst in each reactor). The total reactor volume was loaded with the same catalysts that were used for generating the kinetic data. The feedstock was a similar AR coming from another batch of 13°API heavy crude oil, as also shown in Table 13.4 (AR2).

The catalyst life test was conducted in the fixed-performance mode by increasing periodically reaction temperature to compensate for the activity loss. Unlike most of the residue upgrading processes which typically operate under HDS isoperformance, the target in this process was to keep the API gravity of the product constant, as this parameter is a good measure of the overall character of oils.

The evaluation was carried out under the following operating conditions: LHSV of 0.25 h^{-1}, initial temperature of 380°C, pressure of 9.81 MPa, and H_2/oil ratio of 891 std m^3/m^3. The mass balance runs were performed consecutively by recovering the products every 12 h. Inter-reactor samples were taken every 24 h in order to monitor the behavior of the first reactor. The sample size was kept at less than ~2% of the total feed rate so as to reduce the disturbance of the system. This small amount was sufficient enough to carry out some analyses, particularly to determine metals content in the products for estimating the metals uptake in the first reactor.

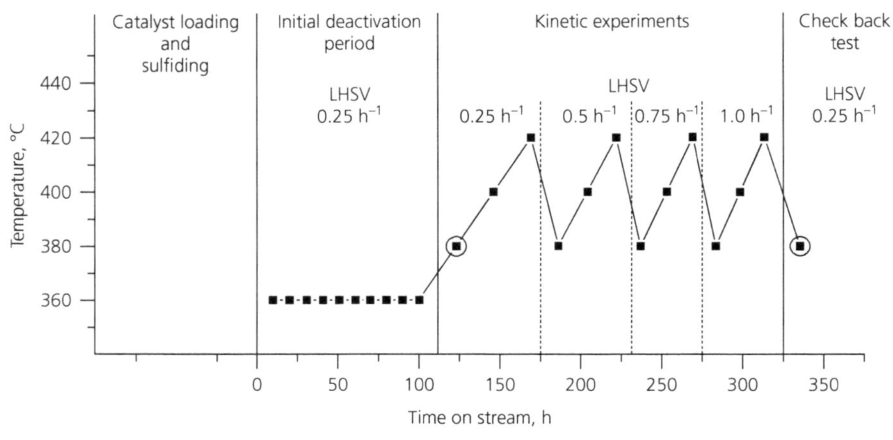

Figure 13.24 Details of the experimental program.

13.5.3 Modeling approach

The mathematical model was constructed on the basis of a three-phase plug-flow reactor model developed by Korsten and Hoffmann [63]. The model incorporates mass transport at the gas–liquid and liquid–solid interfaces and uses correlations to estimate mass-transfer coefficients and fluid properties at process conditions. The feedstock and products are represented by six chemical lumps (S, N, Ni, V, asphaltenes (Asph), and 538°C+ VR), defined by the overall elemental and physical analyses. Thus, the model accounts for the corresponding reactions: HDS, HDN, HDM (nickel (HDNi) and vanadium (HDV) removals), HDAs, and HCR of VR. The gas phase is considered to be constituted of hydrogen, hydrogen sulfide, and the cracking product (CH_4). The reaction term in the mass balance equations is described by apparent kinetic expressions. The reactor model equations were built under the following assumptions:

- The reactor operates in the plug-flow mode.
- Liquid velocity is constant throughout the reactor.
- Liquid vaporization does not take place.
- Pressure is constant.
- There is no heat-transfer resistance between the three phases.
- Intraparticle diffusion is considered in the apparent rate coefficients.
- Catalyst deactivation by coke occurs during the first 100 h on stream and then reaches equilibrium.
- Metals are deposited on the catalyst during the whole cycle.

13.5.3.1 Model equations

The formulation of the model equations is based on the transport of reactants between the gas–liquid–solid phases that takes place in TBRs [51]. Hydrogen, being the main gaseous reactant, is first transferred from the gas phase to the liquid bulk. The reactants in the liquid phase (chemical lumps and dissolved H_2) travel to the catalyst particle in order to react. Products such as H_2S and CH_4 are released to the gas phase passing through the liquid phase, whereas hydrocarbon products return to the liquid.

The change in the molar flow of gaseous compounds along the reactor is equal to the gas–liquid transport rate:

$$\frac{dN_i^G}{dz} = -A_s k_i^L a_L \left(\frac{P}{H_i}\frac{N_i^G}{\sum N_i^G} - C_i^L\right) \tag{13.7}$$

where $i = H_2$, H_2S, and CH_4.

The change in the concentration of gaseous compounds in the liquid phase is attributed to the gas–liquid transport and mass transfer to the solid phase:

$$\frac{dC_i^L}{dz} = \frac{1}{u_L}\left(k_i^L a_L \left(\frac{P}{H_i}\frac{N_i^G}{\sum N_i^G} - C_i^L\right) - k_i^S a_S\left(C_i^L - C_i^S\right)\right) \tag{13.8}$$

where $i = H_2$, H_2S, and CH_4.

The chemical lumps are transferred from the liquid bulk to the catalyst surface:

$$\frac{dC_i^L}{dz} = -\frac{1}{u_L} k_i^S a_S\left(C_i^L - C_i^S\right) \tag{13.9}$$

where i = S, N, Ni, V, Asph, and VR.

The species that travel across the liquid–solid boundary are either consumed or produced by chemical reaction:

$$k_i^S a_S\left(C_i^L - C_i^S\right) = \pm r_j(z,t) \tag{13.10}$$

where $i = H_2$, H_2S, CH_4, S, N, Ni, V, Asph, and VR and r_j represents the local rate of reaction j at the axial position z along the reactor and time t, with j = HDS, HDN, HDNi, HDV, HDAs, and HCR. The "−" sign is for the reactants, whereas the "+" sign is for the products.

It is necessary to include a heat balance to represent the adiabatic operation of the commercial reactor. Given the assumption that the three phases are at the same temperature, the following pseudo-homogeneous energy balance is used:

$$\frac{dT}{dz} = \left(\left(-\Delta H_R\right) r_{HDS}\right)\frac{1}{u_L \rho_L Cp_L + u_G \rho_G Cp_G} \tag{13.11}$$

The heat source term in the energy balance is based on a global heat of reaction (−7820 kJ/kg sulfur) of the AR HDS process [57].

Commercial hydroprocessing reactors generally have several quench injection points. Gas quenching reduces reaction temperature and changes gas composition, which modifies the conditions at the entrance of the next catalyst bed. Quench injection can be represented as the mixing of the quench stream with the gas–liquid bed effluent. The following energy balance can be used to calculate the required quench rate (q) for a certain cold mix temperature (T_{in}) or vice versa:

$$\int_{T_{out}}^{T_{in}} l_{out} Cp_L dT + \int_{T_{out}}^{T_{in}} g_{out} Cp_G dT + \int_{T_Q}^{T_{in}} q Cp_Q dT = 0 \tag{13.12}$$

The actual rate of gas entering the next bed is then obtained by adding the quench rate to the gas effluent of the preceding bed. Gas composition at the entrance of the next bed is adjusted with the individual mass balances of the gaseous species.

A detailed compilation obtained from different literature sources of the correlations used in this model for determining oil properties, gas solubilities, and gas–liquid/liquid–solid mass-transfer coefficients at process conditions has been reported elsewhere [51].

13.5.3.2 Reaction kinetics

The reaction term in the mass balance equations was represented by an apparent power-law kinetic model of nth order:

$$r_j(z,t) = k_j^{app}\phi_j(z,t)\left(C_i^S\right)^{n_j} \tag{13.13}$$

where k_j^{app} is the Arrhenius-type apparent rate coefficient of reaction j, ϕ_j is the deactivation function of reaction j at axial position z and time t, C_i^S is the concentration of the chemical lump i on the catalyst particle, and n_j is the order of reaction

j. In the case of HDS, the inhibitory effect of H_2S adsorption and H_2 concentration was also included:

$$r_{HDS}(z,t) = k_{HDS}^{app}\phi_{HDS}(z,t)\frac{\left(C_S^S\right)^{n_{HDS}}\left(C_{H_2}^S\right)^{0.5}}{\left(1+K_{H_2S}C_{H_2S}^S\right)^2} \quad (13.14)$$

H_2 consumption and H_2S generation were accounted for through overall molar-based stoichiometric coefficients, whereas CH_4 was assumed to be produced exclusively from the cracking of VR.

13.5.3.3 Scale-up of bench reactor data

It is well documented that bench-scale TBRs generally produce kinetic data that cannot be used directly for scale-up and design [53]. The presence of fluid dynamic limitations generates plug-flow deviations and poor wetting of the catalyst bed, affecting overall reactor performance. Because of this, the apparent rate coefficient often increases with reactor scale (i.e., liquid rate) [7]. Thus, the apparent rate constant can be defined as the intrinsic rate constant distorted by diffusional limitations inside the catalyst particle and by incomplete catalyst wetting:

$$k_j^{app} = \eta_0\eta_{CE}k_j^{in} \quad (13.15)$$

where k_j^{in} is the intrinsic rate coefficient and η_0 and η_{CE} represent the catalyst effectiveness factor and external wetting efficiency, respectively. Considering that the experiments were performed with commercial-size catalysts, the effectiveness factor is grouped with k_j^{in} to produce a "particle" rate coefficient (k_j^P):

$$k_j^{app} = \eta_{CE}k_j^P \quad (13.16)$$

k_j^P allows for describing catalyst performance under industrial conditions by avoiding the need to model intraparticle diffusion but only when the catalyst used for the tests is the same as that selected for commercial application. The resulting expression provides a simple relationship between bench-scale and industrial reactor performance with wetting efficiency as the scale parameter.

A useful method to handle this problem is to measure k_j^{app} at various liquid rates, and if possible at two reactor scales, and fit the data to the following empirical equation [74]:

$$\frac{1}{k_j^{app}} = \frac{1}{k_j^P} + \frac{A}{G_L^B} \quad (13.17)$$

where the particle rate constant is approximated in the form of $1/k_j^P$ by extrapolation to infinite superficial liquid mass velocity (G_L) (intercept on the y-axis). For this case, the procedure was applied using the information from the LHSV bench-scale experiments and from a semi-industrial scale test (10 bpd plant).

13.5.3.4 Catalyst deactivation

Deactivation of heavy oil hydroprocessing catalysts is driven by two factors: coking and metal buildup. Coke formation is responsible for the rapid activity decay during the first hours after which apparently reaches equilibrium, whereas metals are accumulated during the whole cycle in a linear fashion [3, 75]. The effect of these two processes on catalyst activity was modeled according to the following empirical expression [73]:

$$\phi_j(z,t) = \frac{1}{\left(1+\alpha_j t\right)^{\beta_j}} - \left(x_{MOC}(z,t)\right)^{\gamma_j} \quad (13.18)$$

The first term of Equation 13.18 is a hyperbolic function that represents the rapid initial activity decline as function of time t. As stated in the list of assumptions, it is considered that this happens during the first 100 h of operation, and the initial deactivation period is over after this time period. The choice of this simple function is based on numerous experimental observations showing that the deactivation curves during start-of-run follow the same pattern even with different types of heavy feeds in a wide range of temperatures [73]. The second term stands for the contribution of metals buildup during the whole cycle. This function is directly linked to the amount of metals-on-catalyst (MOC; x_{MOC}) at axial position z and time t, which allows for establishing a time-evolving axial activity profile.

13.5.4 Simulation of the bench-scale unit

13.5.4.1 Reactor simulation with stable catalyst activity

The rate parameters and stoichiometric coefficients of the reactor model were fitted from the kinetic experiments. The reactor model was applied to simulate the behavior of the bench-scale unit at stable catalyst conditions (after initial deactivation period). Molar concentration profiles of sulfur, nitrogen, and metals in the liquid and solid phases in both reactors (R_1 and R_2) are presented in Figure 13.25. Experimental values are also included for comparison. As observed, the model allows for tracking the evolution of each chemical lump, showing a good agreement with the experimental data. It can be noticed that there is a concentration gradient between the liquid and solid phases which disappears progressively with bed depth. Such behavior is attributed to the mass-transfer resistance at the liquid–solid interface. This resistance is given by the mass-transfer coefficient which depends mainly on the physical properties of the liquid (density and viscosity) and on the superficial liquid mass velocity. In this process, the mass-transfer coefficient is improved along the reactor, because the feed becomes lighter as a result of HCR.

Figure 13.26 shows the molar concentration profiles of H_2 and H_2S in the liquid and solid phases, at a reactor temperature of 380°C. The shape of these profiles is determined by a balance between chemical reaction and gas–liquid mass transfer. From the entrance to a relative reactor length of approximately 0.05 (5% of the total length), H_2 concentration drops quickly, whereas H_2S concentration increases substantially due to the high reaction rates in that section of the reactor. In the rest of the reactor length, H_2 concentrates progressively in the liquid, while H_2S is released from the liquid to the gas phase, and therefore, it is considered that this section is governed by gas–liquid equilibrium. In

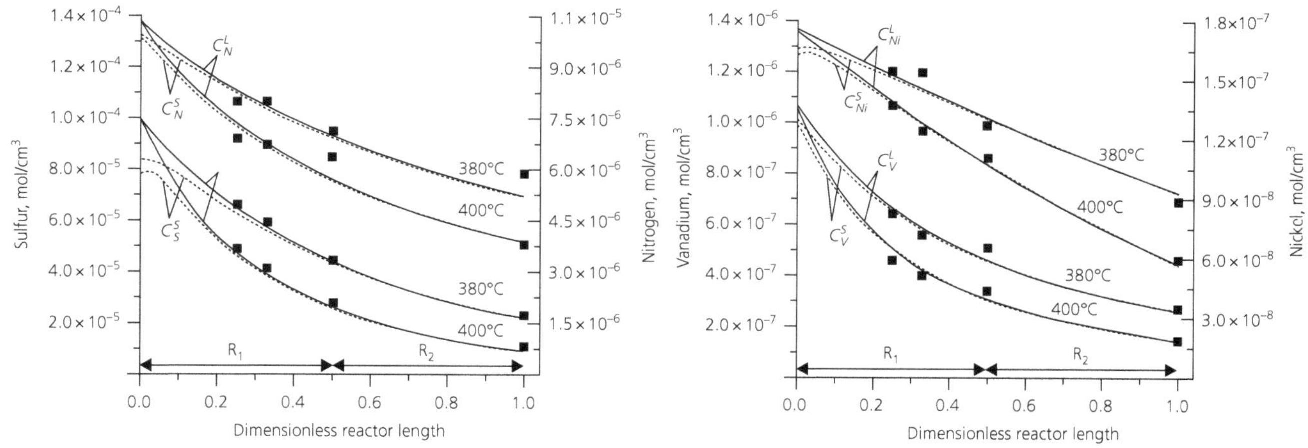

Figure 13.25 Sulfur, nitrogen, and metals molar concentration profiles. Simulated: (—) liquid phase, (---) solid phase; (■) experimental. (Source: Alvarez and Ancheyta [51]. Reproduced with permission of Elsevier.)

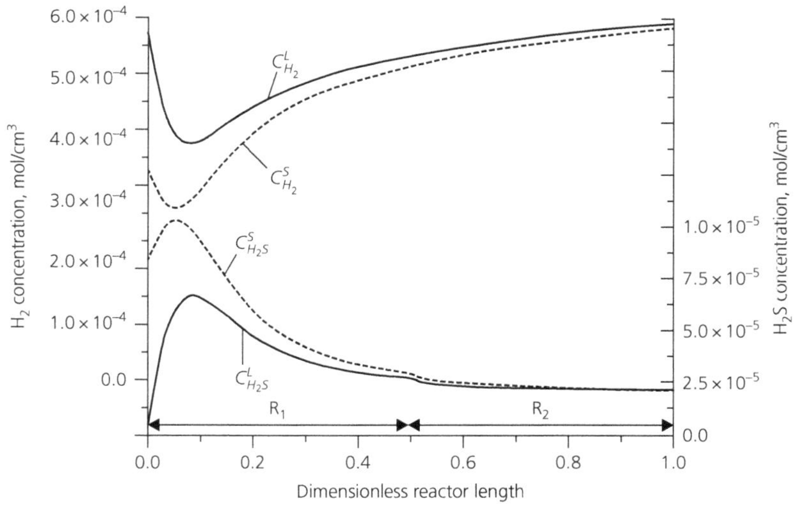

Figure 13.26 Hydrogen and hydrogen sulfide molar concentration profiles. Simulated: (—) liquid phase, (---) solid phase.
(Source: Alvarez and Ancheyta [51]. Reproduced with permission of Elsevier.)

this case, there is also an evolving liquid–solid mass-transfer resistance that disappears as the reaction proceeds.

The evolution of the gas-phase composition at 380°C is depicted in Figure 13.27. Experimental data from chromatographic analyses of the gas phase at the exit of the reaction system are also included to verify the model. It is observed that H_2 partial pressure decreases rapidly as a result of H_2 consumption. H_2S and CH_4 generation causes their partial pressures to increase along the reactor length. At a relative reactor length of 0.5, which corresponds to the inter-reactor zone, there is a notorious jump in H_2 partial pressure and a fall in H_2S and CH_4 pressures. This is caused by the inter-reactor hydrogen stream described in the experiments. Such a stream enriches the gas phase with H_2 and also has the advantage of diluting HDT reaction inhibitors such as H_2S and NH_3 toward the end of the reaction system. This type of gas composition profile is typical of industrial HDT reactors in which multiple hydrogen quench streams are employed.

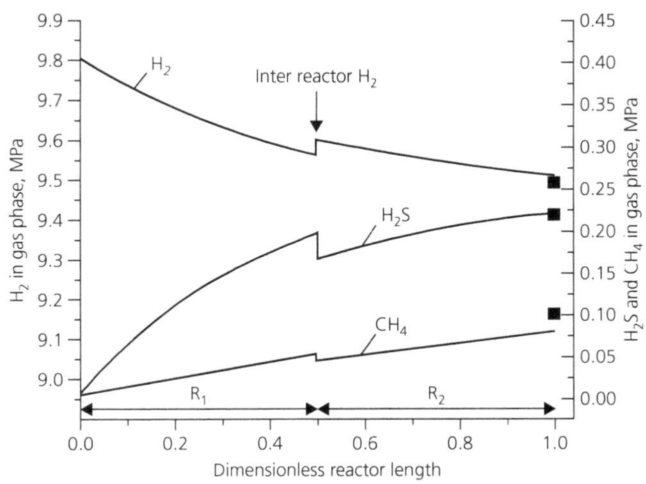

Figure 13.27 Evolution of the gas-phase composition. (—) Simulated. (■) Experimental.
(Source: Alvarez and Ancheyta [51]. Reproduced with permission of Elsevier.)

Figure 13.28 shows the axial profiles of superficial gas velocity (u_G) and H_2/oil ratio (relative to that of the entrance of the first reactor: 890 std m^3/m^3). Experimental values of the H_2/oil ratio at the exit of the reaction system, determined on the basis of hydrogen consumption, are included for comparison. In general, the behavior of both variables resembles that of H_2 partial pressure. This is a result of the H_2 consumption along the reactors, which decreases the H_2/oil ratio quite significantly and produces volumetric contraction of the gas reducing its superficial velocity. As expected, H_2/oil ratio and superficial gas velocity decrease more rapidly at higher temperatures as a result of higher H_2 consumption. On the other hand, the contribution of H_2S and CH_4 to the total gas volumetric flow rate is negligible compared with that of H_2. Detecting such behavior is only possible by formulating the mass balances for the gas phase in terms of molar flows. Most models consider constant gas velocity and therefore do not consider any changes in the gas rate. In this case, the hydrogen stream injected between reactors generates a step increase in both H_2/oil ratio and superficial gas velocity.

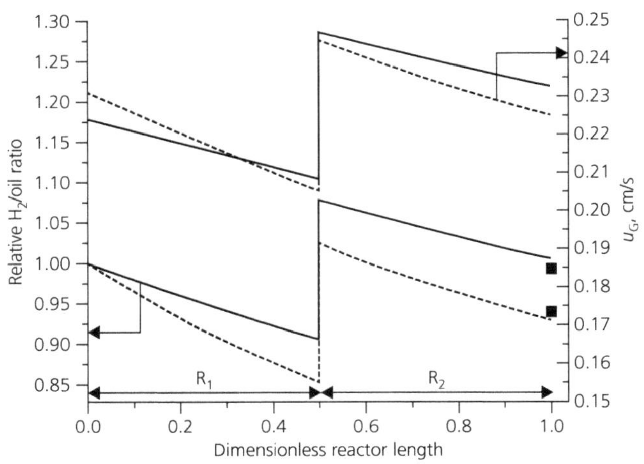

Figure 13.28 H_2/oil ratio and superficial gas velocity profiles. Simulated: (—) 380°C, (---) 400°C; (■) experimental.
(Source: Alvarez and Ancheyta [51]. Reproduced with permission of Elsevier.)

In general, the model predictions are in good agreement with the experimental observations. The model is capable of predicting the evolution of the various chemical lumps along the reactor at various temperatures as well as the gas yields and H_2 consumption.

13.5.4.2 Reactor simulation with time-varying catalyst activity

The data from the catalyst aging test were employed to fit the deactivation functions in order to simulate the performance of the process during time-on-stream. Figure 13.29 illustrates the levels of HDS, HDM, and HCR of VR during the run. It also shows the reaction temperature increase program to keep the upgraded oil quality at 23°API. The model predicts the performance of the process during the entire run reasonably well. The deactivation curves followed different patterns under the fixed-performance mode, for instance, HDS and HDNi decreased progressively with time-on-stream, whereas HDV and HCR gave a better response to the temperature increase. Such behavior is caused by diverse factors such as the properties of the catalysts, activation energies, feed type, and the thermal effects. The discontinuities at around 1500 h and after 2500 h of time-on-stream are attributable to pressure drop-related problems caused by fouling. In those cases, reaction temperature was decreased and the heavy feed was switched to light gas oil in order to flush the unit. Once the pressure drop decreased, the feed was switched back to residue and temperature was increased until the target performance was achieved.

The HDM performance of the first reactor (R_1), along with the global HDM level ($R_1 + R_2$), is presented in Figure 13.30.

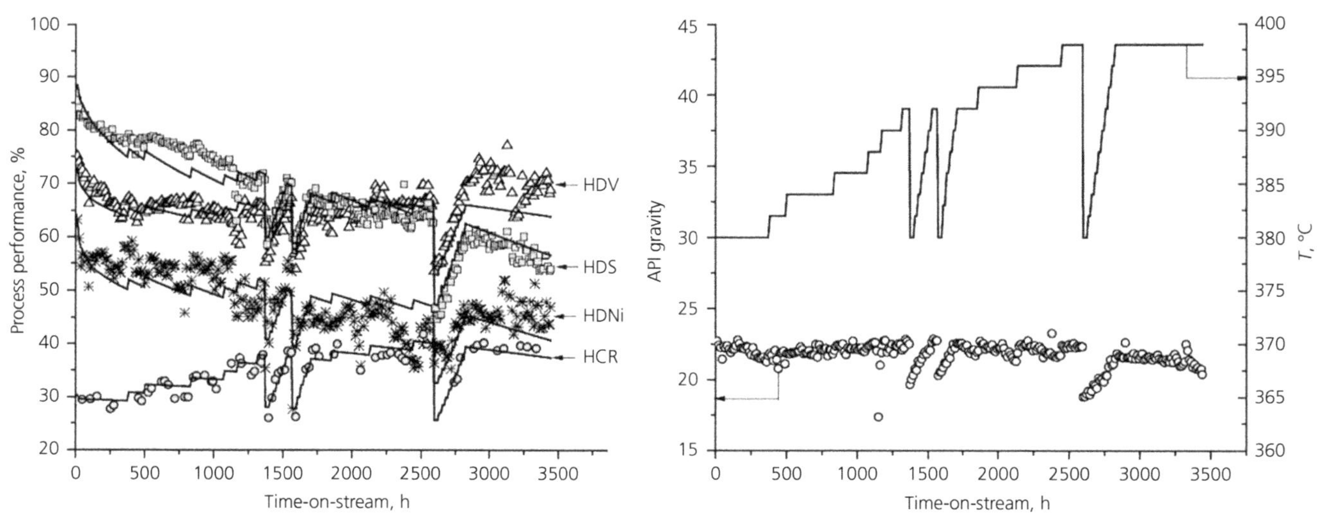

Figure 13.29 Process performance during time-on-stream: (symbols) experimental, (—) simulated.
(Source: Alvarez et al. [73]. Reproduced with permission of Elsevier.)

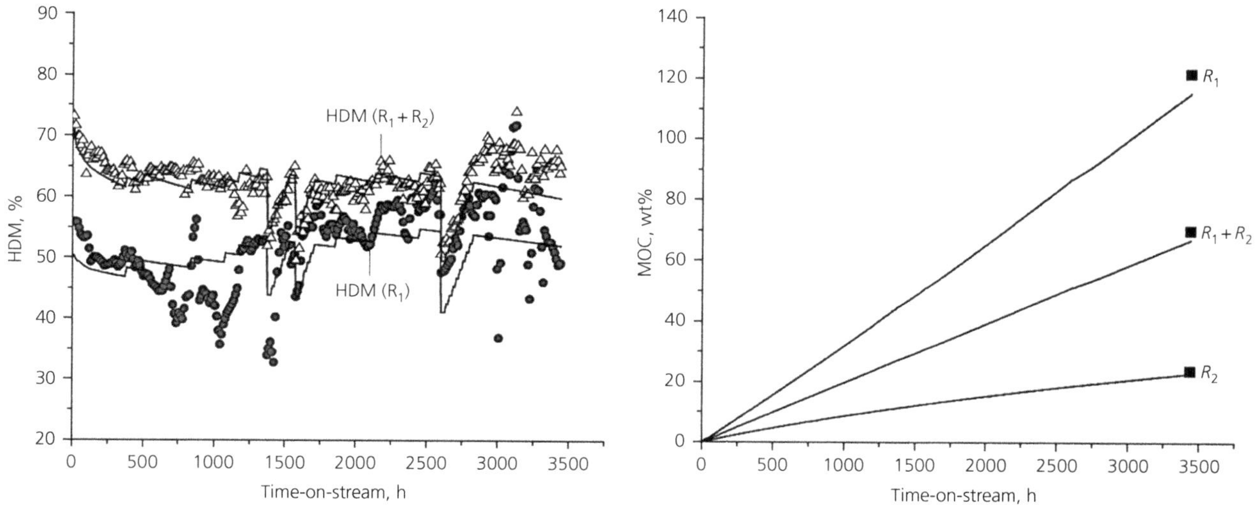

Figure 13.30 HDM performance and MOC: (symbols) experimental, (—) simulated. The experimental values (■) of MOC at the end of the run were estimated based on the metals balance between feed and products.
(Source: Alvarez et al. [73]. Reproduced with permission of Elsevier.)

As mentioned in the description of experiments, the first reactor was monitored with the inter-reactor samples. Evidently, the performance data of the first reactor are widely scattered, but the model is still able to capture the general trend. This is because inter-reactor samples were very small (2% of the effluent) and therefore not completely representative for an accurate kinetic analysis. However, those samples were acceptable enough for a rough estimation of the metals uptake in the first reactor and to verify the model capability. As for the overall HDM, the model works sufficiently well since parameter estimation was carried out using the overall aging data. The same figure shows the evolution of the MOC in both reactors. It is clear that MOC accumulation follows a linear behavior. There is a good match between the model results and the observed MOC, which was estimated from the metals balances between the feed and products. It is noteworthy that metals buildup in the first reactor is as high as approximately 120 wt% (on fresh catalyst basis), whereas it is only approximately 20 wt% in the second reactor, which indicates that the front-end catalyst is protecting the downstream catalysts.

The model is also capable of describing the evolution of the axial MOC profiles, as shown in Figure 13.31. It can be observed that initially (500 h) the axial MOC profile is relatively flat. As time-on-stream increases, the MOC profile acquires the typical descending shape, particularly in the first reactor. Clearly, it can be noticed that the front-end HDM catalyst, located in the first 30% of the total reactor volume (0–0.3), exhibits the highest metals accumulation, followed by the mid-end catalyst, which is distributed in the remaining space of the first reactor and the first part of the second reactor (0.3–0.6). Such a layered catalyst arrangement reduces significantly the metals content in the stream entering to the highly active HDS catalyst bed located in the tail end of the second reactor (0.6–1.0).

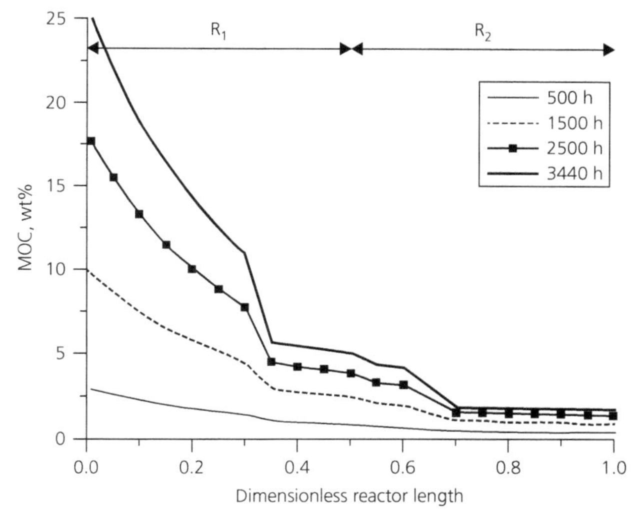

Figure 13.31 Simulated axial MOC profiles. The MOC units are referred to the total amount of fresh catalyst in the reactors.
(Source: Alvarez et al. [73]. Reproduced with permission of Elsevier.)

13.5.5 Scale-up of bench-unit data

In TBR processes it is highly desirable that the whole catalyst bed contributes to the overall conversion. For this to happen, every catalyst particle needs to be covered by a flowing film of liquid [76]. However, bench-scale reactors usually deviate from this ideal behavior as a result of low liquid velocities that cause incomplete catalyst wetting. In such cases, the liquid flows preferentially through certain parts of the bed leaving some of the catalyst particles partially wetted and others completely dry. Therefore, in the scaling up of kinetic data obtained from bench TBRs, it is essential to estimate wetting efficiency in order to determine the extent of catalyst utilization and thus to correct kinetic data.

The scale effect on catalyst wetting efficiency is analyzed in Figure 13.32. The theoretical curve was obtained from Equation 13.17 using the bench- and semi-industrial scale data at 380°C. The shape of the curve indicates that wetting efficiency is a weak function of superficial liquid mass velocity in this system. In fact, the values of wetting efficiency for all experiments fall in the range of commercial operation (0.7–1.0); for instance, at the lowest LHSV, this parameter is already higher than 0.7 (more than 70% catalyst utilization), whereas for the semi-industrial test it is very close to 1, as this plant is roughly 230 times larger in terms of processing capacity that the bench-scale unit operating. Apparently, this indicates that the semi-industrial scale reactor behaves ideally in terms of catalyst utilization, and therefore, this test was considered representative of commercial operation.

Figure 13.33 shows a comparison of semi-industrial and bench-scale performance at LHSV = 0.2 h^{-1} and 380°C. It is evident that the semi-industrial scale outperforms the bench scale as a result of the effect of liquid flow rate on catalyst utilization. The use of wetting efficiency as the scale-up parameter allows accounting for such an effect on reactor performance. From the previous analysis, it was determined that a wetting efficiency of 0.7 correlates sufficiently well bench-scale and semi-industrial reactor performances.

13.5.6 Simulation of the commercial unit

13.5.6.1 Reactor design and simulation with stable catalyst activity

It is well known that there are certain differences between industrial and pilot reactors. The most important one is that industrial reactors operate in the adiabatic mode and the gas recycle stream is distributed along the reactor length to provide quenching to interbed effluents, whereas bench-scale reactors are isothermal, and the gas is supplied generally at the entrance of the reactor only. Therefore, when designing new units, the temperature and H_2/oil ratio profiles of the industrial reactor configuration must be close enough to the conditions previously established during bench-scale experimentation. This configuration is obtained by adjusting inlet reactor temperatures, setting the permitted delta-T per bed (this means estimating the number of catalytic beds and their respective lengths), and finding a proper distribution of the amount of gas that is withdrawn from the recycle loop for quenching.

The reactor model was applied to design and simulate the commercial-scale reactors of the heavy oil upgrading process, keeping in mind the aforementioned criteria. It was considered that the reactors operate at the following average conditions: overall LHSV of 0.25 h^{-1}, temperature of 380°C, H_2/oil ratio of 890 std m^3/m^3, and pressure of 9.81 MPa. The main target was to construct an arrangement in which temperature and H_2/oil ratio profiles matched the required average conditions. This was done through a sequential analysis of several reactor configurations that evolved from one to another, identifying the flaws of a given design and then improving them to meet the established criteria.

Figure 13.34 presents one possible reactor configuration and the simulation of reactor temperature, H_2/oil ratio, and conversion of the chemical lumps. In order to limit the sharp temperature rise caused by the hydroprocessing reactions, the total catalyst volume was divided into six catalyst beds. R_1 required four beds as a result of the large heat release in this section (~72°C), whereas R_2 required only two beds. Bed inlet temperatures and delta-Ts for each reactor were adjusted to be more or less equal in order to match the average temperature

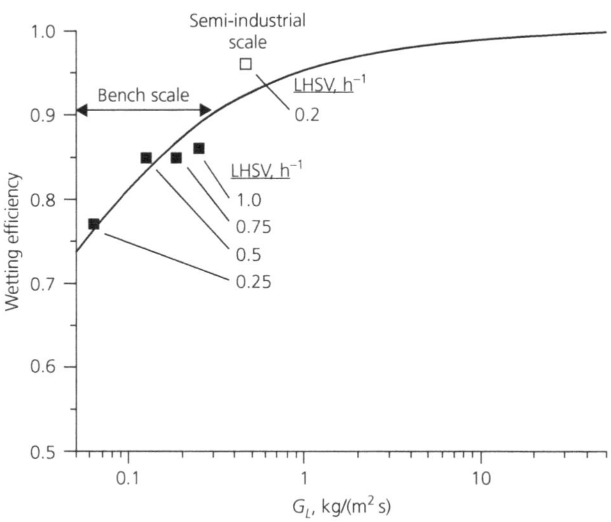

Figure 13.32 Catalyst wetting efficiency as function of superficial liquid mass velocity: (■) bench scale, (□) semi-industrial scale, (—) simulated. (Source: Alvarez et al. [73]. Reproduced with permission of Elsevier.)

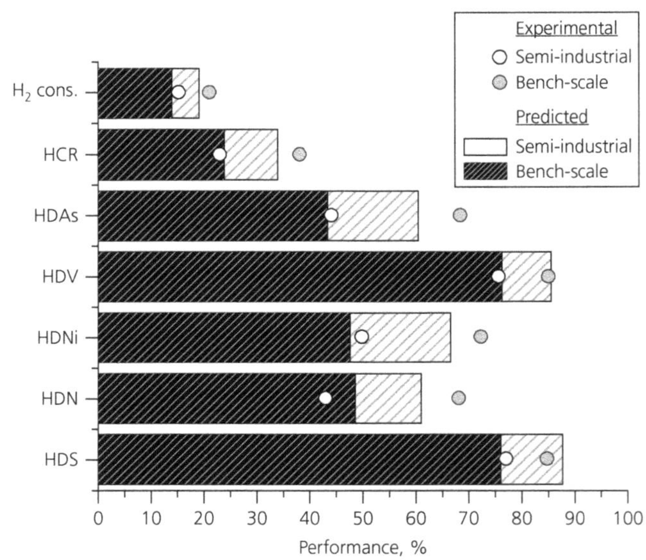

Figure 13.33 Comparison of semi-industrial and bench-scale performances. (Source: Alvarez et al. [73]. Reproduced with permission of Elsevier.)

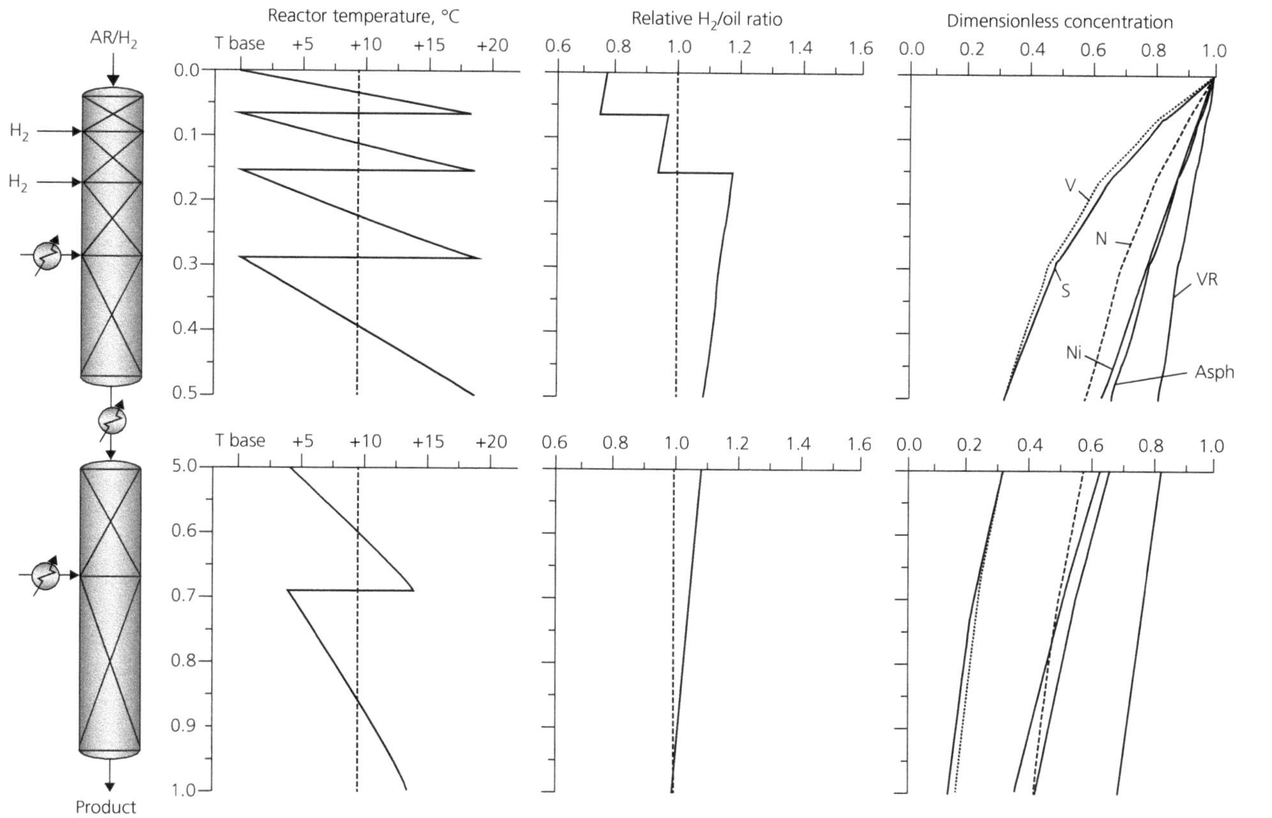

Figure 13.34 Simulation of the proposed industrial reactor configuration. Evolution of reactor temperature, H_2/oil ratio, and concentration of the chemical lumps. (---) Average temperature and H_2/oil ratio.
(Source: Alvarez and Ancheyta [51]. Reproduced with permission of Elsevier.)

(dotted line) in every catalyst bed. Equal delta-Ts per bed produce a catalyst bed arrangement of increasing depths. It can also be noticed from Figure 13.34 that there are only two hydrogen quenches injected at specific positions in R_1 and three heat exchangers distributed along R_1 and R_2. This specific arrangement results from adjusting the amount of quench gas coming from the gas recycle stream to match the average H_2/oil ratio (dotted line).

Figure 13.35 shows the average H_2/oil ratio in both reactors as function of the number of hydrogen quenches. The analysis indicates that two hydrogen quenches is the limit to keep the design H_2/oil ratio (890 std m^3/m^3). If the number of quenches is increased (3, 4, or 5), the average H_2/oil ratio in R_1 begins to drop below the design value, because every extra quench stream reduces the amount of recycled gas entering through the top of R_1. This is not particularly desirable in R_1, as lower H_2/oil ratios will have an adverse effect on catalyst cycle life. From this analysis, it is then concluded that, in this case, it is not possible to quench the total heat release with hydrogen and at the same time keep the H_2/oil ratio at the design average in both reactors. This is the reason why there are three interbed heat exchangers in the proposed reactor configuration. Another option to maintain the design value in R_1 (case 5+) could be increasing the gas recycle rate by 43%; however, this would eventually affect process economics due to elevated compression costs.

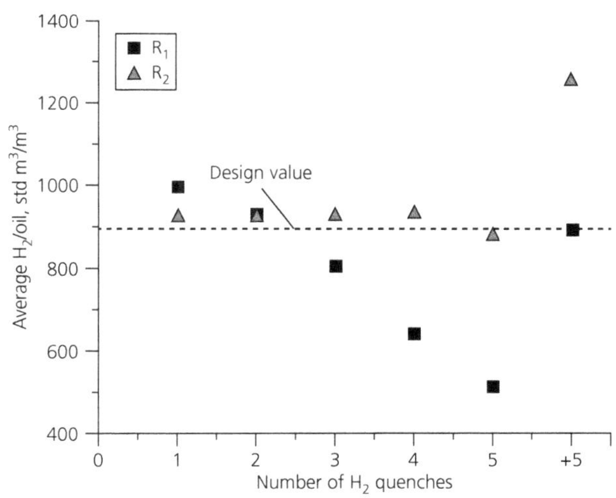

Figure 13.35 Average H_2/oil ratio as function of the number of hydrogen quenches.
(Source: Alvarez et al. [72]. Reproduced with permission of Elsevier.)

13.5.6.2 Commercial reactor simulation with time-varying catalyst activity

The performance model was also applied to analyze the behavior of the industrial-scale process during time-on-stream. In commercial practice, the gradual loss in catalytic activity is offset by increasing the reaction temperature. Shutdown occurs upon the attainment at any position in the reactor of the maximum allowable temperature (MAT), which is established by the metallurgical limits of the reactor vessel. This implies that a wrong reactor design can lead to premature shutdown due to the early appearance of high-temperature zones along the catalyst beds. This factor must be taken into account in reactor design in order to maximize cycle length. The main target of this exercise was to verify if the proposed reactor configuration extends the time to reach the permissible temperature limit (420°C) at any point of the reactor length or, at least, if it equals the cycle length obtained during the aging test (5.2 months).

Figure 13.36 illustrates the time-evolving temperature profiles along the total reactor length. The main feature of the simulation is that the temperature profile moves upward in time because the feed temperature is periodically increased to keep the API gravity of the product constant. The reactor design composed of multiple catalyst beds of increasing lengths allows a relatively uniform temperature distribution along the reactors. In other words, the temperature levels in all catalyst beds are very similar, and there are no excess temperature zones. With such an arrangement, it is possible to operate 4500 h (~6.3 months) without reaching MAT at any point along the reactor length.

It is observed also from Figure 13.36 that the reactors are started up (1 h) at low feed temperature (~350°C). A common safety measure for plant start-up states that the reactor must be started at the lowest possible temperature in order to keep the reactor thermally stable [77]. This rule is based on the fact that the fresh catalyst is more prone to hot spot formation.

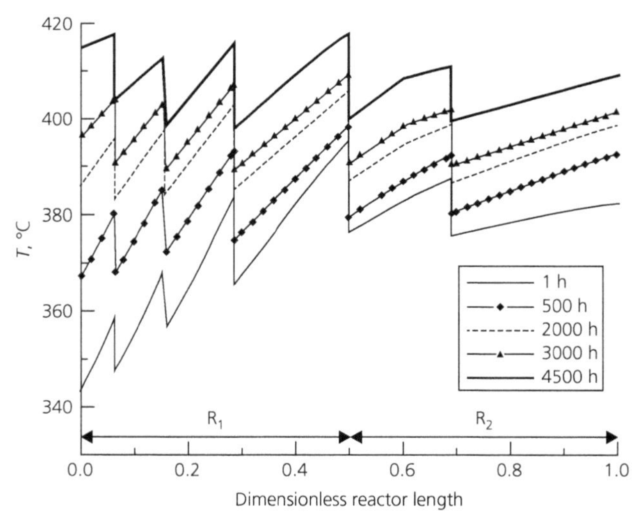

Figure 13.36 Evolution of the axial temperature profiles. (Source: Alvarez et al. [73]. Reproduced with permission of Elsevier.)

In this context, a poor reactor design would lead to the presence of large axial temperature gradients. An example of a wrong design would be the case of a reduced number of catalyst beds in R_1 (2 or 3) of equal length. The resulting bed delta-Ts would produce high-temperature zones at the bed outlets of the reactor. The run would have to be terminated earlier because the bed outlet temperatures would be above the permissible limit too soon. In that case, a substantial amount of catalyst would be underused. The presence of high-temperature zones near the end of R_1 would also shift metals deposition toward the mid- and tail-end catalysts, which are intolerant to metals. This could also induce premature thermal cracking leading to excessive coking and fouling problems.

The conclusion of this analysis is that an appropriate reactor configuration is essential for maximizing cycle length and making better use of the catalyst inventory. For highly exothermic processes, the key is to limit the total heat release to smaller portions (i.e., small delta-T per bed) in order achieve a more even temperature distribution along the reactor. To meet this goal, it is preferable to use multiple beds of increasing length rather than equal bed lengths because the heat release decreases with bed depth. Quench rate must also be carefully selected to obtain the desired temperature profile.

Nomenclature

A	parameter of Equation 13.17
a_L	gas–liquid interfacial area (cm^{-1})
AR	atmospheric residue
A_S	cross-sectional area of the reactor (cm^2)
a_S	liquid–solid interfacial area (cm^{-1})
Asph	asphaltene
B	parameter of Equation 13.17
C_i^G	molar concentration of compound i in gas phase (mol/cm^3)
C_i^L	molar concentration of compound i in liquid phase (mol/cm^3)
C_i^S	molar concentration of compound i in solid phase (mol/cm^3)
Cp_G	gas heat capacity (J/g/K)
Cp_L	liquid heat capacity (J/g/K)
G	gas mass rate (g/s)
G_L	superficial liquid mass velocity (kg/m^2/s)
H_2/oil	hydrogen-to-oil ratio (std m^3/m^3)
HCR	hydrocracking
HDAs	hydrodeasphaltenization
HDM	hydrodemetallization
HDN	hydrodenitrogenation
HDNi	hydrodenickelation
HDS	hydrodesulfurization
HDV	hydrodevanadization
H_i	Henry's law constant for compound i (MPa cm^3/mol)
k_j^{app}	apparent rate coefficient of reaction j

k_j^{in}	intrinsic rate coefficient of reaction j
k_j^p	particle rate coefficient of reaction j
k_i^L	gas–liquid mass-transfer coefficient for compound i (cm/s)
k_i^S	liquid–solid mass-transfer coefficient for compound i (cm/s)
K_{H_2S}	adsorption equilibrium constant for H_2S (cm^3/mol)
l	liquid mass rate (g/s)
MAT	maximum allowable temperature
MOC	metals-on-catalyst
N	nitrogen
Ni	nickel
N_i^G	molar flow of gaseous compound i (mol/s)
n_j	order of reaction j
P	total pressure (MPa)
Q	quench
q	quench fluid mass rate (g/s)
r_j	rate of reaction j (mol cm^3/s)
R_1	first reactor
R_2	second reactor
S	sulfur
t	time
T	temperature
u_G	superficial gas velocity (cm/s)
u_L	superficial liquid velocity (cm/s)
V	vanadium
VR	vacuum residue
x_{MOC}	fractional concentration of metals-on-catalyst
z	axial reactor length

Greek letters

α_j	parameter of Equation 13.18
β_j	parameter of Equation 13.18
γ_j	parameter of Equation 13.18
ΔH_R	overall heat of reaction (kJ/(kg sulfur))
η_0	effectiveness factor
η_{CE}	external catalyst wetting efficiency
ρ_G	gas density at process conditions (g/cm^3)
ρ_L	liquid density at process conditions (g/cm^3)
ϕ_j	deactivation function of reaction j

Subscripts

0	reactor inlet
in	inlet to the following catalytic bed
out	outlet of the previous catalytic bed
Q	quench stream

Non-SI units

API	American Petroleum Institute gravity
Bpcd	barrels per calendar day
Bpd	barrels per day
Mbpd	thousand barrels per day

References

1 Rana MS, Sámano V, Ancheyta J, Diaz JAI. A review of recent advances on process technologies for upgrading heavy oils and residua. Fuel 2007;8:1216–1231.

2 Furimsky E. Selection of catalysts and reactors for hydroprocessing. Appl. Catal. A Gen. 1998;171:177–206.

3 Sie ST. Consequences of catalyst deactivation for process design and operation. Appl. Catal. A Gen. 2001;212:129–151.

4 Mederos FS, Elizalde I, Ancheyta J. Steady-state and dynamic reactor models for hydrotreatment of fractions: A review. Catal. Rev. Sci. Eng. 2009;51:485–607.

5 Ancheyta J, Rana MS, Furimsky E. Hydroprocessing of heavy petroleum feeds: Tutorial. Catal. Today 2005;109:3–15.

6 Ancheyta J. Reactors for hydroprocessing. In: Ancheyta J, Speight JG, editors. Hydroprocessing of heavy oils and residua. 1st ed. Boca Raton, FL: CRC Press/Taylor & Francis; 2007. p 71–120.

7 Satterfield CN. Trickle bed reactors. AIChE J. 1975;21:209–228.

8 Marafi A, Fukase S, Al-Marri M, Stanislaus A. A comparative study of the effect of catalyst type on hydrotreating kinetics of Kuwaiti atmospheric residue. Energy Fuels 2003;17:661–668.

9 Martínez J, Sánchez JL, Ancheyta J, Ruiz RS. A review of process aspects and modeling of ebullated bed reactors for hydrocracking of heavy oils. Catal. Rev. Sci. Eng. 2010;52:60–105.

10 Oil and Gas Journal Surveys. Worldwide refinery-capacities as of January 1, 2011. Oil and Gas Journal. http://www.ogj.com/ogj-survey-downloads.html. Accessed December 6, 2010.

11 Sau M, Basak K, Manna U, Santra M, Verma RP. Effect of organic nitrogen compounds on hydrotreating and hydrocracking reactions. Catal. Today 2005;109:112–119.

12 Dahl IM, Tangstad E, Mostad HB, Andersen K. Effect of hydrotreating on catalytic cracking of a VGO. Energy Fuels 1996;10:85–90.

13 O'Connor P, Brevoord E, Pouwels AC, Wijngaards NJ. Catalyst deactivation in fluid catalytic cracking: A review of mechanisms and testing methods. In: O'Connor P, Takatsuka T, Woolery GL, editors. Deactivation and testing of hydrocarbon-processing catalysts. ACS Symposium Series. Vol. 634, chap 10. Washington, DC: American Chemical Society (ACS) Publications; 1996. p 147–158.

14 Scheuerman GL, Johnson DR, Reynolds BE, Bachtel RW, Threlkel RS. Advances in Chevron RDS technology for heavy oil upgrading flexibility. Fuel Process. Technol. 1993;35:39–54.

15 Kressmann S, Morel F, Harlé V, Kasztelan S. Recent developments in fixed-bed catalytic residue upgrading. Catal. Today 1998;43: 203–215.

16 Ancheyta J, Betancourt G, Marroquín G, Centeno G, Muñoz JAD, Alonso F, inventors. Instituto Mexicano del Petroleo, assignee. Process for the catalytic hydrotreatment of heavy hydrocarbons of petroleum. US patent 7,651,604 B2. January 26, 2010.

17 Speight JG. New approaches to hydroprocessing. Catal. Today 2004;98:55–60.

18 International Energy Agency. 2010. Key world energy statistics. Available at: http://www.iea.org/publications/freepublications. Accessed August 25, 2011.

19 Ancheyta J, Speight J. Heavy oils and residua. In: Ancheyta J, Speight JG, editors. Hydroprocessing of heavy oils and residua. 1st ed. Boca Raton, FL: CRC Press/Taylor & Francis; 2007. p 1–14.

20 Enerdata. 2011. Global energy statistical yearbook. Available at: http://www.enerdata.net/enerdatauk/press-and-publication/publications-analysis.php. Accessed August 27, 2011.

21 Pemex. 2011. Pemex statistical annuary. Available at: http://www.pemex.com/acerca/informes_publicaciones/Paginas/default.aspx. Accessed August 27, 2011.

22 Ancheyta J, Speight J. Hydroprocessing chemistry. In: Ancheyta J, Speight JG, editors. Hydroprocessing of heavy oils and residua. 1st ed. Boca Raton, FL: CRC Press/Taylor & Francis; 2007. p 35–50.

23 Speight JG. The desulfurization of heavy oils and residua. 2nd ed. New York: Marcel Dekker; 2000.

24 Girgis MJ, Gates BC. Reactivities, reaction networks, and kinetics in high-pressure catalytic hydroprocessing. Ind. Eng. Chem. Res. 1991;30:2021–2058.

25 Mochida I, Choi K-H. Current progress in catalysts and catalysis for hydrotreating. In: Hsu CS, Robinson PR, editors. Practical advances in petroleum processing. Vol. 1. New York: Springer; 2006. p 257–296.

26 Vrinat ML. The kinetics of the hydrodesulfurization process: A review. Appl. Catal. 1983;6:137–158.

27 Furimsky E, Massoth FE. Hydrodenitrogenation of petroleum. Catal. Rev. Sci. Eng. 2005;47:297–489.

28 Gioia F, Lee V. Effect of hydrogen partial pressure on catalytic hydrodenitrogenation of quinoline. Ind. Eng. Chem. Process. Des. Dev. 1986;25:918–925.

29 Dolbear GE. Hydrocracking: Reactions, catalysts, and processes. In: Speight JG, editor. Petroleum chemistry and refining. Washington, DC: Taylor & Francis; 1998. chap 7. p 175–197.

30 Ho TC. Hydrodenitrogenation catalysis. Catal. Rev. Sci. Eng. 1988;30:117–160.

31 Furimsky E. Chemistry of catalytic hydrodeoxygenation. Catal. Rev. Sci. Eng. 1983;25:421–458.

32 Janssens JP, Elst G, Schrikkema EG, van Langeveld AD, Sie ST, Moulijn JA. Development of a mechanistic picture of the hydrodemetallization reaction of metallo-tetraphenylporphyrin on a molecular level. Recl. Trav. Chim. Pays-Bas 1996;115:465–473.

33 Beuther H, Schmid BK. Reaction mechanisms and rates in residue hydrodesulfurization. Sixth World Petroleum Congress; June 1963; Frankfurt. Section 3:297–310.

34 Kobayashi S, Kushiyama S, Aizawa R, Koinuma Y, Inoue K, Shimizu Y, Egi K. Kinetic study on the hydrotreating of heavy oil. 2. Effect of catalyst pore size. Ind. Eng. Chem. Res. 1987;26:2245–2250.

35 Ali SA. Thermodynamics of hydroprocessing reactions. In: Ancheyta J, Speight JG, editors. Hydroprocessing of heavy oils and residua. 1st ed. Boca Raton, FL: CRC Press/Taylor & Francis; 2007. p 51–70.

36 Robinson PR, Dolbear GE. Hydrotreating and hydrocracking: Fundamentals. In: Hsu CS, Robinson PR, editors. Practical advances in petroleum processing. Vol. 1. New York: Springer; 2006. p 177–218.

37 Ancheyta J, Centeno G, Trejo F, Marroquín G. Changes in asphaltene properties during hydrotreating of heavy crudes. Energy Fuels 2003;17:1233–1238.

38 Mullins OC. Sulfur and nitrogen structures in asphaltenes and related materials quantified by Xanes spectroscopy. In: Sheu EY, Mullins OC, editors. Asphaltenes: Fundamentals and applications. New York: Plenum Press; 1995. p 53–96.

39 Gauthier T, Danial-Fortain P, Merdrignac I, Guibard I, Quoineaud A-A. Studies on the evolution of asphaltene structure during hydroconversion of petroleum residues. Catal. Today 2008; 130:429–438.

40 Marafi A, Kam E, Stanislaus A. A kinetic study on non-catalytic reactions in hydroprocessing Boscan crude oil. Fuel 2008;87:2131–2140.

41 Klein MT, Hou G, Bertolacini R, Broadbelt LJ, Kumar A. Molecular modeling in heavy hydrocarbon conversions. Boca Raton, FL: CRC Press/Taylor & Francis; 2006.

42 Ho TC. Kinetic modeling of large-scale reaction systems. Catal. Rev. Sci. Eng. 2008;50:287–378.

43 Verstraete JJ, Le Lannic K, Guibard I. Modeling fixed-bed residue hydrotreating processes. Chem. Eng. Sci. 2007;62:5402–5408.

44 Froment GF. Modeling in the development of hydrotreatment processes. Catal. Today 2004;98:43–54.

45 Ghosh P, Andrews AT, Quann RJ, Halbert TR. Detailed kinetic model for the hydro-desulfurization of FCC naphtha. Energy Fuels 2009;23:5743–5759.

46 López-García C, Hudebine D, Schweitzer J-M, Verstraete JJ, Ferré D. In-depth modeling of gasoil hydrotreating: From feedstock reconstruction to reactor stability analysis. Catal. Today 2010;150:279–299.

47 Baltanas MA, Vanraemdonck KK, Froment GF, Mohedas SR. Fundamental kinetic modeling of hydroisomerization and hydrocracking on noble-metal-loaded faujasites. 1. Rate parameters for hydroisomerization. Ind. Eng. Chem. Res. 1989;28:899–910.

48 Kumar H, Froment G. Mechanistic kinetic modeling of the hydrocracking of complex feedstocks, such as vacuum gas oils. Ind. Eng. Chem. Res. 2007;46:5881–5897.

49 Ancheyta J. Modeling and simulation of catalytic reactors for petroleum refining. Hoboken, NJ: John Wiley & Sons, Inc.; 2011.

50 Ho TC, Aris R. On apparent second-order kinetics. AIChE J. 1987;33:1050–1051.

51 Alvarez A, Ancheyta J. Modeling residue hydroprocessing in a multi-fixed-bed reactor system. Appl. Catal. A Gen. 2008;351:148–158.

52 Gray MR, Ayasse AR, Chan EW, Veljkovic M. Kinetics of hydrodesulfurization of thiophenic and sulfide sulfur in Athabasca bitumen. Energy Fuels 1995;9:500–506.

53 Bej SK. Performance evaluation of hydroprocessing catalysts—A review of experimental techniques. Energy Fuels 2002;16:774–784.

54 Jaffe SB. Kinetics of heat release in petroleum hydrogenation. Ind. Eng. Chem. Process. Des. Dev. 1974;13:34–39.

55 Döhler W, Rupp M. Comparison of performance of an industrial VGO-treater with reactor model predictions. Chem. Eng. Technol. 1987;10:349–352.

56 Mohanty S, Saraf DN, Kunzru D. Modeling of a hydrocracking reactor. Fuel Process. Technol. 1991;29:1–17.

57 Shah YT, Paraskos JA. Criteria for axial dispersion effects in adiabatic trickle bed hydroprocessing reactors. Chem. Eng. Sci. 1975;30:1169–1176.

58 Gruia A. Recent advances in hydrocracking. In: Hsu CS, Robinson PR, editors. Practical advances in petroleum processing. Vol. 1. New York: Springer; 2006. p 219–256.

59 Kundu A, Nigam KDP, Duquenne AM, Delmas H. Recent developments in hydroprocessing reactors. Rev. Chem. Eng. 2003; 19:531–605.

60 Robinson PR, Dolbear GE. Commercial hydrotreating and hydrocracking. In: Ancheyta J, Speight JG, editors. Hydroprocessing of heavy oils and residua. 1st ed. Boca Raton, FL: CRC Press/Taylor & Francis; 2007. p 281–312.

61 Hoekstra G. The effects of gas-to-oil rate in ultra low sulfur diesel hydrotreating. Catal. Today 2007;127:99–102.

62 Al-Dahhan MH, Larachi F, Dudukovic MP, Laurent A. High-pressure trickle-bed reactors. Ind. Eng. Chem. Res. 1997;36:3292–3314.

63 Korsten H, Hoffmann U. Three-phase reactor model for hydrotreating in pilot trickle-bed reactors. AIChE J. 1996;42:1350–1360.

64 Chou T. Causes of fouling in hydroprocessing units. Pet. Technol. Q. 2004;Q4:79–85.
65 Alvarez A, Ramírez S, Ancheyta J, Rodríguez LM. Key role of reactor internals in hydroprocessing of oil fractions. Energy Fuels 2007;21:1731–1740.
66 Alvarez A, Ancheyta J, Muñoz JAD. Comparison of quench systems in commercial fixed-bed hydroprocessing reactors. Energy Fuels 2007;21:1133–1144.
67 Ouwerkerk CED, Bratland ES, Hagan AP, Kikkert BLJP, Zonnevylle MC. Performance optimization of fixed bed processes. Pet. Technol. Q. 1999;2:21–30.
68 Gosselink JW. Sulfide catalysts in refineries. Cattech 1998;2: 127–144.
69 Robinson PR. Petroleum processing overview. In: Hsu CS, Robinson PR, editors. Practical advances in petroleum processing. Vol. 1. New York: Springer; 2006. p 1–78.
70 Hallie H. Experience reveals best presulfiding techniques for HDS and HDN catalysts. Oil Gas J. 1982;80:69–74.
71 Marroquín G, Ancheyta J, Díaz JAI. On the effect of reaction conditions on liquid phase sulfiding of a NiMo HDS catalyst. Catal. Today 2004;98:75–81.
72 Alvarez A, Ancheyta J, Muñoz JAD. Modeling, simulation and analysis of heavy oil hydroprocessing in fixed-bed reactors employing liquid quench streams. Appl. Catal. A Gen. 2009;361:1–12.
73 Alvarez A, Ancheyta J, Centeno G, Marroquín G. A modeling study on the effect of reactor configuration on the cycle length of heavy oil fixed-bed hydroprocessing. Fuel 2011;90:3551–3560.
74 Bondi A. Handling kinetics from trickle-phase reactors. Chem. Technol. 1971;1:185–188.
75 Furimsky E, Massoth FE. Deactivation of hydroprocessing catalysts. Catal. Today 1999;52:381–495.
76 Sie ST. Scale effects in laboratory and pilot-plant reactors for trickle-flow processes. Rev. Inst. Fr. Petrol. 1999;46:501–515.
77 Yan TY. Dynamics of a trickle-bed hydrocracker with a quenching system. Can. J. Chem. Eng. 1980;58:259–266.

CHAPTER 14
Catalytic reactors for fuel processing

Gunther Kolb
Decentralized and Mobile Energy Technology Department, Fraunhofer ICT-IMM, Mainz, Germany

Abstract

Fuel processing is the conversion of fossil and regenerative fuels to hydrogen-containing gas mixtures. The chemical conversion is performed in most cases in the gas phase, normally heterogeneously catalyzed in the presence of a solid catalyst. The first step of the conversion procedure is named reforming. It has been established in large-scale industrial processes for many decades. The industrial applications most commonly use natural gas as feedstock. The product of the natural gas reforming process is synthesis gas, a mixture of carbon monoxide and hydrogen, which is then used for numerous processes in large-scale chemical production which are not subject of this section but rather the technology, which provides a hydrogen-containing gas mixture, named reformate, which is a feed suitable for a fuel cell. The fuel cell then converts hydrogen to electrical energy. Depending on the fuel cell type, removal of carbon monoxide might be required, which is achieved by water-gas shift and preferential oxidation reactions performed downstream the reformer. The reactor types suitable for mobile and decentralized applications are, apart from conventional fixed-bed reactors, ceramic and metallic monoliths, (microchannel) plate heat exchangers, and membrane reactors which allow the combination of membrane separation with the reactions mentioned previously.

14.1 Introduction—The basic reactions of fuel processing

Steam reforming is the gas phase conversion of energy carriers such as hydrocarbons and alcohols described by the general formula $C_xH_yO_z$ with steam to a mixture of carbon monoxide and hydrogen according to the following reaction:

$$C_xH_yO_z + (x-z)H_2O \rightarrow xCO + \left(x-z+\frac{y}{2}\right)H_2 \qquad (14.1)$$

The product mixture of the reaction is named reformate. The reaction is endothermic and thus requires heat supply. Besides hydrogen and carbon monoxide, the reformate usually contains significant amounts of unconverted steam, to a lower extent some unconverted fuel and carbon dioxide, the latter being formed by the consecutive water–gas shift (WGS) reaction:

$$CO + H_2O \rightarrow CO_2 + H_2, \quad \Delta H^0_{298} = -40.4\ \text{kJ/mol} \qquad (14.2)$$

The WGS reaction increases the hydrogen concentration of the reformate. This reaction is usually fast enough at the elevated temperatures of hydrocarbon reforming to achieve thermodynamic equilibrium. Owing to its exothermic character, higher reaction temperatures favor the reverse reaction.

Partial oxidation is the conversion of fuels under oxygen-deficient atmosphere:

$$C_xH_yO_z + \frac{(x-z)}{2}(O_2 + 3.76N_2) \rightarrow xCO + \frac{y}{2}H_2 + 3.76\frac{(x-z)}{2}N_2 \qquad (14.3)$$

Two reaction mechanisms do exist in literature for partial oxidation. One of them proposes that the reaction starts with catalytic combustion followed by reactions of lower rate, namely, steam reforming, CO_2 reforming, and WGS. This mechanism is supported by the fact that water is found as primary product of partial oxidation and autothermal reforming in many cases. The other mechanism proposes direct partial oxidation at very short residence times. The reaction is significantly faster than steam reforming and usually performed in the diffusion-limited regime.

Oxidative steam reforming is the general term for the operation of a steam reformer to which a certain amount of air is fed additionally:

$$\begin{aligned} &C_xH_yO_z + n(O_2 + 3.76N_2) + (x-2n-z)H_2O \rightarrow \\ &xCO + \left(x-2n-z+\frac{y}{2}\right)H_2 + 3.76nN_2 \end{aligned} \qquad (14.4)$$

Air addition may be performed in very limited amounts in order to prevent coke formation at the catalyst similar to the "air bleed" which is added to the anode feed of fuel cells running on reformate. In case the amount of oxygen is increased to a degree, where the energy generation by partial oxidation

Multiphase Catalytic Reactors: Theory, Design, Manufacturing, and Applications, First Edition. Edited by Zeynep Ilsen Önsan and Ahmet Kerim Avci.

reaction balances the energy consumption of steam reforming, the overall reaction is theoretically self-sustaining or autothermal. However, this is not the case in a practical system because heat losses need to be compensated. Usually an optimum atomic oxygen/carbon (*O*/*C*) ratio exists for each fuel under thermally neutral conditions to achieve optimum efficiency.

The removal of small amounts of carbon monoxide present in reformate of hydrocarbon and ethanol fuel processors downstream the WGS reactor is commonly performed by the preferential oxidation with air (PrOx):

$$CO + 0.5O_2 \rightarrow CO_2, \quad \Delta H^0_{298} = -283 \text{ kJ/mol} \tag{14.5}$$

The reaction is accompanied by the undesired side reaction of part of the hydrogen present in the reaction mixture:

$$H_2 + 0.5O_2 \rightarrow H_2O, \quad \Delta H^0_{298} = -242 \text{ kJ/mol} \tag{14.6}$$

Usually preferential oxidation requires a minimum of excess air corresponding to an *O*/*CO* ratio or λ-value between 1.5 and 2.0. Under these conditions and full conversion of carbon monoxide, between 0.5 and 1.0 mol of hydrogen is lost for each mole of carbon monoxide converted.

14.2 Theoretical aspects, advantages, and drawbacks of fixed beds versus monoliths, microreactors, and membrane reactors

Conventional fixed catalyst beds, while simple in design and cheap in fabrication, suffer in many applications such as steam reforming from low utilization of the catalyst pellet, channeling of the gas flow through the catalyst bed, and catalyst losses by attrition especially in mobile systems. These problems, along with poor heat transfer in the catalyst bed, make the scale-up of fixed beds a demanding issue. On the other hand, structured reactors such as monoliths and heat exchangers, which are coated with catalyst, are limited by their capacity for taking up the catalyst. Consequently, more active and stable catalysts are required for structured reactors [1], also because the catalyst removal is usually not feasible in such reactors. This may lead to a switch of catalyst technology from low-cost materials to precious metal formulations in many cases. The higher price of these materials needs to be counterbalanced by the lower catalyst mass required.

The most prominent applications of ceramic and metallic monoliths are in the field of automotive exhaust gas treatment, which is not the topic of the current text but rather their application in the field of fuel processing for distributed hydrogen production. Different to ceramic and metallic monoliths developed for automotive exhaust treatment purposes, which nowadays carry channels in the microscale and are actually "microreactors" by definition, microreactors according to the definition chosen here are plate heat exchangers carrying channels in the microscale.

A common feature of monolithic reactors and microreactors are therefore the gas flow in small channels which creates different reactor properties when compared to conventional fixed-bed technology. The flow regime is laminar, the path length for heat and mass transfer is very small, and they have a high surface-to-volume ratio, which makes surface effects dominate over volumetric effects. The share of wall material is higher compared to fixed beds, and consequently the heat transfer in the wall material contributes significantly to the overall heat transfer.

The pressure drop over a channel is calculated as follows:

$$\frac{dp}{dz} = -f\frac{2\rho U^2}{D_h} \tag{14.7}$$

A correlation for the friction factor *f* has been derived by Shah et al. [2]:

$$f = \frac{24}{Re} - \left(1 - 1.3553\alpha + 1.9467\alpha^2 - 1.7012\alpha^3 + 0.9564\alpha^4 - 0.2537\alpha^5\right) \tag{14.8}$$

Monoliths and microreactors carry a multitude, normally thousands, of channels of identical dimensions which are operated in parallel. The high number of channels and their dimensions in the range from 400 to 1000 μm create a pressure drop which is in the order of a few mbar to a few tens of mbar.

Owing to the high degree of parallelization, the scale-up of the microreactor is relatively simple as long as the equipartition of the fluid flow through the channels is achieved.

In practical microchannel systems the *Nu* number is

$$Nu = \frac{kD_h}{\lambda} \tag{14.9}$$

This varies in many cases in the range between 3 and 7 for straight channels involving gaseous medium at low pressure drop. Despite these relatively low numbers, the small diffusion paths allow for fast heat transfer. For liquids much higher *Nu* numbers are achievable owing to the higher heat conductivity of the medium.

It is obvious that straight microchannels are the simplest feasible design for microreactors and that other geometries such as sinusoidal channels or fin geometries do exist, which have the potential for increasing the heat transfer considerably. Hardt et al. [3] demonstrated that *Nu* numbers exceeding 100 can be achieved for air as medium in micro fin arrangements but usually at a much larger penalty of increased pressure drop. Keeping the system pressure drop low is a critical issue for practical fuel processing systems because compression of gases requires energy. The parasitic power losses originating from compressors are known to reduce the efficiency of fuel cell systems considerably [4].

Membrane reactors (MRs) for fuel processing combine the unit operation of membrane separation with catalytic reactions such as reforming and WGS. The membrane separation process is usually performed by hydrogen removal from the reformate by application of membranes made of ceramics or palladium and palladium alloys, while polymeric membranes are less convenient for systems of smaller than industrial scale, because several separation steps are required owing to their relatively low selectivity of the separation process. In MRs the equilibrium of the chemical reactions involved is shifted in a favorable direction, because hydrogen product is removed from the reaction system. Therefore product compositions exceeding the equilibrium of the original feed composition can be achieved.

The hydrogen separation by palladium involves several elementary steps, which include the solution of hydrogen and its diffusion as atomic hydrogen through the membrane bulk material. The introduction of silver into the palladium membrane increases its lifetime but also its cost compared to copper [5]. Generally, the membranes need to be protected against rapid temperature and pressure changes, and they are subject to poisoning by carbon monoxide, hydrogen sulfide, and unsaturated hydrocarbons [5, 6]. The hydrogen flux through the membrane J_{H_2} is proportional to both the diffusion coefficient D_{H_2} of hydrogen in palladium and to Sievert's solubility constant K_s of the hydrogen/palladium system:

$$J_{H_2} \propto D_{H_2} K_S \tag{14.10}$$

$$N_{H_2} = \frac{QA((p_{H_2,1})^n - (p_{H_2,2})^n)}{l} \tag{14.11}$$

The exponent n for the partial pressure is 0.5 provided that the bulk phase diffusion of atomic hydrogen is the only rate-limiting step. In this case Equation 14.10 is named Sievert's law. The hydrogen separation factor s is sometimes used to specify membrane quality. It is defined as

$$s = \frac{\dot{n}_1 \Delta p_1}{\dot{n}_2 \Delta p_2} \tag{14.12}$$

Membrane separation of reformate is usually operated at elevated pressures as the driving force for the permeation process. Thus steam reforming of liquid fuels is the preferred process, because only liquid pumps are required instead of large compressors for air and/or feed pressurization, which create unacceptable parasitic losses. Another measure to increase the hydrogen partial pressure difference between permeate and retentate is to use sweep gas on the retentate side. Because the hydrogen requires humidification for low-temperature PEM fuel cells to prevent membrane dry out, steam is the preferred sweep gas [7]. Yu et al. [8] demonstrated by numerical calculations that the hydrogen production is also higher in the case of steam instead of nitrogen applied as sweep gas.

In order to further minimize the operating pressure required especially for systems of smaller scale, the membrane thickness is reduced to a few micrometers in certain applications. Ceramic supports ensure the mechanical stability of these thin membranes. At such low membrane thicknesses leakage of species other than hydrogen occurs through membrane imperfections ("pinholes").

14.3 Reactor design and fabrication

14.3.1 Fixed-bed reactors

Fixed-bed reactors for distributed fuel processing applications are normally comprised of an insulated stainless steel vessel or tube containing a mesh at the reactor inlet and outlet, respectively, to maintain the catalyst within the reactor. When heat has to be removed from or introduced into the catalyst bed, small tubes are usually introduced into the vessel, which serve as heating or cooling sources via hot or cold fluid flows. The reactor design is then equivalent to that of a shell and tube heat exchanger. However, monolithic reactors and plate heat exchangers are more suitable than fixed beds for the rapid start-up and transient operation requirements of fuel processors of the smaller scale [1].

14.3.2 Monolithic reactors

The fabrication of ceramic monoliths is performed mainly via extrusion techniques [9] resulting in usually elliptic or square-shaped monoliths. Cell densities as high as 1600 cells per square inch (cpsi) are achieved [9], which corresponds to a channel width of about 500 μm of the mostly rectangular or hexagonal channels. Typical key features are 75% porosity and a geometric surface area of 2.8 m^2/l for 400 cpsi for ceramic monoliths [10]. The thermal shock resistance amounts to 800°C and higher.

Owing to the low surface area of the monoliths, a catalyst carrier such as alumina or ceria is deposited onto them usually by wash coating [11]. The ceramic construction material is well compatible with these carriers, does not migrate into the catalyst coatings, and the precious active metal species of the catalysts do not migrate into the monolith bodies [1]. The most widely used construction material for ceramic monolith carriers is cordierite [9] with alumina being an alternative.

Metallic monoliths have numerous advantages over ceramic honeycombs, such as higher mechanical stability, lower wall thickness, and higher heat conductivity of the wall material, while the maximum cell densities are equivalent to those of the ceramic monoliths [9]. The geometric surface area and porosity (exceeding 90%) of metallic monoliths are higher, with the latter resulting in a lower pressure drop at similar cell densities [1]. Electrically heated metallic monoliths have been developed to reduce the start-up time demand. The production of metallic monoliths is performed by rolling up of a pair of flat and corrugated metal foil. The construction material of metallic monoliths are alloys of iron, about 15–20 wt.% chromium, and 5 wt.% aluminum (FeCr alloy), which form a thin alumina layer (0.5 μm) on their surface upon temperature treatment exceeding 850°C [9]. The native alumina layer serves then as an adhesion layer for the catalyst coating and protects the material from corrosion.

The catalyst coatings in ceramic and metallic monoliths are not always evenly distributed, which affects the performance of the monoliths concerning flow equipartition and temperature profile, because the heat conduction is also affected [12].

14.3.3 Microreactors

Microstructured plate heat exchangers are stacked arrangements with a multitude of parallel minichannels and high surface-to-volume ratios in the range of 200 m^2/m^3. The preferred construction material is stainless steel. Wet chemical etching, initially developed for silicon micromachining, is suited for mass production of microchannels. For many metals etching is a cheap and well-established technology, and covers a wide range of channel depths from about less than 100 μm up to 1000 μm and more, which is the channel size usually applied in microstructured reactors for fuel processing applications. However, for applications in the kilowatt range, large microstructured foils are required and costs get to be a critical issue when etching is applied. Embossing is an inexpensive technique for manufacturing metal foils highly suitable for mass production [13, 14]. Even microstructures down to a few micrometer structure size can be achieved by embossing [15, 16].

Sealing by gaskets increases the thermal mass of the devices considerably, which increases start-up time demand. Irreversible sealing techniques make typically use of elevated temperatures for which compatibility to the plate material and its coatings has to be considered. For chemical reactors, the main issue is the thermal stability of the catalyst coating, if being filled in or attached before the sealing, which is the current mainly applied method. Catalyst deactivation may arise from about 300°C to more than 800°C depending on the catalyst formulation which can pose a serious limit to the applicability of the sealing method. When the technique is not compatible, the catalyst or the catalyst coating has to be inserted into the device after the sealing procedure. Laser welding is a viable technique for the sealing of plate heat exchanger/reactors. The cost of the welding procedure is mainly dictated by the power of the laser applied. The spatially limited energy input protects incorporated catalyst from damage. On the contrary, diffusion bonding requires high temperatures with no spatial restriction and also high vacuum. The material is compressed and heated to temperatures close to the melting point, which generates a quasi boundary-free single workpiece. Brazing techniques are widely applied for the sealing of compact plate heat exchangers. However, this is not applicable to devices coated with catalysts, because the brazing lots often contain heavy metals such as cadmium and tin which are poisonous to catalysts [1]. Thus, the brazing step should be followed by a thorough cleaning step before applying catalyst coating. If brazing is done with catalyst-coated plates, the same temperature considerations, as given previously with regard to the catalyst coatings, hold for the melting temperature of brazing lots.

Correct positioning of the microstructured plates to form stacks or other arrangements for the assembly of the reactors is crucial, since a small distortion while bonding may lead to severe deviations from the ideal microchannel shape. Alignment techniques can be based on simple mechanical methods (e.g., use of alignment pins), edge catches in a specially designed assembling device, or optical methods.

In order to coat metallic surfaces with catalyst, a pretreatment to improve the adherence is required [11]. Besides mechanical roughening, chemical and thermal pretreatment are applied frequently. Once the surface is pretreated, the coating slurry needs to be prepared. The most prominent method is to prepare a dispersion of finished catalyst including gelation steps if necessary. The catalyst carrier or the catalyst itself is mixed with binder such as polyvinyl alcohol or methylhydroxyethyl cellulose [17], acid, and solvent, usually water. A lower particle size improves adhesion [9, 18]. It has been demonstrated that the slurry viscosity determines the thickness of the coating. The viscosity itself is determined by the concentration of particles, pH value, and on surfactant addition [19].

The amount of catalyst material, which can be coated onto a microstructured plate exceeds 40 g/m^2. A screen printing method has been developed by the group of the author of this chapter to introduce the catalyst suspension into the microchannels. Alternative but less commonly applied techniques are (i) spray coating [11], which requires a reduction of the viscosity of the slurry or sol, (ii) flame spray deposition [20], and (iii) electrophoretic deposition [21]. After the deposition usually drying and calcinations steps follow, the latter being a heat treatment in air or in other gases for a defined duration. Normally the dried samples are not immediately put into a hot furnace but rather heated up gradually. The final temperature of calcination needs to ensure that organic materials such as binders are completely removed.

14.3.4 Membrane reactors

MRs combine chemical conversion with a membrane separation step. Within the scope of fuel processing, MRs are usually either combined with the reforming step or with WGS. Thin metallic membranes may be produced by techniques such as cold rolling [22]. However, the fabrication of thin palladium membranes onto ceramic surfaces is more complicated and requires methods such as spray pyrolysis chemical vapor deposition and sputtering, but most frequently electroless plating is applied [23]. Palladium particles are produced from palladium solution containing amine complexes of palladium in the presence of reducing agents. Ceramic surfaces such as α-alumina are first sensitized in acidic tin chloride, and then palladium is seeded from acidic palladium ammonia chloride [23].

14.4 Reformers

The light-off temperature, which is required for ignition of fuels over a catalyst, affects the start-up strategy and time demand of autothermal reformers (ATRs) or partial oxidation reactors. The autoignition temperature frequently named light-off temperature under conditions of homogeneous partial oxidation decreases with increasing carbon number and is higher than the light-off temperature over a catalyst. While homogeneous autoignition occurs at 600°C for methane, only 200°C is required for alkanes heavier than *n*-decane [24]. Surface ignition at O/C ratio of 1.0 occurs over platinum at 430°C in case of methane, while 175°C is required for butane.

The catalyst technology for reforming can be roughly separated into nonnoble metal-based catalysts, the most prominent being nickel and noble metal-based catalyst, where rhodium shows the best performance, but also the average highest price. Hickman and Schmidt [25] were one of the first authors proposing partial oxidation of methane over noble metal catalysts at short contact times in the range of a few milliseconds. Platinum and rhodium coated onto porous alumina foam monoliths were tested. Rhodium showed superior performance, namely, higher activity and hydrogen selectivity compared to platinum, which was confirmed by Schmidt et al. [26]. Over rhodium catalyst, the reaction temperature exceeded 900°C, and an *O*/*C* ratio higher than 1.2 was required to achieve full conversion at 10 ms contact time. Huff et al. [27] investigated the performance of platinum, nickel, iridium, and rhodium catalysts deposited onto porous ceramic foams for partial oxidation of methane. Rhodium and nickel showed the highest activity and hydrogen selectivity. However, nickel catalysts suffered frequently from coke formation on their surface even in the case of steam reforming of alcohol fuels such as ethanol [28].

Silberova et al. [29] investigated partial oxidation and oxidative steam reforming of propane over rhodium catalyst containing only 0.01 wt.% rhodium supported by alumina, which was coated onto ceramic foams. Under conditions of partial oxidation (*O*/*C* ratio of 1.3), full conversion of the feed could be achieved for 9 h duration. Hot spots around 900°C were measured in the foam. Under conditions of autothermal reforming, higher reaction temperatures were required to achieve full conversion. The catalyst suffered from deactivation on the short term, which was attributed to the presence of steam in the feed, causing sintering and migration of the rhodium particles. However, the residence time in the foam was rather high with 100 ms. Apart from methane, the observed side products were ethylene, propylene, and even small amounts of acetylene [30].

Ferrandon et al. [31] investigated the effect of the catalyst support on the performance of rhodium catalysts wash coated onto cordierite monoliths for autothermal reforming of gasoline. The samples contained 2 wt.% rhodium supported by either gadolinium/ceria or lanthanum-stabilized alumina, which showed higher activity and superior selectivity. Only 30 ppm of light hydrocarbons ($C > 1$) were detected. However, this originated from the higher surface area and rhodium dispersion of the latter sample. Both samples showed stable performance for more than 50 h test duration.

Qi et al. [32] tested autothermal reforming of n-octane over a ruthenium catalyst, which was composed of 0.5 wt.% ruthenium stabilized by ceria and potassium on γ-alumina. It showed full conversion of *n*-octane for 800 h. However, the selectivity moved from carbon dioxide and methane toward carbon monoxide and light hydrocarbons, which has to be regarded as an indication of catalyst degradation during long-term tests despite the fact that full conversion was achieved. After 800 h the catalyst consequently showed incomplete conversion. Tests performed on the spent catalyst revealed losses of specific surface area and of 33 wt.% of the noble metal.

Dreyer et al. [33] investigated the oxidative steam reforming of diesel surrogates over wash-coated ceramic monoliths carrying 5 wt.% rhodium supported by γ-alumina. The conversion of *n*-decane dropped with increasing *S*/*C* ratio because the reaction temperature decreased significantly. The beneficial effect of higher steam addition on catalyst activity could not be observed because the feed was only preheated to about 250°C, which is not sufficient to maintain high reaction temperatures during autothermal reforming. However, the product spectrum, which was observed, was very similar to the observations of many other groups. Hydrogen and carbon dioxide selectivity was increased by an increasing *S*/*C* ratio, while carbon monoxide selectivity decreased owing to the WGS shift equilibrium. Ethylene and propylene formation could be completely suppressed at *S*/*C* ratios higher than 2.0 even when the *O*/*C* ratio was as low as 0.66. Hexadecane turned out to be more difficult to be converted by the reforming process than *n*-decane in presence of steam.

Karatzas et al. [34] investigated the performance of rhodium catalyst containing 3 wt.% rhodium, 10 wt.% ceria, 10 wt.% lanthana on alumina carrier for the autothermal reforming of *n*-tetradecane, low sulfur diesel containing 6 ppm sulfur, and Fischer–Tropsch diesel over cordierite monolithic reactors. At temperatures exceeding 740°C, full conversion of the feed was achieved.

A first indication of catalyst deactivation in reforming of heavier hydrocarbons is the increased formation of light hydrocarbons, usually ethylene, which, apart from methane, is formed owing to the methanation activity of the catalysts, usually even up to the concentration which is equivalent to the equilibrium of the methanation reaction. Yoon et al. [35] attributed the preferred ethylene formation to homogeneous decomposition of paraffinic feed rather than aromatic, which was proven by reforming experiments without catalyst for paraffinic and aromatic feed. Figure 14.1a shows the linear increase of light hydrocarbons during initial catalyst deactivation in the ATR of Kang et al. [36] over Pt catalyst supported by Gd-doped Ce. With the course of time, the initially linear increase changes to a rather exponential shape as shown in Figure 14.1b. Pors et al. [37] reported increasing release of unconverted benzene, 1,3-butadiene, and propylene during a 1000 h durability test of their ATR operated with German Aral Ultimate diesel. This fuel contained 86 wt.% alkanes and 13.4 wt.% monoaromatics, while the content of diaromatics amounted to 0.5 wt.%.

In order to break up long-chain hydrocarbons, adiabatic prereforming is performed in the industrial scale, which is, however, rarely adopted in fuel processor prototypes of the smaller scale. Chen et al. [38] performed adiabatic prereforming of light distillate ($C_{6.4}H_{14.8}$) with addition of small amount of hydrogen to the feed in a laboratory-scale fuel processor composed of prereformer, ATR, two-stage WGS, and membrane separation. The prereforming catalyst was composed of 50 wt.% NiO, 6 wt.% MgO, and 10 wt.% La_2O_3 and showed 500 h durability when operated at *S*/*C* of 2.4 and H_2/*C* of 0.1. In order to come up with a solution for the light hydrocarbon release of their platinum diesel ATR catalyst, Yoon et al. [39] proposed postreforming in a second reformer stage over ruthenium catalyst supported by Gd-doped CeO_2 at temperatures exceeding 600°C.

The fuel injection turns out to be a critical issue for autothermal reforming of heavy hydrocarbon fuels such as diesel and kerosene. In particular the contact of air with the fuel at elevated

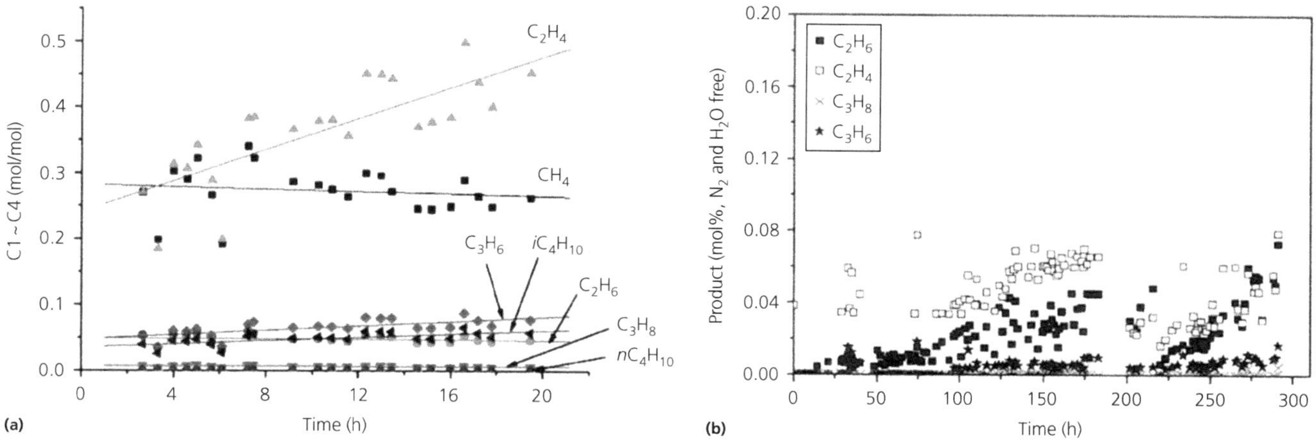

Figure 14.1 Ethylene formation versus time on stream; (a) autothermal reforming of synthetic diesel ($O/C = 1.12$, $S/C = 1.25$ [36]). (Source: Kang et al. [36]. Reproduced with permission of Elsevier.) (b) autothermal reforming of commercial diesel (S/C=3, O/C=1.6 [35]). (Source: Yoon et al. [35]. Reproduced with permission of Elsevier.)

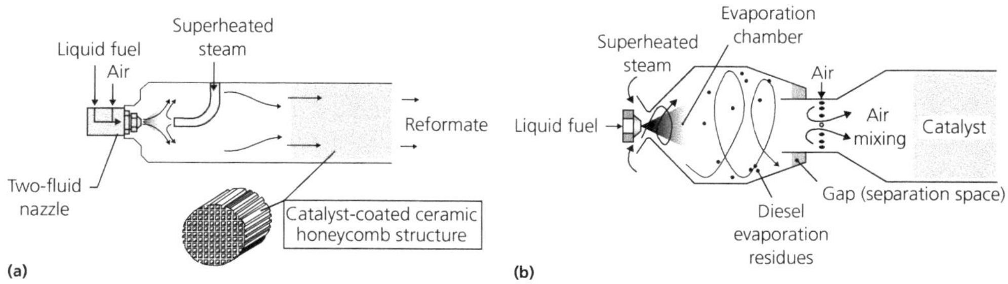

Figure 14.2 Different feed injection systems; (a) ATR-5, (b) ATR-8 [37]. (Source: Pors et al. [37]. Reproduced with permission of John Wiley & Sons, Inc.)

temperatures might lead to coke formation, which then tends to block the feed nozzles of the reformer. Kang et al. [36] reported the successful application of an ultrasonic injector for diesel fuel atomization in their ATR. The diesel fuel was atomized to a droplet size of about 40 µm by ultrasound and then mixed with the air feed, which also prevented the diesel from contacting the wall. Steam was evaporated and mixed to the feed downstream, and the complete feed mixture then entered an inert mixing zone of zirconia balls before it reached the catalyst fixed bed (Pt on Gd-doped CeO_2). The ultrasonic treatment, which consumed only about 20 W electric power, improved the fuel conversion significantly.

Pors et al. [37] presented different, stepwise improved fuel injection systems as part of the diesel ATR development of Jülich research center. While air was mixed with the fuel and steam added downstream in the ATR-5 reactor (Figure 14.2a), superheated steam was mixed with the fuel, which was dispersed in an atomization nozzle in ATR-7 and the air added downstream. Further improvement by adding a cyclone-like separation of nonevaporated residues in ATR-8 (Figure 14.2b) led to an increased durability of the injection system, which was proven experimentally by a 1000 h durability test. Later Pasel et al. [40] reported 2000 h durability of the ATR-8 reactor for Jet-A1 aviation fuel.

O'Connell et al. [41] reported stable evaporation of their fuel injection system for an oxidative steam reformer, which worked similar to ATR-8 by mixing superheated steam and fuel, while the air was added downstream, right in front of the stack of their plate heat exchanger/reactor.

Nilsson et al. [42] reported development work on their injection system, which was based upon atomization of the fuel followed by addition of a preheated mixture of air and steam as shown in Figure 14.3a. This created conditions known as cool flame technology (see in the following paragraph). In this reactor this lead to autoignition and release of unconverted hydrocarbons [42]. An improved design described by Lindstroem et al. [43], which had a much lower residence time of the feed mixture in the inlet mixing chamber as shown in Figure 14.3b, led to much improved performance of the reformer (<15 ppm unconverted diesel in the reformate at temperatures exceeding 730°C). It is also important to preheat the diesel feed before atomization, because the evaporation procedure is then accelerated [44].

The cool flame technology for diesel fuels was adopted by the German Oel und Wärmeinstitut (OWI) for the prereforming of diesel fuels for fuel cell applications. It generates a homogeneous mixture of fuel, air, and steam [45], which avoids the formation of carbonaceous deposits. The cool flame is based upon

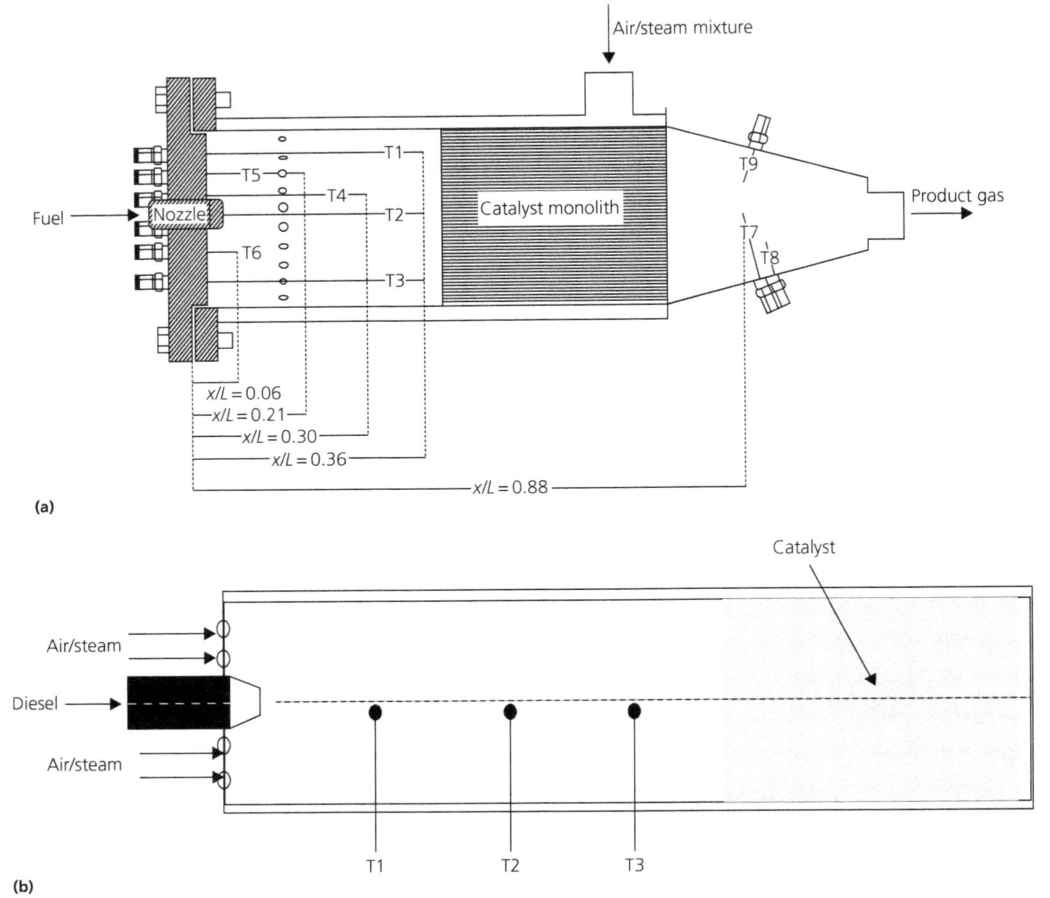

Figure 14.3 Different feed injection systems; (a) first generation [42]. (Source: Nilsson et al. [42]. Reproduced with permission of Elsevier.) (b) improved injection system without recirculation zone [43].
(Source: Lindström et al. [43]. Reproduced with permission of Elsevier.)

homogeneous autoignition of the fuel when atomized into a preheated air stream. Partial oxidation of the fuel takes place, which shortens the chain length of the hydrocarbons [46]. A large variety of products including olefins, alcohols, acids, peroxides, aldehydes, and carbon monoxide are formed [47]. The pale blue color of the cool flame originates from chemiluminescence of electronically exited formaldehyde [47].

Finally catalytic partial dehydrogenation of hydrocarbon fuels should be mentioned as an alternative to complete conversion in reforming processes. Platinum catalysts eventually doped with tin seem to be well suited for this reaction. However, stability issues still require further investigations as indicated by the work of Lucarelli et al. [48].

14.4.1 Fixed-bed reformers

A prominent example of a fixed-bed reactor applied for autothermal reforming is the HotSpot™ fuel processor developed by Johnson Matthey. Platinum and chromium catalyst were introduced as fixed bed. The reformer could be started from ambient temperature when methanol was used as feedstock [49, 50]. The reactor was preheated initially by methanol combustion and then operated under autothermal conditions.

A fixed-bed reformer for catalytic partial oxidation of methane was developed by Recupero et al. [51]. It was designed for an electrical power equivalent of 5 kW_{el}. The reactor was operated at low system pressure between 1 and 3 bar. The *O/C* ratio was ranged between 1.0 and 1.1, while the reaction temperature ranged between 800 and 900°C. The tubular reactor had a volume of 637 cm^3, and the catalyst had to be diluted to a ratio of 1 : 3 with inert material to suppress hot spot formation. The commercial catalyst (CRG-F from British Gas) was operated at 80 000 h^{-1} gas hourly space velocity (*GHSV*), which corresponds to a weight hourly space velocity (*WHSV*) of about 125 l/(h g_{cat}), which is an order of magnitude lower than the values which can be achieved over catalyst coatings in monoliths and plate heat exchangers. This inferior performance originates from the poor utilization of the relatively big catalyst particles owing to mass transfer limitations as described previously. While about 97% methane conversion and 99% selectivity toward hydrogen and carbon monoxide was achieved for 100 h test duration, the selectivity toward carbon deposits exceeded the yields expected from the thermodynamic equilibrium. It was attributed to the Boudouard and methane cracking reactions. A hot spot of 200 K was observed in the reactor which reached a peak temperature of almost 930°C.

Peters et al. [52] used tubular reactors designed for a 50 kW methanol fuel processor, which carried copper/zinc oxide catalyst. The reactor was heated externally by superheated steam at 65 bar to achieve the desired reaction temperature between 260 and 280°C. Up to 50 K temperature drop over the catalyst bed was observed, which originated from the limited heat transfer through the catalyst bed. Owing to the deactivation of the catalyst during 1000 h time on stream, this temperature gradient decreased gradually. Emonts et al. [53] reported that the carbon monoxide concentration determined at the reformer outlet ranged between 1.6 and 2.1 vol.% in the dry reformate at 260 and 280°C reactor temperature, respectively.

Moon et al. [54] operated a fixed-bed reformer under conditions of steam-supported partial oxidation of isooctane and reformulated naphtha (boiling range from 153 to 208°C) over commercial naphtha reforming catalyst from ICI at reaction temperatures between 550 and 700°C. *S/C* ratio of 3.0 was chosen for most experiments, and the *O/C* ratio was set to 1.0 according to the stoichiometry of partial oxidation. *GHSV* was about 8800 h^{-1}. When increasing the *S/C* ratio from 0.5 to 3.0 at a constant *O/C* ratio of 1.0, the methane content in the reformate could be reduced from about 10 vol.% (dry basis) at $S/C = 0.5$ to less than 1% at $S/C = 3.0$ owing to the thermodynamic equilibrium of methanation. The reactor showed stable performance for more than 24 h when fed with sulfur-free (<5 ppm) isooctane under standard conditions. It degraded significantly within a few hours when 100 ppm sulfur was added to the isooctane feed. The deactivation led to an increased methane and carbon dioxide formation, while the carbon monoxide formation decreased [54]. This likely originated from increasing selectivity toward combustion instead of partial oxidation. Steam reforming catalysts usually show the opposite trend when deactivation occurs because they tend to lose their WGS selectivity.

Lindström et al. [55] developed a fixed-bed autothermal methanol reformer designed for a 5 kW_{el} fuel cell operated with copper/zinc oxide catalyst doped with zirconia. The system was started without preheating from ambient temperature by methanol combustion in a start-up burner, which was operated at sixfold air surplus to avoid excessive temperature excursions. Because significant selectivity toward carbon monoxide was observed for the autothermal reforming process, a WGS stage became mandatory [55].

Lattner and Harold [56] performed autothermal reforming of methanol in a relatively big fixed-bed reactor carrying 380 g BASF alumina-supported copper/zinc oxide catalyst modified with zirconia. The *O/C* ratio was set to 0.22 while the *S/C* ratio varied from 0.8 to 1.5. The axial temperature profile of the reactor, which had a length of 50 cm, was rather flat, the hot spot temperature did not exceed 280°C which was achieved by the air distribution system through porous ceramic membrane tubes. More than 95% conversion was achieved. Very low carbon dioxide formation was observed for this reactor; only 0.4 vol.% was found in the reformate. However, the *WHSV* calculated from the data of Lattner and Harold [56] reveals a low value of only 6 l/(h g_{cat}) for the highest *GHSV* of 10 000 h^{-1} reported.

Wang et al. [57] investigated autothermal reforming of liquefied petroleum gas in a fixed-bed reactor over different nickel-based catalysts. The fixed bed had a length of 45.7 cm and a diameter of 1.9 cm. The hydrocarbon feed was either pure propane or liquefied petroleum gas from a fresh (first time filled) and alternatively from a reused bottle, which usually leads to a higher content of higher hydrocarbons and of sulfur odorant. The commercial feedstock contained propane, butane, propylene, ethane, and 40 ppm ethyl mercaptan apart from some nitrogen. The reforming reaction was carried out at 800°C, *S/C* ratio between 1.75 and 1.80, and *O/C* ratio between 0.9 and 1.1. Almost complete conversion of the feed was achieved. However, the catalyst deactivated owing to the presence of sulfur in the feedstock, which was indicated by an increasing content of methane and ethylene in the reformate. The catalyst degradation was faster in the case of the refilled tank owing to the sulfur accumulation in the tank.

Lee et al. [58] from Samsung reported development of a fixed-bed natural gas reformer coupled to a WGS reactor with an electrical power equivalent of 1 kW. The steam reformer was placed in the center of the subsystem, while the annular WGS fixed bed surrounded the reformer separated by an insulation layer. Commercial ruthenium catalyst served for steam reforming, while a copper-based catalyst was used for WGS. A natural gas burner supplied the energy needed by steam reformer. The reformer was operated between 850 and 930°C, while the shift reactor worked between 480 and 530°C. The carbon monoxide content of the reformate was reduced to 0.7 vol.% downstream the shift reactor despite its high operating temperature, because the reformer was operated at high *S/C* ratio between 3 and 5; the water surplus affected the equilibrium of the WGS reaction positively. At full load, the efficiency of this subsystem was 78% which decreased to 72% at 25% load.

Moon et al. [59] presented a breadboard fixed-bed fuel processor for isooctane, which was composed of an ATR and high- and low-temperature shift reactors. The fuel processor was applied for testing different catalysts. A $NiO/CaO/Al_2O_3$ catalyst performed equivalent to a $Ni/Fe/MgO/Al_2O_3$ catalyst for the autothermal reforming reaction.

14.4.2 Monolithic reformers

Rampe et al. [60] developed a monolithic autothermal propane reformer with feed gas preheating functions. The reactor had up to 1.7 kW thermal power output while the volume of the monolith amounted to 283 cm^3. It was operated at high *O/C* ratio of 1.33 and *S/C* ratio of 1.0. Consequently the conditions of the reformer were rather partial oxidation supported by steam, which prevented coke formation and served only to a minor extend as feedstock for steam reforming. However, under these conditions, only 14% of the steam feed was converted, not by steam reforming but rather by WGS (assuming thermodynamic equilibrium of WGS). The hydrogen content of the reformate was only 29 vol.% on a dry basis owing to the operating conditions chosen. Platinum catalyst was applied by Rampe et al. [60], which is known to be a good oxidation catalyst. Even under oxygen-deficient atmosphere ($O/C < 1$), no substantial steam reforming activity was observed in the reactor. Up to 75% reformer efficiency (based upon hydrogen and carbon monoxide formation) could be achieved with the reactor. Preheating the air feed improved the efficiency by about 4%. Propane conversion did not exceed 90%.

Docter et al. [61] developed an ATR for gasoline with an electrical power equivalent of 10 kW_{el}. It was composed of a mixing zone with fuel, air and water injection, and a metallic monolith of 0.5 l volume coated with catalyst. The monolith was heated by electricity at the inlet section and operated at a very high O/C ratio of 1, which is the stoichiometry of partial oxidation. Steam was added to the feed at S/C ratio of 1.5. These operating conditions resulted a low hydrogen content of about 27 vol.%, which was determined for the reformate. The reactor could be turned down by a ratio of 1 : 10 within 2 s while operating temperatures decreased from 800°C to about 660°C. The efficiency of the reactor was still in the range of 80% at more than 2 kW power output.

In order to avoid the hot spot formation frequently observed in fixed catalyst beds, metallic instead of ceramic monoliths are frequently applied for partial oxidation and steam reforming [62]. However, steam reforming would then require external heating. Ryu et al. [63] compared fixed-bed reactors containing nickel catalyst for methane steam reforming to metallic Fecralloy monoliths coated with the same catalyst. Despite about six time lower catalyst mass in the monolithic reactor, 10% higher conversion was achieved, despite the lower temperature in the monolith (Figure 14.4). The authors incorporated their monoliths also into the tubes of a shell and tube heat exchanger to introduce heat transfer capabilities into a monolithic reactor system. However, this approach seems unpractical from a fabrication point of view.

A different approach was taken by Vita et al. [64] which structured their Pt/CeO_2 catalyst itself and wound it up to a jellyroll-like monolith shown in Figure 14.5a. This reactor outperformed a metallic monolith and a fixed bed of the same catalyst in propane oxidative steam reforming, especially at low residence times (Figure 14.5b).

Beretta et al. [65] decreased the hot spot formation of their ceramic monoliths for the partial oxidation of methane by decreasing the cell density from 400 to 115 cpsi, which increased the channel size consequently. Therefore mass transfer limitations occurred, which reduced the oxygen consumption at the channel inlet and pushed the oxidation reaction toward the reactor center; the hot spot was reduced from 900 to 750°C by these means (Figure 14.6).

Fichtner et al. [66] used a monolithic microchannel reactor for partial oxidation of methane. The reaction was carried out at 1000°C temperature, 25 bar pressure and residence times in the order of few milliseconds. The adiabatic hot spot formation was calculated to be 2320°C. This excessive hot spot was expected to be reduced in the metallic honeycomb by axial heat transfer from the oxidation to the steam reforming reaction zones.

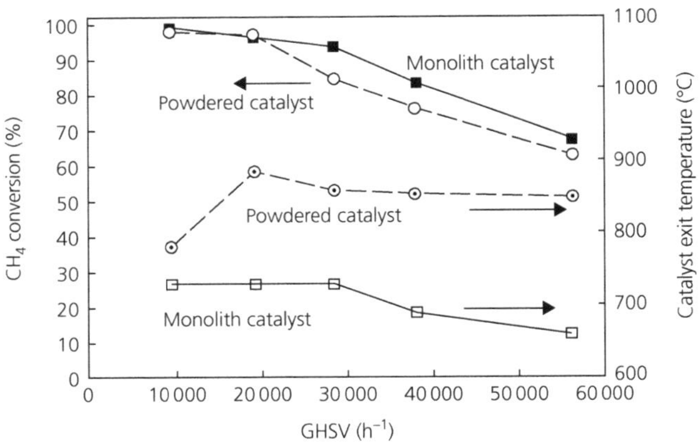

Figure 14.4 Methane conversion and reactor exit temperature versus *GHSV* for a fixed bed compared to a metallic monolith ($S/C = 3$) [63]. (Source: Ryu et al. [63]. Reproduced with permission of Elsevier.)

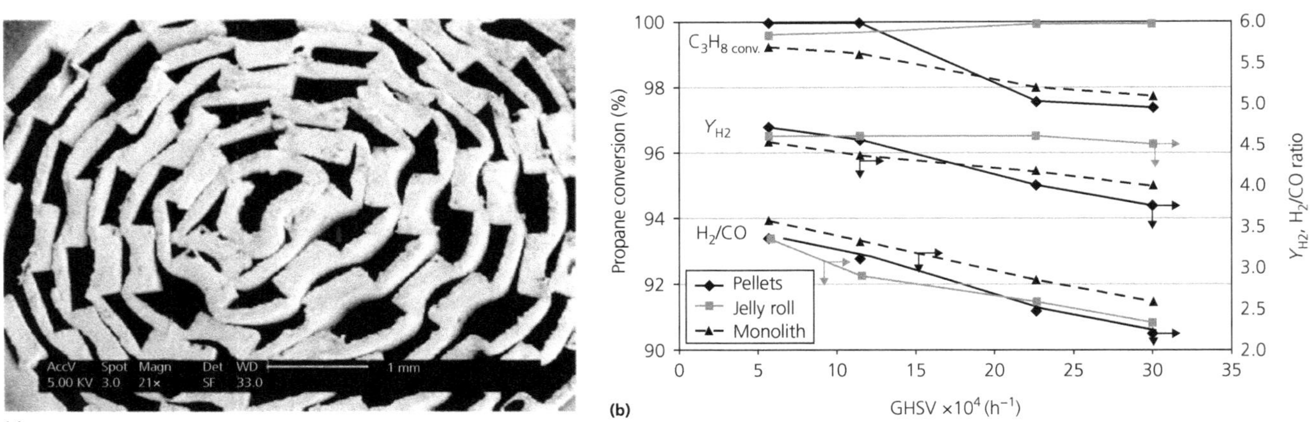

(a) (b)

Figure 14.5 (a) Jellyroll monolith prepared from platinum/ceria catalyst, (b) comparison of propane conversion, hydrogen yield and H_2/CO ratio for fixed bed, jelly roll, and monolithic reactors as determined for different *GHSV* [64]. (Source: Vita et al. [64]. Reproduced with permission of Elsevier.)

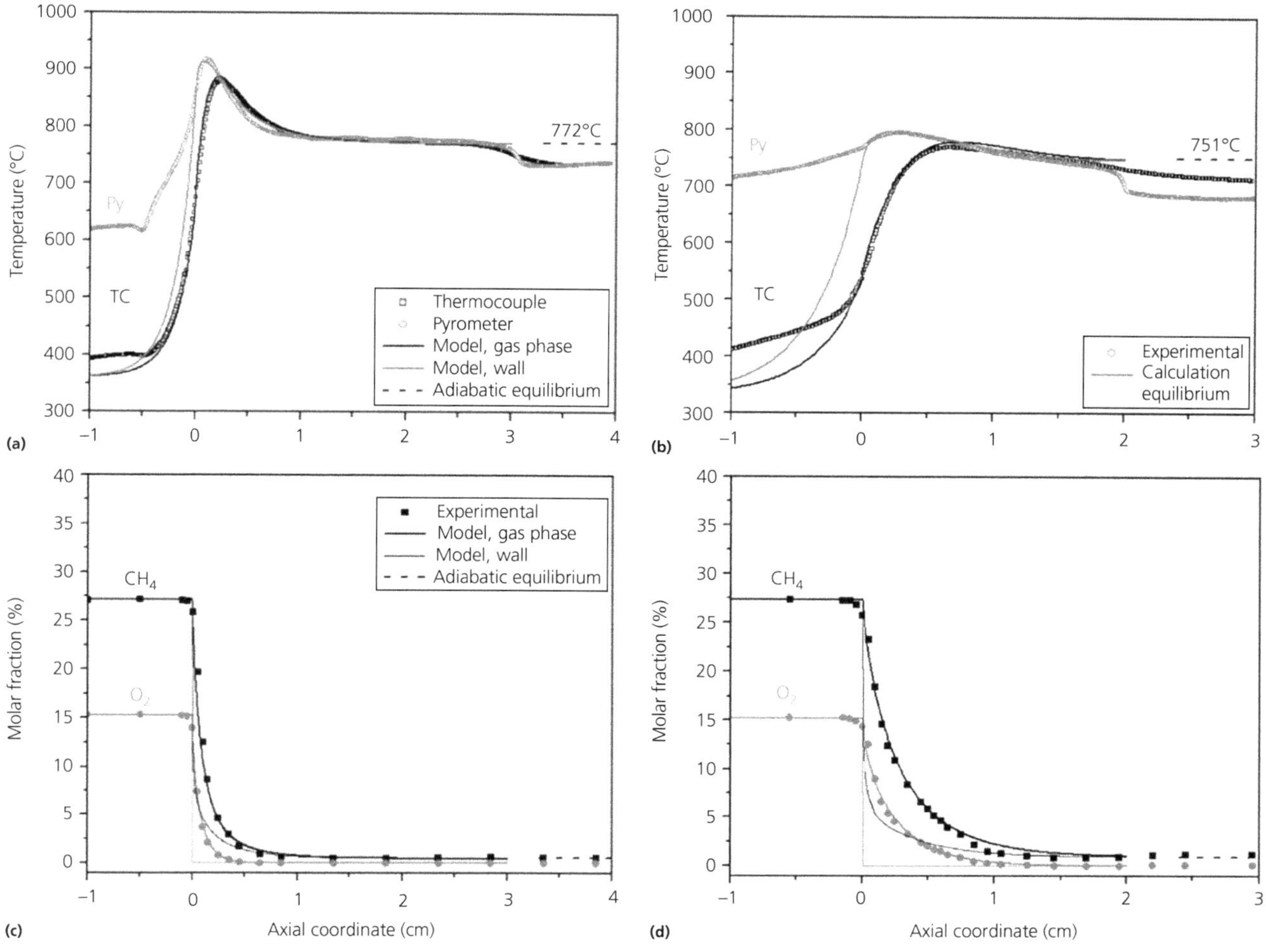

Figure 14.6 Hot spot formation and methane and oxygen concentration profiles in two ceramic monoliths of different cell density for partial oxidation of methane at $O/C = 0.54$ and at a feed inlet temperature of 362°C; (a) temperature profile for 400 cpsi, (b) temperature profile for 115 cpsi, (c) concentration profiles for 400 cpsi, (d) concentration profiles for 115 cpsi [65].
(Source: Beretta et al. [65]. Reproduced with permission of Elsevier.)

Rhodium was chosen as construction material for the reactor, which served as active catalyst species at the same time. Rhodium has a high thermal conductivity of 120 W/(m K). Twenty three foils carrying 28 channels each of which was sealed by electron beam and laser welding. The stack of foils formed a honeycomb which was pressure resistant up to 30 bar. The maximum operating temperature of the reactor was 1200°C. The feed was preheated to 300°C and then fed to the reactor. The experiments were carried out between ambient pressure and 25 bar at O/C ratio 1.0. After ignition of the reaction between 550 and 700°C, 1000°C reaction temperature was then achieved within 1 min, and mainly carbon monoxide and hydrogen were formed. Only 62% conversion of methane but 98% conversion of oxygen was achieved at 1190°C. The performance of the reactor deteriorated when the system pressure was increased. By-product and even soot formation then occurred downstream the reactor.

Jung et al. [67] investigated the partial oxidation of methane over noble metal catalysts coated onto metallic monoliths, which were preheated by electricity. Over palladium catalyst supported by alumina, autoignition of the reaction took place at 270°C and at an O/C ratio of 1.1. However a higher O/C ratio of 1.4 was required to achieve 98.9% methane conversion because the reactor exit temperature of 700°C was too low for full conversion at low O/C ratios.

Microlith-packed metal gauze technology was applied by Castaldi et al. [68] for gasoline reforming. The ATR was composed of a stack of four microlith screens coated with either platinum catalyst supported by lanthanum-stabilized alumina carrier [68] or rhodium on ceria–zirconia carrier [69]. Ceramic seals were put between the screens. The diameter of the screens was 40 mm, while the length of the entire stack including seals and features for sampling and temperature measurement was only between 12 and 40 mm, depending on the number of gauze elements integrated [69]. The isooctane feed rate supplied to this tiny reactor was very high with a thermal equivalent of 3.4 kW [68], which corresponded to gas hourly space velocities between 30 000 and 120 000 h^{-1}. Isooctane was used as surrogate for gasoline. The S/C ratio ranged between 0.5 and 2.1, while the O/C ratio was set between 0.65 and 1.1. With a very low preheating temperature of 200°C of the reactor and of steam, air, and fuel feed, light-off of the platinum catalyst was achieved at S/C of 1.5 and O/C of 0.62. The reactor front (first screen) reached the operating temperature of 800°C within 10 s. The reactor temperature dropped to 650°C toward the outlet, while only 49% conversion of the fuel was achieved. When increasing the O/C ratio to 0.98 and the S/C ratio to 1.5, 94% conversion could be achieved over the platinum catalyst. Owing to the very low residence times applied, only 0.5 vol.% of methane was formed in the reactor under these conditions, because the methanation reaction could not reach thermodynamic equilibrium. No evidence for coke formation was observed over the reactor for 5 h duration. When increasing the bed length to 38 mm, the fuel conversion could be increased to 97% at S/C of 1.24 and O/C of 0.89 for the platinum catalyst. The peak temperatures of the catalyst bed were somewhat lower for the rhodium catalyst and even 98% conversion could be achieved at S/C of 1.53 and O/C of 1.07, while only 0.2 vol.% of methane were found on a dry basis, which is attributed to the higher selectivity toward steam reforming of the rhodium catalyst compared to platinum according to the authors [69]. Then the O/C ratio was increased at constant S/C ratio of 1.5 for the rhodium catalyst, which further increased the conversion. This demonstrates that almost no steam reforming took place in the reactor owing to the low residence times applied but rather partial oxidation accompanied by some WGS.

Catillon et al. [70] investigated the performance of copper/zinc oxide catalyst coated onto copper foams for methanol steam reforming. Significant improvement of the heat transfer by the copper and consequently higher catalyst activity was achieved compared to fixed catalyst beds.

Liu et al. [71] described the performance of their autothermal diesel reformer operated with dodecane and hexadecane. It was composed of four ceramic monoliths switched in series with 600 cpsi, each of which had a diameter of 3.8 cm and a length of 2.54 cm. Less than 0.5 vol.% of methane was detected in the reformate typically, while higher hydrocarbons only appeared in concentrations of few hundred parts per million. The thermal power equivalent of fuel fed to the reformer was in the range up to 5 kW. At a S/C ratio of 2.0 and an O/C ratio of 0.74, the temperature in the four monoliths decreased from 800 (measured downstream the first monolith) to 700°C downstream the fourth monolith. The radial temperature profile did not exceed 30 K [71]. Homogenous reactions upstream the first monolith occurred as assumed by the authors. The reformate composition was determined downstream the four monoliths. The hydrogen and carbon oxide concentration increased over the first three reactors, while methane formation was observed in the fourth monolith, which consumed some of the hydrogen produced in the first three monoliths upstream. The hydrogen concentration in the reformate deviated from the thermodynamic equilibrium, because light hydrocarbons were produced. The highest efficiency of the reformer was determined for a S/C ratio of 2.0 and an O/C ratio of 0.84. When the S/C ratio was increased from 1.1 to 2.8 at constant O/C ratio of 0.68, the temperature of the first reactor was decreased by not more than 40 K. However, the carbon monoxide concentration was lowered by the steam excess, according to the equilibrium of WGS. Coke formation was mainly observed downstream the reactors in the zones of lower temperature [71]. About 1% of the feed was converted to coke according to an estimation of the authors. They assumed that it originated from the Boudouard reaction.

Horng [72] developed a monolithic autothermal methanol reformer. The ceramic monolith was coated with mixed platinum and copper/zinc oxide catalyst. The monolith had 117 mm diameter and 50 mm length, while the whole reactor was more than 510 mm long. Glow plugs were used for the start-up. The reformer could be heated to 200°C at the outlet within 220 s by preheating 70 l/min air with an electric heating with a power output of 960 W. 14 ml/min liquid methanol was injected into this preheated air flow, which corresponded to an O/C ratio of 3.75, a value exceeding the stoichiometric composition for total methanol combustion (O/C ratio of 3.0). Part of the methanol was already converted at the glow plugs. After 220 s, the monolith had a temperature of more than 500°C at the inlet, while the outlet was still rather cold at 80°C. The feed composition was then switched to autothermal conditions. The hydrogen content of the reformate increased to 35 vol.% within 100 s, after 200 s stable conditions were achieved. The carbon monoxide content of the reformate was in the range of 5 vol.%. Decreasing the heating power to 480 W under otherwise same conditions did not change the start-up time demand significantly. These experiments demonstrate that it is feasible to preheat a methanol reformer by catalytic methanol combustion from ambient by simply injecting methanol in a preheated air flow. The preheating is required to avoid excessive cooling of the air feed by methanol evaporation.

Lenz et al. [73] described the development of a 3 kW monolithic steam-supported partial oxidation reactor for jet fuel, which was developed to supply a solid oxide fuel cell (SOFC). The prototype reactor was composed of a ceramic honeycomb monolith (400 cpsi) operated between 950°C at the reactor inlet and 700°C at the reactor outlet [74]. The radial temperature gradient amounted to 50 K which was attributed to inhomogeneous mixing at the reactor inlet. The feed composition corresponded to S/C ratio of 1.75 and O/C ratio of 1.0 at 50 000 h^{-1} $GHSV$. Under these conditions, about 12 vol.% of each carbon monoxide and carbon dioxide were detected in the reformate, while methane was below the detection limit. Later, Lenz et al. [74] described a combination of three monolithic reactors coated with platinum/rhodium catalyst switched in series for jet fuel autothermal reforming. An optimum S/C ratio of 1.5 and an optimum O/C ratio of 0.83 were determined. Under these conditions 78.5% efficiency at 50 000 h^{-1} $GHSV$ was achieved. The conversion did not exceed 92.5%. In the product of these reactors, the authors detected about 0.5 vol.% methane, while the concentration of all higher hydrocarbons up to C_6 did not exceed 0.2 vol.%. Increasing the space velocity did increase the hydrocarbons content in the reformate and further decreased conversion. Interestingly, the steam formation exceeded the thermodynamic equilibrium at the reactor inlet of the first reactor but approached the thermodynamic equilibrium downstream. This indicated that total oxidation was predominant at the reactor inlet while steam reforming consumed the water which had been produced initially. About 1.14 kW heat losses were calculated for the reactor, which emphasized the need for sufficient insulation especially at temperatures exceeding 600°C. When the feed was switched from sulfur-free jet fuel to jet fuel containing 300 ppm sulfur, the conversion dropped quickly and then deteriorated slowly with time on stream.

Qi et al. [75] reported the autothermal reforming of gasoline over cordierite monoliths coated with 0.3 wt.%Rh/3 wt.% MgO/20 wt.% CeO_2/-ZrO_2. The monolith showed superior performance compared to a fixed catalyst bed for autothermal reforming of *n*-octane at an *O*/*C* ratio of 0.76 and *S*/*C* of 2.0. An ATR of 1 kW size equivalent coated with the catalyst showed stable performance for about 30 h, when deactivation of the catalyst started. The hot spot temperature in the reactor was in the range of 900°C, which was assumed to be the origin of catalyst deactivation by sintering.

Nilsson et al. [42] reported autothermal reforming of diesel, gasoline, dimethyl ether, ethanol, E85 (ethanol containing at least 15 wt.% gasoline), and methanol in their monolithic ATR over rhodium catalyst containing 1 wt.% Rh, 10 wt.% CeO_2, and 5 wt.% La_2O_3. While almost full (>99%) conversion was achieved for gasoline and E85, lowest conversion was observed for methanol followed by dimethyl ether, which was attributed to methanol and dimethyl ether decomposition mainly occurring over the catalyst.

Karatzas et al. [34] performed autothermal reforming of tetradecane, low sulfur, and Fischer–Tropsch diesel in a monolithic reformer over rhodium/ceria/lanthana catalyst. The reformer had a thermal power output of 14 kW. It was composed of an inert zirconia-coated alumina foam for feed distribution at the reactor inlet and two 400 cpsi cordierite monoliths coated with the catalyst switched in series. At an *O*/*C* ratio of 0.45, a *S*/*C* ratio of 2.5 and temperatures exceeding 740°C, full conversion of the low sulfur feed was achieved, while the formation of the by-product ethylene was between 100 and 200 ppm. As shown in Figure 14.7, an increasing *S*/*C* ratio suppresses ethylene formation. The catalyst showed stable performance for 40 h duration. Karatzas et al. [44] determined experimentally as shown in Figure 14.8 that the efficiency of their ATR increased with increasing fuel inlet temperature and *O*/*C* ratio.

Fast preheating is possible in case of ceramic monoliths are applied for autothermal reforming. Lindström et al. [43] reported 6 min start-up time demand of their autothermal diesel reformer, which was preheated by a homogeneous diesel burner through a heat exchanger via air (Figure 14.9).

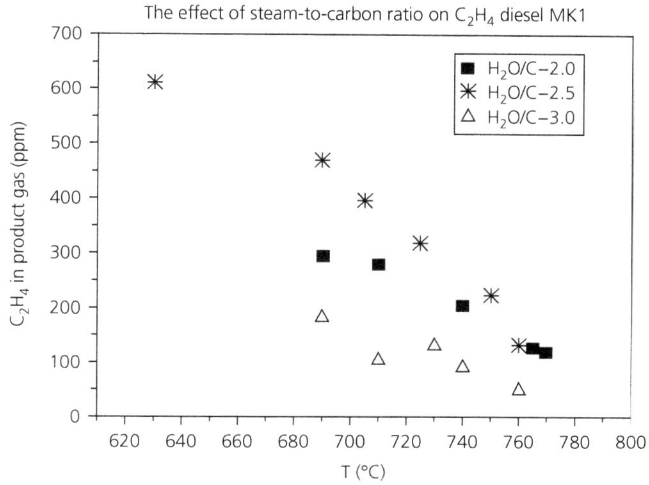

Figure 14.7 Ethylene formation versus temperature at different S/C ratio for autothermal reforming of low sulfur diesel (*O*/*C* = 0.45) [34].
(Source: Karatzas et al. [34]. Reproduced with permission of Elsevier.)

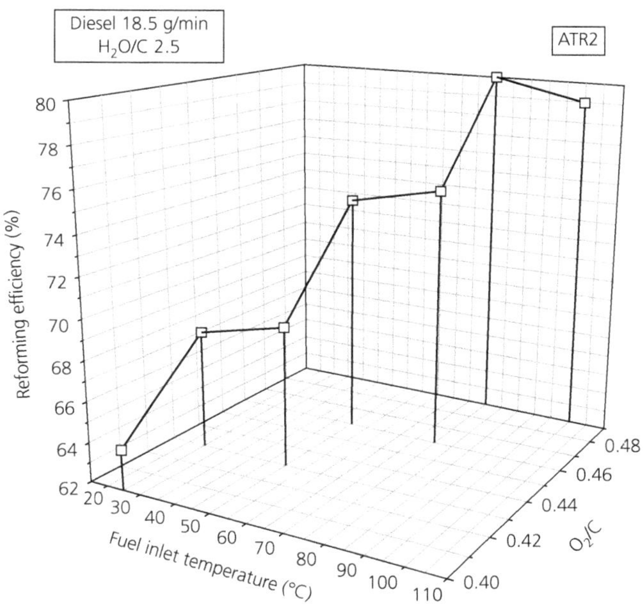

Figure 14.8 Effect of feed inlet temperature and *O*/*C* ratio (here shown as O_2/*C* ratio) on reforming efficiency (*O*/*C* = 2.5, *GHSV* = 9 900–10 800 h^{-1}) [44].
(Source: Karatzas et al. [44]. Reproduced with permission of Elsevier.)

Creaser et al. [76] modeled autothermal diesel reforming in a ceramic monolithic reactor and verified the modeling results with experimental data as shown in Figure 14.10. The model revealed that axial heat conduction plays an important role even in ceramic monoliths. The catalyst temperature was found to be 25°C hotter than the gas phase at the reactor inlet according to the calculations. At the positions of highest reaction rates, the catalyst utilization was as low as 20%. Transport limitations in the washcoat were assumed to be the root cause. The

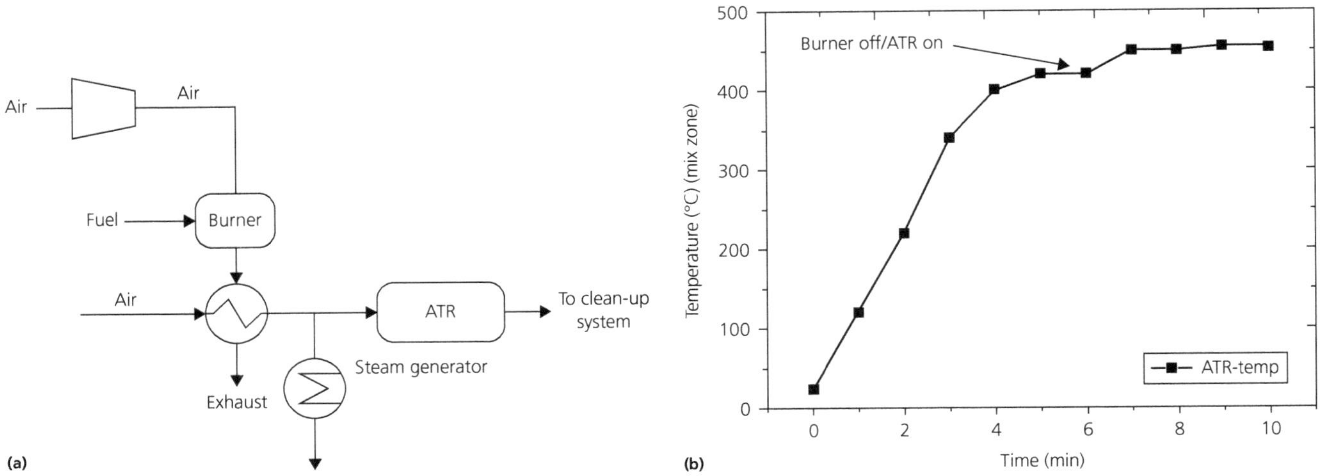

Figure 14.9 (a) System start-up concept, (b) reformer start-up time [43]. (Source: Lindström et al. [43]. Reproduced with permission of Elsevier.)

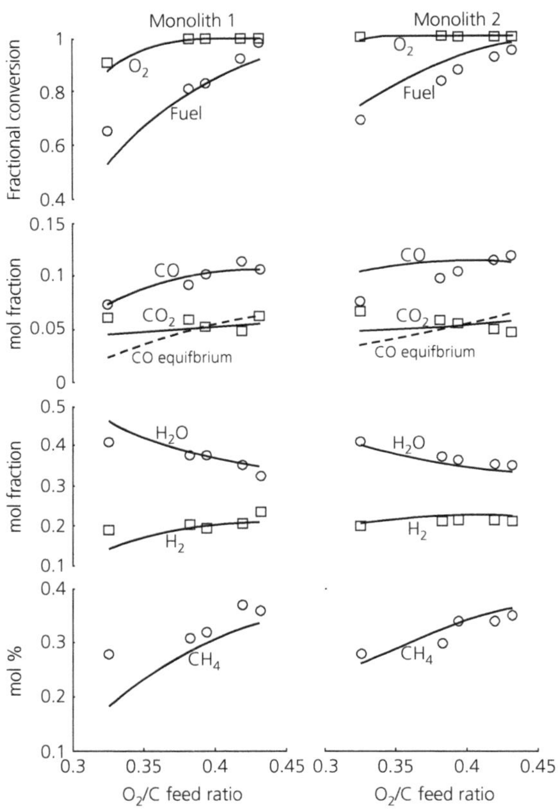

Figure 14.10 Species concentration versus O/C ratio (here expressed as O_2/C) for an autothermal diesel reformer at $S/C = 2.3$. Symbols represent experimental data, the lines represent simulation results [76]. (Source: Creaser et al. [76]. Reproduced with permission of Elsevier.)

multitude of diesel steam reforming reactions were described as lumped model equations, which were first order with respect to the fuel and zero order with respect to steam. The rate of the WGS reaction was only minor, while the equilibrium of WGS was by far exceeded. Partial oxidation was not foreseen in the reaction scheme but only total oxidation, which was also described by a lumped model and first order for oxygen and −0.2 order for the fuel [76].

14.4.3 Plate heat exchangers and microstructured reformers

When catalyst is introduced into the channels or onto the walls of both flow paths of a plate heat exchanger with channels of meso- or microscale, the plate heat exchanger turns into a catalytic heat exchanger/reactor also named catalytic wall reactor. Catalytic (wall) reactors improve the temperature management of exothermic reactions as discussed for many examples in the following paragraphs. The design allows for the coupling of endothermic reactions (such as steam reforming) and exothermic reactions (such as catalytic combustion), which are then separated by merely a few hundred micrometers metal foil between both micro fixed beds or coatings, respectively, creating low-temperature gradients between both sides especially when coatings are applied [77].

In order to avoid hot spot formation, Schildhauer and Geissler [78] proposed plate heat exchanger/reactors for autothermal methanol reforming. The flow of the reformate is redirected into the opposite direction according to this concept, which reduces the initial hot spot formation of the fast partial oxidation reaction by the heat uptake of consecutive steam reforming. Eigenberger et al. [79–82] investigated both theoretically and practically heat exchanger/reactors for steam reforming in the mesoscale. The reactors were initially made from specially prepared ceramic monoliths, which were transformed into heat exchangers. Later on, structured and welded steel foils were applied.

Different flow arrangements exist for heat exchangers, namely, cross flow, countercurrent, and cocurrent flow. The main disadvantages of the cross-flow design are uneven temperature distributions, which also deteriorate the gas composition as demonstrated by numerical calculations for methanol steam reforming by Pan and Wang [83]. Frauhammer et al. [84] proved by numerical calculations that a counterflow arrangement is also not suited for coupling endothermic and exothermic reactions and a cocurrent flow arrangement should be chosen [85]. Catalytic combustion of unconverted hydrogen and residual CO contained in the fuel cell anode off-gas is a well suited heat source for steam reforming of all kind of fuels. The temperature profile in a cocurrently operated microstructured plate heat exchanger/reactor, which is operated as a coupled steam reformer/catalytic afterburner depends on the reactor geometry and size. It is either isothermal [86] or slightly declining over the reactor length axis as indicated by simulations of Petrachi et al. [87] and proven experimentally by O'Connell et al. [41]. This profile originates from the higher rate of oxidation reactions compared to steam reforming.

Figure 14.11 shows the results of numerical simulations of Grote et al. [88] for a diesel steam reformer coupled to a catalytic diesel burner. The temperature is still low at the reactor inlet owing to the energy consumption of steam reforming, while a hot spot is formed downstream by the combustion reaction followed by a declining temperature profile. However, a cross-flow arrangement (not shown here) revealed even higher temperature gradients in the range of 400 K compared to about 100 K for cocurrent flow.

The channels of most plate heat exchanger/reactors are switched in parallel, which reduces the pressure drop compared to alternative flow patterns such as serpentine flow fields. However, flow equipartition is crucial for parallel flow arrangements. It is achieved by perforated plates [89] when a whole stack of plates is fed in parallel from the plate front. Such pinhole plates create additional pressure drop. In case the feed gas is distributed to each plate first and then by a dedicated inlet section to each channel of the plate, a sophisticated geometry of this inlet section [90] helps to achieve flow equipartition. An alternative is the variation of the channel width over the reactor length axis [91].

Reuse et al. [92], Park et al. [93], Kim et al. [94], and Kundu et al. [95] reported the development of combined methanol steam reformers/catalytic burners designed as plate heat exchanger/reactors in the low power range. An early development of an integrated heat exchanger/reactor for methanol steam reforming was carried out by Hermann et al. [96] from GM/OPEL. A 5 kW combined methanol steam reformer/catalytic combustor was built. Instead of channels, fins served as mechanical support and improved the heat transfer. Experimental results were determined at partial load of the device (1–2 kW for the lower heating value (LHV) of the hydrogen produced). At *S/C* ratio of 1.5 and a pressure of 3 bar, full conversion of the methanol was achieved and 0.9 m^3/h hydrogen was produced. Fitzgerald et al. [97] developed a microchannel isooctane heat exchanger/steam reformer, which was heated by combustion gas and produced enough hydrogen to power a 500 W PEM fuel cell. At ambient pressure, a temperature of 650°C, 2.3 ms residence time, and a S/C ratio of 6, up to 95% conversion were achieved at 90% hydrogen selectivity. Whyatt et al. [98] developed a system composed of evaporators, heat exchangers, and four integrated reformers/cross-flow heat exchangers switched in series for isooctane steam reforming. The reformers achieved more than 98.6% conversion at 750°C reaction temperature and a *S/C* ratio of 3. The hydrogen produced was sufficient to power a PEM fuel cell with 13.7 kW electric output. Because the automotive drive train was the final application behind the development work, the required start-up time demand was very short in the range below 1 min. Therefore Whyatt et al. [99] redesigned their system completely. The energy for start-up and continuous operation was provided by homogeneous fuel combustion rather than catalytic combustion, which supplied the reformer and the evaporator, placed right after the reformer, with energy. The target start-up time demand of 60 s could be achieved with this device; however, CO cleanup reactors were not yet incorporated into this system.

Catalysts from Süd Chemie were chosen for methanol steam reforming by Cremers et al. [100]. Because the catalyst activity was relatively low, a micro fixed-bed reactor was built with integrated heat exchanging capabilities. The reactor contained 60 micro fixed-bed passages, which took up 15.9 g of catalyst and 62 heating passages. The reactor was designed to produce

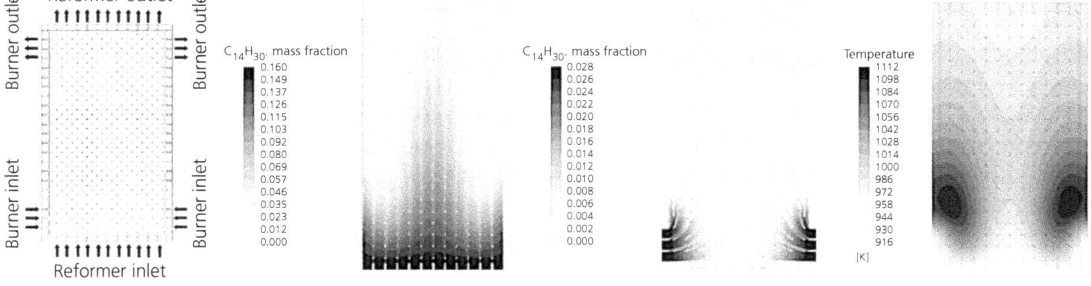

Figure 14.11 Simulation of a cocurrently operated plate heat exchanger for diesel steam reforming coupled to a catalytic diesel burner. Left: flow arrangement, second left: fuel mass fraction in the reformer, third left: fuel mass fraction in the burner, right: temperature in the plate.
(Source: Grote et al. [88]. Reproduced with permission of Elsevier.)

hydrogen for a fuel cell with 500 W electrical power output. Heating oil, preheated by an external burner served as the heat source for the reformer. More than 90% methanol conversion was achieved at 250°C reaction temperature. During the first 4 h of operation, 15% of the initial activity of the catalyst was lost, but then activity remained stable for another 4 h. The catalyst could be regenerated by oxidation and subsequent reduction. A turndown ratio of 1 : 5 could be realized without significant changes of product composition.

Cremers et al. and Pfeifer et al. [100–102] developed reactors with electric power equivalent of 500 W for methanol and methane steam reforming. Pan and Wang [83] developed a cross-flow plate-fin reformer for methanol steam reforming with integrated fuel and water evaporation, steam reforming, and catalytic combustion, which suffered from nonisothermal temperature distribution compensated by low operating temperature of the reactor. The system was then scaled-up by a factor of 14 [103] and operated for 1000 h. The methanol conversion decreased from 100% to about 93% within this time while the carbon monoxide concentration remained below 2 vol.%. The reformate recycle to the afterburner could be reduced to 30%, because heat losses were less dominant for this larger device.

A microstructured diesel steam reformer integrated with a catalytic afterburner was developed by Kolb et al. [104]. It was operated at reaction temperatures exceeding 800°C. The reactor was coated with catalyst from Johnson Matthey Fuel Cells. The anode off-gas and the air supply were fed separately to the burner where they were mixed internally. Full conversion of the diesel fuel was achieved with this reactor, which had a power equivalent of 2 kW thermal energy of the hydrogen produced. The deactivation of the diesel steam reforming catalyst was initially indicated by increasing release of light hydrocarbons into the reformate. Then O'Connell et al. [41] reported the development and experimental evaluation of a reactor based on microchannel technology for the reforming of diesel fuel with 13 kW thermal output of the hydrogen produced. Diesel oxidative steam reforming was performed for 38 h at reaction temperatures exceeding 750°C and *S*/*C* ratios as low as 3.2. Over 98% total diesel conversion was observed at all times, usually only traces of unconverted diesel were detected.

Grote et al. [88] presented an integrated heat exchanger for diesel steam reforming coupled to a diesel burner designed for a thermal input of 10 kW of the diesel fuel, which had a cocurrent flow arrangement. Instead of microchannels, a fin structure was used in the reactor. The reactor was coupled to a superheater for the steam feed by cooling the reformate product. However, the reactor showed a severely declining temperature profile over its length axis of more than 200 K. The formation of light hydrocarbon by-products other than methane was suppressed by increasing reaction temperature and *S*/*C* ratios exceeding 3.5. Lee et al. [105] developed and operated an integrated micro ethanol reformer combined with an integrated catalytic burner for ethanol and an evaporator. The reformer produced enough hydrogen to power a 450 W fuel cell. Micro fixed beds were used and platinum/alumina catalyst served for ethanol combustion. Wichert et al. [106] reported long-term experiments on a coupled steam reformer/catalytic burner for LPG based upon microchannel technology. The reactor was operated, similar to most reactors described previously, in a cocurrent flow arrangement with propane as surrogate for LPG. Long-term stability tests for 1060 h duration were performed at a *S*/*C* ratio of 4.0 and about 750°C reaction temperature with 29 start-up and shutdown cycles. Complete conversion of propane was observed for the full test duration. The gas composition of the reformate agreed well with the thermodynamic equilibrium of propane reforming, WGS, and methanation reactions. The stainless steel construction material of the reactor was not affected by exposure to the operating temperature above 700°C for 1060 h. Neither the reaction temperature nor the start/stop cycles impaired the mechanical integrity of the catalyst coatings as proven after disassembly of the reactor.

14.4.4 Membrane reformers

Besides methane methanol is the fuel that is most frequently processed in membrane reformers. Lin et al. [107] observed much higher methanol conversion for methanol steam reforming in a MR compared to a conventional fixed bed. Then Lin and Rei [108] designed a tubular MR for methanol steam reforming with a central palladium membrane tube, an annular catalyst bed for methanol steam reforming carrying copper-based catalyst which surrounded the membrane, and a second concentric annular bed containing palladium/alumina catalyst for combustion of the retentate, which allowed self-sustaining operation of the reactor. The GHSV in the reformer fixed beds was always below 15 h^{-1}. Up to 85% of the hydrogen could be recovered from the reformate. An energy balance revealed that 74% hydrogen recovery by the membrane separation still left sufficient hydrogen in the retentate to supply the feed preheating and evaporation processes and the steam reforming reaction with energy by combustion of this retentate hydrogen. Later, the same authors [109] reported stable operation of the (commercial) reforming catalyst and of the palladium membrane in the reactor shown in Figure 14.12 at a reactor temperature of 350°C, 15 bar pressure and 95% methanol conversion, and a hydrogen flux of 3.5 $m^3/(m^2\,h)$ for a duration of 900 h. The authors [110] then designed bigger reactor, the palladium membrane of it being prepared by electroless plating. Its permselectivity for H_2/N_2 was in the range of 200–1400, while the permeance varied from 3 to 7 $m^3/(m^2\,h\,atm^{0.5})$. Ten palladium tubes were incorporated into the reactor. At system pressures exceeding 10 bar, more than 95% hydrogen recovery could be achieved with the membrane. For a membrane separator feed containing 82 vol.% hydrogen, 3.9 vol.% methane, and higher hydrocarbons, the purity of the separated hydrogen amounted to 99.6 vol.%, and 0.14 vol.% methane was detected apart from higher hydrocarbons. This separator was operated stably for a duration of more than 1000 h.

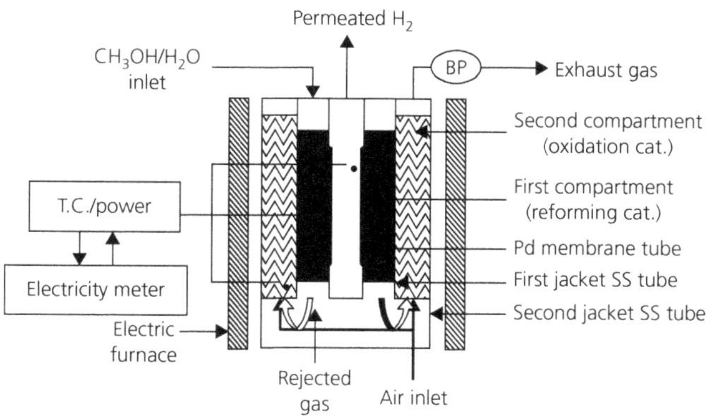

Figure 14.12 Scheme of the membrane reactor for methanol reforming [109]. (Source: Lin and Rei [109]. Reproduced with permission of Elsevier.)

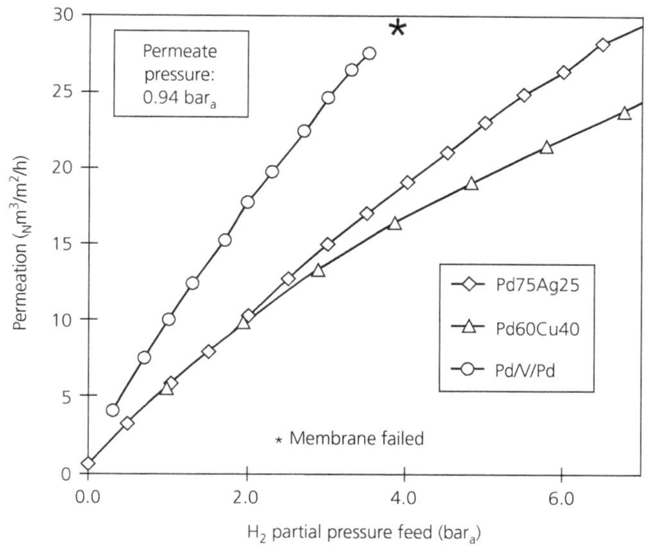

Figure 14.13 Conversion in a membrane reactor applying different membrane types. For comparison, thermodynamic equilibrium conversion and conversion as achieved in a fixed catalyst bed are added [113]. (Source: Wieland et al. [113]. Reproduced with permission of Elsevier.)

Kikuchi [111] described a natural gas MR, which had been developed and operated by Tokyo Gas and Mitsubishi Heavy Industries to supply PEM fuel cells with hydrogen. It was composed of a central burner surrounded by a catalyst bed filled with commercial nickel catalyst. Into the catalyst bed 24 supported palladium membrane tubes were inserted. The membranes had been prepared by electroless plating and were 20 μm thick. Steam was used as sweep gas for the permeate. The reactor carried 14.5 kg catalyst. It was operated at 6.2 bar pressure, *S/C* ratio of 2.4, and 550°C reaction temperature. The conversion of the natural gas was close to 100%, while the equilibrium conversion was only 30% under the operating conditions. The retentate composition was 6 vol.% hydrogen, 1 vol.% carbon monoxide, 91 vol.% carbon dioxide, and 2 vol.% methane.

Basile and Putzuro [112] prepared a palladium membrane by electroless plating palladium with a thickness of 70 μm onto titania tubes and incorporated them into MRs for the partial oxidation of methane. The H_2/N_2 separation factor of these membranes was still quite low (between 3 and 4.7), while carbon deposition on the membrane was negligible for operation up to 400°C.

Wieland et al. [113] combined different membranes with a fixed catalyst bed for methanol steam reforming in a plate-type reactor. Figure 14.13 shows the conversion determined in the MR with different membranes. It is compared to the thermodynamic equilibrium conversion and the conversion achieved in a fixed catalyst bed without membrane. The conversion was higher for the MRs and exceeded the thermodynamic equilibrium of the feed at more than 20 bar pressure. The space velocity, at which the fixed catalyst bed was operated, was rather low. The same authors also performed numerical calculations to enlighten the effect of lower membrane thickness on the performance of their reactor.

Kurungot et al. [114] developed a silica membrane and incorporated it into a catalytic MR for the partial oxidation of methane. Rhodium catalyst on γ-alumina carrier containing 1 wt.% rhodium was coated with 9 μm thickness onto an α-alumina tube. Then the silica membrane, which was only 1.5 μm thick, was coated onto the catalyst. Hydrogen and methane permeation through the membrane were measured revealing increased permeability for hydrogen with increasing temperature, while methane permeability remained at a lower level. Between 100 and 525°C, the separation factor increased from 7.5 to 31. The hydrothermal stability of the membrane was verified at 525°C for 8 h with feed composed of 18 vol.% hydrogen, 18 vol.% methane, and 74 vol.% steam. The separation factor decreased from 31 to 26 with course of the experiment. Partial oxidation of methane was then carried out at an *O/C* ratio of 1.0 and different *S/C* ratios. The reaction was performed under atmospheric pressure and nitrogen sweep gas flow on the permeate side. The results obtained with the membrane were compared to the performance without catalytic membrane in the temperature range between 400 and 575°C. At 525°C reaction temperature, *S/C* of 3.5, and a *WHSV* of 240 l/(h g_{cat}), equilibrium conversion could be exceeded by 37% owing to the hydrogen removal from the reformate. Lowering the *S/C* ratio from 3.5 to 2.5 under these conditions decreased the conversion considerably from 58.7 to 34.2%. At higher *WHSV* carbon formation was suspected, because the membrane color changed from yellow to gray.

Basile et al. [115] prepared a MR for methanol oxidative steam reforming shown schematically in Figure 14.14. A pinhole-free PdAg membrane containing 23 wt.% Ag served for the membrane separation, while sweep gas was used for permeate removal from the membrane outer surface to enhance the mass transport through the membrane. Commercial Cu/ZnO catalyst was applied for methanol reforming. Owing to the low operating

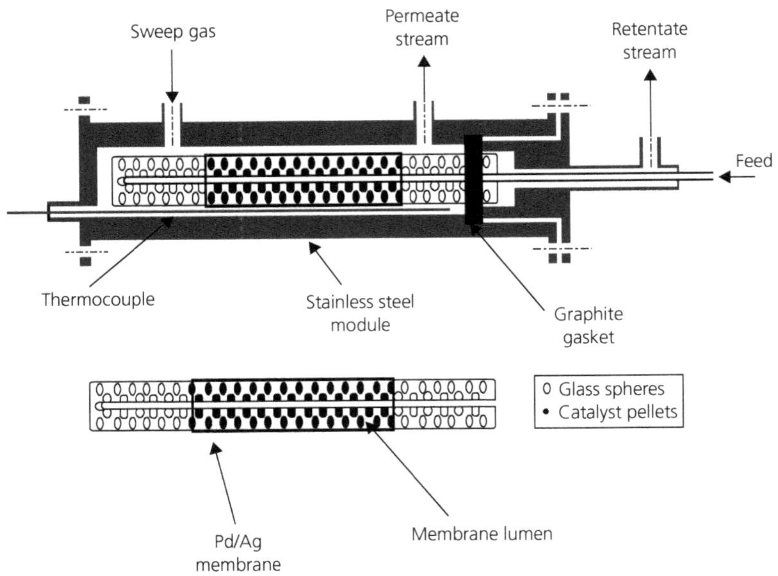

Figure 14.14 Scheme of the membrane reactor for methanol oxidative steam reforming [115].
(Source: Basile et al. [115]. Reproduced with permission of Elsevier.)

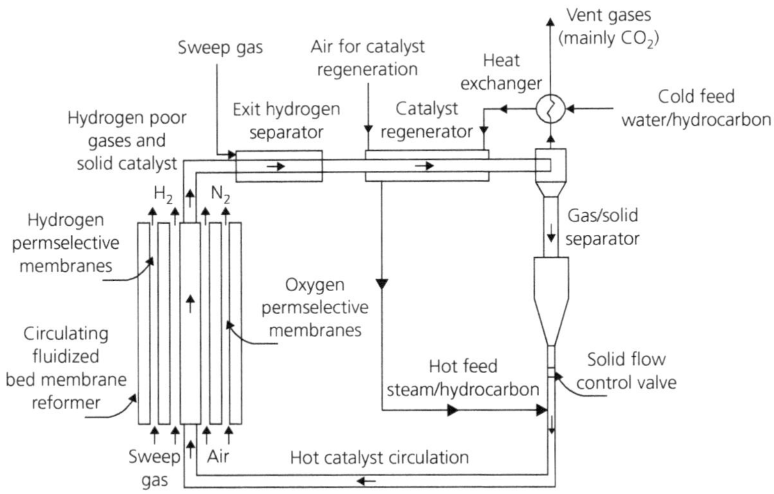

Figure 14.15 Scheme of the fluidized bed membrane reactor [117].
(Source: Chen and Elnashaie [117]. Reproduced with permission of Elsevier.)

temperature required for this catalyst type, not more than 25% of the hydrogen produced by the reforming reaction could be recovered in the permeate. The catalyst showed rapid deactivation during 5 h test duration despite a high *S/C* ratio of 2.7 applied. Then Iulianelli et al. [116] switched to a Cu/Zn/Mg catalyst formulation, which was operated under conditions of pure steam reforming. Up to 94% of the hydrogen could be recovered at a feed pressure of 3.5 bar. Operation of the sweep gas flow in co- and countercurrent flow revealed advantages for the countercurrent flow arrangement.

Chen and Elnashaie [117] proposed a circulating fluidized bed membrane reformer–regenerator, which allows continuous catalyst regeneration, hydrogen separation through a hydrogen permselective membrane, use of sweep gas, and oxygen addition to the feed through an oxygen selective membrane as shown in Figure 14.15.

A different type of MR for the partial oxidation of methane was presented by Ikeguchi et al. [118]. While air was fed to the membrane on the retentate side, methane was fed to the permeate side. This way oxygen ions permeated through the membrane and reacted with methane on the permeate side over 1 wt. % rhodium catalyst supported on magnesia. The membrane was about 800 μm thick. The oxygen flux through the membrane was about 5.5 l oxygen through 100 cm^2 membrane area per hour in absence of methane feed on the permeate side. This value increased to 5.2 l/h when 5 vol.% methane was fed to the permeate side. At 900°C reaction temperature, 30–40% of the methane feed could be converted with high selectivity toward carbon monoxide and hydrogen. The catalyst activity suffered from carbon formation.

An electrochemical MR was presented by Yamaguchi et al. [119]. Proton conducting ceramic electrolyte and a

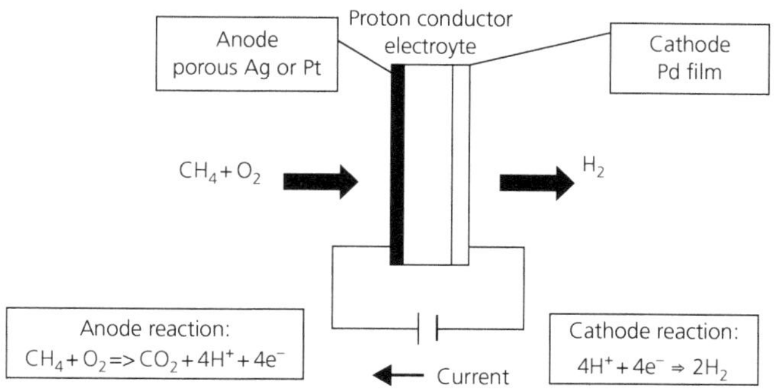

Figure 14.16 Scheme of the electrochemical membrane reactor for partial oxidation of methane [119].
(Source: Yamaguchi et al. [119]. Reproduced with permission of Elsevier.)

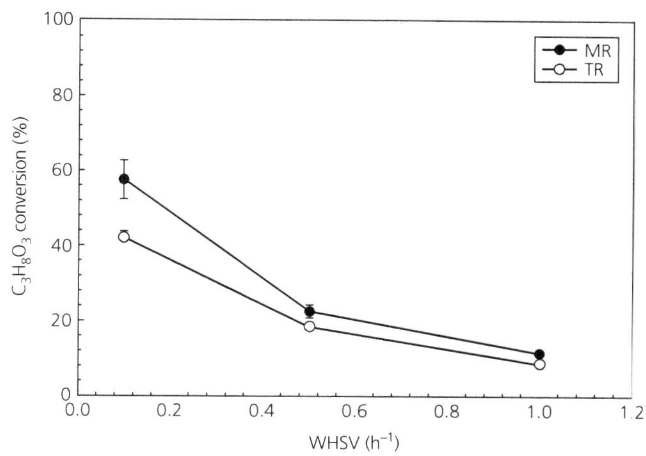

Figure 14.17 Glycerol conversion at different *WHSV* in a membrane reactor (MR) and in a traditional reactor (TR) [125].
(Source: Iulianelli et al. [125]. Reproduced with permission of Elsevier.)

hydrogen-permeable membrane cathode were combined (Figure 14.16). Methane and air were fed to a silver anode and the protons formed by the electrochemical oxidation reaction were transported through the ceramics to the cathode, where hydrogen was formed. The hydrogen permeating through the membrane cathode had high purity. The proton flux through the electrolyte increased with increasing voltage and reaction temperature, which was varied between 400 and 650°C. Anode polarization caused current losses.

An oxygen permeable membrane ($YBa_2Cu_3O_{7-x}$) was applied by Hu et al. [120] for partial oxidation of methane. The membrane served as an oxygen source by diffusion of environmental oxygen into the reactor, which was fed with methane and diluent only. By these means dilution of the product with nitrogen from the air feed could be avoided. Without any catalyst 28% conversion and 95% CO selectivity could be achieved already at a reactor temperature of 900°C. Over Ni/ZrO_2 catalyst nearly 100% methane conversion and 95% CO selectivity were achieved in their small testing reactor. The oxygen permeation flux reached 1.5 ml/cm^2 at 900°C reaction temperature. However, the membrane suffered from stability issues, which were attributed to copper reduction under the experimental conditions applied. Kusakabe et al. [121] prepared yttria-stabilized zirconia membranes on α-alumina support tubes by a sol–gel procedure. This material was investigated as alternative to silica, which is also highly selective for hydrogen permeation, but degraded in the presence of steam. The membrane was then impregnated with platinum and rhodium as catalyst for methane steam reforming. The membrane was 1 μm thick, while the platinum and rhodium loading was 12 and 9 wt.%, respectively. At 500°C the permeance for hydrogen was as high as 10^{-6} mol/(m^2 s Pa). Higher conversion could be achieved in the MR compared to a fixed-bed reactor, but the permeate contained methane and carbon oxides as well.

In recent years, fuels of higher molecular weight and reforming reactions, which have higher tendency toward coke formation such as dry reforming, have been the subject of investigations in the field of membrane reformer development. Mundschau et al. [122] presented a catalytic membrane reformer for diesel fuel, which was composed of yttria-stabilized zirconia for distributed air introduction into fuel feed reaching into the catalytic fixed bed. The latter contained either iron or cobalt perovskite catalysts ($La_{0.5}Sr_{0.5}CoO_{3-\delta}$ or $La_{0.5}Sr_{0.5}FeO_{3-\delta}$). However, the catalyst produced significant amounts of methane impairing the hydrogen yield.

Faroldi et al. [123] performed dry reforming of methane in a Pd MR over catalyst containing 0.6 wt.% ruthenium. At a CO_2/CH_4 ratio of 1.0, a hydrogen permeation flux of 5.68×10^{-7} mol/(s m^2) and a hydrogen recovery of 80% were achieved. When sweep gas was dosed through the central membrane tube, up to 55% conversion were achieved in the reactor. Calabro et al. [124] proposed a pressure swing reformer, which is composed of a fixed catalyst bed for reforming and a membrane separation coupled to a methanation for CO removal downstream. The fixed catalyst bed is periodically switched between reforming operation and combustion of the retentate of the membrane separation process along with the fuel cell anode off-gas by the oxygen depleted air of the cathode off-gas. Pd40Cu membrane was proposed for the separation process. According to the authors, such a membrane can deliver 750 l/min hydrogen at 3.5 bar pressure through 2 m^2 membrane area of 2 μm thickness. Iulianelli et al. [125] investigated glycerol steam reforming in a pinhole-free PdAg MR operated with 0.5 wt.% Ru/Al_2O_3 catalyst. Owing to the maximum operating temperature of the membrane of 450°C, the reaction was performed at 400°C, which is a rather low value for glycerol steam reforming. At $S/C = 2$, a pressure of 5 bar and a low *WHSV* of 0.1 h^{-1}, 60% conversion could be achieved, which was, however, higher that the value found for a conventional fixed-bed reactor (40%, Figure 14.17). Hydrogen recovery was 60% under these conditions. However, the membrane permeability suffered from carbon deposition, which could be removed only partially by hydrogen treatment at 400°C.

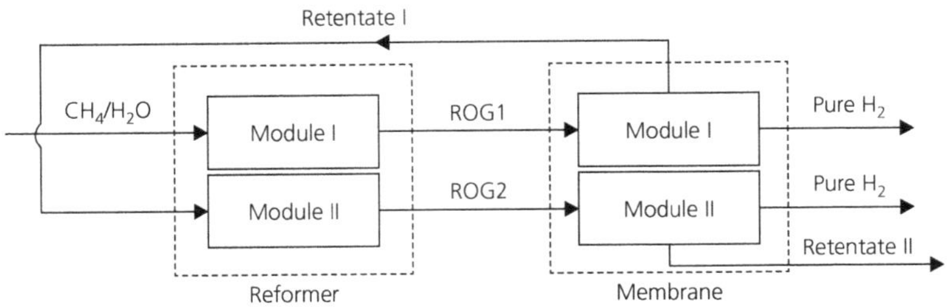

Figure 14.18 Schematic of the staged-separation membrane reactor [126].
(Source: Li et al. [126]. Reproduced with permission of Elsevier.)

In a MR, the membrane separation and the reforming process take place in a single device at the same temperature. Li et al. [126] proposed to separate both processes and to run them at their optimum operating temperature. The authors claim savings of membrane material for palladium-based membranes of up to 84%, because lower membrane thickness is required for the lower temperatures of the membrane separation process, for example, compared to the reaction temperature of methane reforming. Li et al. [126] proposed a two-staged reformer/separator process shown in Figure 14.18, which is, however, still a two-stage reformer followed by separate conventional membrane separation according to the opinion of the author of this section.

14.5 Water-gas shift reactors

The moderately exothermic WGS reaction is limited by its thermodynamic equilibrium. The adiabatic temperature rise of the exothermic reaction requires a separation of the WGS conversion process into two consecutive steps, namely, high-temperature and low-temperature WGS in adiabatic fixed beds or monoliths. An intercooler or water injection is usually switched between both reactors. High-temperature WGS is operated between 375 and 450°C and low-temperature WGS between 200 and 300°C [4]. Depending on the feed composition, the product of high-temperature WGS product contains between 2 and 3 vol.% carbon monoxide, while the product of low-temperature WGS product contains between 0.05 vol.% for the industrial scale [127]. Higher values between 0.3 and 1 vol.% are common for applications of the smaller scale.

When fixed beds of industrial WGS catalysts are applied in small fuel processors, the shift stages dominate the overall system volume and weight by up to 50% owing to the low catalyst utilization [128]. As an alternative to catalysts developed for the industrial scale, precious metal catalysts coatings show at least an order of magnitude higher activity. Certainly the most prominent noble metal catalyst formulation is platinum/ceria [4].

14.5.1 Monolithic reactors

Packed metal gauze technology, which is in fact a monolithic reactor design, was used by Castaldi et al. [68] and Roychoudhury et al. [69] for the WGS reaction. Reformate surrogate containing 3.6 vol.% carbon monoxide was fed to a single-stage WGS reactor, which was operated between 220 and 300°C reaction temperature. At a space velocity around 5000 h^{-1}, 50% conversion was achieved, which corresponded to the thermodynamic equilibrium. Dokupil et al. [129] operated monolithic WGS reactors at a *GHSV* of 20 000 h^{-1} for medium temperature shift and 10 000 h^{-1} for low-temperature shift. The adiabatic temperature rise in the medium temperature reactor operated at 300°C amounted to 68 K, while only 15 K were determined in the low-temperature reactor operated at 280°C. Pasel et al. [130] analyzed the performance of two-stage monolithic WGS reactors coupled to an ATR. Besides the evident conclusion that intermediate water injection is beneficial for the performance of the second reactor owing to the shift of the equilibrium of the reaction and lower reactor inlet temperature, the authors also concluded that adiabatic operation of WGS reactors results in inferior performance compared to isothermal operation, which leads to the introduction of cooling functions into differently design (heat exchanger) reactors, which will be described in the following section. van Dijk et al. [131] proposed quasi-isothermal operation of an adiabatic isothermal WGS reactor. The isothermal behavior was achieved by the high heat conductivity of the aluminum honeycomb monoliths applied. Chen et al. [132] developed a monolithic ATR for ethanol applying $Ir/La_2O_3/ZrO_2$ catalyst supported by ceramic foams. At a *S/C* ratio of 1.5 and an *O/C* ratio of 0.83, 650°C reaction temperature was required to achieve full ethanol conversion and complete suppression of acetaldehyde formation. The reactor was tested under conditions of repeated start and stop, and under continuous operation for a duration of 13 h. However, increased formation of acetaldehyde indicated rapid catalyst deactivation.

14.5.2 Plate heat exchangers and microstructured water-gas shift reactors

The WGS reaction requires a countercurrent cooling concept to achieve an optimum temperature profile, which creates high reaction rate at high temperature at the reactor inlet and a decreasing temperature toward the reactor outlet, which shifts the equilibrium of the reaction toward high conversion. Zalc and Löffler [133] proposed to use a plate heat exchanger to achieve such a temperature profile. The authors calculated that an optimum temperature profile significantly improves the carbon monoxide conversion compared to isothermal and adiabatic reactor operation. The low catalyst utilization in

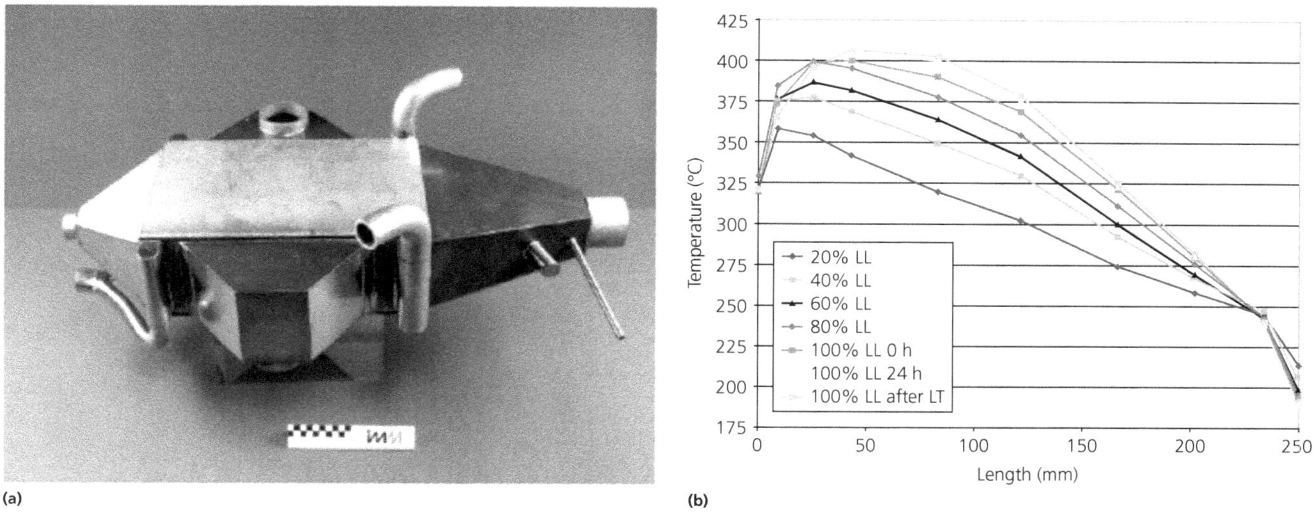

Figure 14.19 (a) Integrated water–gas shift reactor/heat exchanger designed for 5 kW fuel cell system, (b) internal temperature profiles determined during operation of this reactor; the different profiles correspond to different load (100% load is the 5 kW equivalent). The values shown for 100% load were determined at the fresh reactor (0 h) after 24 h under operation (24 h) and after an accidental exposure to reformate of too low temperature for the water–gas shift reaction (after LT) [141].
(Source: O'Connell et al. [141]. Reproduced with permission of Elsevier.)

fixed-bed WGS reactors is overcome by wall-coated microchannel reactors. Noble metal-based catalysts allow the operation of wall-coated monoliths or plate heat exchangers at high volume hourly space velocity (*VHSV*) in the range of 180 l/(h g_{cat}). Microchannel plate heat exchanger technology reduced the required catalyst mass by 50% compared to a fixed bed or monolithic two-staged, adiabatic design [1, 134, 135]. Baier and Kolb [136] showed that a reduction in the flow rates (turndown) improves the performance of such a WGS heat exchanger/reactor. The same authors also calculated the optimum channel size to avoid diffusion limitations. For a channel height of 200 μm, no diffusion limitations occurred, while severe diffusion limitations were observed for channel height of 800 μm, which is in agreement with earlier experimental work performed by Pasel et al. [137]. However, a trade-off is required between catalyst utilization and pressure drop in a practical system [138]. Increasing the channel height from 200 to 800 μm doubles the reactor length required to achieve the same degree of carbon monoxide conversion.

A microchannel heat exchanger/reactor for WGS in the kilowatt size range was described by Kolb et al. [139]. The reactor still had a three-stage cross-flow design for the sake of easier fabrication. Platinum/ceria catalyst was wash coated onto the metal plates, which were sealed by laser welding afterward. The reactor was tested separately and showed equilibrium conversion under the experimental conditions of partial load. Dubrovskiy et al. [140] designed a microstructured reactor/heat exchanger for the WGS reaction containing flat, perforated Mo plates and Mo wires with a diameter of 250 μm and a length of 100 mm coated with porous Mo_2C. The optimum feed temperature was determined to be 400°C and the optimum temperature profile was adjusted in the reactor. The power equivalent of the

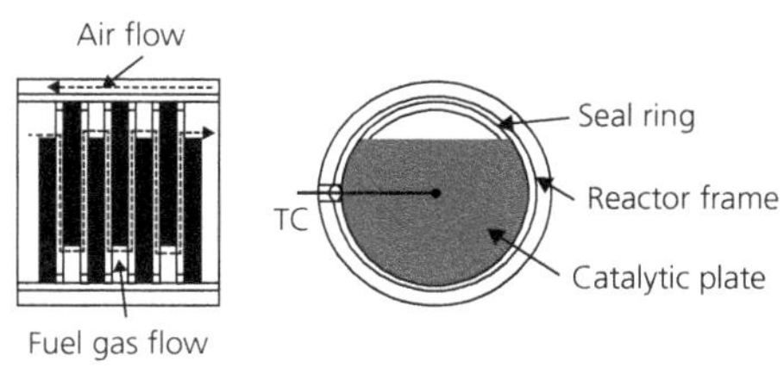

Figure 14.20 Thermal conductive catalyst plates integrated into a water–gas shift reactor. The dotted line indicates the flow paths of reformate and air [143].
(Source: Baronskaya et al. [143]. Reproduced with permission of Elsevier.)

reactor amounted to 45 W. The start-up time demand was only 2 min owing to the low thermal mass of the reactor.

Kolb et al. and O'Connell et al. developed WGS heat exchanger/reactors in the kilowatt scale [104, 141, 142]. A WGS reactor with 5 kW power equivalent is shown in Figure 14.19a. It contains a straight flow path for the reformate, inlets, and outlets from both sides for the cooling gas and cross-flow channels for preheating the reactor. Typical temperature profiles determined experimentally for the counter currently operated plate heat exchanger/reactor are shown in Figure 14.19b. They show a temperature peak at the reactor inlet, owing to the high initial heat of reaction, while the reactor temperature decreased toward the outlet of the reactor. The content of carbon monoxide in the reformate surrogate could be reduced from 10.6 to 0.7 vol.%, which corresponded to 95% conversion. Lower temperatures at the shell of the reactor, especially in the inlet section, were attributed to heat losses to the environment.

Baronskaya et al. [143] described an innovative concept for a heat exchanger/reactor for WGS. As shown in Figure 14.20, sintered plates of Cu/Zn/Zr catalyst mixed with metallic powder

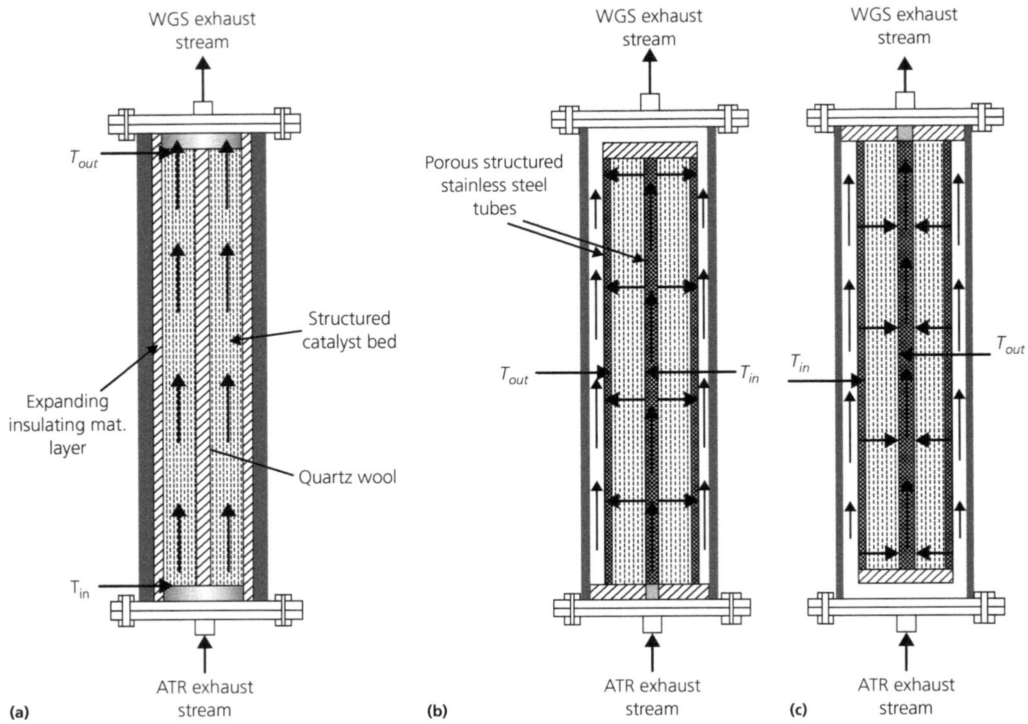

Figure 14.21 (a) Axial, (b) centrifugal, and (c) centripetal flow arrangements of the water–gas shift reactors [144]. (Source: Palma et al. [144]. Reproduced with permission of Elsevier.)

were stacked with spacers and externally cooled by an air flow. By these means, a temperature gradient from 380 to 250°C and more than 90% conversion at a feed concentration of 11.9 vol.% of carbon monoxide could be achieved in a prototype reactor at a *GHSV* of 6300 h^{-1}.

Palma et al. [144] developed WGS reactors in the kilowatt scale, which allowed gradual addition of the reformate through porous sintered stainless steel tubes into ceramic foams (SiC/Al_2O_3) coated with noble metal-based catalyst. Different flow arrangements through the foams were investigated (Figure 14.21), which revealed better performance for radial flow.

14.5.3 Water-gas shift in membrane reactors

Brunetti et al. [145] investigated, by numerical simulations, the potential of MRs for substituting conventional fixed-bed technology. They revealed that a MR has the potential of replacing two-stage adiabatic beds with intermediate water injection similar to the plate heat exchanger technology described previously. Central PdAg membrane tubes were assumed for the reactor positioned in an annular bed of FeCr catalyst with sweep gas flow. Figure 14.22 shows the conversion feasible in a MR compared to the traditional two-stage reactor concept. Figure 14.23 shows the volume savings feasible by applying membrane technology at different reaction temperatures and pressures. At 15 bar the MR requires only 9–12% of the volume of the traditional reactors, because higher pressure obviously increases hydrogen removal by membrane separation and shifts the equilibrium of the WGS reaction. However, higher pressure is a critical issue for decentralized fuel processors of smaller scale because it increases parasitic power losses of the system.

Barbieri et al. [146] developed a MR for the WGS reaction. A palladium/silver film containing 23 wt.% silver, which had a thickness between 1 and 1.5 μm, was prepared by sputtering and coated onto a porous stainless steel support. This preparation method generated a much higher ratio of pore size to film thickness compared to conventional methods. Tubular membranes of 13 mm outer diameter, 10 to 20 mm length, were fabricated. Commercial Cu-based catalyst from Haldor-Topsøe was introduced into the fixed bed. At reaction temperatures between 260 and 300°C, and a GHSV of 2085 h^{-1}, the thermodynamic equilibrium conversion could be exceeded by 5–10% by the membrane technology.

14.6 Carbon monoxide fine cleanup: Preferential oxidation and selective methanation

Both the preferential oxidation of carbon monoxide and the hydrogen oxidation, which occurs in parallel as an undesired side reaction, are highly exothermic. Preferential oxidation

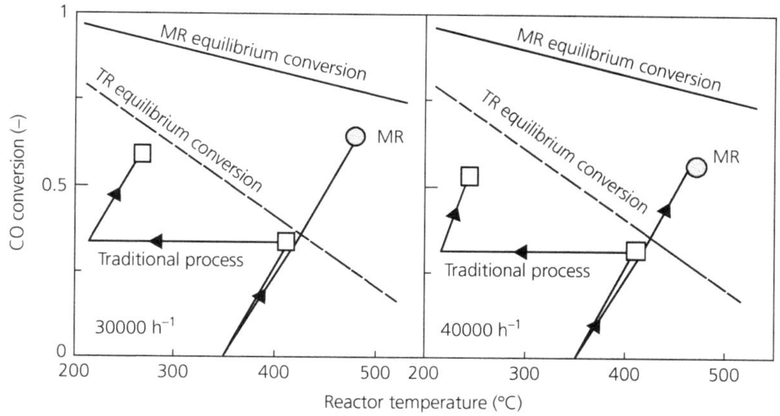

Figure 14.22 CO conversion as a function of reaction temperature for a membrane reactor (MR) and the traditional process at different *GHSV* for 350°C feed temperature and a reaction pressure of 15 bar [145].
(Source: Brunetti et al. [145]. Reproduced with permission of John Wiley & Sons, Inc.)

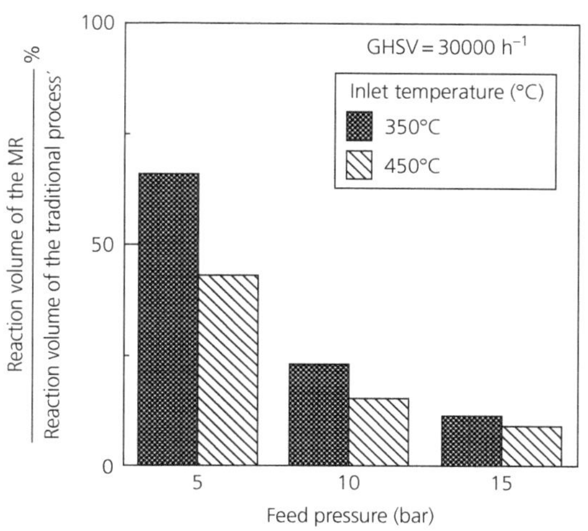

Figure 14.23 Volume reduction by application of membrane technology for water–gas shift for different feed pressure and reaction temperature [145]. (Source: Brunetti et al. [145]. Reproduced with permission of John Wiley & Sons, Inc.)

requires excess air corresponding to an *O*/*CO* ratio which is equivalent to the λ-value ($\lambda = 1$ stands for a stoichiometric ratio) between 1.5 and 2.0, which means that under full conversion of carbon monoxide between 0.5 and 1.0 mol of hydrogen is lost for each mole carbon monoxide converted.

Formation of carbon monoxide over the catalyst by the reverse water–gas shift reaction (RWGS) in an oxygen-deficient atmosphere is frequently observed especially under conditions of partial load, because most catalysts for preferential oxidation of carbon monoxide have some activity for WGS and its reverse reaction. Therefore oversizing the reactor bears the danger of impaired conversion and the same applies for partial load of the reactor unfortunately. Because the concentration of carbon monoxide that is tolerated by low-temperature fuel cells is usually in the range below 100 ppm or less, even low catalytic activity for reverse shift becomes an issue.

Numerical simulation of the preferential oxidation of carbon monoxide was performed by Ouyang et al. [147], which described the behavior of their reactor very well. A simplified mechanism of the reaction network was formulated by the authors. Since the carbon monoxide oxidation requires OH species, its rate increases when hydrogen is present. The simulations showed that the catalyst surface was almost completely covered by carbon monoxide to a certain channel or reactor length (or residence time). The coverage with carbon monoxide decreased abruptly, when the carbon monoxide was completely consumed. At this point, the concentration of adsorbed oxygen and hydrogen increased by almost an order of magnitude, leading to consumption of hydrogen by water formation. The water formation was then limited only by the oxygen present in the gas phase. This means that high selectivity toward carbon dioxide is only feasible at low carbon monoxide conversion over platinum catalyst, which could be verified experimentally by Ouyang et al. [147]. The desorption of carbon monoxide increased with rising temperature over the platinum catalyst. Mass transfer limitations are dominant at higher conversion of carbon monoxide in fixed beds and in microchannels [147], while heat transfer limitations were to be expected only in fixed catalyst beds according to these simulations.

Selective methanation of carbon monoxide does not require air addition to the reformate but suffers not only from competing CO_2 methanation [148],

$$4H_2 + CO_2 \rightarrow 2H_2O + CH_4, \quad \Delta H^0_{298} = -165\ \mathrm{kJ/mol} \qquad (14.13)$$

but also from temperature management problems [149], which can only be solved by multistaged operation when conventional adiabatic fixed beds or monoliths are applied.

The most critical issue is that because the concentration of carbon dioxide is much higher in the reformate compared to that of the carbon monoxide, the catalyst has to be very selective for carbon monoxide methanation. The operating window of methanation catalysts is relatively small, in the temperature range around 250–300°C, because a trade-off is required between sufficient activity and selectivity. Well above 250°C all methanation catalysts tend to be selective for carbon dioxide methanation.

When oxygen is added to the feed, some catalysts convert carbon monoxide to carbon dioxide with very high selectivity while the methane formation starts not before the oxygen is completely consumed [150]. Therefore preferential oxidation and selective methanation could be operated in a combined manner. Care has to be taken when choosing the catalyst for preferential oxidation of carbon monoxide, because some formulations are very sensitive to light hydrocarbon contained in the reformate as reported by Kraaij et al. [151] who observed rapid deactivation.

14.6.1 Fixed-bed reactors

Lee et al. [152] developed two-stage tubular preferential oxidation reactors for a 10 kW_{el} fuel cell system. $Pt/Ru/Al_2O_3$ catalyst was used as fixed bed, which converted 1 vol.% carbon monoxide in the feed to values below 100 ppm between 100 and 140°C temperature at an *O/CO* ratio of 4.0 and a *WHSV* of 40 l/(h g_{cat}). While a single reactor stage with 1 kW_{el} power equivalent reduced the carbon monoxide concentration to less than 200 ppm at an *O/CO* equal to 3.2, the reactor exit temperature increased to 240°C. The introduction of a second-stage reactor reduced the carbon monoxide further below 10 ppm at *O/CO* 2.0 and 4.0 in the first and second stage, respectively. The amount of air fed to the second stage was lower, because only 1000 ppm carbon monoxide had to be converted. Dynamic load changes from 100 to 35% and back to 100% proved that the carbon monoxide in the reformate effluent could be maintained below 100 ppm. One hundred twenty eight parallel tubes with 1.27 mm inner diameter were switched in series to gain a power equivalent of 10 kW_{el}. A cooling stage was introduced between the reactors. The dimensions of the reactor were rather low with 200 mm diameter and 200 mm length. Load changes from 100 to 65% and back were demonstrated for the 10 kW_{el} reactors, proving that carbon monoxide could be maintained at values below 100 ppm. The reactors were brought to operating temperature within 3 min by heating the reactor tubes with reformate of 150°C temperature and oxygen addition at *O/CO* = 3.0. By these means the carbon monoxide could be reduced below 20 ppm within 3 min.

Pan and Wang [153] switched four adiabatic preferential oxidation reactors in series downstream of a methanol reformer/evaporator. Heat exchangers were installed after each reactor. The four reactors were operated at the same inlet temperature of 150°C, and the *O/CO* ratio in the feed increased from 1.6 to 3 to minimize heat formation in the first reactors. Despite these measures, a temperature rise of 121 K was observed in the first reactor while 82 K was observed in the second. While only 50% carbon monoxide conversion was achieved in the first reactor, full conversion was observed after the last stage. The combined steam reformer/cleanup system was operated for 24 h duration. The carbon monoxide content of the reformate could be maintained below 40 ppm.

14.6.2 Monolithic reactors

Zhou et al. [154] reported the operation of a four-stage monolithic preferential oxidation reactor system in a 5 kW methanol reformer for 14 h test duration. Carbon monoxide was reduced to less than 50 ppm; however, four stages of oxygen addition were required.

Dokupil et al. [129] described a monolithic preferential oxidation reactor operated at 100°C temperature and 15 000 h^{-1} *GHSV*. It carried an integrated heat exchanger to improve its thermal management.

Metal foams as alternative to metallic or ceramic monoliths were used by Chin et al. [155] for preferential oxidation. Foams of different pore density (40 pores per inch or 20 pores per inch) were studied. The foams either contained 4 or 12 wt.% of the catalyst, which was composed of 5 wt.% platinum and 0.5 wt.% iron on alumina. Higher conversion was achieved for smaller pore density and lower catalyst loading on the foam. Conversion decreased when the reaction temperature was increased from 100°C and full conversion could not be achieved. This originated from hot spot formation despite the low *O/CO* ratio between 1.0 and 2.0. With ceramic monoliths of comparable size, the authors achieved comparable results. Ahluwalia et al. [156] performed preferential oxidation of carbon monoxide with cordierite monoliths. Commercial catalyst from Engelhard was used for the experiments, which showed light-off at ambient temperature. Reformate surrogate was fed to the reactors, which was composed of 48 vol.% hydrogen, 31 vol.% nitrogen, and different amounts of carbon oxides and steam. The feed temperature ranged between 85 and 100°C. For 1.3 vol.% carbon monoxide in the feed and a low *O/CO* ratio of 1.0, 90 K adiabatic temperature rise was observed in the monolith, while it exceeded 170 K for *O/CO* = 2.0. The carbon monoxide selectivity was almost independent of the carbon monoxide concentration in the feed but rather depended on the *O/CO* ratio for a specific *GHSV*, while the presence of steam had little effect on the carbon monoxide selectivity, which obviously was an inherent property of the catalyst under investigation. Modeling of the reaction system applying experimental results revealed that the carbon monoxide concentration could not be reduced below 10 ppm in a single-stage reactor when carbon monoxide concentration in the reformate feed exceeded 1.05 vol.%. Consequently a concept for two-stage system was developed. About 85% of the carbon monoxide had to be converted in the first reactor. The optimum *O/CO* ratio was in the range of 1.2 for the first reactor, while the second reactor required a higher value of 2.4. By addition of a third stage, the performance of the chain of monoliths could be further improved and even 3.5 vol.% carbon monoxide or more in the feed could be converted completely according to these calculations. The hydrogen loss was low with 1 vol.%. However, the attractive performance of a three-staged system was counterbalanced by a rather complex set-up.

Packed metal gauze technology (named microlith see also Section 4.2) was tested by Castaldi et al. [68] and Roychoudhury et al. [69] for preferential oxidation with surrogate of WGS product, which contained 5000 ppm carbon monoxide. The *O/CO* ratio of the feed was set to 2.4. The adiabatic temperature rise of the reformate was calculated to 45°C under these conditions, while the reactor was operated at about 250°C reaction temperature for 500 h. In a single reactor stage at a high *GHSV* of 150 000 h^{-1}, about 93% conversion was achieved, which corresponds to 350 ppm carbon monoxide at the reactor outlet. Therefore a second-stage reactor was required, which was operated at a higher *O/CO* ratio of 8.0 and even at a higher *GHSV* of 220 000 h^{-1}. The remaining carbon monoxide was converted to a concentration below 20 ppm also under load change conditions.

The local overheating of adiabatic fixed catalyst beds during preferential oxidation of carbon monoxide was demonstrated by Ouyang et al. [157], which observed a temperature rise of almost 70 K at 180°C reaction temperature in the center of their reactor. As a consequence, multiple fixed beds are required for adiabatic operation. While Giroux et al. [1] and Kim et al. [158] proposed two adiabatic stages, Pan and Wang [153] switched four adiabatic preferential oxidation reactors in series. The temperature rise in the first reactor was as high as 121 K. Cruz et al. [159] applied external cooling by cooling air via cooling fins of their monolithic CO-PrOx reactor.

The aforementioned examples demonstrate that the thermal management of monolithic reactors for the preferential oxidation reaction is crucial and requires either up multistage solutions especially during load changes [160] or integrated heat exchange capabilities, which is a dedicated feature of the plate heat exchangers described in the following section.

14.6.3 Plate heat exchangers and microstructured reactors

Zalc and Löffler [133] proposed the application of plate heat exchanger technology for the preferential oxidation of carbon monoxide to improve the heat management. The optimum operating temperature of state-of-the-art preferential oxidation catalysts is in the range of 100°C, which favors evaporation cooling by water, preferably in a cocurrent flow arrangement, for the heat management of plate heat exchanger/reactors.

Dudfield et al. [161] compared a plate and fin heat exchanger (0.25 l reactor volume) with a shell and tube heat exchanger filled with catalyst microspheres (0.25 l reactor volume) and a heat exchanger containing steel granules, which were first sintered to form a porous structure and then wash coated with catalyst (0.25 l reactor volume). The plate and fin design was realized as a sandwich and made of aluminum, while the other devices were made of stainless steel. All reactors were cooled by oil. As to be expected, the pressure drop was the lowest for the plate and fin reactor, 30 times higher for the sintered structure and 8 times higher for the shell and tube heat exchanger. The performance of the reactors was compared at 150°C temperature, 10–175 l/min flow rate of reformate surrogate, and 1–17.5 l/min air feed flow rate. The thermal management of the sintered porous structure created severe hot spots despite the integrated heat exchange functionalities. The two remaining reactors showed comparable performance slightly in favor of the plate and fin design. Hot spots were limited to 20 K.

In order to meet the requirements of a 20 kW fuel processor, which was the final target of their work, Dudfield et al. [161] scaled their reactors up to a dual stage design of 4 l total volume. Each reactor was 108 mm high, 108 mm wide, and 171 mm long; had a volume of 1.85 l and 2.5 kg weight; and carried 8.5 g catalyst. The air feed was split in a ratio of 2/1 between the first and second reactor. When the reactors were operated at a temperature of 160°C and a feed flow rate of 200 l/min, the carbon monoxide content could be decreased from 2.0 vol.% to values below 15 ppm at an *O/CO* ratio of 5.0 [162]. The reactors were operated at full load (20 kW equivalent power output) for approximately 100 h without apparent deactivation [161].

A silicon microreactor for preferential oxidation was designed by Srinivas et al. [163], which was 6 cm × 6 cm wide and long, while the flow path was only 400 μm high. Instead of microchannels, pillars were chosen for the flow distribution in the reactor. The reactor was coated with 2 wt.% platinum/alumina catalyst with a thickness of 10 μm. Tests were performed at an *O/CO* ratio of 2.0 and a high *VHSV* of 120 l/(h g_{cat}). Not more than 90% conversion of carbon monoxide could be achieved in the reactor at 210°C reaction temperature, while similar results were obtained for a small fixed catalyst bed.

Cominos et al. [164] developed a microchannel reactor with integrated heat exchange for the preferential oxidation of carbon monoxide for a fuel processor/fuel cell system with 100 W electrical power output. The reactor was split into three parts, namely, two heat exchangers and the reactor itself, thermally decoupled by insulation material. Fuel cell anode off-gas was foreseen as coolant flowing consecutively through the three devices. The reactor itself contained 19 plates each with 82 microchannels coated with catalyst and sealed by graphite gaskets. Ten plates carrying 75 microchannels each served for the coolant flow within the reactor. This reactor concept was then optimized for size and weight by replacing gaskets with a laser-welded sealing [165]. A length/width ratio of 3 was chosen for the heat exchangers and the reactor in order to minimize axial heat transfer, which would impair the heat exchanger performance.

Ouyang et al. [147] studied the preferential oxidation of carbon monoxide in silicon reactors of the smallest scale fabricated by photolithography and deep reactive ion etching. The reactors had two gas inlets for reformate and air, a premixer, a single reaction channel, and an outlet zone where the product flow was cooled. The channels were sealed by anodic bonding with a Pyrex glass plate. Full conversion of carbon monoxide was achieved between 170 and 300°C reaction temperature.

Lopez et al. [166] described the operation of their mesoscaled folded-plate reactor which was operated with water cooling in a cocurrent flow arrangement, with four stages of air addition (Figure 14.24). Their Au catalyst was operated well below 100°C at an *O/CO* ratio of 3. The CO could be reduced to values below 100 ppm and the reactor had 0.4–0.6 kW electric power equivalent.

A microstructured plate heat exchanger for preferential oxidation in the kilowatt size range was developed by Kolb et al. [139]. The reactor had three-stage cross-flow design for the sake of easier fabrication. Platinum catalyst supported by alumina and zeolite A was coated onto the metal plates, which were sealed by laser welding. The reactor was tested separately with a feed flow rate of 185 l/min, which contained 0.4 vol.% carbon monoxide. The reactor converted more than 90% of the carbon monoxide at 206°C reaction temperature and *O*/*CO* ratio of 4.1 at *WHSV* ranging between 48 l/(h g_{cat}) and 98 l/(h g_{cat}).

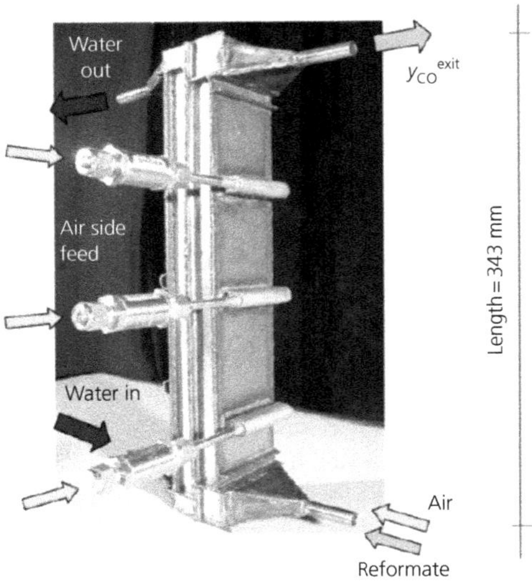

Figure 14.24 Cocurrently liquid-cooled folded-plate reactor for preferential oxidation [166].
(Source: Lopez et al. [166]. Reproduced with permission of Elsevier.)

Kolb et al. [167] reported the design, fabrication, and testing of another preferential oxidation reactor on the kilowatt scale (Figure 14.25a). Evaporation cooling was used here in a cocurrent flow arrangement. The reformate entered the preferential oxidation reactor from both sides, while the water was circulated in a front distribution chamber at the reactor inlet. Similar to the inlet, the purified reformate left the reactor at both sides, while the superheated steam was gathered in a single outlet manifold.

Stable operation and narrow reactor temperature ranges of 10 K could be adjusted inside the reactor at least for lower λ values, that is, *O*/*CO* = 2. Figure 14.25b shows a plot of the CO concentration as determined in the PrOx reactor product over the reactor temperature (average values were chosen) for different experimental conditions. The carbon monoxide concentration in the product was not affected by the *O*/*CO* ratio but depended only on the reactor temperature. Therefore a higher surplus of air (higher *O*/*CO* value) did not improve the reactor performance. The lowest CO values (below detection limit of 5 ppm) were measured for the lowest *O*/*CO* value of 2, because a lower reactor temperature could be achieved owing

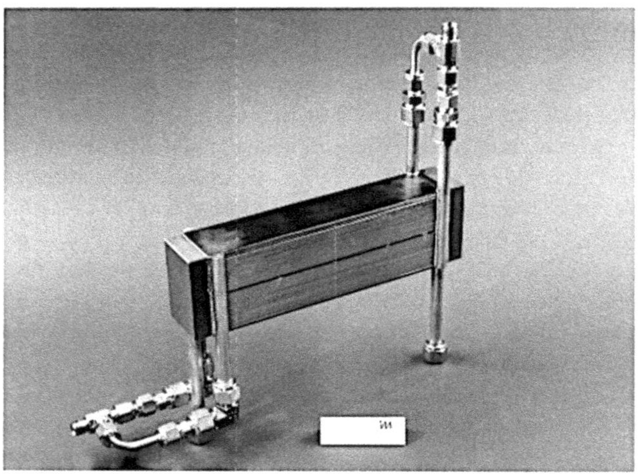

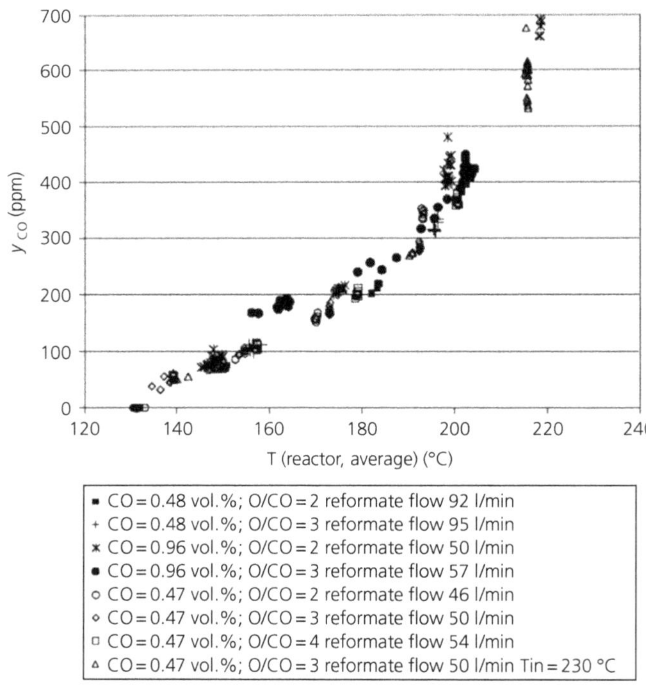

Figure 14.25 (a) Cocurrently operated microreactor for preferential oxidation cooled by water evaporation, (b) CO concentration in the off-gas of the reactor at different total reformate flow rates, CO inlet concentrations and *O*/*CO* values [167].
(Source: Kolb et al. [167]. Reproduced with permission of Elsevier.)

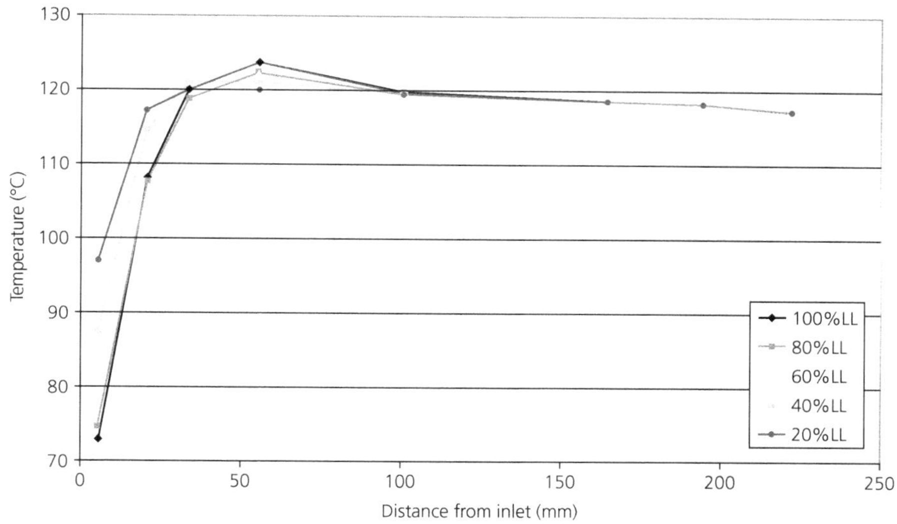

Figure 14.26 Temperature profile of a cocurrently cooled preferential oxidation reactor with 5 $kW_{el,net}$ power equivalent. (Source: Courtesy of Fraunhofer ICT-IMM.)

to the lower heat generation. Higher temperatures favored the reverse WGS reaction leading to higher CO content of the purified reformate as described previously.

O'Connell et al. [141] developed a 5 kW_{el} one-stage preferential oxidation reactor, which was designed and evaluated for the CO cleanup of surrogate diesel reformate. Both partial load operation and load changes could be carried out without significant overshoots of carbon monoxide. Figure 14.26 shows a temperature profile of the reactor. For a CO content in the feed of 1.0 vol.%, the carbon monoxide content was reduced to values below 50 ppm. The reactor was then operated in combination with a 5 kW_{el} WGS reactor switched upstream [142]. Load changes for both reactors could also be carried out without significant overshoots of carbon monoxide.

14.7 Examples of complete fuel processors

Selected examples of complete fuel processors are summarized in the following sections. They document the high complexity of monolith-based systems, which is significantly reduced when integrated plate heat exchanger/reactors are applied. Recent developments in the laboratories of the author's group have revealed even higher compactness by stacking the rectangular plate heat exchanger/reactors. This reduces the system size compared to conventional technology by an order of magnitude.

14.7.1 Monolithic fuel processors

Adachi et al. [168] developed a model for a natural gas fuel processor composed of an ATR designed as metallic foam monolith coated with catalyst and two-stage WGS reactors also designed as foam monoliths followed by two-stage ceramic monoliths for the preferential oxidation of carbon monoxide as shown in Figure 14.27. Figure 14.28 shows the course of temperature and gas composition of feed and reformate as calculated for the different reactors and heat exchangers of the system, while the start-up energy demand of the components is shown in Figure 14.29. Interestingly the start-up energy demand of the ATR is in the same range as the energy demand of both WGS reactors, which partially originates from the lower operating temperature of the latter reactors. The design was tested in a breadboard arrangement. Hexaaluminate was coated on a first section of the ATR to reduce the hot spot in the second section of the reactor, which was coated with noble metal catalyst.

A 3 kW_{el} autothermal gasoline fuel processor was developed by Severin et al. [169]. The fuel was injected into the mixing chamber upstream the ATR. The reformate product of the ATR was cooled in a heat exchanger by the steam and water feed. Two WGS reactors were placed downstream. A heat exchanger installed between them cooled the reformate with part of the reformer air feed. Downstream the WGS followed a preferential oxidation reactor. Finally a heat exchanger cooled the purified reformate and water was removed by condensation before it was fed into the fuel cell. The remaining hydrogen in the fuel cell off-gas was combusted in an afterburner. The afterburner off-gases were used to preheat the air feed of the reformer in a heat exchanger downstream. The reactor temperatures were controlled by water injection into the cooling gas flows of the heat exchangers. All reactors had monolithic design. The ATR was operated at temperatures between 600 and 800°C. According to stoichiometry and thermodynamics, the highest efficiency of the reactor was achieved at the lowest O/C ratio of 0.75. The high- and low-temperature WGS reactors were operated around 450 and 300°C, respectively. At partial load of the system, the operating temperature of the shift reactors could be decreased, which increased the conversion according to the thermodynamic equilibrium. The preferential oxidation reactor reduced the carbon monoxide concentration to approximately 160 ppm. The start-up time demand was approximately 30 min.

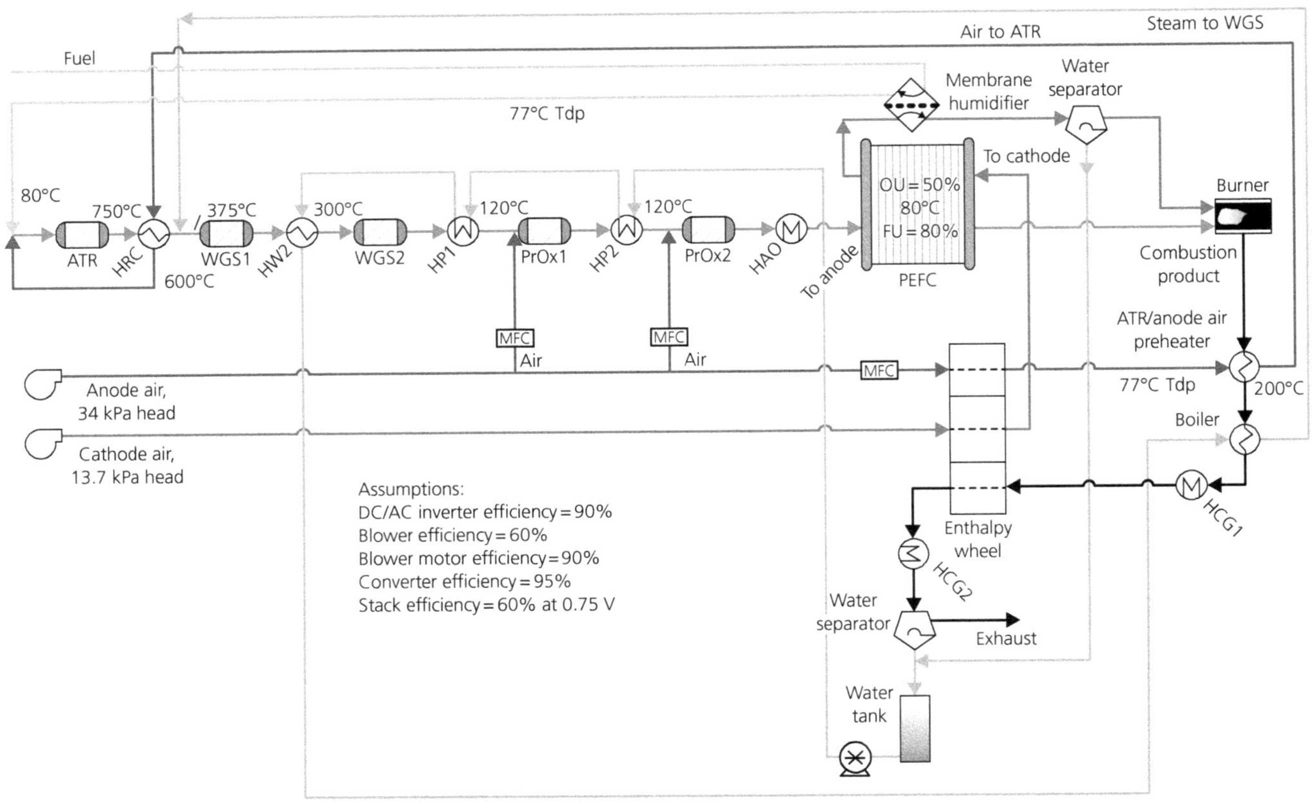

Figure 14.27 Flow scheme of a natural gas fuel processor/fuel cell system [168].
(Source: Adachi et al. [168]. Reproduced with permission of Elsevier.)

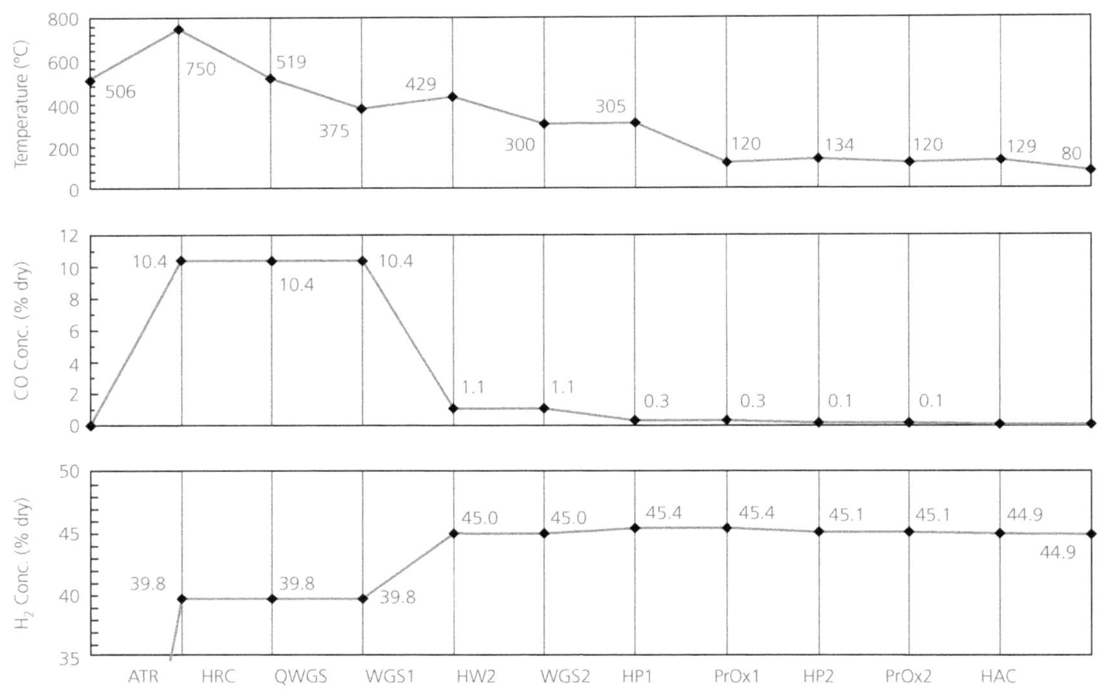

Figure 14.28 Gas temperature, carbon monoxide, and hydrogen content in the gas phase at different components of the natural gas fuel processor/fuel cell system [168].
(Source: Adachi et al. [168]. Reproduced with permission of Elsevier.)

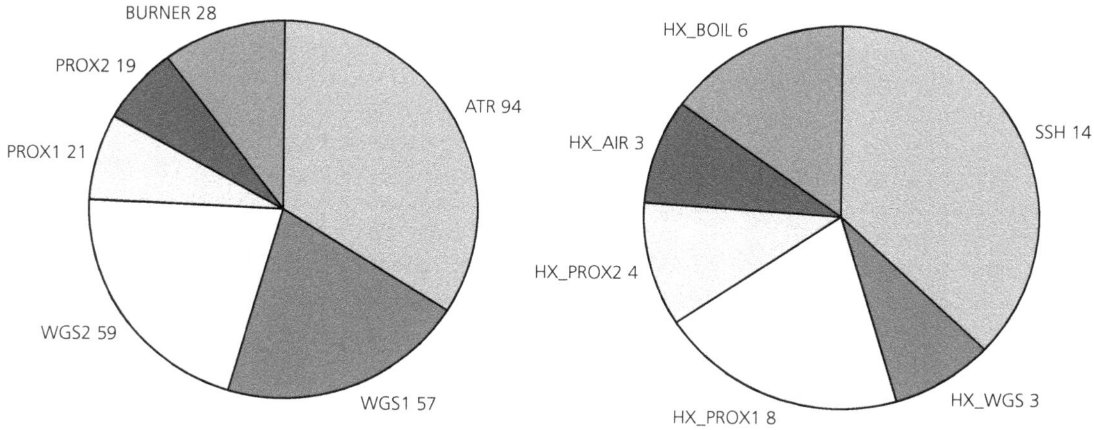

Figure 14.29 Start-up energy demand of the different components of the natural gas fuel processor/fuel cell system [168]. (Source: Adachi et al. [168]. Reproduced with permission of Elsevier.)

Main origin of the start-up time demand was the subsequent heating of a consecutive chain of reactors. Electrical preheating of the inlet section of the reformer monolith reduced the time demand for preheating. The prototype fuel processor had a weight of 18 kg and a volume of 40 l (without insulation).

Lee et al. [170] presented a natural gas fuel processor for the thermal power range up to 16 kW, which was developed by Argonne National Laboratory together with H2fuel LLC Company. Natural gas containing between 20 and 50 ppm sulfur was fed to the monolithic ATR containing commercial catalyst, which was operated between 450 and 750°C. More than 98% methane conversion was achieved and the reformate contained about 40 vol.% hydrogen and 10 vol.% carbon monoxide on a dry basis. The reformate was desulfurized at 350°C by zinc oxide downstream the reformer. Precious metal catalysts were used in annular fixed beds for high- and low-temperature WGS, which surrounded the reformer. The reformer was started by electric heating. The fuel processor was operated for more than 2000 h with stable performance at different load. The efficiency, which was defined as the ratio of the LHV of the hydrogen in the product of the WGS reactor to the LHV of the natural gas feed, was in the ranged between 80% at 20% load and 90% at full load.

Koenig et al. [171] described a 2 kW_{el} combined heat and power (CHP) PEM fuel cell system. The fuel processor was based upon methane steam reforming, WGS, and preferential oxidation of carbon monoxide. As shown in the Sankey diagram (Figure 14.30), about 40% of the thermal energy of the methane fuel was lost to the environment by heat losses and through the thermal energy still contained in hot product and exhaust gases, which emphasizes the importance of efficient insulation and heat recovery in such a system.

Lindermeir et al. [172] described the development work of the German company Webasto for an SOFC-based CHP system. Partial oxidation of methane was chosen as reforming technology. In a two-staged process, part of the fuel was completely oxidized generating carbon dioxide and steam followed by a second-stage fuel addition, which then lead to conditions of partial oxidation. The temperature in the monolithic reformer, where the partial oxidation took place, ranged between 950 and 800°C. The system was operated for 9 h, while the start-up time demand amounted to 3 h.

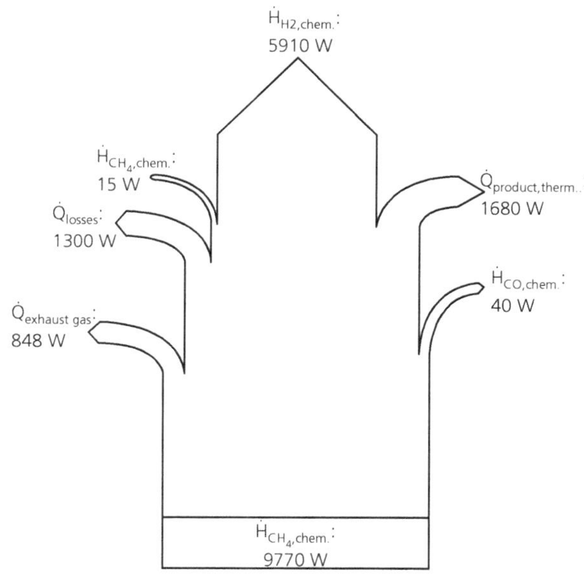

Figure 14.30 Sankey diagram of the 2 kW_{el} CHP PEM fuel cell/methane fuel processor system [171].
(Source: Koenig et al. [171]. Reproduced with permission of Elsevier.)

14.7.2 Plate heat exchanger fuel processors on the meso- and microscale

Dudfield et al. [161] combined a 20 kW methanol reformer with two oil-cooled reactors for the preferential oxidation of carbon monoxide switched in series. The remaining concentration of carbon monoxide in the product was lower than 10 ppm for more than 2 h at a feed concentration of 1.6% carbon monoxide.

Because the reformer was a combination of steam reformer and catalytic burner in plate and fin design, this was regarded as an early and impressive demonstration of the capabilities of the integrated heat exchanger design for fuel processors in the kilowatt range.

Yoshida et al. [173] designed an integrated methanol fuel processor from silicon and Pyrex glass substrates for a power equivalent of 10 W. It contained functional layers for steam reforming, evaporation, and combustion. Commercial Cu/ZnO catalyst served for reforming and the Pt/TiO_2 combustion catalyst was prepared by a sol–gel method. A power density of 2.1 W/cm^3 was determined for the device.

Kolb et al. [86] developed an integrated microstructured fuel processor with an electrical net power output of 100 W. The fuel processor (Figure 14.31) worked very stably under normal operating conditions both under full and partial load. A very narrow temperature profile of only 3 K was achieved in the reformer, which was crucial for the Pd/ZnO catalyst technology applied. Methanol conversion was always complete (>99.9%). At $O/C = 0.25$, $S/C = 1.7$, and 100% load, the carbon monoxide content of the reformate was 1.8 vol.% and the hydrogen content about 50%. The reformate composition did not change significantly at 50% load, only the carbon monoxide content decreased to 1.6 vol.%. The fuel processor was coupled to a high-temperature fuel cell. The high-temperature PEM fuel cell stack generated an electrical power output of 103 W (14.7 V at 7 A) when operated with the reformate from the fuel processor.

An advanced fuel cell/microfuel processor systems is the VeGA system developed by Truma Gerätetechnik GmbH, and Institut für Mikrotechnik Mainz GmbH (IMM) has been described by Wichert et al. [106]. The power output of the systems amounts to about 300 W, while 50 W are consumed by the balance of plant components, which leaves 250 W for the consumer. The fuel processor and a fully integrated and automated system are shown in Figure 14.32a and b, respectively. The fuel processor is composed of an integrated microstructured evaporator and microstructured reformer both integrated with microstructured catalytic burners, heat exchangers (not shown here), and microstructured WGS. Performance data of one of these complete LPG fuel processors which had been operated up to 3500 h in combination with high-temperature PEM fuel cell stacks were reported. Two hundred of these VeGA systems had been fabricated since 2008 and at the time of writing are currently being tested in field trials.

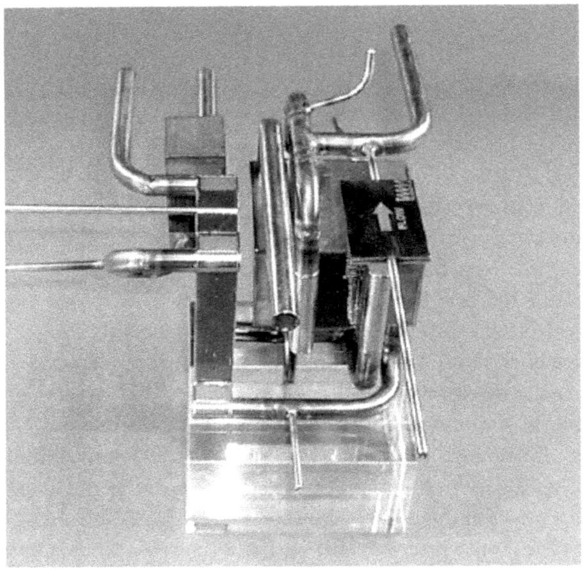

Figure 14.31 Integrated methanol fuel processor with 100 W power equivalent [16, 86].
(Source: Kolb et al. [86]. Reproduced with permission of Wiley-VCH GmbH & Co. KGaA.)

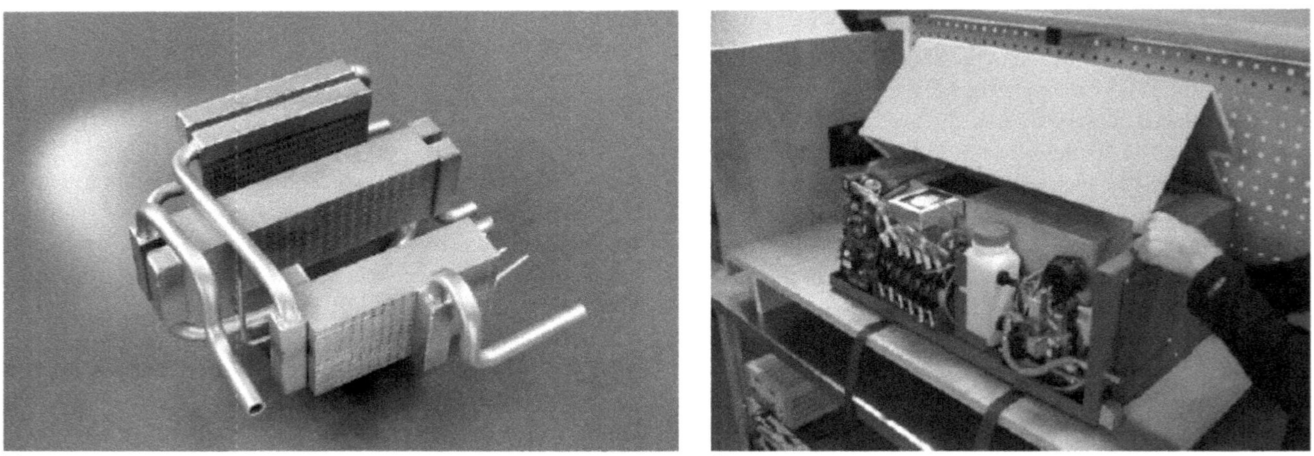

(a) (b)

Figure 14.32 (a) Fuel processor of the VeGA system [16]. (Source: O'Connell et al. [16]. Reproduced with permission of Elsevier.)
(b) 250 W_{el} fuel cell/fuel processor system VeGA developed by a cooperation of TRUMA and IMM [106].
(Source: Wichert et al. [106]. Reproduced with permission of Elsevier.)

Nomenclature

Δp_i	partial pressure difference, Pa
α	channel aspect ratio (ratio of channel width to channel depth), —
λ	heat conductivity of the material, W/(m K)
ρ	density of the fluid, kg/m^3
A	channel cross-sectional area, m^2
A_M	membrane surface area, m^2
D_h	hydraulic diameter (= $4A/P$), m
D_{H_2}	diffusion coefficient of hydrogen in palladium, m^2/s
dp/dz	pressure gradient, Pa/m
f	friction factor, —
$GHSV$	gas hourly space velocity, $L_{feed}/(\text{s}\ L_{reaction\ volume})$
J_{H_2}	hydrogen flux through the membrane, mol/s
k	heat transfer coefficient, W/(m^{2}K)
K_s	Sievert's solubility constant of the hydrogen/palladium system, —
L	membrane thickness, m
$\dot{n}_i$	moles of species i transported through the membrane, mol/s
O/C	molar oxygen to carbon ratio, —
P	channel perimeter, m
Q	permeability coefficient, mol/(m^2 s kPa^n)
s	hydrogen separation factor, —
S/C	molar steam to carbon ratio, —
U	mean flow velocity, m/s
$VHSV$	volume hourly space velocity, $L_{feed}/(\text{h}\ g_{cat})$
$WHSV$	weight hourly space velocity, $kg_{feed}/(\text{s}\ kg_{cat})$

References

1 Giroux, T, Hwang, S, Liu, Y, Ruettinger, W, Shore, L. Monolithic structures as alternatives to particulate catalysts for the reforming of hydrocarbons for hydrogen generation. Appl. Catal. Environ. 2005;56:95–110.

2 Shah, RK, London, AL. Laminar Flow Forced Convection in Ducts. London: Academic Press; 1978.

3 Hardt, S, Ehrfeld, W, Hessel, V, Vanden Bussche, KM. Strategies for size reduction of microreactors by heat transfer enhancement effects. Chem. Eng. Commun. 2003;190(4):540–559.

4 Kolb, G. Fuel Processing for Fuel Cells. Weinheim: Wiley-VCH; 2008.

5 Ledjeff-Hey, K, Formanski, V, Kalk, T, Roes, J. Compact hydrogen production systems for solid polymer fuel cells. J. Power Sources 1998;71:199–207.

6 Qi, A, Peppley, B, Kunal, K. Integrated fuel processors for fuel cell applications: A review. Fuel Process. Technol. 2007;88:3–22.

7 Lattner, JR, Harold, MP. Comparison of conventional and membrane reactor fuel processors for hydrocarbon-based PEM fuel cell systems. Int. J. Hydrog. Energy 2004;29:393–417.

8 Yu, W, Ohmori, T, Yamamoto, T, Endo, A, Nakaiwa, M, Itoh, N. Optimal design and operation of methane steam reforming in a porous ceramic membrane reactor for hydrogen production. Chem. Eng. J. 2007;62:5627–5631.

9 Avila, P, Montes, M, Miro, EE. Monolithic reactors for environmental applications. A review on preparation techniques. Chem. Eng. J. 2005;109:11–36.

10 Tomasic, V, Jovic, F. State-of-the-art in the monolithic catalysts/reactors. Appl. Catal. A Gen. 2006;311:112–121.

11 Meille, V. Review of methods to deposit catalysts on structured surfaces. Appl. Catal. A Gen. 2006;315:1–17.

12 Hayes, RE, Rojas, A, Mmbaga, J. The effective thermal conductivity of monolith honeycomb structures. Catal. Today 2009;147S: S113–S119.

13 Shiu, PP, Knopf, GK, Ostojic, M, Nikumb, S. Rapid fabrication of tooling for microfluidic devices via laser micromachining and hot embossing. J. Micromech. Microeng. 2008;18:2–12.

14 Theis, HE. Handbook of Metalforming Processes. Boca Raton: CRC Press; 1999.

15 DeVries, WR. Analysis of Material Removal Processes. New York: Springer; 1992.

16 O'Connell, M, Kolb, G, Schelhaas, K-P, Wichert, M, Tiemann, D, Pennemann, H, Zapf, R. Towards mass production of microstructured fuel processors for application in future distributed energy generation systems: A review of recent progress at IMM. Chem. Eng. Res. Des. 2012;90:11–18.

17 Germani, G, Stefanescu, A, Schuurman, Y, van Veen, AC. Preparation and characterization of porous alumina-based catalyst coatings in microchannels. Chem. Eng. Sci. 2007;62 (18–20):5084–5091.

18 Agrafiotis, C, Tsetsekou, A. The effect of powder characteristics on washcoat quality. Part I: Alumina washcoats. J. Eur. Ceram. Soc. 2000;20:815–824.

19 Cristiani, C, Valentini, M, Merazzi, M, Neglia, S, Forzatti, P. Effect of aging time on chemical and rheological evolution in γ-Al_2O_3 slurries for dip-coating. Catal. Today 2005;105:492–498.

20 Thybo, S, Jensen, S, Johansen, J, Johannesen, T, Hansen, O, Quaade, UJ. Flame spry deposition of porous catalysts on surfaces and in microsystems. J. Catal. 2004;223:271–277.

21 Wunsch, R, Fichtner, M, Görke, O, Haas-Santo, K, Schubert, K. Process of applying Al_2O_3 coatings in microchannels of completely manufactured microstructured reactors. Chem. Eng. Technol. 2002;25(7):700–703.

22 Tosti, S, Bettinali, L, Violante, V. Rolled thin Pd and Pd-Ag membranes for hydrogen separation and production. Int. J. Hydrog. Energy 2000;25:319–325.

23 Li, A, Liang, W, Hughes, R. Fabrication of dense palladium composite membranes for hydrogen separation. Catal. Today 2000;56:45–51.

24 Williams, KA, Schmidt, LD. Catalytic autoignition of higher alkane partial oxidation on rhodium coated foams. Appl. Catal. A Gen. 2006;299:30–45.

25 Hickman, DA, Schmidt, LD. Production of syngas by direct catalytic oxidation of methane. Science 1993;259:343–347.

26 Schmidt, LD, Huff, M. Partial oxidation of CH_4 and C_2H_6 over noble metal-coated monoliths. Catal. Today 1994;21:443–454.

27 Huff, M, Torniainen, PM, Hickman, DA, Schmidt, LD. Partial oxidation of CH_4, C_2H_6 and C_3H_8 on monoliths at short contact times, In: Curry-Hyde, HE, Howe, RF, editors. Natural Gas Conversion II. Burlington: Elsevier Ltd.; 1994, pp. 315–320.

28 Rodrigues, CP, Schmal, M. Nickel-alumina washcoating on monoliths for the partial oxidation of ethanol to hydrogen production. Int. J. Hydrog. Energy 2011;36:10709–10718.

29 Silberova, B, Venvik, HJ, Walmsley, JC, Holmen, A. Small-scale hydrogen production from propane. Catal. Today 2005;100:457–462.
30 Silberova, B, Venvik, HJ, Holmen, A. Production of hydrogen by short contact time partial oxidation and oxidative steam reforming of propane. Catal. Today 2005;99:69–76.
31 Ferrandon, M, Krause, T. Role of the oxide support on the performance of rhodium catalysts for the autothermal reforming of gasoline and gasoline surrogates to hydrogen. Appl. Catal. A Gen. 2006;311:135–145.
32 Qi, A, Wang, S, Fu, G, Wu, D. Autothermal reforming of n-octane on Ru-based catalysts. Appl. Catal. A Gen. 2005;293:71–82.
33 Dreyer, BJ, Lee, IC, Krummenacher, JJ, Schmidt, LD. Autothermal steam reforming of higher hydrocarbons: n-decane, n-hexadecane and JP-8. Appl. Catal. A Gen. 2006;307:184–194.
34 Karatzas, X, Creaser, D, Grant, A, Dawody, J, Petterson, LJ. Hydrogen generation from n-tetradecane, low-sulfur and Fischer-Tropsch diesel over Rh supported on alumina doped with ceria/lanthana. Catal. Today 2011;164:190–197.
35 Yoon, S, Kang, I, Bae, J. Suppression of ethylene-induced carbon deposition in diesel autothermal reforming. Int. J. Hydrog. Energy 2009;34:1844–1851.
36 Kang, I, Bae, J, Yoon, S, Yoo, Y. Performance improvement of diesel autothermal reformer by applying ultrasonic injector for effective fuel delivery. J. Power Sources 2007;172:845–852.
37 Pors, Z, Pasel, J, Tschauder, A, Dahl, R, Peters, R, Stolten, D. Optimised mixture formation for diesel fuel processing. Fuel Cells 2008;8:129–137.
38 Chen, Y-H, Xu, H, Wang, Y, Jin, X, Xiong, G. Hydrogen production from liquid hydrocarbon fuels for PEMFC operation. Fuel Process. Technol. 2006;87:971–978.
39 Yoon, S, Bae, J. A diesel fuel processor for stable operation of solid oxide fuel cell system: I. Introduction to post-reforming for the diesel fuel processor. Catal. Today 2010;156:49–57.
40 Pasel, J, Meissner, J, Pors, Z, Samsun, RC, Tschauder, A, Peters, R. Autothermal reforming of commercial Jet A-1 on a 5 kW_e scale. Int. J. Hydrog. Energy 2007;32:4847–4858.
41 O'Connell, M, Kolb, G, Schelhaas, KP, Schürer, J, Tiemann, D, Ziogas, A, Hessel, V. Development and evaluation of a microreactor for the reforming of diesel fuel in the kW range. Int. J. Hydrog. Energy 2009;34:6290–6303.
42 Nilsson, M, Karatzas, X, Lindström, B, Petterson, LJ. Assessing the adaptability to varying fuel supply of an autothermal reformer. Chem. Eng. J. 2008;142:309–314.
43 Lindström, B, Karlsson, JAJ, Ekdunge, P, De Verdier, L, Häggendal, B, Dawody, J, Nilsson, M, Petterson, LJ. Diesel fuel reformer for automotive fuel cell applications. Int. J. Hydrog. Energy 2009;34:3367–3381.
44 Karatzas, X, Nilsson, M, Dawody, J, Lindström, B, Petterson, LJ. Characterization and optimization of an autothermal diesel and jet fuel reformer for 5 kW_e mobile fuel cell applications. Chem. Eng. J. 2010;156:366–379.
45 Hartmann, L, Lucka, K, Köhne, H. Mixture preparation by cool flames for diesel-reforming technologies. J. Power Sources 2003;118:286–297.
46 Mator da Silva, J, Hermann, I, Mengel, C, Lucka, K, Köhne, H. Autothermal reforming of gasoline using a cool flame vaporizer. AIChE J. 2004;50(5):1042–1050.
47 Naidja, A, Krishna, CR, Butcher, T, Mahajan, D. Cool flame partial oxidation and its role in combustion and reforming of fuels for fuel cell systems. Prog. Energy Combust. Sci. 2003;29:155–191.
48 Lucarelli, C, Albonetti, S, Vaccari, A, Resini, C, Taillades, G, Roziere, J, Liew, K-E, Ohnesorge, A, Wolff, C, Gabellini, I, Wails, D. On-board H_2 generation by catalytic dehydrogenation of hydrocarbon mixtures. Catal. Today 2011;175:504–508.
49 Dicks, AL. Hydrogen generation from natural gas for the fuel cell systems of tomorrow. J. Power Sources 1996;61:113–124.
50 Jenkins, JW. Catalytic generation of hydrogen from hydrocarbons. US Patent 4,897,253, January 30, 1990.
51 Recupero, V, Pino, L, Di Leonardo, R, Lagana, M, Maggio, G. Hydrogen generator via catalytic partial oxidation of methane for fuel cells. J. Power Sources 1998;71:208–214.
52 Peters, R, Düsterwald, HG, Höhlein, B. Investigation of a methanol reformer concept considering the particular impact of dynamics and long-term stability for use in a fuel-cell powered passenger car. J. Power Sources 2000;86:507–514.
53 Emonts, B, Hansen, JB, Jörgensen, SL, Höhlein, B, Peters, R. Compact methanol reformer test for fuel cell powered light-duty vehicles. J. Power Sources 1998;71:288–293.
54 Moon, DJ, Sreekumar, K, Lee, SD, Lee, BG, Kim, HS. Studies on gasoline fuel processor system for fuel-cell powered vehicles application. Appl. Catal. A Gen. 2001;215:1–9.
55 Lindström, B, Petterson, LJ. Development of a methanol fuelled reformer for fuel cell applications. J. Power Sources 2003;118:71–78.
56 Lattner, JR, Harold, MP. Autothermal reforming of methanol: Experiments and modeling. Catal. Today 2007;120:78–89.
57 Wang, W, Turn, SQ, Keffer, V, Douette, A. Study of process data in autothermal reforming of LPG using multivariate data analysis. Chem. Eng. J. 2007;129:11–19.
58 Lee, D, Lee, HC, Lee, KH, Kim, S. A compact and highly efficient natural gas fuel processor for 1-kW residential polymer electrolyte membrane fuel cells. J. Power Sources 2007;165:337–341.
59 Moon, DJ, Ryu, JW, Yoo, KS, Sung, DJ, Lee, SD. Development of iso-octane fuel processor system for fuel cell applications. Catal. Today 2008;138:222–227.
60 Rampe, T, Heinzel, A, Vogel, B. Hydrogen generation from biogenic and fossil fuels by autothermal reforming. J. Power Sources 2000;86:536–541.
61 Docter, A, Konrad, G, Lamm, A. Reformer für Benzin und benzinähnliche Kraftstoffe. VDI-Berichte 2000;84:399–411.
62 Roh, H-S, Lee, DK, Koo, KY, Jung, UH, Yoon, WL. Natural gas steam reforming for hydrogen production over metal monolith catalyst with efficient heat transfer. Int. J. Hydrog. Energy 2010;35:1613–1619.
63 Ryu, J-H, Lee, K-Y, La, H, Kim, H-J, Yang, J-I, Jung, H. Ni catalyst wash-coated on metal monolith with enhanced heat-transfer capability for steam reforming. J. Power Sources 2007;171:499–505.
64 Vita, A, Pino, L, Cipiti, F, Lagana, M, Recupero, V. Structured reactors as alternative to pellets catalyst for propane oxidative steam reforming. Int. J. Hydrog. Energy 2010;35:9810–9817.
65 Beretta, A, Donazzi, A, Livio, D, Maestri, M, Groppi, G, Tronico, E, Forzatti, P. Optimal design of a CH_4 CPO-reformer with honeycomb catalyst: Combined effect of catalyst load and channel size on the surface temperature profile. Catal. Today 2011;171:79–83.

66 Fichtner, M, Mayer, J, Wolf, D, Schubert, K. Microstructured rhodium catalysts for the partial oxidation of methane to syngas under pressure. Ind. Eng. Chem. Res. 2001;40(16):3475–3483.
67 Jung, H, Yoon, WL, Lee, H, Park, JS, Shin, JS, La, H, Lee, JD. Fast start-up reactor for partial oxidation of methane with electrically heated metallic monolith catalyst. J. Power Sources 2003;124:76–80.
68 Castaldi, M, Lyubovski, M, LaPierre, R, Pfefferle, WC, Roychoudhury, S. Performance of microlith based catalytic reactors for an isooctane reforming system. Society of Automotive engineers 2003; SAE 2003-01-1366:1707–1714.
69 Roychoudhury, S, Lyubovski, M, Shabbir, A. Microlith catalytic reactors for reforming iso-octane-based fuels into hydrogen. J. Power Sources 2005;152:75–86.
70 Catillon, S, Louis, C, Topin, F, Vicente, J, Rouget, R. Improvement of methanol steam reformer for H2 production by addition of copper foam in both the evaporator and the catalytic reactors. Chem. Eng. Trans. 2004;4:111–116.
71 Liu, D-J, Kaun, TD, Liao, H-K, Ahmed, S. Characterization of kilowatt-scale autothermal reformer for production of hydrogen from heavy hydrocarbons. Int. J. Hydrog. Energy 2004;29:1035–1046.
72 Horng, R-F. Transient behaviour of a small methanol reformer for fuel cell during hydrogen production after cold start. Energy Convers. Manag. 2005;46:1193–1207.
73 Lenz, B, Full, J, Siewek, C. Reforming of jet fuel for fuel cell APU's in commercial aircraft. In: Proceedings of the Hydrogen Expo Conference; August 31–September 1, 2005; Hamburg, Germany; 2005. Published on CD.
74 Lenz, B, Aicher, T. Catalytic autothermal reforming of jet fuel. J. Power Sources 2005;149:44–52.
75 Qi, A, Wang, S, Ni, C, Wu, D. Autothermal reforming of gasoline on Rh-based monolithic catalysts. Int. J. Hydrog. Energy 2007;32:981–991.
76 Creaser, D, Karatzas, X, Lundberg, B, Petterson, LJ, Dawody, J. Modeling study of 5 kW_e-scale autothermal diesel fuel reformer. Appl. Catal. A Gen. 2011;404:129–140.
77 Vaccaro, S, Malangone, L, Ciambelli, P. Micro-scale catalytic reactor for syngas production. Ind. Eng. Chem. Res. 2010;49:10924–10933.
78 Schildhauer, TJ, Geissler, K. Reactor concept for improved heat integration in autothermal methanol reforming. Int. J. Hydrog. Energy 2007;32:1806–1810.
79 Gritsch, A, Kolios, G, Eigenberger, G. Reaktorkonzepte zur autothermen Führung endothermer Hochtemperaturreaktionen. Chem. Ing. Tech. 2004;76(6):722–725.
80 Kolios, G, Frauhammer, J, Eigenberger, G. Efficient reactor concepts for coupling of endothermic and exothermic reactions. Chem. Eng. Sci. 2002;57:1505–1510.
81 von Hippel, L, Arntz, D, Frauhammer, J, Eigenberger, G, Friedrich, G. Reaktorkopf für einen monolithischen Gleich- oder Gegenstromreaktor; DE 19653989, June 25, 1998.
82 Frauhammer, J, Friedrich, G, Kolios, G, Klingel, T, Eigenberger, G, von Hippel, L, Arntz, D. Flow distribution concepts for new type monolithic co- or countercurrent reactors. Chem. Eng. Technol. 1999;22:1012–1016.
83 Pan, L, Wang, S. Modeling of a compact plate-fin reformer for methanol steam reforming in fuel cell systems. Chem. Eng. J. 2005;108:51–58.
84 Frauhammer, J, Eigenberger, G, von Hippel, L, Arntz, D. A new reactor concept for endothermic high-temperature reactions. Chem. Eng. Sci. 1999;54(15/16):3661–3670.
85 Anzola, AM, Bruschi, YM, Lopez, E, Schbib, NS, Pedernera, MN, Borio, DO. Heat supply and hydrogen yield in an ethanol microreformer. Ind. Eng. Chem. Res. 2011;50:2698–2705.
86 Kolb, G, Schelhaas, K-P, Wichert, M, Burfeind, J, Hesske, C, Bandlamudi, G. Development of a microstructured methanol fuel processor coupled to a high temperature PEM fuel cell. Chem. Eng. Technol. 2009;32(11):1739–1747.
87 Petrachi, GA, Negro, G, Specchia, S, Saracco, G, Maffetone, PL, Specchia, V. Combining catalytic combustion and steam reforming in a novel multifunctional reactor for on-board hydrogen production from middle distillates. Ind. Eng. Chem. Res. 2005;44:9422–9430.
88 Grote, M, Maximi, M, Yang, Z, Engelhardt, P, Köhne, H, Lucka, K, Brenner, M. Experimental and computational investigations of a compact steam reformer for fuel oil and diesel fuel. J. Power Sources 2011;196:9027–9035.
89 Kolb, G, Baier, T, Schürer, J, Tiemann, D, Ziogas, A, Ehwald, H, Alphonse, P. A micro-structured 5 kW complete fuel processor for iso-octane as hydrogen supply system for mobile auxiliary power units. Part I—Development of the autothermal catalyst and reactor. Chem. Eng. J. 2008;137(1–3):653–663.
90 Commenge, JM, Falk, L, Corriou, JP, Matlosz, M. Optimal design for flow uniformity in microchannel reactors. AIChE J. 2000;48 (2):345–358.
91 Hao, Y, Du, X, Yang, L, Shen, Y, Yang, Y. Numerical simulation of configuration and catalyst-layer effects on micro-channel steam reforming of methanol. Int. J. Hydrog. Energy 2011;36:15611–15621.
92 Reuse, P, Renken, A, Haas-Santo, K, Görke, O, Schubert, K. Hydrogen production for fuel cell application in an autothermal microchannel reactor. Chem. Eng. J. 2004;101(1–3):133–141.
93 Park, G-G, Seo, DJ, Park, S-H, Yoon, Y-G, Kim, C-S, Yoon, W-L. Development of microchannel methanol steam reformer. Chem. Eng. J. 2004;101(1–3):87–92.
94 Kim, T, Kwon, S. Design, fabrication and testing of a catalytic microreactor for hydrogen production. J. Micromech. Microeng. 2006;16:1760–1768.
95 Kundu, A, Jang, JH, Lee, HR, Kim, S-H, Gil, JH, Jung, CR, Oh, YS. MEMS-based micro-fuel processor for application in a cell phone. J. Power Sources 2006;162:572–578.
96 Hermann, I, Lindner, M, Winkelmann, H, Düsterwald, HG. Microreaction technology in fuel processing for fuel cell vehicles. In: Proceedings of the VDE World Microtechnologies Congress, MICRO.tec, Hannover, September 25–27, 2000; Berlin: VDE Verlag; 2000. p. 447–453.
97 Fitzgerald, SP, Wegeng, RS, Tonkovich, ALY, Wang, Y, Freeman, HD, Marco, JL, Roberts, GL, VanderWiel, DP. A compact steam reforming reactor for use in an automotive fuel processor. In: Proceedings of the 4th International Conference on Microreaction Technology, IMRET 4; March 5–9, 2000; Atlanta: AIChE Topical Conference Proceedings; 2000. p. 358–363.
98 Whyatt, GA, TeGrotenhuis, WE, Wegeng, RS, Pederson, LR. Demonstration of energy efficient steam reforming in microchannels for automotive fuel processing. In: Matlosz, M, Ehrfeld, W, Baselt, JP, editors. Proceedings of the Microreaction Technology—IMRET5; May 27–30, 2001; Strasbourg: Springer; 2001. p. 303–312.
99 Whyatt, A, Fischer, CM, Davis, JM. Development of a rapid-start on-board automotive steam reformer. In: Proceedings of the AIChE

Spring Meeting; April 25–29, 2004. Conference; New Orleans: published on CD; 2004.

100 Cremers, C, Stummer, M, Stimming, U, Find, J, Lercher, JA, Kurtz, O, Crämer, K, Haas-Santo, K, Görke, O, Schubert, K. Micro-structured reactors for coupled steam-reforming and catalytic combustion of methane. In: Proceedings of the 7th International Conference on Microreaction Technology, IMRET 7; September 7–10, 2003; Lausanne, Switzerland; Frankfurt: DECHEMA; 2003. p. 100.

101 Cremers, C, Dehlsen, J, Stimming, U, Reuse, P, Renken, A, Haas-Santo, K, Görke, O, Schubert, K. Micro-structured-reactor-system for the steam reforming of methanol. In: Proceedings of the 7th International Conference on Microreaction Technology, IMRET 7; September 7–10, 2003; Lausanne, Switzerland; Frankfurt: DECHEMA; 2003. p. 56.

102 Pfeifer, P, Bohn, L, Görke, O, Haas-Santo, K, Schubert, K. Micro-structured components for hydrogen production from various hydrocarbons. Chem. Eng. Technol. 2005;28(4):474–476.

103 Pan, L, Wang, S. Methanol steam reforming in a compact plate-fin reformer for fuel cell systems. Int. J. Hydrog. Energy 2005;30:973–979.

104 Kolb, G, Schürer, J, Tiemann, D, Wichert, M, Zapf, R, Hessel, V, Löwe, H. Fuel processing in integrated microstructured heat-exchanger reactors. J. Power Sources 2007;171(1):198–204.

105 Lee, Y-K, Kim, K-S, Ahn, J-G, Son, I-H, Shin, WC. Hydrogen production from ethanol over Co/ZnO catalyst in a multi-layered reformer. Int. J. Hydrog. Energy 2010;35:1147–1151.

106 Wichert, M, Men, Y, O'Connell, M, Tiemann, D, Zapf, R, Kolb, G, Butschek, S, Frank, R, Schiegl, A. Self-sustained operation and durability test of a 300 W-class micro-structured LPG fuel processor. Int. J. Hydrog. Energy 2011;36:3496–3504.

107 Lin, Y-M, Lee, G-L, Rei, M-H. An integrated purification and production of hydrogen with a palladium membrane-catalytic reactor. Catal. Today 1998;44:343–349.

108 Lin, Y-M, Rei, M-H. Process development for generating high purity hydrogen by using supported palladium membrane reactor as steam reformer. Int. J. Hydrog. Energy 2000;25:211–219.

109 Lin, Y-M, Rei, M-H. Study on the hydrogen production from methanol steam reforming in supported palladium membrane reactor. Catal. Today 2001;67:77–84.

110 Lin, Y-M, Rei, M-H. Separation of hydrogen from the gas mixture out of catalytic reformer by using supported palladium membrane. Sep. Purif. Technol. 2001;25:87–95.

111 Kikuchi, E. Membrane reactor application to hydrogen production. Catal. Today 2000;56:97–101.

112 Basile, A, Patzuro, L. An experimental study of multilayered composite palladium membrane reactors for partial oxidation of methane to syngas. Catal. Today 2001;67:55–64.

113 Wieland, S, Melin, T, Lamm, A. Membrane reactors for hydrogen production. Chem. Eng. Sci. 2002;57:1571–1576.

114 Kurungot, S, Yamaguchi, T, Nakao, S-I. Rh/γ-Al_2O_3 catalytic layer integrated with sol-gel synthesized microporous silica membrane for compact membrane reactor applications. Catal. Lett. 2003;86 (4):273–278.

115 Basile, A, Gallucci, F, Patzuro, L. Hydrogen production from methanol by oxidative steam reforming carried out in membrane reactor. Catal. Today 2005;104:251–259.

116 Iulianelli, A, Longo, T, Basile, A. Methanol steam reforming reaction in a Pd-Ag membrane reactor for CO-free hydrogen production. Int. J. Hydrog. Energy 2008;33:5583–5588.

117 Chen, Z, Elnashaie, SSEH. Steady-state modelling and bifurcation behavior of circulating fluidized bed membrane reformer-regenerator for the production of hydrogen for fuel cells from heptane. Chem. Eng. Sci. 2004;59:3965–3979.

118 Ikeguchi, M, Mimura, T, Sekine, Y, Kikuchi, E, Matsukata, M. Reaction and oxygen permeation studies in $Sm_{0.4}Ba_{0.6}Fe_{0.8}Co_{0.2}O_{3-\delta}$ membrane reactor for partial oxidation of methane to syngas. Appl. Catal. A Gen. 2005;290:212–220.

119 Yamaguchi, S, Yamamoto, S, Tsuchiya, B., Nagata, S, ShiShido, T. Construction of fuel reformer using proton conducting oxides electrolyte and hydrogen-permeable metal membrane cathode. J. Power Sources 2005;145:712–715.

120 Hu, J, Xing, T, Jia, Q, Hao, H, Yang, D, Guo, Y, Hu, X. Methane partial oxidation to syngas in $YBa_2Cu_3O_{7-x}$ membrane reactor. Appl. Catal. A Gen. 2006;306:29–33.

121 Kusakabe, K, Fumio, S, Eda, T, Oda, M, Sotowa, K-I. Hydrogen production in zirconia membrane reactors for use in PEM fuel cells. Int. J. Hydrog. Energy 2005;30:989–994.

122 Mundschau, MV, Burk, CG, Gribble Jr., DA. Diesel fuel reforming using catalytic membrane reactors. Catal. Today 2008;136:190–205.

123 Faroldi, B, Carrara, C, Lombardo, EA, Cornaglia, LM. Production of ultrapure hydrogen in a Pd-Ag membrane reactor using Ru/La_2O_3 catalysts. Appl. Catal. A Gen. 2007;319:38–46.

124 Calabro, DC, Partridge, RD, Berlowitz, PJ, Carstensen, B, Deckmann, HW, DaPrato, PL, Hershkowitz, F, Socha, RF. A dual functional staged hydrogen purifier for an integrated fuel processor-fuel cell power system. Catal. Today 2007;129:380–390.

125 Iulianelli, A, Seelam, PK, Liguori, S, Longo, T, Keiski, R, Calabro, V, Basile, A. Hydrogen production for PEM fuel cell by gas phase reforming of glycerol as byproduct of bio-diesel. The use of a Pd-Ag membrane reactor at middle reaction temperature. Int. J. Hydrog. Energy 2011;36:3827–3834.

126 Li, A, Lim, CJ, Grace, JR. Staged-separation membrane reactor for methane steam reforming. Chem. Eng. J. 2008;138:452–459.

127 Joensen, F, Rostrup-Nielsen, JR. Conversion of hydrocarbons and alcohols for fuel cells. J. Power Sources 2002;105:195–201.

128 Ghenciu, AF. Review of fuel processing catalysts for hydrogen production in PEM fuel cell systems. Curr. Opin. Solid State Mater. Sci. 2002;6:389–399.

129 Dokupil, M, Spitta, C, Mathiak, J, Beckhaus, P, Heinzel, A. Compact propane fuel processor for auxiliary power unit application. J. Power Sources 2006;157:906–913.

130 Pasel, J, Samsun, RC, Schmitt, D, Peters, R, Stolten, D. Test of a water-gas shift reactor on a 3 kW_e-scale—Design points for high- and low-temperature water-gas shift. J. Power Sources 2005;152:189–195.

131 van Dijk, HAJ, Boon, J, Nyquist, RN, van den Brink, RW. Development of a single stage heat integrated water-gas shift reactor for fuel processing. Chem. Eng. J. 2010;159:182–189.

132 Chen, H, Yu, H, Yang, G, Peng, F, Wang, H, Yang, J. Auto-thermal ethanol micro-reformer with a structural Ir/La_2O_3/ZrO_2 catalyst for hydrogen production. Chem. Eng. J. 2011;167:322–327.

133 Zalc, JM, Löffler, DG. Fuel processing for PEM fuel cells: Transport and kinetic issues of system design. J. Power Sources 2002;111:58–64.

134 TeGrotenhuis, WE, King, DL, Brooks, KP, Holladay, BJ, Wegeng, RS. Optimizing microchannel reactors by trading-off equilibrium and reaction kinetics through temperature management. In: Proceedings of the 6th International Conference on Microreaction Technology, IMRET 6; March 11–14, 2002; New Orleans: AIChE Publication No. 164; 2002. p. 18–28.
135 Kim, G-Y, Mayor, JR, Ni, J. Parametric study of microreactor design for water gas shift reactor using an integrated reaction and heat exchange model. Chem. Eng. J. 2005;110:1–10.
136 Baier, T, Kolb, G. Temperature control of the water-gas shift reaction in microstructured reactors. Chem. Eng. Sci. 2007;62 (17):4602–4611.
137 Pasel, J, Cremer, P, Stalling, J, Wegner, B, Peters, R, Stolten, D. Comparison of two different reactor concepts for the water-gas shift reaction. In: Proceedings of the Fuel Cell Seminar; November 18–21, 2002; Palm Springs, CA; 2002. p. 607–610.
138 Zhai, X, Ding, S, Cheng, Y, Jin, Y, Cheng, Y. CFD simulation with detailed chemistry of steam reforming of methane for hydrogen production in an integrated micro-reactor. Int. J. Hydrog. Energy 2010;35:5383–5392.
139 Kolb, G, Baier, T, Schürer, J, Tiemann, D, Ziogas, A, Specchia, S, Galetti, E, Germani, G, Schuurman, Y. A micro-structured 5 kW complete fuel processor for iso-octane as hydrogen supply system for mobile auxiliary power units. Part II—Development of water-gas shift and preferential oxidation reactors and assembly of the fuel processor. Chem. Eng. J. 2008;138(3):474–489.
140 Dubrovskiy, AR, Rebrov, EV, Kuznetsov, SA, Schouten, JC. A microstructured reactor/heat-exchanger for the water-gas shift reaction operated in the 533-673 K range. Catal. Today 2009;147S:S198–S203.
141 O'Connell, M, Kolb, G, Schelhaas, KP, Schuerer, J, Tiemann, D, Ziogas, A, Hessel, V. The development and evaluation of microstructured reactors for the water-gas shift and preferential oxidation reactions in the 5 kW range. Int. J. Hydrog. Energy 2010;35:2317–2327.
142 O'Connell, M, Kolb, G, Schelhaas, K-P, Schuerer, J, Tiemann, D, Keller, S, Reinhard, D, Hessel, V. Investigation on the combined operation of water gas shift and preferential oxidation reactor system on the kW scale. Ind. Eng. Chem. Res. 2010;49:10917–10923.
143 Baronskaya, NA, Minyokova, TP, Sipatrov, AG, Demeshkina, MP, Khassin, AA, Dimov, SV, Kozlov, SP, Kuznetsov, SP, Terentiev, VY, Khristolyubov, AP, Britzitskiy, OF, Yurieva, TM. Compact reactor for water-gas shift reaction over thermal-conducting catalysts. Chem. Eng. J. 2007;134:195–199.
144 Palma, V, Palo, E, Ciambelli, P. Structured catalytic substrates with radial configurations for the intensification of the WGS stage in H_2 production. Catal. Today 2009;147S:S107–S112.
145 Brunetti, A, Caravella, A, Drioli, E, Barbieri, G. Process intensification by membrane reactors: High-temperature water gas shift reaction as single stage for syngas upgrading. Chem. Eng. Technol. 2012;35(7):1238–1248.
146 Barbieri, G, Bernardo, P, Mattia, R, Drioli, E, Bredesen, R, Klette, H. Pd-based membrane reactor for water-gas shift reaction. Chem. Eng. Trans. 2004;4:55–60.
147 Ouyang, X, Bednarova, L, Besser, RS, Ho, P. Preferential oxidation (PrOx) in a thin-film catalytic microreactor: advantages and limitations. AIChE J. 2005;51(6):1758–1771.
148 Galletti, C, Specchia, S, Specchia, V. CO selective methanation in H2-rich gas for fuel cell applications: Microchannel reactor performance with Ru-based catalysts. Chem. Eng. J. 2011;167:616–621.
149 Xu, G, Chen, X, Zhang, Z-G. Temperature-staged methanation: An alternative method to purify hydrogen-rich fuel gas for PEFC. Chem. Eng. J. 2006;121:97–107.
150 Men, Y, Kolb, G, Zapf, R, Hessel, V, Löwe, H. Selective methanation of carbon oxides in a microchannel reactor—Primary screening and impact of additives. Catal. Today 2007;125:81–87.
151 Kraaij, GJ, Specchia, S, Bollito, G, Mutri, L, Wails, D. Biodiesel fuel processor for APU applications. Int. J. Hydrog. Energy 2009;34:4495–4499.
152 Lee, SH, Han, J, Lee, K-Y. Development of 10kWe preferential oxidation system for fuel cell vehicles. J. Power Sources 2002;109:394–402.
153 Pan, L, Wang, S. A compact integrated fuel-processing system for proton exchange membrane fuel cells. Int. J. Hydrog. Energy 2006;31:447–454.
154 Zhou, S, Yuan, Z, Wang, S. Selective CO oxidation with real methanol reformate over monolithic Pt group catalysts: PEMFC applications. Int. J. Hydrog. Energy 2006;31:924–933.
155 Chin, P, Sun, X, Roberts, GW, Spivey, JJ. Preferential oxidation of carbon monoxide with iron-promoted platinum catalysts supported on metal foams. Appl. Catal. A Gen. 2006;302:22–31.
156 Ahluwalia, RK, Zhang, Q, Chmielewski, DJ, Lauzze, KC, Inbody, MA. Performance of CO preferential oxidation reactor with noble-metal catalyst coated on ceramic monolith for on-board fuel processing applications. Catal. Today 2005;99:271–283.
157 Ouyang, X, Besser, RS. Effect of reactor heat transfer limitations on CO preferential oxidation. J. Power Sources 2005;141:39–46.
158 Kim, K-Y, Han, J, Nam, SW, Lim, T-H, Lee, H-I. Preferential oxidation of CO over CuO/CeO_2 and $Pt\text{-}Co/Al_2O_3$ catalysts in micro-channel reactors. Catal. Today 2008;131:431–436.
159 Cruz, S, Sanz, O, Poyato, R, Laguna, OH, Echave, FJ, Almeida, LC, Centeno, MA, Arzamendi, G, Gandia, LM, Souza-Aguiar, EF, Montes, M, Odriozola, JA. Design and testing of a microchannel reactor for the PROX reaction. Chem. Eng. J. 2011;167:634–642.
160 Srinivas, S, Gulari, E. Preferential CO oxidation in a two-stage packed-bed reactor: Optimization of oxygen split ratio and evaluation of system robustness. Catal. Commun. 2006;7:819–826.
161 Dudfield, CD, Chen, R, Adock, PL. A carbon monoxide PROX reactor for PEM fuel cell automotive application. Int. J. Hydrog. Energy 2001;26:763–775.
162 Dudfield, CD, Chen, R, Adcock, PL. A compact CO selective oxidation reactor for solid polymer fuel cells powered vehicle applications. J. Power Sources 2000;86:214–222.
163 Srinivas, S, Dhingra, A, Im, H, Gulari, E. A scalable silicon microreactor for preferential Co oxidation: Performance comparison with a tubular packed-bed microreactor. Appl. Catal. A Gen. 2004;274:285–293.
164 Cominos, V, Hessel, V, Hofmann, C, Kolb, G, Zapf, R, Ziogas, A, Delsman, E, Schouten, J. Selective oxidation of carbon monoxide in a hydrogen-rich fuel cell feed using a catalyst coated microstructured reactor. Catal. Today 2005;110(1–2):140–153.
165 Delsman, RE, de Croon, MHJM, Pierik, A, Kramer, GJ, Cobden, PD, Hofmann, C, Cominos, V, Schouten, JC. Design and operation

of a preferential oxidation microdevice for a portable fuel processor. Chem. Eng. Sci. 2004;59:4795–4802.

166 Lopez, E, Kolios, G, Eigenberger, G. Preferential Oxidation in a folded-plate reactor. Chem. Eng. Sci. 2007;62:5598–5601.

167 Kolb, G, Hofmann, C, O'Connell, M, Schürer, J. Micro-structured reactors for diesel steam reforming, water-gas shift and preferential oxidation in the kilowatt power range. Catal. Today 2009;147S: S176–S184.

168 Adachi, H, Ahmed, S, Lee, SHD, Papadias, D, Ahluwalia, RK, Bendert, JC, Kanner, SA, Yamazaki, Y. A natural gas fuel processor for a residential fuel cell system. J. Power Sources 2009;188:244–255.

169 Severin, C, Pischinger, S, Ogrzewalla, J. Compact gasoline fuel processor for passenger vehicle APU. J. Power Sources 2005;145:675–682.

170 Lee, SHD, Applegate, DV, Ahmed, S, Calderone, SG, Harvey, TL. Hydrogen from natural gas: Part I—Autothermal reforming in an integrated fuel processor. Int. J. Hydrog. Energy 2005;30:829–842.

171 Koenig, P, Weber, A, Lewald, N, Aicher, T, Joerissen, L, Ivers-Tiffee, E, Szolak, R, Brnedel, M, Kaczerowski, J. Testing and model-aided analysis of a 2 kW_{el} PEMFC CHP system. J. Power Sources 2005;145:327–335.

172 Lindermeir, A, Kah, S, Kavurucu, S, Mühlner, M. On-board diesel fuel processing for an SOFC-APU—Technical challenges for catalysis and reactor design. Appl. Catal. Environ. 2007;70:488–497.

173 Yoshida, K, Tanaka, S, Hiraki, H, Esashi, M. A micro fuel reformer integrated with a combustor and a microchannel evaporator. J. Micromech. Microeng. 2006;16:191–197.

CHAPTER 15

Modeling of the catalytic deoxygenation of fatty acids in a packed bed reactor

Teuvo Kilpiö, Päivi Mäki-Arvela, Tapio Salmi and Dmitry Yu. Murzin
Process Chemistry Centre, Åbo Akademi University, Turku/Åbo, Finland

Abstract

Catalytic deoxygenation is a promising pathway for converting vegetable oil compounds into diesel-like fuel components. Operation of a packed bed reactor was demonstrated for neat stearic acid deoxygenation over Pd supported on mesoporous and microporous carbon. The process was studied in detail with the aid of mathematical modeling. The model was used extensively for simulations and sensitivity analysis, and it was applied for the estimation of the governing kinetic parameters. The model was based on solving the dynamic mass balances with the method of lines. The model contained accumulation, convection, axial dispersion, and reaction terms. Catalyst deactivation was included in the model. The balances were written for the main components (stearic acid, *n*-heptadecane, and *n*-heptadecene). In most of the simulations of this highly selective system, the focal point of interest was on the main reaction leading to heptadecane formation. Catalyst deactivation was modeled using a concept of final activity and taking into account the fact that the number of active sites decreased due to coking with increasing time on stream. The deactivation originated mainly from the reactant. The extent of pore diffusion was evaluated by calculating for the most severe conditions (maximum productivity near feed entrance), the concentration profiles inside the particles, and the effectiveness factor. Even under these circumstances, only extremely small values of the effective diffusivity could make the pore diffusion significant. A model containing the rate constants of the main reaction, side reaction, coking, and the final activity as the adjustable parameters provided an adequate description of the experimental data. The parameter sensitivity study showed that the coking rate exhibited a profound effect on the reaction time at a certain conversion level until a more stable operation was reached. Scale-up criteria to pilot size were also discussed and a pilot simulation study was made.

15.1 Introduction

The research in the field of liquid biofuels has been very active during recent years due to the concern about the future utilization and available reserves of fossil fuel sources. Liquid biofuels comprise bioalcohols [1], fatty acid esters [2], and diesel-like long-chain hydrocarbons produced via hydrodeoxygenation over NiMo and CoMo catalysts [3–5] or catalytic deoxygenation over noble metal catalysts [6–21]. A drawback in the utilization of the first-generation biofuels is their lower energy density due to the presence of oxygen. Especially, biodiesel can have lower lubricity than conventional diesel causing corrosion [22]. Therefore, from the chemical point of view, oxygen-free, long-chain hydrocarbons are preferable as fuels in conventional engines.

Hydrodeoxygenation of fatty acids and vegetable oils has been performed typically in the temperature range of 533–653 K [4, 5] either over sulfided or non-sulfided NiMo catalysts supported on γ-Al_2O_3. The active catalysts often contain also tungsten. The products from the hydrodeoxygenation of fatty acids contain the same carbon number as the original feedstock, opposite to catalytic deoxygenation products of fatty acids.

Selective catalytic deoxygenation of fatty acids and their derivatives in the liquid phase has been demonstrated over Pd/C and Pt/C catalysts at temperature and pressure ranges of 543–633 K and 6–20 bar, respectively [7, 11]. Several feedstocks such as saturated and unsaturated fatty acids and esters as well as triglycerides have been used. Furthermore, catalyst properties and reaction conditions [9, 18] have been optimized, and the possibility of using either batch [20] or continuous reactors [15, 16, 21] has been demonstrated.

In the catalytic deoxygenation of stearic acid, the main liquid-phase product was *n*-heptadecane. Traces of *n*-heptadecene were formed, too. Catalyst deactivation was observed due to the formation of unsaturated *n*-heptadecene, which in turn undergoes cyclization to C17 aromatic compounds. These kinds of compounds can easily form coke on the catalyst surface. The catalyst deactivation was more prominent as the high initial stearic acid concentrations were used. Neither palladium leaching nor sintering was detected [15].

Despite extensive experimental research, there are practically no studies, which quantitatively describe behavior of packed bed

Multiphase Catalytic Reactors: Theory, Design, Manufacturing, and Applications, First Edition. Edited by Zeynep Ilsen Önsan and Ahmet Kerim Avci.

reactors in deoxygenation of fatty acids by taking reaction kinetics and catalyst deactivation into account. The aim of this work was to model a continuous packed bed reactor in catalytic deoxygenation of stearic acid to produce C_{17} linear hydrocarbons. A model based on the numerical solution of the dynamic mass balances including accumulation, convection, axial dispersion, and reaction terms is used for studying the deoxygenation of stearic acid. Furthermore, the extent of the pore diffusion in catalyst particles is investigated. Specific features of the model are the concept of final activity and decreasing number of active sites with increasing time on stream due to coking. According to our knowledge, catalytic deoxygenation of fatty acids in a continuous reactor is modeled for the first time in the current work.

15.2 Experimental data for stearic acid deoxygenation

The simplified reaction scheme for decarboxylation and decarbonylation of stearic acid is given in Figure 15.1. According to this scheme, two parallel liquid-phase reactions, the main reaction producing heptadecane and the side reaction leading to heptadecene, are involved.

The experiments were carried out in a continuous laboratory-scale packed bed reactor (Table 15.1).

According to the analysis, stearic acid transformation in the heterogeneous catalytic packed bed reactor produced heptadecane and heptadecene as main and by-products, respectively, at $p = 10$ and 20 bar and $T = 633$ K [15]. The yield, defined as ((produced heptadecane)/(fed stearic acid)) was initially high for both catalysts (ca. 90%). Pd/C (Aldrich) proved out to be better suited catalyst for this special application than Pd (on mesoporous carbon Sibunit). Less deactivation took place over the Pd/C (Aldrich) catalyst. At the end of the experiment, the yield over Pd/C remained still around 40%, while with the other catalyst, the yield went down to ca. 10%. In the Pd (Sibunit) case, the yield of the side reaction (decarbonylation) was at most ca. 2%, while for Pd/C (Aldrich), it was at most ca. 10%, which decreased then rapidly as a consequence of deactivation [15].

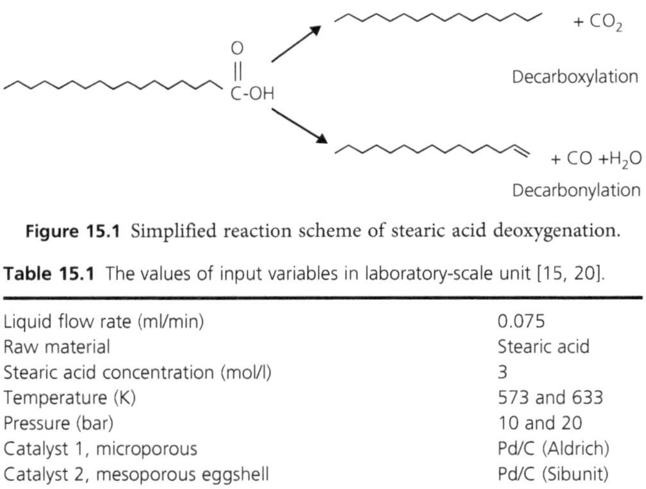

Figure 15.1 Simplified reaction scheme of stearic acid deoxygenation.

Table 15.1 The values of input variables in laboratory-scale unit [15, 20].

Liquid flow rate (ml/min)	0.075
Raw material	Stearic acid
Stearic acid concentration (mol/l)	3
Temperature (K)	573 and 633
Pressure (bar)	10 and 20
Catalyst 1, microporous	Pd/C (Aldrich)
Catalyst 2, mesoporous eggshell	Pd/C (Sibunit)
Bed length (m)	0.15
Bed diameter (m)	0.0159
Catalyst mass (g)	10
Catalyst particle size (mm)	max. 1.5 (Aldrich)
Catalyst particle size (mm)	1.5 (Sibunit)
Metal loading (wt% Pd)	5
Pore volume/mass (ml/g_{cat})	1.23 (Sibunit)
Pore volume/mass (ml/g_{cat})	0.431 (Aldrich)
Thickness eggshell (μm)	50 (Sibunit)
Surface area of catalyst (m^2/g_{cat})	1214 (Aldrich)
Surface area of catalyst (m^2/g_{cat})	504 (Sibunit)

The presence of CO_2 and CO and H_2 was confirmed in the gas phase. Since the main reaction was dominant over the side reaction, one could have expected to have a high CO_2 content in the exit gas (Figure 15.1). However, a high content of carbon as CO was observed. The reaction mechanism on the catalyst surface can explain the differences in the CO and CO_2 concentrations: alkene and formic acid [22] can be produced on the surface, and formic acid can further decompose to CO and H_2O or—alternatively—to CO_2 and H_2. The alkene can either desorb from the surface (decarbonylation product) or be hydrogenated to alkane (decarboxylation product). Thus CO to CO_2 ratio does not necessarily correspond to the overall reaction scheme (Figure 15.1). Another reason, which can partly explain the gas analysis, is much higher solubility of CO_2 than CO in the liquid phase.

The experimental data was used as the basis for the modeling efforts. Some simplifications had to be made and a major one was to concentrate the modeling on the liquid-phase balances.

15.3 Assumptions

The model was developed for the liquid-phase reactions, which were assumed to be irreversible and, thus, were treated independently from the gas-phase reactions. The model included the main reaction and side reaction as shown in Figure 15.1. The main reaction was assumed to follow Langmuir kinetics (kinetics on ideal surfaces), which was also reported for a similar system in a batch reactor [23]. Such kinetics has been found to agree well with the experimental batch data. The side reaction, formation of heptadecene, took place to a much smaller extent and was modeled with a similar kind of reaction kinetics. The decline of the catalyst activity was assumed to take place as a consequence of the decrease in the number of active sites due to coking and was modeled identically for both products. The adsorption terms in the rate expressions for both reactions were identical, since they were based on site occupation.

The reactor was modeled as an axial dispersion unit. The most common single criterion for the extent of axial dispersion is the Peclet (*Pe*) number. A study highlighting the importance of the effect of *Pe* number and the effect of the extent of axial dispersion on the productivity was included. Sensitivity of the model parameters to possible inaccuracy in the value of *Pe* number was also studied. The value of *Pe* number that was used in simulation studies was obtained from empirical correlations presented in the literature [24–26].

Table 15.2 Model equations.

Collection of general equations upon which the modeling work was based on:

Stearic acid balance: $$\frac{\partial C_{St,L}}{\partial t} = \frac{1}{\varepsilon_L}\left(-w_L\frac{\partial C_{St,L}}{\partial l} + D_a\frac{\partial^2 C_{St,L}}{\partial l^2} - \varepsilon_L r'_{St1} - \varepsilon_L r'_{St2}\right) \quad (15.1)$$

Heptadecane balance: $$\frac{\partial C_{HDA,L}}{\partial t} = \frac{1}{\varepsilon_L}\left(-w_L\frac{\partial C_{HDA,L}}{\partial l} + D_a\frac{\partial^2 C_{HDA,L}}{\partial l^2} + \varepsilon_L r'_{St1}\right) \quad (15.2)$$

Heptadecene balance: $$\frac{\partial C_{HDE,L}}{\partial t} = \frac{1}{\varepsilon_L}\left(-w_L\frac{\partial C_{HDE,L}}{\partial l} + D_a\frac{\partial^2 C_{HDE,L}}{\partial l^2} + \varepsilon_L r'_{St2}\right) \quad (15.3)$$

Rate of the main reaction:
$$r_A = r'_{St1} = \frac{p_1 a C_{St,L}}{(1 + p_2 C_{St,L} + p_3 C_{HDE,L})} \quad (15.4)$$

Adsorption parameters:
$$p_1 = k_{St}K_{St},\ p_2 = K_{St},\ p_3 = K_{HDE} \quad (15.5\text{–}15.7)$$

Activity:
$$a = 1 - \theta_{C^\cdot} \quad (15.8)$$

Deactivation:
$$-\frac{da}{dt} = \frac{(a - a_f)p_4 C_{St,L}}{(1 + p_2 C_{St,L} + p_3 C_{HDE,L})} \quad (15.9)$$

Coking rate:
$$r_{C^\cdot} = \frac{p_4 a C_{St,L}}{(1 + p_2 C_{St,L} + p_3 C_{HDE,L})} \quad (15.10)$$

Coking parameter:
$$p_4 = k_{C^\cdot} K_{St} \quad (15.11)$$

Rate of side reaction:
$$r_B = r'_{St2} = \frac{p_5 a C_{St,L}}{(1 + p_2 C_{St,L} + p_3 C_{HDE,L})} \quad (15.12)$$

Step-by-step concentration inside particle:
$$f_{x+1} = \frac{sx^{s-1}\dfrac{f_{x-1}}{2\Delta x} + x^s\dfrac{2f_x - f_{x-1}}{\Delta x^2} + \dfrac{x^s R^2}{D_e}\left(\dfrac{p_1 f_x C_{St,S}}{1 + p_2 f_x C_{St,S} + p_3 C_{HDE,S}}\right)}{\left(\dfrac{sx^{s-1}}{2\Delta x} + \dfrac{x^s}{\Delta x^2}\right)} \quad (15.13)$$

Effectiveness factor:
$$\eta_e = \frac{\sum_{x=1}^{N-1} r_{A,x}V_x}{r_{A,S}V_p} = \frac{\sum_{x=1}^{N-1} \dfrac{p_1 f_x C_{St,S}}{(1 + p_2 f_x C_{St,S} + p_3 C_{HDE,S})}\dfrac{4}{3}\pi\left({r_{i+1}}^3 - {r_i}^3\right)}{\dfrac{p_1 C_{St,S}}{(1 + p_2 C_{St,S} + p_3 C_{HDE,S})}\dfrac{4}{3}\pi R^3} \quad (15.14)$$

Dimensionless concentration, size and shape factor, and diffusivity of liquid mixture:
$$f_x = \frac{C_{St,x}}{C_{St,S}} \quad (15.15) \qquad x = \frac{r}{R} \quad (15.16) \qquad s = \frac{A_p}{V_p}R - 1 \quad (15.17) \qquad D_{AB} = \left(D^0_{AB}\right)^{x_A}\left(D^0_{BA}\right)^{x_B}\Gamma_{AB} \quad (15.18)$$

Thermodynamic correction for diffusivity, Wilke–Chang equation, and viscosity:
$$\Gamma_{AB} = 1 + d\ \ln^0\gamma_A / d\ \ln^0 x_A \approx 1 \quad (15.19)$$
$$D_{AB} = \frac{7.4 \times 10^{-12}\sqrt{\varphi M_B}(T/K)}{(\mu_B/cP)\left(V_A/(cm^3/mol)\right)^{0.6}} \quad (15.20)$$
$$\log_{10}(\mu_L) = A + \frac{B}{T} + CT + DT^2 \quad (15.21)$$

Stearic acid: $A = -3.5929$, $B = 1.3465 \times 10^3$, $C = 2.9104 \times 10^{-3}$, $D = -2.7617 \times 10^{-6}$
Heptadecane: $A = -8.1307$, $B = 1.5791 \times 10^3$, $C = 1.4949 \times 10^{-2}$, $D = -1.1987 \times 10^{-5}$

Liquid holdup:
$$\varepsilon_{L,tot} = k_\varepsilon\left(\frac{d_p}{d_{pipe}}\right)^{0.33} Re_L^{0.14} \quad (15.22)$$

Pressure drop (single phase):
$$\frac{dp}{dl} = -\frac{f\rho_L {w_L}^2}{d_p} \quad (15.23)$$

Ergun equation for friction factor:
$$f = \frac{(1-\varepsilon_p)^2}{{\varepsilon_p}^3}\frac{150}{\varphi d_p \dot{m}_L/\mu_L} + 1.75\frac{(1-\varepsilon_p)}{{\varepsilon_p}^3} \quad (15.24)$$

Correlations for evaluating axial dispersion coefficient (Fu and Tan [24], Ebach and White [25]):
$$\frac{w_L d_p}{\varepsilon_L D_a} = \frac{1.4 * 10^{-4}}{{d_p}^{0.75}\varepsilon_p}\left(\frac{9\pi(1-\varepsilon_p)^2}{16{\varepsilon_p}^3}\right)^{0.25} \quad (15.25)$$
$$\frac{D_a\rho_L}{\mu_L} = 13.5\left(\frac{\rho_L d_p w_L}{\varepsilon_p \mu_L}\right)^{1.06} \quad (15.26)$$

Constant temperature and pressure were assumed and checked to prevail in the reactor. The liquid holdup was assumed to be constant, which was in alignment with the studies for the low superficial velocities [27]. At low superficial velocities, studies have shown that the liquid holdup is very weakly proportional to the liquid *Re* number and independent of the gas superficial velocity. The produced gas was assumed to have a low residence time in the reactor and, consequently, a minor effect on the liquid holdup.

15.4 Model equations

Dynamic mass balances of the liquid-phase compounds, for stearic acid, heptadecane, and heptadecene, are given in Table 15.2 (Eqs. 15.1–15.3).

All the balances have accumulation, convection, axial dispersion, and reaction terms. The equations include liquid holdup, ε_L, and superficial liquid velocity, w_L. Langmuir-type rate equation, for the main reaction, Equation 15.4, included also an activity correction term *a*. K_{St} and K_{HDE} in Equations 15.5–15.7 indicate the adsorption parameters for stearic acid and heptadecene, respectively. Equation 15.4 corresponds to a monomolecular transformation of stearic acid via the adsorption of the reactant to the main product. Adsorption terms for stearic acid and heptadecene were used, since both of these compounds contain functional groups enabling adsorption on the active sites of the catalyst. Reaction rates were assumed not to be limited by heptadecane adsorption. Thus, the adsorption term of heptadecane was neglected. In line with the experimental observations indicating catalyst deactivation, Equation 15.4 (Table 15.2) was modified to incorporate the decrease in catalyst activity. In particular, the activity was assumed to decrease due to coking that reduces the number of active sites (Eq. 15.8 with θ_{C^*} denoting the coverage for coke). Therefore, the decrease in the catalytic activity can be expressed by the form given in Equation 15.9. The equation contains the term a_f, which describes the final activity. A similar equation could have been derived with a concept of partially reversible coking, but the deactivation was modeled as a stearic acid concentration-dependent irreversible coke formation reaction. The rate of coking is described by Equation 15.10, while k_{C^*} in Equation 15.11 indicates the reaction rate constant for coking.

The importance of the pore diffusion in catalyst particles was tested by calculating first the concentration profile inside particles with the Hoyos method [28] (Eq. 15.13) and by calculating the effectiveness factor from the calculated concentration profiles (Eq. 15.14). The pore diffusion study was done only for the Pd/C (Aldrich) catalyst, since the other catalyst was of eggshell type in which the active sites were located in the vicinity of the particle outer surface with the active layer thickness being only 50 μm. The Hoyos method [28] uses the dimensionless concentration and size and shape factors as given in Equations 15.15–15.17, respectively. The extent of pore diffusion was checked by calculating the concentration profiles inside the particles and the resulting effectiveness factor for the most severe conditions (maximum productivity close to the feed entrance). The effective diffusivity was selected to be one tenth of the liquid diffusivity. Mass transfer of the coproduct, heptadecene, was assumed to be sufficient to guarantee its even distribution within the particles. Since the liquid-phase concentrations were high, the liquid diffusivities were calculated using the Vignes expression [29] (Eq. 15.18). D_{AB} indicates the diffusivity for components A in B and D^0_{AB} and D^0_{BA} stand for infinite dilution diffusivities for A in B and B in A, respectively. The exponents x_A and x_B denote the molar fractions of components. The thermodynamic correction for activity, Γ_{AB} (Eq. 15.19), was not applied since the mixture was rather nonpolar. Infinite dilution liquid diffusivities were calculated by using the Wilke–Chang expression (Eq. 15.20). Viscosities of fatty acids and their derivatives are discussed in literature [30–32]. The viscosity of stearic acid at the operation temperature was calculated by using Equation 15.21 for which the parameter values are given in the literature [32]. Viscosity of heptadecane was known to be somewhat lower than the viscosity of stearic acid [32]. The densities of all liquid components were obtained by linear extrapolation of literature data [33].

Pore diffusion calculation was intended to be included in the reactor model, but the preliminary calculations using the Hoyos method for single spherical particles showed that it does not play any significant role here. Only unrealistically small values of effective diffusivity could make the pore diffusion significant enough. When having low superficial velocities for both gas and liquid, the liquid holdup was known to be very weakly dependent on liquid Reynolds number (Eq. 15.22) and practically independent of superficial velocity of gas. Therefore, it could be treated as constant and is taken as 0.4, which is used in the calculation of reaction rates. The liquid holdup estimate was based on the empty space in the packing consisting of equal-sized spheres. The pressure drop was calculated by assuming the presence of only liquid phase (Eq. 15.23). The friction factor was obtained by applying Ergun equation (Eq. 15.24). The axial dispersion coefficients were obtained by using empirical equations given in the literature [26] for the special cases of low superficial liquid velocity (Eqs. 15.25 and 15.26). Values of operating conditions and estimated physical properties are given in Table 15.3.

Table 15.3 Physical properties and variables of the model system.

T (K)	633
p (bar)	20
$\mu_{L,St}$ (cP)	0.18
$\mu_{L,HDA}$ (cP)	0.10
$D^0_{St,HDE}$ (m^2/s)	2.24×10^{-8}
$D^0_{HDE,St}$ (m^2/s)	1.27×10^{-8}
ρ_L (kg/m^3)	670
ε_L ()	0.4
τ (min) (based on liquid holdup)	141

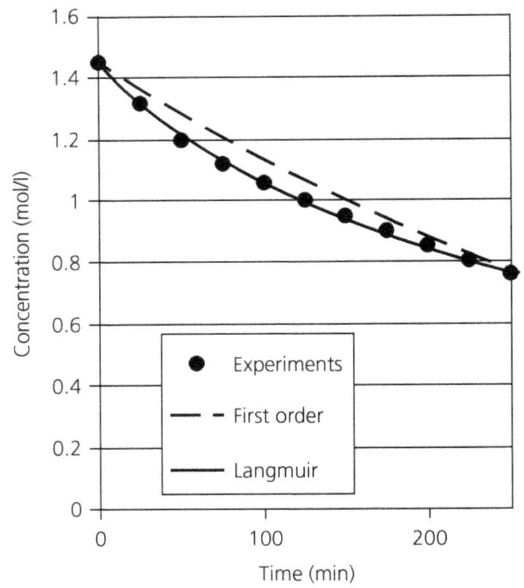

Figure 15.2 The experimental, the first-order power law, and the Langmuir batch reaction curves for stearic acid [20]. Parameters: $p_2 = 0.011$ l/mol, $p_3 = 25$ l/mol.

15.5 Evaluation of the adsorption parameters

The values of the adsorption parameters used in this study were obtained as a result of the minimization of the sum of least squares deviation of calculated and experimental concentration profiles in a batch experiment [20]. The experimental concentration curve is given along with the calculated profiles (based on either the first-order or Langmuir kinetics) in Figure 15.2.

The first-order kinetics serves as a reference since it corresponds to the situation of weak adsorption. The concentration curve based on the Langmuir model followed the experimental trend more precisely and was, therefore, used in further modeling of the packed bed reactor. The adsorption parameters were treated as constants in all studies.

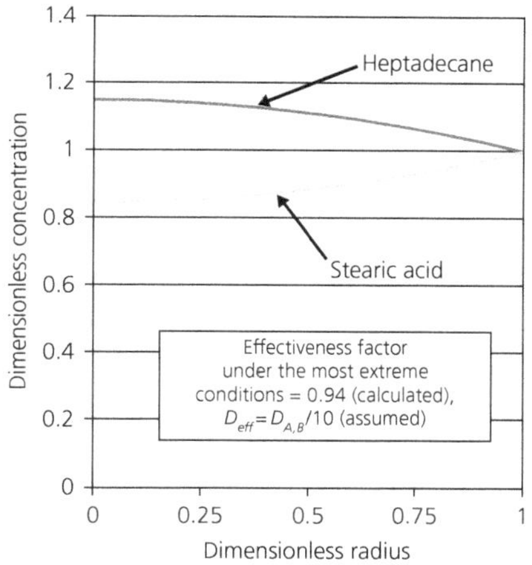

Figure 15.3 Dimensionless initial concentration of stearic acid and heptadecane inside spherical catalyst particles located near the reactor inlet.

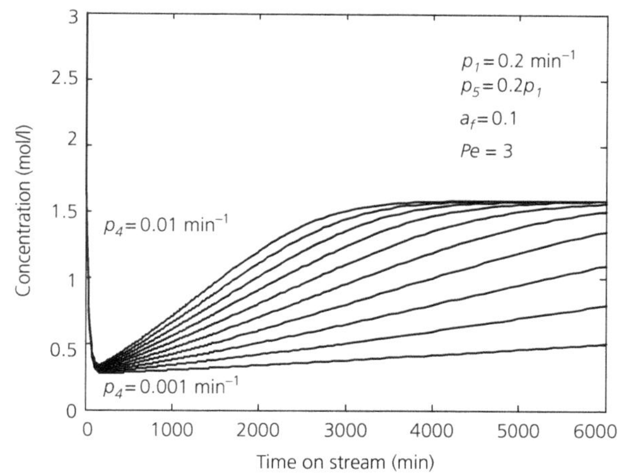

Figure 15.5 Effect of the rate constant of coking on stearic acid concentration.

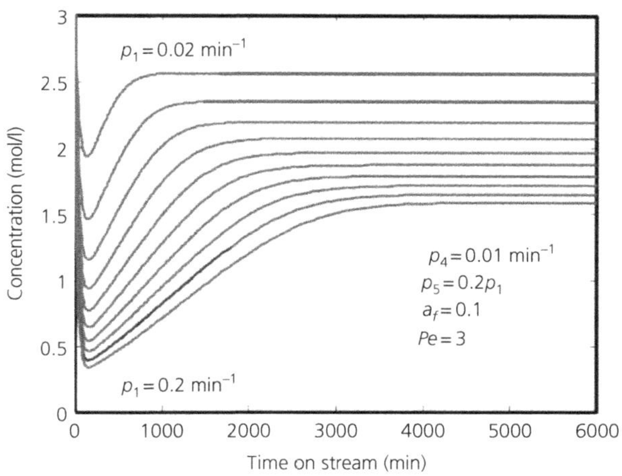

Figure 15.4 Effect of the rate constant of the main reaction on stearic acid concentration.

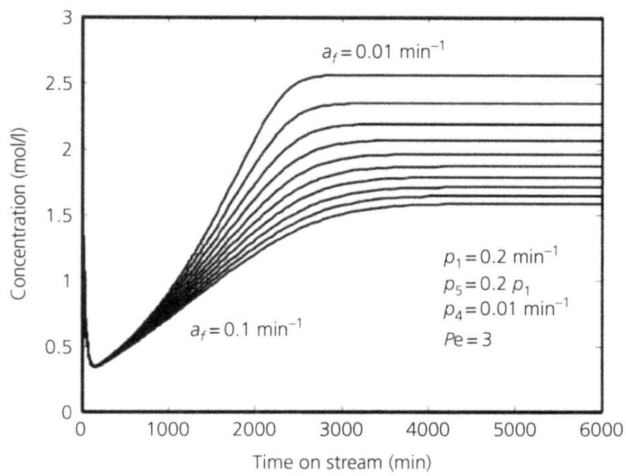

Figure 15.6 Effect of the final activity on stearic acid concentration.

15.6 Particle diffusion study

Particle diffusion was studied with a separate code written in MS Excel™ utilizing the step-by-step routine and gradient-based optimization (Eq. 15.13). The particles were assumed to be spherical. The calculations were made with the highest reaction rate present in the reactor. In practice this meant a place close to feed point location in the beginning of the experiment. Results of calculations are shown in Figure 15.3, where the dimensionless concentrations ($C_i/C_{i,S}$) of stearic acid and heptadecane are illustrated.

The calculation was carried out for Pd/C (Aldrich) catalyst, since the other one was an eggshell catalyst where the active metal was located mainly within the short zone close to the outer surface. Figure 15.3 shows that with the highest reaction rate, the dimensionless concentration of stearic acid dropped in the middle down to 0.935. Since the particle shape was spherical, the volume of the segment closer to the outer surface is larger than other segments, this being among the reasons why the effectiveness factor became very close to 1. As a conclusion it can be stated that the pore diffusion did not retard the main reaction. The effect of pore diffusion on the rate of the side reaction was even less severe, since the severity directly depends on the reaction rate.

15.7 Parameter sensitivity studies

The model parameters used for the reactor simulations were rate parameter of the main and side reactions, p_1 and p_5, respectively, rate parameter of coking, p_4, and final activity, a_f. Sensitivity study of these parameters was carried out by using the model with both the main and side reactions in the presence of coking. *Pe* was set to be 3, a value obtained by applying empirical equations. First, only one parameter at time was changed while maintaining the others constant and the concentration of the stearic acid was observed. The effects of the changes in p_1, p_4, and a_f on stearic acid concentration are presented in Figures 15.4, 15.5, and 15.6, respectively. The parameter values are listed in

Table 15.4 Parameter values of the sensitivity study.

Study/values	p_1 (l/min) (main reaction)	p_4 (l/min) (coking)	p_5 (l/min) (side reaction)	a_f ( ) (final activity)
p_1 (l/min) (main reaction)	0.02–0.2	0.01	$0.2p_1$	0.1
p_4 (l/min) (coking)	0.1	0.001–0.01	0.02	0.1
a_f ( ) (final activity)	0.1	0.01	0.02	0.01–0.1

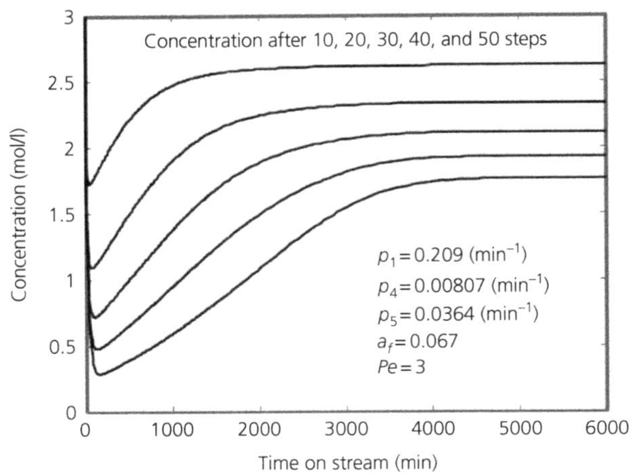

Figure 15.7 Stearic acid profiles inside the reactor.

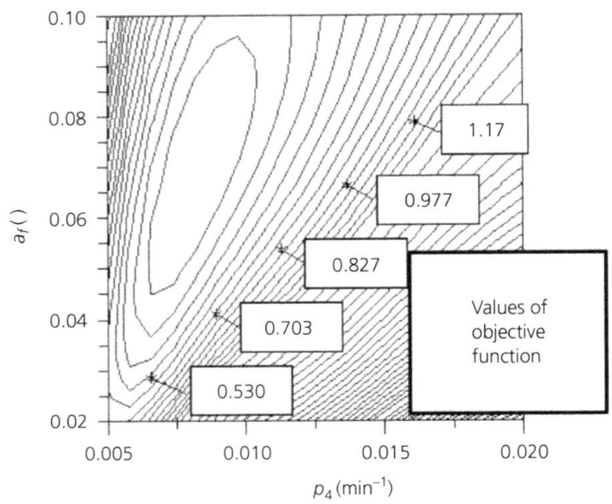

Figure 15.9 Correlation between the deactivation rate and the final activity ($Pe = 3$).

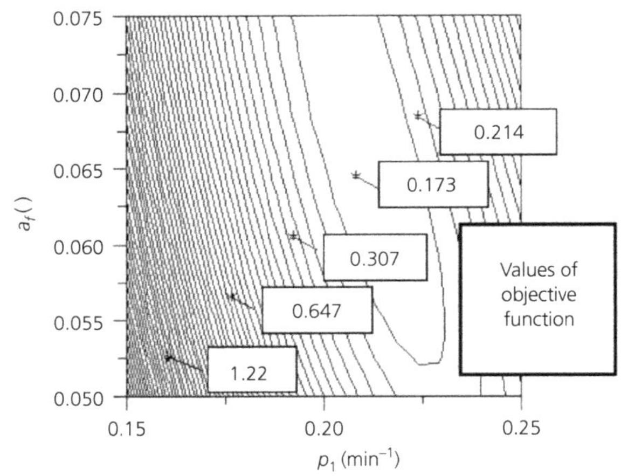

Figure 15.8 Correlation between the main reaction rate and the final activity ($Pe = 3$).

Table 15.4. The presented sensitivity study was made for the Pd/C (Aldrich) case.

Figure 15.4 shows how the changing of the rate parameter of the main reaction had a direct effect on the conversion of stearic acid. In order to keep the selectivity of the main reaction constant, the reaction rate constant of the side reaction was adjusted with the same factor used for changing the rate constant of the main reaction. Having the parameter as 0.2 min^{-1} made the stearic acid concentration to decrease down to 0.3 mol/l. This implied a 90% maximum yield before the deactivation started to dominate and finally decrease the yield down to less than 50%. Having the rate constant as 0.02 min^{-1} gave only 33% maximum yield and less than 20% final yield. When looking closely at the concentration curves, one can note that the reaction rate constant did also change the rate, at which the final yield was reached. This is expected, as the reaction and deactivation are phenomena that compete with each other. The time to reach the final yield with the reaction rate parameter 0.2 min^{-1} was around 5000 min, while it was around 1000 min with 0.02 min^{-1}. A change in the coking rate constant, p_4, influenced most strongly the speed at which the final concentration was reached, Figure 15.5. When the parameter value was increased from 0.001 to 0.01, the time decreased down to ca. 4000 min. The coking parameter also influenced the achieved maximum conversion, which gave further evidence that the reaction rate and coking parameter are coupled. The final activity, Figure 15.6, determined mainly the steady-state stearic acid concentration although it also changed slightly the time to reach it. Figure 15.7 illustrates how the concentration trends were developed inside the reactor for the Pd/C (Aldrich) case. Since the rate of the main reaction was dependent on stearic acid concentration, the reaction rates were higher near the feed entrance and changes in the concentration were significant. The catalyst deactivation was also reactant dependent, which initiated coking from the feed entrance to progress further with time.

15.8 Parameter identification studies

The 2D contour plots obtained using parameter value ranges around the optimum values for the Pd/C (Aldrich) case are illustrated in Figures 15.8, 15.9, and 15.10.

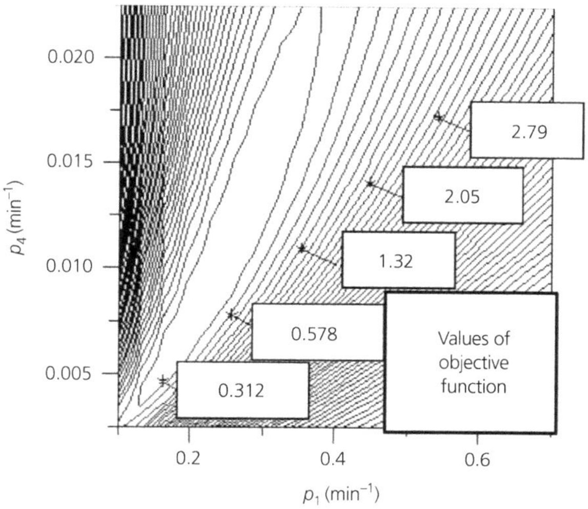

Figure 15.10 Correlation between the main and coking reaction rate parameters ($Pe = 3$).

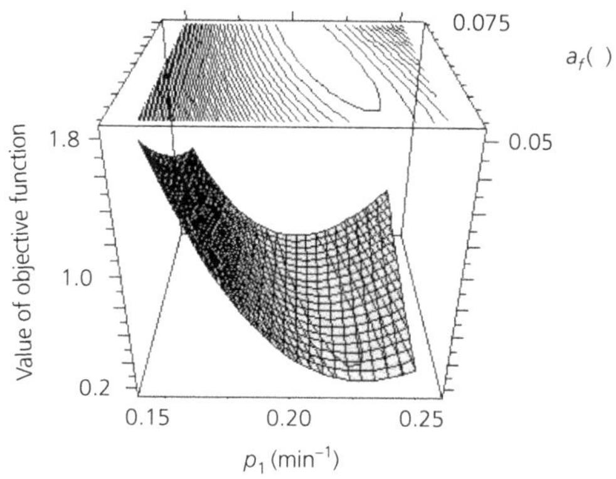

Figure 15.11 Identification between the rate parameter of the main reaction and the final activity.

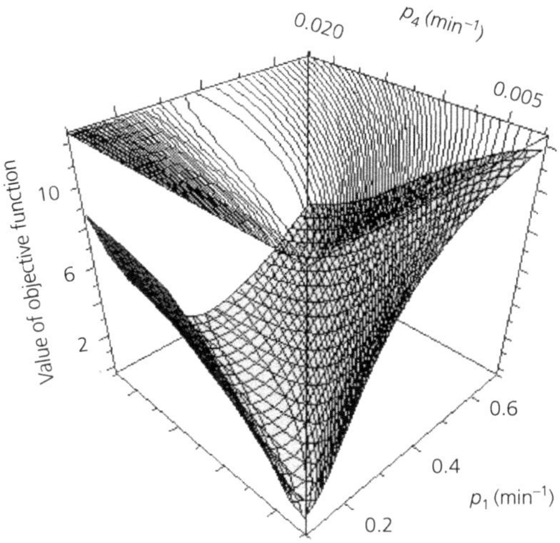

Figure 15.12 Identification between reaction rate parameters of the main reaction and coking.

The plots present the target function (scaled sum of the least squares deviation between observed and calculated concentrations as a function of two parameters), while keeping the third parameter and selectivity constant. The aim was to show how well identifiable the parameters were. Figure 15.8 demonstrates how the effect of the rate constant of the main reaction could be seen to differ from the effect of final activity. This was because the final activity did not influence strongly the productivity at the early period of the experiments, but it dominated later on. The separation of the effect of deactivation rate from the effect of final activity, Figure 15.9, was also clear, since the former strongly affected the time required to reach the final activity while the other adjusted the level at which no further deactivation was visible with time on stream. Among the other ones, the correlation between the rate parameters of the main reaction and deactivation was the strongest, Figure 15.10. Both influenced the reaction rate from the beginning and the mutual correlation of the parameters is revealed. Many parametric combinations produced almost identical values of the target function although the correct order of magnitude of the parameters was detectable. Data from 484 simulations were used to produce each individual contour plot. Each figure presents the norm (measure of the sum of least squares deviation) as a function of two parameters while keeping selectivity and the third parameter constant. Equal weighing factors were used in the computations.

A three-dimensional presentation of the identification of the parameters is given in Figure 15.11 for the rate parameter of the main reaction and the final activity, while Figure 15.12 displays this for the main reaction and the coking rate parameters. These outcomes provide an overview of the difficulty of the parameter optimization task.

15.9 Studies concerning the deviation from ideal plug flow conditions

The Peclet number is regarded as a parameter to define the extent of axial dispersion. It can successfully explain the two extremes of continuous operation: plug flow (high *Pe* number) and ideal mixing of continuous stirred tank. The operation of a real reactor is between these two extremes. Plug flow is usually the more productive, desired flow pattern. In a packed bed, the non-idealities of the flow arise from the wall and packing effects. Liquid tends to flow as rivulets preferring some paths while avoiding others. In practice, this generates some backmixing leading to a residence time distribution (RTD). Modeling the whole flow structure as only axial dispersion is a significant simplification of the complicated fluid dynamics, but it is a preferred way used extensively. Figure 15.13 shows how much the *Pe* number influenced the current reaction system and the consumption of stearic acid. It clearly indicates how much the productivity drops as plug flow conditions are not met. *Pe* number exceeding 100 approaches to the ideal plug flow productivity.

An estimate of the *Pe* number can be obtained from an RTD measurement, that is, variance of residence times in the experiment. Empirical correlation giving the *Pe* number as a function of liquid *Re* (and possibly other dimensionless numbers) is also

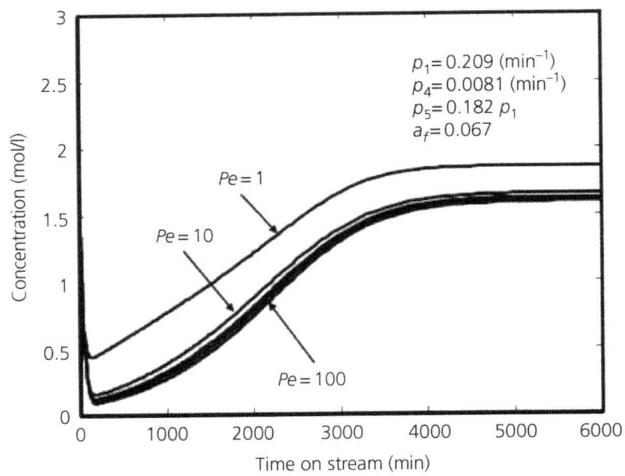

Figure 15.13 The effect of Peclet number on stearic acid consumption (catalyst: (Pd/C (Aldrich))).

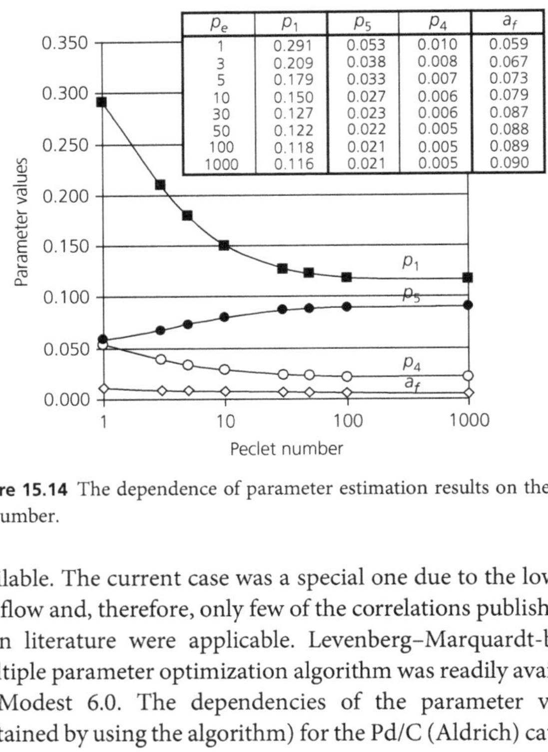

Figure 15.14 The dependence of parameter estimation results on the *Pe* number.

available. The current case was a special one due to the low liquid flow and, therefore, only few of the correlations published in open literature were applicable. Levenberg–Marquardt-based multiple parameter optimization algorithm was readily available in Modest 6.0. The dependencies of the parameter values (obtained by using the algorithm) for the Pd/C (Aldrich) catalyst on the used *Pe* number are shown in Figure 15.14. The parameters included are the rate constants of the main reaction, p_1, side reaction p_5 and coking, p_4, and the final activity, a_f. An estimate for the *Pe* number for this case was 3, which is a rather low value probably caused by diffusion effects in the reactor. Since the accuracy of the empirical equations was not precisely known for this application, Figure 15.14 was prepared. It shows how the values of the parameters (given as a table inset in Figure 15.14) depend on the *Pe* number. Since the parameters were interdependent, all the parameter values do depend on *Pe* number.

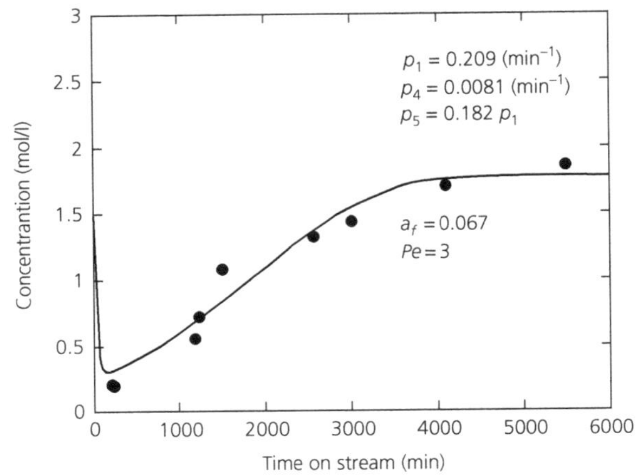

Figure 15.15 Estimated and experimentally observed concentrations of stearic acid (catalyst: (Pd/C (Aldrich))).

15.10 Parameter estimation results

The final parameter estimation was carried out after obtaining a clear view on the effects of various parameters. The results for Pd/C (Aldrich) are presented first, since this catalyst gave a much higher final yield. The estimation was carried out by using a gradient-based Levenberg–Marquardt optimizer in a relatively narrow range of parameter values. The obtained concentration curve for stearic acid is depicted in Figure 15.15 together with the experimental points. The results were obtained with the following parameter values: $p_1 = 0.209\ \text{min}^{-1}$, $p_4 = 0.0081\ \text{min}^{-1}$, $p_5 = 0.182p_1$, and $a_f = 0.067$ for Pd/C (Aldrich). Optimization was carried out without using weighting factors in the least squares optimization routine for any experimental point. As a consequence, the model slightly overpredicted the value of the stearic acid concentration at the first two points.

The Pd/C (Sibunit) case was also studied with the aid of modeling. Since this catalyst lost its activity much faster than Pd/C (Aldrich), the case was of limited importance. The raw material (stearic acid) was consumed over Pd/C (Sibunit) catalyst giving a high initial yield for the main reaction for roughly 250 min, after which the catalyst activity declined rapidly. There were many experimental points indicating this dramatic change. Eventually, the final yield for Pd on mesoporous carbon was much lower than that obtained over the Pd/C (Aldrich). The same model for the deactivation was applied again, and it could adequately describe the experimental observations. Figure 15.16 illustrates the results and gives the parameter values for the case.

15.11 Scale-up considerations

The overall strategy for the scale-up of this study was to select the diameter and the length of the reactor, size and shape of catalyst particles, and the superficial velocity of liquid in such a way that the effects of nonideal flow together with mass and heat

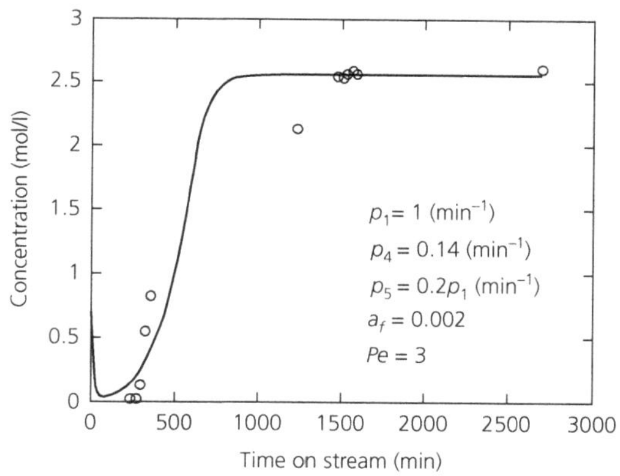

Figure 15.16 Estimated and experimentally observed concentrations of stearic acid (catalyst: (Pd/C (Sibunit))).

transfer in pilot scale were minimized. Since packed bed reactors are the most common reactors for continuous industrial operation, there are thorough review articles available that served as the basis for the selections [34–36].

There are several factors that are negligible in laboratory-scale packed bed operation but may become important during scale-up. Fluid dynamics may change due to channeling and wall effects. In small-scale operation, it is much easier to distribute the feed and avoid channeling. Such wall effects, on the other hand, play a less important role in large-scale operation. Wall effects can be minimized by having a small particle to reactor diameter ratio, d_p/d. However, it is difficult to avoid channeling since liquid tends to favor some paths at the expense of other ones, and thus the liquid flows dynamically in some parts while it stays more stagnant in other locations, typically in pockets between particles. In the current model, the fluid dynamics were lumped into a single parameter, axial dispersion coefficient, or *Pe* number, and, therefore, it will most probably change in scale-up. A collection of empirical equations for predicting the axial dispersion coefficient and the *Pe* number has been presented [26]. The gas velocity is commonly believed to have only a minor effect on the *Pe* number of the liquid. These equations are generally valid only for a limited range of liquid-phase Reynolds number. In most of the equations, the *Pe* number is a function of the liquid Reynolds number and some other dimensionless numbers. Basically, an order of magnitude estimate for the *Pe* number could then be obtained for scale-up.

In the decarboxylation of fatty acids, gases are formed as coproducts. Since the superficial liquid velocity was very low in laboratory scale to provide a long enough residence time, gases were assumed to be removed in a way that the system could be modeled as a single-phase liquid only reactor. In a larger scale, multiple gas withdrawal points may be required to perform this or otherwise gas holdup in a long reactor should be updated as a function of the reactor length and the progress of the reactions.

Pressure drop is usually negligible in a short laboratory-scale reactor with low superficial velocities. In the scale-up, the reactor becomes much longer than in the laboratory scale and pressure drop becomes unavoidable as the gas and liquid superficial velocities are increased.

The advantages and disadvantages of various particle sizes and shapes were also considered in scale-up [34–36]. Larger particles are often selected because small particles in a long reactor generate an unacceptably high pressure drop, which becomes more significant at higher superficial velocities. However, the use of small particles offers the benefits of having better flow distribution and less channeling. They are also less prone to lose effectiveness due to pore diffusion resistance. If pore diffusion is becoming an issue of importance, one can improve the situation by a clever particle shape selection. Particles such as miniliths, wagon wheels, or various types of hollow cylinders can provide more than 10 times larger outer surface area-to-volume ratio than simple spheres or cylinders. The active sites of the catalyst can also be distributed closer to the outer surfaces of particles during catalyst preparation. The particle void fraction and consequently the liquid holdup were other parameters that are influenced by the particle shape selection. The liquid holdup specifies the active volume in the liquid-phase reactor. The third parameter that is influenced by particle shape is the fluid dynamics and consequently the *Pe* number. For the pilot case, the catalyst particle size and shape were decided to be maintained the same as in the laboratory-scale experiments. The decision was justified by still having the overall pressure drop reasonably low.

In the current case, the desired reaction demanded a long residence time for achieving a high product yield. In scale-up, there are two options how to obtain long residence time; either the superficial velocities can be kept at rather low levels or a continuous circulation reactor can be used. Here the selection was to have a single long reactor and increase the flow in a way that a reasonable residence time was maintained. The reactor length was set to be 5 m. The length of laboratory reactor was 15 cm; thus the length of the pilot unit was 33-fold. Generally, the main reason for aiming at higher velocities in a pilot reactor is to improve the *Pe* number and suppress the external heat and mass transfer resistances. The *Pe* number is directly proportional to the superficial velocity and the length of the reactor and, therefore, its increase was intended and expected.

Radial heat transfer becomes a more demanding task as the reactor diameter is increased simply because the surface area for heat transfer increases with the second power of the diameter while the reactor volume increases as the third power of diameter. In industrial applications, this problem can be solved by selecting a multitubular reactor concept and by selecting support materials or bed dilution materials with higher heat conductivities. The heat generation is typically the largest near the feed entrance, and, therefore, cocurrent heat transfer can be applied there if the heat transfer becomes an issue. For the pilot simulation, the reactor diameter was selected to be 5 cm, which is large enough for avoiding the wall effects while still small enough to avoid the generation of severe radial temperature gradients.

Catalyst deactivation was present in the laboratory-scale experiments even for the better catalyst, although the

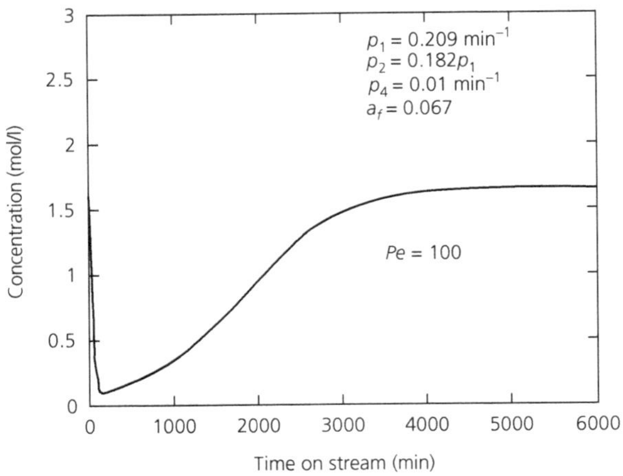

Figure 15.17 Simulated stearic acid concentration change with time in the pilot reactor ($T = 633°C$, $p = 20$ bar, residence time, and active metal concentration same as in the laboratory scale).

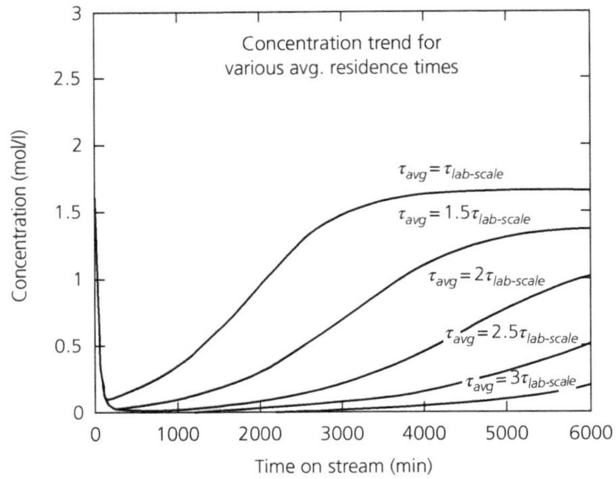

Figure 15.18 Simulated stearic acid concentration change with time in the pilot reactor ($T = 633°C$, $p = 10$ bar, active metal loading same as in the laboratory-scale unit).

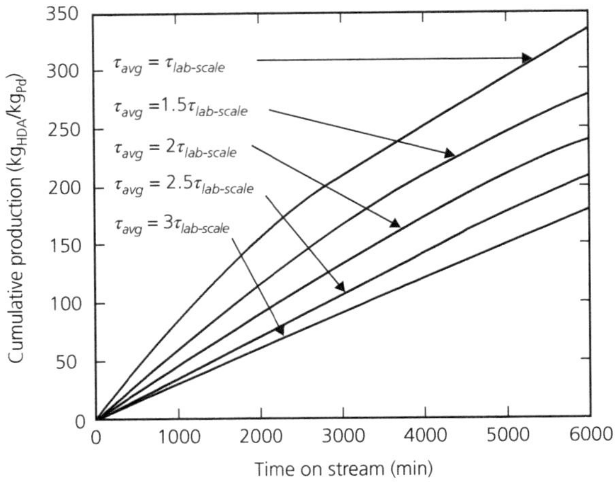

Figure 15.19 Cumulative production of the pilot reactor.

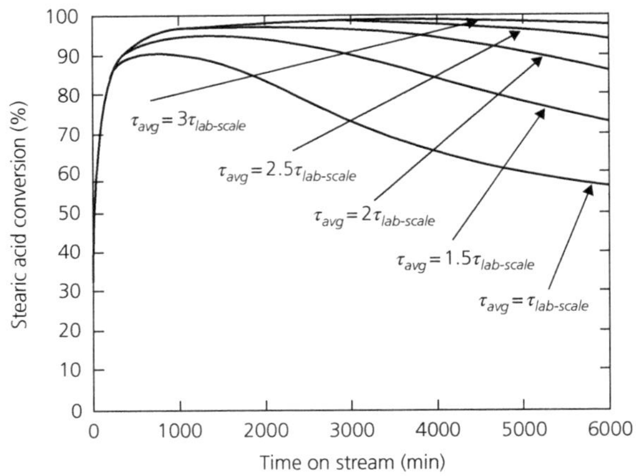

Figure 15.20 Stearic acid conversion (%) as a function of time on stream.

deactivation leveled almost to steady state. Deactivation took place most rapidly near the reactor inlet and progressed then gradually inside the reactor. This implied that by changing the flow periodically from downstream to upstream, in a similar fashion as originally proposed by Matros and Bunimovich [37], one can get more production than by operating in the same flow direction all the time. Catalyst regeneration is outside the scope of this modeling study, but it is a key issue for further experimental studies of this system, if high productivity cannot be maintained better by further catalyst development.

As the superficial velocity of liquid was adjusted upward to provide exactly the same mean residence time for the pilot case as was used for the laboratory-scale experiments, the liquid pressure drop became 1 kPa in the 5 m reactor. This was only 0.1% of total pressure for the lower pressure case and was negligible.

In order to obtain similar yields, the first simulations were performed so that the mean residence time was kept the same. Under these circumstances, it became evident that the scale-up by eliminating avoidable phenomena led to the situation where parameters changed in scale-up were the reactor length and diameter, the superficial liquid velocity, and *Pe* number. In scale-up, the diameter was changed from 1.5 to 5 cm and the length from 15 cm to 5 m. This changed the length-to-diameter ratio from 10 to 100. Together with the increased superficial velocity, this improved the *Pe* number and plug flow characteristics of the reactor. The *Pe* number was calculated to be ca. 3 for the laboratory-scale reactor using empirical equation presented in the literature. With the same equations, it was between 70 and 100 for the pilot scale reactor. A sample simulation of the pilot reactor is presented in Figure 15.17.

In the next step, the residence time of the pilot reactor was varied. By increasing the residence time, the deactivation can be delayed but at the expense of productivity. Figure 15.18 illustrates how much the deactivation was delayed by residence time adjustment. Figure 15.19 reveals how the cumulative production of heptadecane was influenced by the adjustments. The most desirable situation in continuous production would be to have the unit operating at high yields avoiding expensive separation steps. Within the time frame of 6000 min, it is possible to obtain a high product yield by using a residence time that is three times higher than the residence time used in the laboratory-scale experiments.

Figure 15.20 illustrates how the collected overall product is changing as a function of the mean residence time. If the residence time is maintained high, the fraction of reactant (stearic acid) in the collected product remains low. With decreased residence times, increasing amounts of the reactant appears at the reactor outlet.

15.12 Conclusions

Experimental data of stearic acid decarboxylation in a laboratory-scale fixed bed reactor for formation of heptadecane were evaluated studied with the aid of mathematical modeling. Reaction kinetics, catalyst deactivation, and axial dispersion were the central elements of the model. The effect of internal mass transfer resistance in catalyst pores was found negligible due to the slow reaction rates. The model was used for an extensive sensitivity study and parameter estimation. With optimized parameters, the model was able to describe the experimentally observed trends adequately. A reactor scale-up study was made by selecting the reactor geometry (diameter and length of the reactor, size and the shape of the catalyst particles) and operating conditions (superficial liquid velocity, temperature, and pressure) in such a way that nonideal flow and mass and heat transfer phenomena in pilot scale were avoided.

Acknowledgments

This work was part of the activities at the Åbo Akademi University Process Chemistry Centre. Academy of Finland is gratefully acknowledged for financial support.

Nomenclature

a	activity (—)
a, b	Ergun equation parameters (—)
a_f	final activity (—)
A_p	surface area of a particle (m^2)
$C_{HDA,L}$, $C_{HDA,S}$	concentration of heptadecane in liquid and at surface (mol/l)
$C_{HDE,L}$, $C_{HDE,S}$	concentration of heptadecene in liquid and at surface (mol/l)
$C_{St,L}$, $C_{St,x}C_{St,S}$	concentration of stearic acid in liquid, after step x and at surface (mol/l)
D_a	axial dispersion coefficient for liquid (m^2/min)
D_{AB}, D^0_{AB}, D^0_{BA}	diffusivity of component A in component B, infinite dilution diffusivities for components A in B and B in A (m^2/s)
d_p, d_{pipe}, d_K	particle diameter, pipe diameter, and Krischer–Kast diameter (m)
f	friction factor (—)
f_{x-1}, f_x, f_{x+1}	dimensionless concentration before step x, at step x and after step x (—)
k_ε	factor in liquid holdup expression (—)
k_{St}, k_{C^*}	reaction rate constant of main reaction and coking (mol/(l min))
K_{St}, K_{HDA}	adsorption factors for stearic acid and heptadecane (l/mol)
l	axial position (m)
M_B	molar mass of solvent (kg/kmol)
$\dot{m}_L$	mass flux of liquid (kg/(m^{2}s))
N	number of volume elements (—)
p	pressure (Pa)
p_1, p_2, p_3, p_4, p_5	model parameters: main reaction rate parameter (l/min), stearic acid adsorption parameter (l/mol), heptadecene adsorption parameter (l/mol) coking rate parameter (l/min), and side reaction rate parameter (l/min)
$\frac{dp}{dl}$	pressure gradient (Pa/m)
R	particle radius (m)
r	radial location (m)
r_A, $r_{A,x}$, $r_{A,S}$	reaction rate of component A, stearic acid; rate in position x; and rate at catalyst particle surface (mol/(l min))
r_{i+1} r_i	radial location after step $i+1$ and i (m)
r'_{St1}	activity corrected reaction rate for the main reaction (mol/(l min))
r'_{St2}	activity corrected reaction rate for the side reaction (mol/(l min))
Re_L	Reynolds number of liquid (—), $\rho w_L d_p/\mu_L$
s	shape factor (—)
t	time (s)
V_A	molar volume of solute (cm^3/mol)
V_p, V_x	volume of catalyst particle and volume of the element x (m^3)
w_L	superficial liquid velocity (m/min) in balances, (m/s) in dimensionless numbers
x, x_A, x_B	mole fraction of component A and B (—)

Greek letters

Δx	dimensionless length (—)
γ_A	thermodynamic activity of component A (—)
Γ_{AB}	thermodynamic effectivity factor for diffusion (—)
ε_L, $\varepsilon_{L,tot}$, ε_p	holdup of liquid, total holdup of liquid, and void fraction of packed bed (—)
η_e	effectiveness factor for main reaction (—)

$\theta_C{}^*$	occupation degree for coke (—)
μ_L, μ_B	dynamic viscosity of liquid (Pas) and of solvent (cP)
ρ_L, ρ_G	density of liquid and gas (kg/m^3)
ϕ	particle sphericity (—), association factor in Wilke–Chang equation (—)

References

1 Chamy R, Illanes A, Aroca G, Nunez L. Acid hydrolysis of sugar beet pulp as pretreatment for fermentation. Bioresour. Technol. 1994; 50: 149.

2 Li Y, Zhang X, Sun L, Zhang J, Xu H. Fatty acid methyl ester synthesis catalyzed by solid superacid catalyst $SO_4{}^{2-}/ZrO_2–TiO_2/La^{3+}$. Appl. Energy 2010; 87: 156.

3 Laurent E, Delmon B. Study of the hydrodeoxygenation of carbonyl, carboxylic and guaiacyl groups over sulfided CoMo/(-Al_2O_3) and NiMo/(-Al_2O_3) catalysts. I. Catalytic reaction schemes. Appl. Catal. A 1994; 109: 77.

4 Gomes JR. Vegetable oil hydroconversion process. US Patent 2006,018,620. December 1, 2005.

5 Priecel P, Kubicka D, Capek L, Bastl Z, Rynek P. The role of Ni species in the deoxygenation of rapeseed oil over NiMO-alumina catalysts. Appl. Catal. A Gen. 2011; 397: 127.

6 Murzin D, Kubičková I, Snåre M, Mäki-Arvela P Myllyoja J. Method for the manufacture of hydrocarbons. US Patent 7, 491,858. February 17, 2009.

7 Kubičková I, Snåre M, Mäki-Arvela P, Eränen K, Murzin DY. Hydrocarbons for diesel fuel via decarboxylation of vegetable oils. Catal. Today 2005; 106: 197.

8 Snåre M, Kubičková I, Mäki-Arvela P, Eränen K, Murzin DY. Continuous deoxygenation of ethyl stearate—A model reaction for production of diesel fuel hydrocarbons. Catal. Org. React. 2006; 115: 415.

9 Snåre M, Kubicková I, Mäki-Arvela P, Eränen K, Wärnå J, Murzin DY. Production of diesel fuel from renewable feeds: Kinetics of ethyl stearate decarboxylation. Chem. Eng. J. 2007; 134: 29.

10 Snåre M, Kubicková I, Mäki-Arvela P, Chichova D, Eränen K, Murzin DY. Catalytic deoxygenation of unsaturated renewable feedstocks for production of diesel fuel hydrocarbons. Fuel 2008; 87: 933.

11 Snåre M, Kubicková I, Mäki-Arvela P, Eränen K, Murzin DY. Heterogeneous catalytic deoxygenation of stearic acid for production of biodiesel. Ind. Eng. Chem. Res. 2006; 45: 5708.

12 Simakova I, Simakova O, Mäki-Arvela P, Murzin D. Decarboxylation of fatty acids over Pd supported on mesoporous carbon. Catal. Today 2010; 150: 28.

13 Simakova I, Simakova O, Mäki-Arvela P, Simakov A, Estrada M, Murzin DY. Deoxygenation of stearic acid over supported Pd catalysts: Effect of metal dispersion. Appl. Catal. A Gen. 2009; 355: 100.

14 Simakova I, Rozmysłowicz B, Simakova O, Mäki-Arvela P, Simakov A, Murzin DY. Catalytic deoxygenation of C18 fatty acids over mesoporous Pd/C catalyst for synthesis of biofuels. Top. Catal. 2011; 54: 460.

15 Lestari S, Mäki-Arvela P, Bernas H, Simakova O, Sjöholm R, Beltramini J, Max LG, Myllyoja J, Simakova I, Murzin DY. Catalytic deoxygenation of stearic acid in a continuous reactor over a mesoporous carbon supported Pd catalyst. Energy Fuels 2009; 23: 3842.

16 Bernas H, Eränen K, Simakova I, Leino AR, Kordás K, Myllyoja J, Mäki-Arvela P, Salmi T, Murzin DY. Deoxygenation of fatty acids in a fixed bed reactor. Fuel 2010; 89: 2033.

17 Lestari S, Mäki-Arvela P, Simakova I, Beltramini J, Max LG, Murzin DY. Catalytic deoxygenation of stearic acid and palmitic acid in a semibatch mode. Catal. Lett. 2009; 130: 48.

18 Mäki-Arvela P, Kubicková I, Eränen K, Snåre M, Murzin D. Catalytic decarboxylation of fatty acids and their derivatives. Energy Fuels 2007; 21: 30.

19 Rozmysłowicz B, Mäki-Arvela P, Lestari S, Simakova OA, Eränen K, Simakova I, Murzin DY, Salmi T. Catalytic deoxygenation of tall oil fatty acids over a palladium-mesoporous carbon catalyst: A new source of biofuels. Top. Catal. 2010; 53: 1277.

20 Snåre M. Development of next generation biodiesel technology [Doctoral dissertation], Turku/Åbo: Åbo Akademi University; 2006.

21 Mäki-Arvela P, Snåre M, Eränen K, Myllyoja J, Murzin DY. Continuous decarboxylation of lauric acid over Pd/C catalyst. Fuel 2008; 87: 3543.

22 Fazal MA, Haseeb AS, Ma A, Masjuki HH. Biodiesel feasibility study: An evaluation of material compatibility, performance, emission and engine durability. Renew. Sustain. Energy Rev. 2011; 15: 1314.

23 Salmi T, Mikkola JP, Wärnå J. Chemical reaction engineering and reactor technology. Boca Raton, FL: CRC Press; 2009.

24 Fu MS, Tan CS. Liquid hold-up and axial dispersion in trickle bed reactors. Chem. Eng. Sci. 1996; 51: 5357.

25 Ebach EA, White RR. Mixing of fluids flowing through beds of packed solids. AIChE J. 1958; 4: 161.

26 Piche S, Larachi F, Iliuta I, Grandjean B. Improving the prediction of liquid back-mixing in trickle-bed reactors using a neural network approach. J. Chem. Technol. Biotechnol. 2002; 77: 989.

27 Lange R, Schubert M, Bauer T. Liquid hold-up in trickle bed reactors at very low liquid Reynolds numbers. Ind. Eng. Chem. Res. 2005; 44: 6504.

28 Hoyos B, Cadavid JG, Rangel H. Formulation and numerical calculation of non-isothermal effectiveness factor for finite cylindrical catalysts with bidimensional diffusion. Lat. Am. Appl. Res. 2004; 34: 17.

29 Vignes A. Diffusion in binary solutions. Ind. Eng. Chem. Fundam. 1966; 5: 189.

30 Assen E, Rytter E, Øye H. Viscosity of n-hydrocarbons and their mixtures. Ind. Eng. Chem. Res. 1990; 29: 1635.

31 Noureddini H, Teoh B, Clements LD. Viscosities of vegetable oils and fatty acids. J. Am. Oil Chem. Soc. 1992; 69: 1189.

32 Yaws CL. The Yaws' handbook of physical properties for hydrocarbons and chemicals. Houston, TX: Gulf Publishing Company; 2005.

33 Noureddini H, Teoh BC, Davis CL. Densities of vegetable oils and fatty acids. J. Am. Oil Chem. Soc. 1992; 69: 1184.

34 Pangarkar K, Schildhauer T, van Ommen J, Nijenhuis J, Kapteijn F, Moulijn J. Structured packings for multiphase catalytic reactors. Ind. Eng. Chem. Res. 2008; 47: 3720.

35 Krishna R, Sie T. Strategies for multiphase reactor selection. Chem. Eng. Sci. 1994; 49: 4029.

36 Mederos F, Ancheyta J, Chen J. Review on criteria to ensure ideal behaviors in trickle-bed reactors. Appl. Catal. A Gen. 2009; 355: 1–19.

37 Matros YS, Bunimovich GA. Reverse-flow operation in fixed bed catalytic reactors. Catal. Rev. Sci. Eng. 1996; 38: 1.

Index

Multiphase Catalytic Reactors: Theory, Design, Manufacturing, and Applications, First Edition. Edited by Zeynep Ilsen Önsan and Ahmet Kerim Avci.

www.ingramcontent.com/pod-product-compliance
Ingram Content Group UK Ltd.
Pitfield, Milton Keynes, MK11 3LW, UK
UKHW050952290926
15470UKWH00005B/1771

9 781118 115763